Lecture Notes in Control and Information Sciences

Volume 497

This series reports new developments in the fields of control and information sciences—quickly, informally and at a high level. The type of material considered for publication includes:

1. Preliminary drafts of monographs and advanced textbooks
2. Lectures on a new field, or presenting a new angle on a classical field
3. Research reports
4. Reports of meetings, provided they are

   (a) of exceptional interest and
   (b) devoted to a specific topic. The timeliness of subject material is very important.

Indexed by EI-Compendex, SCOPUS, Ulrich's, MathSciNet, Current Index to Statistics, Current Mathematical Publications, Mathematical Reviews, IngentaConnect, MetaPress and Springerlink.

Vasile Drăgan · Samir Aberkane · Ioan Lucian Popa

Robust Control of Jump Linear Stochastic Systems

Applications to Sampled-Data Control

Springer

Vasile Drăgan
Institute of Mathematics "Simion Stoilov"
of the Romanian Academy
Bucharest, Romania

Academy of Romanian Scientists
Bucharest, Romania

Ioan Lucian Popa
Department of Computing, Mathematics
and Electronics
1 Decembrie 1918 University
Alba Iulia, Romania

Samir Aberkane
Université de Lorraine, CNRS, CRAN
Nancy, France

ISSN 0170-8643 ISSN 1610-7411 (electronic)
Lecture Notes in Control and Information Sciences
ISBN 978-3-031-84072-2 ISBN 978-3-031-84070-8 (eBook)
https://doi.org/10.1007/978-3-031-84070-8

Mathematics Subject Classification: 93BXX, 93B05, 93B36, 93Cxx, 93C05, 93C57, 93EXX, 93E03, 93E15, 93E20, 49N25

This Springer imprint is published by the registered company Springer Nature Switzerland AG
The registered company address is: Gewerbestrasse 11, 6330 Cham, Switzerland

To my children, Alina and Cătălin and their sons

In memory of my father, Makhlouf. To my daughters, Zélie and Félicie

To my sons, Horea and Dacian and my wife Maria-Aurora

Preface

This monograph contains recent developments in the theory of linear stochastic systems controlled by impulses. The systems under consideration are simultaneously affected by multiplicative white noise perturbations and Markovian jumping.

The goal is to offer a comprehensive, thorough and self-sufficient exploration of the subject. We provide solutions to various theoretical and practical aspects in connection with this class of stochastic systems including stability analysis, optimal control, robust stabilization, minimization of the effect of some exogenous random disturbances through the behavior of the considered systems, and the estimation of some remote signals generated by dynamical systems modeled by jump stochastic linear differential equations. Motivated on the one hand by the recent progress obtained in connection with the deterministic dynamical systems controlled by impulses, and on the other hand by the remarkable developments from the recent decades done in relation to the control of linear stochastic systems, we decided to write this book dedicated to some important classes of stochastic systems controlled by impulses. We are convinced that this approach enlarges the area of possible applications in connection with different aspects of the control of physical, engineering, biological processes corrupted by random phenomena.

The appropriate mathematical framework for a clear formulation and a rigorous solution of the control problems for stochastic dynamical systems controlled by impulses is the theory of jump stochastic differential equations. Roughly speaking, a dynamical system described by jump stochastic differential equations can have some abrupt changes of its trajectories at certain time instances and it cannot be considered as having an evolution in continuous time in the usual sense. These abrupt changes of its trajectories are named impulsive phenomena, and they can be produced as an effect of some impulsive controls. The main idea of impulsive control is that the control action is only allowed at some discrete-time instants, inducing instantaneous changes on the trajectories of the controlled system. As in the deterministic case, the control by impulses of some stochastic processes may reduce the cost of the control process by reducing the amount of transmitted information.

In this context we distinguish two main objectives of this book. The first one is to develop a mathematical theory of time-varying jump linear stochastic systems including both white noise perturbations and Markovian jumping. The second one is to develop analysis and design methods for control problems for linear stochastic systems controlled by impulses such as linear quadratic optimal control, robust stabilization with respect to different classes of parametric uncertainties modeled by white noise perturbations with unknown intensity, robust estimation/filtering of some remote signals where the robustness is considered with respect to both stochastic parameter uncertainties and exogenous random phenomena affecting the dynamical system. From this perspective, the contents of the book may be divided into two main parts. The first five chapters are dedicated to the development of self-contained theoretical foundations for the description, analysis and control of time-varying dynamical systems described by impulsive Itô-type differential equations with Markovian jump perturbations. To the best of our knowledge, the majority of the results reported in these chapters are original and new. The objective is to provide the reader with the necessary theoretical background in order to address the second part of the book.

After some prerequisites concerning measure and probability theory which will be used in the next chapters of the book, Chap. 1 describes the main types of jump stochastic linear differential equations encountered in the next developments. The behavior of the trajectories of such systems of stochastic differential equations with impulses in the neighborhood of the impulse time instances is analyzed. In the last part of Chap. 1, we show how a stochastic sampled-data control problem can be equivalently converted into a control problem for a stochastic dynamical system controlled by impulses.

Chapter 2 is entirely dedicated to the problem of exponential stability of a class of deterministic jump linear differential equations. Both forward and backward jump linear differential equations defined on a Banach space are analyzed, and an accompanying discrete-time linear equation is associated. It is shown that the exponential stability of the zero solution of a forward jump linear differential equation is equivalent to the exponential stability of the zero solution of its accompanying discrete-time linear equation. The main purpose of this chapter is to provide characterizations of the exponential stability of the zero solution of a forward jump linear differential equation with positive evolution on an ordered Hilbert space. These characterizations are given in terms of the existence of some globally defined and bounded solution of some suitable forward or backward non-homogeneous jump differential equations or in terms of feasibility of some forward or backward affine differential inequalities with jumps. The last section of Chap. 2 is devoted to Lyapunov-type jump linear differential equations. This class of jump linear differential equations will be used in the next chapter to characterize the exponential stability in mean square of some large classes of jump stochastic linear differential equations. It is worth to notice that unlike the case of a forward or backward time-invariant non-homogeneous ordinary linear differential equation whose bounded (on the whole real axis) solution, if any, is constant, in the case of a forward or backward time-invariant non-homogeneous jump linear differential equation, its unique bounded (on the real axis) solution is not constant. Indeed, such a solution is a periodic function with the period equal to the length of the dwell time. This fact has important consequences in the derivation of some criteria for exponential stability in mean square for jump stochastic linear differential equations with constant coefficients.

In the first part of Chap. 3 we provide criteria for exponential stability in mean square of a large class of jump stochastic linear differential equations with state multiplicative white noise perturbations and/or Markov jump perturbations. These criteria are directly obtained from the results derived in Sects. 2.4 and 2.5. In the last part of this chapter, we highlight several useful properties of the solutions of non-homogeneous jump stochastic linear differential equation which will be often involved in the next chapters. In this context a fundamental auxiliary result was proved (see Lemma 3.1) which analyzes the behavior of a quadratic form along the trajectories of a non-homogeneous jump stochastic linear differential equation in the neighborhood of a discontinuity of the matrix of the quadratic form. The utility of this result will be highlighted in the development from the next chapters.

In Chap. 4 we deal with the stabilizability and detectability properties of a jump linear stochastic system. For this type of systems, we introduce the concepts of mean-square stabilizability by linear state feedback and of mean-square stabilizability by impulses in a state feedback form. Based on the criteria for exponential stability in mean square derived in Chap. 3, we are able to propose necessary and sufficient conditions for these types of mean-square stabilizability. In the case of sampled-data stochastic controlled systems we introduce the concept of stabilizability by sampled state measurements and provide a set of conditions which guarantee this kind of stabilizability property. The mean-square detectability of a jump linear stochastic system is also defined in this chapter and criteria which allow us to decide if a jump linear stochastic system is mean-square detectable or not are provided. In the last section of this chapter we extend to this general framework the well-known duality property between stabilizability and detectability of a linear time-invariant finite-dimensional deterministic system. This duality will be used in Sect. 5.6 to obtain the existence of the bounded and stabilizing solution of a forward jump matrix Riccati differential equation.

Chapter 5 is dedicated to some classes of jump Riccati differential equations arising in connection with different control problems for jump linear stochastic systems. Our attention is focused on the maximal, minimal and stabilizing solutions of these equations. Necessary and sufficient conditions for the existence of these kinds of solutions are given. In our approach, in this monograph, we have considered a general class of backward jump nonlinear differential equation that contains as particular cases different kind of jump matrix Riccati differential equations arising in stochastic control. In the last part of Chap. 5, we propose iterative procedures for the numerical computation of the maximal and stabilizing solutions of this jump Riccati differential equations. Their utility will be proven by illustrative examples given in the other chapters.

In the second part of this book we illustrate the applicability of the theoretical results derived in the first chapters in the solution of control problems for jump linear stochastic systems controlled by impulses. Chapter 6 is devoted to the linear quadratic optimal control problem for jump linear stochastic systems controlled by impulses. In the case when the control process is on infinite time horizon, we consider both the free end state and the fixed end state cases. Here, we particularly highlighted the role of the minimal solution of the stochastic jump matrix Riccati differential equations in the case of the free end state linear quadratic optimal control problem as well as the role of the maximal solution of this type of Riccati equations in the case of the fixed end state linear quadratic optimal control problem.

In Chap. 7, we introduce the concept of H_2-norm of a jump linear stochastic system. This norm measures the effect of some exogenous disturbances of additive white noise type on some outputs of the considered system. We provide explicit formulae of the state-space representation of the controllers which achieve the minimal value of the H_2-norm. The problem of the filtering/estimation of a remote signal generated by a dynamical system modeled by jump linear stochastic differential equations subject to additive and multiplicative white noise perturbations and Markovian jumping is studied in Chap. 8. Chapter 9 is dedicated to the problem of robust stability and robust stabilization of a jump linear stochastic system affected by random parametric uncertainties with unknown intensity. The level of robustness of the stability is measured based on the concept of stability radius. To obtain an estimate of a lower bound for the stability radius as well as for the designing of the corresponding robust stabilizing control, we use the scaling perturbations technique introduced by D. Hinrichsen and A. J. Pritchard in the case of the linear time-invariant finite-dimensional deterministic systems.

Each chapter opens with a brief abstract that outlines its content and concludes with a "Notes and References" section, where additional material related to the topic is explored.

This monograph is based on original recent results obtained by the authors. A part, which refers to control problems for jump stochastic linear systems affected by multiplicative white noise perturbations, was published in prestigious control journals. All the other parts are published here for the first time. We are convinced that the material from this book does not emphasize the whole potential of the proposed control methods for jump stochastic systems that are simultaneously affected by multiplicative white noise perturbations and Markov jumping, but it gives all the theoretical and numerical details the readers need to develop their own applications.

The target audience of the book includes researchers in theoretical and applied mathematics, research engineers, as well as graduate students interested in stochastic modeling and control. Some parts of this monograph could be useful even for undergraduate students.

We should like to express our gratitude to Mrs. Viorica Drăgan for her hard work and patience in the realization of this manuscript.

Finally, the authors sincerely appreciate the ongoing support and guidance provided by the publishers throughout this project.

Mihailesti, Romania
Nancy, France
Alba Iulia, Romania
October 2024

Vasile Drăgan
Samir Aberkane
Ioan Lucian Popa

Contents

Abbreviations

AS	Asymptotic stable (Definition 2.1 (b))
ASMS	Asymptotically stable in mean square (Definition 3.1)
BGLJLDE	Backward generalized Lyapunov-type jump linear differential equation (Sect. 7.2.1)
BJLDE	Backward jump linear differential equation (Sect. 2.1.4)
BJLDI	Backward linear differential inequality with jumps (Theorem 2.10)
BJMRDE	Generalized backward jump matrix Riccati differential equation (Sect. 5.1.2)
DTLE	Discrete-time linear equation (Sect. 2.1.2)
ES	Exponentially stable (Definition 2.1 (a))
ESMS	Exponential stability in the mean square (Definition 3.1)
ESMS-C	Exponentially stable in mean square with conditioning (Definition 3.1)
FGLJLDE	Forward generalized Lyapunov-type jump linear differential equation (Sect. 7.2.1)
FJLDE	Forward jump linear differential equation (Eq. 2.4)
FJLDI	Forward linear differential inequality with jumps (Theorem 2.10)
GBJMRDE	Generalized backward jump matrix Riccati differential equations (Sect. 5.1.1)
GLDE	Generalized Lyapunov differential equation (Sect. 2.5.2)
GLILDE	Generalized Lyapunov linear differential equations with impulses (Sect. 2.5.2)
GLJLDE	Generalized Lyapunov linear differential equations with jumps (Sect. 2.5.2)
GLJMLDE	Generalized Lyapunov jump matrix linear differential equation (Sect. 5.1.4)
GUMSAS	Global uniform mean-square asymptotic stable (Definition 9.2)
ICLSS	Impulsive controlled linear stochastic system (Sect. 4.1)

ICLSSMJ	Impulsive controlled linear stochastic systems with Markovian jumping (Sect. 1.2.2)
JSLDE	Jump stochastic linear differential equations (Sect. 3.1)
L_2-SSMS	L_2 Stochastically stable in mean square (Definition 3.1)
L_2-SSMS-C	L_2 Stochastically stable in mean square with conditioning (Definition 3.1)
LMDI	Linear matrix differential inequality (Proposition 4.4)
LMI	Linear matrix inequality (Example 3.1)
SESMS	Strong exponentially stable in mean square (Definition 3.1)
SJMRDE	Stochastic jump matrix Riccati differential equation (Sect. 5.1.2)
UMSS	Uniform mean-square stable (Definition 9.2)
$\mathscr{L}^2$−GUMSS	$\mathscr{L}^2$−global uniform mean-square stable (Definition 9.1)
$\mathscr{L}^2$−GUMSS-C	$\mathscr{L}^2$−global uniform mean-square stable with conditioning (Definition 9.1)

Chapter 1
Preliminaries to Probability Theory and Jump Stochastic Linear Differential Equations

Abstract This first chapter collects in its first part, for the readers convenience, some definitions and fundamental results concerning the measure theory and the stochastic processes theory which are needed in the following developments of the book. All the presented results will be given without proof, such results being classical, and appropriate references will be added for the interested readers. The second part of the chapter is devoted to the presentation of the different types of stochastic differential equations with jumps that will be studied in the rest of the book. Fundamental existence and unicity results of the solutions of such equations will be also stated and proved (Propositions 1.1, 1.2 and 1.3).

1.1 Several Preliminaries to Measure Theory, Stochastic Processes and Stochastic Integrals

In this section, we present the definitions of the main mathematical notions and concepts which will be involved in the developments in this monograph.

The basic results related to the notions defined will be presented without proofs. For the reader convenience we shall provide bibliographical references where the proof may be found.

1.1.1 Measure Theory: Integration

Let Ω be a non-empty set and 2^{Ω} be the set of all its subsets.

Definition 1.1 ([1–3]) A collection of subsets $\mathfrak{F} \subset 2^{\Omega}$ is named σ-algebra on Ω if the following axioms hold:

(a) $\Omega \in \mathfrak{F}$;
(b) if $M \in \mathfrak{F}$ then $\Omega \setminus M \in \mathfrak{F}$;
(c) if $\{M_i\}_{i \in \mathbb{Z}_+}$ is a sequence of subsets from $\mathfrak{F}$ then $\bigcup_{i=0}^{\infty} M_i \in \mathfrak{F}$.

V. Drăgan et al., *Robust Control of Jump Linear Stochastic Systems*, Lecture Notes in Control and Information Sciences 497, https://doi.org/10.1007/978-3-031-84070-8_1

A pair $(\Omega, \mathfrak{F})$ where $\mathfrak{F}$ is a σ-algebra on Ω will be named *measurable space*.

If $\mathfrak{M} \subset 2^{\Omega}$ then $\sigma(\mathfrak{M})$ denotes the smallest σ-algebra containing all subsets from $\mathfrak{M}$. Often $\sigma(\mathfrak{M})$ will be names σ-algebra generated by $\mathfrak{M}$. When $\Omega = \mathbb{R}$ or $\Omega = \mathscr{I}$, where $\mathscr{I} \subset \mathbb{R}$ is an interval, we shall denote $\mathfrak{B}(\mathbb{R})$ or $\mathfrak{B}(\mathscr{I})$, respectively, the σ-algebra of Borel subsets of $\mathbb{R}$, or $\mathscr{I}$, respectively, that is the smallest σ-algebra containing all open subsets from $\mathbb{R}$, or $\mathscr{I}$, respectively.

If $(\Omega_1, \mathfrak{F}_1)$ and $(\Omega_2, \mathfrak{F}_2)$ are two measurable spaces, then $\mathfrak{F}_1 \times \mathfrak{F}_2$ stands for the smallest σ-algebra on $\Omega_1 \times \Omega_2$ containing all subsets $M_1 \times M_2$ where $M_1 \in \mathfrak{F}_1$ and $M_2 \in \mathfrak{F}_2$.

Definition 1.2 ([1–3]) Let $(\Omega, \mathfrak{F})$ be a given measurable space. A function $f : \Omega \to \mathbb{R}$ is named real valued measurable function with respect to σ-algebra $\mathfrak{F}$ or a measurable function (for short) if for any $M \in \mathfrak{B}(\mathbb{R})$ the set $\{\omega \in \Omega \mid f(\omega) \in M\}$ lies in $\mathfrak{F}$.

One may check (see Chap. 5.1 [2]) that $f : \Omega \to \mathbb{R}$ is a measurable function with respect to σ-algebra $\mathfrak{F}$ if and only if $f^{-1}((-\infty, a)) \in \mathfrak{F}$, for any $a \in \mathbb{R}$.

Definition 1.3 ([2]) For a given measurable space $(\Omega, \mathfrak{F})$, a measurable function $f : \Omega \to \mathbb{R}$ is named **step function** if it takes a finite number of distinct values.

Remark 1.1

(i) The simplest example of step function is the indicator function χ_M of a set M from $\mathfrak{F}$, that is

$$\chi_M(\omega) = \begin{cases} 1 & \text{if } \omega \in M \\ 0 & \text{if } \omega \in \Omega \setminus M. \end{cases}$$

(ii) Directly from Definitions 1.2 and 1.3 one obtains that for a given measurable space $(\Omega, \mathfrak{F})$, a measurable function $f : \Omega \to \mathbb{R}$ is a step function if and only if it is of the form

$$f = \sum_{i=1}^{N} x_i \chi_{M_i} \tag{1.1}$$

where $x_i \in \mathbb{R}$, $1 \le i \le N$ are the distinct values of f and $M_i = \{\omega \in \Omega \mid f(\omega) = x_i\}$.

Definition 1.4 ([1])

(i) If $(\Omega, \mathfrak{F})$ is a given measurable space, a function $\mu : \mathfrak{F} \to [0, \infty]$ is a measure on $(\Omega, \mathfrak{F})$ if:

(a) $\mu(\emptyset) = 0$;
(b) if $M_k \in \mathfrak{F}$, $k \in \mathbb{Z}_+$ are such that $M_i \cap M_j = \emptyset$ whenever $i \ne j$, then

$$\mu\left(\bigcup_{k=0}^{\infty} M_k\right) = \sum_{k=0}^{\infty} \mu(M_k).$$

In this case the triple $(\Omega, \mathfrak{F}, \mu)$ will be names *space with measure.*

(ii) If $\mu(\Omega) = 1$ then μ is a probability on $\mathfrak{F}$ and the triple $(\Omega, \mathfrak{F}, \mu)$ is named *probability space.* In this case, the subset M from $\mathfrak{F}$ is named *event* and the functions $f : \Omega \to \mathbb{R}$ which are measurable with respect to σ-algebra $\mathfrak{F}$ will be named *random variables.*

Note 1.1 In the sequel, we shall denote $(\Omega, \mathfrak{F}, \mathscr{P})$ instead of $(\Omega, \mathfrak{F}, \mu)$ whenever we refer to a probability space.

The proof of the next theorem may be found for example in Vestrup [2].

Theorem 1.1 *Let $(\Omega, \mathfrak{F}, \mu)$ be a space with measure and $f : \Omega \to \mathbb{R}$ be a measurable function such that $f(\omega) \geq 0$, for all $\omega \in \Omega$. Under these conditions there exists a sequence of step functions $\{f_n\}_{n\geq 1}$, $f_n : \Omega \to \mathbb{R}$ which have the properties:*

(i) $0 \leq f_n(\omega) \leq f_{n+1}(\omega)$ *for all* $\omega \in \Omega$;
(ii) $0 \leq \mathbb{I}(f_n) \leq \mathbb{I}(f_{n+1})$ *for all* $n \geq 1$ *where*

$$\mathbb{I}(f_n) \triangleq \sum_{i=1}^{N_n} x_i^{(n)} \mu(M_i^{(n)}) \tag{1.2}$$

(with the convention $0 \cdot \infty = 0$*), if*

$$f_n = \sum_{i=1}^{N_n} x_i^{(n)} \chi_{M_i^{(n)}}. \tag{1.3}$$

(iii) $\lim_{n\to\infty} f_n(\omega) = f(\omega)$ *for all* $\omega \in \Omega$.

It is worth mentioning that during the proof of the preceding theorem, explicit formulae of the values $x_i^{(n)}$ of the function $f_n(\omega)$ are provided (see, e.g., [2]).

Definition 1.5 Let $f : \Omega \to \mathbb{R}$ be a measurable function on the space with measure $(\Omega, \mathfrak{F}, \mu)$ such that $f(\omega) \geq 0$, for all $\omega \in \Omega$.

(a) If f is a step function we define

$$\int_\Omega f \, \mathrm{d}\mu \triangleq \sum_{i=1}^{N} x_i \mu(M_i) \tag{1.4}$$

(with the convention $0 \cdot \infty = 0$), where the number $x_i \geq 0$ and the set $M_i \in \mathfrak{F}$ are those which appear in the representation of type (1.1) of the step function f.

(b) If f is a measurable function with non-negative values we define

$$\int_\Omega f \mathrm{d}\mu \triangleq \lim_{n\to\infty} \mathbb{I}(f_n), \tag{1.5}$$

$\mathbb{I}(f_n)$ being defined as in (1.2) for the sequence of step functions f_n whose existence is guaranteed by Theorem 1.1.

(c) If $f : \Omega \to \mathbb{R}$ is a measurable function we define

$$\int_{\Omega} f \, \mathrm{d}\mu \triangleq \int_{\Omega} f^+ \, \mathrm{d}\mu - \int_{\Omega} f^- \, \mathrm{d}\mu \tag{1.6}$$

provided that at least one of $\int_\Omega f^+ \, \mathrm{d}\mu$ or $\int_\Omega f^- \, \mathrm{d}\mu$ to be finite, where

$$f^+ := \frac{1}{2}(|f| + f)$$

and

$$f^- = \frac{1}{2}(|f| - f).$$

(d) The measurable function $f : \Omega \to \mathbb{R}$ is integrable if $\int_{\Omega} f \, \mathrm{d}\mu$ defined by (1.6) is finite.

It is worth noticing that the value of the limit from (1.5) does not depend upon by the choice of the sequence of step functions $\{f_n\}$ which is satisfying

$$\lim_{n\to\infty} f_n(\omega) = f(\omega), \quad \omega \in \Omega.$$

One proofs (see [4]) that the measurable function $f : \Omega \to \mathbb{R}$ is integrable if and only if

$$\int_{\Omega} |f| \, \mathrm{d}\mu < \infty. \tag{1.7}$$

If $1 \le p < \infty$ we denote $\mathscr{L}^p(\Omega, \mathbb{R})$ the linear space of all measurable functions $f : \Omega \to \mathbb{R}$ with the property that

$$\int_{\Omega} |f|^p \, \mathrm{d}\mu < \infty. \tag{1.8}$$

Based on (1.7) and (1.8) we may conclude that a measurable function f lies in $\mathscr{L}^p(\Omega, \mathbb{R})$ if and only if f^p is integrable.

The linear space $\mathscr{L}^p(\Omega, \mathbb{R})$ is a Banach space with respect to the norm

$$\|f\|_{\mathscr{L}^p} \triangleq \left(\int_{\Omega} |f|^p \, \mathrm{d}\mu \right)^{1/p}. \tag{1.9}$$

Remark 1.2

(a) Bearing in mind the terminology introduced in Definition 1.4(b) for the case when the space with measure $(\Omega, \mathfrak{F}, \mu)$ is a probability space, we further display the terminology used in the stochastic context for the concept of integrability and the space of integrable functions. So, in the case of a random variable $f : \Omega \to \mathbb{R}$ we write $\mathbb{E}[f]$ instead of $\int_\Omega f \, \mathrm{d}\mu$, and it will be named the expectation of the random variable f. In this case (1.6) becomes

$$\mathbb{E}[f] = \mathbb{E}[f^+] - \mathbb{E}[f^-] \tag{1.10}$$

provided that at least one of the terms from right hand side to be finite.

(b) A random variable f is integrable if and only if

$$\mathbb{E}[f] < \infty. \tag{1.11}$$

(c) If $(\Omega, \mathfrak{F}, \mathscr{P})$ is a probability space, then $\mathscr{L}^p(\Omega, \mathbb{R})$ denotes the linear space of random variables $f : \Omega \to \mathbb{R}$ which are satisfying

$$\mathbb{E}[|f|^p] < \infty.$$

The norm (1.9) becomes

$$\|f\|_{\mathscr{L}^p} = \left(\mathbb{E}[|f|^p]\right)^{1/p}. \tag{1.12}$$

(d) A vector valued function $f : \Omega \to \mathbb{R}^m$ is a m-dimensional random vector if each of its components f_j, $1 \le j \le m$ is a random variable. If $1 \le p < \infty$ we denote $\mathscr{L}^p(\Omega, \mathbb{R}^m)$ the linear space of m-dimensional random vectors $f : \Omega \to \mathbb{R}^m$ with the property that each of its component f_j, $1 \le j \le m$ lies in $\mathscr{L}^p(\Omega, \mathbb{R})$.

1.1.2 *Convergence of the Sequences of Measurable Functions*

Here we recall several kinds of convergence of sequences of measurable functions defined on a given space with measure $(\Omega, \mathfrak{F}, \mu)$. In the case when $\mu(\Omega) = 1$ these definitions are applicable to the sequences of random variables.

Definition 1.6 Let $\{f_n\}_{n\ge1}$ be a sequence of measurable functions on the space with measure $(\Omega, \mathfrak{F}, \mu)$. Then:

(a) the sequence $\{f_n\}_{n\ge1}$ is **almost everywhere convergent** to a measurable function $f : \Omega \to \mathbb{R}$ if there exists a subset $\Omega_0 \subset \Omega$ with $\mu(\Omega_0) = 0$ such that

$$\lim_{n\to\infty} f_n(\omega) = f(\omega) \tag{1.13}$$

for all $\omega \in \Omega \setminus \Omega_0$.

$$\lambda(A) := \int_{\Omega} \chi_A x \, d\mathscr{P} = \int_{A} x \, d\mathscr{P}.$$

Applying Radon–Nikodym theorem (see Theorem 1.2.6 Drăgan et al. [6]) in the case of $(\Omega, \mathfrak{G}, \lambda)$ we deduce that there exists an unique $(mod\ \mathscr{P})$ integrable function $y : \Omega \to \mathbb{R}$ measurable with respect to the σ-algebra $\mathfrak{G}$, and satisfying

$$\int_{A} y \, d\mathscr{P} = \int_{A} x \, d\mathscr{P}, \quad \text{for all} \quad A \in \mathfrak{G}.$$

The random variable with the properties stated before will be named conditional expectation with respect to the σ-algebra $\mathfrak{G}$ of the random variable x, and it will be denoted $\mathbb{E}[x \mid \mathfrak{G}]$. If $x, x_1, x_2, \ldots, x_m$ are random variables then we shall write $\mathbb{E}[x \mid (x_1, x_2, \ldots, x_m)]$ instead of $\mathbb{E}[x \mid \sigma[x_1, x_2, \ldots, x_m]]$.

If x is an integrable random variable on $A \in \mathfrak{F}$ with $\mathscr{P}(A) > 0$, then

$$\mathbb{E}[x \mid A] := \mathbb{E}[x \mid \chi_A] = \frac{1}{\mathscr{P}(A)} \int_{A} x \, d\mathscr{P} \quad a.s.$$

The main properties of the conditional expectation used in the next chapters are listed below.

Theorem 1.7 ([4]) *Let x, y be integrable random variables on the probability space $(\Omega, \mathfrak{F}, \mathscr{P})$ and $\mathfrak{G}, \mathfrak{H} \subset \mathfrak{F}$ be two σ-algebras. The following hold:*

(i) $\mathbb{E}[\mathbb{E}[x \mid \mathfrak{G}]] = \mathbb{E}[x]$;
(ii) $\mathbb{E}[\mathbb{E}[x \mid \mathfrak{G}] \mid \mathfrak{H}] = \mathbb{E}[x \mid \mathfrak{H}]$ *a.s. if* $\mathfrak{H} \subset \mathfrak{G}$;
(iii) $\mathbb{E}[xy \mid \mathfrak{G}] = y\mathbb{E}[x \mid \mathfrak{G}]$ *if y is measurable with respect to* $\mathfrak{G}$;
(iv) *if x is independent of σ-algebra $\mathfrak{G}$, then* $\mathbb{E}[x \mid \mathfrak{G}] = \mathbb{E}[x]$.

1.1.5 Stochastic Processes

Let $(\Omega, \mathfrak{F}, \mathscr{P})$ be a given probability space and $\mathscr{I} \subset \mathbb{R}$ be an interval.

Definition 1.8 A function $(t, \omega) \to x(t, \omega) : \mathscr{I} \times \Omega \to \mathbb{R}^m$ with the property that for each $t \in \mathscr{I}$, $x(t, \cdot)$ is a random vector will be named stochastic process.

Usually, the dependence with respect to the argument ω of a stochastic process is suppressed, being implicit. That is why, we shall write $\mathbf{x} = \{x(t)\}_{t \in \mathscr{I}}$ when we refer to a stochastic process for which the parameter t takes values in $\mathscr{I}$.

Definition 1.9 We say that a stochastic process $\mathbf{x} = \{x(t)\}_{t \in \mathscr{I}}$ is:

(a) **continuous** if for almost all $\omega \in \Omega$, the function $t \to (t, \omega) : \mathscr{I} \to \mathbb{R}^m$ is a continuous function;

(b) **right continuous** if for almost all $\omega \in \Omega$, the function $t \to (t, \omega) : \mathscr{J} \to \mathbb{R}^m$ is right continuous;
(c) **continuous in probability** if for all $t_0 \in \mathscr{J}$ and for any $\{t_n\}_{n\geq 1} \subset \mathscr{J}$ such that $\lim_{n\to\infty} t_n = t_0$ it follows that $x(t_n) \xrightarrow{\mathscr{P}} x(t_0)$;
(d) **measurable stochastic process** if $(t, \omega) \to x(t, \omega) : \mathscr{J} \times \Omega \to \mathbb{R}^m$ is a measurable vector on the measurable space $(\mathscr{J} \times \Omega, \mathfrak{B}(\mathscr{J}) \times \mathfrak{F})$, $\mathfrak{B}(\mathscr{J})$ being the σ-algebra of Borel subset of $\mathscr{J}$.

Let $\mathfrak{M} = \{\mathscr{M}_t\}_{t\in\mathscr{J}}$ be a family of σ-algebras $\mathscr{M}_t \subset \mathfrak{F}$, $t \in \mathscr{J}$, such that $\mathscr{M}_{t_1} \subset \mathscr{M}_{t_2}$, if $t_1 \leq t_2$.

Definition 1.10 We shall say that the stochastic process $\mathbf{x} = \{x(t)\}_{t\in\mathscr{J}}$ is named non-anticipative with respect to the family of the σ-algebras $\mathfrak{M}$, if

(a) $\{x(t)\}_{t\in\mathscr{J}}$ is a measurable stochastic process;
(b) for each $t \in \mathscr{J}$, $x(t, \cdot)$ is $\mathscr{M}_t$-measurable

In the case when (b) holds, the stochastic process $\{x(t)\}_{t\in\mathscr{J}}$ is named **adapted** to the family of the σ-algebras $\mathfrak{M}$.

If $1 \leq p < \infty$, $\mathscr{L}^p(\mathscr{J} \times \Omega, \mathbb{R}^m)$ denotes the linear space of the measurable stochastic processes $x : \mathscr{J} \times \Omega \to \mathbb{R}^m$ with the property that $\mathbb{E}\left[\int_{\mathscr{J}} |x(t)|^p \, \mathrm{d}t\right] < \infty$.

$\mathscr{L}^p_{\mathfrak{M}}(\mathscr{J} \times \Omega, \mathbb{R}^m)$ denotes the linear subspace of the measurable stochastic processes $\mathbf{x} = \{x(t)\}_{t\in\mathscr{J}} \in \mathscr{L}^p(\mathscr{J} \times \Omega, \mathbb{R}^m)$ which is adapted to the family of the σ-algebras $\mathfrak{M}$.

Theorem 1.8 (see Theorem 1.6.1 [6]) *If $\mathfrak{M} = \{\mathscr{M}_t\}_{t\in\mathscr{J}}$ is a family of σ-algebras. If for all t $\mathscr{M}_t$ contains all subsets $\mathscr{A} \in \mathfrak{F}$ with $\mathscr{P}(\mathscr{A}) = 0$ then $\mathscr{L}^p_{\mathfrak{M}}(\mathscr{J} \times \Omega, \mathbb{R}^m)$ is a closed subspace in $\mathscr{L}^p(\mathscr{J} \times \Omega, \mathbb{R}^m)$.*

Corollary 1.2 *If $\mathfrak{M} = \{\mathscr{M}_t\}_{t\in\mathscr{J}}$ is an increasing family of σ-algebras and if for any $t \in \mathscr{J}$, σ-algebra $\mathscr{M}_t$ contains all subsets $\mathscr{A} \in \mathfrak{F}$ with $\mathscr{P}(\mathscr{A}) = 0$ then $\mathscr{L}^2_{\mathfrak{M}}(\mathscr{J} \times \Omega, \mathbb{R}^m)$ is a Hilbert space with respect to the inner product*

$$\langle \mathbf{x}, \mathbf{y} \rangle_{\mathscr{L}^2} \triangleq \mathbb{E}\left[\int_{\mathscr{J}} y^\top(t) x(t) \, \mathrm{d}t\right]$$

for all $\mathbf{x} = \{x(t)\}_{t\in\mathscr{J}}$, $\mathbf{y} = \{y(t)\}_{t\in\mathscr{J}}$ from $\mathscr{L}^2_{\mathfrak{M}}(\mathscr{J} \times \Omega, \mathbb{R}^m)$.

Often in the sequel, we shall denote $\mathscr{L}^2_{\mathfrak{M}}(\mathscr{J}, \mathbb{R}^m)$ instead of $\mathscr{L}^2_{\mathfrak{M}}(\mathscr{J} \times \Omega, \mathbb{R}^m)$ because the dependence with respect to the variable Ω of the stochastic process is implicit.

Let $\mathfrak{T} = \{\tau_k\}_{k\geq 0} \subset \mathbb{R}_+$ such that $0 = \tau_0 < \tau_1 < \tau_2 < \ldots < \tau_k < \tau_{k+1} < \ldots$ be a partition of the semiaxis $\mathbb{R}_+$.

Let $\mathfrak{M} = \{\mathscr{M}_t\}_{t\in\mathbb{R}_+}$ be a non-decreasing family of σ-algebras such that for each $t \in \mathbb{R}_+$, $\mathscr{M}_t \subset \mathfrak{F}$ and contains all subsets $\mathscr{A} \in \mathfrak{F}$ with $\mathscr{P}(\mathscr{A}) = 0$.

Definition 1.11

(a) If $0 \le \kappa_1 \le \kappa_2$ are natural numbers, $\ell^2_{\mathfrak{M}}\{\kappa_1, \kappa_2; \mathbb{R}^m\}$ stands for the linear space of sequences $\mathbf{z} = \{z(k)\}_{\kappa_1\le k\le\kappa_2}$ of random vectors with the property that for each k, $z(k) : \Omega \to \mathbb{R}^m$ is such that $\sigma[z(k)] \subset \mathscr{M}_{\tau_k}$ and $\mathbb{E}[|z(k)|^2] < \infty$.
(b) $\ell^2_{\mathfrak{M}}\{\mathbb{Z}_+; \mathbb{R}^m\}$ denotes the linear space of the sequences $\mathbf{z} = \{z(k)\}_{k\in\mathbb{Z}_+}$ of random vectors having the properties: $z(k) : \Omega \to \mathbb{R}^m$, $\sigma(z_k) \subset \mathscr{M}_{\tau_k}$ for all $k \in \mathbb{Z}_+$ and $\sum_{k=0}^{\infty} \mathbb{E}[|z(k)|^2] < \infty$.

One may check that $\ell^2_{\mathfrak{M}}\{\mathbb{Z}_+; \mathbb{R}^m\}$ is a Hilbert space with respect to the inner product

$$\langle \mathbf{z}_1, \mathbf{z}_2 \rangle \stackrel{\Delta}{=} \sum_{k=0}^{\infty} \mathbb{E}[z_1^{\mathsf{T}}(k) z_2(k)],$$

for all $\mathbf{z}_j = \{z_j(k)\}_{k\in\mathbb{Z}_+} \in \ell^2_{\mathfrak{M}}\{\mathbb{Z}_+; \mathbb{R}^m\}$, $j = 1, 2$.

1.1.6 Wiener Processes and Markov Processes

Throughout in this subsection $\mathscr{J} = \mathbb{R}_+$.

A. **Wiener processes.**

Definition 1.12 A Brownian motion or one-dimensional Wiener process $\{\beta(t)\}_{t\in\mathbb{R}_+}$ is a stochastic process with the properties:

(a) $\beta(0) = 0$ a.s.;
(b) $\{\beta(t)\}_{t\in\mathbb{R}_+}$ is a stochastic process with independent increments, that means, for any sequence $\{t_k\}_{k\ge 0}$ with $0 \le t_0 < t_1 < \ldots < t_j < \ldots$ the random variables $\beta(t_{k+1}) - \beta(t_k)$, $k \ge 0$, are independent;
(c) if $0 \le s < t$ then $\beta(t) - \beta(s)$ is a random variable with normal distribution

$$\mathbb{E}[\beta(t) - \beta(s)] = \mu(t - s)$$

and

$$\mathbb{E}[(\beta(t) - \beta(s))^2] = \lambda(t - s),$$

μ and λ being real constants with $\lambda > 0$.

If $\mu = 0$ and $\lambda = 1$, $\{\beta(t)\}_{t\ge 0}$ is named **standard Brownian motion** or **one-dimensional standard Wiener process.**

If $\{\beta(t)\}_{t\geq 0}$ is a Brownian motion, $\{\widetilde{\beta}(t)\}_{t\geq 0}$ defined by

$$\widetilde{\beta}(t) \triangleq \frac{1}{\sqrt{\lambda}}(\beta(t) - t\mu), \quad t \geq 0,$$

is a standard Brownian motion or one-dimensional standard Wiener process.

Definition 1.13

(a) A r-dimensional Wiener process is a stochastic process $\{w(t)\}_{t\geq 0}$ with $w(t) = (w_1(t)\ w_2(t)\ \dots\ w_r(t))^\top$, where for which $1 \leq i \leq r$, $\{w_i(t)\}t \geq 0$ is a Brownian motion with $\mu_i = 0$ and $\lambda_i > 0$;
(b) a r-dimensional stochastic process $\{w(t)\}_{t\geq 0}$ ($w(t) = (w_1(t)\ w_2(t)\ \dots\ w_r(t))^\top$) is named r-dimensional **standard Wiener process** if for each $1 \leq i \leq r$, $\{w_i(t)\}_{t\geq 0}$ is one-dimensional standard Wiener process and the σ-algebras $\sigma[w_i(t);\ t \geq 0]$, $1 \leq i \leq r$, are independent.

Remark 1.3

(a) If $\{w(t)\}_{t\geq 0}$ is a r-dimensional Wiener process in the sense of Definition 1.13 (a), then $\mathbb{E}[w(t) - w(s)] = 0$ and $\mathbb{E}[(w(t) - w(s))(w(t) - w(s))^\top] = (t - s)\Lambda$, for all $t \geq s \geq 0$, where $\Lambda > 0$.
(b) If $\{w(t)\}_{t\geq 0}$ is a r-dimensional standard Wiener process, then $\mathbb{E}[w(t) - w(s)] = 0$ and $\mathbb{E}[(w(t) - w(s))(w(t) - w(s))^\top] = I_r(t - s)$, for all $t \geq s \geq 0$, where I_r being the identity matrix of size $r \times r$.
(c) In the developments from the next chapters the Wiener process in the sense of Definition 1.13(a) will be used to model exogenous disturbances as well as in modeling some random parametric uncertainties. Λ is viewed as a measure of the intensity of those disturbances which will be taken into account in the analysis of the robustness of stability and in synthesis of some control laws which achieve robust stability of the controlled system under consideration.

For each $t \geq 0$ we denote $\widetilde{\mathscr{F}}_t \subset \mathfrak{F}$ the smallest σ-algebra containing all subsets $\mathscr{A} \in \mathfrak{F}$ with $\mathscr{P}(\mathscr{A}) = 0$ and with respect to which all random vectors $\{w(s)\}_{t\geq s\geq 0}$ are measurable.

We set $\mathscr{U}_t = \sigma[w(t + h) - w(t);\ h \geq 0]$. One proves (see, e.g., Theorem 1.7.1 from Drăgan [6]) that for each $t \geq 0$ $\widetilde{\mathscr{F}}_t$ is independent of $\mathscr{U}_t$.

B. Standard homogeneous Markov process.

Definition 1.14 A family of matrices $\{P(t)\}_{t>0}$ where $P(t) = (p_{ij}(t))_{1\leq i,j\leq N}$ is named transition semigroup if:

(i) for each $t > 0$, $P(t)$ is a stochastic matrix, that is $0 \leq p_{ij}(t) \leq 1$ and $\sum_{k=1}^{N} p_{ik}(t) = 1$, for all $i, j \in \mathscr{N} = \{1, 2, \dots, N\}$;
(ii) $P(t + s) = P(t)P(s)$, for all $t > 0$, $s > 0$.

More details regarding semigroup theory may be found in Menaldi [7].

Definition 1.15 A stochastic process $\{\eta(t)\}_{t\geq 0}$ is a **standard homogeneous Markov process** with the state space the finite set $\mathcal{N}$ and the transition semigroup $\{P(t)\}_{t>0}$ $(P(t) = (p_{ij}(t))_{i,j\in\mathcal{N}})$ if:

(i) $\eta(t) : \Omega \to \mathcal{N}$, for all $t \geq 0$;
(ii) $\mathscr{P}\{\eta(t+h) = j \mid \eta(s),\ t \geq s\} = p_{\eta(t)j}(h)$ a.s. for all $t \geq 0$, $h > 0$, $1 \leq j \leq N$;
(iii) $\lim_{t\to 0^+} P(t) = I_N$;
(iv) $\{\eta(t)\}_{t\geq 0}$ is a right continuous stochastic process.

From the previous definition one sees that a standard homogeneous Markov process is a triple $(\{\eta(t)\}_{t\geq 0}, \{P(t)\}_{t>0}, \mathcal{N})$ where $\{P(t)\}_{t>0}$ is a given transition semigroup, $\mathcal{N}$ is a given finite set and $\{\eta(t)\}_{t\geq 0}$ is any stochastic process satisfying condition (i)–(iv) from above.

The main properties of a standard homogeneous Markov process are listed below. The proof may be found in G. Ciucu [8], Chung [9], and Doob [10].

Theorem 1.9 *A standard homogeneous Markov process* $(\{\eta(t)\}_{t\geq 0}, \{P(t)\}_{t>0}, \mathcal{N})$ *has the properties:*

(i) $\mathscr{P}(\{\eta(t+h) = j \mid \eta(t) = i\} = p_{ij}(h)$, *for all* $i, j \in \mathcal{N}$, $h > 0$, $t \geq 0$ *if* $\mathscr{P}\{\eta(t) = i\} > 0$;
(ii) if x *is a bounded random variable measurable with respect to the* σ*-algebra* $\sigma[\eta(s),\ s \geq t]$ *then* $\mathbb{E}[x \mid \eta(u), t \geq u] = \mathbb{E}[x \mid \eta(t)]$ *a.s.;*
(iii) $\{\eta(t)\}_{t\geq 0}$ *is a stochastic process which is continuous in probability;*
(iv) $p_{ii}(t) > 0$, *for all* $i \in \mathcal{N}$, $t > 0$;
(v) $\lim_{t\to\infty} P(t)$ *is well defined;*
(vi) there exists a constant matrix $Q \in \mathbb{R}^{N\times N}$ *such that* $P(t) = e^{Qt}$, $t > 0$ *and its elements* q_{ij} *have the properties:* $q_{ij} \geq 0$ *if* $i \neq j$ *and* $\sum_{k=1}^{N} q_{ik} = 0$, *for all* $i, j \in \mathcal{N}$.

1.1.7 Stochastic Integrals

In this monograph we shall use stochastic integrals of Itô type for stochastic processes which are non-anticipative with respect to a family of σ-algebras generated by a triple of stochastic processes defined by a given probability space $(\Omega, \mathfrak{F}, \mathscr{P})$: a r_1-dimensional standard Wiener process $\{w(t)\}_{t\geq 0}$, a standard homogeneous Markov process $\{\eta(t)\}_{t\geq 0}$, and a sequence $\{w_d(k)\}_{k\geq 0}$ of r_2-dimensional independent random vectors.

Along of the family of σ-algebras $\widetilde{\mathscr{F}} = \{\widetilde{\mathscr{F}}_t\}_{t\geq 0} \subset \mathfrak{F}$ generated by the r_1-dimensional standard Wiener process $\{w(t)\}t \geq 0$ we consider the families of σ-algebras $\mathscr{G} = \{\mathscr{G}_t\}_{t\geq 0}$ and $\widehat{\mathscr{F}} = \{\widehat{\mathscr{F}}_t\}_{t\geq 0}$ where for each $t \geq 0$, $\mathscr{G}_t = \sigma[\eta(s);\ t \geq s \geq 0]$ and $\widehat{\mathscr{F}}_t = \sigma[w_d(k);\ t > \tau_k]$.

In the definition of the σ-algebras $\{\widehat{\mathscr{F}}_t\}_{t\geq 0}$, a given partition $\mathfrak{T} = \{\tau_k,\ k \in \mathbb{Z}_+\}$ of the semiaxis $\mathbb{R}_+$ is involved. Assume that for each $t \geq 0$, σ-algebras $\widetilde{\mathscr{F}}$, $\widehat{\mathscr{F}}$, $\mathscr{G}_t$ are mutually independent.

Throughout in this book $\mathfrak{H} = \{\mathscr{H}_t\}_{t\geq 0}$, where for each $t \geq 0$, $\mathscr{H}_t \subset \mathfrak{F}$ stands for the smallest σ-algebra which contains the σ-algebras $\widetilde{\mathscr{F}}_t$, $\widehat{\mathscr{F}}_t$ and $\mathscr{G}_t$ augmented with all subsets $\mathscr{A} \in \mathfrak{F}$ with $\mathscr{P}(\mathscr{A}) = 0$. Following the line of the proof of Theorem 1.9.1 from Drăgan et al. [6] one can prove that for each $t \geq 0$, the σ-algebra $\sigma[w(t+h) - w(t);\ h > 0]$ is independent of the σ-algebra $\mathscr{H}_t$. If $[a, b] \subset [0, \infty)$, then the family of the σ-algebra $\mathfrak{H} = \{\mathscr{H}_t\}_{t\in[a,b]}$ has the properties used in the theory of the stochastic integrals of Itô type.

So, based on the developments from Friedman [11] (see also G. Ciucu [8], I. Ghihman [12], A. Lipser [13], Neveu [14], H. Schneider [15]), one obtains that if $f \in \mathscr{L}^2_{\mathfrak{H}}([a, b]; \mathbb{R}^{r_1})$ one can define the stochastic integral of Itô type $\int_a^b f(t)dw(t)$. A list of the usual properties of Itô integrals may be found in Sect. 1.9 from Drăgan et al. [6].

1.2 Systems of Stochastic Linear Differential Equations Controlled by Impulses

In this section we make a brief overview of the main types of systems of stochastic linear differential equations controlled by impulses which will be involved in the developments of the next chapters of the book.

1.2.1 *Systems of Linear Differential Equations of Itô Type Controlled by Impulses*

Let us consider the controlled systems having the state-space representation of the form:

$$\mathrm{d}x(t) = (A_0(t)x(t) + G_0(t)v(t))\mathrm{d}t + \sum_{j=1}^{r_1}(A_j(t)x(t) + G_j(t)v(t))\mathrm{d}w_j(t),$$

$$\tau_k < t \leq \tau_{k+1} \tag{1.16a}$$

$$x(\tau_k^+) = A_{d0}(\tau_k)x(\tau_k) + G_{d0}(\tau_k)v_d(k) + B_{d0}(\tau_k)u(k)$$
$$+ \sum_{j=1}^{r_2}(A_j(\tau_k)x(\tau_k) + G_{dj}(\tau_k)v_d(k) + B_{dj}(\tau_k)u(k))w_{dj}(k), \quad k \in \mathbb{Z}_+ \tag{1.16b}$$

$$y(\tau_k) = C_0(\tau_k)x(\tau_k) + \sum_{j=1}^{r_2} w_{dj}(k)C_j(\tau_k)x(\tau_k) + D(\tau_k)v_d(k), \quad k \in \mathbb{Z}_+ \tag{1.17}$$

$$z(t) = C_z(t)x(t), \quad t \in \mathbb{R}_+ \tag{1.18a}$$

$$z_d(\tau_k) = C_{dz}(\tau_k)x(\tau_k) + D_{dz}(\tau_k)u(k), \quad k \in \mathbb{Z}_+. \tag{1.18b}$$

The real numbers τ_k, $k \in \mathbb{Z}_+$, are such that

$$0 = \tau_0 < \tau_1 < \ldots < \tau_k < \tau_{k+1} < \ldots$$

determining a partition of the real semiaxis $\mathbb{R}_+$.
Regarding the stochastic processes $\{w_j(t)\}_{t\geq 0}$ and $\{w_{dj}(k)\}_{k\in\mathbb{Z}_+}$ involved in (1.16) and (1.17) we make the assumption:

Assumption 1.1

(a) $\{w(t)\}_{t\geq 0}$ $(w(t) = (w_1(t)\ w_2(t)\ \ldots\ w_{r_1}(t))^\top)$ is a r_1-dimensional standard Wiener process defined on a given probability space $(\Omega, \mathfrak{F}, \mathscr{P})$.
(b) $\{w_d(k)\}_{k\in\mathbb{Z}_+}$ $(w_d(k) = (w_{d1}(k)\ w_{d2}(k)\ \ldots\ w_{dr_2}(k))^\top)$ is a sequence of r_2-dimensional independent random vectors defined on the same probability space $(\Omega, \mathfrak{F}, \mathscr{P})$ with zero mean and $\mathbb{E}[w_d(k)w_d^\top(k)] = I_{r_2}$, $k \in \mathbb{Z}_+$.
(c) $\{w(t)\}_{t\geq 0}$, $\{w_d(k)\}_{k\in\mathbb{Z}_+}$ are independent stochastic processes.

In (1.16)–(1.18) $x(t) \in \mathbb{R}^n$ denotes the vector of the state parameters at time t and $u(k) \in \mathbb{R}^{n_u}$ are vectors of control parameters at instance time τ_k, $k \in \mathbb{Z}_+$. The sequence $\{y(\tau_k)\}_{k\in\mathbb{Z}_+} \subset \mathbb{R}^{n_y}$ denotes the sequence of measurements available to feed a controller. In (1.18), $z(t) \in \mathbb{R}^{m_z}$, $t \geq 0$, is the continuous-time controlled output, while $z_d(\tau_k)$, $k \in \mathbb{Z}_+$ is the discrete-time controlled output.
In (1.16), $v(t) \in \mathbb{R}^{n_v}$, $t \in \mathbb{R}_+$ and $v_d(k) \in \mathbb{R}^{n_{v_d}}$, $k \in \mathbb{Z}_+$ are stochastic processes which model some exogenous disturbances which affect the behavior of the controlled system. For each $t > 0$, we denote $\mathscr{F}_t \subset \mathfrak{F}$ the σ-algebra generated by the random vectors $w(s)$, $0 \leq s \leq t$ and $w_d(k)$, $k \geq 0$ such that $\tau_k < t$. For $t = 0$, $\mathscr{F}_0 \subset \mathfrak{F}$ is the trivial σ-algebra formed by Ω and $\emptyset$. For each $t \geq 0$ the σ-algebra $\mathscr{F}_t$ is augmented by all subsets $\mathfrak{A} \in \mathfrak{F}$ with $\mathscr{P}(\mathfrak{A}) = 0$.
Regarding the solutions of the stochastic differential equation (1.16) we prove:

Proposition 1.1 *Assume:*

(a) Assumption 1.1 is fulfilled;
(b) $t \to (A_j(t), G_j(t)) : \mathbb{R}_+ \to \mathbb{R}^{n\times n} \times \mathbb{R}^{n\times n_v}$, $0 \leq j \leq r_1$ *are bounded and continuous matrix-valued functions;*
(c) $\{A_{dj}(\tau_k)\}_{k\in\mathbb{Z}_+} \subset \mathbb{R}^{n\times n}$, $\{B_{dj}(\tau_k)\}_{k\in\mathbb{Z}_+} \subset \mathbb{R}^{n\times n_u}$, $\{G_{dj}(\tau_k)\}_{k\in\mathbb{Z}_+} \subset \mathbb{R}^{n\times n_{v_d}}$, $0 \leq j \leq r_2$ *are bounded matrix-valued sequences;*
(d) $v(\cdot) \in \mathscr{L}^2_{\mathscr{F}}([0, T], \mathbb{R}^{n_v})$ *for all* $T > 0$ *while for each* $k \in \mathbb{Z}_+$ $v_d(k) : \Omega \to \mathbb{R}^{n_{v_d}}$ *is* $\mathscr{F}_{\tau_k}$*-measurable such that* $\mathbb{E}[|v_d(k)|^2] < \infty$;
(e) $\lim_{k\to\infty} \tau_k = +\infty$.

Let $t_0 \in \mathbb{R}_+$ *and* $k_0 \in \mathbb{Z}_+$ *be such that* $t_0 \in (\tau_{k_0-1}, \tau_{k_0}]$. *Under these conditions, for each random vector* $x_0 : \Omega \to \mathbb{R}^n$ *which is* $\mathscr{F}_{t_0}$*-measurable with* $\mathbb{E}[|x_0|^2] < \infty$ *and for each sequence* $\mathbf{u}_{k_0} = \{u(k)\}_{k\geq k_0} \subset \mathbb{R}^{n_u}$ *with the property that for each* $k \geq k_0$,

$u(k)$ is $\mathscr{F}_{\tau_k}$-measurable and $\mathbb{E}[|u(k)|^2] < \infty$, the system of stochastic linear differential equations (1.16) has a unique solution $x(t) = x(t; t_0, x_0, \mathbf{u}_{k_0})$ with the properties:

(i) $x(\cdot)$ is almost surely (a.s.) continuous in any $t \geq t_0, t \neq \tau_k$ and is left continuous in $t = \tau_k$, $k \geq k_0$;
(ii) $x(\cdot; t_0, x_0, \mathbf{u}_{k_0}) \in \mathscr{L}^2_{\mathscr{F}}([t_0, T], \mathbb{R}^n)$, for all $T > t_0$;
(iii) $x(t_0) = x_0$ and $\lim_{t \searrow \tau_k} x(t) = x(\tau_k^+)$ for all $k \geq k_0$.

Proof Since (1.16a) is a non-homogeneous stochastic linear differential equation with the forcing term $v(\cdot)$ satisfying the assumption (d) we may infer that this equation has a unique solution $x(t) = x(t; t_0, x_0)$, $t \geq t_0$ which has the properties:

(α) $t \to x(t)$ is a.s. continuous in any $t \geq t_0$;
(β) $x(\cdot) \in \mathscr{L}^2_{\mathscr{F}}([t_0, T], \mathbb{R}^n)$ for all $T > t_0$;
(γ) $x(t_0) = x_0$.

The value $x(\tau_{k_0})$ of the solution $x(\cdot; t_0, x_0)$ is used in Eq. (1.16b) (written for k replaced by k_0) to compute the value of $x(\tau_{k_0}^+)$. One sees that under the considered assumptions $x(\tau_{k_0}^+)$ is $(\mathscr{F}_{\tau_{k_0}} \vee \sigma[w_d(k_0)])$-measurable and $\mathbb{E}[|x(\tau_{k_0}^+)|^2] < \infty$. Taking $x(\tau_{k_0}^+)$ as an initial condition at time $t = \tau_{k_0}$ we deduce that (1.16a) has a unique solution $x(t; \tau_{k_0}, x(\tau_{k_0}^+))$, $t \geq \tau_{k_0}$ which has the properties:

(α') $t \to x(t; \tau_{k_0}, x(\tau_{k_0}^+))$ is a.s. continuous in any $t \geq \tau_{k_0}$;
(β') $x(\cdot; \tau_{k_0}, x(\tau_{k_0}^+)) \in \mathscr{L}^2_{\mathscr{F}}([\tau_{k_0}, T], \mathbb{R}^n)$ for all $T > \tau_{k_0}$;
(γ') $x(\tau_{k_0}; \tau_{k_0}, x(\tau_{k_0}^+)) = x(\tau_{k_0}^+)$.

The value $x(\tau_{k_0+1}; \tau_{k_0}, x(\tau_{k_0}^+))$ is then used in Eq. (1.16b) written for k replaced by $k_0 + 1$ to compute $x(\tau_{k_0+1}^+)$. Under the considered assumptions $x(\tau_{k_0+1}^+)$ is a random vector which is $(\mathscr{F}_{\tau_{k_0+1}} \vee \sigma[w_d(k_0 + 1)])$-measurable and $\mathbb{E}[|x(\tau_{k_0+1}^+)|^2] < \infty$. Taking $x(\tau_{k_0+1}^+)$ as initial value at $t = \tau_{k_0+1}$ one considers $x(t; \tau_{k_0+1}, x(\tau_{k_0+1}^+))$, $t \geq \tau_{k_0+1}$, the unique solution of (1.16a) which has the properties similar to (α') – (γ'), restricted to the interval $[\tau_{k_0+1}, \infty)$. Proceeding in this way we may construct step by step the solution $x(\cdot; t_0, x_0, \mathbf{u}_{k_0})$ of (1.16) as:

$$x(t; t_0, x_0, \mathbf{u}_{k_0}) := \begin{cases} x(t; t_0, x_0), & \text{if } \; t \in [t_0, \tau_{k_0}) \\ x(t; \tau_k, x(\tau_k^+)), & \text{if } \; t \in (\tau_k, \tau_{k+1}], \quad k \geq k_0. \end{cases}$$

The dependence with respect to the terms of the sequence $\mathbf{u}_{k_0}$ is realized by $x(\tau_k^+)$ via (1.16b). Thus the proof is complete.

Remark 1.4 In (1.16)–(1.18) the controls $u(k)$ act as impulses at some instances time τ_k, $k = 0, 1, \ldots$. In the sequel, the controlled systems of type (1.16)–(1.18) will be named **impulsive controlled linear stochastic systems** (ICLSS) of Itô type and τ_k will be named impulsive time instances.

1.2.2 Impulsive Controlled Linear Stochastic Systems Subject to Jump Markov Perturbations

We consider the controlled linear stochastic system having the state-space representation described by

$$\dot{x}(t) = A(t, \eta(t))x(t) + G(t, \eta(t))v(t), \quad \tau_k < t \leq \tau_{k+1} \tag{1.19a}$$

$$x(\tau_k^+) = A_d(\tau_k, \eta(\tau_k))x(\tau_k) + G_d(\tau_k, \eta(\tau_k))v_d(k) + B_d(\tau_k, \eta(\tau_k))u(k), \quad k \in \mathbb{Z}_+ \tag{1.19b}$$

$$y(\tau_k) = C(\tau_k, \eta(\tau_k))x(\tau_k) + D(\tau_k, \eta(\tau_k))v_d(k), \quad k \in \mathbb{Z}_+ \tag{1.20}$$

$$z(t) = C_z(t, \eta(t))x(t), \quad t \in \mathbb{R}_+ \tag{1.21a}$$

$$z_d(\tau_k) = C_{dz}(\tau_k, \eta(\tau_k))x(\tau_k) + D_{dz}(\tau_k, \eta(\tau_k))u(k), \quad k \in \mathbb{Z}_+. \tag{1.21b}$$

In (1.19)–(1.21), $x(t) \in \mathbb{R}^n$, $y(\tau_k) \in \mathbb{R}^{n_y}$, $(z(t), z_d(\tau_k)) \in \mathbb{R}^{n_z} \times \mathbb{R}^{n_{z_d}}$, $u(k) \in \mathbb{R}^{n_u}$ are state parameters, the measured output, the controlled outputs and the control parameters, respectively, and $\{\eta(t)\}_{t\geq 0}$ is a standard homogeneous Markov process defined on a probability space $(\Omega, \mathfrak{F}, \mathscr{P})$ taking values in the finite set $\mathcal{N} = \{1, 2, \ldots, N\}$ and having the transition semigroup $P(t) = e^{Qt}$, $t \geq 0$. The elements q_{ij} of the generator matrix Q satisfy the following conditions:

$$q_{ij} \geq 0 \text{ if } i \neq j, \qquad \sum_{l=1}^{N} q_{il} = 0, \text{ for all } i, j \in \mathcal{N}. \tag{1.22}$$

For each $t \geq 0$, we denote $\mathscr{G}_t$ the σ-algebra generated by the random variables $\eta(s)$, $0 \leq s \leq t$ and augmented by all sets $\mathfrak{A} \in \mathfrak{F}$ with $\mathscr{P}(\mathfrak{A}) = 0$. Whenever $\eta(t) = i \in \mathcal{N}$ we shall write $A(t, i)$, $G(t, i)$ and so on, instead of $A(t, \eta(t))$, $G(t, \eta(t))$ and so on. Regarding the trajectories of the system (1.19a) we have:

Proposition 1.2 *Assume:*

(a) $t \to (A(t, i), G(t, i)) : \mathbb{R}_+ \to \mathbb{R}^{n\times n} \times \mathbb{R}^{n\times n_v}$, $i \in \mathcal{N}$ *are bounded and continuous matrix-valued functions;*

(b) $\{A_d(\tau_k, i)\}_{k\in\mathbb{Z}_+} \subset \mathbb{R}^{n\times n}$, $\{G_d(\tau_k, i)\}_{k\in\mathbb{Z}_+} \subset \mathbb{R}^{n\times n_{v_d}}$, $\{B_d(\tau_k, i)\}_{k\in\mathbb{Z}_+} \subset \mathbb{R}^{n\times n_u}$, $i \in \mathcal{N}$, *are bounded matrix-valued sequences;*

(c) $v(\cdot) \in \mathscr{L}^2_{\mathscr{G}}([0, T], \mathbb{R}^{n_v})$ *for all* $T > 0$ *while for each* $k \in \mathbb{Z}_+$, $v_d(k) : \Omega \to \mathbb{R}^{n_{v_d}}$ *is a random vector,* $\mathscr{G}_{\tau_k}$*-measurable with* $\mathbb{E}[|v_d(k)|^2] < \infty$;

(d) $\lim_{k\to\infty} \tau_k = \infty$.

Let $t_o \in \mathbb{R}_+$ *and* $k_o \in \mathbb{Z}_+$ *be such that* $t_0 \in (\tau_{k_0-1}, \tau_{k_0}]$. *Under these conditions for each random vector* $x_0 : \Omega \to \mathbb{R}^n$ *which is* $\mathscr{G}_{t_0}$*-measurable and satisfies* $\mathbb{E}[|x_0|^2] <$

∞ and for each sequence $\mathbf{u}_{k_0} = \{u(k)\}_{k \geq k_0} \subset \mathbb{R}^{n_u}$ with the properties that for each $k \geq k_0$, $u(k)$ is a random vector $\mathcal{G}_{\tau_k}$-measurable such that $\mathbb{E}[|u(k)|^2] < \infty$, the system of linear differential equations (1.19) has a unique solution $x(t) = x(t; t_0, x_0, \mathbf{u}_{k_0})$, $t \geq t_0$, which has the properties:

(i) $x(\cdot)$ is a.s. continuous in any $t \geq t_0$, $t \neq \tau_k$ and it is left continuous in each $t = \tau_k$, $k \geq k_0$;
(ii) $x(\cdot) \in \mathcal{L}^2_{\mathcal{G}}([t_0, T], \mathbb{R}^n)$ for all $T > t_0$;
(iii) $x(t_0) = x_0$ and for each $k \geq k_0$, $\lim_{t \searrow \tau_k} x(t) = x(\tau_k^+)$.

Proof The proof follows the same line as the proof of Proposition 1.1. In this case, for each $k \geq k_0$, $x(\tau_{k_0}^+)$ is the initial condition of the restriction of the solution $x(t; t_0, x_0, \mathbf{u}_{k_0})$ to the interval $(\tau_k, \tau_{k+1}]$.

Remark 1.5

(a) Since in the system (1.19)–(1.21) the controls $u(k)$ act as some impulses at discrete-time instances τ_k, $k = 0, 1, 2, \ldots$ we shall call this kind of systems impulsive controlled linear stochastic systems with Markovian jumping (ICLSSMJ).
(b) The evolution of the system (1.19) is affected by two kinds of switching:

(α) random switching generated by the transition from the current mode i_1 to another mode i_2 of the Markov process $(\eta(t))$;
(β) switching at impulsive time instances τ_k, $k = 0, 1, \ldots$.

The jumps generated by the transition from the current mode i_1 to another mode $i_2 \in \mathcal{N}$ do not produce discontinuities of the trajectories, while the jumps taking place at the impulsive time instances τ_k may produce discontinuities of the trajectories of the system (1.19) according to (1.19b).

1.2.3 General Class of Impulsive Controlled Linear Stochastic Systems

Let us consider the linear system controlled by impulses having the state-space representation described by:

$$\begin{aligned} dx(t) &= (A_0(t, \eta(t))x(t) + G_0(t, \eta(t))v(t))\,dt \\ &\quad + \sum_{j=1}^{r_1} (A_j(t, \eta(t))x(t) + G_j(t, \eta(t))v(t))dw_j(t), \quad \tau_k < t \leq \tau_{k+1} \end{aligned} \tag{1.23a}$$

$$x(\tau_k^+) = A_{d0}(k, \eta(\tau_k))x(\tau_k) + G_{d0}(k, \eta(\tau_k))v_d(k) + B_{d0}(k, \eta(\tau_k))u(k)$$
$$+ \sum_{j=1}^{r_2} w_{dj}(k)(A_{dj}(k, \eta(\tau_k))x(\tau_k) + G_{dj}(k, \eta_k)v_d(k)$$
$$+ B_{dj}(k, \eta(\tau_k))u(k)), \quad k \in \mathbb{Z}_+ \tag{1.23b}$$

$$y(\tau_k) = C_0(k, \eta(\tau_k))x(\tau_k) + \sum_{j=1}^{r_2} w_{dj}(k)C_j(k, \tau_k)x(\tau_k)$$
$$+ D(k, \eta(\tau_k))v_d(k), \tag{1.24}$$

$$z(t) = C_z(t, \eta(t))x(t), \quad t \in \mathbb{R}_+ \tag{1.25a}$$
$$z_d(\tau_k) = C_{dz}(k, \eta(\tau_k))x(\tau_k) + D_{dz}(k, \eta(\tau_k))u(k), \quad k \in \mathbb{Z}_+. \tag{1.25b}$$

As in the previous cases, $x(t) \in \mathbb{R}^n$, $y(\tau_k) \in \mathbb{R}^{n_y}$, $(z(t), z_d(\tau_k)) \in \mathbb{R}^{n_z} \times \mathbb{R}^{n_{zd}}$, $u(k) \in \mathbb{R}^{n_u}$ are the state parameters, the measured output, the controlled outputs (sometimes named quality outputs) and control parameters, respectively.

The stochastic processes $\{w(t)\}_{t\in\mathbb{R}_+}$, $\{w_d(k)\}_{k\in\mathbb{Z}_+}$ and $\{\eta(t)\}_{t\in\mathbb{R}_+}$ which affect the controlled system modeled by (1.23)–(1.25) are satisfying the assumption:

Assumption 1.2

(a) $\{w(t)\}_{t\geq 0}$ $(w(t) = (w_1(t)\ w_2(t)\ \dots\ w_{r_1}(t))^\top)$ is a r_1-dimensional standard Wiener process defined on a given probability space $(\Omega, \mathfrak{F}, \mathcal{P})$;

(b) $\{w_d(k)\}_{k\in\mathbb{Z}_+}$ $(w_d(k) = (w_{d1}(k)\ w_{d2}(k)\ \dots\ w_{dr_2}(k))^\top)$ is a sequence of r_2-dimensional independent random vectors defined on the same probability space $(\Omega, \mathfrak{F}, \mathcal{P})$ with zero mean and covariance matrix $\mathbb{E}[w_d(k)w_d^\top(k)] = I_{r_2}$ for all $k \in \mathbb{Z}_+$;

(c) $\{\eta(t)\}_{t\geq 0}$ is a homogeneous standard Markov process defined on the probability space $(\Omega, \mathfrak{F}, \mathcal{P})$ taking values in the finite set $\mathcal{N} = \{1, 2, \dots, N\}$ and having the transition semigroup $P(t) = e^{Qt}$, $t \geq 0$. The elements q_{ij} of the generator matrix Q satisfy a condition of type (1.22);

(d) $\{w(t)\}_{t\geq 0}$, $\{w_d(k)\}_{k\in\mathbb{Z}_+}$ and $\{\eta(t)\}_{t\in\mathbb{R}_+}$ are mutually independent stochastic processes.

For each $t \geq 0$ we consider $\mathcal{F}_t \subset \mathfrak{F}$ the σ-algebra generated by the random vectors $w(s)$, $0 \leq s \leq t$ and $w_d(k)$, $k \geq 0$ such that $\tau_k < t$ and $\mathcal{G}_t \subset \mathfrak{F}$ be the σ-algebra generated by the random variables $\eta(s)$, $0 \leq s \leq t$. We assume that both $\mathcal{F}_t$ and $\mathcal{G}_t$ contain all sets $\mathbb{M} \in \mathfrak{F}$ such that $\mathcal{P}(\mathbb{M}) = 0$.

With this notation, Assumption 1.2 is fulfilled if and only if Assumption 1.1 is fulfilled and for each $t \geq 0$, the σ-algebra $\mathcal{F}_t$ is independent of the σ-algebra $\mathcal{G}_t$. We set $\mathcal{H}_t = \mathcal{F}_t \vee \mathcal{G}_t$, i.e., $\mathcal{H}_t \subset \mathfrak{F}$ is the smallest σ-algebra which contains the σ-algebras $\mathcal{F}_t$ and $\mathcal{G}_t$.

The next result provides information about the properties of the trajectories of the system (1.23).

Proposition 1.3 *Assume:*

(a) Assumption 1.2 is fulfilled;
(b) $t \to (A_j(t,i), G_j(t,i)) : \mathbb{R}_+ \to \mathbb{R}^{n\times n} \times \mathbb{R}^{n\times n_v}, \quad 0 \le j \le r_1, \quad i \in \mathcal{N}$ *are bounded and continuous matrix-valued functions;*
(c) $\{A_{dj}(\tau_k,i)\}_{k\in\mathbb{Z}_+} \subset \mathbb{R}^{n\times n}, \quad \{G_{dj}(\tau_k,i)\}_{k\in\mathbb{Z}_+} \subset \mathbb{R}^{n\times n_{v_d}}, \quad \{B_{dj}(\tau_k,i)\}_{k\in\mathbb{Z}_+} \subset \mathbb{R}^{n\times n_u}$, $0 \le j \le r_2$, $i \in \mathcal{N}$ *are bounded matrix-valued sequences;*
(d) $v(\cdot) \in \mathcal{L}^2_{\mathcal{H}}([0,T],\mathbb{R}^{n_v})$ *for all* $T > 0$ *while for each* $k \in \mathbb{Z}_+$, $v_d(k) : \Omega \to \mathbb{R}^{n_{v_d}}$ *is a random vector which is* $\mathcal{H}_{\tau_k}$*-measurable and* $\mathbb{E}[|v_d(k)|^2] < \infty$;
(e) $\lim\limits_{k\to\infty} \tau_k = +\infty$.

Let $t_0 \in \mathbb{R}_+$ *and* $k_0 \in \mathbb{Z}_+$ *be such that* $\tau_{k_0-1} < t_0 < \tau_{k_0}$. *Under these conditions, for each random vector* $x_0 : \Omega \to \mathbb{R}^n$ *which is* $\mathcal{H}_{t_0}$*-measurable and* $\mathbb{E}[|x_0|^2] < \infty$, *and for each sequence* $\mathbf{u}_{k_0} = \{u(k)\}_{k\ge k_0} \subset \mathbb{R}^{n_u}$ *with the property that for any* $k \ge k_0$, $u(k)$ *is a random vector* $\mathcal{H}_{\tau_k}$*-measurable and* $\mathbb{E}[|u(k)|^2] < \infty$, *the stochastic linear differential equation (1.23) has a unique solution* $x(t) := x(t; t_0, x_0, \mathbf{u}_{k_0})$ *with the properties:*

(i) $x(\cdot)$ *is a.s. continuous in any* $t \ne \tau_k$ *and it is left continuous in* $t = \tau_k$ *for any* $k \ge k_0$;
(ii) $x(\cdot) \in \mathcal{L}^2_{\mathcal{H}}([t_0,T],\mathbb{R}^n)$, *for all* $T > t_0$;
(iii) $x(t_0) = x_0$ *and for each* $k \ge k_0$, $\lim\limits_{\substack{t\to\tau_k \\ t>\tau_k}} x(t) = x(\tau_k^+)$.

Proof May be done following step by step the proof of Proposition 1.1.

1.3 Sampled-Data Linear Stochastic Controlled Systems

Let us consider a linear stochastic controlled system having the state-space representation described by:

$$\begin{aligned} dx(t) &= (A_0(t,\eta(t))x(t) + G_0(t,\eta(t))v(t) + B_0(t,\eta(t))u(t))dt \qquad (1.26a)\\ &\quad + \sum_{j=1}^{r_1}(A_j(t,\eta(t))x(t) + G_j(t,\eta(t))v(t) + B_j(t,\eta(t))u(t))dw_j(t),\\ &\quad t \ge t_0,\ x(t_0) = x_0,\\ y(t) &= C_0(t,\eta(t))x(t) \qquad (1.26b)\\ z(t) &= \begin{pmatrix} C_z(t,\eta(t))x(t) \\ D_z(t,\eta(t))u(t) \end{pmatrix} \qquad (1.26c) \end{aligned}$$

where $x(t) \in \mathbb{R}^n$ denotes the vector of the state parameters at time t, $u(t) \in \mathbb{R}^{n_u}$ stands for the vector of control parameters, $y(t) \in \mathbb{R}^{n_y}$ is the vector of measurements available to activate a controller with some imposed performance specifications and $z(t) \in \mathbb{R}^{n_z} \times \mathbb{R}^{\hat{n}_z}$ is a controlled output sometimes called quality output, $v(t) \in \mathbb{R}^{n_v}$ models different exogenous disturbances which affect the considered controlled system. In (1.26), $\{w_j(t)\}_{t\geq 0}$, $1 \leq j \leq r_1$, $\{\eta(t)\}_{t\geq 0}$ are stochastic processes as in the case of the system (1.23).

We assume that the measured values are only available at discrete-time instances τ_k, $k = 0, 1, \ldots$. It is expected that the measurements not to be perfect. Hence, the measured values $y(\tau_k)$ are affected by some random perturbations due to the reliability of the measurements devices as well as by the external disturbances due to the environment. The effect of these perturbations is modeled by a stochastic process which affects the measured output values. So, the inputs of the controller will be of the form

$$\tilde{y}(k) = y(\tau_k) + \nu(k), \quad k = 0, 1, \ldots .$$

In the developments from this monograph, we consider that the random perturbations which affect the measurements are of the form

$$\nu(k) = \sum_{j=1}^{r_2} w_{dj}(k) C_j(k, \eta(\tau_k)) x(\tau_k) + D(k, \eta(\tau_k)) v_d(k)$$

where $\{w_{dj}(k)\}_{k\in\mathbb{Z}_+}$, $1 \leq j \leq r_2$ are sequences of independent random vectors and $v_d(k) : \Omega \to \mathbb{R}^{n_{v_d}}$ are random vectors measurable with respect to $\mathscr{H}_{\tau_k}$ with finite second moments. Thus, when we talk about a sampled-data controlled system, the measured output will be of the form

$$\begin{aligned} y(\tau_k) = C_0(k, \eta(\tau_k)) x(\tau_k) + \sum_{j=1}^{r_2} w_{dj}(k) C_j(k, \eta(\tau_k)) x(\tau_k) \\ + D(k, \eta(\tau_k)) v_d(k). \end{aligned} \tag{1.27}$$

For the uniformity of the writing we denote $C_0(k, \eta(\tau_k))$ instead of $C_0(\tau_k, \eta(\tau_k))$.

To control the system (1.26a) using measurements of type (1.27) we shall use discrete-time controllers of the form:

$$x_c(k+1) = \mathscr{A}_c(k) x_c(k) + \mathscr{B}_c(k) y(\tau_k) \tag{1.28a}$$
$$y_c(k) = \mathscr{C}_c(k) x_c(k) + \mathscr{D}_c(k) y(\tau_k) \tag{1.28b}$$

or in the form of a jump linear system:

$$\dot{x}_c(t) = A_c(t) x_c(t), \quad \tau_k < t \leq \tau_{k+1} \tag{1.29a}$$
$$x_c(\tau_k^+) = \mathscr{A}_d(k) x_c(\tau_k) + \mathscr{B}_d(k) y(\tau_k) \tag{1.29b}$$
$$y_c(k) = \mathscr{C}_c(k) x_c(\tau_k) + \mathscr{D}_c(k) y(\tau_k), \quad k \in \mathbb{Z}_+. \tag{1.29c}$$

Since in both cases (1.28) and (1.29), $y_c(k)$ are discrete-time quantities while the control $u(t)$ is a continuous-time function, in order to use the outputs of a controller (1.28) or (1.29) to control a system of type (1.26a) it is necessary to insert an operator to convert the discrete-time signals generated by the controller into continuous-time signals which are used to control the system (1.26a).

Among such operators, we shall use in this monograph the simplest one, namely a "zero-order hold" (ZOH), associated to a partition of the semiaxis $\mathbb{R}_+$, generated by the sequence of the sampling time instances $\{\tau_k\}_{k\in\mathbb{Z}_+}$. In this way, we obtain

$$u(t) = u(k), \quad \tau_k < t \leq \tau_{k+1}, \tag{1.30a}$$

$$u(k) = y_c(k), \quad k \in \mathbb{Z}_+, \tag{1.30b}$$

$y_c(k)$ being the output of a controller (1.28) or (1.29), respectively. From (1.30) is clear that the generated controls are piecewise constant functions.

Substituting (1.30a) in (1.26a) and (1.26c) we obtain

$$\begin{aligned} dx(t) &= (A_0(t,\eta(t))x(t) + G_0(t,\eta(t))v(t) + B_0(t,\eta(t))u(k))dt \\ &\quad + \sum_{j=1}^{r_1}(A_j(t,\eta(t))x(t) + G_j(t,\eta(t))v(t) + B_j(t,\eta(t))u(k))dw_j(t), \\ &\quad \tau_k < t \leq \tau_{k+1}, \end{aligned} \tag{1.31a}$$

$$z(t) = \begin{pmatrix} C_z(t,\eta(t))x(t) \\ D_z(\tau_k,\eta(\tau_k))u(k) \end{pmatrix}, \quad \tau_k < t \leq \tau_{k+1}, \quad k = 0, 1, \ldots . \tag{1.31b}$$

Let $\phi : \mathbb{R}_+ \to \mathbb{R}^{n_u}$ be defined by

$$\phi(t) = u(k), \quad t \in (\tau_k, \tau_{k+1}], \ k \in \mathbb{Z}_+. \tag{1.32}$$

It is obvious that $\phi(\cdot)$ is differentiable in any $t \in \mathbb{R}_+$, $t \neq \tau_k$ and we have $\dot{\phi}(t) = 0$. Also we have

$$\phi(\tau_k^+) = \lim_{\substack{t\to\tau_k \\ t>\tau_k}} \phi(t) = u(k) \ \text{ for all } \ k \in \mathbb{Z}_+.$$

We set $\mathbf{x}(t) = (x^\top(t) \ \ \phi^\top(t))^\top$. By direct calculation one obtains that if $x(\cdot)$ is a trajectory of (1.31a) determined by the control $\{u(k)\}_{k\in\mathbb{Z}_+}$ and $\phi(\cdot)$ is defined via (1.32) based on the same vectors $u(k)$, then $\mathbf{x}(\cdot)$ is a solution of the following system of stochastic differential equations controlled by impulses:

$$\begin{aligned} d\mathbf{x}(t) &= (\mathcal{A}_0(t,\eta(t))\mathbf{x}(t) + G_0(t,\eta(t))v(t))dt \\ &\quad + \sum_{j=1}^{r_1}(\mathcal{A}_j(t,\eta(t))\mathbf{x}(t) + G_j(t,\eta(t))v(t))dw_j(t), \quad \tau_k < t \leq \tau_{k+1}, \end{aligned} \tag{1.33a}$$

$$\mathbf{x}(\tau_k^+) = \mathcal{A}_{d0}\mathbf{x}(\tau_k) + \mathcal{B}_{d0}u(k), \quad k \in \mathbb{Z}_+ \tag{1.33b}$$

$$z(t) = \mathscr{C}_z(t, \eta(t))\mathbf{x}(t), \quad t \in \mathbb{R}_+, \tag{1.34a}$$

$$z_d(\tau_k) = \mathscr{D}_{dz}(k, \eta(\tau_k))u(k), \quad k \in \mathbb{Z}_+, \tag{1.34b}$$

where we denoted

$$\mathscr{A}_j(t,i) = \begin{pmatrix} A_j(t,i) & B_j(t,i) \\ 0_{n_u n} & 0_{n_u} \end{pmatrix}, \quad 0 \leq j \leq r_1, \quad 1 \leq i \leq N, \quad t \in \mathbb{R}_+, \tag{1.35a}$$

$$\mathscr{A}_{d0} = \begin{pmatrix} I_n & 0_{nn_u} \\ 0_{n_u n} & 0_{n_u} \end{pmatrix}, \quad \mathscr{B}_{d0} = \begin{pmatrix} 0_{nn_u} \\ I_{n_u} \end{pmatrix}, \tag{1.35b}$$

$$\mathscr{C}_z(t,i) = (C_z(t,i) \;\; 0_{n_z n_u}), \quad 1 \leq i \leq N, \quad t \in \mathbb{R}_+, \tag{1.35c}$$

$$\mathscr{D}_{dz}(k,i) = D_z(\tau_k, i), \quad 1 \leq i \leq N, \quad t \in \mathbb{R}_+. \tag{1.35d}$$

One sees that the system (1.33) is a system of stochastic linear differential equations controlled by impulses of type (1.23) and (1.34) is a quality output of type (1.25). By direct calculations involving (1.35) one obtains that if $\mathbf{x}(t)$ is a solution of (1.33) determined by the control $\{u(k)\}_{k \in \mathbb{Z}_+}$, then $x(t)$ formed by the first n components of $\mathbf{x}(t)$ is a trajectory of (1.31a) determined by the same control $\{u(k)\}_{k \in \mathbb{Z}_+}$.

In the next chapters we shall use systems controlled by impulses of type (1.34), (1.35) to obtain solutions of different optimal control problems as well as optimal filtering problems for sampled-data controlled systems.

1.4 Notes and References

The results enclosed in Sect. 1.1.1 for which the proofs are omitted are well known and can be found in the monographs about measure theory and stochastic process theory. The results enclosed in Sects. 1.2 and 1.3 are new.

References

1. Rudin, W.: Real & Complex Analysis. McGraw-Hill (1987)
2. Vestrup, E.M.: The Theory of Measures and Integration. Wiley (2003)
3. Billingsley, P.: Probability and Measure (Anniversary ed.). Wiley (2012)
4. Bauer, H.: Measure and Integration Theory. Walter de Gruyter (2001)
5. Bartle, R.G.: The Elements of Integration and Lebesgue Measure. Wiley (1995)
6. Drăgan, V., Morozan, T., Stoica, A.M.: Mathematical Methods in Robust Control of Linear Stochastic Systems. Springer, New York (2013)
7. Menaldi, J.L.: Stochastic Differential Equations with Jumps, p. 77. Mathematics Faculty Research Publications (2008). https://digitalcommons.wayne.edu/mathfrp/77
8. Ciucu, G., Tudor, C.: Teoria Probalităţilor şi Aplicaţii (Probability Theory and Applications). Editura Ştiinţifică şi Enciclopedică, Bucureşti (1983)
9. Chung, K.L.: Markov Chains with Stationary Transition Probabilities. Springer, Berlin, Heidelberg (1960)

10. Doob, J.L.: Stochastic Processes. Wiley, New York (1967)
11. Friedman, A.: Stochastic Differential Equations and Applications, vol. 1. Academic (1975)
12. Ghihman, I., Skorokhod, A.: Introduction to the Theory of Random Processes (translated from Russian). W.B. Saunders Co., Philadelphia (1969)
13. Lipser, A., Shiryaev, A.: Statistics of Random Processes. Nauka, Moscow (1977); Springer, Berlin (1974)
14. Neveu, J.: Integrales stochastiques et applications. Notes de cours multigraphiées (1972)
15. Schneider, H., Vidyasagar, M.: Cross-positive matrices. SIAM J. Numer. Anal. **7**(4), 508–519 (1970)

Chapter 2
Stability of Jump Linear Differential Equations

Abstract In this chapter the problem of exponential stability of the zero state equilibrium of a class of jump linear differential equations on a real ordered Banach space is investigated. These equations are defined by a special class of jump operator valued functions. The results presented here could be viewed, in some sense, as an extension to the *non-continuous* case of the results developed in this chapter from Drăgan et al. in mathematical methods in robust control of linear stochastic systems (2013) [1]. The differential equations addressed in Drăgan et al. in mathematical methods in robust control of linear stochastic systems (2013) [1] (and named differential equations with positive evolution) are defined by a special class of strongly continuous operator valued functions. These differential equations are natural extensions to the time-varying case of the linear differential equations with constant coefficients on an ordered Banach space defined by a linear and bounded operator with positive semigroup. In the present chapter, discontinuities of the operator valued functions are allowed at some time instances called *jump time instances*. The jump linear differential equations with positive evolution studied in this chapter contain as special cases Lyapunov-type jump differential equations arising in a natural way in connection to the problem of exponential stability in the mean-square sense of a large class of jump stochastic linear differential equations. Properties of forward/backward jump operator valued functions on a Banach space are emphasized in Sect. 2.1. Section 2.2 is devoted to the characterization of the exponential stability and asymptotic stability of forward jump linear differential equations. In the main result of this section, the stability of the forward jump linear equations is equivalently formulated as the stability of a corresponding class of discrete-time linear equations (Theorem 2.1). The periodic case as well as the time-invariant case are also treated in this section. Sections 2.3 and 2.4 are devoted to the case of jump linear differential equations on a Hilbert space. The fundamental results obtained in these sections are then exploited in the last part of this chapter in order to obtain criteria for the exponential stability of a class of generalized Lyapunov linear differential equations with jumps on a Hilbert space as well as for the existence of some bounded solutions on the whole real semiaxis of the corresponding non-homogeneous generalized Lyapunov differential equations with jumps.

V. Drăgan et al., *Robust Control of Jump Linear Stochastic Systems*, Lecture Notes in Control and Information Sciences 497, https://doi.org/10.1007/978-3-031-84070-8_2

2.1 Jump Linear Differential Equations on a Banach Space

2.1.1 Forward Jump Linear Differential Equations

Let $(\mathscr{X}, \|\cdot\|)$ be a real Banach space. We denote $\mathscr{B}(\mathscr{X})$ the set of all linear and bounded operators $\mathscr{L} : \mathscr{X} \to \mathscr{X}$. It is known that $\mathscr{B}(\mathscr{X})$ is a Banach space with respect to the norm

$$\|\mathscr{L}\| = \sup_{\|x\|=1} \|\mathscr{L}x\|. \tag{2.1}$$

Let $t \to \mathscr{L}(t) : \mathbb{R}_+ \to \mathscr{B}(\mathscr{X})$ be a continuous valued function.[1]

Let $\mathfrak{T} = \{\tau_0, \tau_1, \ldots, \tau_k, \tau_{k+1}, \ldots\}$ be a partition of $\mathbb{R}_+$ such that: $0 = \tau_0 < \tau_1 < \ldots < \tau_k < \tau_{k+1} < \ldots$. For each $\tau_k \in \mathfrak{T}$ we consider an operator $\mathscr{I}(\tau_k) \in \mathscr{B}(\mathscr{X})$. Employing the triple $(\mathscr{L}(\cdot), \mathscr{I}(\cdot), \mathfrak{T})$ we introduce the following linear differential equation on $\mathscr{X}$

$$\dot{x}(t) = \mathscr{L}(t)x(t), \qquad \tau_k < t \leq \tau_{k+1} \tag{2.2a}$$

$$x(\tau_k^+) = \mathscr{I}(\tau_k)x(\tau_k), \qquad k \in \mathbb{Z}_+. \tag{2.2b}$$

A differential equation of type (2.2) will be named forward jump linear differential equation (FJLDE) and the terms τ_k of the sequence $\mathfrak{T}$ will be called jump time instances.

To clarify the meaning of the adjective "forward jump" attached to the linear differential Eq. (2.2), let us consider $(t_0, x_0) \in \mathbb{R}_+ \times \mathscr{X}$ and let us denote $x(\cdot; t_0, x_0)$ the solution of (2.2) which starts from x_0 at the initial time instance t_0.

If $t_0 \in [\tau_{k_0}, \tau_{k_0+1})$ then $x(\cdot; t_0, x_0)$ has a possible discontinuity in each $\tau_k \geq \tau_{k_0+1}$. More precisely, $x(\cdot; t_0, x_0)$ regarded as a solution of the ordinary differential equation

$$\dot{x}(t) = \mathscr{L}(t)x(t)$$

is a continuous function on the whole interval $[t_0, \infty)$. Its value $x(\tau_{k_0+1}; t_0, x_0)$ is used in (2.2b) (written for k replaced by $k_0 + 1$) to compute $x(\tau_{k_0+1}^+)$ which will serve as an initial value of the restriction of the solution $x(\cdot; t_0, x_0)$ to the interval $[\tau_{k_0+1}, \tau_{k_0+2}]$. Hence, for $t \in [\tau_{k_0+1}, \tau_{k_0+2}]$

$$x(t; t_0, x_0) = x(t; \tau_{k_0+1}, x(\tau_{k_0+1}^+)).$$

Since $x(\cdot; \tau_{k_0+1}, x(\tau_{k_0+1}^+))$ is continuous on the interval $[\tau_{k_0+1}, \tau_{k_0+2}]$, we have:

$$\lim_{\substack{t \to \tau_{k_0+2} \\ t < \tau_{k_0+2}}} x(t; t_0, x_0) = \lim_{\substack{t \to \tau_{k_0+2} \\ t < \tau_{k_0+2}}} x(t; \tau_{k_0+1}, x(\tau_{k_0+1}^+)) := x(\tau_{k_0+2}; t_0, x_0).$$

[1] Here continuity must be understand with respect to the topology induced by the norm $\|\cdot\|$ defined on $\mathscr{B}(\mathscr{X})$.

Then, $x(\tau_{k_0+2}; t_0, x_0)$ is used in (2.2b) (written for k replaced by $k_0 + 2$) to compute $x(\tau^+_{k_0+2})$ which will serve as initial value of the restriction of the solution $x(\cdot; t_0, x_0)$ to the interval $[\tau_{k_0+2}, \tau_{k_0+3}]$. This means that for $t \in [\tau_{k_0+2}, \tau_{k_0+3}]$ we have $x(t; t_0, x_0) = x(t; \tau_{k_0+2}, x(\tau^+_{k_0+2}))$. Further, $x(\tau_{k_0+3}; t_0, x_0)$ is used in (2.2b) (written for k replaced by $k_0 + 3$) to compute $x(\tau^+_{k_0+3})$ and the described procedure may continue to compute $x(\cdot; t_0, x_0)$ on all intervals $[\tau_k, \tau_{k+1}]$, $k \geq k_0 + 1$. Thus we obtain:

Proposition 2.1 *For each $(t_0, x_0) \in \mathbb{R}_+ \times \mathscr{X}$, the FJLDE (2.2) has a unique solution $x(t; t_0, x_0)$, $t \geq t_0$, which is a differentiable function on each interval $(\tau_k, \tau_{k+1}) \cap [t_0, \infty)$, left continuous in any τ_k having possible discontinuities at time instances $\tau_k \in \mathfrak{T} \cap [t_0, \infty)$ and satisfies $x(t_0; t_0, x_0) = x_0$.*

Note 2.1 (a) To highlight the left continuity at time instances τ_k of the solutions of a FJLDE we write $\tau_k < t \leq \tau_{k+1}$ to describe the intervals of evolution of the dynamics in continuous time of such equations.

(b) It is worth mentioning that unlike the ordinary linear differential equation

$$\dot{x}(t) = \mathscr{L}(t)x(t), \quad t \in \mathbb{R}_+,$$

the solutions $x(\cdot; t_0, x_0)$ of a FJLDE of type (2.2) can be extended only for $t \geq t_0$. This justify the adjective "forward jump" associated to a linear differential equation of type (2.2).

Let $T(t, t_0)$ be the linear evolution operator on $\mathscr{X}$ defined by the ordinary linear differential equation

$$\dot{x}(t) = \mathscr{L}(t)x(t), \quad t \in \mathbb{R}_+.$$

For a precise definition of a linear evolution operator we refer to Chap. 3 in Daleckiĭ and Kreĭn [2].

Remark 2.1 The main properties of a linear evolution operator which will be used in the following are:

(a) for each $s \in \mathbb{R}_+$, the function $t \to T(t, s)$ is differentiable and satisfies the following differential equation on $\mathscr{B}(\mathscr{X})$:

$$\frac{\mathrm{d}}{\mathrm{d}t}T(t, s) = \mathscr{L}(t)T(t, s);$$

(b) for each $t \in \mathbb{R}_+$, the function $s \to T(t, s)$ is differentiable and satisfies the following linear differential equation on $\mathscr{B}(\mathscr{X})$:

$$\frac{\mathrm{d}}{\mathrm{d}s}T(t, s) = -T(t, s)\mathscr{L}(s);$$

(c) $T(t, t) = I_{\mathscr{X}}$ for all $t \in \mathbb{R}_+$, $I_{\mathscr{X}}$ being the identity operator on $\mathscr{X}$;

(d)

$$\|T(t,s)\| \le e^{\lambda|t-s|}, \tag{2.3}$$

for all $t, s \in \mathbb{R}_+$, where $\lambda = \sup_{t\in\mathbb{R}_+} \|\mathscr{L}(t)\|$;

(e) if $\mathscr{L}(t) = \mathscr{L}$ for all $t \in \mathbb{R}_+$, then

$$T(t,s) = e^{\mathscr{L}(t-s)} \triangleq \sum_{k=0}^{\infty} \frac{(t-s)^k}{k!}\mathscr{L}^k;$$

(f) if there exists $\theta > 0$ such that $\mathscr{L}(t+\theta) = \mathscr{L}(t)$ for all $t \in \mathbb{R}_+$, then $T(t + j\theta, s + j\theta) = T(t,s)$, for all $t, s \in \mathbb{R}_+$, $j \in \mathbb{Z}$.

The solution of the linear differential equation

$$\dot{x}(t) = \mathscr{L}(t)x(t)$$

satisfies

$$x(t) = T(t,t_0)x(t_0), \quad \text{for all } t, t_0 \in \mathbb{R}_+.$$

The solution $x(t; t_0, x_0)$ of the FJLDE (2.2) has the representation

$$x(t; t_0, x_0) = \mathbf{T}(t, t_0)x_0 \tag{2.4}$$

for all $t \ge t_0 \ge 0$, $x_0 \in \mathscr{X}$, where we denoted

$$\mathbf{T}(t,t_0) = T(t,\tau_k)\mathscr{I}(\tau_k)\prod_{l=0}^{k-k_0-1} T(\tau_{k-l},\tau_{k-l-1})\mathscr{I}(\tau_{k-l-1})T(\tau_{k_0},t_0) \tag{2.5}$$

if $t \in [\tau_k, \tau_{k+1}]$ and $t_0 \in (\tau_{k_0-1}, \tau_{k_0}]$, $k > k_0$,

$$\mathbf{T}(t,t_0) = T(t,\tau_k)\mathscr{I}(\tau_k)T(\tau_k,t_0) \tag{2.6}$$

if $\tau_{k-1} < t_0 \le \tau_k \le t \le \tau_{k+1}$, $k \in \mathbb{Z}_+$,

$$\mathbf{T}(t,t_0) = T(t,t_0) \tag{2.7}$$

if $\tau_k < t_0 \le t \le \tau_{k+1}$. Moreover, if $t \in (\tau_k, \tau_{k+1})$ then $t \to \mathbf{T}(t, t_0)$ is differentiable and satisfies

$$\frac{\partial}{\partial t}\mathbf{T}(t,t_0) = \mathscr{L}(t)\mathbf{T}(t,t_0)$$

and if $t_0 \in (\tau_{k_0-1}, \tau_{k_0})$ then $t_0 \to \mathbf{T}(t, t_0)$ is differentiable and satisfies

$$\frac{\partial}{\partial t_0}\mathbf{T}(t, t_0) = -\mathbf{T}(t, t_0)\mathscr{L}(t_0).$$

We set

$$\mathbf{T}(t, \tau_{k_0-1}^{+}) \stackrel{\Delta}{=} \lim_{\substack{t_0 \to \tau_{k_0-1} \\ t_0 > \tau_{k_0-1}}} \mathbf{T}(t, t_0). \tag{2.8}$$

This is obtained directly from (2.5)–(2.7).

2.1.2 Forward Jump Linear Discrete-Time Equations

Now we consider the discrete-time linear equation (DTLE) associated to (2.2):

$$z_{k+1} = \mathscr{L}_d(k) z_k \tag{2.9}$$

where

$$\mathscr{L}_d(k) \stackrel{\Delta}{=} T(\tau_{k+1}, \tau_k)\mathscr{I}(\tau_k). \tag{2.10}$$

One sees that $\mathscr{L}_d(k) \in \mathscr{B}(\mathscr{X})$, for all $k \in \mathbb{Z}_+$.

Lemma 2.1 *There exists a one to one correspondence between the set of the solutions of FJLDE (2.2) and the set of the solutions of DTLE (2.9) and (2.10).*

Proof Let $\tilde{x}(\cdot)$ be an arbitrary solution of (2.2). From (2.2a) we obtain that

$$\tilde{x}(\tau_{k+1}) = T(\tau_{k+1}, \tau_k)x(\tau_k^{+}).$$

Then, from (2.2b) and (2.10) we obtain that the sequence $\{\tilde{x}(\tau_k)\}_{k \geq k_0}$ solves (2.9). Conversely, if $\{z_k\}_{k \geq k_0}$ is a solution of the DTLE (2.9) and (2.10) we set $\tilde{x}(t) = T(t, \tau_k)\mathscr{I}(\tau_k)z_k, \quad \tau_k < t \leq \tau_{k+1}, \quad k \geq k_0$. One sees that $\tilde{x}(t)$ solves (2.2a). From $\tilde{x}(\tau_{k+1}) = T(\tau_{k+1}, \tau_k)\mathscr{I}(\tau_k)z_k$ together with (2.9) and (2.10) we deduce that $\tilde{x}(\tau_{k+1}) = z_{k+1}$, for all $k \geq k_0$. This allows us to deduce that

$$\tilde{x}(\tau_k^{+}) = \lim_{\substack{t \to \tau_k \\ t \geq \tau_k}} \tilde{x}(t) = \mathscr{I}(\tau_k)\tilde{x}(\tau_k).$$

Hence, $\tilde{x}(\cdot)$ is a solution of the FJLDE (2.2). Thus the proof is complete.

We set

$$\mathbf{T}_d(k, k_0) = \begin{cases} \mathscr{L}_d(k-1)\mathscr{L}_d(k-2)\cdots\mathscr{L}_d(k_0), & \text{if} \quad k > k_0 \geq 0 \\ I_{\mathscr{X}}, \quad \text{if} \quad k = k_0. \end{cases} \tag{2.11}$$

$\mathbf{T}_d(k, k_0)$ will be named linear evolution operator defined by the DTLE (2.9) and (2.10). The solution of (2.9) satisfies

$$z_k = \mathbf{T}_d(k, k_0) z_{k_0}, \quad \text{for all} \quad k \geq k_0.$$

From (2.5) and (2.11) we obtain

$$\mathbf{T}(t, t_0) = T(t, \tau_k)\mathscr{I}(\tau_k)\mathbf{T}_d(k, k_0)T(\tau_{k_0}, t_0), \tag{2.12}$$

if $t \in [\tau_k, \tau_{k+1}]$ and $t_0 \in (\tau_{k_0-1}, \tau_{k_0}]$. From (2.3), (2.10) and (2.11) one gets

$$\|\mathbf{T}_d(k, k_0)\| \leq \sup_{k_0 \leq l \leq k-1} \|\mathscr{I}(\tau_l)\| e^{\lambda(\tau_k - \tau_{k_0})}, \tag{2.13}$$

for all $k > k_0 \geq 0$.

2.1.3 A Non-homogeneous FJLDE

Let us consider the affine (non-homogeneous) jump differential equation

$$\dot{x}(t) = \mathscr{L}(t)x(t) + f(t), \qquad \tau_k < t \leq \tau_{k+1} \tag{2.14a}$$
$$x(\tau_k^+) = \mathscr{I}(\tau_k)x(\tau_k) + g_k, \qquad k \in \mathbb{Z}_+, \tag{2.14b}$$

where $\mathscr{L}(\cdot)$, $\mathscr{I}(\cdot)$ are as in (2.2) and $t \to f(t) : \mathbb{R}_+ \to \mathscr{X}$ is a continuous vector valued function and $g_k \in \mathscr{X}$, $k \in \mathbb{Z}_+$. Regarding the representation of the solutions of the non-homogeneous FJLDE (2.14) we have:

Lemma 2.2 *A solution $x(\cdot; t_o, x_0)$ of the non-homogeneous FJLDE (2.14) which starts from $x_0 \in \mathscr{X}$ at the initial time instance t_0 has the representation*

$$x(t; t_0, x_0) = \mathbf{T}(t, t_0)x_0 + \sum_{l=k_0}^{k} \mathbf{T}(t, \tau_l^+)g_l + \int_{t_0}^{t} \mathbf{T}(t, s)f(s)ds, \tag{2.15}$$

for all $t \geq t_0$ such that $t \in [\tau_k, \tau_{k+1}]$ and $t_0 \in (\tau_{k_0-1}, \tau_{k_0}]$.

Proof We set $z_k := x(\tau_k; t_0, x_0)$. One sees that the sequence $\{z_k\}_{k \geq k_0}$ solves the affine DTLE

$$z_{k+1} = \mathscr{L}_d(k)z_k + \zeta_k \tag{2.16}$$

where $\mathscr{L}_d(k)$ is the linear operator defined in (2.10) and $\zeta_k := T(\tau_{k+1}, \tau_k)g_k + \int_{z_k}^{\tau_{k+1}} T(\tau_{k+1}, s)f(s)\text{d}s$. Using the representation formula of the solutions of the DTLE (2.16) we obtain $x(\tau_k; t_0, x_0) = \mathbf{T}_d(k, k_0)x(\tau_{k_0}; t_0, x_0) + \sum_{l=k_0}^{k-1} \mathbf{T}_d(k, l+1)\mathbf{T}(\tau_{l+1}, \tau_l)g_l + \sum_{l=k_0}^{k-1} \mathbf{T}_d(k, l+1) \int_{\tau_l}^{\tau_{l+1}} T(\tau_{l+1}, s)f(s)\text{d}s$. On the other hand from $x(t; t_0, x_0) = T(t, \tau_k)\mathscr{I}(\tau_k)x(\tau_k; t_0, x_0) + T(t, \tau_k)g_k$we get

$$x(t; t_0, x_0) = T(t, \tau_k)\mathscr{I}(\tau_k)\mathbf{T}_d(k, k_0)x(\tau_{k_0}; t_0, x_0) + \sum_{l=k_0}^{k} \mathbf{T}(t, \tau_l^+)g_l + \int_{\tau_{k_0}}^{t} \mathbf{T}(t, s)f(s)ds. \tag{2.17}$$

Now, if we take into account that $t_0 \in (\tau_{k_0-1}, \tau_{k_0}]$ we obtain from (2.14a) that

$$x(\tau_{k_0}; t_0, x_0) = T(\tau_{k_0}, t_0)x_0 + \int_{t_0}^{\tau_{k_0}} T(\tau_{k_0}, s)f(s)ds.$$

Substituting this formula in (2.17) and taking into account again (2.12) we obtain (2.15). Thus the proof ends.

2.1.4 Backward Jump Linear Differential Equations

The quintuple $(\mathscr{L}(\cdot), f(\cdot), \mathscr{I}(\cdot), \{g_k\}_{k\in\mathbb{Z}_+}, \mathfrak{T})$ involved in (2.14) can define also the following kind of jump linear differential equation

$$-\dot{y}(t) = \mathscr{L}(t)y(t) + f(t), \qquad \tau_k \le t < \tau_{k+1}, \tag{2.18a}$$

$$y(\tau_k^-) = \mathscr{I}(\tau_k)y(\tau_k) + g_k, \qquad k \in \mathbb{Z}_+. \tag{2.18b}$$

If $(t_0, y_0) \in (0, \infty) \times \mathscr{X}$ we denote $y(t; t_0, y_0)$ the solution of the jump differential equation which satisfies the initial condition $y(t_0; t_0, y_0) = y_0$. If $t_0 \in [\tau_{k_0}, \tau_{k_0+1})$ then $y(\tau_{k_0}; t_0, y_0)$ is used in (2.18b) written for $k = k_0$ to compute $y(\tau_{k_0}^-; t_0, y_0)$. This will be used as terminal value at τ_{k_0} imposed to the solution $y(\cdot; t_0, y_0)$ on the interval $(\tau_{k_0-1}, \tau_{k_0}]$. Repeating this finding, on each interval $[\tau_{k-1}, \tau_k]$ with $k = k_0 - 1,\ k_0 - 2, \cdots,\ 1$ we may conclude that the solution $y(\cdot; t_0, y_0)$ may be defined on the whole interval $[0, t_0]$. Thus we obtain:

Proposition 2.2 *Under the considered assumptions, for each $(t_0, y_0) \in (0, \infty) \times \mathscr{X}$ the non-homogeneous (affine) jump linear differential equation (2.18) has a unique solution $y(\cdot; t_0, y_0)$ starting from y_0 as the initial time t_0. This solution is defined on the whole interval $[0, t_0]$ being a differentiable function on each interval $(\tau_{k-1}, \tau_k) \cap [0, t_0]$, right continuous in any $\tau_k \in \mathfrak{T} \cap [0, t_0)$ with possible discontinuities in the impulse times τ_k.*

Note 2.2 (a) Unlike the ordinary differential equation

$$-\dot{y}(t) = \mathscr{L}(t)y(t) + f(t), \quad t \in \mathbb{R}_+,$$

whose solutions are defined on the whole semiaxis $\mathbb{R}_+$, the solution $y(\cdot; t_0, y_0)$ of the non-homogeneous jump linear differential equation of (2.18) can be extended only on the interval $[0, \tau_{k_0+1}]$ where τ_{k_0+1} is such that $\tau_{k_0} \le t < \tau_{k_0+1}$. For these reasons, the jump differential equation (2.18) will be named backward jump linear differential equation (BJLDE).

(b) To highlight the right continuity at time instances τ_k of the solutions of a BJLDE we write $\tau_k \le t < \tau_{k+1}$ to describe the intervals of evolution of the dynamics in continuous time of such equations.

Let us denote $T^a(t, t_0)$ the linear evolution operator on $\mathscr{X}$ defined by the ordinary linear differential equation

$$\dot{y}(t) = -\mathscr{L}(t)y(t), \qquad t \in \mathbb{R}_+.$$

According to the terminology used in Chap. 2 from Drăgan et al. [1], $T^a(t, t_0)$ will be named anticausal evolution operator associated to the operator valued function $\mathscr{L}(\cdot)$. The solutions of the differential equation

$$-\dot{y}(t) = \mathscr{L}(t)y(t)$$

satisfy the equality

$$y(t) = T^a(t, t_0)y(t_0), \quad \text{for all } t, t_0 \in \mathbb{R}_+.$$

If $y(t; t_0, y_0)$ is the solution of the BJLDE (2.18) with $g_k = 0$, $k \in \mathbb{Z}_+$ and $f(t) = 0$, $t \in \mathbb{R}_+$, then we have the representation formula

$$y(t; t_0, y_0) = \mathbf{T}^a(t, t_0)y_0,$$

where we have denoted

$$\begin{aligned}\mathbf{T}^a(t, t_0) =& T^a(t, \tau_k)\mathscr{I}(\tau_k)T^a(\tau_k, \tau_{k+1})\mathscr{I}(\tau_{k+1})T^a(\tau_{k+1}, \tau_{k+2})\mathscr{I}(\tau_{k+2}) \\ & \cdot \ldots \cdot \mathscr{I}(\tau_{k_0-1})T^a(\tau_{k_0-1}, \tau_{k_0})\mathscr{I}(\tau_{k_0})T^a(\tau_{k_0}, t_0)\end{aligned} \tag{2.19}$$

if $t \in [\tau_{k-1}, \tau_k]$ and $t_0 \in [\tau_{k_0}, \tau_{k_0+1})$, $k < k_0$,

$$\mathbf{T}^a(t, t_0) = T^a(t, t_0), \tag{2.20}$$

if $t, t_0 \in [\tau_{k-1}, \tau_k]$. By definition $\mathbf{T}^a(t, \tau_{k_0+1}^-) = \lim_{\substack{t_0 \to \tau_{k_0+1} \\ t_0 < \tau_{k_0+1}}} \mathbf{T}^a(t, t_0)$. This is computed directly from (2.19) and (2.20), respectively. Now, we may obtain the representation formula of the solution with given initial values of a BJLDE (2.18).

Lemma 2.3 *For each $(t_0, x_0) \in (0, \infty) \times \mathscr{X}$, the solution $y(\cdot; t_0, x_0)$ of the BJLDE (2.18) which starts from y_0 at the initial time t_0 has the representation formula*

$$y(t; t_0, y_0) = \mathbf{T}^a(t, t_0)y_0 + \sum_{l=k}^{k_0} \mathbf{T}^a(t, \tau_l^-)g_l + \int_t^{t_0} \mathbf{T}^a(t, s)f(s)ds \tag{2.21}$$

if $t \in [\tau_{k-1}, \tau_k]$, $t_0 \in [\tau_{k_0}, \tau_{k_0+1})$, $k \le k_0$ *and*

$$y(t; t_0, y_0) = \mathbf{T}^a(t, t_0)y_0 + \int_t^{t_0} \mathbf{T}^a(t, s)f(s)ds \tag{2.22}$$

if $\tau_{k-1} \le t \le t_0 < \tau_k$, $k \in \mathbb{Z}_+$.

Proof Let $k_0 \in \mathbb{Z}_+$ be such that $t_0 \in [\tau_{k_0}, \tau_{k_0+1})$. For each $k \le k_0$ we obtain from (2.18a) that

$$y(\tau_k; t_0, y_0) = T^a(\tau_k, \tau_{k+1})y(\tau_{k+1}^-; t_0, y_0) + \int_{\tau_k}^{\tau_{k+1}} T^a(\tau_k, s)f(s)\mathrm{d}s.$$

Substituting this equality in (2.18b) we obtain that the sequence $\{y(\tau_k^-; t_0, y_0)\}_{k\le k_0}$ solves the following backward discrete-time linear equation:

$$z_k = \mathcal{L}_d^a(k)z_{k+1} + \zeta_k \tag{2.23}$$

where we have denoted

$$\mathcal{L}_d^a(k) \triangleq \mathcal{I}(\tau_k)T^a(\tau_k, \tau_{k+1})$$

and

$$\zeta_k \triangleq g_k + \mathcal{I}(\tau_k) \int_{\tau_k}^{\tau_{k+1}} T^a(\tau_k, s)f(s)\mathrm{d}s.$$

We set

$$\mathbf{T}_d^a(k, k_0)y \triangleq \begin{cases} \mathcal{L}_d^a(k)\mathcal{L}_d^a(k+1)\cdots\mathcal{L}_d^a(k_0-1), & \text{if} \quad k < k_0 \\ y, \quad \text{if} \quad k = k_0. \end{cases}$$

This allows us to rewrite (2.19) in the compact form:

$$\mathbf{T}^a(t, t_0) = T^a(t, \tau_k)\mathbf{T}_d^a(k, k_0)\mathcal{I}(\tau_{k_0})T^a(\tau_{k_0}, t_0) \tag{2.24}$$

if $t \in [\tau_k, \tau_{k+1}]$. The solution of the DTLE (2.23) has the representation

$$z_k = \mathbf{T}_d^a(k, k_0) + \sum_{l=k}^{k_0-1} \mathbf{T}_d^a(k, l)\zeta_l.$$

For $z_k = y(\tau_k^-; t_0, y_0)$ we get

$$\begin{aligned} y(t; t_0, y_0) &= T^a(t, \tau_k) y(\tau_k^-; t_0, y_0) \\ &= T^a(t, \tau_k)\mathbf{T}_d^a(k, k_0) y(\tau_{k_0}^-; t_0, y_0) + \sum_{l=k}^{k_0-1} T^a(t, \tau_k)\mathbf{T}_d^a(k, l) g_l \\ &+ \sum_{l=k}^{k_0-1} \int_{\tau_l}^{\tau_{l+1}} T^a(t, \tau_k)\mathbf{T}_d^a(k, l)\mathscr{I}(\tau_l) T^a(\tau_l, s) f(s) ds + \int_t^{\tau_k} T^a(t, s) f(s) ds. \end{aligned} \tag{2.25}$$

Substituting in (2.25) $y(\tau_{k_0}^-; t_0, y_0)$ by $\mathscr{I}(\tau_{k_0}) T^a(\tau_{k_0}, t_0) y_0 + g_{k_o}$ we obtain via (2.24) that (2.25) coincides to (2.21). The representation formula (2.22) is obtained directly from (2.18a). Thus the proof ends.

In Sect. 2.3 we shall see that in the case when $\mathscr{X}$ is an ordered Hilbert space, a kind of BJLDE associated to a FJLDE of the form (2.2) is used to derive some criteria for exponential stability.

Note 2.3 The results proved in this section as well as in the other sections of this chapter, still hold if $t \to \mathscr{L}(t) : \mathbb{R}_+ \to \mathscr{B}(\mathscr{X})$ is a bounded and piecewise continuous operator valued function having lateral limits in any $t \in \mathbb{R}_+$ and $t \to f(t) : \mathbb{R}_+ \to \mathscr{X}$ is a bounded and piecewise continuous vector valued function having lateral limits in any $t \in \mathbb{R}_+$.

2.2 Exponential Stability and Asymptotic Stability of Forward Jump Linear Differential Equations

Regarding the partition $\mathfrak{T}$ of the semiaxis $\mathbb{R}_+$, it is worth mentioning that the lengths $\tau_{k+1} - \tau_k$ of the intervals between two impulses are not necessary equal. However, in our approach, they obey the following assumption:

Assumption 2.1

$$0 < \underline{h} \le \tau_{k+1} - \tau_k \le \bar{h}, \quad \text{for all } k \in \mathbb{Z}_+, \tag{2.26}$$

where $\underline{h}$ and $\bar{h}$ are positive constants not depending upon k.

It should be noted that if (2.26) is fulfilled, then $\lim_{k \to \infty} \tau_k = +\infty$.

2.2.1 Definitions and Several General Results

Definition 2.1 We say that the zero solution of the FJLDE (2.2) is:

(a) exponentially stable (ES) if there exists $\beta \geq 1,\ \alpha > 0$, such that the solution $x(\cdot; t_0, x_0)$ of (2.2) satisfies

$$\|x(t; t_0, x_0)\| \leq \beta e^{-\alpha(t-t_0)} \|x_0\| \tag{2.27}$$

for all $t \geq t_0 \geq 0$ and $x_0 \in \mathscr{X}$;

(b) asymptotic stable (AS) if

$$\lim_{t\to\infty} \|x(t; t_0, x_0)\| = 0, \tag{2.28}$$

for all $(t_0, x_0) \in \mathbb{R}_+ \times \mathscr{X}$.

Remark 2.2

(a) Due to the linear character of the equation (2.2), both the exponential stability and asymptotic stability are not associated to a certain solution, they are in fact, properties of the whole equation. That is why, in many cases we shall say that "the FJLDE (2.2) is ES or AS, respectively," instead of "the zero solution of the FJLDE (2.2) is ES or AS, respectively."
(b) From (2.27) and (2.28) one sees that the FJLDE (2.2) is AS if it is ES. The converse implication is true only in some special cases.
(c) From (2.4) and (2.27) ((2.28 respectively), one obtains that the FJLDE (2.2) is:

   (i) exponentially stable if and only if

$$\|\mathbf{T}(t, t_0)\| \leq \beta e^{-\alpha(t-t_0)}, \tag{2.29}$$

   for all $t \geq t_0$ where $\alpha > 0,\ \beta \geq 1$ are constants not depending upon t and t_0;

   (ii) asymptotically stable if and only if

$$\lim_{t\to\infty} \|\mathbf{T}(t, t_0)\| = 0, \tag{2.30}$$

   for all $t_0 \in \mathbb{R}_+$.

The next result establishes a relation between the properties of exponential stability and asymptotic stability, respectively, of the FJLDE (2.2) and the exponential stability and asymptotic stability, respectively for the DTLE (2.9) and (2.10). First, we recall the definitions of this concept for a discrete-time linear equation.

Definition 2.2 We say that the DTLE

$$z_{k+1} = \hat{\mathscr{L}}_d(k) z_k$$

(where $\hat{\mathscr{L}}_d \in \mathscr{B}(\mathscr{X})$) is:

(i) exponentially stable (ES) if its solution $z(k; k_0, z_{k_0})$ satisfy

$$\|z(k; k_0, z_{k_0})\| \leq \gamma \delta^{k-k_0} \|z_{k_0}\|, \tag{2.31}$$

for all $k \geq k_0 \in \mathbb{Z}_+$ and $z_{k_0} \in \mathscr{X}$, where $\gamma \geq 1$ and $\delta \in (0, 1)$ do not depend upon k, k_0, z_{k_0};

(ii) asymptotically stable (AS) if $\lim_{k\to\infty} \|z(k; k_0, z_{k_0})\| = 0$, for all $(k_0, z_{k_0}) \in \mathbb{Z}_+ \times \mathscr{X}$.

Now we prove:

Theorem 2.1 *If the impulsive time instances satisfy the Assumption 2.1 then the following equivalences hold:*

(i) the FJLDE (2.2) is ES if and only if the DTLE (2.9) and (2.10) is ES;
(ii) the FJLDE (2.2) is AS if and only if the DTLE (2.9) and (2.10) is AS.

Proof *(i) The necessity.* We show that the DTLE (2.9) and (2.10) is ES if the FJLDE (2.2) is ES. Let $z(k; k_0, z_{k_0})$, $k \geq k_0$ be an arbitrary solution of the DTLE (2.9) and (2.10). We denote $z_k = z(k; k_0, z_{k_0})$ and define

$$\tilde{x}(t) = T(t; \tau_k)\mathscr{I}(\tau_k)z_k, \tag{2.32}$$

$t \in [\tau_k, \tau_{k+1}]$, $k \geq k_0$. As in the proof of Lemma 2.1 one shows that $\tilde{x}(t)$ is the solution of the FJLDE (2.2) corresponding to the initial condition (τ_{k_0}, z_{k_0}), i.e., $\tilde{x}(t) = \tilde{x}(t; \tau_{k_0}, z_{k_0})$. Additionally, from (2.10) and (2.32) we deduce that

$$\tilde{x}(\tau_k; \tau_{k_0}, z_{k_0}) = z_k, \quad \text{for all } \; k \geq k_0. \tag{2.33}$$

Thus we get

$$\|z_k\| = \|\tilde{x}(\tau_k; \tau_{k_0}, z_{k_0}\| \leq \beta e^{-\alpha(\tau_k - \tau_{k_0})} \|z_{k_0}\|, \tag{2.34}$$

for all $k \geq k_0 \in \mathbb{Z}_+$ where $\beta \geq 1$, $\alpha > 0$ do not depend upon k, k_0, z_{k_0}.

On the other hand (2.26) allows us to deduce that

$$e^{-\alpha(\tau_k - \tau_{k_0})} = \prod_{l=k_0}^{k-1} e^{-\alpha(\tau_{l+1} - \tau_l)} \leq e^{-\alpha \underline{h}(k-k_0)}$$

for all $k \geq k_0 \in \mathbb{Z}_+$. So, (2.34) yields

$$\|z(k; k_0, z_{k_0})\| = \|z_k\| \leq \beta e^{-\alpha \underline{h}(k-k_0)} \|z_{k_0}\|,$$

for all $k \geq k_0 \geq 0$, $z_{k_0} \in \mathscr{X}$. Hence, the solutions of the DTLE (2.9) and (2.10) satisfy (2.31) with $\gamma = \beta$ and $\delta = e^{-\alpha \underline{h}} \in (0, 1)$. Thus, we have proved that the DTLE (2.9) and (2.10) is ES if the FJLDE (2.2) is ES.

We prove now the *sufficiency of (i)*. To this end, we have to show that the FJLDE (2.2) is ES if the DTLE (2.9) and (2.10) is ES. Let $x(t) = x(t; t_0, x_0)$, $t \geq t_0 \geq 0$ be an arbitrary solution of (2.2). Let $k_0 \in \mathbb{Z}_+$ be such that $t_0 \in (\tau_{k_0-1}, \tau_{k_0}]$. As in the proof of Lemma 2.1 one shows that $z_k = z(k; k_0, x(\tau_{k_0})) \triangleq x(\tau_k; \tau_{k_0}, x(\tau_{k_0}))$ solves the DTLE (2.9) and (2.10). Hence,

$$\|x(\tau_k; \tau_{k_0}, x(\tau_{k_0}))\| = \|z(k; k_0, x(\tau_{k_0}))\| \leq \gamma \delta^{k-k_0} \|x(\tau_{k_0})\|, \tag{2.35}$$

for all $k \geq k_0 \in \mathbb{Z}_+$, where $\gamma \geq 1$, $\delta \in (0, 1)$. Employing again (2.26) we may infer that $\delta \leq \delta^{\bar{h}^{-1}(\tau_{l+1}-\tau_l)}$ which leads to $\delta^{k-k_0} \leq \delta^{\bar{b}^{-1}(\tau_k-\tau_{k_0})}$, for all $k \geq k_0$. So, we obtain from (2.35) that

$$\|x(\tau_k; \tau_{k_0}, x(\tau_{k_0}))\| \leq \gamma e^{-\alpha(\tau_k-\tau_{k_0})}, \tag{2.36}$$

for all $k \geq k_0$, where $\alpha \triangleq -\bar{h}^{-1} \log \delta$. For $t \in [\tau_k, \tau_{k+1}]$, we have

$$x(t; \tau_{k_0}, x(\tau_{k_0})) = T(t, \tau_k) x(\tau_k; \tau_{k_0}, x(\tau_{k_0})).$$

Based on (2.3) and (2.36) we deduce:

$$\|x(t; \tau_{k_0}, x(\tau_{k_0}))\| \leq \gamma e^{\lambda(\tau_{k+1}-\tau_k)} e^{-\alpha(\tau_k-\tau_{k_0})} \|x(\tau_{k_0})\| \leq \gamma e^{\lambda \bar{h}} e^{-\alpha(\tau_k-\tau_{k_0})} \|x(\tau_{k_0})\|$$

and obtain

$$\|x(t; \tau_{k_0}, x(\tau_{k_0}))\| \leq \beta_1 e^{-\alpha(t-\tau_{k_0})} \|x(\tau_{k_0})\|, \tag{2.37}$$

for all $t \in (\tau_{k_0}, \infty)$, $\beta_1 \triangleq \gamma e^{(\lambda+\alpha)\bar{h}}$ does not depend upon t and k_0. On the other hand,

$$\|x(\tau_{k_0})\| \leq \|T(\tau_{k_0}, t_0)\| \cdot \|x_0\| \leq e^{\lambda(\tau_{k_0}-t_0)} \|x_0\| \leq e^{\lambda \bar{h}} \|x_0\|. \tag{2.38}$$

This leads to

$$\|x(\tau_{k_0})\| \leq \beta_2 e^{-\alpha(t_{k_0}-t_0)} \|x_0\|,$$

where $\beta_2 = e^{(\lambda+\alpha)\bar{h}}$. Since $x(t; t_0, x_0) = x(t; \tau_{k_0}, x(\tau_{k_0}))$, for all $t \geq \tau_{k_0}$, we deduce via (2.37) and (2.38) that

$$\|x(t; t_0, x_0)\| \leq \beta e^{-\alpha(t-t_0)} \|x_o\|, \quad \text{for all} \quad t \geq \tau_{k_0}, \tag{2.39}$$

where we have denoted $\beta = \gamma e^{2(\lambda+\alpha)\bar{h}}$. Finally, let us remark that (2.39) remains true for $t \in [t_0, \tau_{k_0}]$, too, because in this case we have

$$\|x(t; t_0, x_0)\| \leq \|T(t, t_0)\| \cdot \|x_0\| \leq e^{(\lambda+\alpha)\bar{h}} e^{-\alpha(t-t_0)} \|x_0\|$$

and $e^{(\lambda+\alpha)\bar{h}} \leq \beta$. Thus the proof of the equivalence *(i)* is complete.

The necessity part from *(ii)* follows immediately from (2.33). It remains to prove the sufficiency part of *(ii)*. Let $x(t; t_0, x_0)$ be an arbitrary solution of (2.2). We have to show that $\lim_{t\to\infty} \|x(t; t_0, x_0)\| = 0$ if the DTLE (2.9) and (2.10) is AS. Let $k_0 \in \mathbb{Z}_+$ be such that $t_0 \in (\tau_{k_0-1}, \tau_{k_0}]$. We set $x(\tau_{k_0}) \overset{\Delta}{=} x(\tau_{k_0}; t_0, x_0)$. As in the proof of Lemma 2.1 one shows that the sequence $\{x(\tau_k; \tau_{k_0}, x(\tau_{k_0}))\}_{k\geq k_0}$ is a solution of the DTLE (2.9) and (2.10). Hence $\lim_{k\to\infty} \|x(\tau_k; \tau_{k_0}, x(\tau_{k_0}))\| = 0$. On the other hand from $x(t; t_0, x_0) = x(t; \tau_{k_0}, x(\tau_{k_0}))$, for all $t \geq \tau_{k_0}$ we obtain

$$x(t; t_0, x_0) = T(t; \tau_k)x(\tau_k; \tau_{k_0}, x(\tau_{k_0})).$$

Based on (2.26) we may deduce that

$$\|x(t; t_0, x_0)\| \leq e^{\lambda\bar{h}} \|x(\tau_k; \tau_{k_0}, x(\tau_{k_0}))\|$$

for all $t \in [\tau_k, \tau_{k+1}]$, $k \geq k_0$. Hence,

$$0 \leq \lim_{t\to\infty} \|x(t; t_0, x_0)\| \leq e^{\lambda\bar{h}} \lim_{k\to\infty} \|x(\tau_k; \tau_{k_0}, x(\tau_{k_0}))\| = 0.$$

Thus the proof ends.

The utility of the result proved in Theorem 2.1 will be revealed later in order to derive some criteria for exponential stability and asymptotic stability of a FJLDE of type (2.2). In what follows, we first prove a result which emphasize some properties of the solutions of an affine FJLDE of type (2.14).

Theorem 2.2 *Assume:*

(a) the FJLDE (2.2) is ES;
(b) $\{g_k\}_{k\in\mathbb{Z}_+} \in \ell^2(\mathbb{Z}_+, \mathscr{X})$ *and* $f(\cdot) \in L^2([a, b], \mathscr{X})$.

Under these conditions, for each $(t_0, x_0) \in \mathbb{R}_+ \times \mathscr{X}$ *the solution* $x(\cdot; t_0, x_0)$ *of the affine FJLDE (2.14) has the properties:*

(i) $x(\cdot; t_0, x_0) \in L^2([t_0, \infty), \mathscr{X})$;
(ii) $x(\cdot; t_0, x_0)$ *is a bounded function on* $[t_0, \infty)$;
(iii) $\lim_{t\to\infty} x(t; t_0, x_0) = 0$.

Proof From (2.15), (2.29) it follows that it is sufficient to prove the properties from the statement for $x_0 = 0$. Let $k_o \in \mathbb{Z}_+$ be such that $t_0 \in [\tau_{k_0-1}, \tau_{k_0}]$ and $k_t \in \mathbb{Z}_+$ be such that $t \in [\tau_{k_t}, \tau_{k_t+1}]$. Let

$$x_1(t) \overset{\Delta}{=} \begin{cases} \sum_{l=k_0}^{k_t} \mathbf{T}(t, \tau_l^+)g_l, & \text{for } t \geq \tau_{k_0} \\ 0, & \text{for } t \in [t_0, \tau_{k_0}) \end{cases} \tag{2.40a}$$

$$x_2(t) \triangleq \int_{t_0}^{t} \mathbf{T}(t,s)f(s)\mathrm{d}s. \tag{2.40b}$$

From (2.15) and (2.40) we have $x(t;t_0,0) = x_1(t) + x_2(t)$, for all $t \geq t_0$. To prove *(i)* we have to show that $\int_{t_0}^{\infty} \|x_k(t)\|^2 \mathrm{d}t < \infty, \quad k = 1,2$. From (2.29) together with the definition of $\mathbf{T}(t,\tau_l^+)$ given in (2.8) we obtain via Cauchy–Schwarz inequality that

$$\|x_1(t)\|^2 \leq \beta^2 \sum_{l=k_0}^{k_t} e^{-\alpha(t-\tau_l)} \sum_{l=k_0}^{k_t} e^{-\alpha(t-\tau_l)} \|g_l\|^2. \tag{2.41}$$

Reasoning as in the proof of Theorem 2.1 we obtain via (2.26) that

$$\sum_{l=k_0}^{k_t} e^{-\alpha(t-\tau_l)} \leq \sum_{l=k_0}^{k_t} e^{-\alpha(\tau_{k_t}-\tau_l)} \leq \sum_{l=k_0}^{k_t} e^{-\alpha\underline{h}^{-1}(k_t-l)} \leq c_0,$$

for all $t \geq \tau_{k_0}$ where $c_0 = \frac{e^{\alpha\underline{h}^{-1}}}{e^{\alpha\underline{h}^{-1}}-1}$ does not depend upon t. This allows us to deduce from (2.41) that $\int_{\tau_{k_0}}^{\infty} \|x_1(t)\|^2 \mathrm{d}t \leq \beta^2 c_0 \sum_{l=k_0}^{\infty} \int_{\tau_l}^{\infty} e^{-\alpha(t-\tau_l)}\mathrm{d}t \|g_l\|^2$ which yields $\int_{t_0}^{\infty} \|x_1(t)\|^2 \mathrm{d}t \leq \beta^2 c_0 \alpha^{-1} \sum_{l=0}^{\infty} \|g_l\|^2$. Hence, $x_1(\cdot) \in L^2([t_0,\infty),\mathscr{X})$. Further, (2.40b) and (2.29) lead to

$$\|x_2(t)\|^2 \leq \beta^2 \int_{t_0}^{t} e^{-\alpha(t-s)}\mathrm{d}s \cdot \int_{t_0}^{t} e^{-\alpha(t-s)} \|f(s)\|^2 \mathrm{d}s \leq \frac{\beta^2}{\alpha} \int_{t_0}^{t} e^{-\alpha(t-s)} \|f(s)\|^2 \mathrm{d}s.$$

Thus

$$\int_{t_0}^{\infty} \|x_2(t)\|^2 \mathrm{d}t \leq \frac{\beta^2}{\alpha} \int_{t_0}^{\infty}\int_{t_0}^{t} e^{-\alpha(t-s)} \|f(s)\|^2 \mathrm{d}s\mathrm{d}t = \frac{\beta^2}{\alpha} \int_{t_0}^{\infty} \|f(s)\|^2 \left(\int_{s}^{\infty} e^{-\alpha(t-s)}\mathrm{d}s \right) \mathrm{d}t$$
$$\leq \left(\frac{\beta}{\alpha}\right)^2 \int_{t_0}^{t} \|f(s)\|^2 \mathrm{d}s.$$

So we have shown that $x_2(\cdot) \in L^2([t_0,\infty),\mathscr{X})$. Finally, we may conclude that $x(\cdot;t_0,0) \in L^2([t_0,\infty),\mathscr{X})$ which confirm the validity of *(i)*.
To prove that *(ii)* holds, let us remark that $\|x_1(t)\|^2 \leq \beta \sum_{l=k_0}^{k_t} e^{-2\alpha(t-\tau_l)} \sum_{l=k_0}^{k_t} \|g_l\|^2$. Proceeding as in the first part of the proof one shows that $\sum_{l=k_0}^{k_t} e^{-2\alpha(t-\tau_l)} \leq \tilde{c}_0$, for all $t \geq \tau_{k_0}$ where $\tilde{c}_0 = \frac{e^{2\alpha\underline{h}^{-1}}}{e^{2\alpha\underline{h}^{-1}}-1}$ does not depend upon t. Hence,

$$\|x_1(t)\| \leq \beta\sqrt{\tilde{c}_0}\left(\sum_{l=0}^{\infty}\|g_l\|^2\right)^{1/2}, \tag{2.42}$$

for all $t \geq t_0$. Further, (2.40b) yields

$$\|x_2(t)\| \leq \frac{\beta}{\sqrt{2\alpha}}\left(\int_0^{\infty}\|f(s)\|^2 ds\right)^{1/2}, \tag{2.43}$$

for all $t \geq t_0$. From (2.42) and (2.43) together with (2.29) we may conclude that $x(\cdot; t_0, x_0)$ is bounded on $[t_0, \infty)$ because $x(t; t_0, x_0) = \mathbf{T}(t, t_0)x_0 + x_1(t) + x_2(t)$, for all $t \geq t_0$. To prove *(iii)*, let us remark that $\{x(\tau_k; t_0, x_0)\}_{k\geq k_0}$ solves the affine DTLE:

$$z_{k+1} = \mathscr{L}_d(k)z_k + \zeta_k, \tag{2.44}$$

where $\mathscr{L}_d(k)$ is the linear operator defined in (2.10) and $\zeta_k = T(\tau_{k+1}, \tau_k)g_k + \int_{\tau_k}^{\tau_{k+1}} T(\tau_{k+1}, s)f(s)ds$, for all $k \geq k_0$, where $k_0 \in \mathbb{Z}_+$ is such that $t_0 \in (\tau_{k_0-1}, \tau_{k_0}]$. We show that

$$\sum_{k=k_0}^{\infty}\|\zeta_k\|^2 < \infty. \tag{2.45}$$

We have $\|\zeta_k\|^2 \leq 2\|T(\tau_{k+1}, \tau_k)\|^2\|g_k\|^2 + 2\int_{\tau_k}^{\tau_{k+1}}\|T(\tau_{k+1}, s)\|^2\|f(s)\|^2ds$. Further, (2.3), (2.26) yield $\|\zeta_k\|^2 \leq 2e^{2\lambda\bar{h}}\|g_k\|^2 + 2e^{2\lambda\bar{h}}\int_{\tau_k}^{\tau_{k+1}}\|f(s)\|^2ds$. We get

$$\sum_{k=k_0}^{\infty}\|\zeta_k\|^2 \leq \gamma\sum_{k=k_0}^{\infty}\|g_k\|^2 + \gamma\int_{t_0}^{\infty}\|f(s)\|^2ds < \infty,$$

where $\gamma = 2e^{2\lambda\bar{h}}$. Thus we have shown that (2.45) holds. Hence,

$$\lim_{k\to\infty}\|\zeta_k\|^2 = 0. \tag{2.46}$$

From (2.44) we obtain $x(\tau_k; t_0, x_0) = \mathbf{T}_d(t, k_0)x(\tau_{k_0}; t_0, x_0) + \sum_{l=k_0}^{k-1}\mathbf{T}_d(k, l+1)\zeta_l$, for all $k \geq k_0 + 1$. Based on the equivalence proved in *(i)* from Theorem 2.1 we deduce that if the FJLDE (2.2) is ES then there exist $\gamma \geq 1$, $\delta \in (0, 1)$ such that

$$\|\mathbf{T}_d(k, k_0)\| \leq \gamma\delta^{k-k_0},$$

for all $k \geq k_0$. Hence,

$$\|x(\tau_k; t_0, x_0)\| \leq \gamma\delta^{k-k_0}\|x(\tau_{k_0}; t_0, x_0)\| + \gamma\sum_{l=k_0}^{k-1}\delta^{k-l-1}\|\zeta_l\|, \tag{2.47}$$

for all $k \geq k_0 + 1$. We show that

$$\lim_{k\to\infty} \sum_{l=k_0}^{k-1} \delta^{k-l-1} \|\zeta_l\| = 0. \tag{2.48}$$

To this end, we shall use the Cezaro–Stolz rule for the convergence of the sequence of real numbers (see, e.g., Muresan [3]). We set $\sum_{l=k_0}^{k-1} \delta^{k-l-1} \|\zeta_l\| \stackrel{\Delta}{=} \frac{u_k}{v_k}$, where $u_k = \sum_{k=k_0}^{k-1} \delta^{-l-1} \|\zeta_l\|$ and $v_k = \delta^{-k}$, $k \geq k_0 + 1$. One sees that $\lim\limits_{k\to\infty} v_k = \infty$. Employing (2.46) we deduce that

$$\lim_{k\to\infty} \frac{u_{k+1} - u_k}{v_{k+1} - v_k} = 0.$$

Hence, (2.48) holds. From (2.47) and (2.48) we may conclude that

$$\lim_{k\to\infty} \|x(\tau_k; t_0, x_0)\| = 0. \tag{2.49}$$

For $t \geq t_0$ we write $x(t; t_0, x_0) = T(t, \tau_{k_t})x(\tau_{k_t}; t_0, x_0)$ where $k_t \in \mathbb{Z}_+$ is such that $\tau_{k_t} \leq t < \tau_{k_t+1}$. We have $\|x(t; t_0, x_0)\| \leq e^{\lambda\bar{h}} \|x(\tau_{k_t}; t_0, x_0\|$. Invoking (2.49) we may infer that $\lim\limits_{t\to\infty} \|x(t; t_0, x_0)\| = 0$. Thus the proof is complete.

Remark 2.3 The boundedness property of the solution $x(\cdot; t_0, x_0)$ stated in *(ii)* from Theorem 2.2 remains true also in the case when the assumption *(b)* from the mentioned theorem is replaced by $\sup\limits_{k\in\mathbb{Z}_+} \|g_k\| < \infty$ and $\sup\limits_{t\geq t_0} \|f(t)\| < \infty$. Indeed, if $x_k(\cdot)$, $k = 1, 2$, are the functions defined in (2.40) we have $\|x_1(t)\| \leq c_0 \sup\limits_{l\in\mathbb{Z}_+} \|g_l\|$ and $\|x_2(t)\| \leq \frac{\beta}{\alpha} \sup\limits_{t\geq t_0} \|f(t)\|$, where $c_0 = \frac{e^{\alpha \underline{h}^{-1}}}{e^{\alpha \underline{h}^{-1}} - 1}$.

2.2.2 *The Periodic Case*

In this subsection we assume that the triple $(\mathscr{L}(\cdot), \mathscr{I}(\cdot), \mathfrak{T})$ which describes the FJLDE (2.2) obeys the following assumption:

Assumption 2.2 (a) $t \to \mathscr{L}(t) : \mathbb{R}_+ \to \mathscr{B}(\mathscr{X})$ is a continuous operator valued function with respect to the norm (2.1) which is periodic of period $\theta_c > 0$;

(b) $\tau_k = kh$, for all $k \in \mathbb{Z}_+$, where $h > 0$ is fixed;

(c) there exists $\theta_d \in \mathbb{Z}_+$, $\theta_d \geq 1$ such that $\mathscr{I}((k + \theta_d)h) = \mathscr{I}(kh)$, for all $k \in \mathbb{Z}_+$;

(d) $\dfrac{\theta_c}{h\theta_d}$ is a rational number.

Lemma 2.4 *Under the Assumption 2.2 if $\mathscr{L}_d(k)$, $k \in \mathbb{Z}_+$ are the linear operators defined in (2.10) then we have:*

$$\mathscr{L}_d(k + \tilde{\mathbf{k}}) = \mathscr{L}_d(k), \ \text{for all} \ k \in \mathbb{Z}_+$$

where

$$\tilde{\mathbf{k}} = \tilde{p}\theta_d, \tag{2.50}$$

$\tilde{p}$ being determined from the equality

$$\frac{\theta_c}{h\theta_d} = \frac{\tilde{p}}{\tilde{q}} \tag{2.51}$$

in which $\tilde{p}$, $\tilde{q}$ are coprime natural numbers.

Proof Under the Assumption 2.2 we obtain from (2.10) and (2.51) that

$$\begin{aligned}\mathscr{L}_d(k + \tilde{\mathbf{k}}) &= T((k + \tilde{\mathbf{k}} + 1)h, (k + \tilde{\mathbf{k}})h)\mathscr{I}((k + \tilde{\mathbf{k}})h) \\ &= T((k + 1)h + \tilde{q}\theta_c, kh + \tilde{q}\theta_c)\mathscr{I}((k + \tilde{p}\theta_d)h).\end{aligned}$$

Further, invoking the property *(f)* from the Remark 2.1 together with the equality *(c)* from Assumption 2.2 we get

$$\mathscr{L}_d(k + \tilde{\mathbf{k}}) = T((k + 1)h, kh)\mathscr{I}(kh) = \mathscr{L}_d(k),$$

for all $k \in \mathbb{Z}_+$, which ends the proof.

According to the terminology used in the case of a discrete-time linear equation with periodic coefficients, the linear operator $\mathbf{T}_d(\tilde{\mathbf{k}}, 0)$ will be named **the monodromy operator** defined by the DTLE (2.9) and (2.10). Regarding the exponential stability of the DTLE (2.9) and (2.10) we prove:

Proposition 2.3 *Under the Assumption 2.2 the following are equivalent:*

(i) the DTLE (2.9) and (2.10) is ES;
(ii) the spectral radius of the monodromy operator $\mathbf{T}_d(\tilde{\mathbf{k}}, 0)$ satisfies

$$\rho(\mathbf{T}_d(\tilde{\mathbf{k}}, 0)) < 1.$$

Proof First we prove the implication *(i)* ⇒ *(ii)*. Let $x_0 \in \mathscr{X}$ be arbitrary and $x(t) \triangleq x(k; 0, x_0)$ be the solution of the DTLE (2.9) and (2.10) which satisfies $x(0; 0, x_0) = x_0$. For $k = l\tilde{\mathbf{k}}$ we have the representation formula $x(l\tilde{\mathbf{k}}; 0, x_0) = \mathbf{T}_d^l(\tilde{\mathbf{k}}, 0)x_0$. Since *(i)* holds, we obtain via (2.31) that $\|\mathbf{T}_d^l(\tilde{\mathbf{k}}, 0)x_0\| \leq \gamma\delta^{l\tilde{\mathbf{k}}}\|x_0\|$, for all $l \in \mathbb{Z}_+$, where $\gamma \geq 1$ and $\delta \in (0, 1)$. Hence, $\|\mathbf{T}_d^l(\tilde{\mathbf{k}}, 0)\| \leq \gamma\delta^{l\tilde{\mathbf{k}}}$. From the definition of the spectral radius of a bounded linear operator, we deduce that

$$\rho(\mathbf{T}_d(\tilde{\mathbf{k}}, 0)) = \lim_{l\to\infty} \sup \|\mathbf{T}_d^l(\tilde{\mathbf{k}}, 0)\|^{1/l} \le \delta^{\tilde{\mathbf{k}}} < 1.$$

Thus we have shown that *(ii)* is true if *(i)* holds.
Let us prove now the converse implication. If *(ii)* is true, we may chose $\tilde{\delta}$ such that

$$\rho(\mathbf{T}_d(\tilde{\mathbf{k}}, 0)) < \tilde{\delta} < 1.$$

From the definition of the spectral radius, we deduce that there exists $l_0 \ge 1$ such that $\|\mathbf{T}_d^l(\tilde{\mathbf{k}}, 0)\| \le \tilde{\delta}^l$, for all $l \ge l_0$. This allows us to deduce that

$$\|\mathbf{T}_d^l(\tilde{\mathbf{k}}, 0)\| \le \tilde{\gamma}\tilde{\delta}^l, \tag{2.52}$$

for all $l \ge 0$, where $\tilde{\gamma} \triangleq \max\{1, \sup_{0\le j\le l_0} (\tilde{\delta}^{-j}\|\mathbf{T}_d(l\tilde{\mathbf{k}}, 0)\|)\}$. Let $k \ge k_0 \ge 0$ be arbitrary. Let $m = \left[\frac{k}{\tilde{\mathbf{k}}}\right]$ and $n = \left[\frac{k_0}{\tilde{\mathbf{k}}}\right]$. Assume firstly that $m \ge n+1$. In this case we may write

$$\mathbf{T}_d(k, k_0) = \mathbf{T}_d(k, m\tilde{\mathbf{k}}) \cdot \mathbf{T}_d(m\tilde{\mathbf{k}}, (m+1)\tilde{\mathbf{k}}) \cdot \mathbf{T}_d((m+1)\tilde{\mathbf{k}}, k_0).$$

Employing (2.13) and (2.52) we deduce that

$$\|\mathbf{T}_d(k, k_0)\| \le \beta^2\tilde{\gamma}e^{2\tilde{\mathbf{k}}\lambda h}\tilde{\delta}^{m-n-1}$$

where we have denoted $\beta = \max_{0\le j\le \tilde{\mathbf{k}}-1} \{\|\mathscr{I}(jh)\|\}$. Based on the properties of the integer part of a real number, we obtain

$$\|\mathbf{T}_d(k, k_0)\| \le \hat{\gamma}\hat{\delta}^{k-k_0} \tag{2.53}$$

for all $k \ge k_0 \ge 0$ such that $\left[\frac{k}{\tilde{\mathbf{k}}}\right] - \left[\frac{k_0}{\tilde{\mathbf{k}}}\right] \ge 1$, where we have denoted $\hat{\gamma} = \beta^2\tilde{\gamma}e^{2\tilde{\mathbf{k}}\lambda h}\tilde{\delta}^{-2\tilde{\mathbf{k}}}$ and $\hat{\delta} = \tilde{\delta}^{1/\tilde{\mathbf{k}}} \in (0, 1)$. If $\left[\frac{k}{\tilde{\mathbf{k}}}\right] - \left[\frac{k_0}{\tilde{\mathbf{k}}}\right] < 1$, it follows that $0 \le k_0 < k \le \tilde{\mathbf{k}}$. So, (2.53) leads to

$$\|\mathbf{T}_d(k, k_0)\| \le \gamma_1\hat{\delta}^{k-k_0},$$

for all $k \ge k_0 \ge 0$, with $\gamma_1 = \max\{\hat{\delta}^{-\tilde{\mathbf{k}}}, \hat{\gamma}\}$. Thus the proof is complete.

Combining the results proved in *(i)* of Theorem 2.1 and the Proposition 2.3, we obtain:

Corollary 2.1 *Under the Assumption 2.2 the following are equivalent:*

(i) the FJLDE (2.2) is ES;
(ii) the spectral radius of the monodromy operator $\mathbf{T}_d(\tilde{\mathbf{k}}, 0)$ *defined by the DTLE (2.9) and (2.10) satisfies* $\rho(\mathbf{T}_d(\tilde{\mathbf{k}}, 0)) < 1$.

Let us remark that if the Assumption 2.2 holds, then the FJLDE (2.2) may be defined on the whole real axis. In this case, it takes the form:

$$\dot{x}(t) = \mathscr{L}(t)x(t), \qquad kh < t \leq (k+1)h \tag{2.54a}$$
$$x(kh^+) = \mathscr{I}(kh)x(kh), \qquad k \in \mathbb{Z}. \tag{2.54b}$$

Here $\mathscr{L}(\cdot)$ extends to the whole real axis, setting $\mathscr{L}(t) \triangleq \mathscr{L}\left(t - \left[\frac{t}{\theta_c}\right]\theta_c\right)$ if $t < 0$. Similarly, $\mathscr{I}(kh)$ is defined for $k < 0$ by $\mathscr{I}(kh) \triangleq \mathscr{I}\left(\left(k - \left[\frac{k}{\theta_d}\right]\theta_d\right)h\right)$.

Remark 2.4 Under the Assumption 2.2 the following hold:

(a) the linear evolution operator $T(t,s)$ defined by the ordinary linear differential equation $\dot{x}(t) = \mathscr{L}(t)x(t), \quad t \in \mathbb{R}$, is defined for any $t, s \in \mathbb{R}$. In this case, the equalities from Remark 2.1 *(f)* become: $T(t + j\theta_c, s + j\theta_c) = T(t,s)$, for all $t, s \in \mathbb{R},\ j \in \mathbb{Z}$;

(b) as we already saw in the proof of Lemma 2.4, the equalities (2.10) become:

$$\mathscr{L}_d(k) = T((k+1)h, kh)\mathscr{I}(kh), \tag{2.55}$$

if the Assumption 2.2 is fulfilled. From *(a)* we may infer that (2.55) makes sense for $k \in \mathbb{Z}$. Under the same arguments as in the proof of Lemma 2.4 one shows that

$$\mathscr{L}_d(k + \tilde{\mathbf{k}}) = \mathscr{L}_d(k), \quad \text{for all } \ k \in \mathbb{Z}. \tag{2.56}$$

From (2.11) and (2.56) we deduce that:

$$\mathbf{T}_d(k + j\tilde{\mathbf{k}}, k_0 + j\tilde{\mathbf{k}}) = \mathbf{T}(k, k_0), \tag{2.57}$$

for all $k, k_0, j \in \mathbb{Z}$;

(c) the equalities (2.12) become:

$$\mathbf{T}(t,t_0) = T\left(t, \left[\frac{t}{h}\right]h\right)\mathscr{I}\left(\left[\frac{t}{h}\right]h\right)\mathbf{T}_d\left(\left[\frac{t}{h}\right], \left[\frac{t_0}{h}\right]+1\right)T\left(\left(\left[\frac{t_0}{h}\right]+1\right)h, t_0\right)$$

if $\left[\frac{t_0}{h}\right] < \left[\frac{t}{h}\right]$ and $\mathbf{T}(t,t_0) = T(t,t_0)$, if $\left[\frac{t_0}{h}\right] = \left[\frac{t}{h}\right]$ for all $t \geq t_0,\ t, t_0 \in \mathbb{R}$;

(d)

$$\mathbf{T}(t + j\tilde{\theta}, t_0 + j\tilde{\theta}) = \mathbf{T}(t, t_0), \tag{2.58}$$

for all $t \geq t_0,\ t, t_0 \in \mathbb{R},\ j \in \mathbb{Z}$, where

$$\tilde{\theta} = \tilde{q}\theta_c = \tilde{p}h\theta_d, \tag{2.59}$$

$\tilde{p}, \tilde{q}$ being coprime natural numbers which are satisfying (2.51);

(e) from (2.58) we may deduce that if (2.29) hold for $t \geq t_0 \geq 0$, then it still holds for any $t \geq t_0,\ t, t_0 \in \mathbb{R}$. Indeed

$$\|\mathbf{T}(t, t_0)\| = \|\mathbf{T}\left(t - \left[\frac{t_0}{\tilde{\theta}}\right]\tilde{\theta}, t_0 - \left[\frac{t_0}{\tilde{\theta}}\right]\tilde{\theta}\right)\| \leq \beta e^{-\alpha(t-t_0)},$$

for all $t \geq t_0$, $t, t_0 \in \mathbb{R}$;

(f) based on (2.57) one shows that

$$\|\mathbf{T}_d(k, k_0)\| \leq \gamma\delta^{k-k_0} \tag{2.60}$$

holds for any $k \geq k_0$, $k, k_0 \in \mathbb{Z}$ if it is true for $k \geq k_0 \geq 0$.

Let us consider now the non-homogeneous FJLDE (2.14) under the condition that the triple $(\mathscr{L}(\cdot), \mathscr{I}(\cdot), \mathfrak{T})$ satisfies the Assumption 2.2 and the forcing terms g_k, $f(\cdot)$ satisfy the following assumption:

Assumption 2.3 $\{g_k\}_{k\in\mathbb{Z}} \subset \mathscr{X}$ is a periodic sequence of period θ_d, while $t \to f(t)$ is a continuous vector valued function which is periodic with period θ_c, θ_c, θ_d being the same as in the Assumption 2.2.

From the Remark 2.3 we deduce that under the Assumption 2.3, all solutions $x(\cdot; t_0, x_0)$ of (2.14) are bounded if the FJLDE (2.2) is ES.

Further, we show that under the Assumptions 2.2 and 2.3 among the solutions of the non-homogeneous FJLDE (2.14), there exists a periodic solution if the FJLDE (2.2) is ES. First, let us remark that under the Assumption 2.2 *(b),* the representation formula (2.15) becomes:

$$x(t; t_0, x_0) = \mathbf{T}(t, t_0)x_0 + \sum_{l=\left[\frac{t_0}{h}\right]+1}^{\left[\frac{t}{h}\right]} \mathbf{T}(t, lh^+)g_l + \int_{t_0}^{t} \mathbf{T}(t, s)f(s)\mathrm{d}s, \text{ if } \left[\frac{t_0}{h}\right] < \left[\frac{t}{h}\right] \tag{2.61a}$$

$$x(t; t_0, x_0) = T(t, t_0)x_0 + \int_{t_0}^{t} T(t, s)f(s)\mathrm{d}s, \text{ if } \left[\frac{t_0}{h}\right] = \left[\frac{t}{h}\right]. \tag{2.61b}$$

If the periodicity property stated in the Assumptions 2.2 and 2.3 are satisfied, then the non-homogeneous FJLDE (2.14) may be defined on the whole real axis.
To this end, one can be take $f(t) = f\left(t - \left[\frac{t}{\theta_c}\right]\theta_c\right)$ if $t < 0$ and $g_k = g_{k-\left[\frac{k}{\theta_d}\right]\theta_d}$, for $k \in \mathbb{Z}$, $k < 0$. In this case, (2.14) becomes:

$$\dot{x}(t) = \mathscr{L}(t)x(t) + f(t), \qquad kh < t \leq (k+1)h \tag{2.62a}$$

$$x(kh^+) = \mathscr{I}(kh)x(kh) + g_k, \qquad k \in \mathbb{Z}. \tag{2.62b}$$

As in the proof of Lemma 2.2 one shows that for each $(t_0, x_0) \in \mathbb{R} \times \mathscr{X}$ the solution $x(\cdot; t_0, x_0)$ of (2.62) has the representation formula of type (2.61).

The next result shows that under the exponential stability of the FJLDE (2.54), the non-homogeneous FJLDE (2.62) has a unique solution which is bounded on the whole real axis.

Theorem 2.3 *Assume:*

(a) the Assumptions 2.2 and 2.3 are fulfilled;
(b) the FJLDE (2.54) is ES.

Under these conditions the non-homogeneous FJLDE (2.62) has a unique solution which is bounded on $\mathbb{R}$*. This solution is left continuous in each* t *and is described by:*

$$\tilde{x}(t) = \sum_{l=-\infty}^{[\frac{t}{h}]} \mathbf{T}(t, lh^+)g_l + \int_{-\infty}^{t} \mathbf{T}(t, s)f(s)ds, \tag{2.63}$$

for all $t \in \mathbb{R}$ *and has the properties:*

(i) $\tilde{x}(\cdot)$ *is a periodic function of period* $\tilde{\theta}$*, where* $\tilde{\theta}$ *was defined in (2.59);*
(ii) for any $(t_0, x_0) \in \mathbb{R} \times \mathcal{X}$ *the solution* $x(\cdot; t_0, x_0)$ *of (2.62) satisfies:*

$$\|x(t; t_0, x_0) - \tilde{x}(t)\| \leq \beta e^{-\alpha(t-t_0)}\|x_0 - \tilde{x}(t_0)\|,$$

for all $t \geq t_0$*, where* $\beta \geq 1$*,* $\alpha > 0$ *do not depend upon* t*,* t_0*,* x_0*.*

Proof We divide the proof into several steps.

Step 1. We show that both the sum and the integral from the right hand side of (2.63) are convergent. According to the Remark 2.2 *(c)* and the Remark 2.4 *(c)*, it follows that if the FJLDE (2.54) is ES, then there exists $\beta \geq 1$, $\alpha > 0$, such that $\|\mathbf{T}(t, s)\| \leq \beta e^{-\alpha(t-s)}$, for all $t \geq s$, $t, s \in \mathbb{R}$. From $\mathbf{T}(t, lh^+) = \lim_{\substack{s \to lh \\ s > lh}} \mathbf{T}(t, s)$ we obtain that $\|\mathbf{T}(t, lh^+)\| \leq \beta e^{-\alpha(t-lh)}$, for all $t \geq lh$. Let $k = \left[\frac{t}{h}\right]$. We have

$$\left\|\sum_{l=-\infty}^{[\frac{t}{h}]} \mathbf{T}(t, lh^+)g_l\right\| \leq c_g\beta \sum_{l=-\infty}^{k} e^{-\alpha(t-lh)} \leq c_g\beta \sum_{l=-\infty}^{k} e^{-\alpha(k-l)h}$$

which yields

$$\left\|\sum_{l=-\infty}^{[\frac{t}{h}]} \mathbf{T}(t, lh^+)g_l\right\| \leq \beta_1 c_g, \tag{2.64}$$

for all $t \in \mathbb{R}$ where $c_g := \max_{0 \leq j \leq \theta_d} \|g_j\|$, $\beta_1 = \beta\frac{e^{\alpha h}}{e^{\alpha h}-1}$. Also we have

$$\left\|\int_{-\infty}^{t} \mathbf{T}(t, s)f(s)\mathrm{d}s\right\| \leq \frac{\beta}{\alpha}c_f, \tag{2.65}$$

for all $t \in \mathbb{R}$, where $c_f := \max_{0 \le s \le \theta_c} \|f(s)\|$. So, (2.64) and (2.65) confirm the convergence in norm of the sum and of the integral from the right hand side of (2.63).

Step 2. Let us consider the non-homogeneous DTLE

$$z_{k+1} = \mathscr{L}_d(k)x_k + \zeta_k \tag{2.66}$$

$k \in \mathbb{Z}$, where $\mathscr{L}_d(k)$ is defined in (2.55) and

$$\zeta_k := T((k+1)h, kh)g_k + \int_{kh}^{(k+1)h} T((k+1)h, s)f(s)\mathrm{d}s. \tag{2.67}$$

From (2.56) we infer that $\mathscr{L}_d(k)$ is a periodic sequence of period $\tilde{\mathbf{k}}$.

We show now, that $\{\zeta_k\}_{k\in\mathbb{Z}} \subset \mathscr{X}$ is also a periodic sequence of period $\tilde{\mathbf{k}}$. Involving (2.59) together with $\tilde{\mathbf{k}} = \tilde{p}\theta_d$ we get

$$\begin{aligned}\zeta_{k+\tilde{\mathbf{k}}} &= T((k+\tilde{\mathbf{k}}+1)h, (k+\tilde{\mathbf{k}})h)g_{k+\tilde{\mathbf{k}}} + \int_{k+\tilde{\mathbf{k}}h}^{(k+\tilde{\mathbf{k}}+1)h} T((k+\tilde{\mathbf{k}}+1)h, s)f(s)\mathrm{d}s \\ &= T((k+1)h+\tilde{\theta}, kh+\tilde{\theta})g_{k+\tilde{p}\theta_d} + \int_{kh+\tilde{\theta}}^{(k+1)h+\tilde{\theta}} T((k+1)h+\tilde{\theta}, s)f(s)\mathrm{d}s.\end{aligned} \tag{2.68}$$

Employing (2.58) together with the periodicity properties of the sequence $\{g_k\}_{k\in\mathbb{Z}}$ and of the function $f(\cdot)$ stated in the Assumption 2.3, we obtain (after a change of the variable of integration) that the last part from (2.68) reduces to (2.67). This confirms that $\{\theta_k\}_{k\in\mathbb{Z}}$ is a periodic sequence of period $\tilde{\mathbf{k}}$. On the other hand, from *(i)* of Theorems 2.1 and (2.56), we deduce that there exist $\gamma \ge 1$, $\delta \in (0, 1)$ such that (2.60) holds for any $k \ge k_0$, $k, k_0 \in \mathbb{Z}$. For each $k \in \mathbb{Z}$ we define

$$\hat{z}_k = \sum_{l=-\infty}^{k-1} \mathbf{T}_d(k, l+1)\zeta_l. \tag{2.69}$$

Based on (2.60) and the boundedness of the sequence $\{\zeta_l\}_{l\in\mathbb{Z}}$ we deduce that for each $k \in \mathbb{Z}$ the limit $\lim_{n\to-\infty} \sum_{l=n}^{k-1} \mathbf{T}_d(k, l+1)\zeta_k$ is well defined in $(\mathscr{X}, \|\cdot\|)$ and it provides the value of $\hat{z}_k$ via (2.69). By direct calculation, involving (2.11) and (2.69) one obtains that $\{\hat{z}_k\}_{k\in\mathbb{Z}}$ solves (2.66). Employing (2.57) and the periodicity property of the sequence $\{\zeta_l\}_{l\in\mathbb{Z}}$ we get

$$\hat{z}_{k+\tilde{\mathbf{k}}} = \sum_{l=-\infty}^{k+\tilde{\mathbf{k}}-1} \mathbf{T}_d(k+\tilde{\mathbf{k}}, l+1)\zeta_l = \sum_{l=-\infty}^{k-1} \mathbf{T}_d(k+\tilde{\mathbf{k}}, l+1+\tilde{\mathbf{k}})\zeta_{l+\tilde{\mathbf{k}}}$$
$$= \sum_{l=-\infty}^{k-1} \mathbf{T}_d(k, l+1)\zeta_l = \hat{z}_k,$$

for all $k \in \mathbb{Z}$. So, $\{\hat{z}_k\}_{k\in\mathbb{Z}}$ is a periodic solution of period $\tilde{\mathbf{k}}$ of (2.66).

Let $\{\check{z}_k\}_{k\in\mathbb{Z}}$ be an arbitrary bounded solution of the DTLE (2.66). For an arbitrary $n \le k$, $n, k \in \mathbb{Z}$ we have $\check{z}_k = \mathbf{T}_d(k, n)\check{z}_n + \sum_{l=n}^{k-1} \mathbf{T}_d(k, l+1)\zeta_l$. Letting $n \to -\infty$, we obtain via (2.60) and the boundedness of the sequence $\{\check{z}_k\}_{k\in\mathbb{Z}}$, that $\check{z}_k = \sum_{l=-\infty}^{k-1} \mathbf{T}_d(k, l+1)\zeta_l = \hat{z}_k$. Thus, we have shown that $\{\hat{z}_k\}_{k\in\mathbb{Z}}$ defined in (2.69) is the unique bounded on $\mathbb{Z}$ solution of the non-homogeneous DTLE (2.66).

Step 3. Let $k \in \mathbb{Z}$ be arbitrary, but fixed. For $t \in (kh, (k+1)h]$, we define

$$\hat{x}(t) \triangleq T(t, kh)\left(\mathscr{I}(kh)\hat{z}_k + g_k\right) + \int_{kh}^{t} T(t, s)f(s)\mathrm{d}s. \tag{2.70}$$

From (2.70) one obtains that

$$\frac{\mathrm{d}}{\mathrm{d}t}\hat{x}(t) = \mathscr{L}(t)\hat{x}(t) + f(t), \quad kh < t \le (k+1)h. \tag{2.71}$$

Writing (2.70) for $t \in ((k-1)h, kh]$ we obtain via (2.11) and (2.69) that $\hat{x}(kh) = \hat{z}_k$, for all $k \in \mathbb{Z}$. This allows us to rewrite (2.70) as:

$$\hat{x}(t) = T(t, kh)\left(\mathscr{I}(kh)\hat{x}(kh) + g_k\right) + \int_{kh}^{t} T(t, s)f(s)\mathrm{d}s.$$

Hence,

$$\hat{x}(kh^+) = \lim_{\substack{t\to kh\\ t>kh}} \hat{x}(t) = \mathscr{I}(kh)\hat{x}(kh) + g_k, \tag{2.72}$$

for all $k \in \mathbb{Z}$. From (2.71) and (2.72) we may conclude that $t \to \hat{x}(t) : \mathbb{R} \to \mathscr{X}$ is a solution of the non-homogeneous FJLDE (2.62). Now we show that $\hat{x}(\cdot)$ is a periodic function of period $\tilde{\theta}$ where $\tilde{\theta}$ was defined in (2.59). To this end, we take into account that, if $t \in (kh, (k+1)h]$, then $t + \tilde{\theta} \in ((k+\tilde{p}\theta_d)h, (k+\tilde{p}\theta_d+1)h]$. In this case (2.70) yields

$$\hat{x}(t+\tilde{\theta}) = T(t+\tilde{\theta}, kh+\tilde{\theta})(\mathscr{I}((k+\tilde{p}\theta_d)h)\hat{z}_{k+\tilde{\mathbf{k}}} + g_{k+\tilde{\mathbf{k}}}) + \int_{kh+\tilde{\theta}}^{t+\tilde{\theta}} T(t+\tilde{\theta}, s)f(s)\mathrm{d}s. \tag{2.73}$$

Employing (2.58) together with the periodicity property of the sequences $\{\mathscr{I}(kh)\}_{k\in\mathbb{Z}}$, $\{\hat{z}_k\}_{k\in\mathbb{Z}}$ and $\{g_k\}_{k\in\mathbb{Z}}$ as well as of the function $f(\cdot)$ we may infer that the right hand side of (2.73) will be just the right hand side of (2.70). Thus we obtained that $\hat{x}(t+\tilde{\theta}) = \hat{x}(t)$, for all $t \in \mathbb{R}$ which confirms the periodicity property of the solution $\hat{x}(\cdot)$. Further, we show that $\hat{x}(\cdot)$ is the unique bounded on $\mathbb{R}$ solution of (2.62). Let $t \to \check{x}(t) : \mathbb{R} \to \mathscr{X}$ be another bounded on $\mathbb{R}$ solution of (2.62). One obtains that the sequence $\{\check{x}(kh)\}_{k\in\mathbb{Z}}$ is a bounded on $\mathbb{Z}$ solution of the non-homogeneous DTLE (2.66). In Step 2 we have shown that the unique bounded on $\mathbb{Z}$ solution of (2.66) is $\{\hat{z}_k\}_{k\in\mathbb{Z}}$ defined by (2.69). Hence $\check{x}(kh) = \hat{z}_k$, for all $k \in \mathbb{Z}$. Integrating (2.62a) satisfied by $\check{x}(\cdot)$ we obtain via (2.62b) that

$$\check{x}(t) = T(t, kh)\left(\mathscr{I}(kh)\check{x}(kh) + g_k\right) + \int_{kh}^{t} T(t,s) f(s)\mathrm{d}s, \tag{2.74}$$

for all $kh < t \le (k+1)h$, $k \in \mathbb{Z}$. From (2.70) and (2.74) we deduce that $\check{x}(t) = \hat{x}(t)$, for all $t \in \mathbb{R}$, because $\check{x}(kh) = \hat{z}_k$, for all $k \in \mathbb{Z}$. Substituting (2.69) in (2.70) one obtains by direct calculation, involving (2.67) and the equality from Remark 2.4 (c) that $\hat{x}(\cdot)$ defined via (2.70) coincides to $\tilde{x}(\cdot)$ defined by (2.63). Thus we have proved in an indirect way that $\tilde{x}(\cdot)$ defined by (2.63) satisfies (i). Finally, let us remark that if $x(\cdot; t_0, x_0)$ is an arbitrary solution of (2.62) then

$$y(t) \overset{\Delta}{=} x(t; t_0, x_0) - \tilde{x}(t), \ t \ge t_0,$$

is a solution of FJLDE (2.54). So, the exponential decreasing of the deviation of $x(\cdot; t_0, x_0)$ from $\tilde{x}(t)$ stated in (ii) is a direct consequence of the assumption (b). Thus the proof ends.

The result proved in Theorem 2.3 will be used in Chap. 7 in connection with the computation of the performance achieved by the filter.

2.2.3 The Time-Invariant Case

For the beginning we define the time-invariant character of a FJLDE (2.2).

Definition 2.3 We say that the FJLDE (2.2) is in the time-invariant case if the following conditions are simultaneously satisfied:

(a) $\mathscr{L}(t) = \mathscr{L} \in \mathscr{B}(\mathscr{X})$, for all $t \in \mathbb{R}_+$;
(b) $\mathscr{I}(\tau_k) = \mathscr{I} \in \mathscr{B}(\mathscr{X})$, for all $k \in \mathbb{Z}_+$;
(c) $\tau_k = kh$, for all $k \in \mathbb{Z}_+$, with $h > 0$ fixed.

In this case (2.2) takes the form

$$\dot{x}(t) = \mathscr{L}x(t), \qquad kh < t \le (k+1)h \tag{2.75a}$$

$$x(kh^+) = \mathscr{I}x(kh), \qquad k \in \mathbb{Z}. \tag{2.75b}$$

According to the Remark 2.1 (e) the linear operator (2.10) associated to the system (2.75) takes the form

$$\mathscr{L}_d = e^{\mathscr{L}h}\mathscr{I}. \tag{2.76}$$

It is obvious that $\mathscr{L}_d \in \mathscr{B}(\mathscr{X})$.

The time-invariant analogous of the result proved in Corollary 2.1 is:

Corollary 2.2 *For the FJLDE (2.75), the following are equivalent:*

(i) the system (2.75) is ES;
(ii) the spectral radius of the linear operator introduced by (2.76) satisfies

$$\rho(\mathscr{L}_d) < 1.$$

Proof follows immediately employing Theorem 2.1 applied in the case of the DTLE: $z_{k+1} = \mathscr{L}_d z_k, \; k \in \mathbb{Z}$.

Let us consider now, the special case of a FJLDE (2.2) for which the conditions (a) and (b) from Definition 2.3 are satisfied, but condition (c) is violated. In this case the system (2.2) becomes

$$\dot{x}(t) = \mathscr{L}x(t), \qquad \tau_k < t \le \tau_{k+1} \tag{2.77a}$$
$$x(\tau_k^+) = \mathscr{I}x(\tau_k), \qquad k \in \mathbb{Z}_+. \tag{2.77b}$$

The accompanying DTLE of type (2.9) is defined by the operators

$$\mathscr{L}_d(k) = e^{\mathscr{L}(\tau_{k+1}-\tau_k)}\mathscr{I}, \;\; k \in \mathbb{Z}_+. \tag{2.78}$$

From (2.9) and (2.78), one sees that in the case of the FJLDE of type (2.77), the accompanying discrete-time equation is a time-varying discrete-time linear equation.

From Theorem 2.1 we deduce that even if the coefficients of (2.77) are constant, both the exponential stability and asymptotic stability must be approached as in the time-varying case. This fact will become clearer in the next section.

Let us consider the non-homogeneous system of forward jump linear differential equation

$$\dot{x}(t) = \mathscr{L}x(t) + f(t), \qquad kh < t \le (k+1)h \tag{2.79a}$$
$$x(kh^+) = \mathscr{I}x(kh) + g_k, \qquad k \in \mathbb{Z}, \tag{2.79b}$$

where $\mathscr{L}, \mathscr{I} \in \mathscr{B}(\mathscr{X})$ and the forcing terms $\{g_k\}_{k\in\mathbb{Z}} \subset \mathscr{X}$, $f : \mathbb{R} \to \mathscr{X}$ satisfy the Assumption 2.3.

Remark 2.5 A special case of (2.79) is that where $g_k = g \in \mathscr{X}$, for all $k \in \mathbb{Z}$ and $f(t) = f \in \mathscr{X}$, for all $t \in \mathbb{R}$. In this special case, we may take $\theta_d = 1$ and $\theta_c = h$, and we have $\frac{\theta_c}{\theta_d h} = 1$. It is worth noticing that the choose of $\theta_c = h$ is based on the fact that a constant function may be viewed as a periodic function with arbitrary period.

In the case of the non-homogeneous FJLDE (2.79) we obtain the following result which may be viewed as a time-invariant counterpart of that proved in Theorem 2.3.

Theorem 2.4 *If the spectral radius of the linear and bounded operator defined by (2.76) satisfies $\rho(\mathcal{L}_d) < 1$, then the following hold:*

- (i) *if $\{g_k\}_{k\in\mathbb{Z}} \subset \mathcal{X}$ is a periodic sequence of period $\theta_d \geq 1$ and $t \to f(t) : \mathcal{X} \to \mathcal{X}$ is a periodic continuous function of period $\theta_c > 0$ and if $\frac{\theta_c}{h\theta_d}$ is a rational number, then the non-homogeneous FJLDE (2.79) has a unique bounded on $\mathbb{R}$ solution. Additionally, this solution is a left continuous function in each t and it is periodic function of period $\tilde{\theta}$, $\tilde{\theta}$ being defined as in (2.59);*
- (ii) *if $g_k = g \in \mathcal{X}$, for all $k \in \mathbb{Z}$ and $f(t) = f \in \mathcal{X}$, for all $t \in \mathbb{R}$, then the unique bounded on $\mathbb{R}$ solution of the non-homogeneous FJLDE (2.79) is a periodic function of period h.*

Proof (i) follows from Theorem 2.3, because (2.79) is a special case of (2.62). The validity of the statement (ii) follows immediately from (i) together with the Remark 2.5.

2.3 Linear Differential Equation with Jumps on a Hilbert Space

In this section $(\mathcal{X}, \langle\cdot,\cdot\rangle)$ is a real Hilbert space. We denote $|\cdot|$ the norm on $\mathcal{X}$ induced by the inner product $\langle\cdot,\cdot\rangle$, i.e.,

$$|x| = (\langle x, x\rangle)^{1/2} \tag{2.80}$$

for all $x \in \mathcal{X}$. The norm of type (2.1) induced by (2.80) will be denoted $\|\cdot\|_2$. In this framework beside the FJLDE (2.2), the triple $(\mathcal{L}(\cdot), \mathcal{I}(\cdot), \mathfrak{T})$ defines the following BJLDE

$$-\dot{y}(t) = \mathcal{L}^*(t)y(t), \qquad \tau_k \leq t < \tau_{k+1} \tag{2.81a}$$

$$y(\tau_k^-) = \mathcal{I}^*(\tau_k)y(\tau_k), \qquad k \in \mathbb{Z}_+. \tag{2.81b}$$

As usual, the superscript * is used to denote the adjoint of a linear bounded operator on $\mathcal{X}$. Based on Remark 2.1 (b) we may infer that the anticausal linear evolution operator on $\mathcal{X}$ defined by the operator valued function $\mathcal{L}^*(\cdot)$ is

$$T^a(t, s) = T^*(s, t), \quad \text{for all } s \geq t \geq 0$$

$T^*(\cdot,\cdot)$ being the adjoint of the linear evolution operator $T(\cdot,\cdot)$ on $\mathscr{X}$, defined by the ordinary linear differential equation

$$\dot{x}(t) = \mathscr{L}(t)x(t), \quad t \in \mathbb{R}_+.$$

Let $y(\cdot; t_0, y_0)$ be a solution of the BJLDE (2.81). Denoting $z_k := y(\tau_k^-; t_0, y_0)$, we obtain that $\{z_k\}_{0\le k\le k_0}$ solves the following backward DTLE:

$$z_k = \mathscr{L}_d^*(k)z_{k+1} \tag{2.82}$$

where k_0 is such that $t_0 \in [\tau_{k_0}, \tau_{k_0+1})$ and

$$\mathscr{L}_d^*(k) = \mathscr{I}^*(\tau_k)T^*(\tau_{k+1}, \tau_k). \tag{2.83}$$

One sees that the operator from the right hand side of (2.83) is just the adjoint of the operator of the right hand side of (2.10). This explains the notation adopted for the operators which define the DTLE (2.82). The anticausal evolution operator defined on $\mathscr{X}$ by the DTLE (2.82) is

$$\mathbf{T}_d^*(l,k) = \begin{cases} \mathscr{L}_d^*(k)\mathscr{L}_d^*(k+1)\cdots\mathscr{L}_d^*(l-1), & \text{if } \; l > k \\ I_{\mathscr{X}}, \quad \text{if } \; l = k. \end{cases}$$

The solutions $y(\cdot; t_0, y_0)$ of (2.81) have the representation:

$$y(t; t_0, y_0) = \mathbf{T}^*(t_0, t)y_0,$$

for all $0 \le t \le t_0$, $y_0 \in \mathscr{X}$, where

$$\mathbf{T}^*(t_0, t) = T^*(\tau_k, t)\mathbf{T}_d^*(k_0, k)\mathscr{I}^*(\tau_{k_0})T^*(t_0, \tau_{k_0}) \tag{2.84}$$

if $t \in [\tau_k, \tau_{k+1}]$ and $t_0 \in [\tau_{k_0}, \tau_{k_0+1})$, $k < k_0$ and

$$\mathbf{T}^*(t_0, t) = T^*(t_0, t), \tag{2.85}$$

if $\tau_k \le t \le t_0 < \tau_{k+1}$. In the sequel, (2.81) will be named dual-FJLDE associated to the FJLDE (2.2).

Let us consider the non-homogeneous BJLDE

$$-\dot{y}(t) = \mathscr{L}^*(t)y(t) + f(t), \qquad \tau_k \le t < \tau_{k+1} \tag{2.86a}$$

$$y(\tau_k^-) = \mathscr{I}^*(\tau_k)y(\tau_k) + g_k, \qquad k \in \mathbb{Z}_+. \tag{2.86b}$$

The next result will play an important role in the derivation of some criteria for exponential stability of a FJLDE of type (2.2) which defines a positive evolution on an ordered Hilbert space.

Theorem 2.5 *Under the Assumption 2.1, if the FJLDE (2.2) is ES then the following hold:*

(*i*) *for each bounded sequence* $\{g_k\}_{k\in\mathbb{Z}_+} \subset \mathcal{X}$ *and for each bounded and piecewise continuous vector valued function* $f(\cdot) : \mathbb{R}_+ \to \mathcal{X}$ *having lateral limits in each* $t \in \mathbb{R}_+$ *the non-homogeneous BJLDE (2.86) has a unique bounded on* $\mathbb{R}_+$ *solution. This solution is a right continuous function and has the representation formula:*

$$\tilde{y}(t) = \sum_{l=k}^{\infty} \mathbf{T}^*(\tau_l^-, t)g_l + \int_t^{\infty} \mathbf{T}^*(s, t)f(s)ds \tag{2.87}$$

for all $t \in \mathbb{R}_+$, *where* k *is such that* $t \in [\tau_{k-1}, \tau_k)$;

(*ii*) *if the triple* $(\mathcal{L}(\cdot), \mathcal{I}(\cdot), \mathfrak{T})$ *satisfies the Assumption 2.2 and the forcing terms* $\{g_k\}_{k\in\mathbb{Z}_+}$ *and* $f(\cdot)$ *satisfy the Assumption 2.3 then,* $\tilde{y}(\cdot)$ *is a periodic function of period* $\tilde{\theta}$, *with* $\tilde{\theta}$ *defined in (2.59);*

(*iii*) *if the FJLDE (2.2) is time invariant (in the sense of Definition 2.3) and if* $g_k = g \in \mathcal{X}$, *for all* $k \in \mathbb{Z}_+$ *and* $f(t) = f \in \mathcal{X}$, *for all* $t \in \mathbb{R}_+$, *then the unique bounded on* $\mathbb{R}_+$ *solution of the BJLDE (2.86) is a periodic function of period* h *and it is right continuous in any* $t \in \mathbb{R}_+$.

Proof Let us consider the non-homogeneous backward DTLE:

$$z_k = \mathcal{L}_d^*(k)z_{k+1} + \hat{\zeta}_k, \tag{2.88}$$

where

$$\hat{\zeta}_k \triangleq g_k + \mathcal{I}^*(\tau_k) \int_{\tau_k}^{\tau_{k+1}} T^*(s, \tau_k)f(s)\mathrm{d}s, \quad k \in \mathbb{Z}_+. \tag{2.89}$$

Writing (2.29) for $\tau_{k-1} < t_0 \le \tau_k \le t \le \tau_{k+1}$ we deduce that $\|\mathcal{I}(\tau_k)\|_2 \le \beta$, for all $k \in \mathbb{Z}_+$ if the FJLDE (2.2) is ES. Since $\|\mathcal{I}^*(\tau_k)\|_2 = \|\mathcal{I}(\tau_k)\|_2$ we may conclude that under the considered assumptions the sequence $\{\hat{\zeta}_k\}_{k\in\mathbb{Z}_+}$ defined in (2.89) is bounded. For each $k \in \mathbb{Z}_+$ we define

$$\hat{z}_k = \sum_{l=k}^{\infty} \mathbf{T}_d^*(l, k)\hat{\zeta}_l. \tag{2.90}$$

Since FJLDE (2.2) is assumed to be ES, we deduce via (*i*) of Theorem 2.1 that there exist $\gamma > 0$, $\delta \in (0, 1)$ such that $\|\mathbf{T}_d^*(l, k)\|_2 \le \gamma\delta^{l-k}$ for all $l \ge k \ge 0$. This allows us to infer that the sum from the right hand side of (2.90) is convergent in norm and defines a solution of the DTLE (2.88). Additionally $|z_k| \le \hat{\gamma}$, for all $k \in \mathbb{Z}_+$, where $\hat{\gamma} > 0$ is a constant not depending upon $k \in \mathbb{Z}_+$. Applying Theorem 2.7 (*i*) from Drăgan et al. [4] in the case of the DTLE (2.88) we deduce that $\{\hat{z}_k\}_{k\in\mathbb{Z}_+}$ is the unique bounded solution of (2.88). Let $t \to \hat{y}(t) : \mathbb{R}_+ \to \mathcal{X}$ be defined by

$$\hat{y}(t) = T^*(\tau_k, t)\hat{z}_k + \int_t^{\tau_k} T^*(s,t)f(s)\mathrm{d}s, \tag{2.91}$$

if $t \in [\tau_{k-1}, \tau_k)$, $k \in \mathbb{Z}_+$. Invoking Remark 2.1 (b) we deduce that $\hat{y}(\cdot)$ is differentiable in any $t \neq \tau_k$, $k \in \mathbb{Z}_+$ and satisfies (2.86a). Furthermore (2.91) together with the exponential stability of (2.2) lead to $|\hat{y}(t)| \leq \hat{c}$, for all $t \in \mathbb{R}$, where $\hat{c}$ does not depend upon t. From (2.91) we obtain that $\hat{y}(\tau_k^-) = \lim_{\substack{t\to\tau_k \\ t<\tau_k}} \hat{y}(t) = \hat{z}_k$, for all $k \in \mathbb{Z}_+$. This allows us to rewrite (2.88) satisfied by $\{\hat{z}_k\}_{k\in\mathbb{Z}_+}$ as:

$$\begin{aligned}\hat{y}(\tau_k^-) &= \mathscr{I}^*(\tau_k)\left(T^*(\tau_{k+1},\tau_k)\hat{y}(\tau_{k+1}^-) + \int_{\tau_k}^{\tau_{k+1}} T^*(s,\tau_k)f(s)\mathrm{d}s\right) + g_k \\ &= \mathscr{I}^*(\tau_k)\hat{y}(\tau_k) + g_k.\end{aligned}$$

This confirms the fact that $\hat{y}(\cdot)$ defined in (2.91) satisfies (2.86b). So, we have obtained that $\hat{y}(\cdot)$ defined in (2.91) is a bounded on $\mathbb{R}_+$ solution of (2.86).

Let us now show that it is the unique bounded on $\mathbb{R}_+$ solution of (2.86). Let $\check{y}(\cdot)$ be another bounded on $\mathbb{R}_+$ solution of the non-homogeneous BJLDE (2.86). By direct computation one obtains that the sequence $\{\check{y}(\tau_k^-)\}_{k\in\mathbb{Z}_+}$ is a bounded on $\mathbb{Z}_+$ solution of the DTLE (2.88)–(2.89). From the uniqueness of the bounded on $\mathbb{Z}_+$ solution of the DTLE (2.88)–(2.89) it follows that $\check{y}(\tau_k^-) = \hat{z}_k$, for all $k \in \mathbb{Z}_+$. This allows us to conclude via (2.90)–(2.91) that $\check{y}(t)$ coincides with $\hat{y}(t)$ on each interval $[\tau_{k-1}, \tau_k]$, $k \geq 1$, which confirms the uniqueness of the bounded on $\mathbb{R}_+$ solution of (2.86). Substituting (2.90) in (2.91) we obtain based on (2.89) and (2.84) written for t_0 replaced by s that $\hat{y}(t)$ has the representation formula described by the right hand side of (2.87). This means that $\hat{y}(t) = \tilde{y}(t)$, for all $t \in \mathbb{R}_+$. So the proof of (i) is complete.

To prove (ii), let us remark that if the triple $(\mathscr{L}(\cdot), \mathscr{I}(\cdot), \mathfrak{T})$ satisfies the Assumption 2.2, we obtain via Lemma 2.4 that the sequence $\{\mathscr{L}_d(k)\}_{k\in\mathbb{Z}_+}$ is a periodic sequence of period $\tilde{\mathbf{k}}$, where $\tilde{\mathbf{k}}$ is defined in (2.50). Under the Assumptions 2.2–2.3, (2.89) takes the form

$$\hat{\zeta}_k = g_k + \mathscr{I}^*(kh)\int_{kh}^{(k+1)h} T^*(s,kh)f(s)\mathrm{d}s.$$

Employing (2.58) and (2.59) together with the periodic property of the sequence $\{g_k\}_{k\in\mathbb{Z}_+}$ and of the function $f(\cdot)$, we deduce that $\hat{\zeta}_k = \hat{\zeta}_{k+\tilde{\mathbf{k}}}$, for all $k \in \mathbb{Z}_+$. Applying Theorem 2.5 (ii) from Drăgan et al. [4] in the case of the backward DTLE (2.88)–(2.89) we deduce that the unique bounded on $\mathbb{Z}_+$ solution of this equation defined via (2.90) is a periodic sequence of period $\tilde{\mathbf{k}}$. Further, invoking again (2.58)–(2.59) we obtain that $\hat{y}(\cdot)$ defined by (2.91) is a periodic function of period $\tilde{\theta}$. The conclusion follows from the equality $\tilde{y}(t) = \hat{y}(t)$, for all $t \in \mathbb{R}_+$.

(iii) If the FJLDE (2.2) is time invariant, then it can be viewed as a periodic system with $\theta_c = h$ and $\theta_d = 1$. In this case (2.50) is satisfied with $\tilde{p} = \tilde{q} = 1$. From (ii), it follows that the unique bounded on $\mathbb{R}_+$ solution of (2.86) is a periodic function of period $\tilde{\theta} = h$. Thus the proof ends.

2.4 The Case of Forward Jump Linear Differential Equations that Define a Positive Evolution on an Ordered Hilbert Space

In this section we consider a special class of FJLDEs of type (2.2) that define a positive evolution on an ordered Hilbert space. For this kind of differential equations we shall derive a set of criteria for exponential stability. They will be used in the next chapter to obtain criteria for exponential stability in the mean-square sense of a large class of systems of stochastic linear differential equations with impulses.

2.4.1 Several Preliminary Issues

Let $\mathscr{X}$ be a real Hilbert space and $\mathscr{X}^+ \subset \mathscr{X}$ be a convex cone. The convex cone $\mathscr{X}^+$ induces the order relation $\succcurlyeq$ on $\mathscr{X}$. This means that if $x, y \in \mathscr{X}$ we have $x \succcurlyeq y$ if $x - y \in \mathscr{X}^+$. By convention $x \preccurlyeq y$ if and only if $y - x \in \mathscr{X}^+$.

In the next definition we recall the main types of convex cones involved in this section.

Definition 2.4 (i) A convex cone $\mathscr{X}^+$ is a **pointed cone** if whenever $x \in \mathscr{X}^+$ and $-x \in \mathscr{X}^+$ it follows that $x = 0$.

(ii) A convex cone $\mathscr{X}^+$ is a **closed cone** if it is a closed set in the topology induced by the norm (2.80).

(iii) A convex cone $\mathscr{X}^+$ is a **solid cone** if its interior $Int\ \mathscr{X}^+$ is a non-empty set.

(iv) $\mathscr{X}^+$ is a **normal convex cone** if there exists a constant $b > 0$ with the property that for any $x, y \in \mathscr{X}^+$ such that $x \preccurlyeq y$ it follows that $|x| \leq b|y|$.

(v) We say that the norm $|\cdot|$ is monotone with respect to the convex cone $\mathscr{X}^+$ if for any $x, y \in \mathscr{X}^+$ such that $x \preccurlyeq y$ it follows that $|x| \leq |y|$.

For other types of cones and useful properties we refer to Krasnosel'skiy [5], Krasnosel'skiy [6] and Damm [7]. We recall that for each $\xi \in Int\ \mathscr{X}^+$ the Minkovski functional $x \to |x|_\xi : \mathscr{X} \to \mathbb{R}_+$ is defined by

$$|x|_\xi = \inf\{t > 0 | -\xi \leq \frac{1}{t}x \leq \xi\}. \tag{2.92}$$

The next result is a consequence of the developments from Drăgan et al. [4] Sect. 2.1.2.

Proposition 2.4 *If $\mathscr{X}^+ \subset \mathscr{X}$ is a convex closed, solid, normal cone such that $\mathscr{X}^+ \neq \mathscr{X}$, then for each $\xi \in Int\ \mathscr{X}^+$, the Minkovski functional defined by (2.92) is a norm on $\mathscr{X}$ equivalent to the norm $|\cdot|$ defined in (2.80), more precisely, there exist $c_k > 0$, $k = 1, 2$ (possible depending on ξ) such that*

$$c_1|x| \leq |x|_\xi \leq c_2|x|, \quad \textit{for all} \quad x \in \mathscr{X}. \tag{2.93}$$

If $T \in \mathscr{B}(\mathscr{X})$ we denote $\|T\|_\xi$ the norm of type (2.1) induced by the norm $|x|_\xi$.

Remark 2.6 (a) From (2.93) one obtains that there exist positive constants (possibly depending on ξ), $\tilde{c}_k$, $k = 1, 2$, such that

$$\tilde{c}_1\|T\|_\xi \leq \|T\|_2 \leq \tilde{c}_2\|T\|_\xi, \quad \text{for all} \quad T \in \mathscr{B}(\mathscr{X}).$$

(b) The equality $\|T\|_2 = \|T^*\|_2$ which holds for any $T \in \mathscr{B}(\mathscr{X})$ is not true when $\|\cdot\|_2$ is replaced by $\|\cdot\|_\xi$. However, one may prove that there exist positive constants $\hat{c}_k$, $k = 1, 2$, (possibly depending on $\xi \in Int\ \mathscr{X}$) such that

$$\hat{c}_1\|T\|_\xi \leq \|T^*\|_\xi \leq \hat{c}_2\|T\|_\xi, \quad \text{for all} \quad T \in \mathscr{B}(\mathscr{X}).$$

Definition 2.5 We say that a linear operator $T : \mathscr{X} \to \mathscr{X}$ is a positive operator and we shall write $T \geq 0$, if $T\mathscr{X}^+ \subset \mathscr{X}^+$.

Corollary 2.3 *If $T \in \mathscr{B}(\mathscr{X})$ is a positive operator, then*

$$\|T\|_\xi = |T\xi|_\xi \quad \textit{for all} \quad \xi \in Int\ \mathscr{X}^+.$$

Proof Follows from (2.1) written for $|\cdot|_\xi$ and $\|\cdot\|_\xi$ together with the statements (vii)–(ix) from Theorem 2.1 from Drăgan et al. [4].

Definition 2.6 (a) We say that the linear differential equation

$$\dot{x}(t) = \mathscr{L}(t)x(t) \tag{2.94}$$

defines a positive evolution on the ordered space $(\mathscr{X}, \mathscr{X}^+)$ if the corresponding linear evolution operator $T(t, t_0)$ satisfies $T(t, t_0)\mathscr{X}^+ \subset \mathscr{X}^+$, for all $t \geq t_0 \geq 0$.

(b) We say that the FJLDE (2.2) defines a positive evolution on the interval $\mathscr{I} \subset \mathbb{R}_+$ with respect to the ordered Hilbert space $(\mathscr{X}, \mathscr{X}^+)$ if

$$x(t; t_0, x_0) \in \mathscr{X}^+, \quad \text{for all} \quad t \geq t_0, \ t, t_0 \in \mathscr{I} \quad \text{whenever} \quad x_0 \in \mathscr{X}^+. \tag{2.95}$$

If $\mathscr{I} = \mathbb{R}_+$, we shall say that the FJLDE (2.2) generates a positive evolution on the ordered Hilbert space $(\mathscr{X}, \mathscr{X}^+)$ if (2.95) holds for any $t \geq t_0 \geq 0$.

The next result will play an important role in the derivation of the main results of this section.

Proposition 2.5 *The following hold:*

(i) the FJLDE (2.2) defines a positive evolution on the interval $\mathcal{I}$ *with respect to the ordered Hilbert space* $(\mathcal{X}, \mathcal{X}^+)$ *if and only if the linear evolution operator defined in (2.5)–(2.7) satisfies* $\mathbf{T}(t, t_0)\mathcal{X}^+ \subset \mathcal{X}^+ \ \forall \ t \geq t_0, \ t, t_0 \in \mathcal{I}$;

(ii) if the FJLDE (2.2) defines a positive evolution on the ordered Hilbert space $(\mathcal{X}, \mathcal{X}^+)$ *then for any* $k \in \mathbb{Z}_+$ *the following are true:*

(a) $\mathcal{I}(\tau_k)\mathcal{X}^+ \subset \mathcal{X}^+$;

(b) $T(t, t_0)\mathcal{X}^+ \subset \mathcal{X}^+$ *for all* $\tau_k \leq t_0 \leq t \leq \tau_{k+1}$, $T(t, t_0)$ *being the linear evolution operator on* $\mathcal{X}$ *defined by the linear differential equation (2.94);*

(iii) if the FJLDE (2.2) defines a positive evolution on the ordered Hilbert space $(\mathcal{X}, \mathcal{X}^+)$ *then the linear operators* $\mathcal{L}_d(k)$ *defined in (2.10) are positive operators.*

Proof (i) follows immediately from (2.4) and the Definition 2.6 (ii). To prove (ii), let $k \in \mathbb{Z}_+$ be arbitrary but fixed. If the FJLDE (2.2) defines a positive evolution on the ordered space $(\mathcal{X}, \mathcal{X}^+)$ if follows that $x(t; \tau_k, x_0) \in \mathcal{X}^+$ for all $t \geq \tau_k$ if $x_0 \in \mathcal{X}^+$. We have $x(t; \tau_k, x_0) = T(t, \tau_k)\mathcal{I}(\tau_k)x_0$ if $t \in (\tau_k, \tau_{k+1}]$. Hence,

$$T(t, \tau_k)\mathcal{I}(\tau_k)x_0 \in \mathcal{X}^+ \text{ if } t \in (\tau_k, \tau_{k+1}].$$

This allows us to deduce that $\mathcal{I}(\tau_k)x_0 = \lim\limits_{\substack{t \to \tau_k \\ t > \tau_k}} T(t, \tau_k)\mathcal{I}(\tau_k)x_0 \in \mathcal{X}^+$ because $\mathcal{X}^+$ is a closed convex cone. Since x_0 is arbitrary in $\mathcal{X}^+$, we may infer that the property (a) holds. Further, from (2.4) and (2.7) we deduce that $x(t; t_0, x_0) = T(t, t_0)x_0$ for all $\tau_k < t_0 \leq t \leq \tau_{k+1}$. This means that

$$T(t, t_0)x_0 \in \mathcal{X}^+ \text{ for all } \tau_k < t_0 \leq t \leq \tau_{k+1} \text{ and for all } x_0 \in \mathcal{X}^+. \tag{2.96}$$

One obtains that (2.96) remains true if $t_0 \to \tau_k$, $t_0 > \tau_k$ as well as $t \to \tau_{k+1}$, $t < \tau_{k+1}$, because $\mathcal{X}^+$ is a closed subset. This confirms the assertion (b) of (ii).

(iii) is obvious because for each $k \in \mathbb{Z}_+$ the linear operator $\mathcal{L}_d(k)$ is obtained by composing two positive operators. Thus the proof ends.

Now, we introduce the following subset associated to the convex cone $\mathcal{X}^+$

$$(\mathcal{X}^+)^* = \{x \in \mathcal{X} \,|\, \langle x, y \rangle \geq 0 \text{ for all } y \in \mathcal{X}^+\}.$$

One checks that $(\mathcal{X}^+)^*$ is also a convex cone. It is called the dual cone of $\mathcal{X}^+$.

Definition 2.7 The cone $\mathcal{X}^+$ is named **self-dual** if $(\mathcal{X}^+)^* = \mathcal{X}^+$.

The following result collects several properties of a self-dual convex cone.

Proposition 2.6 *For a self-dual convex cone the following hold:*

(i) $\mathscr{X}^+$ *is a pointed convex cone;*
(ii) *the norm* $|\cdot|$ *defined by (2.80) is monotone with respect to the cone* $\mathscr{X}^+$*;*
(iii) $\mathscr{X}^+$ *is a normal convex cone;*
(iv) *for any* $T \in \mathscr{B}(\mathscr{X})$ *the following are equivalent:*

(a) T *is a positive operator;*
(b) *the adjoint operator* T^* *is a positive operator.*

Proof (i) If both x and $-x$ lie in $\mathscr{X}^+$, then $\langle x, -x\rangle \geq 0$ which yields $|x|^2 \leq 0$ which is equivalent with $x = 0$.

(ii) Let $x, y \in \mathscr{X}^+$ be arbitrary, such that $x \succcurlyeq y$. So, both $x - y$ and $x + y$ lie in $\mathscr{X}^+$. This leads to $\langle x - y, x + y\rangle \geq 0$ which is equivalent to $|x|^2 \geq |y|^2$.

(iii) From Definition 2.4 and (ii) above, it follows that $\mathscr{X}^+$ is a normal convex cone with the constant $b = 1$.

(iv) To prove the implication $(a) \Rightarrow (b)$ we have to show that $T^*y \in \mathscr{X}^+$ for any $y \in \mathscr{X}^+$. Let $x \in \mathscr{X}^+$ be arbitrary. We have

$$\langle x, T^*y\rangle = \langle Tx, y\rangle \geq 0$$

because $Tx \in \mathscr{X}^+$ which is a self-dual cone. The implication $(b) \Rightarrow (a)$ is obtained in a similar way.

2.4.2 Criteria for Exponential Stability of a FJLDE that Defines a Positive Evolution on an Ordered Hilbert Space

Let us remark that based on Remark 2.6 (a) the property of the exponential stability of a FJLDE (2.2) defined on an ordered Hilbert space may be defined with respect to any of the Minkowski norms $|\cdot|_\xi$, $\xi \in Int\ \mathscr{X}^+$ if the order relation is induced by a closed, solid, normal, convex cone.

The next result shows that in the case of a FJLDE of type (2.2) which defines a positive evolution on the ordered Hilbert space $(\mathscr{X}, \mathscr{X}^+)$, it is sufficient that only one solution starting from the interior of the convex cone to have an exponential decrease in order to be sure that all its solutions have an exponential decrease with the same decrease rate.

Proposition 2.7 *Assume that* $\mathscr{X}^+ \subset \mathscr{X}$ *is a closed, solid, normal, convex cone. If the FJLDE (2.2) defines a positive evolution on the ordered space* $(\mathscr{X}, \mathscr{X}^+)$ *then the following are equivalent:*

(i) the FJLDE (2.2) is ES;
(ii) there exists $\xi \in Int\ \mathscr{X}^+$ such that

$$|x(t; t_0, \xi)|_\xi \leq \beta e^{-\alpha(t-t_0)} \text{ for all } t \geq t_0 \geq 0, \tag{2.97}$$

where $\beta \geq 1$, $\alpha > 0$ do not depend upon t and t_0.

Proof The implication $(i) \Rightarrow (ii)$ follows immediately from the definition of the exponential stability together with (2.93) and from the fact that $|\xi|_\xi = 1$ (see Theorem 2.1 (ix) from Drăgan et al. [4]).
Now, we prove $(ii) \Rightarrow (i)$. From (2.4) and (2.97) we get

$$|\mathbf{T}(t, t_0)\xi|_\xi \leq \beta e^{-\alpha(t-t_0)}, \quad \text{for all } t \geq t_0 \geq 0.$$

Employing Proposition 2.5 (i) and Corollary 2.3 we further obtain

$$\|\mathbf{T}(t, t_0)\|_\xi \leq \beta e^{-\alpha(t-t_0)}, \quad \text{for all } t \geq t_0 \geq 0.$$

The conclusion is obtained from Remark 2.2 (c). Thus the proof ends.

Definition 2.8 (a) We say that a sequence $\{g_k\}_{k\in\mathbb{Z}_+} \subset \mathscr{X}^+$ is **uniform positive,** and we shall write $g_k \ggcurly 0$, $k \in \mathbb{Z}_+$ if there exists $\xi \in Int\ \mathscr{X}^+$ and a scalar $\varepsilon > 0$ such that $g_k \succeq \varepsilon\xi$, for all $k \in \mathbb{Z}_+$.
(b) We say that a vector valued function $t \to f(t) : \mathbb{R}_+ \to \mathscr{X}^+$ is uniform positive and we shall write $f(t) \ggcurly 0$, $t \in \mathbb{R}_+$, if there exists $\xi \in Int\ \mathscr{X}^+$ and a scalar $\varepsilon > 0$ such that $f(t) \succcurlyeq \varepsilon\xi$, for all $t \in \mathbb{R}_+$.

Remark 2.7

(a) The uniform positive property is not related to a certain vector $\xi \in Int\ \mathscr{X}^+$. Indeed, if $\zeta \in Int\ \mathscr{X}^+$ is an arbitrary vector, then there exists $\gamma > 0$ such that $0 \preccurlyeq \gamma\zeta \preccurlyeq \xi$. Hence, $g_k \succcurlyeq \hat{\varepsilon}\zeta$ for all $k \in \mathbb{Z}_+$, where $\hat{\varepsilon} = \varepsilon\gamma$.
(b) A sequence $\{g_k\}_{k\in\mathbb{Z}_+}$ will be named uniform negative and we shall write $g_k \llcurly 0$, $k \in \mathbb{Z}_+$ if $\{-g_k\}_{k\in\mathbb{Z}_+}$ is uniform positive. Analogously, a function $t \to h(t) : \mathbb{R}_+ \to \mathscr{X}$ is uniform negative and we shall write $h(t) \ll 0$, $t \in \mathbb{R}_+$ if $-h(t)$ is uniform positive.

The next result will be used in the proof of the main result of this section.

Lemma 2.5 *Assume:*

(a) $\mathscr{X}$ is a real Hilbert space ordered by the ordering relation induced by the solid, closed, convex cone $\mathscr{X}^+$ such that $\mathscr{X}^+ \neq \mathscr{X}$;
(b) $t \to \mathscr{L} : \mathbb{R}_+ \to \mathscr{B}(\mathscr{X})$ is a bounded and continuous operator valued function.

Let $T(s, t)$, $s \geq t \geq 0$ be the linear evolution operator on $\mathscr{X}$ defined by the linear differential equation

$$\dot{x}(t) = \mathscr{L}(t)x(t). \tag{2.98}$$

Under the considered assumptions, for each $\xi \in Int\, \mathscr{X}^+$ there exist the constants $\delta_k > 0$, $k = 1, 2$, depending upon ξ, such that:

(i) $T^*(s,t)\xi \succcurlyeq \frac{1}{2}\xi$ for all $t \leq s \leq t + \delta_1$;
(ii) $T(s,t)\xi \succcurlyeq \frac{1}{2}\xi$ for all $t \leq s \leq t + \delta_2$.

Proof To prove (i) let us consider $y_s(t) \stackrel{\Delta}{=} T^*(s,t)\xi$, $t \leq s$. Employing Remark 2.1 (i) we obtain $\dot{y}_s(t) = -\mathscr{L}^*(t)y_s(t)$. This allows us to deduce that

$$y_s(t) = \xi + \int_t^s \mathscr{L}^*(\tau)y_s(\tau)d\tau \quad \text{for all } \ 0 \leq t \leq s. \tag{2.99}$$

This yields $|y_s(t)| \leq |\xi| + \int_t^s \|\mathscr{L}^*(\tau)\|_2 |y_s(\tau)| d\tau$. Let $\lambda = \sup_{\tau \in \mathbb{R}_+} \|\mathscr{L}(\tau)\|_2$. Taking into account that $\|\mathscr{L}(\tau)\|_2 = \|\mathscr{L}^*(\tau)\|_2$, we obtain via Gronwall's Lemma (see, e.g., Gronwall [8] or Amann [9]) that

$$|y_s(t)| \leq |\xi| e^{\lambda(s-t)} \quad \text{for all } \ 0 \leq t \leq s. \tag{2.100}$$

Further, (2.99) and (2.100) lead to $|y_s(t) - \xi| \leq |\xi| \left(e^{\lambda(s-t)} - 1\right)$. Invoking (2.93) we deduce that

$$|y_s(t) - \xi|_\xi \leq c_2 |\xi| \left(e^{\lambda(s-t)} - 1\right) \quad \text{for all } \ 0 \leq t \leq s < \infty.$$

Choose $\delta_1 > 0$ such that $c_2 |\xi| \left(e^{\lambda \delta_1} - 1\right) < \frac{1}{2}$. With this choice we may write

$$|y_s(t) - \xi|_\xi \leq \frac{1}{2},$$

for all $t, s \in \mathbb{R}_+$ such that $t \leq s \leq t + \delta_1$. This is equivalent to

$$|2(y_s(t) - \xi)|_\xi \leq 1, \quad \text{for all } \ 0 \leq t \leq s \leq t + \delta_1.$$

We deduce that $-\xi \preccurlyeq 2(y_s(t) - \xi) \preccurlyeq \xi$, for all $t, s \in \mathbb{R}_+$ such that $t \leq s \leq t + \delta_1$. Thus we may conclude that $T^*(s,t)\xi \succcurlyeq \frac{1}{2}\xi$, for all $t, s \in \mathbb{R}_+$ with $0 \leq s - t \leq \delta_1$, which confirms (i).

To prove (ii), we consider $x_t(s) := T(s,t)\xi$, $s \geq t \geq 0$ and repeat step by step the reasons from above. Thus the proof ends.

The next result emphasizes several additional properties of the bounded on $\mathbb{R}_+$ solution of a non-homogeneous BJLDE of type (2.86).

Theorem 2.6 *Assume:*

(a) $\mathscr{X}$ is a real Hilbert space ordered by an order relation induced by the closed, solid, self-dual convex cone $\mathscr{X}^+$;

(b) the FJLDE (2.2) defines a positive evolution on the ordered space $(\mathscr{X}, \mathscr{X}^+)$;
(c) the impulsive time instants sequence $\{\tau_k\}_{k\in\mathbb{Z}_+}$ *satisfies the Assumption 2.1.*

Under these assumptions, the following are true if the FJLDE (2.2) is ES:

(i) for each bounded sequence $\{g_k\}_{k\in\mathbb{Z}_+} \subset \mathscr{X}^+$ *and each bounded and piecewise continuous vector valued function* $t \to f(t) : \mathbb{R}_+ \to \mathscr{X}$ *with lateral limits in each t satisfying the property that* $f(t) \succeq 0$, *for all* $t \in \mathbb{R}_+$ *the unique bounded on* $\mathbb{R}_+$ *solution* $\tilde{y}(\cdot)$ *of the non-homogeneous BJLDE (2.86) has the property that* $\tilde{y}(t) \succeq 0$ *for all* $t \in \mathbb{R}_+$;
(ii) if the bounded sequence $\{g_k\}_{k\in\mathbb{Z}_+} \subset Int\ \mathscr{X}$ *is uniform positive and the function* $f(\cdot)$ *has the properties from (i) above, then the unique bounded on* $\mathbb{R}_+$ *solution of the non-homogeneous BJLDE (2.86) is uniform positive.*

Proof From Theorem 2.5, we know that for each bounded sequence $\{g_k\}_{k\in\mathbb{Z}_+} \subset \mathscr{X}$ and for each bounded and piecewise continuous function with lateral limits in each t, the BJLDE (2.86) has a unique bounded on $\mathbb{R}_+$ solution $\tilde{y}(\cdot)$. We have to show that if $g_k \in \mathscr{X}^+$, for each $k \in \mathbb{Z}_+$ and $f(t) \in \mathscr{X}^+$, for all $t \in \mathbb{R}_+$, then $\tilde{y}(t) \in \mathscr{X}^+$, for all $t \in \mathbb{R}_+$. This fact follows immediately from the representation formula (2.87) of the solution $\tilde{y}(\cdot)$ together with the Proposition 2.5 (i) and Proposition 2.6 (iv). Thus we have obtained that (i) holds.

To prove that (ii) is also true, let us remark that from (2.86) written for the unique bounded solution $\tilde{y}(\cdot)$ we obtain that the sequence $\{\tilde{y}(\tau_k^-)\}_{k\in\mathbb{Z}_+}$ is a bounded solution of the DTLE (2.88)–(2.89). From (2.89) one sees that $\hat{\zeta}_k \gg 0$, $k \in \mathbb{Z}_+$ if $g_k \gg 0$, $k \in \mathbb{Z}_+$. Applying Theorem 2.1 (i) we deduce that the accompanying DTLE (2.9) is ES if FJLDE (2.2) is ES. From Theorem 2.5 (iv) in Drăgan et al. [4] applied to the non-homogeneous DTLE (2.88)–(2.89) we deduce that the unique bounded on $\mathbb{Z}_+$ solution of this backward DTLE is uniform positive. Hence, for $\xi \in Int\ \mathscr{X}^+$, there exists $\varepsilon > 0$ such that

$$\tilde{y}(\tau_k^-) \succeq \varepsilon\xi, \quad \text{for all} \ \ k \in \mathbb{Z}_+. \tag{2.101}$$

From (2.86a) written for the solution $\tilde{y}(\cdot)$ we obtain that

$$\tilde{y}(t) = T^*(\tau_{k+1}, t)\tilde{y}(\tau_{k+1}^-) + \int_t^{\tau_{k+1}} T^*(s,t)f(s)ds, \quad \tau_k \le t \le \tau_{k+1}, \quad k \in \mathbb{Z}_+.$$

Invoking Proposition 2.5 (ii) and Proposition 2.6 (iv) we infer that for any $\tau_k \le t \le s \le \tau_{k+1}$, $k \in \mathbb{Z}_+$ we have $\int_t^{\tau_{k+1}} T^*(s,t)f(s)ds \succeq 0$ because $\mathscr{X}^+$ is an closed convex cone. Hence, $\tilde{y}(t) \succeq T^*(\tau_{k+1}, t)\tilde{y}(\tau_{k+1}^-)$, for all $\tau_k \le t \le \tau_{k+1}$, $k \in \mathbb{Z}_+$. Let $\delta_1 > 0$ be the constant provided by Lemma 2.5 (i). Dividing the interval $[t, \tau_{k+1}]$ into a union of disjoint intervals with a length less or equal to δ_1, we can deduce (via (2.101)) that $\tilde{y}(t) \succeq \hat{\varepsilon}\xi$, for all $t \in \mathbb{R}_+$, where $\hat{\varepsilon} > 0$ is a constant depending upon the ratio $\frac{\bar{h}}{\delta_2}$. Thus the proof is complete.

The next result provides a list of necessary and sufficient conditions for the exponential stability of a FJLDE that defines a positive evolution on an ordered Hilbert space.

Theorem 2.7 *Under the assumptions from Theorem 2.6, the following are equivalent:*

(i) the FJLDE (2.2) is ES;

(ii) for any bounded and uniform positive sequence $\{g_k\}_{k\in\mathbb{Z}_+} \subset Int\ \mathscr{X}^+$ *and any bounded piecewise continuous vector valued function* $t \to f(t) : \mathbb{R}_+ \to \mathscr{X}$ *which has lateral limits in each* $t \in \mathbb{R}_+$ *and takes values in* $\mathscr{X}^+$*, the non-homogeneous BJLDE*

$$-\dot{y}(t) = \mathscr{L}^*(t)y(t) + f(t), \qquad \tau_k \le t < \tau_{k+1}, \tag{2.102a}$$

$$y(\tau_k^-) = \mathscr{I}^*(\tau_k)y(\tau_k) + g_k, \qquad k \in \mathbb{Z}_+, \tag{2.102b}$$

has a bounded and uniform positive solution $y(\cdot) : \mathbb{R}_+ \to \mathscr{X}$*;*

(iii) there exists a bounded and uniform positive sequence $\{\tilde{g}_k\}_{k\in\mathbb{Z}_+} \subset Int\ \mathscr{X}^+$ *and a bounded and continuous on each interval* (τ_k, τ_{k+1}) *vector valued function* $t \to \tilde{f}(t) : \mathbb{R}_+ \to \mathscr{X}$ *which has lateral limits in any* τ_k *and takes values in* $\mathscr{X}^+$*, such that the corresponding non-homogeneous BJLDE of type (2.102) has a bounded solution* $\tilde{y}(\cdot) : \mathbb{R}_+ \to \mathscr{X}^+$*;*

(iv) there exists a bounded and uniform positive function $t \to z(t) : \mathbb{R}_+ \to \mathscr{X}$ *which is differentiable on each interval* (τ_k, τ_{k+1}) *with bounded derivative and both* $z(\cdot)$*,* $\dot{z}(\cdot)$ *have lateral limits in any* τ_k *and satisfy the following linear differential inequality with jumps*

$$\dot{z}(t) + \mathscr{L}^*(t)z(t) \preccurlyeq 0, \qquad \tau_k \le t < \tau_{k+1}, \tag{2.103a}$$

$$\mathscr{I}^*(\tau_k)z(\tau_k) - z(\tau_k^-) \ll 0, \qquad k \in \mathbb{Z}_+. \tag{2.103b}$$

Proof The implication $(i) \Rightarrow (ii)$ follows from Theorem 2.6 (ii), while the implication $(ii) \Rightarrow (iii)$ is obvious. Now, we prove $(iii) \Rightarrow (i)$. If (iii) is true, then from (2.102b) written for $\tilde{g}_k$ instead of g_k and $\tilde{y}(\cdot)$ instead of $y(\cdot)$, we conclude via Proposition 2.5 (ii) that $\tilde{y}(\tau_k^-) \gg 0$, $k \in \mathbb{Z}_+$. By direct calculation one obtains that the sequence $\{\tilde{y}(\tau_k^-)\}_{k\in\mathbb{Z}_+}$ is a bounded and uniform positive solution of the following non-homogeneous backward DTLE

$$z_k = \mathscr{L}_d^*(k)z_{k+1} + \tilde{\zeta}_k, \tag{2.104}$$

where $\tilde{\zeta}_k := \tilde{g}_k + \mathscr{I}^*(\tau_k)\int_{\tau_k}^{\tau_{k+1}} T^*(s, \tau_k)f(s)ds$, $\mathscr{L}_d^*(k)$ being the adjoint of the linear operator $\mathscr{L}_d(k)$ defined as in (2.10). Under the considered assumptions $\{\tilde{\zeta}_k\}_{k\in\mathbb{Z}_+}$ is a bounded and uniform positive sequence from $Int\ \mathscr{X}^+$. Employing $(iv) \Rightarrow (i)$ from Theorem 2.4 from Drăgan et al. [4] we deduce that the DTLE (2.9) and (2.10) is ES. Finally, from the equivalence (i) from Theorem 2.1 we may conclude that the FJLDE is ES which confirms that the implication $(iii) \Rightarrow (i)$ is true. The validity of the implication $(iii) \Rightarrow (iv)$ is obvious because any bounded and uniform positive solution on $\mathbb{R}_+$ of (2.102) is also a solution of (2.103).

Finally we have to prove the implication $(iv) \Rightarrow (iii)$. If $t \to \tilde{z}(t) : \mathbb{R}_+ \to \mathscr{X}$ is a solution of (2.103) we set

$$\tilde{f}(t) \stackrel{\Delta}{=} \dot{\tilde{z}}(t) - \mathscr{L}^*(t)\tilde{z}(t), \quad t \in \mathbb{R}_+$$

and

$$\tilde{g}_k \stackrel{\Delta}{=} \tilde{z}(\tau_k^-) - \mathscr{I}^*(\tau_k)\tilde{z}(\tau_k).$$

With this choice one obtains that $\tilde{z}(\cdot)$ is a bounded and uniform positive solution of the following non-homogeneous BJLDE

$$\begin{aligned} -\dot{\tilde{z}}(t) &= \mathscr{L}^*(t)\tilde{z}(t) + \tilde{f}(t), & \tau_k \le t < \tau_{k+1}, \\ \tilde{z}(\tau_k^-) &= \mathscr{I}^*(\tau_k)\tilde{z}(\tau_k) + \tilde{g}_k, & k \in \mathbb{Z}_+, \end{aligned}$$

which shows that (iii) is satisfied. Thus the proof ends.

Remark 2.8 From the previous theorem, one sees that only the uniform positivity of the sequence $\{g_k\}_{k\in\mathbb{Z}_+}$ which acts at the impulsive time instances τ_k is essential to characterize the exponential stability of a FJLDE which defines a positive evolution on an ordered Hilbert space.

2.4.3 The Periodic Case

We begin this section by showing that in the case of the FJLDE (2.2) with periodic coefficients which defines a positive evolution on an ordered Hilbert space, its exponential stability is equivalent to the asymptotic stability. In the developments in this subsection, we shall refer to the version (2.54) of the FJLDE (2.2).

Theorem 2.8 *Assume:*

(a) $\mathscr{X}$ is a real Hilbert space ordered by the order relation induced by a close, solid, self-dual, convex cone $\mathscr{X}^+$;
(b) the triplet $(\mathscr{L}(\cdot), \mathscr{I}(\cdot), \mathfrak{T})$ satisfies the Assumption 2.2;
(c) the FJLDE (2.54) defines a positive evolution on the ordered space $(\mathscr{X}, \mathscr{X}^+)$.

Under these assumptions, the following are equivalent:

(i) the FJLDE (2.54) is ES;
(ii) the FJLDE (2.54) is AS;
(iii) the spectral radius of the monodromy operator $\mathbf{T}_d(\tilde{\mathbf{k}}, 0)$ defined by the DTLE (2.9) and (2.10) satisfies $\rho(\mathbf{T}_d(\tilde{\mathbf{k}}, 0)) < 1$, $\tilde{\mathbf{k}}$ being the period defined in Lemma 2.4.

Proof The equivalence $(i) \Leftrightarrow (iii)$ follows from Corollary 2.1. The implication $(i) \Rightarrow (ii)$ follows directly from the definition of the exponential stability and of the asymptotic stability of the FJLDE (2.54).

It remains to prove the implication $(ii) \Rightarrow (iii)$. We set $\tilde{\rho} := \rho(\mathbf{T}_d(\tilde{\mathbf{k}}, 0))$. Let us assume by contradiction that $\tilde{\rho} \geq 1$. Applying Theorem 9.11 from MA Krasnosel'skij [10] (see also Appendix 2.6 from Schaefer [11] or Theorem 2.6 (i) from Damm [12]) in the case of the linear operator $\mathcal{T} = \mathbf{T}_d^*(\tilde{\mathbf{k}}, 0)$ we deduce that there exists $\tilde{y} \in \mathcal{X}^+$, $\tilde{y} \neq 0$, such that $\mathbf{T}_d(\tilde{\mathbf{k}}, 0)\tilde{y} = \tilde{\rho}\tilde{y}$. Let $z(k; 0, \tilde{y})$ be the solution of the DTLE (2.9) which satisfies $z(0; 0, \tilde{y}) = \tilde{y}$. One checks that $z(j\tilde{\mathbf{k}}; 0, \tilde{y}) = \tilde{\rho}^j \tilde{y}$. Hence, $z(k; 0, \tilde{y})$ cannot tends to 0, when $k \to \infty$. So, the DTLE (2.9) and (2.10) is not AS. In these conditions, Theorem 2.1 (ii) shows that the FJLDE (2.54) is not AS which contradicts the assertion (ii) from the statement. Therefore, $\tilde{\rho} < 1$. Thus the proof is complete.

Now, we prove a result which highlights some interesting properties of the unique bounded on $\mathbb{R}$ solution of a non-homogeneous FJLDE of type (2.62).

Theorem 2.9 *Under the assumptions of Theorem 2.8, the following hold if the FJLDE (2.54) is ES:*

(i) for each sequence $\{g_k\}_{k\in\mathbb{Z}} \subset \mathcal{X}^+$ which is periodic of period $\theta_d \geq 1$ and for each piecewise continuous vector valued function $f(\cdot) : \mathbb{R} \to \mathcal{X}^+$ with the properties:

(a) it has finite lateral limits in any $t \in \mathbb{R}$;
(b) it is a periodic function of period $\theta_c > 0$;

(θ_d, θ_c being the same as in the Assumption 2.2), then the unique bounded solution $\tilde{x}(\cdot) : \mathbb{R} \to \mathcal{X}^+$ of the corresponding non-homogeneous FJLDE (2.62) is a periodic function of period $\tilde{\theta}$ with $\tilde{\theta}$ defined in (2.59);

(ii) if the periodic sequence of period $\theta_d \geq 1$, $\{g_k\}_{k\in\mathbb{Z}}$ lies in the interior of $\mathcal{X}^+$ and the vector valued function $f(\cdot)$ has the properties from (i) above, then the unique bounded on $\mathbb{R}$ solution of the non-homogeneous FJLDE (2.62) is a periodic function of period $\tilde{\theta}$ and satisfies $\tilde{x} \gg 0$, for all $t \in \mathbb{R}$.

Proof (i) follows immediately from Proposition 2.5 (i) and the representation formula (2.63). Now we prove (ii). From (2.62a) written for $\tilde{x}(\cdot)$ we have

$$\tilde{x}(t) = T(t, kh)\mathcal{J}(kh)\tilde{x}(kh) + \int_{kh}^{t} T(t, s)f(s)\mathrm{d}s + T(t, kh)g_k,$$

for all $kh < t \leq (k+1)h$, $k \in \mathbb{Z}$. Employing Proposition 2.5 (ii) we deduce that $T(t, kh)\mathcal{J}(kh)\tilde{x}(kh) \succeq 0$ because $\tilde{x}(kh) \succeq 0$, for all $k \in \mathbb{Z}$. Also, $\int_{kh}^{t} T(t, s)f(s)\mathrm{d}s \succeq 0$ because $\mathcal{X}^+$ is a closed convex cone. Hence, $\tilde{x}(t) \succeq T(t, kh)g_k$. Invoking Lemma 2.5 (ii) we may conclude that

$$T(t, kh)g_k \succcurlyeq \left(\frac{1}{2}\right)^{\left[\frac{h}{\delta_2}\right]+1} g_k, \quad \text{for all } t \in [kh, (k+1)h], \quad k \in \mathbb{Z}.$$

Since g_k is a periodic sequence in $Int\ \mathscr{X}^+$ we may deduce that there exists $\xi \in Int\ \mathscr{X}^+$, $\varepsilon > 0$ such that

$$\tilde{x}(t) \succcurlyeq \left(\frac{1}{2}\right)^{\left[\frac{h}{\delta_2}\right]+1} \varepsilon\xi, \quad \text{for all } t \in \mathbb{R}.$$

This completes the proof.

The next result provides a list of necessary and sufficient conditions for the exponential stability of a FJLDE of type (2.54) which is the version of (2.2) under the Assumption 2.2.

Theorem 2.10 *Under the assumptions of Theorem 2.8, the following are equivalent:*

(i) the FJLDE (2.54) is ES;

(ii) for any sequence $\{g_k\}_{k\in\mathbb{Z}_+} \subset Int\ \mathscr{X}^+$ which is periodic with period $\theta_d \geq 1$ and for any piecewise continuous vector valued function $t \to f(t) : \mathbb{R}_+ \to \mathscr{X}^+$ which has finite lateral limits in any $t \in \mathbb{R}_+$ and it is periodic with period $\theta_c > 0$ (with θ_d and θ_c being the same as in the Assumption 2.2) the BJLDE (2.102) with $\tau_k = kh$ has a unique bounded and uniform positive solution $t \to y(t) : \mathbb{R}_+ \to Int\ \mathscr{X}^+$ which is periodic of period $\tilde{\theta}$ with $\tilde{\theta}$ defined in (2.59);

(iii) there exists a sequence $\{\tilde{g}_k\}_{k\in\mathbb{Z}_+} \subset Int\ \mathscr{X}^+$ periodic of period θ_d and a function $t \to \tilde{f}(t) : \mathbb{R}_+ \to \mathscr{X}^+$ with the properties from (ii) such that the corresponding non-homogeneous BJLDE of type (2.102) with $\tau_k = kh$ has a bounded solution $t \to \tilde{y}(t) : \mathbb{R}_+ \to \mathscr{X}^+$ which is a periodic function of period $\tilde{\theta}$;

(iv) there exists a function $t \to z(t) : \mathbb{R}_+ \to Int\ \mathscr{X}^+$ which is a C^1 function on each interval $(kh, (k+1)h)$, right continuous in each kh and periodic of period $\tilde{\theta}$ satisfying a backward linear differential inequality with jumps (BJLDI) of type (2.103) with $\tau_k = kh$;

(v) for any sequence $\{g_k\}_{k\in\mathbb{Z}} \subset Int\ \mathscr{X}^+$ which is periodic of period θ_d and any function $t \to f(t) : \mathbb{R} \to \mathscr{X}^+$ piecewise continuous with finite lateral limits in any $t \in \mathbb{R}$ and periodic of period θ_c, (θ_d and θ_c being the same as in Assumption 2.2), the non-homogeneous FJLDE:

$$\dot{x}(t) = \mathscr{L}(t)x(t) + f(t), \qquad kh < t \leq (k+1)h, \tag{2.105a}$$

$$x(kh^+) = \mathscr{I}(kh)x(kh) + g_k, \qquad k \in \mathbb{Z} \tag{2.105b}$$

has a solution $t \to \tilde{x}(t) : \mathbb{R} \to Int\ \mathscr{X}^+$ which is periodic of period $\tilde{\theta}$;

(vi) there exists a sequence $\{\tilde{g}_k\}_{k\in\mathbb{Z}} \subset Int\ \mathscr{X}^+$ periodic of period $\theta_d \geq 1$ and a continuous on each interval $(kh, (k+1)h)$ function, $t \to \tilde{f}(t) : \mathbb{R} \to \mathscr{X}^+$ which has lateral limits in any kh and is periodic of period $\theta_c > 0$ with the

property that the corresponding non-homogeneous FJLDE of type (2.105) has a periodic solution of period $\tilde{\theta}$, $t \to \tilde{x}(t) : \mathbb{R} \to \mathcal{X}^+$;

(vii) there exists a function $t \to \mathbf{z}(t) : \mathbb{R} \to Int\ \mathcal{X}^+$ *which is a* C^1*−function on any interval* $(kh, (k+1)h)$, *left continuous in each* kh, *periodic of period* $\tilde{\theta}$ *which solves the following forward linear differential inequality with jumps (FJLDI):*

$$\dot{\mathbf{z}}(t) - \mathcal{L}(t)\mathbf{z}(t) \succcurlyeq 0, \qquad kh < t \leq (k+1)h, \tag{2.106a}$$

$$\mathbf{z}(kh^+) - \mathcal{J}(kh)\mathbf{z}(kh) \gg 0, \qquad k \in \mathbb{Z}. \tag{2.106b}$$

Proof The equivalences $(i) \Leftrightarrow (ii) \Leftrightarrow (iii) \Leftrightarrow (iv)$ follow from Theorem 2.6 and Theorem 2.5 (ii). The equivalences $(i) \Leftrightarrow (v) \Leftrightarrow (vi) \Leftrightarrow (vii)$ can be proved in an analogous way as in Theorem 2.6. We take into account that based on Theorem 2.3 (ii) the unique bounded solution of the non-homogeneous FJLDE (2.105) is periodic of period $\tilde{\theta}$.

Remark 2.9 Criteria as those from $(v) to (vii)$ (from the previous theorem) described based on a non-homogeneous FJLDE can be obtained only in the periodic case, because in this case, the coefficients of the FJLDE (2.2) can be extended in a natural way by periodicity to the whole real axis.

2.4.4 The Time-Invariant Case

In this subsection we consider the case when the FJLDE (2.2) is time invariant in the sense of Definition 2.3. In this case, it takes the form of (2.75). The analogous of Theorem 2.8 from above is given below.

Theorem 2.11 *Assume:*

(a) $\mathcal{X}$ *is a real Hilbert space ordered by the order relation induced by a closed, solid, self-dual, convex cone* $\mathcal{X}^+$;

(b) any solution $x(\cdot; t_0, x_0)$ *of the FJLDE (2.75) lies in* $\mathcal{X}^+$ *if* $x_0 \in \mathcal{X}^+$.

Under these assumptions, the following are equivalent:

(i) the FJLDE (2.75) is ES;
(ii) the FJLDE (2.75) is AS;
(iii) the spectral radius of the linear operator $e^{\mathcal{L}h}\mathcal{J}$ *satisfies* $\rho(e^{\mathcal{L}h}\mathcal{J}) < 1$.

Proof This is analogous to the proof of the Theorem 2.8. One takes into account that in the time-invariant case $\tilde{\mathbf{k}} = 1$ and $\mathbf{T}_d(\tilde{\mathbf{k}}, 0) = e^{\mathcal{L}h}\mathcal{J}$.

The next result collects a list of necessary and sufficient conditions for the exponential stability of the FJLDE (2.75) which defines a positive evolution on an ordered Hilbert space.

Theorem 2.12 *Under the assumptions from Theorem 2.11 the following are equivalent:*

(i) the FJLDE (2.75) is ES;

(ii) for any $g \in Int\ \mathscr{X}^+$ and any $f \in \mathscr{X}^+$ the non-homogeneous BJLDE

$$-\dot{y}(t) = \mathscr{L}^* y(t) + f, \qquad kh \le t < (k+1)h \tag{2.107a}$$

$$y(kh^-) = \mathscr{I}^* y(kh) + g, \qquad k \in \mathbb{Z} \tag{2.107b}$$

has a unique solution $t \to y(t) : \mathbb{R} \to Int\ \mathscr{X}^+$ which is a periodic function of period h;

(iii) there exist $\tilde{g} \in Int\ \mathscr{X}^+$ and $\tilde{f} \in \mathscr{X}^+$ with the property that the corresponding non-homogeneous BJLDE of type (2.107) has a unique solution $t \to \tilde{y}(t) : \mathbb{R} \to \mathscr{X}^+$ which is a periodic function of period h;

(iv) there exists a function $t \to z(t) : \mathbb{R} \to Int\ \mathscr{X}^+$ which is a C^1-function on each interval $(kh, (k+1)h)$, $k \in \mathbb{Z}$, right continuous together with its derivative in each kh, periodic of period h and satisfying the following $BJLDI$:

$$\mathscr{L}^* z(t) + \dot{z}(t) \preccurlyeq 0, \qquad kh < t < (k+1)h \tag{2.108a}$$

$$\mathscr{I}^* z(kh) - z(kh^-) \ll 0, \qquad k \in \mathbb{Z}; \tag{2.108b}$$

(v) for any $g \in Int\ \mathscr{X}^+$ and any $f \in \mathscr{X}^+$ the non-homogeneous FJLDE

$$\dot{x}(t) = \mathscr{L} x(t) + f, \qquad kh < t \le (k+1)h \tag{2.109a}$$

$$x(kh^+) = \mathscr{I} x(kh) + g, \qquad k \in \mathbb{Z} \tag{2.109b}$$

has a unique solution $t \to x(t) : \mathbb{R} \to Int\ \mathscr{X}^+$ and additionally this solution is periodic of period h;

(vi) there exist $\tilde{g} \in Int\ \mathscr{X}^+$ and $\tilde{f} \in \mathscr{X}^+$ with the property that the corresponding non-homogeneous FJLDE (2.109) has a unique solution $t \to \tilde{x}(t) : \mathbb{R} \to \mathscr{X}^+$ which is periodic of period h;

(vii) there exists a function $t \to \mathbf{z}(t) : \mathbb{R} \to Int\ \mathscr{X}^+$ which is a C^1-function on each interval $(kh, (k+1)h)$, left continuous together with its derivative in every kh, $k \in \mathbb{Z}$ and solves the following FJLDI:

$$\mathscr{L}\mathbf{z}(t) - \dot{\mathbf{z}}(t) \preccurlyeq 0, \qquad kh < t < (k+1)h, \tag{2.110a}$$

$$\mathscr{I}\mathbf{z}(kh) - \mathbf{z}(kh^+) \lll 0, \qquad k \in \mathbb{Z}. \tag{2.110b}$$

Proof The results are obtained from Theorem 2.10 in the special case $\theta_d = 1$, $\theta_c = \tilde{\theta} = h$.

2.5 Generalized Lyapunov Differential Equations with Jumps

In what follows, we will exploit the fundamental results obtained in the previous sections in order to obtain criteria for the exponential stability of a class of generalized Lyapunov linear differential equations with jumps on a Hilbert space as well as for the existence of some bounded solutions on the whole real semiaxis of the corresponding non-homogeneous generalized Lyapunov differential equations with jumps. These equations will play an important role in the characterization of the exponential stability in mean square of stochastic linear differential equation with jumps.

2.5.1 *Hilbert Spaces of Symmetric Matrices*

Let $\mathcal{S}_n \subset \mathbb{R}^{n \times n}$ be the linear subspace of the symmetric matrices of size $n \times n$. If $N \geq 1$ is a fixed natural number, we denote $\mathcal{S}_n^N := \underbrace{\mathcal{S}_n \times \mathcal{S}_n \times \cdots \mathcal{S}_n}_{N \text{ times}}$. One checks that $\mathcal{S}_n^N$ is a real vector space. Its elements $\mathbf{X}$ are finite sequences of symmetric matrices, that is $\mathbf{X} = (X_1, X_2, \ldots, X_N)$. On $\mathcal{S}_n^N$ we introduce the inner product

$$\langle \mathbf{X}, \mathbf{Y} \rangle = \sum_{i=1}^{N} Tr[X_i Y_i], \tag{2.111}$$

for all $\mathbf{X} = (X_1, X_2, \ldots, X_N)$ and $\mathbf{Y} = (Y_1, Y_2, \ldots, Y_N) \in \mathcal{S}_n^N$. In this way $\mathcal{S}_n^N$ can be organized as a real Hilbert space. Let

$$\mathcal{S}_n^{N+} = \{\mathbf{X} \in \mathcal{S}_n^N | \mathbf{X} = (X_1, X_2, \ldots, X_N),\ X_i \geq 0,\ 1 \leq i \leq N\}.$$

Here $X_i \geq 0$ means that X_i is a positive semidefinite matrix. One shows that $\mathcal{S}_n^{N+}$ is a closed, solid, self-dual, convex cone (see Example 2.1.3, Remark 2.1.4 from Drăgan et al. [1]). Its interior $Int\ \mathcal{S}_n^{N+}$ consists of all elements $\mathbf{X} = (X_1, X_2, \ldots, X_N) \in \mathcal{S}_n^N$ such that $X_i > 0$, for all $1 \leq i \leq N$.

We denote $\succcurlyeq$ the order relation on $\mathcal{S}_n^N$ induced by the convex cone $\mathcal{S}_n^{N+}$. Adapting (2.92) to the case of the ordered Hilbert space $(\mathcal{S}_n^N, \mathcal{S}_n^{N+})$ we obtain that the Minkovski norm associated to $\xi = \mathbf{I}_n^N \stackrel{\Delta}{=} (I_n, I_n, \ldots, I_n) \in Int\ \mathcal{S}_n^{N+}$, we obtain that

$$|\mathbf{X}|_\xi = \sup_{1 \leq i \leq N} \|X_i\| \stackrel{\Delta}{=} \|\mathbf{X}\|_\infty, \quad \text{for all } \mathbf{X} \in \mathcal{S}_n^N. \tag{2.112}$$

In (2.112), $\|X_i\|$ is the spectral norm of the symmetric matrix X_i, that is $\|X_i\| = \max\{|\lambda|;\ \lambda \in Spec\ X_i\}$.

For a linear operator $\mathscr{L} : \mathscr{S}_n^N \to \mathscr{S}_n^N$ we denote $\|\mathscr{L}\|_2$ and $\|\mathscr{L}\|_\infty$, respectively, the operator norms induced by the norm $|\cdot|$ (defined by the inner product (2.111)) and by the norm $|\cdot|_\xi$ defined in (2.112), respectively.

From Corollary 2.3 applied in the special case of the ordered space $(\mathscr{S}_n^N, \mathscr{S}_n^{N+})$ we obtain:

Corollary 2.4 *If $\mathscr{L} : \mathscr{S}_n^N \to \mathscr{S}_n^N$ is a positive operator then $\|\mathscr{L}\|_\infty = \|\mathscr{L}\mathbf{I}_n^N\|_\infty$.*

2.5.2 Generalized Lyapunov Linear Differential Equations with Jumps

For each $t \in \mathbb{R}_+$ we consider the linear operator $\mathscr{L}(t) : \mathscr{S}_n^N \to \mathscr{S}_n^N$ defined as follows:

$$\mathscr{L}(t)[\mathbf{X}] = (\mathscr{L}_1(t)[\mathbf{X}], \mathscr{L}_2(t)[\mathbf{X}], \dots, \mathscr{L}_N(t)[\mathbf{X}]) \tag{2.113a}$$

$$\mathscr{L}_i(t)[\mathbf{X}] = A_0(t,i)X_i + X_i A_0^\top(t,i) + \sum_{l=1}^{r_1} A_l(t,i)X_i A_l^\top(t,i) + \sum_{j=1}^{N} q_{ji}X_j,$$
$$1 \le i \le N, \quad \text{for all } \mathbf{X} = (X_1, X_2, \dots, X_N) \in \mathscr{S}_n^N. \tag{2.113b}$$

In (2.113), $t \to A_l(t,i) : \mathbb{R}_+ \to \mathbb{R}^{n\times n}$, $1 \le i \le N$, $1 \le l \le r_1$ are bounded and continuous matrix-valued functions and the real numbers q_{ji} satisfy the condition (1.22).

Based on the definition of the adjoint of a linear operator with respect to the inner product, one obtains that:

$$\mathscr{L}^*(t)[\mathbf{X}] = (\mathscr{L}_1^*(t)[\mathbf{X}], \mathscr{L}_2^*(t)[\mathbf{X}], \dots, \mathscr{L}_N^*(t)[\mathbf{X}]) \tag{2.114a}$$

$$\mathscr{L}_i^*(t)[\mathbf{X}] = A_0^\top(t,i)X_i + X_i A_0(t,i) + \sum_{l=1}^{r_1} A_l^\top(t,i)X_i A_l(t,i) + \sum_{j=1}^{N} q_{ij}X_j,$$
$$1 \le i \le N, \quad \text{for all } \mathbf{X} \in \mathscr{S}_n^N. \tag{2.114b}$$

The linear operators of type (2.113) are known as Lyapunov-type operators associated to stochastic linear differential equations of the from (1.23a) and were widely studied in Chaps. 2 and 3 from Drăgan et al. [1]. Beside the operators (2.113) we consider the linear operators $\mathscr{J}(\tau_k) : \mathscr{S}_n^N \to \mathscr{S}_n^N$ defined by

$$\mathscr{J}(\tau_k)[\mathbf{X}] = (\mathscr{J}_1(\tau_k)[\mathbf{X}], \mathscr{J}_2(\tau_k)[\mathbf{X}], \dots, \mathscr{J}_N(\tau_k)[\mathbf{X}]) \tag{2.115a}$$

$$\mathscr{J}_i(\tau_k)[\mathbf{X}] = \sum_{l=0}^{r_2} A_{dl}(\tau_k,i)X_i A_{dl}^\top(\tau_k,i), \quad 1 \le i \le N, \quad \mathbf{X} \in \mathscr{S}_n^N, \tag{2.115b}$$

where $\tau_k \in \mathbb{R}_+$, $k \in \mathbb{Z}_+$ are given real numbers. The adjoint of the linear operator (2.115) with respect to the inner product (2.111) is described by

$$\mathscr{I}^*(\tau_k)[\mathbf{X}] = (\mathscr{I}_1^*(\tau_k)[\mathbf{X}], \mathscr{I}_2^*(\tau_k)[\mathbf{X}], \ldots, \mathscr{I}_N^*(\tau_k)[\mathbf{X}]) \tag{2.116a}$$

$$\mathscr{I}_i^*(\tau_k)[\mathbf{X}] = \sum_{l=0}^{r_2} A_{dl}^\top(\tau_k, i) X_i A_{dl}(\tau_k, i), \quad 1 \le i \le N, \quad \mathbf{X} = (X_1, X_2, \ldots, X_N) \in \mathscr{S}_n^N. \tag{2.116b}$$

In (2.115) and (2.116), $A_{dl}(\tau_k, i) \in \mathbb{R}^{n \times n}$ are given matrices. Employing the linear operators $\mathscr{L}(t)$ defined in (2.113) and $\mathscr{I}(\tau_k)$ described in (2.115) we may introduce the following linear differential equations with jumps on the Hilbert space $\mathscr{S}_n^N$:

$$\dot{\mathbf{X}}(t) = \mathscr{L}(t)[\mathbf{X}(t)], \qquad \tau_k < t \le \tau_{k+1} \tag{2.117a}$$

$$\mathbf{X}(\tau_k^+) = \mathscr{I}(\tau_k)[\mathbf{X}(\tau_k)], \qquad k \in \mathbb{Z}_+. \tag{2.117b}$$

The sequence of impulsive time instants $\{\tau_k\}_{k \in \mathbb{Z}_+}$ satisfies the Assumption 2.1.

In this monograph, the differential equations of the form (2.117) will be named *generalized Lyapunov linear differential equations with jumps (or with impulses)* (GLJLDE) (or GLILDE).

One sees that (2.117) is a FJLDE of type (2.2) defined on the Hilbert space $\mathscr{X} = \mathscr{S}_n^N$. Applying Proposition 2.1 in the special case of (2.117) we deduce that for each $(t_0, \mathbf{X}_0) \in \mathbb{R}_+ \times \mathscr{S}_n^N$, the GLJLDE (2.117) has a unique solution $t \to \mathbf{X}(t; t_0, \mathbf{X}_0) : [t_0, \infty) \to \mathscr{S}_n^N$ satisfying $\mathbf{X}(t_0; t_0, \mathbf{X}_0) = \mathbf{X}_0$. This solution is differentiable in any $t \neq \tau_k$ and has possible discontinuities in the impulsive time instants $\tau_k \in [t_0, \infty)$. We denote $\mathbf{T}(t, t_0)$, $t \ge t_0 \ge 0$ the linear evolution operator on $\mathscr{S}_n^N$ defined by (2.117). We deduce

$$\mathbf{T}(t, t_0)\mathbf{X}_0 = \mathbf{X}(t; t_0, \mathbf{X}_0), \quad \text{for all } t \ge t_0 \ge 0 \text{ and } \mathbf{X}_0 \in \mathscr{S}_n^N.$$

Formally, $\mathbf{T}(t, t_0)$ is described as in (2.5)–(2.7) where $\mathscr{I}(\tau_k)$ are defined as in (2.115) and $T(t, s)$ is the linear evolution operator on $\mathscr{S}_n^N$ defined by the generalized Lyapunov differential equation (GLDE):

$$\dot{\mathbf{X}}(t) = \mathscr{L}(t)[\mathbf{X}(t)], \quad t \in \mathbb{R}_+. \tag{2.118}$$

Remark 2.10 Applying Theorem 2.6.1 from Drăgan et al. [1] in the case of the linear differential Equation (2.118) we deduce that

$$T(t, s)\mathscr{S}_n^{N+} \subset \mathscr{S}_n^{N+}, \quad \text{for all } t \ge s \ge 0. \tag{2.119}$$

Combining (2.118) to (2.119) we may conclude via (2.5)–(2.7) that

$$\mathbf{T}(t, t_0)\mathcal{S}_n^{N+} \subset \mathcal{S}_n^{N+}, \quad \text{for all } t \geq t_0 \geq 0.$$

Finally, from Proposition 2.5 (i) applied in the case of (2.117) we deduce that a generalized Lyapunov linear differential equation with jumps of type (2.117) defines a positive evolution on the ordered space $(\mathcal{S}_n^N, \mathcal{S}_n^{N+})$.

Definition 2.9 We say that the GLJLDE (2.117) defines an exponentially stable evolution on $\mathcal{S}_n^N$ or, equivalently the GLJLDE (2.117) is ES if there exist $\beta \geq 1$, $\alpha > 0$ such that

$$\|\mathbf{X}(t; t_0, \mathbf{X}_0)\| \leq \beta e^{-\alpha(t-t_0)}\|\mathbf{X}_0\|, \quad \text{for all } t \geq t_0 \geq 0, \quad \mathbf{X}_0 \in \mathcal{S}_n^N. \tag{2.120}$$

In (2.120), $\|\cdot\|$ can be any norm on $\mathcal{S}_n^N$. A particular choice of the norm corresponds to a particular constant β, because all norms on the vector space $\mathcal{S}_n^N$ are equivalent.
From Remark 2.10 it follows that we may apply the results proved in Sect. 2.4 to obtain criteria for exponential stability of the GLJLDE (2.117) as well as for the existence of some bounded solutions on the whole real semiaxis of the corresponding non-homogeneous generalized Lyapunov differential equations with jumps.

The next result is obtained directly applying Theorem 2.5 and Theorem 2.6.

Theorem 2.13 *Assume:*

(a) the impulsive time instants sequence $\{\tau_k\}_{k\in\mathbb{Z}_+}$ *satisfies the Assumption 2.1;*
(b) the GLJLDE (2.117) defines an exponentially stable evolution.

Under these conditions, the following hold:

(i) for each bounded and piecewise continuous function $t \to \mathbf{H}(t) : \mathbb{R}_+ \to \mathcal{S}_n^N$ *having lateral limits in each* $t \in \mathbb{R}_+$ *and for each bounded sequence* $\{\mathbf{\Gamma}_k\}_{k\in\mathbb{Z}_+} \subset \mathcal{S}_n^N$ *the non-homogeneous backward generalized Lyapunov differential equation with jumps*

$$\dot{Y}(t, i) = \mathcal{L}_i^*(t)[\mathbf{Y}(t)] + H(t, i), \qquad \tau_k \leq t < \tau_{k+1} \tag{2.121a}$$
$$Y(\tau_k^-, i) = \mathcal{I}_i^*(\tau_k)[\mathbf{Y}(\tau_k)] + \Gamma_k(i), \qquad k \in \mathbb{Z}_+ \tag{2.121b}$$

$1 \leq i \leq N$, *has a unique solution* $\tilde{\mathbf{Y}}(t) = (\tilde{Y}(t, 1), \tilde{Y}(t, 2), \ldots, \tilde{Y}(t, N))$, $t \in \mathbb{R}_+$ *with the property that* $\sup_{t\in\mathbb{R}_+} \|\tilde{Y}(t, i)\| < \infty$, $1 \leq i \leq N$;

(ii) if $\tau_k = kh$, $k \in \mathbb{Z}_+$ *and if there exists a real number* $\theta_c > 0$ *and an integer* $\theta_d \geq 1$ *with the properties:*

(α) $\frac{\theta_c}{h\theta_d}$ *is a rational number;*
(β) $A_l(t + \theta_c, i) = A_l(t, i)$, $1 \leq l \leq r_1$, $H(t + \theta_c, i) = H(t, i)$, $t \in \mathbb{R}_+$, $i \in \{1, 2, \ldots, N\}$;

(γ) $A_{dl}((k+\theta_d)h, i) = A_{dl}(kh, i)$, $0 \le l \le r_2$, $\Gamma_{k+\theta_d}(i) = \Gamma_k(i)$, *for all* $k \in \mathbb{Z}_+$, $1 \le i \le N$,

then, the unique bounded solution of (2.121) is a periodic function of period $\tilde{\theta}$ *where* $\tilde{\theta}$ *is computed as in (2.59) based on* θ_c, θ_d, h;

(*iii*) *if* $\tau_k = kh$, $k \in \mathbb{Z}_+$ *and* $A_l(t, i) = A_l(i)$, $0 \le l \le r_1$, $H(t, i) = H(i)$, *for all* $t \in \mathbb{R}_+$, $A_{dl}(kh, i) = A_{dl}(i)$, $0 \le l \le r_2$, $\Gamma_k(i) = \Gamma(i)$, *for all* $k \in \mathbb{Z}_+$, $1 \le i \le N$, *then the unique bounded on* $\mathbb{R}_+$ *solution of (2.121) is a differentiable function on each interval* $(kh, (k+1)h)$, *right continuous in* kh, $k \in \mathbb{Z}_+$ *and periodic of period* h;

(*iv*) *if* $H(t, i) \ge 0$, $t \in \mathbb{R}_+$, $\Gamma_k(i) \ge 0$, *for all* $k \in \mathbb{Z}_+$, $1 \le i \le N$, *then the unique bounded on* $\mathbb{R}_+$ *solution of (2.121),* $\tilde{\mathbf{Y}}(\cdot)$ *lies in* $\mathcal{S}_n^{N+}$, *that is, it satisfies*

$$0 \le \tilde{Y}(t, i) \le \gamma I_n, \quad \textit{for all} \;\; t \in \mathbb{R}_+, \quad 1 \le i \le N;$$

(*v*) *if* $H(t, i) \ge 0$, $t \in \mathbb{R}_+$ *and* $\Gamma_k(i) \ge \delta I_n$, *for all* $k \in \mathbb{Z}_+$, $1 \le i \le N$, *then the unique bounded on* $\mathbb{R}_+$ *solution of (2.121) is uniform positive, that is, there exist* $\gamma_k > 0$, $k = 1, 2$, *such that*

$$0 \le \gamma_1 I_n \le \tilde{Y}(t, i) \le \gamma_2 I_n, \quad \textit{for all} \;\; t \in \mathbb{R}_+, \quad 1 \le i \le N.$$

Specializing the results proved in Theorems 2.3, 2.4 and 2.9 to the case of the GLJLDE of type (2.117) we obtain:

Theorem 2.14 *Assume:*

(*a*) $\tau_k = kh$, *for all* $k \in \mathbb{Z}$, $h > 0$ *is fixed;*
(*b*) *the GLJLDE (2.117) defines an exponentially stable evolution.*

Under these conditions the following hold:

(*i*) *if there exist a real number* $\theta_c > 0$ *and an integer* $\theta_d \ge 1$ *such that* $\frac{\theta_c}{h\theta_d}$ *is a rational number, and* $A_l(\cdot, i)$, $0 \le l \le r_1$, $1 \le i \le N$ *are periodic functions of period* θ_c *and* $\{A_{dl}(kh, i)\}_{k \in \mathbb{Z}}$, $0 \le l \le r_2$, $0 \le i \le N$ *are periodic sequences of period* θ_d, *then, for each function* $t \to \mathbf{H}(t) : \mathbb{R} \to \mathcal{S}_n^N$ *which is continuous on each interval* $(kh, (k+1)h)$ *with finite lateral limits in each* kh, $k \in \mathbb{Z}$ *and it is periodic with period* θ_c *and for each sequence* $\{\mathbf{\Gamma}_k\}_{k \in \mathbb{Z}} \subset \mathcal{S}_n^N$ *periodic of period* θ_d, *the unique bounded on* $\mathbb{R}$ *solution of the non-homogeneous GLJLDE*

$$\dot{X}(t, i) = \mathcal{L}_i(t)[\mathbf{X}(t)] + H(t, i), \qquad kh < t \le (k+1)h \tag{2.122a}$$

$$X(kh^+, i) = \mathcal{I}_i(kh)[\mathbf{X}(kh)] + \Gamma_k(i), \qquad k \in \mathbb{Z}, \tag{2.122b}$$

$1 \le i \le N$, *is a periodic function of period* $\tilde{\theta}$ *where* $\tilde{\theta}$ *is computed as in (2.59) based on the numbers* θ_c, θ_d, h;

(*ii*) *if* $A_l(t, i) = A_l(i)$, $0 \le l \le r_1$, $H(t, i) = H(i)$, $A_{dl}(kh, i) = A_{dl}(i)$, $\Gamma_k(i) = \Gamma(i)$, $0 \le l \le r_2$, $1 \le i \le N$, $k \in \mathbb{Z}$, *then the unique bounded on* $\mathbb{R}$ *solution of*

(2.122) is a C^1*-function on each interval* $(kh, (k+1)h)$*, is left continuous in each* kh*,* $k \in \mathbb{Z}$ *and it is a periodic function of period* h*;*

(iii) if $A_l(\cdot, i)$*,* $A_{dl}(\cdot, i)$*,* $H(\cdot, i)$*,* $\Gamma_k(i)$ *satisfy conditions from* (i) *and additionally* $H(t,i) \geq 0$ *for all* $t \in \mathbb{R}$ *and the periodic sequence* $\{\mathbf{\Gamma}_k\}_{k\in\mathbb{Z}}$ *has the property that* $\Gamma_k(i) \geq 0$*, then the unique bounded on* $\mathbb{R}$ *solution* $\tilde{\mathbf{X}}(\cdot) = (\tilde{X}(\cdot, 1), \tilde{X}(\cdot, 2), \ldots, \tilde{X}(\cdot, N))$ *of (2.122) lie in* $\mathcal{S}_n^{N+}$*;*

(iv) if $\mathbf{H}(\cdot)$ *is as in* (iii)*, but the periodic sequence of period* θ_d $\{\mathbf{\Gamma}_k\}_{k\in\mathbb{Z}} \subset Int\ \mathcal{S}_n^{N+}$*, then the unique bounded on* $\mathbb{R}$ *solution of (2.122) is uniform positive function.*

2.6 Notes and References

For a rigorous treatment of the theory of stability of differential equations with positive evolution on a Banach/Hilbert space with application to the problem of exponential stability of a class of generalized Lyapunov linear differential equations (which are related to the study of stability of Itô-type stochastic systems as well as of Itô-type stochastic systems with Markovian jumps) one can refer to Damm [7]; Dragan et al. [13]; Drăgan et al. [1].
To the best of the authors knowledge, a similar treatment for jump linear differential equations related to a Lyapunov theory for linear stochastic impulsive systems with Markovian jumps has not yet been addressed in the literature. Hence, this chapter can be viewed as one of the first attempts for a rigorous characterization of such mathematical objects.

References

1. Drăgan, V., Morozan, T., Stoica, A.M.: Mathematical Methods in Robust Control of Linear Stochastic Systems. Springer-Verlag, New York (2013)
2. Daleckiĭ, J.L., Kreĭn, M.G.: Stability of solutions of differential equations in banach space, vol. 43. Transl. Math. Monogr. Am. Math, Soc (1974)
3. Muresan, M.: A Concrete Approach to Classical Analysis. Springer-Verlag, New York (2009)
4. Drăgan, V., Morozan, T., Stoica, A.M.: Mathematical Methods in Robust Control of Discrete-Time Linear Stochastic Systems. Springer-Verlag, New York (2010)
5. Krasnosel'skiy, M.A.: Positive Solutions of Operator Equations (Russian). Fizmatgiz, Moskow (1962)
6. Krasnosel'skiy, M.A., Lifshits, J.A., Sobolev, A.V.: Positive Linear Systems: The Method of Positive Operators, Sigma Series in Applied Mathematics, vol. 5. Heldermann, Berlin (1989)
7. Damm, T.: Rational Matrix Equations in Stochastic Control, vol. 297. Springer-Verlag, Berlin Heidelberg (2004)
8. Gronwall, T.H.: Note on the derivatives with respect to a parameter of the solutions of a system of differential equations. Ann. Math. **20**(4), 292–296 (1919)
9. Amann, H.: Ordinary Differential Equations: An Introduction to Nonlinear Analysis. Walter De Gruyter Inc, Berlin (1990)
10. Krasnosel'skij, M.A., AS, Lifshits, J.A.: Positive Linear Systems: The Method of Positive Operators. Sigma Series in Applied Mathematics, vol. 5, Heldermann Verlag, Berlin (1989)

11. Schaefer, H.H.: Topological Vector Spaces. Springer, Berlin, Heidelberg, New York (1971)
12. Damm, T., D.H.: Newton's method for concave operators with resolvent positive derivatives in ordered banach spaces. Linear Algebra Appl **363**, 43–64 (2003)
13. Dragan, V., Damm, T., Freiling, G., Morozan, T.: Differential equations with positive evolutions and some applications. Results Math. **48**, 206–236 (2005)

Chapter 3
Mean-Square Stability of Systems Described by Impulsive Stochastic Linear Differential Equations

Abstract This chapter addresses the problem of mean-square stability of systems described by impulsive stochastic linear differential equations. We consider a general class of stochastic impulsive differential equations that encompasses impulsive stochastic linear differential equations of Itô type as well as impulsive linear differential equations subject to Markovian jumping as particular cases. In Sect. 3.2, we first introduce a Lyapunov-type jump differential equations that belong to the category of GLJLDEs introduced in Chap. 2. Such an equation is associated to the considered class of impulsive stochastic linear differential equations, and it plays a key role in characterizing mean-square stability properties for such systems. In Sect. 3.3, several mean-square stability concepts are defined. The different connections/relations between these stability notions are then resumed in Propositions 3.2 and 3.3 and Theorem 3.1, respectively. A particular emphasis is then put on the so-called exponential stability in the mean-square sense (ESMS). This emphasis is motivated by the fact that ESMS benefits from a Lyapunov-type characterization. Hence, in Sect. 3.4, Lyapunov criteria for ESMS are given in Theorem 3.2 for the general time-varying case, as well as for the time-periodic case, Theorem 3.3 and the time-invariant case, Theorem 3.4, respectively. The main theoretical support used in this section is the theory of jump linear differential equations with positive evolution developed in Chap. 2. Finally, in Sect. 3.5, we shall study some useful properties of the solutions of a class of non-homogeneous jump stochastic linear differential equations often named *affine* jump stochastic linear differential equations.

3.1 Systems Described by Impulsive Stochastic Linear Differential Equations

In this chapter we study different kinds of stability in the mean-square sense of systems modeled by impulsive stochastic linear differential equations of the form

$$dx(t) = A_0(t, \eta(t))x(t)dt + \sum_{l=1}^{r_1} A_l(t, \eta(t))x(t)dw_l(t), \qquad \tau_k < t \le \tau_{k+1} \tag{3.1a}$$

V. Drăgan et al., *Robust Control of Jump Linear Stochastic Systems*, Lecture Notes in Control and Information Sciences 497, https://doi.org/10.1007/978-3-031-84070-8_3

$$x(\tau_k^+) = A_{d0}(\tau_k, \eta(\tau_k))x(\tau_k) + \sum_{l=1}^{r_2} w_{dl}(k)A_{dl}(\tau_k, \eta(\tau_k))x(\tau_k), \qquad k \in \mathbb{Z}_+, \tag{3.1b}$$

where $x(t) \in \mathbb{R}^n$ are the state parameters and $\{w(t)\}_{t\geq 0}$ ($w(t) = (w_1(t),\ w_2(t),\ \dots,\ w_{r_1}(t))^\top$), $\{w_d(k)\}_{k\in\mathbb{Z}_+}$ $\left(w_d(k) = (w_{d1}(k),\ w_{d2}(k),\ \dots,\ w_{dr_2}(k))^\top\right)$ and $\{\eta(t)\}_{t\geq 0}$ are stochastic processes which were described in Sect. 1.2 and satisfying Assumption 1.2.

In (3.1), for any $i \in \mathcal{N}$, $t \to A_l(t,i) : \mathbb{R}_+ \to \mathbb{R}^{n\times n}, 0 \leq l \leq r_1$ are bounded and continuous matrix-valued functions and $\{A_{dl}(\tau_k, i)\}_{k\in\mathbb{Z}_+} \subset \mathbb{R}^{n\times n}, 0 \leq l \leq r_2$ are bounded matrix-valued sequences.

Note 3.1 The results derived in this chapter still hold if $t \to A_l(t,i) : \mathbb{R}_+ \to \mathbb{R}^{n\times n}, 0 \leq l \leq r_1, i \in \mathcal{N}$ are bounded, piecewise continuous matrix-valued functions having lateral limits in any $t \in \mathbb{R}_+$.

In order to motivate the interest for the study of the stability of systems of type (3.1), let us consider a stochastic linear system controlled by impulses of the form:

$$dx(t) = A_0(t, \eta(t))x(t)dt + \sum_{l=1}^{r_1} A_l(t, \eta(t))x(t)dw_l(t), \qquad \tau_k < t \leq \tau_{k+1} \tag{3.2a}$$

$$x(\tau_k^+) = A_{d0}(\tau_k, \eta(\tau_k))x(\tau_k) + B_0(\tau_k, \eta(\tau_k))u(k), \qquad k \in \mathbb{Z}_+, \tag{3.2b}$$

$$y(\tau_k) = C_0(\tau_k, \eta(\tau_k))x(\tau_k) + \sum_{l=1}^{r_2} w_{dl}(k)C_l(\tau_k, \eta(\tau_k))x(\tau_k), \tag{3.2c}$$

where $u(k) \in \mathbb{R}^{n_u}$ are the control parameters and $y(\tau_k) \in \mathbb{R}^{n_y}$ are the measurements which are used to activate a controller for obtaining the desired behavior of the closed-loop system.

Let us assume that to control the system (3.2) we use controls of the form $u(k) = K(\tau_k, \eta(\tau_k))y(\tau_k)$. Substituting this control in (3.2) we obtain the closed-loop system:

$$dx(t) = A_0(t, \eta(t))x(t)dt + \sum_{l=1}^{r_1} A_l(t, \eta(t))x(t)dw_l(t), \qquad \tau_k < t \leq \tau_{k+1} \tag{3.3a}$$

$$\begin{aligned} x(\tau_k^+) &= \left(A_{d0}(\tau_k, \eta(\tau_k)) + B_0(\tau_k, \eta(\tau_k))K(\tau_k, \eta(\tau_k))C_0(\tau_k, \eta(\tau_k))\right)x(\tau_k) \\ &\quad + \sum_{l=1}^{r_2} w_{dl}(k)B_0(\tau_k, \eta(\tau_k))K(\tau_k, \eta(\tau_k))C_l(\tau_k, \eta(\tau_k))x(\tau_k), \ k \in \mathbb{Z}_+. \end{aligned} \tag{3.3b}$$

It is obvious that (3.3) is a special case of (3.1). Among the controller specifications, stability of the closed-loop system plays a key role. In the stochastic framework, several stability notions could be defined based on the convergent nature of the system trajectories to an equilibrium position. In this chapter, we are interested in the exponential stability in mean square, because in this case there exist criteria expressed by necessary and sufficient conditions or only by sufficient conditions which allow us to design some stabilizing controllers which have to satisfy also additional impose performance specifications. We shall define several types of mean-square stability of the state equilibrium $x(t) = 0$ of the systems of the form (3.1), and we shall provide necessary and sufficient conditions which guarantee the desired kind of stability.

Two important special cases of the system (3.1) will be also analyzed in detail, namely:

I. **Systems modeled by impulsive stochastic linear differential equations of Itô type.**

This case occurs when in the system (3.1) the Markov process $\eta(t)$ has only one state, this means that $\mathcal{N} = \{1\}$. In this case, the system takes the form:

$$dx(t) = A_0(t)x(t)dt + \sum_{l=1}^{r_1} A_l(t)x(t)dw_l(t), \qquad \tau_k < t \leq \tau_{k+1} \tag{3.4a}$$

$$x(\tau_k^+) = A_{d0}(\tau_k)x(\tau_k) + \sum_{l=1}^{r_2} w_{dl}(k)A_{dl}(\tau_k)x(\tau_k), \qquad k \in \mathbb{Z}_+. \tag{3.4b}$$

II. **Systems described by impulsive linear differential equations subject to Markovian jumping.**

If in (3.1) we have $A_l(t, i) = 0$, $1 \leq l \leq r_1$, $t \in \mathbb{R}_+$, $A_{dl}(\tau_k, i) = 0$, $0 \leq l \leq r_2$, $k \in \mathbb{Z}_+$, $i \in \mathcal{N}$, we obtain a system described by impulsive linear differential equations of the form:

$$\dot{x}(t) = A_0(t, \eta(t))x(t), \qquad \tau_k < t \leq \tau_{k+1} \tag{3.5a}$$

$$x(\tau_k^+) = A_d(\tau_k, \eta(\tau_k)))x(\tau_k), \qquad k \in \mathbb{Z}_+. \tag{3.5b}$$

As we already pointed out in Remark 1.5(b) the trajectories of the systems described by impulsive linear differential equations subject to some Markovian perturbations may exhibit discontinuities only to the impulsive time instances τ_k. The switchings due to the transitions of the Markovian process from the mode i_1 to another mode i_2 do not produce discontinuities of the trajectories of the considered system. Hence, we call systems described by impulsive linear differential equations with or without Markovian switching as "jump stochastic linear differential equations" (JSLDE).

3.2 Generalized Lyapunov Differential Equations with Impulses Associated to a JSLDE

Let $x(t) = x(t; t_0, x_0)$ be a solution of the JSLDE (3.1), For each $j \in \mathcal{N}$ we consider

$$\mathbf{X}^{(j)}(t) = (X_1^{(j)}(t), X_2^{(j)}(t), \dots, X_N^{(j)}(t)) \in \mathcal{S}_n^N$$

where

$$X_i^{(j)}(t) \triangleq \mathbb{E}\left[x(t)x^\top(t)\chi_{\{\eta(t)=i\}}|\eta(t_0) = j\right], \tag{3.6}$$

for all $i \in \mathcal{N}$, where $\chi_{\{\eta(t)=i\}}$ is the indicator function of the event $\{\eta(t) = i\}$.

Proposition 3.1 *Under Assumptions 1.2 and 2.1 for each* $(t_0, x_0) \in \mathbb{R}_+ \times \mathbb{R}^n$ *the function* $t \to \mathbf{X}^{(j)}(t) : [t_0, \infty) \to \mathcal{S}_n^N$ *is a solution of a GLJLDE of type (2.117) and has the initial condition*

$$\mathbf{X}^{(j)}(t_0) = (0, \dots, 0, \underbrace{x_0 x_0^\top}_{j\text{th-position}}, 0, \dots, 0).$$

Proof Applying the Itô formula (see for example Theorem 2.11.4 from [1]) to the function $v(t, x, i) \triangleq x^\top H(i)x$ and to the stochastic process $x(t) = x(t; t_0, x_0)$, $t \geq t_0$ where $H(i) \in \mathcal{S}_n$, we obtain:

$$\mathbb{E}[v(t, x(t),\eta(t))|\eta(t_0) = j] = \mathbb{E}[v(\tau, x(\tau), \eta(\tau))|\eta(t_0) = j] + \int_\tau^t \mathbb{E}[x^\top(s)\mathcal{L}^*_{\eta(s)}(s)[\mathbf{H}]x(s)|\eta(t_0) = j]ds,$$

for all $t_0 \leq \tau_k < \tau < t < \tau_{k+1}$, where $\mathcal{L}^*_{\eta(s)}(s)[\cdot]$ is computed as in (2.113) based on the coefficients of Eq. (3.1a) when $\eta(s) = i \in \mathcal{N}$ and $\mathbf{H} = (H(1), H(2), \dots, H(N))$. Letting $\tau \to \tau_k$ and taking into account that $\eta(t)$ has right continuous sampled paths we obtain:

$$\mathbb{E}[v(t, x(t), \eta(t)|\eta(t_0) = j] = \mathbb{E}[v(\tau_k, x(\tau_k^+), \eta(\tau_k)) \mid \eta(t_0) = j] + \int_{\tau_k}^t \mathbb{E}[x^\top(s)\mathcal{L}^*_{\eta(s)}(s)[\mathbf{H}]x(s) \mid \eta(t_0) = j]ds, \tag{3.7}$$

for all $t_0 \leq \tau_k < t < \tau_{k+1}$, $j \in \mathcal{N}$, $k \in \mathbb{Z}_+$, $\mathbf{H} \in \mathcal{S}_n^N$. Let $i_0 \in \mathcal{N}$ be fixed, but arbitrary. We define $\mathbf{H}^{i_0} = (H^{i_0}(1), H^{i_0}(2), \dots, H^{i_0}(N))$ as

$$H^{i_0}(i) = \begin{cases} xx^\top, & \text{if } i = i_0 \\ 0, & \text{if } i \neq i_0, \end{cases}$$

where $x \in \mathbb{R}^n$ is arbitrary but fixed. By direct calculation, involving (2.117b), (2.114b) and (3.6) we obtain:

$$\mathbb{E}[x^\top(s)\mathscr{L}^*_{\eta(s)}(s)[\mathbf{H}^{i_0}]x(s) \mid \eta(t_0) = j] = x^\top \mathscr{L}_{i_0}(s)[\mathbf{X}^{(j)}(s)]x. \tag{3.8}$$

Writing (3.7) for $\mathbf{H}$ replaced by $\mathbf{H}^{i_0}$ we obtain via (3.6) and (3.8) that

$$x^\top X^{(j)}_{i_0}(t)x = x^\top \mathbb{E}[x(\tau_k^+)x^\top(\tau_k^+)\chi_{\{\eta(\tau_k)=i_0\}} \mid \eta(t_0) = j]x$$
$$+ x^\top \int_{\tau_k}^{t} \mathscr{L}_{i_0}(s)[\mathbf{X}^{(j)}(s)]\mathrm{d}s\ x.$$

This equality is equivalent to

$$X^{(j)}_{i_0}(t) = \mathbb{E}[x(\tau_k^+)x^\top(\tau_k^+)\chi_{\{\eta(\tau_k)=i_0\}} \mid \eta(t_0) = j] + \int_{\tau_k}^{t} \mathscr{L}_{i_0}(s)[\mathbf{X}^{(j)}(s)]\mathrm{d}s, \tag{3.9}$$

for all $t_0 \le \tau_k < t < \tau_{k+1}$, $k \in \mathbb{Z}_+$, because x was arbitrarily chosen in $\mathbb{R}^n$. Since the integrant from the right hand side of (3.9) is a continuous function with respect to s, it follows that $X^{(j)}_{i_0}(\cdot)$ is differentiable on each interval (τ_k, τ_{k+1}) and we have

$$\dot{X}^{(j)}_{i_0}(t) = \mathscr{L}_{i_0}(t)[\mathbf{X}^{(j)}(t)], \qquad \tau_k < t \le \tau_{k+1} \tag{3.10a}$$

$$X^{(j)}_{i_0}(\tau_k^+) = \mathbb{E}[x(\tau_k^+)x^\top(\tau_k^+)\chi_{\{\eta(\tau_k)=i_0\}} \mid \eta(t_0) = j], \tag{3.10b}$$

for all $k \in \mathbb{Z}_+$ such that $\tau_k \ge t_0$, $i_0 \in \mathscr{N}$. Further, we show how one can compute the right hand side of (3.10b) using the matrices which appear in (3.1b). To this end we compute $\mathbb{E}[x(\tau_k^+)x^\top(\tau_k^+)\chi_{\{\eta(\tau_k)=i_0\}} \mid \mathscr{G}_{\tau_k}]$. If we take into account that $w_d(k)$ and $x(\tau_k)$ are independent random vectors and that $\sigma[w_d(k)]$ and $\mathscr{G}_{\tau_k}$ are independent σ-algebras we obtain that

$$\begin{aligned}
&\mathbb{E}[x(\tau_k^+)x^\top(\tau_k^+)\chi_{\{\eta(\tau_k)=i_0\}} \mid \mathscr{G}_{\tau_k}] \\
&= \sum_{l=0}^{r_2} \mathbb{E}[A_{dl}(\tau_k, \eta(\tau_k))x(\tau_k)x^\top(\tau_k)A^\top_{dl}(\tau_k, \eta(\tau_k))\chi_{\{\eta(\tau_k)=i_0\}} \mid \mathscr{G}_{\tau_k}] \\
&= \sum_{l=0}^{r_2} A_{dl}(\tau_k, i_0)\mathbb{E}[x(\tau_k)x^\top(\tau_k)\chi_{\{\eta(\tau_k)=i_0\}} \mid \mathscr{G}_{\tau_k}]A^\top_{dl}(\tau_k, i_0).
\end{aligned}$$

For the last equality we took in account that

$$\chi_{\{\eta(\tau_k)=i\}}\chi_{\{\eta(\tau_k)=i_0\}} = \begin{cases} 0, & \text{if } \ i \neq i_0 \\ \chi_{\{\eta(\tau_k)=i_0\}}, & \text{if } \ i = i_0. \end{cases}$$

Since $\sigma[\eta(t_0)] \subset \mathcal{G}_{\tau_k}$ we may conclude that

$$\mathbb{E}[x(\tau_k^+)x^\top(\tau_k^+)\chi_{\{\eta(\tau_k)=i_0\}} \mid \eta(t_0) = j]$$
$$= \sum_{l=0}^{r_2} A_{dl}(\tau_k, i_0)\mathbb{E}[x(\tau_k)x^\top(\tau_k)\chi_{\{\eta(\tau_k)=i_0\}} \mid \eta(t_0) = j]A_{dl}^\top(\tau_k, i_0).$$

Thus we have obtained that (3.10b) becomes

$$X_{i_0}^{(j)}(\tau_k^+) = \sum_{l=0}^{r_2} A_{dl}(\tau_k, i_0)X_{i_0}^{(j)}(\tau_k)A_{dl}^\top(\tau_k, i_0) = \mathcal{I}_{i_0}(\tau_k)[\mathbf{X}^{(j)}(\tau_k)]. \tag{3.11}$$

From (3.10a) and (3.11) we may conclude that for each $j \in \mathcal{N}$, $t \to \mathbf{X}^{(j)}(t) : [t_0, \infty) \to \mathcal{S}_n^N$ is a solution of a GLJLDE of type (2.117). The initial value $\mathbf{X}^{(j)}(t_0)$ claimed in the statement is obtained immediately from (3.6). Thus the proof ends.

Employing (3.10a) and (3.11) beside (2.113a) and (2.115b), respectively, we obtain the following detailed expression of the initial value problem (IVP) solved by $\mathbf{X}^{(j)}(\cdot)$:

$$\dot{X}_i^{(j)}(t) = A_0(t, i)X_i^{(j)}(t) + X_i^{(j)}(t)A_0^\top(t, i)$$
$$+ \sum_{l=1}^{r_1} A_l(t, i)X_i^{(j)}(t)A_l^\top(t, i) + \sum_{\iota=1}^{N} q_{\iota i}X_\iota^{(j)}(t), \quad \tau_k < t \le \tau_{k+1} \tag{3.12a}$$

$$X_i^{(j)}(\tau_k^+) = \sum_{l=0}^{r_2} A_{dl}(\tau_k, i)X_i^{(j)}(\tau_k)A_{dl}^\top(\tau_k, i), \quad k = k_0, k_0 + 1, \ldots \tag{3.12b}$$

$$X_i^{(j)}(t_0) = \begin{cases} 0, & \text{if } i \ne j \\ x_0x_0^\top, & \text{if } i = j, \end{cases} \tag{3.12c}$$

where k_0 is such that $t_0 \in (\tau_{k_0-1}, \tau_{k_0}]$, $j \in \mathcal{N}$. In some cases, instead of $\mathbf{X}^{(j)}(t)$ we will use the function taking values in $\mathcal{S}_n^N : t \to \mathbf{X}(t) = (X_1(t), X_2(t), \ldots, X_N(t))$, defined by

$$X_i(t) := \mathbb{E}[x(t; t_0, x_0)x^\top(t; t_0, x_0)\chi_{\{\eta(t)=i\}}], \tag{3.13}$$

for all $i \in \mathcal{N}$, $t \ge t_0 \ge 0$, $x_0 \in \mathbb{R}^n$. Comparing (3.6) with (3.13) we obtain from Proposition 3.1:

Corollary 3.1 *Under Assumption 1.2, $t \to \mathbf{X}(t) : [t_0, \infty) \to \mathcal{S}_n^N$ is the solution of GLJLDE (3.12a) and (3.12b) having the initial value*

$$\mathbf{X}(t_0) = (\pi_1(t_0)x_0x_0^\top, \pi_2(t_0)x_0x_0^\top, \ldots, \pi_N(t_0)x_0x_0^\top), \tag{3.14}$$

where $\pi(t_0) = (\pi_1(t_0), \pi_2(t_0), \ldots, \pi_N(t_0))$ is the probability distribution of the random variable $\eta(t_0)$, that is $\pi_i(t_0) = \mathcal{P}\{\eta(t_0) = i\}$, $i \in \mathcal{N}$.

Proof From (3.6) and (3.13) one sees that

$$X_i(t) = \sum_{j=1}^{N} \pi_j(t_0) X_i^{(j)}(t)$$

or equivalently

$$\mathbf{X}(t) = \sum_{j=1}^{N} \pi_j(t_0) \mathbf{X}^{(j)}(t).$$

This confirms that $\mathbf{X}(\cdot)$ solves (3.12a), (3.12b) because these are linear equations.

The formula given in (3.14) for the initial state is obtained from

$$\mathbf{X}(t_0) = \sum_{j=1}^{N} \pi_j(t_0) \mathbf{X}^{(j)}(t_0)$$

together with (3.12c).

Now, let us notice that in the case when the JSLDE (3.1) is in the special form of a system modeled by impulsive stochastic linear differential equations of Itô type as in (3.4), then (3.6) coincides with (3.13), because, in this case, the Markov process has only one state, that is $\mathcal{N} = \{1\}$. In this case, the IVP (3.12) takes the form:

$$\dot{X}(t) = A_0(t)X(t) + X(t)A_0^\top(t) + \sum_{l=1}^{r_1} A_l(t)X(t)A_l^\top(t), \quad \tau_k < t \leq \tau_{k+1} \tag{3.15a}$$

$$X(\tau_k^+) = \sum_{l=0}^{r_2} A_{dl}(\tau_k)X(\tau_k)A_{dl}^\top(\tau_k), \quad k = k_0, k_0+1, \dots \tag{3.15b}$$

$$X(t_0) = x_0 x_0^\top \tag{3.15c}$$

where k_0 is such that $\tau_{k_0-1} < t_0 \leq \tau_{k_0}$.

Let $\mathbf{T}(t, t_0)$ be the linear evolution operator defined on $\mathcal{S}_n^N$ by the GLJLDE (3.12a) and (3.12b). If $\mathbf{H} \in \mathcal{S}_n^N$ we have that $\mathbf{T}(t, t_0)[\mathbf{H}] = (\mathbf{T}_1(t, t_0)[\mathbf{H}], \mathbf{T}_2(t, t_0)[\mathbf{H}], \dots, \mathbf{T}_N(t, t_0)[\mathbf{H}])$ where $\mathbf{T}_i(t, t_0)[\mathbf{H}] \in \mathcal{S}_n$, for all $i \in \mathcal{N}$.

Remark 3.1 (a) from Remark 2.10 we may infer that $\mathbf{T}_i(t, t_0)[\mathbf{H}]$ is a positive semidefinite matrix in $\mathcal{S}_n$, if $\mathbf{H} \succcurlyeq 0$;

(b) for each $(t_0, x_0) \in \mathbb{R}_+ \times \mathbb{R}^n$ and for each $j \in \mathcal{N}$ we may write

$$\mathbb{E}[x^\top(t;t_0,x_0)x(t;t_0,x_0)|\eta(t_0)=j] = Tr[\mathbb{E}[x(t;t_0,x_0)x^\top(t;t_0,x_0)\chi_{\{\eta(t_0)=j\}}|\eta(t_0)=j]]$$
$$= \sum_{i=1}^{N} Tr\left[\mathbb{E}[x(t;t_0,x_0)x^\top(t;t_0,x_0)\chi_{\{\eta(t)=i\}}|\eta(t_0)=j]\right] = \sum_{i=1}^{N} Tr[X_i^{(j)}(t)].$$

This leads to

$$\mathbb{E}[|x(t;t_0,x_0)|^2 \mid \eta(t_0)=j] = \sum_{i=1}^{N} Tr[\mathbf{T}_i(t,t_0)[\mathbf{X}^{(j)}(t_0)]], \tag{3.16}$$

where $\mathbf{X}^{(j)}(t_0)$ is defined in (3.12c);

(c) if $x(t;t_0,x_0)$ is a solution of the JSLDE (3.4), then (3.16) takes the form

$$\mathbb{E}[|x(t;t_0,x_0)|^2] = Tr[\mathbf{T}(t,t_0)[x_0x_0^\top]], \tag{3.17}$$

for all $t \geq t_0 \geq 0$, $x_0 \in \mathbb{R}^n$, $\mathbf{T}(t,t_0)$ being now the linear evolution operator defined on $\mathcal{S}_n^1$ by the GLJLDE (3.15a) and (3.15b).

3.3 Several Kinds of Mean-Square Stability of a JSLDE

In this section we define several kinds of stability in mean square of the JSLDE of type (3.1) and we will highlight the relation between them.

Definition 3.1 We say that a JSLDE (3.1) is:

(a) **strong exponentially stable in mean square** (SESMS) if there exist $\beta \geq 1$, $\alpha > 0$ such that

$$\|\mathbf{T}(t,t_0)\|_\infty \leq \beta e^{-\alpha(t-t_0)}, \quad \text{for all } t \geq t_0 \geq 0, \tag{3.18}$$

$\mathbf{T}(t,t_0)$ being the linear evolution operator defined on $\mathcal{S}_n^N$ by the GLJLDE (3.12a) and (3.12b) associated to the JSLDE (3.1);

(b) **exponentially stable in mean square with conditioning** (ESMS-C) if there exist $\beta \geq 1$, $\alpha > 0$ such that the solutions of the JSLDE (3.1) satisfy

$$\mathbb{E}[|x(t;t_0,x_0)|^2 \mid \eta(t_0)=j] \leq \beta e^{-\alpha(t-t_0)}|x_0|^2, \tag{3.19}$$

for all $t \geq t_0 \geq 0$, $x_0 \in \mathbb{R}^n$, $j \in \mathcal{N}$ and for any initial probability distribution $\pi_0 = (\pi_1(0), \pi_2(0), \ldots, \pi_N(0))$;

(c) **exponentially stable in mean square** (ESMS) if there exist $\beta \geq 1$, $\alpha > 0$ such that the solutions of the JSLDE (3.1) satisfy

$$\mathbb{E}[|x(t;t_0,x_0)|^2] \leq \beta e^{-\alpha(t-t_0)}|x_0|^2, \tag{3.20}$$

for all $t \geq t_0 \geq 0$, $x_0 \in \mathbb{R}^n$ and any initial probability distribution π_0 of the Markov process;

(d) L_2 **stochastically stable in mean square with conditioning** (L_2-SSMS-C) if

$$\int_{t_0}^{\infty} \mathbb{E}[|x(t; t_0, x_0)|^2 \mid \eta(t_0) = j]\mathrm{d}t \leq \delta |x_0|^2, \tag{3.21}$$

for all $t_0 \geq 0$, $x_0 \in \mathbb{R}^n$, $j \in \mathcal{N}$ and any initial probability distribution π_0 of the Markov process, where $\delta > 0$ is a constant which does not depend upon t_0, x_0, j, π_0;

(e) L_2 **stochastically stable in mean square** (L_2-SSMS) if

$$\int_{t_0}^{\infty} \mathbb{E}[|x(t; t_0, x_0)|^2]\mathrm{d}t \leq \delta |x_0|^2, \tag{3.22}$$

for all $t_0 \geq 0$, $x_0 \in \mathbb{R}^n$ and for any initial probability distribution π_0 of the Markov process, where $\delta > 0$ is a constant which does not depend upon t_0, x_0, π_0;

(f) **asymptotically stable in mean square** (ASMS) if its solutions satisfy

$$\lim_{t \to \infty} \mathbb{E}[|x(t; t_0, x_0)|^2] = 0,$$

for all $t_0 \geq 0$, $x_0 \in \mathbb{R}^n$ and for any initial probability distribution π_0 of the Markov process.

Remark 3.2 In the case when the JSLDE (3.1) is in the special form of (3.4), then (3.19) coincides with (3.20) and (3.21) coincides with (3.22). Hence, in this case, only the concepts introduced by Definition 3.1(a), (c), (e) and (f) are relevant.

The next result is obtained immediately from Definition 3.1.

Proposition 3.2 *If Assumptions 1.2 and 2.1 hold then the following implications are true:*

(i) if the JSLDE (3.1) is SESMS, then it is ESMS-C;
(ii) if the JSLDE (3.1) is ESMS-C, then it is ESMS;
(iii) if the JSLDE (3.1) is ESMS, then it is ASMS;
(iv) if the JSLDE (3.1) is ESMS-C, then it is L_2*-SSMS-C;*
(v) if the JSLDE (3.1) is ESMS, then it is L_2*-SSMS;*
(vi) if the JSLDE (3.1) is L_2*-SSMS-C, then it is* L_2*-SSMS.*

Proof The implications (ii), (vi) follow from (3.19) and (3.21), respectively, if is taken into account that $\mathbb{E}[\cdot] = \sum_{j=1}^{N} \pi_{t_0}(j)\mathbb{E}[\cdot \mid \eta(t_0) = j]$ with $\sum_{j=1}^{N} \pi_{t_0}(j) = 1$ and $\pi_{t_0}(j) \geq 0$. The implications (iii), (iv) and (v) are obvious.

It remains to prove the implication (i). To this end we shall employ (3.16). First, based on Remark 3.1(a) we may infer that

$$Tr[\mathbf{T}_i(t, t_0)[\mathbf{X}^{(j)}(t_0)]] \leq n\|\mathbf{T}_i(t, t_0)[\mathbf{X}^{(j)}(t_0)]\|,$$

$\|\cdot\|$ being the spectral norm of a symmetric matrix. Involving (2.112) written for $X_i := \mathbf{T}_i(t, t_0)[\mathbf{X}^{(j)}(t_0)]$ we obtain from (3.16) that

$$\mathbb{E}[|x(t; t_0, x_0)|^2 \mid \eta(t_0) = j] \le nN\|\mathbf{T}(t, t_0)[\mathbf{X}^{(j)}(t_0)]\|_\infty$$
$$\le nN\|\mathbf{T}(t, t_0)\|_\infty \|\mathbf{X}^{(j)}(t_0)\|_\infty.$$

Further (2.112) and (3.12c) yield

$$\|\mathbf{X}^{(j)}(t_0)\|_\infty = |x_0|^2.$$

Finally (3.18) allows us to deduce that

$$\mathbb{E}[|x(t; t_0, x_0)|^2 \mid \eta(t_0) = j] \le nN\beta e^{-\alpha(t-t_0)}|x_0|^2, \tag{3.23}$$

for all $t \ge t_0 \ge 0$, $x_0 \in \mathbb{R}^n$, $j \in \mathcal{N}$ and any initial probability distribution π_0 of the Markov process. Thus the proof is complete.

None of the converse implications from Proposition 3.2 is true in the absence of some additional assumptions. In the next two results we show under which condition a part of the implications from Proposition 3.2 becomes equivalent.

Proposition 3.3 *Under Assumption 1.1 the following are equivalent for a JSLDE of type (3.4):*

(i) the JSLDE (3.4) is SESMS;
(ii) the JSLDE (3.4) is ESMS.

Proof The implication $(i) \Rightarrow (ii)$ follows from Proposition 3.2. Now we prove $(ii) \Rightarrow (i)$. Let $\mathbf{T}(t, t_0)$ be the linear evolution operator on $\mathcal{S}_n^1$ defined by the GLJLDE (3.15). Based on the Remark 2.10 we obtain from Corollary 2.4 and (2.112) (in the special case $N = 1$) that $\|\mathbf{T}(t, t_0)\|_\infty = \|\mathbf{T}(t, t_0)[I_n]\|$, fot all $t \ge t_0 \ge 0$. Employing (3.17) we may write

$$\|\mathbf{T}(t, t_0)\|_\infty \le \sum_{i=1}^{n} \|\mathbf{T}(t, t_0)[e_i e_i^\top]\| = \sum_{i=1}^{n} \mathbb{E}[|x(t; t_0, e_i)|^2],$$

where $e_i \in \mathbb{R}^n$ is the ith column of the identity matrix I_n. Finally, (3.20) leads to $\|\mathbf{T}(t, t_0)\|_\infty \le \beta e^{-\alpha(t-t_0)}$, for all $t \ge t_0 \ge 0$ which shows that the JSLDE (3.4) is SESMS. Thus the proof is complete.

To ease the presentation of the next result, let us define the concept of a periodic JSLDE.

Definition 3.2 We say that the JSLDE (3.1) is in the periodic case if the following conditions are simultaneously satisfied:

(a) $\tau_k = kh$, for all $k \in \mathbb{Z}_+$ with $h > 0$ fixed;

(b) there exists a real number $\theta_c > 0$ with the property that $A_l(t + \theta_c, i) = A_l(t, i), \ 0 \leq l \leq r_1,$ for all $(t, i) \in \mathbb{R}_+ \times \mathcal{N}$;
(c) there exists a natural number $\theta_d \geq 1$ with the property that $A_{dl}((k + \theta_d)h, i) = A_{dl}(kh, i),$ for all $0 \leq l \leq r_2, \ (k, i) \in \mathbb{Z}_+ \times \mathcal{N}$;
(d) $\frac{\theta_c}{h\theta_d}$ is a rational number.

Now we prove:

Theorem 3.1 *Assume:*

(a) Assumption 1.2 is fulfilled;
(b) the JSLDE (3.1) is in the periodic case.

Under these conditions the following are equivalent:

(i) the JSLDE (3.1) is SESMS;
(ii) the JSLDE (3.1) is ESMS-C;
(iii) the JSLDE (3.1) is ESMS;
(iv) the JSLDE (3.1) is ASMS.

Proof The implications $(i) \Rightarrow (ii) \Rightarrow (iii) \Rightarrow (iv)$ follow from Proposition 3.2. Now we prove $(iv) \Rightarrow (i)$. Let $\mathbf{T}(t, t_0)$ be the linear evolution operator on $\mathcal{S}_n^N$ defined by the GLJLDE (3.12a) and (3.12b). We have to show that there exist $\beta \geq 1$, $\alpha > 0$ for which (3.18) holds. Based on Remark 2.10 and Corollary 2.4 we may infer that

$$\|\mathbf{T}(t, 0)\|_\infty = \|\mathbf{T}(t, 0)[\mathbf{I}^N]\|_\infty$$

which leads to

$$\|\mathbf{T}(t, 0)\|_\infty \leq \sum_{j=1}^{N} \|\mathbf{T}(t, 0)[\mathbf{I}^{(j)}]\|_\infty \tag{3.24}$$

where $\mathbf{I}^N = (I_n, I_n, \ldots, I_n) \in \mathcal{S}_n^N$ and $\mathbf{I}^{(j)} = (\mathbf{I}_1^{(j)}, \mathbf{I}_2^{(j)}, \ldots, \mathbf{I}_N^{(j)})$ with

$$\mathbf{I}_i^{(j)}(t_0) = \begin{cases} 0, & \text{if } i \neq j \\ I_n, & \text{if } i = j. \end{cases}$$

For each $j \in \mathcal{N}$ we denote $x^{(j)}(t, x_0)$ the solution of the JSLDE:

$$dx(t) = A_0(t, \eta^{(j)}(t))x(t)dt + \sum_{l=1}^{r_1} A_l(t, \eta^{(j)}(t))x(t)dw_l(t), \ kh < t \leq (k+1)h \tag{3.25a}$$

$$x(kh^+) = (A_{d0}(kh, \eta^{(j)}(kh)) + \sum_{l=1}^{r_2} w_{dl}(k)A_{dl}(kh, \eta^{(j)}(kh)))x(kh), \ k \in \mathbb{Z}_+ \tag{3.25b}$$

which verifies $x^{(j)}(0, x_0) = x_0$. In (3.25), $\{\eta^{(j)}(t)\}_{t\geq 0}$ is the version of $\{\eta(t)\}_{t\geq 0}$ when the initial probability distribution is $\pi_0^{(j)} = (\pi_1^{(j)}(0), \pi_2^{(j)}(0), \ldots, \pi_N^{(j)}(0))$ with

$$\pi_i^{(j)}(0) = \begin{cases} 0, & \text{if } i \neq j \\ 1, & \text{if } i = j. \end{cases} \tag{3.26}$$

Since (3.25) is one of the versions of (3.1) and the property of asymptotic stability does not depend upon the initial probability distribution of the Markov process, we deduce that if (iv) holds, then

$$\lim_{t\to\infty} \mathbb{E}[|x^{(j)}(t_0, x_0)|^2] = 0, \tag{3.27}$$

for all $x_0 \in \mathbb{R}^n$ and any $j \in \mathcal{N}$.

Let $e_i \in \mathbb{R}^n$ be the ith column of the identity matrix I_n. Let $x^{(j)}(\cdot; e_i)$ be the solution of the JSLDE (3.25) with the initial condition $x^{(j)}(0, e_i) = e_i$. We set $\mathbf{I}^{(j,i)} = (\mathbf{I}_1^{(j,i)}, \mathbf{I}_2^{(j,i)}, \ldots, \mathbf{I}_N^{(j,i)})$ where

$$\mathbf{I}_l^{(j,i)} = \begin{cases} 0, & \text{if } l \neq j \\ e_i e_i^\top, & \text{if } l = j. \end{cases}$$

One sees that

$$\sum_{i=1}^{n} \mathbf{I}^{(j,i)} = \mathbf{I}^j, \quad \text{for all } 1 \leq j \leq N. \tag{3.28}$$

Further, combining (3.24), (3.14), (3.26) and (3.28) we deduce that

$$\begin{aligned}
\sum_{i=1}^{n} \mathbb{E}[|x^{(j)}(t, e_i)|^2] &= \sum_{i=1}^{n}\sum_{l=1}^{N} Tr[\mathbb{E}[x^{(j)}(t, e_i)(x^{(j)}(t, e_i))^\top \chi_{\{\eta^{(j)}(t)=l\}}]] \\
&= \sum_{i=1}^{n}\sum_{l=1}^{N} Tr[\mathbf{T}_l(t, 0)[\mathbf{I}^{(j,i)}]] = \sum_{l=1}^{N} Tr[\mathbf{T}_l(t, 0)[\mathbf{I}^{(j)}]] \\
&\geq \sum_{l=1}^{N} \|\mathbf{T}_l(t, 0)[\mathbf{I}^{(j)}]\|
\end{aligned} \tag{3.29}$$

where $\|\mathbf{T}_l(t, 0)[\mathbf{I}^{(j)}]\|$ is the spectral norm of the positive semidefinite matrix $\mathbf{T}_l(t, 0)[\mathbf{I}^{(j)}]$. From (2.112) written for $\mathbf{X} \triangleq \mathbf{T}(t, 0)[\mathbf{I}^{(j)}]$ and (3.29) we get

$$\|\mathbf{T}(t, 0)[\mathbf{I}^{(j)}]\|_\infty \leq \sum_{i=1}^{n} \mathbb{E}[|x^{(j)}(t, e_i)|^2].$$

Using (3.27) written for $x_0 = e_i$ we deduce that $\lim_{t\to\infty} \|\mathbf{T}(t, 0)\|_\infty = 0$.

Particularly, we have that $\lim_{k\to\infty} \|\mathbf{T}(kh, 0)\|_\infty = 0$. According to (2.11) and (2.55) written for the special case of a GLJLDE of type (3.12a) and (3.12b) with $\tau_k = kh$, the previous equality may be rewritten in the equivalent form

$$\lim_{k\to\infty} \|\mathbf{T}_d(k, 0)\|_\infty = 0. \tag{3.30}$$

Let $\tilde{p}$ and $\tilde{q}$ be coprime natural numbers satisfying

$$\frac{\theta_c}{h\theta_d} = \frac{\tilde{p}}{\tilde{q}} \tag{3.31a}$$

$$\tilde{\theta} \stackrel{\Delta}{=} h\tilde{p}\theta_d = \tilde{q}\theta_c \tag{3.31b}$$

$$\tilde{\mathbf{k}} \stackrel{\Delta}{=} \tilde{p}\theta_d. \tag{3.31c}$$

Employing (2.55)–(2.57) specialized for the case of the operators (2.113) and (2.115) we may obtain from (3.30) that

$$\lim_{l\to\infty} \|\mathbf{T}_d^{(l)}(\tilde{\mathbf{k}}, 0)\|_\infty = 0.$$

Hence, the eigenvalues of the monodromy operator $\mathbf{T}_d(\tilde{\mathbf{k}}, 0)$ are in the interior of the disk $|\lambda| < 1$. This allows us to conclude that the spectral radius of the monodromy operator $\mathbf{T}_d(\tilde{\mathbf{k}}, 0)$ satisfies $\rho(\mathbf{T}_d(\tilde{\mathbf{k}}, 0)) < 1$. Finally, from Corollary 2.1 specialized to the case of the GLJLDE (3.12a) and (3.12b) we deduce that (3.18) is satisfied. The proof ends.

3.4 Lyapunov-Type Criteria for the Exponential Stability in Mean Square

In this section we shall use some results derived in Sects. 2.4 and 2.5 in order to obtain useful criteria for exponential stability in mean square of a JSLDE of type (3.1) and its particular forms (3.4) and (3.5).

3.4.1 *The General Time-Varying Case*

The result proved in Theorem 2.8, specialized to a GLJLDE of type (2.117), gives:

Theorem 3.2 *Assume:*

(a) Assumption 1.2 is fulfilled;
(b) the impulsive time instants τ_k satisfy a condition of type (2.26).

Under these conditions, the following are equivalent:

(i) *the JSLDE (3.1) is SESMS;*

(ii) *for any bounded and continuous function* $t \to (H(t,1), H(t,2), \dots, H(t,N)) : \mathbb{R}_+ \to \mathcal{S}_n^{N+}$ *and for any uniform positive and bounded sequence* $\{\mathbf{\Gamma}_k\}_{k\in\mathbb{Z}_+} \subset Int\ \mathcal{S}_n^{N+}$*, the non-homogeneous backward GLJLDE:*

$$-\dot{Y}_i(t) = A_0^\top(t,i)Y_i(t) + Y_i(t)A_0(t,i) + \sum_{l=1}^{r_1} A_l^\top(t,i)Y_i(t)A_l(t,i) + \sum_{j=1}^{N} q_{ij}Y_j(t) + H(t,i), \quad \tau_k \le t < \tau_{k+1}, \tag{3.32a}$$

$$Y_i(\tau_k^-) = \sum_{l=0}^{r_2} A_{dl}^\top(\tau_k,i)Y_i(\tau_k)A_{dl}(\tau_k,i) + \Gamma_k(i), \quad k \in \mathbb{Z}_+, \quad i \in \mathcal{N}, \tag{3.32b}$$

has a unique bounded solution $t \to (Y_1(t), Y_2(t), \dots, Y_N(t)) : \mathbb{R}_+ \to \mathcal{S}_n^N$ *which is uniform positive;*

(iii) *there exists a bounded and continuous function* $t \to (\tilde{H}(t,1), \dots, \tilde{H}(t,N)) : \mathbb{R}_+ \to \mathcal{S}_n^{N+}$ *and a bounded and uniform positive sequence* $\{\tilde{\mathbf{\Gamma}}_k\}_{k\in\mathbb{Z}_+} \subset Int\ \mathcal{S}_n^{N+}$ *with the property that the corresponding non-homogeneous backward GLJLDE of type (3.32) has a unique bounded solution* $t \to (\tilde{Y}_1(t), \tilde{Y}_2(t), \dots, \tilde{Y}_N(t)) : \mathbb{R}_+ \to \mathcal{S}_n^{N+}$;

(iv) *there exists a function* $t \to (Z_1(t), Z_2(t), \dots, Z_N(t)) : \mathbb{R}_+ \to \mathcal{S}_n^N$ *with the properties:*

(α) *it is bounded and uniform positive;*

(β) *it is differentiable on each interval* (τ_k, τ_{k+1}) *and right continuous in any* $t = \tau_k$, $k \in \mathbb{Z}_+$;

(γ) *it solves the backward linear differential inequality of Lyapunov type with jumps*

$$\dot{Z}_i(t) + A_0^\top(t,i)Z_i(t) + Z_i(t)A_0(t,i) + \sum_{l=1}^{r_1} A_l^\top(t,i)Z_i(t)A_l(t,i) + \sum_{j=1}^{N} q_{ij}Z_j(t) \le 0, \quad \tau_k \le t < \tau_{k+1}, \tag{3.33a}$$

$$\sum_{l=0}^{r_2} A_{dl}^\top(\tau_k,i)Z_i(\tau_k)A_{dl}(\tau_k,i) - Z_i(\tau_k^-) + \gamma I_n \le 0,, \quad k \in \mathbb{Z}_+, \tag{3.33b}$$

with $\gamma > 0$ *a constant which does not depend upon* k.

Proof It is obtained directly from Theorem 2.8 adapted to the operators defined in (2.114) and (2.116).

In the case of a JSLDE without Markov jump perturbations the result stated in Theorem 3.2 takes the form:

Corollary 3.2 *Assume:*

(a) Assumption 1.1 is fulfilled;
(b) the impulsive time instants τ_k satisfy a condition of type (2.26).

Under these conditions, the following are equivalent:

(i) the JSLDE (3.4) is ESMS;
(ii) for any bounded and continuous function $t \to H(t) : \mathbb{R}_+ \to \mathcal{S}_n^{1+}$ and for any uniform positive sequence $\{\Gamma_k\}_{k \in \mathbb{Z}_+} \subset Int\ \mathcal{S}_n^{1+}$, the homogeneous backward GLJLDE

$$
\begin{aligned}
&-\dot{Y}(t) = A_0^\top(t)Y(t) + Y(t)A_0(t) + \sum_{l=1}^{r_1} A_l^\top(t)Y(t)A_l(t) + H(t),\\
&\tau_k \le t < \tau_{k+1}, \qquad (3.34a)\\
&Y(\tau_k^-) = \sum_{l=0}^{r_2} A_{dl}^\top(\tau_k)Y(\tau_k)A_{dl}(\tau_k) + \Gamma_k, \quad k \in \mathbb{Z}_+, \qquad (3.34b)
\end{aligned}
$$

has a unique bounded and uniform positive solution $t \to Y(t) : \mathbb{R}_+ \to \mathcal{S}_n^1$;
(iii) there exists a bounded and continuous function $t \to \tilde{H}(t) : \mathbb{R}_+ \to \mathcal{S}_n^{1+}$ and a bounded and uniform positive sequence $\{\tilde{\Gamma}_k\}_{k \in \mathbb{Z}_+} \subset Int\ \mathcal{S}_n^{1+}$ with the property that the corresponding non-homogeneous backward GLJLDE of type (3.34) has a unique bounded solution $t \to \tilde{Y}(t) : \mathbb{R}_+ \to \mathcal{S}_n^{1+}$;
(iv) there exists a function $t \to Z(t) : \mathbb{R}_+ \to \mathcal{S}_n^{1+}$ with the properties:

(α') it is bounded and uniform positive;
(β') it is differentiable on each interval (τ_k, τ_{k+1}) and right continuous in each $t = \tau_k$, $k \in \mathbb{Z}_+$;
(γ') it solves the following backward linear differential inequality with jumps

$$
\dot{Z}(t) + A_0^\top(t)Z(t) + Z(t)A_0(t) + \sum_{l=1}^{r_1} A_l^\top(t)Z(t)A_l(t) \le 0,\ \tau_k \le t < \tau_{k+1}, \qquad (3.35a)
$$

$$
\sum_{l=0}^{r_2} A_{dl}^\top(\tau_k)Z(\tau_k)A_{dl}(\tau_k) - Z(\tau_k^-) + \gamma I_n \le 0 \qquad (3.35b)
$$

$\gamma > 0$ is a constant not depending upon k.

Proof It follows immediately from Theorem 3.2 together with Proposition 3.3 which makes possible the replacement of the property of SESMS from the point (i) of the statement of Theorem 3.2 with the property of ESMS of the considered JSLDE involved in the assumption (i) of Corollary 3.2.

Remark 3.3

(a) the presence of the parameter $\gamma > 0$ in (3.33b) and in (3.35b), respectively, is considered to mark the fact that the involved inequalities are uniform with respect to $k \in \mathbb{Z}_+$;
(b) it is worth noticing that in the general time-varying case of a type (3.1) system with $N \geq 2$, the statements (ii)–(iv) from Theorem 3.2 provide necessary and sufficient conditions for strong exponential stability in mean square of a JSLDE and they are only sufficient conditions for the property of exponential stability in mean square. We shall see in the next subsection that in the periodic case, the statements (ii)–(iv) provide necessary and sufficient conditions for the property of exponential stability in mean square for JSLDE of type (3.1) with $N \geq 2$.

3.4.2 The Periodic Case

The results proved in Theorem 2.12 specialized to the case of a GLJLDE (2.117) together with Theorem 3.1 lead to:

Theorem 3.3 *Assume:*

(a) Assumption 1.2 is fulfilled;
(b) the JSLDE (3.1) is in the periodic case (in the sense of Definition 3.2).

Let $\theta_c \in \mathbb{R}$ and $\theta_d \in \mathbb{Z}$ be such that $\theta_c > 0$, $\theta_d \geq 1$ associated to the system (3.1) according to Definition 3.2. We set $\tilde{\theta} \triangleq \tilde{p}h\theta_d = \tilde{q}\theta_c$ where $\tilde{p}$ and $\tilde{q}$ are coprime natural numbers satisfying $\frac{\tilde{p}}{\tilde{q}} = \frac{\theta_c}{h\theta_d}$. Under these conditions, the following are equivalent:

(i) the JSLDE (3.1) is ESMS;
(ii) for any continuous function $t \to (H(t,1), H(t,2), \ldots, H(t,N)) : \mathbb{R} \to \mathcal{S}_n^N$ such that $H(t+\theta_c, i) = H(t,i) \geq 0$, for all $(t,i) \in \mathbb{R} \times \mathcal{N}$ and for any sequence $\{\mathbf{\Gamma}_k\}_{k\in\mathbb{Z}} \subset \operatorname{Int} \mathcal{S}_n^{N+}$, where $\mathbf{\Gamma}_k = (\Gamma_k(1), \Gamma_k(2), \ldots, \Gamma_k(N))$ such that $\Gamma_{k+\theta_d}(i) = \Gamma_k(i) > 0$, for all $(k,i) \in \mathbb{Z} \times \mathcal{N}$, the non-homogeneous backward GLJLDE:

$$
-\dot{Y}_i(t) = A_0^\top(t,i)Y_i(t) + Y_i(t)A_0(t,i) + \sum_{l=1}^{r_1} A_l^\top(t,i)Y_i(t)A_l(t,i)
$$
$$
+\sum_{j=1}^{N} q_{ij}Y_j(t) + H(t,i), \quad kh \leq t < (k+1)h, \tag{3.36a}
$$

$$Y_i(kh^-) = \sum_{l=0}^{r_2} A_{dl}^\top(kh, i) Y_i(kh) A_{dl}(kh, i) + \Gamma_k(i), \quad k \in \mathbb{Z}, \tag{3.36b}$$

has a unique bounded solution $t \to (Y_1(t), Y_2(t), \ldots, Y_N(t))$ *and this solution has the properties*

$$Y_i(t + \tilde{\theta}) = Y_i(t) > 0, \quad \textit{for all} \quad (t, i) \in \mathbb{R} \times \mathcal{N};$$

(*iii*) *there exist a continuous function* $t \to (\tilde{H}(t, 1), \ldots, \tilde{H}(t, N)) : \mathbb{R} \to \mathcal{S}_n^N$ *such that* $\tilde{H}(t + \theta_c, i) = \tilde{H}(t, i) \geq 0$, *for all* $(t, i) \in \mathbb{R} \times \mathcal{N}$ *and a sequence* $\{\tilde{\mathbf{\Gamma}}_k\}_{k \in \mathbb{Z}} \subset Int\ \mathcal{S}_n^{N+}$ *where* $\tilde{\mathbf{\Gamma}}_k = (\tilde{\Gamma}_k(1), \tilde{\Gamma}_k(2), \ldots, \tilde{\Gamma}_k(N))$ *periodic of period* θ_d *with the property that the corresponding non-homogeneous backward GLJLDE of type (3.36) has a unique bounded solution* $t \to (\tilde{Y}_1(t), \tilde{Y}_2(t), \ldots, \tilde{Y}_N(t)) : \mathbb{R} \to \mathcal{S}_n^{N+}$ *and this solution is periodic of period* $\tilde{\theta}$;

(*iv*) *there exists a function* $t \to (Z_1(t), Z_2(t), \ldots, Z_N(t)) : \mathbb{R} \to \mathcal{S}_n^N$ *with the properties:*

(α) *it is bounded and uniform positive on* $\mathbb{R}$;
(β) *it is a periodic function of period* $\tilde{\theta}$;
(γ) *it is differentiable on each interval* $(kh, (k+1)h)$ *and right continuous in each* $t = kh$, $k \in \mathbb{Z}$;
(δ) *it solves the following backward linear differential inequality of Lyapunov type with jumps:*

$$\dot{Z}_i(t) + A_0^\top(t, i) Z_i(t) + Z_i(t) A_0(t, i) + \sum_{l=1}^{r_1} A_l^\top(t, i) Z_i(t) A_l(t, i)$$
$$+ \sum_{j=1}^{N} q_{ij} Z_j(t) \leq 0, \quad kh \leq t < (k+1)h, \tag{3.37a}$$
$$\sum_{l=0}^{r_2} A_{dl}^\top(kh, i) Z_i(kh) A_{dl}(kh, i) - Z_i(kh^-) < 0, \quad k \in \mathbb{Z}; \tag{3.37b}$$

(*v*) *for any continuous function* $t \to (H(t, 1), H(t, 2), \ldots, H(t, N)) : \mathbb{R} \to \mathcal{S}_n^N$ *with the property that* $H(t + \theta_c, i) = H(t, i) \geq 0$, *for all* $(t, i) \in \mathbb{R} \times \mathcal{N}$ *and for any sequence* $\{\mathbf{\Gamma}_k\}_{k \in \mathbb{Z}} \subset Int\ \mathcal{S}_n^{N+}$, *where* $\mathbf{\Gamma}_k = (\Gamma_k(1), \Gamma_k(2), \ldots, \Gamma_k(N))$ *with the property that* $\Gamma_{k+\theta_d}(i) = \Gamma_k(i) > 0$, $k \in \mathbb{Z}$, *the non-homogeneous forward GLJLDE*

$$\dot{X}_i(t) = A_0(t, i) X_i(t) + X_i(t) A_0^\top(t, i) + \sum_{l=1}^{r_1} A_l(t, i) X_i(t) A_l^\top(t, i)$$
$$+ \sum_{j=1}^{N} q_{ji} X_j(t) + H(t, i), \quad kh < t \leq (k+1)h, \tag{3.38a}$$

$$X_i(kh^+) = \sum_{l=0}^{r_2} A_{dl}(kh, i) X_i(kh) A_{dl}^{\top}(kh, i) + \Gamma_k(i), \quad k \in \mathbb{Z}, \tag{3.38b}$$

has a unique bounded solution $t \to (X_1(t), X_2(t), \ldots, X_N(t)) : \mathbb{R} \to Int\ \mathcal{S}_n^{N+}$ and this solution is periodic of period $\tilde{\theta}$;

(vi) *there exist a continuous function $t \to (\tilde{H}(t, 1), \ldots, \tilde{H}(t, N)) : \mathbb{R} \to \mathcal{S}_n^N$ with $\tilde{H}(t + \theta_c, i) = \tilde{H}(t, i) \geq 0$, for all $(t, i) \in \mathbb{R} \times \mathcal{N}$ and a sequence $\{\tilde{\Gamma}_k\}_{k \in \mathbb{Z}} \subset Int\ \mathcal{S}_n^{N+}$ which is periodic of period $\tilde{\theta}_d$ such that the corresponding non-homogeneous forward GLJLDE of type (3.38) has a unique bounded solution $t \to (\tilde{X}_1(t), \ldots, \tilde{X}_N(t)) : \mathbb{R} \to \mathcal{S}_n^{N+}$ which is a periodic function of period $\tilde{\theta}$;*

(vii) *there exists a function $t \to (\tilde{Z}_1(t), \tilde{Z}_2(t), \ldots, \tilde{Z}_N(t)) : \mathbb{R} \to \mathcal{S}_n^N$ with the properties:*

(α') *it is bounded and uniform positive on $\mathbb{R}$;*
(β') *it is a periodic function of period $\tilde{\theta}$;*
(γ') *it is differentiable on each interval $(kh, (k+1)h)$ and left continuous in each $t = kh$, $k \in \mathbb{Z}$;*
(δ') *it solves the following non-homogeneous forward differential inequality of Lyapunov type with jumps:*

$$-\dot{\tilde{Z}}_i(t) + A_0(t, i)\tilde{Z}_i(t) + \tilde{Z}_i(t)A_0^{\top}(t, i) + \sum_{l=1}^{r_1} A_l(t, i)\tilde{Z}_i(t)A_l^{\top}(t, i)$$
$$+ \sum_{j=1}^{N} q_{ji}\tilde{Z}_j(t) \leq 0, \quad kh < t \leq (k+1)h,$$
$$\sum_{l=0}^{r_2} A_{dl}(kh, i)\tilde{Z}_i(kh)A_{dl}^{\top}(kh, i) - \tilde{Z}_i(kh^+) < 0, \quad k \in \mathbb{Z}.$$

When the JSLDE (3.1) is in the special form of (3.4) the result stated in Theorem 3.3 becomes:

Corollary 3.3 *Assume:*

(a) *Assumption 1.1 is fulfilled;*
(b) *the JSLDE (3.4) is in the periodic case (in the sense of Definition 3.2).*

Let $\theta_c \in \mathbb{R}$ and $\theta_d \in \mathbb{Z}$ be such that $\theta_c > 0$, $\theta_d \geq 1$ associated to the system (3.4) according to Definition 3.2. We set $\tilde{\theta} \triangleq \tilde{p}h\theta_d = \tilde{q}\theta_c$ where $\tilde{p}$ and $\tilde{q}$ are coprime natural numbers satisfying $\frac{\tilde{p}}{\tilde{q}} = \frac{\theta_c}{h\theta_d}$. Under these conditions, the following are equivalent:

(i) *the JSLDE (3.4) is ESMS;*
(ii) *for any continuous function $t \to H(t) : \mathbb{R} \to \mathcal{S}_n^1$ with the property $H(t + \theta_c) = H(t) \geq 0$, for all $t \in \mathbb{R}$ and for any sequence $\{\Gamma_k\}_{k \in \mathbb{Z}} \subset Int\ \mathcal{S}_n^{1+}$, which is periodic of period θ_d, the non-homogeneous backward GLJLDE*

$$-\dot{Y}(t) = A_0^\top(t)Y(t) + Y(t)A_0(t) + \sum_{l=1}^{r_1} A_l^\top(t)Y(t)A_l(t) + H(t),$$

$$kh \leq t < (k+1)h, \tag{3.39a}$$

$$Y(kh^-) = \sum_{l=0}^{r_2} A_{dl}^\top(kh)Y(kh)A_{dl}(kh) + \Gamma_k, \quad k \in \mathbb{Z}, \tag{3.39b}$$

has a unique bounded solution $t \to Y(t) : \mathbb{R} \to \ Int\ \mathcal{S}_n^{1+}$ *and this solution is periodic of period* $\tilde{\theta}$;

(*iii*) *there exist a continuous function* $t \to \tilde{H}(t) : \mathbb{R} \to \mathcal{S}_n^1$ *with the property that* $\tilde{H}(t+\theta_c) = \tilde{H}(t) \geq 0$, *for all* $(t) \in \mathbb{R}$ *and a sequence* $\{\tilde{\Gamma}_k\}_{k\in\mathbb{Z}} \subset Int\ \mathcal{S}_n^{1+}$ *which is periodic of period* θ_d *such that the corresponding non-homogeneous backward GLJLDE of type (3.39) has a unique bounded solution* $t \to \tilde{Y}(t) : \mathbb{R} \to \ \mathcal{S}_n^{1+}$ *and additionally this solution is a periodic function of period* $\tilde{\theta}$;

(*iv*) *there exists a function* $t \to Z(t) : \mathbb{R} \to \mathcal{S}_n^1$ *with the properties:*

(α) *it is bounded and uniform positive on* $\mathbb{R}$;
(β) *it is a periodic function of period* $\tilde{\theta}$;
(γ) *it is differentiable on each interval* $(kh, (k+1)h)$ *and it is right continuous in each* $t = kh$, $k \in \mathbb{Z}$;
(δ) *it solves the following non-homogeneous backward differential inequality of Lyapunov type with jumps*

$$\dot{Z}(t) + A_0^\top(t)Z(t) + Z(t)A_0(t) + \sum_{l=1}^{r_1} A_l^\top(t)Z(t)A_l(t) \leq 0, \quad kh \leq t < (k+1)h,$$

$$\sum_{l=0}^{r_2} A_{dl}^\top(kh)Z(kh)A_{dl}(kh) - Z(kh^-) < 0, \quad k \in \mathbb{Z};$$

(*v*) *for any continuous function* $t \to H(t) : \mathbb{R} \to \mathcal{S}_n^1$ *with the property that* $H(t+\theta_c) = H(t) \geq 0$, $t \in \mathbb{R}$ *and for any sequence* $\{\Gamma_k\}_{k\in\mathbb{Z}} \subset \ Int\ \mathcal{S}_n^{1+}$, *which is periodic with period* θ_d, *the non-homogeneous forward GLJLDE*

$$\dot{X}(t) = A_0(t)X(t) + X(t)A_0^\top(t) + \sum_{l=1}^{r_1} A_l(t)X(t)A_l^\top(t) + H(t),$$

$$kh < t \leq (k+1)h, \tag{3.40a}$$

$$X(kh^+) = \sum_{l=0}^{r_2} A_{dl}(kh)X(kh)A_{dl}^\top(k) + \Gamma_k, \quad k \in \mathbb{Z}, \tag{3.40b}$$

has an unique bounded solution $t \to X(t) : \mathbb{R} \to \ Int\ \mathcal{S}_n^{1+}$ *and this solution is a periodic function of period* $\tilde{\theta}$;

(vi) *there exist a continuous function* $t \to \tilde{H}(t) : \mathbb{R} \to \mathcal{S}_n^1$ *with the property that* $\tilde{H}(t + \theta_c) = \tilde{H}(t) \geq 0$, *for all* $t \in \mathbb{R}$ *and a sequence* $\{\tilde{\Gamma}_k\}_{k \in \mathbb{Z}} \subset Int\ \mathcal{S}_n^{1+}$ *which is periodic of period* θ_d *such that the corresponding non-homogeneous forward GLJLDE of type (3.40) has a unique bounded solution* $t \to \tilde{X}(t) : \mathbb{R} \to \mathcal{S}_n^{1+}$ *and this solution is a periodic function of period* $\tilde{\theta}$;

(vii) *there exists a function* $t \to \tilde{Z}(t) : \mathbb{R} \to \mathcal{S}_n^1$ *with the properties:*

(α) *is bounded and uniform positive on* $\mathbb{R}$;
(β) *is a periodic function of period* $\tilde{\theta}$;
(γ) *is differentiable on each interval* $(kh, (k+1)h)$ *and left continuous in each* $t = kh$, $k \in \mathbb{Z}$;
(δ) *solves the following forward differential inequality of Lyapunov type with jumps:*

$$\dot{\tilde{Z}}(t) \geq A_0(t)\tilde{Z}(t) + \tilde{Z}(t)A_0^\top(t) + \sum_{l=1}^{r_1} A_l(t)\tilde{Z}(t)A_l^\top(t),\ kh < t \leq (k+1)h,$$

$$\sum_{l=0}^{r_2} A_{dl}(kh)\tilde{Z}(kh)A_{dl}^\top(k) - \tilde{Z}(kh^+) < 0, \quad k \in \mathbb{Z}.$$

3.4.3 The Time-Invariant Case

We start by defining the concept of time-invariant jump stochastic linear differential equations.

Definition 3.3 We say that a JSLDE (3.1) is in the time-invariant case if the following conditions are simultaneously satisfied:

(i) $A_l(t,i) = A_l(i) \in \mathbb{R}^{n \times n}$, for all $(t,i) \in \mathbb{R}_+ \times \mathcal{N}$, $0 \leq l \leq r_1$;
(ii) $A_{dl}(\tau_k, i) = A_{dl}(i) \in \mathbb{R}^{n \times n}$, for all $(k,i) \in \mathbb{Z}_+ \times \mathcal{N}$, $0 \leq l \leq r_2$;
(iii) $\tau_k = kh$, for all $k \in \mathbb{Z}_+$, where $h > 0$ is given.

In this case, the JSLDE (3.1) takes the form

$$dx(t) = A_0(\eta(t))x(t)dt + \sum_{l=1}^{r_1} A_l(\eta(t))x(t)dw_l(t), \quad kh < t \leq (k+1)h, \tag{3.41a}$$

$$X(kh^+) = A_{d0}(\eta(kh))x(kh) + \sum_{l=1}^{r_2} w_{dl}(k)A_{dl}(\eta(kh))x(kh), \quad k \in \mathbb{Z}. \tag{3.41b}$$

The time-invariant version of the JSLDE (3.4) is:

$$dx(t) = A_0x(t)dt + \sum_{l=1}^{r_1} A_lx(t)dw_l(t), \quad kh < t \leq (k+1)h, \tag{3.42a}$$

$$X(kh^+) = \left(A_{d0} + \sum_{l=1}^{r_2} w_{dl}(k)A_{dl}\right)x(kh), \quad k \in \mathbb{Z}. \tag{3.42b}$$

Since JSLDE of type (3.41) may be regarded as a JSLDE in the periodic case with $\theta_c = h$ and $\theta_d = 1$, we may use the results stated in Theorem 3.3 in order to obtain a set of criteria for exponential stability in the mean-square sense of a JSLDE in the time-invariant case. Thus, we obtain:

Theorem 3.4 *Assume that Assumption 1.2 is fulfilled. Then the following are equivalent:*

(i) the JSLDE (3.41) is ESMS;

(ii) for any $\mathbf{H} = (H(1), H(2), \ldots, H(N) \in \mathcal{S}_n^{N+}$ *and any* $\mathbf{\Gamma} = (\Gamma(1), \Gamma(2), \ldots, \Gamma(N)) \in Int\ \mathcal{S}_n^{N+}$*, the non-homogeneous backward GLJLDE*

$$-\dot{Y}_i(t) = A_0^\top(i)Y_i(t) + Y_i(t)A_0(i) + \sum_{l=1}^{r_1} A_l^\top(i)Y_i(t)A_l(i)$$

$$+\sum_{j=1}^{N} q_{ij}Y_j(t) + H(i), \quad kh \le t < (k+1)h, \tag{3.43a}$$

$$Y_i(kh^-) = \sum_{l=0}^{r_2} A_{dl}^\top(i)Y_i(kh)A_{dl}(i) + \Gamma(i), \quad k \in \mathbb{Z}, \quad i \in \mathcal{N}, \tag{3.43b}$$

has a unique bounded solution $t \to (Y_1(t), Y_2(t), \ldots, Y_N(t)) : \mathbb{R} \to Int\ \mathcal{S}_n^N$ *and this solution is a periodic function of period* h *and satisfies* $Y_i(t) > 0$, $(t, i) \in \mathbb{R} \times \mathcal{N}$;

(iii) there exists $(\tilde{\mathbf{H}}, \tilde{\mathbf{\Gamma}}) \in \mathcal{S}_n^{N+} \times Int\ \mathcal{S}_n^{N+}$, $\tilde{\mathbf{H}} = (\tilde{H}(1), \tilde{H}(2), \ldots, \tilde{H}(N))$, $\tilde{\mathbf{\Gamma}} = (\tilde{\Gamma}(1), \tilde{\Gamma}(2), \ldots, \tilde{\Gamma}(N))$ *with the property that the corresponding non-homogeneous backward GLJLDE (3.43) has a unique bounded solution* $t \to (\tilde{Y}_1(t), \tilde{Y}_2(t), \ldots, \tilde{Y}_N(t)) : \mathbb{R} \to \mathcal{S}_n^{N+}$*. Additionally, this solution is a periodic function with period* h*;*

(iv) there exists a function $t \to (Z_1(t), \ldots, Z_N(t)) : \mathbb{R} \to \mathcal{S}_n^N$ *with the properties:*

(α) it is bounded and uniform positive on $\mathbb{R}$*;*

(β) it is a periodic function of period h*;*

(γ) it is differentiable on each interval $(kh, (k+1)h)$ *and is right continuous in each* $t = kh$, $k \in \mathbb{Z}$*;*

(δ) it solves the following backward linear differential inequalities of Lyapunov type with jumps

$$\dot{Z}_i(t) + A_0^\top(i)Z_i(t) + Z_i(t)A_0(i) + \sum_{l=1}^{r_1} A_l^\top(i)Z_i(t)A_l(i)$$

$$+\sum_{j=1}^{N} q_{ij} Z_j(t) \le 0,$$

$$kh \le t < (k+1)h,$$

$$\sum_{l=0}^{r_2} A_{dl}^{\top}(i) Z_i(kh) A_{dl}(i) - Z_i(kh^-) < 0, \quad k \in \mathbb{Z};$$

(v) *for any* $\mathbf{H} = (H(1), H(2), \ldots, H(N)) \in \mathcal{S}_n^{N+}$ *and any* $\mathbf{\Gamma} = (\Gamma(1), \Gamma(2), \ldots, \Gamma(N)) \in Int\ \mathcal{S}_n^{N+}$ *the non-homogeneous forward GLJLDE*

$$\dot{X}_i(t) = A_0(i) X_i(t) + X_i(t) A_0^{\top}(i) + \sum_{l=1}^{r_1} A_l(i) X_i(t) A_l^{\top}(i)$$
$$+\sum_{j=1}^{N} q_{ji} X_j(t) + H(i), \quad kh < t \le (k+1)h, \tag{3.44a}$$

$$X(kh^+) = \sum_{l=0}^{r_2} A_{dl}(i) X_i(kh) A_{dl}^{\top}(i) + \Gamma(i), \quad k \in \mathbb{Z}, \quad i \in \mathcal{N}, \tag{3.44b}$$

has an unique bounded solution $t \to (X_1(t), X_2(t), \ldots, X_N(t)) : \mathbb{R} \to \mathcal{S}_n^{N+}$. *Additionally, this solution is a periodic function of period h and satisfies* $X_i(t) > 0$, *for all* $(t, i) \in \mathbb{R} \times \mathbb{Z}$;

(vi) *there exists* $(\tilde{\mathbf{H}}, \tilde{\mathbf{\Gamma}}) \in \mathcal{S}_n^{N+} \times Int\ \mathcal{S}_n^{N+}$, $\tilde{\mathbf{H}} = (\tilde{H}(1), \tilde{H}(2), \ldots, \tilde{H}(N))$, $\tilde{\mathbf{\Gamma}} = (\tilde{\Gamma}(1), \tilde{\Gamma}(2), \ldots, \tilde{\Gamma}(N))$ *with the property that the corresponding non-homogeneous forward GLJLDE of type (3.44) has a unique bounded solution* $t \to (\tilde{X}_1(t), \tilde{X}_2(t), \ldots, \tilde{X}_N(t)) : \mathbb{R} \to \mathcal{S}_n^{N+}$. *Additionally, this solution is a periodic function of period h;*

(vii) *there exists a function* $t \to (\tilde{Z}_1(t), \tilde{Z}_2(t), \ldots, \tilde{Z}_N(t)) : \mathbb{R} \to \mathcal{S}_n^{N+}$ *with the properties:*

(α') *it is bounded and uniform positive function on* $\mathbb{R}$;
(β') *it is a periodic function of period h*;
(γ') *it is differentiable on each interval* $(kh, (k+1)h)$ *and it is left continuous in each* $t = kh$, $k \in \mathbb{Z}$;
(δ') *it solves the following forward linear differential inequality of Lyapunov type with jumps*

$$-\dot{\tilde{Z}}_i(t) + A_0(i)\tilde{Z}_i(t) + \tilde{Z}_i(t) A_0^{\top}(i) + \sum_{l=1}^{r_1} A_l(i)\tilde{Z}_i(t) A_l^{\top}(i)$$

$$+\sum_{j=1}^{N} q_{ji} \tilde{Z}_j(t) \le 0, \quad kh < t < (k+1)h,$$

$$\tilde{Z}_i(kh^+) > \sum_{l=1}^{r_2} A_{dl}(i)\tilde{Z}_i(kh)A_{dl}^\top(i), \quad k \in \mathbb{Z}, \quad i \in \mathcal{N}.$$

In the case when the JSLDE (3.41) is in the special form of (3.42), the criteria for exponential stability in the mean-square sense provided in the previous theorem become:

Corollary 3.4 *Under Assumption 1.1, the following are equivalent:*

(i) the JSLDE (3.42) is ESMS;

(ii) for any $H \in \mathcal{S}_n^{1+}$ and $\Gamma \in Int\ \mathcal{S}_n^{1+}$, the non-homogeneous backward GLJLDE

$$-\dot{Y}(t) = A_0^\top Y(t) + Y(t)A_0 + \sum_{l=1}^{r_1} A_l^\top Y(t)A_l + H, \quad kh \le t < (k+1)h, \tag{3.45a}$$

$$Y(kh^-) = \sum_{l=0}^{r_2} A_{dl}^\top Y(kh)A_{dl} + \Gamma, \quad k \in \mathbb{Z}, \tag{3.45b}$$

has a unique bounded solution $t \to Y(t) : \mathbb{R} \to; Int\ \mathcal{S}_n^1$. Additionally, this solution is a periodic function of period h and satisfies $Y(t) > 0$, $t \in \mathbb{R}$;

(iii) there exists $\tilde{H} \in \mathcal{S}_n^{1+}$, $\tilde{\Gamma} \in Int\ \mathcal{S}_n^{1+}$ with the property that the corresponding non-homogeneous backward GLJLDE of type (3.45) has a unique bounded solution $t \to \tilde{Y}(t) : \mathbb{R} \to \mathcal{S}_n^1$ and additionally this solution is a periodic function of period h;

(iv) there exists a function $t \to Z(t) : \mathbb{R} \to \mathcal{S}_n^1$ with the properties:

(α) it is bounded and uniform positive on $\mathbb{R}$;
(β) it is a periodic function of period h;
(γ) it is differentiable on each interval $(kh, (k+1)h)$ and right continuous with finite left lateral limit in each $t = kh$, $k \in \mathbb{Z}$;
(δ) it solves the following backward linear differential inequalities of Lyapunov type with jumps

$$\dot{Z}(t) + A_0^\top Z(t) + Z(t)A_0 + \sum_{l=1}^{r_1} A_l^\top Z(t)A_l \le 0, \quad kh \le t < (k+1)h,$$

$$Z(kh^-) - \sum_{l=0}^{r_2} A_{dl}^\top Z(kh)A_{dl} > 0, \quad k \in \mathbb{Z};$$

(v) for any $H \in \mathcal{S}_n^{1+}$, $\Gamma \in Int\ \mathcal{S}_n^{1+}$ the non-homogeneous forward GLJLDE

$$\dot{X}(t) = A_0X(t) + X(t)A_0^{\top} + \sum_{l=1}^{r_1} A_lX(t)A_l^{\top} + H, \quad kh < t \leq (k+1)h, \tag{3.46a}$$

$$X(kh^+) = \sum_{l=0}^{r_2} A_{dl}X(kh)A_{dl}^{\top} + \Gamma, \quad k \in \mathbb{Z}, \tag{3.46b}$$

has an unique bounded solution $t \to X(t) : \mathbb{R} \to \mathcal{S}_n^1$ *and additionally this solution is a periodic function of period* h *with the property* $X(t) > 0$, *for all* $t \in \mathbb{R}$;

(vi) *there exist* $\tilde{H} \in \mathcal{S}_n^{1+}$, $\tilde{\Gamma} \in Int\ \mathcal{S}_n^{1+}$, *with the property that the corresponding non-homogeneous forward GLJLDE of type (3.46) has a unique bounded solution* $t \to \tilde{X}(t) : \mathbb{R} \to \mathcal{S}_n^1$ *which is a periodic function of period* h *and* $\tilde{X}(t) \geq 0$, *for all* $t \in \mathbb{R}$;

(vii) *there exists a function* $t \to \tilde{Z}(t) : \mathbb{R} \to \mathcal{S}_n^1$ *with the properties:*

(α') *it is bounded and uniform positive function on* $\mathbb{R}$;
(β') *it is a periodic function of period* h;
(γ') *it is differentiable on each interval* $(kh, (k+1)h)$ *and it is left continuous with finite right lateral limit in each* $t = kh$, $k \in \mathbb{Z}$;
(δ') *it solves the following forward linear differential inequality of Lyapunov type with jumps*

$$\dot{\tilde{Z}}(t) \geq A_0\tilde{Z}(t) + \tilde{Z}(t)A_0^{\top} + \sum_{l=1}^{r_1} A_l\tilde{Z}(t)A_l^{\top}, \quad kh < t \leq (k+1)h,$$

$$\tilde{Z}(kh^+) > \sum_{l=0}^{r_2} A_{dl}\tilde{Z}(kh)A_{dl}^{\top}, \quad k \in \mathbb{Z}.$$

Remark 3.4 In the case of a system of impulsive linear differential equations subject to Markovian jumps of type (3.5), criteria for exponential stability in mean square can be obtained from Theorem 3.2 for the general time-varying case, Theorem 3.3 for the periodic case and Theorem 3.4, respectively, for the time-invariant case, taking $A_l(\cdot, i) = 0$ for $1 \leq l \leq r_1$ and $A_{dl}(\cdot, i) = 0$ for $1 \leq l \leq r_2$, $i \in \mathcal{N}$.

Example 3.1 The following example is adapted from [2]. The system parameters are described as follows:

$$A_0 = \begin{bmatrix} -1 & 0 \\ 1 & -2 \end{bmatrix}, \quad A_{d0} = \begin{bmatrix} 2 & 1 \\ 1 & 3 \end{bmatrix}, \quad A_1 = \mu I_2, \quad A_{d1} = \delta I_2$$

for some scalars $\mu, \delta \geq 0$. For the particular case where the matrix A_1 takes the above form, we will show via the equivalence $(i) \leftrightarrow (vi)$ from Corollary 3.4 that the ESMS condition for system (3.42) can be reformulated as an LMI feasibility problem involving a *single* matrix inequality. This simplifies drastically the sum of squares programming used in [2].

Indeed, it follows from the equivalence $(i) \leftrightarrow (vi)$ from Corollary 3.4 that the JSLDE (3.42) with the given coefficients is ESMS if and only if the forward non-homogeneous Lyapunov jump linear differential equation:

$$\dot{X}(t) = A_0 X(t) + X(t) A_0^T + \mu^2 X(t), kh < t \leq (k+1)h \tag{3.47}$$

and

$$X(kh^+) = A_{d,0} X(kh) A_{d,0}^T + \delta^2 X(kh) + I_2, k \in \mathbb{Z} \tag{3.48}$$

has a unique bounded solution $\tilde{X}(\cdot) : \mathbb{R} \to \mathcal{S}_2$. Additionally this solution is a periodic function with period h and it is uniform positive definite. From (3.47) we obtain the following representation of the solution $\tilde{X}(\cdot)$:

$$\tilde{X}(t) = e^{(A_0 + \frac{1}{2}\mu^2 I_2)(t-kh)} \tilde{X}(kh^+) e^{(A_0 + \frac{1}{2}\mu^2 I_2)^T (t-kh)}$$

for all $kh < t \leq (k+1)h, k \in \mathbb{Z}$. Hence

$$\tilde{X}(kh) = e^{(A_0 + \frac{1}{2}\mu^2 I_2)h} \tilde{X}((k-1)h^+) e^{(A_0 + \frac{1}{2}\mu^2 I_2)^T h}.$$

Substituting this equality in (3.48) and taking into account that $\tilde{X}((k-1)h^+) = \tilde{X}(kh^+) > 0$ for all $k \in \mathbb{Z}$ (due to the periodicity property) we obtain that the system under consideration is ESMS if and only if the linear equation:

$$\begin{aligned} \tilde{X} = {} & A_{d,0} e^{(A_0 + \frac{1}{2}\mu^2 I_2)h} \tilde{X} e^{(A_0 + \frac{1}{2}\mu^2 I_2)^T h} A_{d,0}^T + \\ & \delta^2 e^{(A_0 + \frac{1}{2}\mu^2 I_2)h} \tilde{X} e^{(A_0 + \frac{1}{2}\mu^2 I_2)^T h} + I_2 \end{aligned} \tag{3.49}$$

has a solution $\tilde{X} > 0$ or equivalently, if the LMI:

$$\begin{aligned} & A_{d,0} e^{(A_0 + \frac{1}{2}\mu^2 I_2)h} \tilde{X} e^{(A_0 + \frac{1}{2}\mu^2 I_2)^T h} A_{d,0}^T + \\ & \delta^2 e^{(A_0 + \frac{1}{2}\mu^2 I_2)h} \tilde{X} e^{(A_0 + \frac{1}{2}\mu^2 I_2)^T h} - \tilde{X} < 0 \end{aligned} \tag{3.50}$$

has a feasible solution $\tilde{X} > 0$.

We now follow a similar protocol as in [2], i.e., for several values of the parameters μ and δ, we compute the smallest constant dwell time h for which the JLSS (3.42) remains stable. We solve the LMI (3.50) for each case and the results are gathered in Table 3.1.

One can see that the results reported in [2] are recovered here.

Table 3.1 Estimated smallest constant dwell-time h for the system (3.42) for various values for μ and δ

μ/δ	0	0.6	1.2	1.8	2.4
0	1.1350	1.1568	1.2030	1.2732	1.3593
0.3	1.1800	1.2089	1.2577	1.3317	1.4223
0.6	1.3787	1.3992	1.4576	1.5456	1.6528
0.9	1.8770	1.9072	1.9918	2.1181	2.2700
1.2	3.9315	4.0005	4.1932	4.4752	4.8083

3.5 Systems of Non-homogeneous Stochastic Jump Linear Differential Equations

In this section, we shall highlight some useful properties of the solutions of the following non-homogeneous jump stochastic linear differential equations (JSLDE) often named affine JSLDE:

$$dx(t) = [A_0(t, \eta(t))x(t)+ f_0(t)]dt + \sum_{l=1}^{r_1}[A_l(t, \eta(t))x(t) + f_l(t)]dw_l(t),$$

$$\tau_k < t \le \tau_{k+1}, \tag{3.51a}$$

$$x(\tau_k^+) = A_{d0}(\tau_k, \eta(\tau_k))x(\tau_k) + g_0(k) + \sum_{l=1}^{r_2} w_{dl}(k)[A_{dl}(\tau_k, \eta(\tau_k))x(\tau_k) + g_l(k)],$$

$$k \in \mathbb{Z}_+, \tag{3.51b}$$

where $x(t) \in \mathbb{R}^n$ are the state parameters and $\{w(t)\}_{t\ge 0}$ $(w(t) = (w_1(t), \dots, w_{r_1}(t))^\top)$, $\{w_d(k)\}_{k\in\mathbb{Z}_+}$ $(w_d(t) = (w_{d1}(k), \dots, w_{dr_2}(k))^\top)$ and $\{\eta(t)\}_{t\ge 0}$ are stochastic processes which were described in Sect. 1.2 and satisfy Assumption 1.2. The forcing terms $f_l(\cdot)$ and $g_l(k)$ from (3.51) satisfy the assumption:

Assumption 3.1

(i) for each $0 \le l \le r_1$ the stochastic process $\{f_l(t)\}_{t\ge 0}$ lies in $\mathcal{L}^2_{\mathcal{H}}([0, T], \mathbb{R}^n)$ for any $T > 0$;
(ii) for each $0 \le l \le r_2$, $\{g_l(k)\}_{k\in\mathbb{Z}_+}$ lies in $\ell^2_{\mathcal{H}}\{0, \mathbf{k}; \mathbb{R}^n\}$ for any $\mathbf{k} \in \mathbb{Z}_+, \mathbf{k} \ge 1$.

For each $t_0 \in \mathbb{R}_+$ we denote $\mathfrak{X}_{t_0}$ the set of random vectors $x_0 : \Omega \to \mathbb{R}^n$ which are $\mathcal{H}_{t_0}$-measurable and satisfy $\mathbb{E}[|x_0|^2] < \infty$.

Let $\Phi(t, t_0)$, $t \ge t_0 \ge 0$ be the fundamental matrix solution of the SLDE of type (3.1a) obtained taking $f_l(\cdot) = 0$ for $0 \le l \le r_1$ in (3.51a). This means that $t \to \Phi(t, t_0)$ is the solution of the following IVP on the linear space $\mathbb{R}^{n\times n}$:

$$d\Phi(t,t_0) = A_0(t,\eta(t))\Phi(t,t_0)dt + \sum_{l=1}^{r_1} A_l(t,\eta(t))\Phi(t,t_0)dw_l(t), \quad t \geq t_0$$
$$\Phi(t_0,t_0) = I_n. \tag{3.52}$$

Invoking Proposition 1.12.1 from [1], see also [3] Chap. 3, we deduce that for any $t \geq t_0$ the matrix $\Phi(t,t_0)$ is invertible and its inverse is given by

$$\Phi^{-1}(t,t_0) = \tilde{\Phi}^{\top}(t,t_0) \ \ a.s., \quad \text{for all} \ \ t \geq t_0, \tag{3.53}$$

where $t \to \tilde{\Phi}(t,t_0)$ is the solution of the following IVP on the linear space $\mathbb{R}^{n\times n}$:

$$d\tilde{\Phi}(t,t_0) = \left[-A_0^{\top}(t,\eta(t)) + \sum_{l=1}^{r_1}(A_l^{\top}(t,\eta(t)))^2\right]\tilde{\Phi}(t,t_0)dt$$
$$- \sum_{l=1}^{r_1} A_l^{\top}(t,\eta(t))\tilde{\Phi}(t,t_0)dw_l(t), \quad t \geq t_0 \tag{3.54}$$
$$\tilde{\Phi}(t_0,t_0) = I_n\ , \quad \text{for all} \ \ t_0 \geq 0.$$

Applying Corollary 1.2, Chap. 5 in [4] in the case of the IVPs (3.52) and (3.54), respectively, we obtain that for each $t > t_0 \geq 0$ there exist positive constants δ_l, $l = 1, 2$, depending upon T and $T - t_0$ such that:

$$\mathbb{E}\left[\sup_{t_0\leq t\leq T} |\Phi(t,t_0)x_0|^2\right] \leq \delta_1\left(1 + \mathbb{E}[|x_0|^2]\right), \tag{3.55a}$$
$$\mathbb{E}\left[\sup_{t_0\leq t\leq T} |\Phi^{-1}(t,t_0)x_0|^2\right] \leq \delta_2\left(1 + \mathbb{E}[|x_0|^2]\right), \tag{3.55b}$$

when $x_0 \in \mathfrak{X}_{t_0}$. Proceeding as in Sect. 1.12 from [1] (see also [3], Chap. 3) one obtains the following representation formulae of the solution of (3.51a):

$$x(t;t_0,x_0) = \Phi(t,t_0) + \Phi(t,t_0)\int_{t_0}^{t} \Phi^{-1}(s,t_0)\left(f_0(s) - \sum_{l=1}^{r_1} A_l(s,\eta(s))f_l(s)\right)ds$$
$$+ \sum_{l=1}^{r_1}\Phi(t,t_0)\int_{t_0}^{t}\Phi^{-1}(s,t_0)f_l(s)dw_l(s), \quad a.s. \ \ t \in [t_0, \tau_{k_0}] \tag{3.56}$$

and

$$x(t;t_0,x_0) = \Phi(t,\tau_k)x(\tau_k^+)$$
$$+ \Phi(t,\tau_k)\int_{\tau_k}^{t} \phi^{-1}(s,\tau_k)\left(f_0(s) - \sum_{l=1}^{r_1} A_l(s,\eta(s))f_l(s)\right)\mathrm{d}s$$
$$+ \sum_{l=1}^{r_1} \Phi(t,\tau_k)\int_{\tau_k}^{t} \Phi^{-1}(s,\tau_k)f_l(s)\mathrm{d}w_l(s), \quad a.s.\ t \in [\tau_k,\tau_{k+1}], \tag{3.57}$$

$k \geq k_0$, where k_0 is such that $\tau_{k_0-1} < t_0 \leq \tau_{k_0}$ and $x(\tau_k^+)$ are computed as in (3.51b).

The following result highlights several useful properties of the solution $x(\cdot;t_0,x_0)$ of the IVP (3.51).

Proposition 3.4 *Assume:*

(a) Assumptions 1.2, 2.1 and 3.1 are fulfilled;
(b) for each $i \in \mathcal{N}$ the functions $t \to A_l(t,i) : \mathbb{R}_+ \to \mathbb{R}^{n\times n}$, $0 \leq l \leq r_1$ are bounded and continuous matrix-valued functions and $\{A_{dl}(\tau_k,i)\}_{k\in\mathbb{Z}} \subset \mathbb{R}^{n\times n}$, $0 \leq l \leq r_2$, are bounded matrix-valued sequences.

Under these assumptions, for each $t_0 \in \mathbb{R}_+$ and $x_0 \in \mathfrak{X}_{t_0}$, the JSLDE (3.51) has a unique solution $x(t) = x(t;t_0,x_0)$, $t \geq t_0$ having the properties:

(i) $x(\cdot)$ is a.s. continuous in any $t \neq \tau_k$ and it is left continuous in $t = \tau_k$, for all $k \in \mathbb{Z}_+$ for which $\tau_k > t_0$;
(ii) $x(\cdot) \in \mathcal{L}^2_{\mathcal{H}}([t_0,T];\mathbb{R}^n)$, for all $T > t_0$;
(iii) $x(t_0) = x_0$ and $\lim_{\substack{t\to\tau_k \\ t>\tau_k}} x(t) = x(\tau_k^+)$ for all $\tau_k \geq t_0$;
(iv) for each $k \geq k_0$ there exist $\nu_k > 0$ which may depend upon τ_k and $\tau_{k+1} - \tau_k$, such that

$$\mathbb{E}\left[\sup_{t_0\leq t\leq\tau_{k_0}} |x(t;t_0,x_0)|^2\right] \leq \nu_{k_0}\left(1 + \mathbb{E}[|x_0|^2] + \sum_{l=0}^{r_1}\int_{t_0}^{\tau_{k_0}} \mathbb{E}[|f_l(s)|^2]\mathrm{d}s\right)$$

and

$$\mathbb{E}\left[\sup_{\tau_k\leq t\leq\tau_{k+1}} |x(t;t_0,x_0)|^2\right] \leq \nu_k\left(1 + \mathbb{E}[|x(\tau_k^+)|^2] + \sum_{l=0}^{r_1}\int_{\tau_k}^{\tau_{k+1}} \mathbb{E}[|f_l(s)|^2]\mathrm{d}s\right)$$

if $k \geq k_0$.

Proof The validity of (i)–(iii) may be checked directly, employing the representation formulae (3.56)–(3.57). The proof of (iv) may be done by direct calculation involving (3.55)–(3.57) together with Theorem 3.6, Chap. 4 from [4]. The details are omitted.

Whenever we say that $x(t; t_0, x_0)$, $t \geq t_0$ is a solution of the JSLDE (3.51) we have in mind the solution having the properties stipulated in Proposition 3.4.

3.5.1 An Auxiliary Result

In what follows, we will present a new technical result that will play a key role in the next developments. It is based on the property of continuity in probability of a homogeneous standard Markov process with a finite number of states (see, e.g., Theorem 1.8.2 from [1], see also [5–7].

Lemma 3.1 *Assume that the assumptions of Proposition 3.4 are fulfilled. Let $t \to X(t,i) : \mathbb{R}_+ \to \mathcal{S}_n$, $i \in \mathcal{N}$, be continuous functions in any $t \neq \tau_k$ and right continuous with finite lateral limits in $t = \tau_k$, $k \in \mathbb{Z}_+$. Let $t_0 \in \mathbb{R}_+$ and $x_0 \in \mathfrak{X}_{t_0}$ be given and $k_0 \in \mathbb{Z}_+$ such that $\tau_{k_0-1} < t_0 \leq \tau_{k_0}$. Let $x(t; t_0, x_0)$, $t \geq t_0$ be the solution of the JSLDE (3.51) satisfying $x(t_0; t_0, x_0) = x_0$. If $\tau_k > t_0$ we have the equalities:*

$$\begin{aligned}
&\lim_{\substack{t\to\tau_k\\ t>\tau_k}} \mathbb{E}\left[x^\top(t; t_0, x_0)X(t, \eta(t))x(t; t_0, x_0) \mid \eta(t_0) = j\right] \\
&= \mathbb{E}\left[x^\top(\tau_k^+)X(\tau_k, \eta(\tau_k))x(\tau_k^+) \mid \eta(t_0) = j\right], \qquad (3.58a)\\
&\lim_{\substack{t\to\tau_k\\ t<\tau_k}} \mathbb{E}\left[x^\top(t; t_0, x_0)X(t, \eta(t))x(t; t_0, x_0) \mid \eta(t_0) = j\right] \\
&= \mathbb{E}\left[x^\top(\tau_k; t_0, x_0)X(\tau_k^-, \eta(\tau_k))x(\tau_k; t_0, x_0) \mid \eta(t_0) = j\right], \qquad (3.58b)
\end{aligned}$$

for all $j \in \mathcal{N}$, $x(\tau_k^+)$ being computed via (3.51b) based on $x(\tau_k; t_0, x_0)$.

Proof The equality from (3.58a) follows immediately from the Lebesgue-dominant convergence theorem and with the right continuity in τ_k of $X(\cdot, i)$ and of the Markov process $\eta(t)$, together with Proposition 3.4 (iii). We have to prove the equality from (3.58b). To this end, we shall employ the property of continuity in probability of a standard homogeneous Markov process with a finite number of states.

Let $\{t_p\}_{p\geq 1} \subset (t_0, \tau_k)$ be an arbitrary sequence such that

$$\lim_{p\to\infty} t_p = \tau_k. \qquad (3.59)$$

We show that

$$\lim_{p\to\infty} \mathbb{E}[|X_{l_1 l_2}(t_p, \eta(t_p)) - X_{l_1 l_2}(\tau_k^-, \eta(\tau_k))| \mid \eta(t_0) = j] = 0, \qquad (3.60)$$

for all $1 \leq l_1, l_2 \leq n, j \in \mathcal{N}$, where $X_{l_1 l_2}(\cdot, i)$ are the scalar components of the matrices $X(\cdot, i), i \in \mathcal{N}$. First we write

$$\begin{aligned}
&\mathbb{E}[|X_{l_1 l_2}(t_p, \eta(t_p)) - X_{l_1 l_2}(\tau_k^-, \eta(\tau_k))| \mid \eta(t_0) = j] \\
&\leq \mathbb{E}[|X_{l_1 l_2}(t_p, \eta(t_p)) - X_{l_1 l_2}(\tau_k^-, \eta(t_p))| \mid \eta(t_0) = j] \\
&+ \mathbb{E}[|X_{l_1 l_2}(\tau_k^-, \eta(t_p)) - X_{l_1 l_2}(\tau_k^-, \eta(\tau_k))| \mid \eta(t_0) = j].
\end{aligned} \tag{3.61}$$

Further we obtain

$$\begin{aligned}
&\mathbb{E}[|X_{l_1 l_2}(t_p, \eta(t_p)) - X_{l_1 l_2}(\tau_k^-, \eta(t_p))| \mid \eta(t_0) = j] \\
&\leq \sum_{i=1}^{N} |X_{l_1 l_2}(t_p, i) - X_{l_1 l_2}(\tau_k^-, i) \mid \mathbb{E}[\chi_{\{\eta(t_p)=i\}} \mid \eta(t_0) = j],
\end{aligned} \tag{3.62}$$

$\chi_{\{\eta(\cdot)=i\}}$ being the indicator function of the event $\{\eta(\cdot) = i\|$. Hence,

$$0 \leq \mathbb{E}[\chi_{\{\eta(t_p)=i\}} \mid \eta(t_0) = j] \leq 1.$$

This allows us to deduce from (3.62) that

$$\begin{aligned}
0 &\leq \lim_{p \to \infty} \mathbb{E}[|X_{l_1 l_2}(t_p, \eta(t_p)) - X_{l_1 l_2}(\tau_k^-, \eta(t_p))| \mid \eta(t_0) = j] \\
&\leq \lim_{p \to \infty} \sum_{i=1}^{N} |X_{l_1 l_2}(t_p, i) - X_{l_1 l_2}(\tau_k^-, i)| = 0, \quad \text{for all } \ j \in \mathcal{N}.
\end{aligned} \tag{3.63}$$

For the last equality we have used (3.59) together with the definition of the left limit of $X_{l_1 l_2}(\tau_k^-, i)$. Employing Corollary 1.1.4 and Theorem 1.8.2 (vi) from [1] we deduce that the stochastic process $(t, \omega) \to |X_{l_1 l_2}(\tau_k^-, \eta(t, \omega)) - X_{l_1 l_2}(\tau_k^-, \eta(\tau_k, \omega))| : [t_0, \hat{\tau}] \times \Omega \to \mathbb{R}$ is continuous in probability at $t = \tau_k$, when $\hat{\tau} > \tau_k$. Hence, $|X_{l_1 l_2}(\tau_k^-, \eta(t_p, \omega)) - X_{l_1 l_2}(\tau_k^-, \eta(\tau_k))| \xrightarrow{\mathcal{P}} 0$ when $p \to \infty$ for arbitrary sequence $\{t_p\}_{p \geq 1} \subset (t_0, \tau_k)$ satisfying (3.59).

Here and in the sequel, the symbol $\xrightarrow{\mathcal{P}}$ denotes the convergence in probability. Further, Theorem 1.2.4 from [1] applied in the case of the measurable space $(\Omega, \mathfrak{F}, \mathcal{P})$ gives

$$\lim_{p \to \infty} \mathbb{E}[|X_{l_1 l_2}(\tau_k^-, \eta(t_p)) - X_{l_1 l_2}(\tau_k^-, \eta(\tau_k))| \mid \eta(t_0) = j] = 0, \tag{3.64}$$

for all $j \in \mathcal{N}$. Combining (3.61)–(3.64) we deduce that (3.60) holds for any $j \in \mathcal{N}$ and any sequence $\{t_p\}_{p \geq 1} \subset (t_0, \tau_k)$ satisfying (3.59). Applying Theorem 1.1.9 from [1] we may infer that

$$X_{l_1 l_2}(t_p, \eta(t_p)) \xrightarrow{\mathcal{P}} X_{l_1 l_2}(\tau_k^-, \eta(\tau_k)) \tag{3.65}$$

when $p \to \infty$, for any sequence $\{t_p\}_{p\geq 1} \subset (t_0, \tau_k)$ satisfying (3.59).

On the other hand, Proposition 3.4 (i) combined with Theorem 1.1.2 from [1] leads to

$$x_l(t_p; t_0, x_0) \xrightarrow{\mathscr{P}} x_l(\tau_k; t_0, x_0), \quad \text{when} \quad p \to \infty \tag{3.66}$$

for any sequence $\{t_p\}_{p\geq 1} \subset (t_0, \tau_k)$ satisfying (3.59), $x_l(\cdot; t_0, x_0)$, $1 \leq l \leq n$ being the l^{-th} scalar component of the solution $x(\cdot; t_0, x_0)$. Invoking again Corollary 1.1.4 from [1] we may deduce from (3.65) and (3.66) that

$$x^\top(t_p; t_0, x_0)X(t_p, \eta(t_p))x(t_p; t_0, x_0) \xrightarrow{\mathscr{P}} x^\top(\tau_k; t_0, x_0)X(\tau_k^-; \eta(\tau_k))x(\tau_k; t_0, x_0), \tag{3.67}$$

when $p \to \infty$, for all $\{t_p\}_{p\geq 1} \subset (t_0, \tau_k)$ satisfying (3.59). On the other hand, invoking again Proposition 3.4 (i) and (iv) we may deduce that for any sequence $\{t_p\}_{p\geq 1} \subset (t_0, \tau_k)$ satisfying (3.59) we have

$$|x^\top(t_p; t_0, x_0)X(t_p, \eta(t_p))x(t_p; t_0, x_0)| \leq \gamma\tilde{\delta} \quad a.s. \tag{3.68}$$

for all $p \geq p_1$ where $\gamma \geq \sup_{t\in[t_0,\tau_k]} \|X(t, i)\|$ and $\tilde{\delta} > 0$ is such that

$$\sup_{\tau_{k-1}\leq t\leq \tau_k} |x(t; t_0, x_0)|^2 \leq \tilde{\delta} \quad a.s.$$

when $k \geq k_0 + 1$ or

$$\sup_{t_0\leq t\leq \tau_k} |x(t; t_0, x_0)|^2 \leq \tilde{\delta} \quad a.s.$$

when $k = k_0$, $k_0 \in \mathbb{Z}_+$ being such that $\tau_{k_0-1} < t_0 < \tau_{k_0}$. In (3.68) $p_1 \in \mathbb{Z}_+$ is such that $\{t_p\}_{p\geq p_1} \subset (\tau_{k-1}, \tau_k)$ if $k \geq k_0 + 1$ and $p_1 = 1$ if $k = k_0$. Applying Theorem 1.2.3 from [1] we deduce via (3.67) and (3.68) that

$$\begin{aligned}
&\lim_{p\to\infty} \mathbb{E}[x^\top(t_p; t_0, x_0)X(t_p, \eta(t_p))x(t_p; t_0, x_0) \mid \eta(t_0) = j] \\
&= \mathbb{E}[x^\top(\tau_k; t_0, x_0)X(\tau_k^-; \eta(\tau_k))x(\tau_k; t_0, x_0) \mid \eta(t_0) = j],
\end{aligned} \tag{3.69}$$

for all $j \in \mathcal{N}$. This confirms the validity of the equality (3.58b) because (3.69) holds for an arbitrary sequence $\{t_p\}_{p\geq 1} \subset (t_0, \tau_k)$ satisfying (3.59). Thus the proof ends.

3.5.2 On the Behavior of the Solutions of a Non-homogeneous JSLDE

Let us consider the non-homogeneous backward GLJLDE associated to the JSLDE (3.51):

$$-\dot{X}(t,i) = A_0^\top(t,i)X(t,i) + X(t,i)A_0(t,i) + \sum_{l=1}^{r_1} A_l^\top(t,i)X(t,i)A_l(t,i)$$

$$+ \sum_{j=1}^{N} q_{ij}X(t,j) + I_n, \quad \tau_k \le t < \tau_{k+1}, \tag{3.70a}$$

$$X(\tau_k^-, i) = \sum_{l=0}^{r_2} A_{dl}^\top(\tau_k, i)X(\tau_k, i)A_{dl}(\tau_k, i) + I_n, \quad k \in \mathbb{Z}_+, \quad i \in \mathcal{N}. \tag{3.70b}$$

Invoking the equivalence $(i) \Leftrightarrow (ii)$ from Theorem 3.2 we obtain:

Corollary 3.5 *Assume:*

(a) Assumptions 1.2 and 2.1 are fulfilled;
(b) for each $i \in \mathcal{N}$ the matrix-valued functions $t \to A_l(t,i) : \mathbb{R}_+ \to \mathbb{R}^{n\times n}$, $0 \le l \le r_1$ are bounded and continuous while the matrix-valued sequences $\{A_{dl}(\tau_k, i)\}_{k\in\mathbb{Z}_+} \subset \mathbb{R}^{n\times n}$, $0 \le l \le r_2$, are bounded .

Under these conditions, the following are equivalent:

(i) the JSLDE of type (3.1) obtained taking $f_l(\cdot) = 0$, $0 \le l \le r_1$ and $g_l(\cdot) = 0$, $0 \le l \le r_2$ in (3.51) is SESMS;
(ii) the non-homogeneous backward GLJLDE (3.70) has a unique solution $t \to \tilde{\mathbf{X}}(t) = (\tilde{X}(t,1), \tilde{X}(t,2), \dots, \tilde{X}(t,N)) : \mathbb{R}_+ \to \mathcal{S}_n^N$ such that

$$0 < \tilde{c}_1 I_n \le \tilde{X}(t,i) \le \tilde{c}_2 I_n < +\infty, \tag{3.71}$$

for all $(t,i) \in \mathbb{R}_+ \times \mathcal{N}$, $\tilde{c}_l, l1, 2$ being constants not depending upon (t,i).

The next result highlights some useful properties of the solution of the non-homogeneous JSLDE (3.51). They may be viewed as the stochastic counterpart of the result proved in Theorem 2.2 for the solutions of a non-homogeneous FJLDE (2.2).

Theorem 3.5 *Assume:*

(a) the assumptions from Proposition 3.4 are satisfied;
(b) the JSLDE obtained tacking $f_l(\cdot) = 0$, $0 \le l \le r_1$ and $g_j(\cdot) = 0$, $0 \le j \le r_2$ is SESMS.

Let $x(t) := x(t; t_0, x_0)$, $t \ge t_0$ be a solution of the non-homogeneous JSLDE (3.51) with $x_0 \in \mathfrak{X}_{t_0}$. Under the considered assumptions the following hold:

(i) there exist $\beta \geq 1$, $\alpha > 0$ *such that*

$$\mathbb{E}[|x(t)|^2 \mid \eta(t_0) = j] \leq \beta(e^{-\alpha(t-t_0)}\mathbb{E}[|x_0|^2 \mid \eta(t_0) = j]$$

$$+\sum_{l=0}^{r_2}\sum_{k=k_0}^{k_t} e^{-\alpha(t-\tau_k)}\mathbb{E}[|g_l(k)|^2|\eta(t_0)=j]+\sum_{l=0}^{r_1}\int_{t_0}^{t} e^{-\alpha(t-s)}\mathbb{E}[|f_l(s)|^2|\eta(t_0)=j]\mathrm{d}s)$$

for all $t \geq t_{k_0}$ *where* k_0, $k_t \in \mathbb{Z}_+$ *are such that* $t_0 \in (\tau_{k_0-1}, \tau_{k_0}]$ *and* $t \in [\tau_{k_t}, \tau_{k_t+1})$, *and*

$$\mathbb{E}[|x(t)|^2 \mid \eta(t_0) = j] \leq \beta\left(e^{-\alpha(t-t_0)}\mathbb{E}[|x_0|^2 \mid \eta(t_0) = j]\right.$$

$$\left.+\sum_{l=0}^{r_1}\int_{t_0}^{t} e^{-\alpha(t-s)}\mathbb{E}[|f_l(s)|^2 \mid \eta(t_0) = j]\mathrm{d}s\right)$$

$t_0 \leq t \leq \tau_{k_0}$, *for all* $j \in \mathcal{N}$, $x_0 \in \mathfrak{X}_{t_0}$.

Furthermore, if $\{f_l(t)\}_{t\geq 0} \subset \mathcal{L}^2_{\mathcal{H}}(\mathbb{R}_+, \mathbb{R}^n), 0 \leq l \leq r_1$ *and* $\{g_l(k)\}_{k\in\mathbb{Z}_+} \in \ell^2_{\mathcal{H}}\{\mathbb{Z}_+, \mathbb{R}^n\}, 0 \leq l \leq r_2$, *then:*

(ii) there exist $\gamma > 0$ *such that*

$$\int_{t_0}^{\infty} \mathbb{E}[|x(t)|^2 \mid \eta(t_0) = j]\mathrm{d}t \leq \gamma\left(\mathbb{E}[|x_0|^2 \mid \eta(t_0) = j]\right.$$

$$+\sum_{l=0}^{r_2}\sum_{k=k_0}^{\infty} \mathbb{E}[|g_l(k)|^2 \mid \eta(t_0) = j] + \sum_{l=0}^{r_1}\int_{t_0}^{\infty} \mathbb{E}[|f_l(t)|^2 \mid \eta(t_0) = j]\mathrm{d}t)$$

for all $x_0 \in \mathfrak{X}_{t_0}$, $j \in \mathcal{N}$, γ *not depending upon* j *and* x_0;

(iii) $\lim_{t\to\infty} \mathbb{E}[|x(t)|^2 \mid \eta(t_0) = j] = 0$, *for all* $j \in \mathcal{N}$, $x_0 \in \mathfrak{X}_{t_0}$.

Proof Let $t_0 \in \mathbb{R}_+$ and $x_0 \in \mathfrak{X}_{t_0}$ be given. Let $k_0 \in \mathbb{Z}_+$ be such that $t_0 \in (\tau_{k_0-1}, \tau_{k_0}]$. For each $k \geq k_0$ we consider intervals of type $[\hat{\tau}_k, t) \subset (\tau_k, \tau_{k+1})$ and $[\hat{\tau}_0, t] \subset (t_0, \tau_{k_0})$, respectively. Let $\tilde{\mathbf{X}}(\cdot) = (\tilde{X}(\cdot, 1), \tilde{X}(\cdot, 2), \ldots, \tilde{X}(\cdot, N))$ be the unique bounded on $\mathbb{R}_+$ solution of the non-homogeneous backward GLJLDE (3.70). Applying the Itô-type formula (see Chapter 1 from [1]) in the case of the function $v(t, x, i) \triangleq x^\top \tilde{X}(t, i)x$ and of the stochastic process $x(t) = x(t; t_0, x_0)$, $t \geq t_0$, we obtain via (3.70a) that

$$\mathbb{E}[v(t, x(t), \eta(t)) \mid \eta(t_0) = j] = \mathbb{E}[v(\hat{\tau}_k, x(\hat{\tau}_k), \eta(\hat{\tau}_k) \mid \eta(t_0) = j] \tag{3.72}$$

$$+\,\mathbb{E}[\int_{\hat{\tau}_k}^{t} (-|x(s)|^2 + \mathcal{W}(s, x(s), \mathbf{f}(s), \eta(s)))\mathrm{d}s \mid \eta(t_0) = j]$$

if $k \geq k_0$ and $t_0 < \hat{\tau}_0 < t < \tau_{k_0}$ if $k = 0$, where we have denoted

$$\mathscr{W}(s, x(s), \mathbf{f}(s), \eta(s)) \stackrel{\Delta}{=} 2x^\top(s)(\tilde{X}(s, \eta(s)) f_0(s) + \sum_{l=1}^{r_1} A_l^\top(s, \eta(s))\tilde{X}(s, \eta(s)) f_l(s))$$
$$+ \sum_{l=1}^{r_1} f_l^\top(s)\tilde{X}(s, \eta(s)) f_l(s) \tag{3.73}$$

and $\mathbf{f}(s) = (f_0(s), f_1(s), \ldots, f_{r_1}(s))$. Letting $\hat{\tau}_k \to \tau_k$ (for $k \geq 0$) and $\hat{\tau}_0 \to t_0$, respectively (for $k = 0$), in (3.72) we obtain

$$\mathbb{E}[x^\top(t)\tilde{X}(t, \eta(t))x(t) \mid \eta(t_0) = j] = \mathbb{E}[x^\top(\tau_k^+)\tilde{X}(\tau_k, \eta(\tau_k))x(\tau_k^+) \mid \eta(t_0) = j]$$
$$+ \mathbb{E}[\int_{\tau_k}^{t} (-|x(s)|^2 + \mathscr{W}(s, x(s), \mathbf{f}(s), \eta(s)))\mathrm{d}s \mid \eta(t_0) = j] \tag{3.74}$$

for all $j \in \mathscr{N}$, $t \in (\tau_k, \tau_{k+1})$, $k \geq k_0$, and

$$\mathbb{E}[x^\top(t)\tilde{X}(t, \eta(t))x(t) \mid \eta(t_0) = j] = \mathbb{E}[x_0^\top \tilde{X}(t_0, j)x_0 \mid \eta(t_0) = j]$$
$$+ \mathbb{E}\left[\int_{t_0}^{t} (-|x(s)|^2 + \mathscr{W}(s, x(s), \mathbf{f}(s), \eta(s)))\mathrm{d}s \mid \eta(t_0) = j\right] \tag{3.75}$$

for all $j \in \mathscr{N}$, $t \in (t_0, \tau_{k_0})$. From (3.74) and (3.75) we obtain via Theorem 1.2.5 from [1] (see also [8, 9]) that

$$\frac{\mathrm{d}}{\mathrm{d}t}\mathbb{E}\left[x^\top(t)\tilde{X}(t, \eta(t))x(t) \mid \eta(t_0) = j\right] = -\mathbb{E}\left[|x(t)|^2 \mid \eta(t_0) = j\right]$$
$$+ \mathbb{E}\left[\mathscr{W}(t, x(t), \mathbf{f}(t), \eta(t)) \mid \eta(t_0) = j\right] \tag{3.76}$$

a. e. $t \in (t_0, T)$, $t > t_0$ being arbitrary but fixed and

$$\mathbb{E}\left[x^\top(t)\tilde{X}(t, \eta(t))x(t) | \eta(t_0) = j\right]\Big|_{t=\tau_k} = \mathbb{E}\left[x^\top(\tau_k^+)\tilde{X}(\tau_k, \eta(\tau_k))x(\tau_k^+) | \eta(t_0) = j\right],$$
$$k \geq k_0 \tag{3.77a}$$
$$\mathbb{E}\left[x^\top(t)\tilde{X}(t, \eta(t))x(t) \mid \eta(t_0) = j\right]\Big|_{t=t_0} = \mathbb{E}\left[x^\top\tilde{X}(t_0, j)x_0 \mid \eta(t_0) = j\right]. \tag{3.77b}$$

We write:

$$-|x(t)|^2 + \mathscr{W}(t, x(t), \mathbf{f}(t), \eta(t)) = -\frac{1}{2}|x(t)|^2 + \left|\frac{1}{2} - 2\tilde{X}(t, \eta(t)) f_0(t)\right|^2$$

$$-|\frac{1}{2} - 2\sum_{l=1}^{r_1} A_l^\top(t, \eta(t))\tilde{X}(t, \eta(t)) f_l(t)|^2 + \mathscr{W}_1(t, \mathbf{f}(t), \eta(t)) \tag{3.78a}$$

$$\mathscr{W}_1(t, \mathbf{f}(t), \eta(t)) = 4|\tilde{X}(t, \eta(t) f_0(t)|^2 + 4\sum_{l_1, l_2=1}^{r_1} f_{l_1}^\top(t)\tilde{X}(t, \eta(t)) A_{l_1}(t, \eta(t))$$

$$\cdot A_{l_2}^\top(t, \eta(t))\tilde{X}(t, \eta(t)) f_{l_2}(t) + \sum_{l=1}^{r_1} f_l^\top(t)\tilde{X}(t, \eta(t)) f_l(t), \quad t \geq t_0. \tag{3.78b}$$

From (3.78b), we may infer that there exists a constant $\gamma_c > 0$ depending upon $\tilde{c}_2$ and upon $\sup_{t\geq 0} \|A_l(t, i)\|$, such that:

$$0 \leq \mathscr{W}_1(t, \mathbf{f}(t), \eta(t)) \leq \gamma_c \sum_{l=0}^{r_1} |f_l(t)|^2, \quad t \geq t_0. \tag{3.79}$$

Employing (3.76), (3.78) and (3.79) we get:

$$\frac{\mathrm{d}}{\mathrm{d}t}\mathbb{E}[x^\top(t)\tilde{X}(t, \eta(t))x(t) \mid \eta(t_0) = j] \leq -\frac{1}{2}\mathbb{E}[|x(t)|^2 \mid \eta(t_0) = j] + \varphi_j(t) \tag{3.80a}$$

$$\varphi_j(t) := \gamma_c \sum_{l=0}^{r_1} \mathbb{E}[|f_l(t)|^2 \mid \eta(t_0) = j], \quad j \in \mathscr{N}, \quad a.a. \ t \geq t_0. \tag{3.80b}$$

Further, (3.71) and (3.80) yield:

$$\begin{aligned} &\frac{\mathrm{d}}{\mathrm{d}t}\mathbb{E}[x^\top(t)\tilde{X}(t, \eta(t))x(t) \mid \eta(t) = j] \\ &\leq \alpha_1 \mathbb{E}[x^\top(t)\tilde{X}(t, \eta(t)x(t) \mid \eta(t_0) = j] + \varphi_j(t), \end{aligned} \tag{3.81}$$

a.a. $t \geq 0$, $j \in \mathscr{N}$, where $\alpha_1 := \frac{1}{2\tilde{c}_2}$ and $\varphi_j(t)$ is introduced via (3.80b).
Proceeding as in the computation of the right hand side of (3.10b) we can make explicit the right hand side of (3.77a). Thus, substituting (3.51b) in (3.77a) we obtain

$$\begin{aligned} &\mathbb{E}[x^\top(\tau_k^+)\tilde{X}(\tau_k, \eta(\tau_k))x(\tau_k^+) \mid \eta(t_0) = j] \\ &= \sum_{l=0}^{r_2} \mathbb{E}[x^\top(\tau_k)A_{dl}^\top(\tau_k, \eta(\tau_k))\tilde{X}(\tau_k, \eta(\tau_k))A_{dl}(\tau_k, \eta(\tau_k))x(\tau_k) \mid \eta(t_0) = j] \end{aligned}$$

$$+2\sum_{l=0}^{r_2}\mathbb{E}[x^\top(\tau_k)A_{dl}^\top(\tau_k,\eta(\tau_k))\tilde{X}(\tau_k,\eta(\tau_k))g_l(k)\mid\eta(t_0)=j]$$

$$+\sum_{l=0}^{r_2}\mathbb{E}[g_l^\top(k)\tilde{X}(\tau_k,\eta(\tau_k))g_l(k)\mid\eta(t_0)=j].$$

Now, (3.70b) yields

$$\begin{aligned}
&\mathbb{E}[x^\top(\tau_k^+)\tilde{X}(\tau_k,\eta(\tau_k))x(\tau_k^+)\mid\eta(t_0)=j]\\
&=\mathbb{E}[x^\top(\tau_k)\tilde{X}(\tau_k^-,\eta(\tau_k))x(\tau_k)|\,\eta(t_0)=j]-\mathbb{E}[|x(\tau_k)|^2\mid\eta(t_0)=j]\\
&\quad+2\sum_{l=0}^{r_2}\mathbb{E}[x^\top(\tau_k)A_{dl}^\top(\tau_k,\eta(\tau_k))\tilde{X}(\tau_k,\eta(\tau_k))g_l(k)\,|\eta(t_0)=j]\\
&\quad+\sum_{l=0}^{r_2}\mathbb{E}[g_l^\top(k)\tilde{X}(\tau_k,\eta(\tau_k))g_l(k)\mid\eta(t_0)=j].
\end{aligned}\tag{3.82}$$

Involving again the square completion technique we get:

$$\begin{aligned}
&-|x(\tau_k)|^2+\sum_{l=0}^{r_2}x^\top(\tau_k)A_{dl}^\top(\tau_k,\eta(\tau_k))\tilde{X}(\tau_k,\eta(\tau_k))g_l(k)=-\frac{1}{2}|x(\tau_k)|^2\\
&-\sum_{l=0}^{r_2}\left|\frac{1}{\sqrt{2(r_2+1)}}x(\tau_k)-\sqrt{2(r_2+1)}A_{dl}^\top(\tau_k,\eta(\tau_k))\tilde{X}(\tau_k,\eta(\tau_k))g_l(k)\right|^2\\
&+2(r_2+1)\sum_{l=0}^{r_2}|A_{dl}^\top(\tau_k,\eta(\tau_k))\tilde{X}(\tau_k,\eta(\tau_k))g_l(k))|^2.
\end{aligned}$$

Substituting this equality in (3.82), we deduce that

$$\begin{aligned}
&\mathbb{E}[x^\top(\tau_k^+)\tilde{X}(\tau_k,\eta(\tau_k))x(\tau_k^+)\mid\eta(t_0)=j]\\
&\qquad\le\delta_1\mathbb{E}[x^\top(\tau_k)\tilde{X}(\tau_k^-,\eta(\tau_k))x(\tau_k)\mid\eta(t_0)=j]+\zeta_j(k)
\end{aligned}\tag{3.83a}$$

$$\zeta_j(k)\stackrel{\Delta}{=}\gamma_d\sum_{l=0}^{r_2}\mathbb{E}[|g_l(k)|^2\,|\eta(t_0)=j]\tag{3.83b}$$

and $\delta_1=1-\frac{1}{2\tilde{c}_3}\in(0,1)$. Let $t\to z_j(t):[t_0,\infty)\to\mathbb{R}$ be the solution of the IVP:

$$\dot{z}_j(t)=-\alpha_1 z_j(t)+\varphi_j(t),\quad \tau_k<t\le\tau_{k+1}\tag{3.84a}$$

$$z_j(\tau_k^+)=\delta_1 z_j(\tau_k)+\zeta_j(k),\quad k=k_0,k_0+1,\ldots\tag{3.84b}$$

$$z_j(t_0)=\mathbb{E}[x_0^\top\tilde{X}(t_0,j)x_0\mid\eta(t_0)=j].\tag{3.84c}$$

One sees that (3.84) is an IVP defined by an one-dimensional non-homogeneous FJLDE of type (2.14). In this case (2.15) gives

$$z_j(t) = \mathbf{T}(t, t_0)z_j(t_0) + \sum_{k=k_0}^{k_t} \mathbf{T}(t, \tau_k^+)\zeta_j(k) + \int_{t_0}^{t} \mathbf{T}(t, s)\varphi_j(s)\mathrm{d}s \ \text{ if } t > \tau_{k_0} \quad (3.85)$$

and

$$z_j(t) = \mathbf{T}(t, t_0)z_j(t_0) + \int_{t_0}^{t} \mathbf{T}(t, s)\varphi_j(s)\mathrm{d}s \ \text{ if } t_0 \le t \le \tau_{k_0} \quad (3.86)$$

where $\mathbf{T}(t, s)$ is the linear evolution operator on $\mathbb{R}$ defined as in (2.5)–(2.7) specialized to the one-dimensional FJLDE:

$$\dot{z}_j(t) = -\alpha_1 z_j(t), \quad \tau_k < t \le \tau_{k+1} \quad (3.87a)$$
$$z_j(\tau_k^+) = \delta_1 z_j(\tau_k), \quad k \in \mathbb{Z}_+. \quad (3.87b)$$

In this special case, the accompanying DTLE of type (2.9) and (2.10) becomes

$$z_{k+1} = \mathscr{L}_d(k)z_k \quad (3.88a)$$
$$\mathscr{L}_d(k) = \delta_1 e^{-\alpha_1(\tau_{k+1}-\tau_k)}, \quad k \in \mathbb{Z}_+. \quad (3.88b)$$

In this case (2.11) yields

$$\mathbf{T}_d(k, l) = \delta_1^{k-l} e^{-\alpha_1(\tau_k-\tau_l)}, \quad \text{for all } k \ge l \ge k_0. \quad (3.89)$$

We have $0 \le \mathbf{T}_d(k, l) \le \delta^{k-l}$, for all $k \ge l \ge k_0$ where $\delta := \delta_1 e^{-\alpha_1/\underline{h}} \in (0, 1)$. Hence, the DTLE (3.88) is exponentially stable. Based on Theorem 2.1 (i) applied in the special case of the FJLDE (3.87) we deduce that there exist $\tilde{\beta} > 1, \alpha > 0$ such that $0 < \mathbf{T}(t, s) \le \tilde{\beta} e^{-\alpha(t-s)}$, for all $t \ge s \ge t_0$. This allows us to deduce from (3.85) and (3.86) that

$$0 \le z_j(t) \le \tilde{\beta}(e^{-\alpha(t-t_0)}z_j(t_0) + \sum_{k=k_0}^{k_t} e^{-\alpha(t-\tau_k)}\zeta_j(k) + \int_{t_0}^{t} e^{-\alpha(t-s)}\varphi_j(s)\mathrm{d}s) \quad (3.90)$$

if $t \ge \tau_{k_t} \ge \tau_{k_0}$ and

$$0 \le z_j(t) \le \tilde{\beta}(e^{-\alpha(t-t_0)}z_j(t_0) + \int_{t_0}^{t} e^{-\alpha(t-s)}\varphi_j(s)\mathrm{d}s) \quad (3.91)$$

if $t_0 \le t \le \tau_{k_0}$. Let us now show that

$$\mathbb{E}[x^\top(t)\tilde{X}(t,\eta(t))x(t) \mid \eta(t_0)=j] \le z_j(t), \quad \text{for all } t \ge t_0, \quad j \in \mathcal{N}. \tag{3.92}$$

Indeed, from (3.77b), (3.81), (3.84) we deduce that (3.92) hold for $t_0 \le t < \tau_{k_0}$. Then, using Lemma 3.1 we deduce that

$$\mathbb{E}[x^\top(\tau_{k_0})\tilde{X}(\tau_{k_0}^-,\eta(\tau_{k_0}))x(\tau_{k_0}) \mid \eta(t_0)=j] = \lim_{\substack{t\to\tau_{k_0}\\ t<\tau_{k_0}}} \tilde{X}(t,\eta(t))x(t) \mid \eta(t_0)=j]$$
$$\le z_j(\tau_{k_0}), \quad \text{for all } j \in \mathcal{N}. \tag{3.93}$$

From (3.83) and (3.84b) written for $k = k_0$, together with (3.93) we have:

$$\mathbb{E}[x^\top(\tau_{k_0}^+)\tilde{X}(\tau_{k_0},\eta(\tau_{k_0}))x(\tau_{k_0}^+) \mid \eta(t_0)=j] \le z_j(\tau_{k_0}^+).$$

This inequality allows us to infer that (3.92) holds for $\tau_{h_0} \le t < \tau_{k_0+1}$. Invoking again Lemma 3.1, we obtain that

$$\mathbb{E}[x^\top(\tau_{k_0+1})\tilde{X}(\tau_{k_0+1}^-,\eta(\tau_{k_0+1}))x(\tau_{k_0+1}) \mid \eta(t_0)=j] \le z_j(\tau_{k_0+1})$$

which yields via (3.83) and (3.84b) written for $k = k_0 + 1$, that

$$\mathbb{E}[x^\top(\tau_{k_0+1}^+)\tilde{X}(\tau_{k_0+1},\eta(\tau_{k_0+1}))x(\tau_{k_0+1}^+) \mid \eta(t_0)=j] \le z_j(\tau_{k_0+1}^+).$$

Proceeding in this way, it may be shown inductively, for $k \ge k_0$, that (3.92) holds on each interval $[\tau_k, \tau_{k+1})$. So, we may conclude that (3.92) holds for any $t \ge t_0$.

The upper bounds for $\mathbb{E}[|x(t)|^2 \mid \eta(t_0)=j]$ stated in (*i*) from the statement are obtained from (3.71) and (3.90) and (3.92) together with (3.80b) and (3.83b) taking $\beta := \frac{\tilde{\beta}}{\tilde{c}_1}\max\{1,\gamma_c,\gamma_d\}$. In the following we have:

$$\int_{t_0}^{\infty} \mathbb{E}[|x(t)|^2 \mid \eta(t_0)=j]dt \le \frac{\beta}{\alpha}\mathbb{E}[|x_0|^2 \mid \eta(t_0)=j] \tag{3.94}$$
$$+\beta\sum_{l=0}^{r_1}\int_{t_0}^{\infty}\int_{t_0}^{t} e^{-\alpha(t-s)}\mathbb{E}[|f_l(s)|^2 \mid \eta(t_0)=j]ds\, dt$$
$$+\beta\sum_{l=0}^{r_2}\int_{t_0}^{\infty}\left(\sum_{k=k_0}^{k_t} e^{-\alpha(t-\tau_k)}\mathbb{E}[|g_l(k)|^2 \,|\eta(t_0)=j]\right)dt.$$

Let $T > t_0$ be arbitrary but fixed. We have

$$\int_{t_0}^{T}\int_{t_0}^{t} e^{-\alpha(t-s)}\mathbb{E}[|f_l(s)|^2 \mid \eta(t_0)=j]ds\, dt$$

$$= \int_{t_0}^{T}\int_{s}^{T} e^{-\alpha(t-s)}\mathrm{d}t\mathbb{E}[|f_l(s)|^2\ \eta(t_0) = j]\mathrm{d}s \leq \frac{1}{\alpha}\int_{t_0}^{T} \mathbb{E}[|f_l(s)|^2\ |\eta(t_0) = j]\mathrm{d}s.$$

Letting $T \to \infty$ we obtain

$$\int_{t_0}^{\infty}\int_{t_0}^{t} e^{-\alpha(t-s)}\mathbb{E}[|f_l(s)|^2|\eta(t_0)=j]\mathrm{d}s\mathrm{d}t \leq \frac{1}{\alpha}\int_{t_0}^{\infty} \mathbb{E}[|f_l(s)|^2|\eta(t_0)=j]\mathrm{d}s < \infty, \quad (3.95)$$

$0 \leq l \leq r_1$, because $f_l(\cdot) \in \mathcal{L}^2_{\mathcal{H}}(\mathbb{R}_+, \mathbb{R}^n)$. Let $\tau_k > \tau_{k_0}$ be arbitrary but fixed. We may write

$$\int_{t_0}^{\tau_{k+1}} \sum_{\iota=k_0}^{k_t} e^{-\alpha(t-\tau_\iota)}\mathbb{E}[|g_l(\iota)|^2 \mid \eta(t_0) = j]\mathrm{d}t$$
$$= \sum_{\iota=k_0}^{k}\int_{\tau_\iota}^{\tau_{k+1}} e^{-\alpha(t-\tau_\iota)}\mathrm{d}t\mathbb{E}[|g_l(\iota)|^2|\eta(t_0) = j] \leq \frac{1}{\alpha}\sum_{\iota=k_0}^{k} \mathbb{E}[|g_l(\iota)|^2|\eta(t_0) = j].$$

Letting $k \to \infty$ we infer that

$$\int_{t_0}^{\infty} \sum_{\iota=k_0}^{k_t} e^{-\alpha(t-\tau_\iota)}\mathbb{E}[|g_l(\iota)|^2|\eta(t_0) = j]\mathrm{d}t \leq \sum_{\iota=k_0}^{\infty}\mathbb{E}[|g_l(\iota)|^2|\eta(t_0) = j] < \infty \quad (3.96)$$

because $g_l(\cdot) \in \ell^2_{\mathcal{H}}(\mathbb{Z}_+, \mathbb{R}^n)$ and $\lim_{k\to\infty} \tau_k = \infty$. The conclusion in the statement (ii) is obtained from (3.94)–(3.96). Having in mind (3.92) we may infer that to prove (iii) it is sufficient to show that $\lim_{t\to\infty} z_j(t) = 0,\ j \in \mathcal{N}$, when $z_j(\cdot)$ is the solution of the IVP (3.84). One sees that the sequence $\{z_j(\tau_k)\}_{k\geq k_0}$ solves the discrete-time IVP:

$$\begin{aligned} y_j(k+1) &= \mathcal{L}_d(k)y_j(k) + \mu_j(k), \quad k \geq k_0 \\ y_j(k_0) &= z_j(\tau_{k_0}), \end{aligned} \quad (3.97)$$

k_0 being such that $\tau_{k_0-1} < t_0 \leq \tau_{k_0}$, with

$$\mu_j(k) \triangleq e^{-\alpha_1(\tau_{k+1}-\tau_k)}\zeta_j(k) + \int_{\tau_k}^{\tau_{k+1}} e^{-\alpha_1(\tau_{k+1}-s)}\varphi_j(s)\mathrm{d}s, \quad (3.98)$$

$\zeta_j(k)$ are defined in (3.83b), while $\varphi_j(s)$ are defined in (3.80b). From (3.98) one sees that

$$\sum_{k=k_0}^{\infty} \mu_j(k) \leq \sum_{k=k_0}^{\infty} \zeta_j(k) + \int_{k_0}^{\infty} \varphi_j(s)ds < \infty$$

and $\mu_j(k) \geq 0$ for all $k \geq k_0$. Hence,

$$\lim_{k \to \infty} \mu_j(k) = 0, \quad j \in \mathcal{N}. \tag{3.99}$$

From (3.97) we obtain that

$$z_j(\tau_k) = \mathbf{T}_d(k, k_0)z_j(\tau_{k_0}) + \sum_{l=k_0}^{k-1} \mathbf{T}_d(k, l+1)\mu_j(l).$$

Further (3.89) yields

$$z_j(\tau_k) \leq \delta^{k-k_0} z_j(\tau_{k_0}) + \sum_{l=k_0}^{k-1} \delta^{k-l-1} \mu_j(l) \ \text{ for all } \ k \geq k_0.$$

As in the proof of Theorem 2.2 we deduce via (3.99) that

$$\lim_{k \to \infty} z_j(\tau_k) = 0, \tag{3.100}$$

because $\delta \in (0, 1)$ and $z_j(\tau_k) \geq 0$. Let now $t > t_0$ be arbitrary. From (3.84) we deduce that

$$z_j(t) = \delta_1 e^{-\alpha_1(t-\tau_{k_t})} z_j(\tau_{k_t}), \tag{3.101}$$

where $k_t \in \mathbb{Z}_+$ is such that $\tau_{k_t} \leq t \leq \tau_{k_t+1}$. Hence, $\lim_{t \to \infty} \tau_{k_t} = \infty$. Combining (3.100) and (3.101) we may conclude that $\lim_{t \to \infty} z_j(t) = 0$, which ends the proof.

The next result provides information regarding the sequence $\{x(\tau_k; t_0, x_0)\}_{k \in \mathbb{Z}_+}$ of the values at impulse times τ_k of the solution $x(\cdot; t_0, x_0)$ of the non-homogeneous JSLDE (3.51).

Corollary 3.6 *Let $t_0 \in \mathbb{R}_+$ and $k_0 \in \mathbb{Z}_+$ be such that $\tau_{k_0-1} < t \leq \tau_{k_0}$. Let $x_0 \in \mathfrak{X}_{t_0}$ be arbitrary. If the assumptions from Theorem 3.5 are fulfilled, then the values $x(\tau_k) \triangleq x(\tau_k; t_0, x_0)$ of the solution $x(\cdot; t_0, x_0)$ of the JSLDE (3.51) satisfy:*

(i)

$$\mathbb{E}[|x(\tau_k)|^2 \mid \eta(t_0) = j] \leq \beta \left(e^{-\alpha(\tau_k - t_0)} \mathbb{E}[|x_0|^2 \mid \eta(t_0) = j] \right.$$
$$+ \sum_{l=0}^{r_2} \sum_{\iota=k_0}^{k-1} e^{-\alpha(\tau_k - \tau_\iota)} \mathbb{E}[|g_l(\iota)|^2 \mid \eta(t_0) = j]$$

$$+\sum_{l=0}^{r_1}\int_{t_0}^{\tau_k} e^{-\alpha(\tau_k-s)}\mathbb{E}[|f_l(s)|^2 \mid \eta(t_0)=j]\mathrm{d}s\Bigg) \quad \text{for all } k \geq k_0+1 \tag{3.102a}$$

and

$$\mathbb{E}[|x(\tau_{k_0})|^2 \mid \eta(t_0)=j] \leq \beta\Big(e^{-\alpha(\tau_{k_0}-t_0)}\mathbb{E}[|x_0|^2 \mid \eta(t_0)=j]$$
$$+\sum_{l=0}^{r_1}\int_{t_0}^{\tau_{k_0}} e^{-\alpha(\tau_{k_0}-s)}\mathbb{E}[|f_l(s)|^2 \mid \eta(t_0)=j]\mathrm{d}s\Bigg) \quad \text{for } k=k_0 \tag{3.102b}$$

where $\beta \geq 1$, $\alpha > 0$, *are constants not depending upon* t_0, x_0 *and* j.
Furthermore, if $\{f_l(t)\}_{t\geq 0} \subset \mathcal{L}^2_{\mathcal{H}}(\mathbb{R}_+, \mathbb{R}^n)$, $0 \leq l \leq r_1$ *and* $\{g_l(k)\}_{k\in\mathbb{Z}_+} \in \ell^2_{\mathcal{H}}\{\mathbb{Z}_+, \mathbb{R}^n\}$, $0 \leq l \leq r_2$, *then:*

(*ii*) $\{x(\tau_k)\}_{k\geq k_0} \in \ell^2_{\mathcal{H}}(k_0, \infty; \mathbb{R}^n)$ *and we have*

$$\sum_{k=k_0}^{\infty}\mathbb{E}[|x(\tau_k)|^2 \mid \eta(t_0)=j] \leq \hat{\gamma}(\mathbb{E}[|x_0|^2 \mid \eta(t_0)=j]$$
$$+\sum_{l=0}^{r_2}\sum_{k=k_0}^{\infty}\mathbb{E}[|g_l(k)|^2 \mid \eta(t_0)=j] + \sum_{l=0}^{r_1}\int_{t_0}^{\infty}\mathbb{E}[|f_l(s)|^2 \mid \eta(t_0)=j]\mathrm{d}s), \tag{3.103}$$

(*iii*) $\lim_{k\to\infty}\mathbb{E}[|x(\tau_k)|^2 \mid \eta(t_0)=j]=0$.

Proof The inequalities (3.102) are directly obtained from Theorem 3.5 (i) taking $t=\tau_k$. Further, (3.103) is obtained by direct calculation from (3.102). The equality

$$\lim_{k\to\infty}\mathbb{E}[|x(\tau_k)|^2 \mid \eta(t_0)=j]=0 \tag{3.104}$$

is obtained from Theorem 3.5 (iii) using the fact that $\lim_{k\to\infty}\tau_k=\infty$ when Assumption 2.1 is fulfilled.

Remark 3.5 In the absence of assumption (b) in Theorem 3.5 we cannot be sure that the estimates from Theorem 3.5 (i) and Corollary 3.6 (i) are uniform with respect to t and t_0 from $\mathbb{R}_+$. Repeating on each interval of type $[t_0, \tau_{k_0}]$, $[\tau_k, \tau_{k+1}]$ the reasons made in Remark 3.6.1 from [1] one may show that for each compact interval $[t_0, T]$ there exists a constant $\beta > 0$ depending upon T and $T-t_0$ with the property that:

$$\mathbb{E}[|x(t;t_0,x_0)|^2 \mid \eta(t_0)=j] \leq \beta(\mathbb{E}[|x_0|^2 \mid \eta(t_0)=j]$$
$$+\sum_{l=0}^{r_2}\sum_{k=k_0}^{k_T}\mathbb{E}[|g_l(k)|^2 \mid \eta(t_0)=j] + \sum_{l=0}^{r_1}\int_{t_0}^{T}\mathbb{E}[|f_l(s)|^2 \mid \eta(t_0)=j]\mathrm{d}s)$$

for all $t \in [t_0, T]$, $x_0 \in \mathfrak{X}_{t_0}$, $f_l(\cdot) \in \mathcal{L}^2_{\mathcal{H}}(t_0, T; \mathbb{R}^n)$, $0 \leq l \leq r_1$ and for all $g_l(k) : \Omega \to \mathbb{R}^n$, $\mathcal{H}_{\tau_k}$–measurable with

$$\mathbb{E}[|g_l(k)|^2] < \infty, \quad k_0 \leq k \leq k_T, \quad 0 \leq l \leq r_2,$$

where k_0 and $k_T \in \mathbb{Z}_+$ are such that $\tau_{k_0-1} < t_0 \leq \tau_{k_0}$ and $\tau_{k_T} \leq T < \tau_{k_T+1}$.

3.5.3 *Input–Output Operators Defined by Systems Described by Jump Stochastic Linear Differential Equations*

We consider the affine (non-homogeneous) system of jump stochastic linear differential equations:

$$\begin{aligned} dx(t) &= (A_0(t, \eta(t))x(t) + G_0(t, \eta(t))v(t))dt + \sum_{l=1}^{r_1}(A_l(t, \eta(t))x(t) \\ &\quad + G_l(t, \eta(t))v(t))dw_l(t), \qquad \tau_k < t \leq \tau_{k+1} \qquad (3.105a) \end{aligned}$$

$$\begin{aligned} x(\tau_k^+) &= A_{d0}(\tau_k, \eta(\tau_k))x(\tau_k) + G_{d0}(\tau_k, \eta(\tau_k))v_d(k) + \sum_{l=1}^{r_2} w_{dl}(k)(A_{dl}(\tau_k, \eta(\tau_k))x(\tau_k) \\ &\quad + G_{dl}(\tau_k, \eta(\tau_k))v_d(k)), \qquad k \in \mathbb{Z}_+, \qquad (3.105b) \end{aligned}$$

having the outputs

$$z_c(t) = C_z(t, \eta(t))x(t), \quad t \in \mathbb{R}_+, \qquad (3.106a)$$
$$z_d(\tau_k) = C_{dz}(\tau_k, \eta(\tau_k))x(\tau_k) + D_{dz}(\tau_k, \eta(\tau_k))v_d(k), \qquad k \in \mathbb{Z}_+, \qquad (3.106b)$$

where $x(t) \in \mathbb{R}^n$ is the vector of the state parameters, $z_c(t) \in \mathbb{R}^{n_{zc}}$ are the discrete-time outputs at the impulse times τ_k, $v(\cdot) \in \mathcal{L}^2_{\mathcal{H}}(\mathbb{R}_+, \mathbb{R}^{m_v})$ and $v_d(k) : \Omega \to \mathbb{R}^{m_{v_d}}$ are random vectors $\mathcal{H}_{\tau_k}$–measurable with finite second moments. In (3.105), $\{w(t)\}_{t \geq 0}$ ($w(t) = (w_1(t), w_2(t), \ldots, w_{r_1}(t))^\top$), $\{w_d(k)\}_{k \in \mathbb{Z}_+}$ ($w_d(k) = (w_{d1}(k), w_{d2}(k), \ldots, w_{dr_2}(k))^\top$) and $\{\eta(t)\}_{t \geq 0}$ are stochastic processes which satisfy Assumption 1.2. Regarding the coefficients of the system (3.105) and (3.106) we make the assumption:

Assumption 3.2

(a) $t \to (A_l(t, i), G_l(t, i)) : \mathbb{R}_+ \to \mathbb{R}^{n \times n} \times \mathbb{R}^{n \times m_v}$, $0 \leq l \leq r_1$ and $t \to C_z(t, i) : \mathbb{R}_+ \to \mathbb{R}^{n_{zc} \times n}$, $i \in \mathcal{N}$ are bounded and continuous matrix-valued functions;

(b) $\{A_{dl}(\tau_k, i)\}_{k \in \mathbb{Z}_+} \subset \mathbb{R}^{n \times n}$, $\{G_{dl}(\tau_k, i)\}_{k \in \mathbb{Z}_+} \subset \mathbb{R}^{n \times m_{v_d}}$, $0 \leq l \leq r_2$, $\{C_{dz}(\tau_k, i)\}_{k \in \mathbb{Z}_+} \subset \mathbb{R}^{n_{zd} \times n}$, $\{D_{dz}(\tau_k, i)\}_{k \in \mathbb{Z}_+} \subset \mathbb{R}^{n_{zd} \times m_{v_d}}$, $i \in \mathcal{N}$, are bounded matrix-valued sequences.

Taking $f_l(t) \triangleq G_l(t, \eta(t))v(t), t \in \mathbb{R}_+, 0 \le l \le r_1, g_l(k) \triangleq G_{dl}(\tau_k, \eta(\tau_k))v_d(k)$, $k \in \mathbb{Z}_+$, $0 \le l \le r_2$ it becomes clear that the JSLDE (3.105) is a non-homogeneous JSLDE of the form (3.51). So, the results proved in Theorem 3.5 and Corollary 3.6 are applicable to the solutions of (3.105). Thus we obtain:

Corollary 3.7 *Assume:*

(a) Assumptions 1.2 and 3.2 are fulfilled;
(b) the JSLDE obtained taking $v(\cdot) \equiv 0$ and $v_d(\cdot) \equiv 0$ in (3.105) is SESMS.

Let $t_0 \in \mathbb{R}_+$ and $x(\cdot; t_0, 0, v(\cdot), v_d(\cdot))$ be the solution of (3.105) determined by $v(\cdot) \in \mathscr{L}^2_{\mathscr{H}}(t_0, \infty; \mathbb{R}^{m_v})$ and $\{v_d(k)\}_{k \ge k_0} \in \ell^2_{\mathscr{H}}\{k_0, \infty; \mathbb{R}^{m_{v_d}}\}$ having the initial value $x(t_0; t_0, 0, v(\cdot), v_d(\cdot)) = 0$.

Under the considered assumptions $x(\cdot; t_0, 0, v(\cdot), v_d(\cdot)) \in \mathscr{L}^2_{\mathscr{H}}(t_0, \infty; \mathbb{R}^n)$ and $\{x(\tau_k; t_0, 0, v(\cdot), v_d(\cdot))\}_{k \ge k_0} \in \ell^2_{\mathscr{H}}\{k_0, \infty; \mathbb{R}^n\}$. Further, we have

$$\int_{t_0}^{\infty} \mathbb{E}[|x(t; t_0, 0, v(\cdot), v_d(\cdot))|^2]\mathrm{d}t \le \mu \left(\int_{t_0}^{\infty} \mathbb{E}[|v(t)|^2]\mathrm{d}t + \sum_{k=k_0}^{\infty} \mathbb{E}[|v_d(k)|^2] \right) \tag{3.107}$$

and

$$\sum_{k=k_0}^{\infty} \mathbb{E}[|x(\tau_k; t_0, 0, v(\cdot), v_d(\cdot))|^2] \le \hat{\mu} \left(\int_{t_0}^{\infty} \mathbb{E}[|v(t)|^2]\mathrm{d}t + \sum_{k=k_0}^{\infty} \mathbb{E}[|v_d(k)|^2] \right) \tag{3.108}$$

where $k_0 \in \mathbb{Z}_+$ is such that $\tau_{k_0-1} < t_0 \le \tau_{k_0}$ and

$$\mu = \gamma \ \max \left\{ \sum_{l=0}^{r_1} \max_{i \in \mathscr{N}} \left\{ \sup_{t \ge 0} \|G_l(t, i)\|^2 \right\}, \sum_{l=0}^{r_2} \max_{i \in \mathscr{N}} \left\{ \sup_{k \in \mathbb{Z}_+} \|G_{dl}(\tau_k, i)\|^2 \right\} \right\} \tag{3.109}$$

$$\hat{\mu} = \hat{\gamma} \ \max \left\{ \sum_{l=0}^{r_1} \max_{i \in \mathscr{N}} \left\{ \sup_{t \ge 0} \|G_l(t, i)\|^2 \right\}, \sum_{l=0}^{r_2} \max_{i \in \mathscr{N}} \left\{ \sup_{k \in \mathbb{Z}_+} \|G_{dl}(\tau_k, i)\|^2 \right\} \right\} \tag{3.110}$$

with $\gamma > 0$ being the constant provided by Theorem 3.5 (i) while $\hat{\gamma}$ is the constant provided by Corollary 3.6 (i) applied in the case of (3.105).

Now, we consider the linear operator induced by the non-homogeneous JSLDE (3.105) and (3.106) on the Hilbert space $\mathscr{L}^2_{\mathscr{H}}(t_0, \infty; \mathbb{R}^{m_v}) \times \ell^2_{\mathscr{H}}\{k_0, \infty; \mathbb{R}^{m_{v_d}}\}$ defined by:

$$\mathbb{L}_{t_0\infty}(v(\cdot), v_d(\cdot)) = \left(\mathbb{L}_{ct_0\infty}(v(\cdot), v_d(\cdot)), \mathbb{L}_{dt_0\infty}(v(\cdot), v_d(\cdot))\right), \tag{3.111a}$$

$$\mathbb{L}_{ct_0\infty}(v(\cdot), v_d(\cdot))(t) \triangleq C_z(t, \eta(t))x(t; t_0, 0, v(\cdot), v_d(\cdot)), \quad t \in [t_0, \infty) \tag{3.111b}$$

$$\begin{aligned}\mathbb{L}_{dt_0\infty}(v(\cdot), v_d(\cdot))(k) &\triangleq C_{dz}(\tau_k, \eta(\tau_k))x(\tau_k; t_0, 0, v(\cdot), v_d(\cdot)) \\ &\quad + D_{dz}(\tau_k, \eta(\tau_k))v_d(k), \quad k = k_0, k_0+1, \ldots\end{aligned} \tag{3.111c}$$

for all $v(\cdot) \in \mathcal{L}^2_{\mathcal{H}}(t_0, \infty; \mathbb{R}^{m_v})$ *and* $\{v_d(k)\}_{k\geq k_0} \in \ell^2_{\mathcal{H}}\{k_0, \infty; \mathbb{R}^{m_{vd}}\}$ *where* $k_0 \in \mathbb{Z}_+$ *is related by* t_0 *through*

$$\tau_{k_0-1} < t_0 \leq \tau_{k_0}. \tag{3.112}$$

From (3.107)–(3.110) we deduce that $\mathbb{L}_{ct_0\infty}(v(\cdot), v_d(\cdot))(\cdot) \in \mathcal{L}^2_{\mathcal{H}}(t_0, \infty; \mathbb{R}^{n_{zc}})$ *and* $\{\mathbb{L}_{dt_0\infty}(v(\cdot), v_d(\cdot))(k)\}_{k\geq k_0} \in \ell^2_{\mathcal{H}}\{k_0, \infty; \mathbb{R}^{n_{zd}}\}$ *if* $v(\cdot) \in \mathcal{L}^2_{\mathcal{H}}(t_0, \infty; \mathbb{R}^{m_v})$ *and* $\{v_d(k)\}_{k\geq k_0} \in \ell^2_{\mathcal{H}}\{k_0, \infty; \mathbb{R}^{m_{vd}}\}$. *Hence, under the assumptions of Corollary 3.7 together with the linear dependence of the solution of (3.105) with respect to* $(v(\cdot), v_d(\cdot))$ *we may conclude that* $(v(\cdot), v_d(\cdot)) \to \mathbb{L}_{t_0\infty}(v(\cdot), v_d(\cdot)) : \mathcal{L}^2_{\mathcal{H}}(t_0, \infty; \mathbb{R}^{m_v}) \times \ell^2_{\mathcal{H}}\{k_0, \infty; \mathbb{R}^{m_{vd}}\} \to \mathcal{L}^2_{\mathcal{H}}(t_0, \infty; \mathbb{R}^{n_{zc}}) \times \ell^2_{\mathcal{H}}\{k_0, \infty; \mathbb{R}^{n_{zd}}\}$ *is a linear and bounded operator,* k_0 *being related to* t_0 *via (3.112).*

The linear operator $\mathbb{L}_{t_0\infty}$ *will be named input–output operator defined by the system (3.105) and (3.106) on the interval* $[t_0, \infty)$. *In this case the system (3.105) and (3.106) is the state-space representation on the interval* $[t_0, \infty)$ *of* $\mathbb{L}_{t_0\infty}$. *If (3.111b) and (3.111c) are restricted to a compact interval* $[t_0, T]$, *one obtains a linear operator* $\mathbb{L}_{t_0T} : \mathcal{L}^2_{\mathcal{H}}(t_0, T; \mathbb{R}^{m_v}) \times \ell^2_{\mathcal{H}}\{k_0, k_T; \mathbb{R}^{m_{vd}}\} \to \mathcal{L}^2_{\mathcal{H}}(t_0, T; \mathbb{R}^{n_{zc}}) \times \ell^2_{\mathcal{H}}\{k_0, k_T; \mathbb{R}^{n_{zd}}\}$ *where* k_0 *is related to* t_0 *via (3.112) and* $k_T \in \mathbb{Z}_+$ *is related to* T *by*

$$\tau_{k_T} \leq T < \tau_{k_T+1}. \tag{3.113}$$

Based on Remark 3.5 we may conclude that $\mathbb{L}_{t_0T}$ *is a linear and bounded operator. It will be named the input–output operator defined by (3.105) and (3.106) on the interval* $[t_0, T]$ *and (3.105) and (3.106) is a state-space representation of the input–output operator* $\mathbb{L}_{t_0T}$. *The role of the input–output operators in the analysis of the robustness of the stability with respect to different classes of parametric uncertainties will be emphasized in Chap. 9.*

3.6 Notes and References

Stability theory is one of the cornerstones of control theory. Consequently, huge efforts and contributions have been made over the past decades in this field, both in the deterministic and the stochastic frameworks. Giving an exhaustive bibliographical review is not possible, instead we limit ourselves to cite some of the main monographs that addressed problem related to stability of stochastic systems. Readers who are interested on systems described by Itô-type differential equations can refer to [10–17].

To the best of the authors knowledge, ESMS for systems described by differential equations with Markovian jump perturbations has been introduced and studied for the first time in [18]. Since then, great achievements have been obtained in the stability theory of MJLSs making it a solidly grounded research field. Without any intention of being exhaustive, we also cite here some of the main works which have marked decades of contributions [1, 19–22]. The class of Itô differential equations with Markovian switching has been addressed for example in [3, 23, 24].

Impulsive stochastic systems where the continuous-time part consists of an Itô-type linear stochastic differential equation and the discrete-time part is a stochastic difference equation have been also extensively studied in the literature; see, e.g., [25, 26] and the references therein. However, as pointed out by [2], very few address the case where both parts of the impulsive system are affected by noise [27, 28] and, when this is the case, average dwell-time conditions, state-dependent impulse times or stochastically arriving impulses [29, 30] are most of the time only considered. The case of purely time-dependent and deterministic impulses' arrival times has been addressed recently in [2]. In our work, we have adopted a similar time-dependent deterministic *impulse protocol*.

Impulsive Markovian jump stochastic systems have been also addressed in the literature. One of the crucial issues for this class of systems is the description of the behavior of the Markovian process in the neighborhood of the impulse times. Roughly speaking, two main directions appear in the literature. In the first one, the impulsive effect is considered to be induced by the switching of the Markov process; hence the impulse times correspond to the Markov switching instants. A *dwell time* approach has been used in [31–33] while in [34, 35] the authors defined a Markov process that has both continuous and discrete modes. In this context, the impulsive effect is a consequence of the switching of the Markov process into a discrete mode. At the impulse times, a discrete-time approach can then be adopted because of the nature of the considered Markovian process. The aforementioned settings lead to a continuous/discrete formulation of the problem. This is illustrated for example in Theorem 3.2 in [34] or Theorem 5.17 from [35] where the stability results are formulated via both continuous-time and discrete-time generalized Lyapunov-type operators. Note that such framework can not be used in a sampled-data control context.

In the second direction, the impulsive effect on the system state is not a consequence of the Markovian process switching. The impulses are time-dependent and deterministic and hence independent of the Markovian switching instants. This setting allows to consider sampled-data control problems. In this context, the issue that we have raised at the beginning of this discussion became crucial. In [36–39], the authors tackled this problem by using the following assumption: $\eta(\tau_k^-) = \eta(\tau_k)$, $k \in \mathbb{Z}_+$, almost surely. In other words, they assume that the intersection between the set of impulse times and the set the Markovian process switching times is a set with zero measure. In [40], the authors assumed that switching process can be modeled as a discrete-time Markov chain at the impulse times.

In our work, no such assumptions have been imposed. Instead, we have obtained a technical result, Lemma 3.1, that rigorously describes the behavior of the term $\mathbb{E}\left[x^\top(\tau_k)X(\tau_k, \eta(\tau_k))x(\tau_k) \mid \eta(t_0) = j\right]$ in the neighbor of the impulse times τ_k, $k \in \mathbb{Z}_+$. To the best of the author knowledge, it is the first time that such a technical result is obtained in the literature. Using this result leads us to formulate our stability results in terms of solutions to a class of generalized Lyapunov-type differential equations with jumps, as well as the main results from Chaps. 6–9.

References

1. Drăgan, V., Morozan, T., Stoica, A.M.: Mathematical Methods in Robust Control of Linear Stochastic Systems. Springer-Verlag, New York (2013)
2. Briat, C.: Stability analysis and stabilization of stochastic linear impulsive, switched and sampled-data systems under dwell-time constraints. Automatica **74**(1), 279–287 (2016)
3. Mao, X., Yuan, C.: Stochastic Differential Equations with Markovian Switching. Imperial College Press (2006)
4. Friedman, A.: Stochastic Differential Equations and Applications, vol, vol. 1. Academic Press (1975)
5. Chung, K.L.: Markov Chains with Stationary Transition Probabilities. Springer, Berlin, Heidelberg (1960)
6. Doob, J.L.: Stochastic Processes. Wiley, New York (1967)
7. Ghihman, I., AS,: Introduction to the Theory of Random Processes (translated from Russian). W.B. Saunders Co., Philadelphia (1969)
8. Dunford N, Schwartz JT (1988) Linear Operators, Part 1: General Theory. Wiley-Interscience
9. Natanson IP (1961) Theory of Functions of a Real Variable, vol 1. Frederick Ungar; Revised edition
10. Arnold, L.: Stochastic Differential Equations. Theory and Applications, Wiley, New York (1974)
11. Arnold, L.: Random Dynamical Systems. Springer, Berlin (1988)
12. Curtain, R.: Stability of Stochastic Dynamical Systems. Springer (2014)
13. Damm, T.: Rational Matrix Equations in Stochastic Control, vol. 297. Springer-Verlag, Berlin Heidelberg (2004)
14. Khasminskii, R.: Stochastic Stability of Differential Equations. Sijythoff and Noordhoff, Alpen aan den Ryn (1980)
15. Kushner H (1980) Stochastic Stability and Control. (Academic, New York, 1967) 99. G.S. Ladde, V. Lakshmikantham, Random Differential Inequalities (Academic, New York, 1980)
16. Øksendal, B.: Stochastic Differential Equations. Springer-Verlag, Berlin Heidelberg (2003)
17. Willems, J.: in Stochastic Problems in Dynamics. Pitman, London (1977)
18. Kats, I., Krasovskii, N.: On stability of systems with random parameters. Prikl Mat Mekh (in Russian) **24**, 809–823 (1960)
19. Mariton, M.: Jump Linear Systems in Automatic Control. CRC Press (1990)
20. Costa, O.L.V., Fragoso, M.D., Todorov, M.G.: Continuous-Time Markov Jump Linear Systems. Springer-Verlag, Berlin Heidelberg (2013)
21. Boukas EK (2008) Control of singular systems with random abrupt changes. Springer-Science & Buisiness Media
22. Bolzern P, Colaneri P (2015) Positive Markov jump linear systems. Nnow Publishers
23. Yin GG, Zhou C (2009) Hybrid switching diffusions: Properties and Applications. Springer-Science & Buisiness Media
24. Yao, X., Wu, L., Zheng, X.: Filtering and control of stochastic jump hybrid systems. Springer (2016)
25. Hou, L., Michel, A.N.: Moment stability of discontinuous stochastic dynamical systems. IEEE Transactions on Automatic Control **46**, 938–943 (2005)
26. Teel, A.R., Subbaraman, A., Sferlazza, A.: Stability analysis for stochastic hybrid systems: A survey. Automatica **50**, 2435–2456 (2014)
27. Chen, W.H., Wang, J.G., Tang, Y.J., Lu, X.: Robust h_∞ control of uncertain linear impulsive stochastic systems. International Journal of Robust and Nonlinear Control **18**, 1348–1371 (2008)
28. Feng Z, Teo K (2010) Optimal feedback control for stochastic impulsive linear systems subject to poisson processes. in Optimization and Optimal Control, ser. Springer Optimization and Its Applications, A. Chinchuluun, P. M. Pardalos, R. Enkhbat, and I. Tseveendorj, Eds. Springer New York

29. Hespanha J, Teel AR (2006) Stochastic impulsive systems driven by renewal processes. in 17th International Symposium on Mathematical Theory of Networks and Systems, Kyoto, Japan
30. Antunes, D., Hespanha, J.P., Silvestre, C.: Volterra integral approach to impulsive renewal systems: Application to networked control. IEEE Transactions on Automatic Control **57**, 607–619 (2013)
31. Xiong, J., Lam, J., Shu, Z., Mao, X.: Stability analysis of continuous-time switched systems with a random switching signal. IEEE Transactions on Automatic Control **59**, 180–186 (2013)
32. Xiang W, Lam J, Li P (2018) On stability and h_∞ control of switched systems with random switching signals. Automatica p 419–425
33. Wang G, Zhang Q, Yang C, Su C (2016) Stability and stabilization of continuous-time stochastic markovian jump systems with random switching signals. J Franklin Inst p 1339–1357
34. Souza, M., Fioravanti, A.R., Araujo, V.S.P.: Impulsive markov jump linear systems: Stability analysis and h_2 control. Nonlinear Analysis: Hybrid Systems **42**(101), 089 (2021)
35. Briat, C.: Stability analysis and stabilization of linear symmetric matrix-valued continuous, discrete, and impulsive dynamical systems-a unified approach for the stability analysis and the stabilization of linear systems. Nonlinear Analysis: Hybrid Systems **43**(101), 242 (2022)
36. Geromel, J.C.: Differential Linear Matrix Inequalities. Springer, In Sampled-Data Systems Filtering and Control (2023)
37. Geromel, J.C., Gabriel, G.W.: Optimal h_2 state feedback sampled-data control design of markov jump linear systems. Automatica **54**(1), 182–188 (2015)
38. Gabriel, G.W., Geromel, J.C.: Performance evaluation of sampled-data control of markov jump linear systems. Automatica **84**(1), 212–215 (2015)
39. Gabriel, G.W., Conçalves, T.R., Geromel, J.C.: Optimal and robust sampled-data control of markov jump linear systems: A differential lmi approach. IEEE Transactions on Automatic Control **63**(9), 3054–3060 (2018)
40. Hu, L., Shi, P., Huang, B.: Stochastic stability and robust control for sampled-data systems with markovian jump parameters. J Math Anal Appl **313**, 504–517 (2006)

Chapter 4
Structural Properties of Jump Linear Stochastic Systems

Abstract In this chapter we define and characterize some fundamental control theory concepts, namely the stabilizability and the detectability notions, for a general class of systems described by impulsive stochastic linear differential equations. To the best of the authors knowledge, it seems that such structural properties for impulsive stochastic systems have not been considered yet in the literature. In Sect. 4.2, we define the concepts of stabilizability in mean square by linear state feedback as well as the concept of stabilizability in mean square by impulses in the state feedback form of the considered class of jump linear stochastic systems. Then, based on Lyapunov criteria for exponential stability in mean square provided in Sect. 3.4 we derive several necessary and sufficient conditions which will allow us to test the property of mean-square stabilizability by state feedback and mean-square stabilizability by impulses. These results are then specialized in Sect. 4.3 to the case of stochastic sampled-data controlled systems by introducing the concept of mean-square stabilizability by sample states measurements (see Definition 4.5). In Sect. 4.4, we define the concept of mean-square detectability of a jump linear stochastic system and we provide methods which allow us to decide if a given jump linear stochastic system is detectable or not. More precisely, we shall show how we may use the results from Sect. 3.4 in order to obtain necessary and sufficient conditions for mean-square detectability of the stochastic systems under consideration. The detectability result obtained in Sect. 4.4 will then be applied in Sect. 4.5 to the mean-square stability problem. The objective here is to propose a less conservative result than the one obtained in Theorem 3.2. We show that under the detectability assumption a kind of mean-square stability of a jump stochastic linear differential equation of type (3.1) is achieved even in the case when the non-homogeneous backward GLJLDE (3.32) has the forcing term positive semidefinite instead of uniform positive definite and has a positive semidefinite global solution instead of a uniform positive one. Finally, Sect. 4.6 addresses the *Duality* between the concepts of mean-square stabilizability and mean-square detectability defined in Sects. 4.2 and 4.4, respectively.

V. Drăgan et al., *Robust Control of Jump Linear Stochastic Systems*, Lecture Notes in Control and Information Sciences 497, https://doi.org/10.1007/978-3-031-84070-8_4

4.1 Jump Linear Stochastic Systems

In this section we provide a short description of the main categories of jump linear stochastic systems studied in this chapter. First we consider the jump linear stochastic system (JLSS) described by:

$$\begin{aligned} dx(t) &= (A_0(t,\eta(t))x(t) + B_0(t,\eta(t))u_c(t))dt \\ &\quad + \sum_{l=1}^{r_1}(A_l(t,\eta(t))x(t) + B_l(t,\eta(t))u_c(t))dw_l(t), \qquad \tau_k < t \leq \tau_{k+1} \end{aligned} \tag{4.1a}$$

$$\begin{aligned} x(\tau_k^+) &= A_{d0}(\tau_k,\eta(\tau_k))x(\tau_k) + B_{d0}(\tau_k,\eta(\tau_k))u_d(k) \\ &\quad + \sum_{l=1}^{r_2} w_{dl}(k)\left(A_{dl}(\tau_k,\eta(\tau_k))x(\tau_k) + B_{dl}(\tau_k,\eta(\tau_k))u_d(k)\right), \quad k \in \mathbb{Z}_+, \end{aligned} \tag{4.1b}$$

$$dz_c(t) = C_{c0}(t,\eta(t))x(t) + \sum_{l=1}^{r_1} C_{cl}(t,\eta(t))x(t)dw_l(t), \qquad t \in \mathbb{R}_+ \tag{4.2a}$$

$$z_d(\tau_k) = (C_{d0}(\tau_k,\eta(\tau_k)) + \sum_{l=1}^{r_2} w_{dl}(k)C_{dl}(\tau_k,\eta(\tau_k)))x(\tau_k), \quad k \in \mathbb{Z}_+, \tag{4.2b}$$

where $x(t) \in \mathbb{R}^n$ are the state parameters, $u_c(t) \in \mathbb{R}^{m_c}$ are the control parameters acting in continuous time, while $u_d(k) \in \mathbb{R}^{m_d}$ are the control parameters which are acting by impulses only at time instances τ_k, $k \in \mathbb{Z}_+$. In (4.2), $z_c(t) \in \mathbb{R}^{n_{z_c}}$ is an output which evolve in continuous time and $z_d(\tau_k) \in \mathbb{R}^{n_{z_d}}$ is an output which is observed only at discrete time instances τ_k, $k \in \mathbb{Z}_+$. The stochastic processes $\{w(t)\}_{t\geq 0}$ ($w(t) = (w_1(t)\ w_2(t)\ \dots w_{r_1}(t))^\top$), $\{w_d(k)\}_{k\geq 0}$ ($w_d(k) = (w_{d1}(k)\ w_{d2}(k)\ \dots\ w_{dr_2}(k))^\top$) and $\{\eta(t)\}_{t\geq 0}$ involved in (4.1)–(4.2) are those described in Chap. 1 and are satisfying the Assumption 1.2.

If in (4.1b) we have $A_{d0}(\tau_k,i) = I_n$, $B_{d0}(\tau_k,i) = 0$, $A_{dl}(\tau_k,i) = 0$, $B_{dl}(\tau_k,i) = 0$, $1 \leq l \leq r_2$, $i \in \mathcal{N}$, $k \in \mathbb{Z}_+$ then (4.1) reduces to:

$$\begin{aligned} dx(t) &= (A_0(t,\eta(t)x(t) + B_0(t,\eta(t))u_c(t))dt \\ &\quad + \sum_{l=1}^{r_1}(A_l(t,\eta(t))x(t) + B_l(t,\eta(t))u_c(t))dw_l(t), \qquad t \in \mathbb{R}_+ \end{aligned}$$

which is a controlled system described by Itô differential equations with Markovian switching intensively studied in Drăgan et al. [1]. On the other hand, if in (4.1a) we would have $A_l(t,i) = 0$, $B_l(t,i) = 0$, for $t \in \mathbb{R}_+$, $0 \leq l \leq r_1$, $i \in \mathcal{N}$, then $x(t) = x(\tau_k^+)$, $t \in (\tau_k, \tau_{k+1}]$ a.s. Particularly, $x(\tau_{k+1}) = x(\tau_k^+)$ a.s. In this special case, (4.1) becomes:

$$x(\tau_{k+1}) = A_{d0}(\tau_k, \eta(\tau_k))x(\tau_k) + B_{d0}(\tau_k, \eta(\tau_k))u_d(k)$$
$$+ \sum_{l=1}^{r_2} w_{dl}(k)(A_{dl}(\tau_k, \eta(\tau_k))x(\tau_k) + B_{dl}(\tau_k, \eta(\tau_k))u_d(k)), \quad k \in \mathbb{Z}_+,$$

which is a controlled system described by discrete-time linear stochastic equations simultaneously affected by sequences of independent random variables and a Markov chain. This kind of controlled stochastic systems were investigated in Drăgan et al. [2].

Let us remark that if $\{\eta(t)\}_{t\geq 0}$ is a standard homogeneous Markov process with a finite number of states, $\mathcal{N}$, then the sequence of random variables $\{\eta(\tau_k)\}_{k\geq 0}$ is a (possibly non-homogeneous) Markov chain having the set of states $\mathcal{N}$ and the transition probability matrices $\mathbb{P}_k = e^{Q(\tau_{k+1}-\tau_k)}$, $k \geq 0$.

The above discussion shows that the controlled system of type (4.1) includes as special cases the continuous time controlled stochastic linear systems described by Itô differential equations subject to Markovian switching as well as discrete-time controlled linear stochastic systems described by linear equations simultaneously affected by sequences of independent random variables and a Markov chain.

In the present chapter, we assume that the coefficients of the systems (4.1)–(4.2) are satisfying the assumption:

Assumption 4.1

(a) for each $i \in \mathcal{N}$, $t \to (A_l(t,i), B_l(t,i)) : \mathbb{R}_+ \to \mathbb{R}^{n\times n} \times \mathbb{R}^{n\times m_c}$, $t \to C_{cl}(t,i) : \mathbb{R}_+ \to \mathbb{R}^{n_{zc}\times n}$, $0 \leq l \leq r_1$, are bounded and continuous matrix-valued functions;
(b) $\{A_{dl}(\tau_k, i)\}_{k\in\mathbb{Z}_+} \subset \mathbb{R}^{n\times n}$, $\{B_{dl}(\tau_k, i)\}_{k\in\mathbb{Z}_+} \subset \mathbb{R}^{n\times m_d}$, $\{C_{dl}(\tau_k, i)\}_{k\in\mathbb{Z}_+} \subset \mathbb{R}^{n_{zd}\times n}$, $0 \leq l \leq r_2$, are bounded matrix-valued sequences.

Note 4.1 The developments from this chapter still hold if for each $i \in \mathcal{N}$, $t \to (A_l(t,i), B_l(t,i)) : \mathbb{R}_+ \to \mathbb{R}^{n\times n} \times \mathbb{R}^{n\times m_c}$, $t \to C_{cl}(t,i) : \mathbb{R}_+ \to \mathbb{R}^{n_{zc}\times n}$, $0 \leq l \leq r_1$, are bounded and piecewise continuous matrix-valued functions having lateral limits in any $t \in \mathbb{R}_+$.

Often a JLSS of type (4.1) may be identified with a triple of the form

$$((\mathbf{A}(\cdot)|\mathbf{A}_d(\cdot)), (\mathbf{B}(\cdot)|\mathbf{B}_d(\cdot)), Q) \tag{4.3}$$

where

$$\mathbf{A}(\cdot) = (\mathbb{A}_0(\cdot), \mathbb{A}_1(\cdot), \ldots, \mathbb{A}_{r_1}(\cdot)) \tag{4.4a}$$

$$\mathbf{B}(\cdot) = (\mathbb{B}_0(\cdot), \mathbb{B}_1(\cdot), \ldots, \mathbb{B}_{r_1}(\cdot)), \tag{4.4b}$$

$$\mathbf{A}_d(\cdot) = (\mathbb{A}_{d0}(\cdot), \mathbb{A}_{d1}(\cdot), \ldots, \mathbb{A}_{dr_2}(\cdot)) \tag{4.5a}$$

$$\mathbf{B}_d(\cdot) = (\mathbb{B}_{d0}(\cdot), \mathbb{B}_{d1}(\cdot), \ldots, \mathbb{B}_{dr_2}(\cdot)), \tag{4.5b}$$

$$\mathbb{A}_l(\cdot) = (A_l(\cdot, 1), A_l(\cdot, 2), \ldots, A_l(\cdot, N)) \tag{4.6a}$$

$$\mathbb{B}_l(\cdot) = (B_l(\cdot, 1), B_l(\cdot, 2), \ldots, B_l(\cdot, N)), \tag{4.6b}$$

$A_l(\cdot, i)$, $B_l(\cdot, i)$ being matrix-valued functions satisfying the Assumption 4.1 (a),

$$\mathbb{A}_{dl}(\cdot) = (A_{dl}(\cdot, 1), A_{dl}(\cdot, 2), \ldots, A_{dl}(\cdot, N)) \tag{4.7a}$$

$$\mathbb{B}_{dl}(\cdot) = (B_{dl}(\cdot, 1), B_{dl}(\cdot, 2), \ldots, B_{dl}(\cdot, N)), \tag{4.7b}$$

$(A_{dl}(\cdot, i), B_{dl}(\cdot, i))$ are satisfying the Assumption 4.1 (b). $Q \in \mathbb{R}^{N\times N}$ is a matrix whose elements q_{ij} are satisfying conditions (1.22).

The next two definitions are adapted from Definitions 3.2 and 3.3, respectively, to the case of a JLSS of type (4.1)–(4.2).

Definition 4.1 We say that the JLSS (4.1)–(4.2) is time periodic if the following conditions are simultaneously fulfilled:

(a) $\tau_k = kh$, for all $k \in \mathbb{Z}_+$ where $h > 0$ is fixed;
(b) there exists a real number $\theta_c > 0$ with the property that $A_l(t + \theta_c, i) = A_l(t, i)$, $B_l(t + \theta_c, i) = B_l(t, i)$, $C_{cl}(t + \theta_c, i) = C_{cl}(t, i)$, $0 \le l \le r_1$, for all $t \in \mathbb{R}_+$;
(c) there exists an integer $\theta_d \ge 1$ with the property that $A_{dl}((k + \theta_d)h, i) = A_{dl}(kh, i)$, $B_{dl}((k + \theta_d)h, i) = B_{dl}(kh, i)$, $C_{dl}((k + \theta_d)h, i) = C_{dl}(kh, i)$, $0 \le l \le r_2$, for all $k \in \mathbb{Z}_+$, $i \in \mathcal{N}$;
(d) $\frac{\theta_c}{\theta_d h}$ is a rational number.

As in the case of the Definition 3.2, if the property $d)$ from above is satisfied, then there exist two coprime natural numbers $\tilde{p}$, $\tilde{q}$ with the property that

$$\frac{\theta_c}{\theta_d h} = \frac{\tilde{p}}{\tilde{q}}. \tag{4.8}$$

If (4.8) is satisfied we set:

$$\tilde{\theta} \triangleq \tilde{q}\theta_c = \tilde{p}\theta_d h \tag{4.9a}$$

$$\tilde{\mathbf{k}} \triangleq \tilde{p}\theta_d. \tag{4.9b}$$

Definition 4.2 We say that the JLSS (4.1)–(4.2) is in the time-invariant case, if the following conditions are simultaneously satisfied:

(a) $\tau_k = kh$, for all $k \in \mathbb{Z}_+$, where $h > 0$ does not depend upon k;
(b) for any $(t, i) \in \mathbb{R}_+ \times \mathcal{N}$, the following equalities hold: $A_l(t, i) = A_l(i) \in \mathbb{R}^{n\times n}$, $B_l(t, i) = B_l(i) \in \mathbb{R}^{n\times m_c}$ and $C_{cl}(t, i) = C_{cl}(i) \in \mathbb{R}^{n_{z_c}\times n}$, $0 \le l \le r_1$;
(c) for any $(k, i) \in \mathbb{Z}_+ \times \mathcal{N}$ the following hold: $A_{dl}(\tau_k, i) = A_{dl}(i) \in \mathbb{R}^{n\times n}$, $B_{dl}(\tau_k, i) = B_{dl}(i) \in \mathbb{R}^{n\times m_d}$ and $C_{dl}(\tau_k, i) = C_{dl}(i) \in \mathbb{R}^{n_{z_d}\times n}$ for all $0 \le l \le r_2$.

If the conditions from the previous definition are satisfied, the JLSS (4.1)–(4.2) may be rewritten as:

$$\begin{aligned} dx(t) &= (A_0(\eta(t))x(t) + B_0(\eta(t))u_c(t))dt \\ &\quad + \sum_{l=1}^{r_1}(A_l(\eta(t))x(t) + B_l(\eta(t))u_c(t))dw_l(t), \qquad kh < t \le (k+1)h, \end{aligned} \tag{4.10a}$$

$$\begin{aligned} x(kh^+) &= A_{d0}(\eta(kh))x(kh) + B_{d0}(\eta(kh))u_d(k) \\ &\quad + \sum_{l=1}^{r_2} w_{dl}(k)\left(A_{dl}(\eta(kh))x(kh) + B_{dl}(\eta(kh))u_d(k)\right), \quad k \in \mathbb{Z}_+, \end{aligned} \tag{4.10b}$$

$$dz_c(t) = C_{c0}(\eta(t))x(t) + \sum_{l=1}^{r_1} C_{cl}(\eta(t))x(t)dw_l(t), \qquad t \in \mathbb{R}_+ \tag{4.11a}$$

$$z_d(kh) = C_{d0}(\eta(kh))x(kh) + \sum_{l=1}^{r_2} w_{dl}(k)C_{dl}(kh, \eta(kh))x(kh), \quad k \in \mathbb{Z}_+. \tag{4.11b}$$

Beside the JLSS (4.1)–(4.2) and its time-invariant version (4.10)–(4.11) one considers the following two important special cases:

$$\begin{aligned} dx(t) &= (A_0(t)x(t) + B_0(t)u_c(t))dt \\ &\quad + \sum_{l=1}^{r_1}(A_l(t)x(t) + B_l(t)u_c(t))dw_l(t), \qquad \tau_k < t \le \tau_{k+1}, \end{aligned} \tag{4.12a}$$

$$\begin{aligned} x(\tau_k^+) &= A_{d0}(\tau_k)x(\tau_k) + B_{d0}(\tau_k)u_d(k) \\ &\quad + \sum_{l=1}^{r_2} w_{dl}(k)\left(A_{dl}(\tau_k)x(\tau_k) + B_{dl}(\tau_k)u_d(k)\right), \quad k \in \mathbb{Z}_+, \end{aligned} \tag{4.12b}$$

$$dz_c(t) = C_{c0}(t)x(t) + \sum_{l=1}^{r_1} C_{cl}(t)x(t)dw_l(t), \qquad t \in \mathbb{R}_+ \tag{4.13a}$$

$$z_d(\tau_k) = C_{d0}(\tau_k)x(\tau_k) + \sum_{l=1}^{r_2} w_{dl}(k)C_{dl}(\tau_k)x(\tau_k), \quad k \in \mathbb{Z}_+, \tag{4.13b}$$

and

$$\dot{x}(t) = A_0(t, \eta(t))x(t) + B_0(t, \eta(t))u_c(t), \qquad \tau_k < t \le \tau_{k+1}, \tag{4.14a}$$

$$x(\tau_k^+) = A_{d0}(\tau_k, \eta(\tau_k))x(\tau_k) + B_{d0}(\tau_k, \eta(\tau_k))u_d(k), \quad k \in \mathbb{Z}_+, \tag{4.14b}$$

having the outputs $z_c(t) = C_{c0}(t, \eta(t))x(t)$ and $z_d(\tau_k) = C_{d0}(\tau_k, \eta(\tau_k))x(\tau_k)$.

Often a jump linear stochastic system of type (4.12) may be identified to a pair of the form $((\hat{\mathbf{A}}(\cdot)|\hat{\mathbf{A}}_d(\cdot)), (\hat{\mathbf{B}}(\cdot)|\hat{\mathbf{B}}_d(\cdot)))$ where $\hat{\mathbf{A}}(\cdot) = (A_0(\cdot), A_1(\cdot), \ldots, A_{r_1}(\cdot))$, $\hat{\mathbf{B}}(\cdot) = (B_0(\cdot), B_1(\cdot), \ldots, B_{r_1}(\cdot))$, $A_l(\cdot), B_l(\cdot), 0 \le l \le r_1$ being the coefficients of (4.12a). $\hat{\mathbf{A}}_d(\cdot) = (A_{d0}(\cdot), A_{d1}(\cdot), \ldots, A_{dr_2}(\cdot))$, $\hat{\mathbf{B}}_d(\cdot) = (B_{d0}(\cdot), B_{d1}(\cdot), \ldots, B_{dr_2}(\cdot))$, where $A_{dl}(\cdot), B_{dl}(\cdot), 0 \le l \le r_2$ being the coefficients of (4.12b).

The time-invariant version of (4.12)–(4.13) and (4.14) are:

$$dx(t) = (A_0x(t) + B_0u_c(t))dt + \sum_{l=1}^{r_1}(A_lx(t) + B_lu_c(t))dw_l(t), \ kh < t \le (k+1)h, \tag{4.15a}$$

$$x(kh^+) = A_{d0}x(kh) + B_{d0}u_d(k) + \sum_{j=1}^{r_2} w_{dj}(k)\left(A_{dj}x(kh) + B_{dj}u_d(k)\right), \ k \in \mathbb{Z}_+, \tag{4.15b}$$

$$d_c(t) = C_{c0}x(t) + \sum_{l=1}^{r_1} C_{cl}x(t)dw_l(t), \qquad t \in \mathbb{R}_+ \tag{4.16a}$$

$$z_d(kh) = (C_{d0} + \sum_{l=1}^{r_2} w_{dl}C_{dl})x(kh), \quad k \in \mathbb{Z}_+, \tag{4.16b}$$

and

$$\dot{x}(t) = A_0(\eta(t))x(t) + B_0(\eta(t))u_c(t), \qquad kh < t \le (k+1)h, \tag{4.17a}$$

$$x(kh^+) = A_{d0}(\eta(kh))x(kh) + B_{d0}(\eta(kh))u_d(k), \quad k \in \mathbb{Z}_+, \tag{4.17b}$$

having the outputs $z_c(t) = C_{c0}(\eta(t))x(t)$, $t \in \mathbb{R}_+$, $z_d(kh) = C_{d0}(\eta(kh))x(kh)$, $k \in \mathbb{Z}_+$. Let us note that systems of type (4.15)–(4.16) with $A_l = 0$, $B_l = 0$, $1 \le l \le r_1$

and $A_{dj} = 0$, $B_{dj} = 0$, $1 \leq j \leq r_2$ were considered in this chapter in the monograph Ichikawa and Katayama [3].

If in (4.1), we have $B_l(\cdot, i) = 0$, $i \in \mathcal{N}$, $0 \leq l \leq r_1$, then the system is controlled only by controls $u_d(k)$ which are acting only at time instances τ_k. That is why in this case the system (4.1) with $B_l(\cdot, i) = 0$, $i \in \mathcal{N}$, $0 \leq l \leq r_1$ will be named impulsive controlled linear stochastic system (ICLSS).

4.2 Mean-Square Stabilizability of Jump Linear Stochastic Systems

In this section, we define the concepts of stabilizability in mean square by linear state feedback as well as the concept of stabilizability in mean square by impulses in the state feedback form of jump linear stochastic systems described in the previous section. Based on Lyapunov criteria for exponential stability in mean square provided in Sect. 3.4 we derive several necessary and sufficient conditions which will allow us to test the property of mean-square stabilizability by state feedback and mean-square stabilizability by impulses.

4.2.1 *Basic Definitions*

Definition 4.3

(a) We say that the JLSS (4.1) or equivalently the triple $((\mathbf{A}(\cdot)|\mathbf{A}_d(\cdot)), (\mathbf{B}(\cdot)|\mathbf{B}_d(\cdot)), Q)$ is **stabilizable in mean square by linear state feedback** if there exist bounded matrix-valued functions $t \to F_c(t, i) : \mathbb{R}_+ \to \mathbb{R}^{m_c \times n}$ continuous on each interval (τ_k, τ_{k+1}) with finite lateral limits in any τ_k and matrix-valued sequences $\{F_d(\tau_k, i)\}_{k \in \mathbb{Z}_+} \subset \mathbb{R}^{m_d \times n}$, $i \in \mathcal{N}$, with the property that the closed-loop system:

$$\begin{aligned} dx(t) &= (A_0(t, \eta(t)) + B_0(t, \eta(t))F_c(t, \eta(t)))x(t)dt \\ &+ \sum_{l=1}^{r_1}(A_l(t, \eta(t)) + B_l(t, \eta(t))F_c(t, \eta(t)))x(t)dw_l(t), \quad \tau_k < t \leq \tau_{k+1} \end{aligned} \tag{4.18a}$$

$$\begin{aligned} x(\tau_k^+) &= (A_{d0}(\tau_k, \eta(\tau_k)) + B_{d0}(\tau_k, \eta(\tau_k))F_d(\tau_k, \eta(\tau_k)))x(\tau_k) \\ &+ \sum_{l=1}^{r_2} w_{dl}(k)\,(A_{dl}(\tau_k, \eta(\tau_k)) + B_{dl}(\tau_k, \eta(\tau_k))F_d(\tau_k, \eta(\tau_k)))\,x(\tau_k), \ k \in \mathbb{Z}_+, \end{aligned} \tag{4.18b}$$

is SESMS.

(b) We say that the JLSS of type (4.12) or equivalently the pair $((\hat{\mathbf{A}}(\cdot)|\hat{\mathbf{A}}_d(\cdot)), (\hat{\mathbf{B}}(\cdot)|\hat{\mathbf{B}}_d(\cdot)))$ is **stabilizable in mean square by state feedback** if there exist a bounded matrix-valued function $t \to F_c(t) : \mathbb{R}_+ \to \mathbb{R}^{m_c \times n}$ continuous on each interval (τ_k, τ_{k+1}) with finite lateral limits in any τ_k and a bounded matrix-valued sequence $\{F_d(\tau_k)\}_{k \in \mathbb{Z}_+} \subset \mathbb{R}^{m_d \times n}$ with the property that the closed-lops system:

$$\begin{aligned} dx(t) &= (A_0(t) + B_0(t)F_c(t))x(t)dt \\ &\quad + \sum_{l=1}^{r_1}(A_l(t) + B_l(t)F_c(t))x(t)dw_l(t), \qquad \tau_k < t \le \tau_{k+1}, \end{aligned} \tag{4.19a}$$

$$\begin{aligned} x(\tau_k^+) &= (A_{d0}(\tau_k) + B_{d0}(\tau_k)F_d(\tau_k))x(\tau_k) \\ &\quad + \sum_{l=1}^{r_2} w_{dl}(k)\,(A_{dl}(\tau_k) + B_{dl}(\tau_k)F_d(\tau_k))\,x(\tau_k), \quad k \in \mathbb{Z}_+, \end{aligned} \tag{4.19b}$$

is ESMS.

(c) We say that the JLSS (4.14) is **stabilizable in mean square by state feedback** if there exist matrix-valued bounded functions $t \to F_c(t,i) : \mathbb{R}_+ \to \mathbb{R}^{m_c \times n}$ continuous on each interval (τ_k, τ_{k+1}) with finite lateral limits in any τ_k and bounded matrix-valued sequences $\{F_d(\tau_k, i)\}_{k \in \mathbb{Z}_+} \subset \mathbb{R}^{m_d \times n}$ with the property that the closed-loop system:

$$\dot{x}(t) = (A_0(t,\eta(t)) + B_0(t,\eta(t))F_c(t,\eta(t)))x(t), \qquad \tau_k < t \le \tau_{k+1}, \tag{4.20a}$$

$$x(\tau_k^+) = (A_{d0}(\tau_k,\eta(\tau_k)) + B_{d0}(\tau_k,\eta(\tau_k))F_d(\tau_k,\eta(\tau_k)))x(\tau_k), k \in \mathbb{Z}_+, \tag{4.20b}$$

is SESMS.

The pairs $(F_c(\cdot,i), \{F_d(\tau_k,i)\}_{k \in \mathbb{Z}_+})$, $i \in \mathcal{N}$, for which the closed-loop systems (4.18) or (4.20) are SESMS as well as the pair $(F_c(\cdot), \{F_d(\tau_k)\}_{k \in \mathbb{Z}_+})$, for which the closed-loop system (4.19) is ESMS will be called **pairs of stabilizing feedback gains**.

Remark 4.1

(a) In the next chapter we shall show that if the JLSS (4.1) or one of its particular versions (4.12), (4.14) are in the periodic case in the sense of the Definition 4.1, then it is mean-square stabilizable by state feedback if and only if among the pairs of stabilizing feedback gains there are pairs of stabilizing feedback gains of the form $\left(F_c(\cdot,i), \{F_d(kh,i)\}_{k \in \mathbb{Z}_+}\right)$, $i \in \mathcal{N}$ (and $(F_c(\cdot), \{F_d(kh)\}_{k \in \mathbb{Z}_+})$,, respectively) in which $F_c(\cdot,i)$ $(F_c(\cdot)$,, respectively) are matrix-valued functions continuous on each interval $(kh, (k+1)h)$ with finite lateral limits in any kh and are periodic of period $\tilde{\theta}$ and $\{F_d(kh,i)\}_{k \in \mathbb{Z}_+}$ ($\{F_d(kh)\}_{k \in \mathbb{Z}}$,, respectively) are periodic sequences of period $\tilde{\mathbf{k}}$ where $\tilde{\theta}$ and $\tilde{\mathbf{k}}$ were defined in (4.9).

(b) We shall show also in the next chapter that if the JLSS (4.1) or one of its special forms (4.12), (4.14) is time invariant, then, it is stabilizable in mean square by state feedback if and only if among the pairs of stabilizing feedback gains there are stabilizing feedback gains of the form $\left(F_c(\cdot, i), \{F_d(kh, i)\}_{k\in\mathbb{Z}_+}\right)$, $i \in \mathcal{N}$ (and $(F_c(\cdot), \{F_d(kh)\}_{k\in\mathbb{Z}_+})$,, respectively) in which $F_c(\cdot, i)$ ($F_c(\cdot)$,, respectively) are continuous functions on each interval $(kh, (k+1)h)$ with finite lateral limits in any kh and are periodic with period h and the sequences $\{F_d(kh, i)\}_{k\in\mathbb{Z}_+}$, $i \in \mathcal{N}$, ($\{F_d(kh)\}_{k\in\mathbb{Z}_+}$,, respectively) are constant sequences, that is $F_d(kh, i) = F_d(i)$ for all $(k, i) \in \mathbb{Z}_+ \times \mathcal{N}$ ($F_d(kh) = F_d$ for all $k \in \mathbb{Z}_+$,, respectively).

Definition 4.4

(a) We say that the JLSS (4.1) or equivalently the triple $((\mathbf{A}(\cdot)|\mathbf{A}_d(\cdot)), (\mathbf{B}(\cdot)|\mathbf{B}_d(\cdot)), Q)$ is **stabilizable in mean square by impulses in a state feedback form** if it is stabilizable in mean square by linear state feedback and among the pairs of stabilizing feedback gains there are pairs of stabilizing feedback gains of the form $(0, \{F_d(\tau_k, i)\}_{k\in\mathbb{Z}_+})$, $i \in \mathcal{N}$. In this case, the closed-loop system (4.18) becomes:

$$dx(t) = A_0(t, \eta(t))x(t)dt + \sum_{l=1}^{r_1} A_l(t, \eta(t))x(t)dw_l(t), \ \tau_k < t \le \tau_{k+1} \tag{4.21a}$$

$$\begin{aligned} x(\tau_k^+) &= (A_{d0}(\tau_k, \eta(\tau_k)) + B_{d0}(\tau_k, \eta(\tau_k))F_d(\tau_k, \eta(\tau_k)))x(\tau_k) \\ &\quad + \sum_{l=1}^{r_2} w_{dl}(k)\,(A_{dl}(\tau_k, \eta(\tau_k)) + B_{dl}(\tau_k, \eta(\tau_k))F_d(\tau_k, \eta(\tau_k)))\,x(\tau_k), \quad k \in \mathbb{Z}_+, \end{aligned} \tag{4.21b}$$

and it is SESMS.

(b) The JLSS (4.12) or equivalently the pair $((\hat{\mathbf{A}}(\cdot)|\hat{\mathbf{A}}_d(\cdot)), (\hat{\mathbf{B}}(\cdot)|\hat{\mathbf{B}}_d(\cdot)))$ is **stabilizable in mean square by impulses in a state feedback form** if it is stabilizable in mean square by linear state feedback and among the pairs of stabilizing feedback gains there are pairs of the form $(0, \{F_d(\tau_k)\}_{k\in\mathbb{Z}_+})$. In this case the closed-loop system (4.19) takes the form:

$$dx(t) = A_0(t)x(t)dt + \sum_{l=1}^{r_1} A_l(t)x(t)dw_l(t), \qquad \tau_k < t \le \tau_{k+1}, \tag{4.22a}$$

$$\begin{aligned} x(\tau_k^+) &= (A_{d0}(\tau_k) + B_{d0}(\tau_k)F_d(\tau_k))x(\tau_k) \\ &\quad + \sum_{l=1}^{r_2} w_{dl}(k)\,(A_{dl}(\tau_k) + B_{dl}(\tau_k)F_d(\tau_k))\,x(\tau_k), \quad k \in \mathbb{Z}_+, \end{aligned} \tag{4.22b}$$

and it is ESMS.

(c) We say that the JLSS (4.14) is **stabilizable in mean square by impulses in a state feedback form** if it is stabilizable in mean square by linear state feedback and among the stabilizing feedback gains there are pairs of the form

$(0, \{F_d(\tau_k, i)\}_{k \in \mathbb{Z}_+}), i \in \mathcal{N},$. In this case, the closed-loop system (4.20) reduces to:

$$\dot{x}(t) = A_0(t, \eta(t))x(t), \qquad \tau_k < t \leq \tau_{k+1}, \tag{4.23a}$$

$$x(\tau_k^+) = (A_{d0}(\tau_k, \eta(\tau_k)) + B_{d0}(\tau_k, \eta(\tau_k))F_d(\tau_k, \eta(\tau_k)))x(\tau_k), k \in \mathbb{Z}_+, \tag{4.23b}$$

and it is SESMS.

The matrix-valued sequences $\{F_d(\tau_k, i)\}_{k \in \mathbb{Z}_+} \subset \mathbb{R}^{m_d \times n}$, $i \in \mathcal{N}$, for which the closed-loop systems (4.21) and (4.23) are SESMS, as well as the sequences $\{F_d(\tau_k)\}_{k \in \mathbb{Z}_+} \subset \mathbb{R}^{m_d \times n}$ for which the closed-loop system (4.22) is ESMS will be named sequences of stabilizing feedback gains.

Remark 4.2 In (a) and (c) from Definitions 4.3 and 4.4, the stability notion of SESMS has been used in order to be able to apply Theorem 3.2 in the general time-varying case to derive some necessary and sufficient conditions for mean-square stabilizability. According to Proposition 3.3, in the case of the closed-loop systems of type (4.19) and (4.22), the notion of SESMS is equivalent to the ESMS stability notion. That is why in part (b) of the Definitions 4.3 and 4.4 we have required that the closed-loop systems is ESMS. Based on Theorem 3.1, we also deduce that in the periodic case, the property of SESMS of the closed-loop system is equivalent to the property of ESMS of such a system.

4.2.2 Necessary and Sufficient Conditions for Mean-Square Stabilizability of a JLSS

Now we show how we can use the results derived in Sect. 3.4 in order to obtain necessary and sufficient conditions for the stabilizability in the mean-square sense of a JLSS and to provide methods for obtaining some pairs of stabilizing feedback gains, or sequences of stabilizing feedback gains, respectively. Applying Theorem 3.2 in the case of the closed-loop system (4.18) we obtain:

Proposition 4.1 *Under the Assumptions 1.2, 2.1 and 4.1, the following are equivalent:*

(i) the JLSS (4.1) is mean-square stabilizable by linear state feedback;
(ii) there exist functions $t \to (X_i(t), K_i(t)) : \mathbb{R}_+ \to \mathcal{S}_n \times \mathbb{R}^{m_c \times n}$ *and the sequences* $\{Z_i(\tau_k)\}_{k \in \mathbb{Z}_+} \subset \mathcal{S}_n$, $\{V_i(\tau_k)\}_{k \in \mathbb{Z}_+} \subset \mathbb{R}^{m_d \times n}$, $i \in \mathcal{N}$ *with the properties:*

(α) $X_i(\cdot)$ *are differentiable on each interval* (τ_k, τ_{k+1}) *and right continuous in any* $t = \tau_k$, $k \in \mathbb{Z}_+$;
(β) $0 \prec \gamma_1 I_n \preceq X_i(t) \preceq \gamma_2 I_n$; $\|\dot{X}_i(t)\| \leq \gamma_3$, *for all* $t \in \mathbb{R}_+$, $0 \prec \gamma_4 I_n \preceq Z_i(\tau_k) \preceq \gamma_5 I_n$, *for all* $k \in \mathbb{Z}_+$, *for some positive constants* γ_j, $1 \leq j \leq 5$;

(γ) *$K_i(\cdot)$ are piecewise continuous and bounded matrix-valued functions and $\{V_i(\tau_k)\}_{k\in\mathbb{Z}_+}$ are bounded matrix-valued sequences;*

(δ) *$(X_i(\cdot), K_i(\cdot), \{Z_i(\tau_k), V_i(\tau_k)\}_{k\in\mathbb{Z}_+})$, $i \in \mathcal{N}$ solves the following system of linear matrix differential inequalities with jumps:*

$$\begin{pmatrix} -\dot{X}_i(t) + \Theta_1(t, X_i(t), K_i(t), i) & \overset{r_1}{\underset{l=1}{\text{row}}}\{(A_l(t,i)X_i(t) + B_l(t,i)K_i(t))^\top\} & \overset{N}{\underset{\substack{j=1\\ j\neq i}}{\text{row}}}\{\sqrt{q_{ij}}X_i(t)\} \\ * & -\Theta_2(X_i(t)) & 0 \\ * & * & -\Theta_3(\mathbf{X}_{-i}(t)) \end{pmatrix} \prec -\xi_1 I$$

$$\tau_k \le t < \tau_{k+1}, \tag{4.24a}$$

$$X_i(\tau_k^-) = Z_i(\tau_k), \tag{4.24b}$$

$$\begin{pmatrix} -Z_i(\tau_k) & \overset{r_2}{\underset{l=0}{\text{row}}}\{A_{dl}(\tau_k,i)Z_i(\tau_k) + B_{dl}(\tau_k,i)V_i(\tau_k)^\top\} \\ * & -\Theta_4(X_i(\tau_k)) \end{pmatrix} \prec -\xi_2 I,$$

$$k \in \mathbb{Z}_+,\ i \in \mathcal{N}, \tag{4.24c}$$

where

$$\Theta_1(t, X_i(t), K_i(t), i) \overset{\Delta}{=} (A_0(t,i) + \frac{1}{2}q_{ii}I_n)X_i(t) + X_i(t)(A_0(t,i) + \frac{1}{2}q_{ii}I_n)^\top + B_0(t,i)K_i(t) + K_i^\top(t)B_0^\top(t,i),$$

$$\Theta_2(X_i(t)) \overset{\Delta}{=} \overset{r_1}{\underset{j=1}{\text{diag}}}\{X_i(t)\} \in \mathcal{S}_{n_1},\quad n_1 = r_1 n,$$

$$\Theta_3(\mathbf{X}_{-i}(t)) \overset{\Delta}{=} \overset{N}{\underset{\substack{j=1\\ j\neq i}}{\text{diag}}}\{X_j(t)\} \subset \mathcal{S}_{n_2},\quad n_2 = (N-1)n,$$

$$\Theta_4(X_i(\tau_k)) \overset{\Delta}{=} \overset{r_2}{\underset{l=0}{\text{diag}}}\{X_i(\tau_k)\} \in \mathcal{S}_{n_3},\quad n_3 = (r_2+1)n.$$

If $\left(X_i(\cdot), K_i(\cdot), \{Z_i(\tau_k), V_i(\tau_k)\}_{k\in\mathbb{Z}_+}\right)$, $i \in \mathcal{N}$ is a solution of (4.24) then

$$F_c(t,i) \overset{\Delta}{=} K_i(t)X_i^{-1}(t) \qquad F_d(\tau_k,i) \overset{\Delta}{=} V_i(\tau_k)Z_i^{-1}(\tau_k),\quad (t,k) \in \mathbb{R}_+ \times \mathbb{Z}_+,$$

$i \in \mathcal{N}$ provide a pair of stabilizing feedback gains for the JLSS (4.1).

Proof The equivalence $(i) \Leftrightarrow (ii)$ from Theorem 3.2 applied in the case of the closed-loop system (4.18) taking $H_i(t) = I_n$, $t \in \mathbb{R}_+$, $\Gamma_k(i) = I_n$, $k \in \mathbb{Z}_+$, $i \in \mathcal{N}$ allows us to deduce that the JLSS (4.1) is mean-square stabilizable by linear state feedback if and only if there exists matrix-valued functions $t \to F_c(t,i) : \mathbb{R}_+ \to \mathbb{R}^{m_c\times n}$ which are continuous on each interval (τ_k, τ_{k+1}) having finite lateral limits in any τ_k, $k \in \mathbb{Z}_+$ and bounded sequences $\{F_d(\tau_k, i)\}_{k\in\mathbb{Z}_+} \subset \mathbb{R}^{m_d\times n}$, $i \in \mathcal{N}$, such that the non-homogeneous backward GLJLDE:

$$\begin{aligned}
&\dot{Y}_i(t) + (A_0(t,i) + B_0(t,i)F_c(t,i))^\top Y_i(t) + Y_i(t)(A_0(t,i) + B_0(t,i)F_c(t,i)) \\
&\quad + \sum_{l=1}^{r_1}(A_l(t,i) + B_l(t,i)F_c(t,i))^\top Y_i(t)(A_l(t,i) + B_l(t,i)F_c(t,i)) \\
&\quad + \sum_{j=1}^{N} q_{ij}Y_j(t) + I_n = 0, \quad \tau_k \le t < \tau_{k+1} \qquad (4.25a)
\end{aligned}$$

$$\begin{aligned}
Y_i(\tau_k^-) &= \sum_{l=0}^{r_2}(A_{dl}(\tau_k,i) + B_{dl}(\tau_k,i)F_d(\tau_k,i))^\top Y_i(\tau_k) \\
&\quad \times (A_{dl}(\tau_k,i) + B_{dl}(\tau_k,i)F_d(\tau_k,i)) + I_n, \quad k \in \mathbb{Z}_+ \qquad (4.25b)
\end{aligned}$$

has a unique solution $t \to (Y_1(t), Y_2(t), \ldots, Y_N(t)) : \mathbb{R}_+ \to \mathcal{S}_n^N$ with the property that there exist two positive constants c_1, c_2 such that

$$0 \prec c_1 I_n \preceq Y_i(t) \preceq c_2 I_n,$$

for all $(t, i) \in \mathbb{R}_+ \times \mathcal{N}$. Employing the Schur complement technique and defining $X_i(t) := Y_i^{-1}(t)$, $K_i(t) := F_c(t,i)Y_i^{-1}(t)$, $Z_i(\tau_k) := Y_i^{-1}(\tau_k^-)$, $V_i(\tau_k) := F_d(\tau_k, i)Y_i^{-1}(\tau_k^-)$, one shows that the feasibility of (4.25) is equivalent to the solvability of (4.24). Thus the proof ends.

Further, we analyze the problem of mean-square stabilizability by impulses in a state feedback form as well as the mean-square stabilizability by linear state feedback of a time-invariant counterpart of the JLSS (4.1) or (4.12), respectively.

I. Let us first consider the simpler version of a JLSS (4.15):

$$dx(t) = A_0x(t)dt + \sum_{l=1}^{r_1} A_lx(t)dw_l(t), \qquad kh < t \le (k+1)h, \qquad (4.26a)$$

$$x(kh^+) = A_{d0}x(kh) + B_{d0}u_d(k) + \sum_{l=1}^{r_2} w_{dl}(k)(A_{dl}x(kh) + B_{dl}u_d(k)), \; k \in \mathbb{Z}_+. \qquad (4.26b)$$

According to the Definition 4.4 (*b*) and Remark 4.2 (*b*), the JLSS (4.26) is stabilizable in mean square by impulses if there exists a matrix $F_d \in \mathbb{R}^{m_d\times n}$ such that the closed-loop system

$$dx(t) = A_0x(t)dt + \sum_{l=1}^{r_1} A_lx(t)dw_l(t), \qquad kh < t \le (k+1)h, \qquad (4.27a)$$

$$x(kh^+) = [A_{d0} + B_{d0}F_d + \sum_{l=1}^{r_2} w_{dl}(k)(A_{dl} + B_{dl}F_d)]x(kh), \quad k \in \mathbb{Z}_+ \tag{4.27b}$$

is ESMS. Invoking the equivalence $(i) \Leftrightarrow (iv)$ from Corollary 3.4 we deduce that the JSLDE (4.27) is ESMS if and only if the non-homogeneous forward GLJLDE

$$\dot{X}(t) = \hat{\mathscr{L}}[X(t)], \qquad kh < t \leq (k+1)h, \tag{4.28a}$$

$$X(kh^+) = \sum_{l=0}^{r_2}(A_{dl} + B_{dl}F_d)X(kh)(A_{dl} + B_{dl}F_d)^\top + I_n, \quad k \in \mathbb{Z} \tag{4.28b}$$

has a unique bounded solution $t \to \tilde{X}(t) : \mathbb{R} \to \mathscr{S}_n$ and this solution is a periodic function of period h such that $\tilde{X}(t) > 0$, for all $t \in \mathbb{R}$. In (4.28a), the operator $X \to \hat{\mathscr{L}}[X] : \mathscr{S}_n \to \mathscr{S}_n$ is defined by

$$\hat{\mathscr{L}}[X] = A_0X + XA_0^\top + \sum_{l=1}^{r_1} A_lXA_l^\top. \tag{4.29}$$

Now we are in position to prove:

Proposition 4.2 *Under the Assumption 1.1 the following are equivalent:*

i) the JLSS (4.26) is stabilizable in mean square by impulses;
ii) the non-homogeneous forward GLJLDE (4.28) has a unique bounded solution $t \to \tilde{X}(t) : \mathbb{R} \to \mathscr{S}_n$ and this solution is a periodic function of period h satisfying $\tilde{X}(t) > 0$, for all $t \in \mathbb{R}$;
iii) there exist matrices $Z \in \mathscr{S}_n$, $Z > 0$ and $K \in \mathbb{R}^{m_d \times n}$ solving the following LMI:

$$\begin{pmatrix} -Z & \mathscr{W}_1(Z, K) \\ \mathscr{W}_1^\top(Z, K) & -\mathscr{W}_2(Z) \end{pmatrix} < 0 \tag{4.30}$$

where we have denoted

$$\begin{aligned} \mathscr{W}_1(Z, K) &= (A_{d0}e^{\hat{\mathscr{L}}h}[Z] + B_{d0}K \;\; A_{d1}e^{\hat{\mathscr{L}}h}[Z] + B_{d1}K \;\; \ldots \\ &\quad A_{dr_2}e^{\hat{\mathscr{L}}h}[Z] + B_{dr_2}K) \in \mathbb{R}^{n \times \hat{n}} \\ \mathscr{W}_2(Z) &= diag(e^{\hat{\mathscr{L}}h}[Z] \;\; e^{\hat{\mathscr{L}}h}[Z] \;\; \ldots \;\; e^{\hat{\mathscr{L}}h}[Z]) \in \mathscr{S}_{\hat{n}} \\ &\text{with } \hat{n} := (r_2+1)n \text{ and } e^{\hat{\mathscr{L}}h}[Z] := \sum_{j=0}^{\infty} \frac{h^j}{j!}\hat{\mathscr{L}}^j[Z]. \end{aligned} \tag{4.31}$$

If (Z, K) is a solution of the LMI (4.30) then

$$F_d := K(e^{\hat{\mathscr{L}}h}[Z])^{-1} \tag{4.32}$$

achieves the exponential stability in mean square of the corresponding closed-loop JSLDE of type (4.27).

iv) There exists a function $t \to Z(t) : [0,\ h] \to \mathcal{S}_n$ and a matrix $V \in \mathbb{R}^{m_d \times n}$ satisfying the following Lyapunov-type linear differential inequality with jumps:

$$- \dot{Z}(t) + A_0 Z(t) + Z(t) A_0^\top + \sum_{l=1}^{r_1} A_l Z(t) A_l^\top \leq 0,\ kh < t \leq (k+1)h \tag{4.33a}$$

$$\begin{pmatrix} Z(0) - I_n & \mathcal{W}_1(Z(h), V) \\ \mathcal{W}_1^\top(Z(h), V) & \mathcal{W}_2(Z(h)) \end{pmatrix} > 0 \tag{4.33b}$$

where we have denoted

$$\begin{aligned} &\mathcal{W}_1(Z(h), V) = (A_{d0} Z(h) + B_{d0} V \ \ A_{d1} Z(h) + B_{d1} V \ \ \ldots \\ &\qquad A_{dr_2} Z(h) + B_{dr_2} V) \in \mathbb{R}^{n \times \hat{n}} \\ &\mathcal{W}_2(Z) = diag(Z(h) \ \ Z(h) \ \ \ldots \ \ Z(h)) \in \mathcal{S}_{\hat{n}} \\ &with\ \ \hat{n} := (r_2 + 1)n. \end{aligned}$$

If $(Z(\cdot), V)$ is a solution of (4.33) then the control law $u_d(k) = V Z^{-1}(h) x(kh)$, $k \in \mathbb{Z}_+$ stabilizes the JLSS (4.27).

Proof The equivalence $(i) \Leftrightarrow (ii)$ follows directly from the chain of implications $(i) \Rightarrow (v) \Rightarrow (vi) \Rightarrow (i)$ from Corollary 3.4 with $\tilde{H} = \tilde{\Gamma} = I_n$ applied to the closed-loop JSLDE (4.27). If (ii) holds, we take $t \to X(t) : \mathbb{R} \to \mathcal{S}_n$ as the unique bounded positive definite and $h-$periodic solution of (4.28). We have:

$$X(0) = X(h) = e^{\hat{\mathcal{L}} h}[X(0^+)] + \int_0^h e^{\hat{\mathcal{L}}(h-s)}[I_n] \mathrm{d}s.$$

Substituting this representation of $X(0)$ in (4.28b) written for $k = 0$ we obtain

$$X(0^+) = \sum_{l=0}^{r_2} (A_{dl} + B_{dl} F_d) e^{\hat{\mathcal{L}}} [X(0^+)] (A_{dl} + B_{dl} F_d)^\top + \hat{H} \tag{4.34}$$

where

$$\hat{H} \triangleq I_n + \sum_{l=0}^{r_2} (A_{dl} + B_{dl} F_d) \int_0^h e^{\hat{\mathcal{L}}(h-s)}[I_n] ds (A_{dl} + B_{dl} F_d)^\top. \tag{4.35}$$

Further, (4.29) allows us to write

$$\hat{\mathscr{L}}[X] = \hat{\mathscr{L}}_0[X] + \hat{\mathscr{L}}_1[X] \tag{4.36}$$

where

$$\hat{\mathscr{L}}_0[X] := A_0X + XA_0^\top, \tag{4.37a}$$

$$\hat{\mathscr{L}}_1[X] := \sum_{l=1}^{r_1} A_l X A_l^\top, \tag{4.37b}$$

for all $X \in \mathscr{S}_n$. So, (4.36), (4.37) yield: $\hat{\mathscr{L}}[X] \geq \hat{\mathscr{L}}_0[X]$, for all $X \in \mathscr{S}_n^+$. Applying the time-invariant version of part (i) in Theorem 2.2.5 from Drăgan et al. [1] we deduce that

$$e^{\hat{\mathscr{L}}t}[X] \geq e^{\hat{\mathscr{L}}_0 t}[X] \geq 0, \quad \text{for all } t \geq 0,\ X \in \mathscr{S}_n^+. \tag{4.38}$$

From (4.38) written for $X = I_n$ and (4.35), we may infer that $\hat{H} \geq I_n > 0$. Thus (4.34) shows that $X(0^+)$ solves the following LMI:

$$\sum_{l=0}^{r_2} (A_{dl} + B_{dl}F_d)e^{\hat{\mathscr{L}}h}[X(0^+)](A_{dl} + B_{dl}F_d)^\top - X(0^+) < 0. \tag{4.39}$$

On the other hand (4.28b) written for $k = 0$ gives

$$X(0^+) \geq I_n > 0, \tag{4.40}$$

because $X(0) > 0$. Now, from (4.38) and (4.40) we deduce that

$$e^{\hat{\mathscr{L}}h}[X(0^+)] \geq e^{\hat{\mathscr{L}}_0 h}[X(0^+)] = e^{A_0 h}X(0^+)e^{A_0^\top h} > 0$$

because $e^{A_0 h}$ is an invertible matrix. Employing the Schur complement techniques, we obtain that (4.39) is equivalent to

$$\begin{pmatrix} -X(0^+) & \mathscr{W}_1(X(0^+),\ F_d e^{\hat{\mathscr{L}}h}[X(0^+)]) \\ \mathscr{W}_1^\top(X(0^+),\ F_d e^{\hat{\mathscr{L}}h}[X(0^+)]) & -\mathscr{W}_2(X(0^+)) \end{pmatrix} < 0.$$

Hence, (4.30) is satisfied by $(Z, K) = (X(0^+),\ F_d e^{\hat{\mathscr{L}}h}[X(0^+)])$ which means that (iii) holds if (ii) is true.

Let us prove now that $(iii) \Rightarrow (i)$. If there exist $Z \in \mathscr{S}_n$, $Z > 0$ and $K \in \mathbb{R}^{m_d \times n}$ satisfying (4.30), then, the Schur complement techniques allows us to deduce that

$$\sum_{l=0}^{r_2} (A_{dl} + B_{dl}F_d)e^{\hat{\mathscr{L}}h}[Z](A_{dl} + B_{dl}F_d)^\top - Z < 0$$

where

$$F_d := K\left(e^{\hat{\mathscr{L}}h}[Z]\right)^{-1}. \tag{4.41}$$

Let $\Gamma \triangleq Z - \sum_{l=0}^{r_2}(A_{dl} + B_{dl}F_d)e^{\hat{\mathscr{L}}h}[Z](A_{dl} + B_{dl}F_d)^\top$. So, $\Gamma > 0$ and Z solves the equation

$$Z = \sum_{l=0}^{r_2}(A_{dl} + B_{dl}F_d)e^{\hat{\mathscr{L}}h}[Z](A_{dl} + B_{dl}F_d)^\top + \Gamma. \tag{4.42}$$

Let $t \to X(t) : \mathbb{R} \to \mathscr{S}_n$ be defined by:

$$X(t) := e^{\hat{\mathscr{L}}(t-kh)}[Z] \text{ if } t \in (kh, (k+1)h], \quad k \in \mathbb{Z}. \tag{4.43}$$

One sees that $\lim_{\substack{t\to kh\\ t>kh}} X(t) = Z$ and $X(kh) = e^{\hat{\mathscr{L}}h}[Z]$, for all $k \in \mathbb{Z}$.
Hence, (4.42) becomes

$$X(kh^+) = \sum_{l=0}^{r_2}(A_{dl} + B_{dl}F_d)X(kh)(A_{dl} + B_{dl}F_d)^\top + \Gamma, \quad \text{for all } k \in \mathbb{Z}. \tag{4.44}$$

From (4.38) and (4.43) we deduce that

$$X(t) \geq e^{A_0(t-kh)}Ze^{A_0^\top(t-kh)} > 0, \quad \text{for all } kh < t \leq (k+1)h, \ k \in \mathbb{Z}.$$

Employing (4.43) and (4.44) we may conclude that $X(\cdot)$ is a bounded and positive definite solution of the following non-homogeneous forward GLJLDE:

$$\dot{X}(t) = \hat{\mathscr{L}}[X(t)], \quad kh < t \leq (k+1)h \tag{4.45a}$$

$$X(kh^+)=\sum_{l=0}^{r_2}(A_{dl}+B_{dl}F_d)X(kh)(A_{dl}+B_{dl}F_d)^\top+\Gamma, \ k \in \mathbb{Z} \text{ with } \Gamma > 0. \tag{4.45b}$$

Invoking the implication $(vi) \Rightarrow (i)$ from Corollary 3.4 in the case of (4.45) we may conclude that the JSLDE of type (4.27) corresponding to F_d defined via (4.41) is ESMS, which means that the JLSS (4.26) is stabilizable in mean square by impulses defined as

$$u_d(k) = F_d x(kh), \quad k \in \mathbb{Z}_+.$$

Thus the proof is complete.

Remark 4.3 1. From the numerical point of view, the main challenge in solving the LMI (4.30) is to find an efficient procedure for the computation of $e^{\hat{\mathscr{L}}}[Z]$. Such a procedure is described as follows: In order to obtain a description of the dependence $Z \to e^{\hat{\mathcal{L}}}[Z] : \mathcal{S}_n \to \mathcal{S}_n$, we use the fact that $\mathcal{S}_n$ is a $\frac{n(n+1)}{2}$ dimensional vector space. Hence, if $\{\Xi^{\mu\nu} \mid 1 \leq \mu \leq \nu \leq n\}$ is a basis of the vector space $\mathcal{S}_n$, then any element Z of $\mathcal{S}_n$ has the unique representation:

$$Z = \sum_{\mu=1}^{n} \sum_{\nu=\mu}^{n} z_{\mu\nu} \Xi^{\mu\nu}.$$

Hence,

$$e^{\hat{\mathcal{L}}h}[Z] = \sum_{\mu=1}^{n} \sum_{\nu=\mu}^{n} z_{\mu\nu} \tilde{\Xi}^{\mu\nu} \tag{4.46}$$

where

$$\tilde{\Xi}^{\mu\nu} \triangleq e^{\hat{\mathcal{L}}h}[\Xi^{\mu\nu}], \quad 1 \leq \mu \leq \nu \leq n. \tag{4.47}$$

Therefore, before testing the feasibility of the LMI (4.30), one looks for a basis $\Xi^{\mu\nu}$, $1 \leq \mu \leq \nu \leq n$ of the vector space $\mathcal{S}_n$. Then, one computes the matrices $\tilde{\Xi}^{\mu\nu}$ via (4.47). To this end one uses a truncation of the series from (4.31) written for Z replaced successively by $\tilde{\Xi}^{\mu\nu}$. In practice, one may use the canonical basis of the vector space $\mathcal{S}_n$ consisting of the matrices

$$\Xi^{\mu\nu} = \{\Xi_{ij}^{\mu\nu} \mid i, j \in \{1, 2, \ldots, n\}\}$$

where

$$\Xi_{ij}^{\mu\nu} = \begin{cases} 1, & \text{if } (ij) = (\mu\nu) \text{ or } (ji) = (\mu\nu) \\ 0, & \text{in other cases.} \end{cases} \tag{4.48}$$

2. Regarding the numerical aspects of condition iv) from Proposition 4.2, we propose a piecewise linear approximation method. In this method, it is proposed to impose a piecewise linear structure to the matrix-valued function $Z(\cdot)$ involved in the conditions (4.33). Let $\bar{q} \in \mathbb{Z}_+$, $\bar{q} \geq 1$ be fixed. Let $\delta > 0$ be defined by $\delta = \frac{h}{\bar{q}}$. We compute the matrices $Z(q\delta)$, $0 \leq q \leq \bar{q}$ from the following LMIs:

$$-\frac{Z((q+1)\delta) - Z(q\delta)}{\delta} + \mathcal{L}(Z((q+1)\delta)) \leq 0 \tag{4.49}$$

$$-\frac{Z((q+1)\delta) - Z(q\delta)}{\delta} + \mathcal{L}(Z(q\delta)) \leq 0 \tag{4.50}$$

$0 \leq q \leq \bar{q} - 1$, where:

$$\mathcal{L}(Z) \triangleq A_0 Z + Z A_0^\top + \sum_{l=1}^{r_1} A_l Z A_l^\top. \tag{4.51}$$

We define:

$$Z(t) \triangleq \frac{Z((q+1)\delta) - Z(q\delta)}{\delta}(t - q\delta) + Z(q\delta) \tag{4.52}$$

$t \in [q\delta,\ (q+1)\delta]$, $0 \le q \le \bar{q} - 1$. Multiplying (4.49) by $\frac{t-q\delta}{\delta}$ and (4.50) by $1 - \frac{t-q\delta}{\delta}$ for $q\delta < t < (q+1)\delta$, we obtain that $Z(\cdot)$ defined by (4.52) satisfies $-\dot{Z}(t) + \mathscr{L}(Z(t)) \le 0$, $t \in (0,\ h)$, $t \ne q\delta$, $0 \le q \le \bar{q} - 1$. If we take into account that $Z(q\delta) = Z(0)$ when $q = 0$ and $Z(q\delta) = Z(h)$ when $q = \bar{q}$, it follows that the system formed by (4.33b), (4.49), (4.50) forms a system of LMIs which provides via (4.52) a solution $(Z(\cdot),\ V)$ of (4.33).

Example 4.1 Consider the following JLSS:

$$dx(t) = A_0 x(t)dt + A_1 x(t)dw_1(t), \qquad kh < t \le (k+1)h, \tag{4.53a}$$

$$x(kh^+) = A_{d0}x(kh) + B_{d0}u_d(k) + w_{d1}(k)(A_{d1}x(kh) + B_{d1}u_d(k)),\ \ k \in \mathbb{Z}_+ \tag{4.53b}$$

with:

$$A_0 = \begin{bmatrix} 0.4855 & 0.8707 \\ 1.0260 & -0.3818 \end{bmatrix},\ A_1 = \begin{bmatrix} 0.4289 & -0.8999 \\ -0.2991 & 0.6347 \end{bmatrix},\ A_{d0} = \begin{bmatrix} 0.0675 & 0.2917 \\ -0.1871 & 0.9877 \end{bmatrix}$$

$$A_{d1} = \begin{bmatrix} 0.3929 & 0.2798 \\ 0.1946 & 0.0512 \end{bmatrix},\ B_{d0} = \begin{bmatrix} -0.7745 \\ 0.7868 \end{bmatrix},\ B_{d1} = \begin{bmatrix} 1.4089 \\ -0.5341 \end{bmatrix}.$$

The above example has been randomly generated for illustration purpose. We fix $h = 0.1$. We would like to verify the mean-square stabilizability by impulses of the JLSS (4.53). To this end, we apply Proposition 4.2 and we solve the LMI feasibility problem (4.30) by following the procedure described in Remark 4.3. From the numerical point of view, we use the LMI Control Toolbox for use with MATLAB®. We obtain a feasible solution given by:

$$Z = \begin{bmatrix} 31.4890 & 27.3674 \\ 27.3674 & 29.3405 \end{bmatrix},\ K = \begin{bmatrix} -20.4667 & -13.8667 \end{bmatrix} \tag{4.54}$$

confirming the mean-square stabilizability by impulses of the JLSS (4.53). The stabilizing feedback gain is then computed as:

$$F_d = \begin{bmatrix} -0.0574 & -0.2086 \end{bmatrix}. \tag{4.55}$$

Example 4.2 Let us consider the system described in Example 3.1 (an example that has been originally proposed in Briat [4]) with $B_{d0} = \begin{bmatrix} 1 \\ 2 \end{bmatrix}$ and $B_{d1} = \begin{bmatrix} 2 \\ 3 \end{bmatrix}$. Similarly to Example 3.1, we will show that in the case where the matrix A_1 is of the form $A_1 = \mu I_2$, the stabilizability (in mean square by impulses) condition for the JLSS (4.26) can be reformulated as an LMI feasibility problem involving a single matrix inequality. We will use to this end the equivalence $(i) \leftrightarrow (ii)$ from Proposition 4.2. Thus the system under consideration is mean-square stabilizable if and only if there exists a matrix F_d such that the forward LJLDE:

$$\begin{aligned} \dot{X}(t) &= A_0 X(t) + X(t) A_0^T + \mu^2 X(t), \qquad kh < t \le (k+1)h \\ X(kh^+) &= (A_{d0} + B_{d0} F_d) X(kh) (A_{d0} + B_{d0} F_d)^T + \\ &\quad + (\delta I_2 + B_{d1} F_d) X(kh) (\delta I_2 + B_{d1} F_d)^T + I_2, \quad k \in \mathbb{Z} \end{aligned} \tag{4.56}$$

has a unique bounded solution $\bar{X}(\cdot) : \mathbb{R} \to \mathcal{S}_2$ which is a periodic function of period h and it is uniform positive definite matrix. From the first equation of (4.56) we have that:

$$\bar{X}(t) = e^{(A_0 + \frac{1}{2}\mu^2 I_2)(t-kh)} \bar{X}((k-1)h^+) e^{(A_0 + \frac{1}{2}\mu^2 I_2)^T (t-kh)}$$

which leads to

$$\bar{X}(kh) = e^{(A_0 + \frac{1}{2}\mu^2 I_2)h} \tilde{X}((k-1)h^+) e^{(A_0 + \frac{1}{2}\mu^2 I_2)^T h}.$$

Substituting this equality in the second equation of (4.56) we get that the system under consideration is stabilizable in mean square if and only if there exists a matrix F_d with the property that the following LMI has a solution $\hat{X} > 0$:

$$\begin{aligned} &- \hat{X} + (A_{d0} + B_{d0} F_d) e^{(A_0 + \frac{1}{2}\mu^2 I_2)h} \hat{X} e^{(A_0 + \frac{1}{2}\mu^2 I_2)^T h} (A_{d0} + B_{d0} F_d)^T + \\ &(\delta I_2 + B_{d1} F_d) e^{(A_0 + \frac{1}{2}\mu^2 I_2)h} \hat{X} e^{(A_0 + \frac{1}{2}\mu^2 I_2)^T h} (\delta I_2 + B_{d1} F_d)^T < 0 \end{aligned} \tag{4.57}$$

or equivalent if and only if the following LMI has a solution $Z > 0$ and $K \in \mathbb{R}^{1 \times 2}$:

$$\begin{pmatrix} -Z & \mathcal{W}_1(Z, K) \\ \mathcal{W}_1^T(Z, K) & \mathcal{W}_2(Z) \end{pmatrix} < 0 \tag{4.58}$$

where: $\mathcal{W}_1(Z, K) = (A_{d0} e^{(A_0 + \frac{1}{2}\mu^2 I_2)h} Z + B_{d0} K \quad \delta e^{(A_0 + \frac{1}{2}\mu^2 I_2)h} Z + B_{d1} K)$, $\mathcal{W}_2$ $(Z) = -\operatorname{diag}\{Z, Z\}$. If (Z, K) is a solution of (4.58) then the control law

$$u_d(k) = K Z^{-1} e^{-(A_0 + \frac{1}{2}\mu^2 I_2)h} x(kh), \quad k \in \mathbb{Z}_+$$

stabilizes our system.

As a numerical application, we fix $\mu = 0.6$ and $\delta = 1.2$. For $h = 1.5$, we solve the LMI feasibility problem (4.58) and obtain the following stabilizing feedback gain:

$$F_d = \begin{bmatrix} -0.3556 & -0.5889 \end{bmatrix}. \tag{4.59}$$

The evolution of the corresponding closed-loop system state corresponding to a single realization of the stochastic processes is depicted in Fig. 4.1. We have also performed Monte-Carlo simulations (corresponding to different realizations of the stochastic

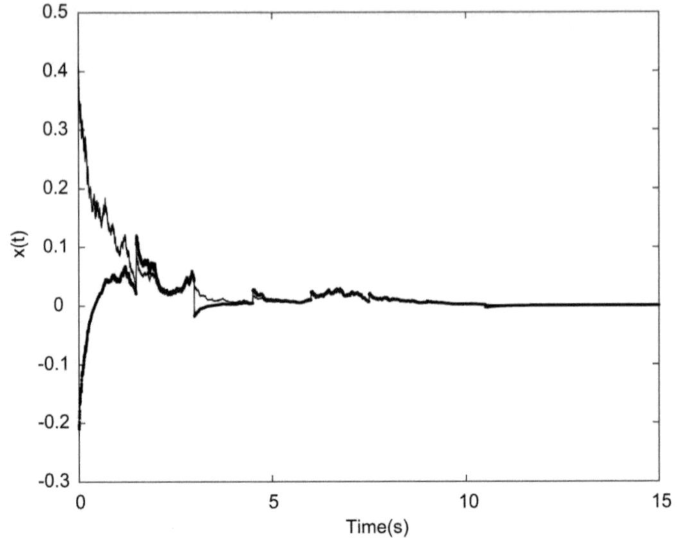

Fig. 4.1 Closed-loop state evolution for $h = 1.5$

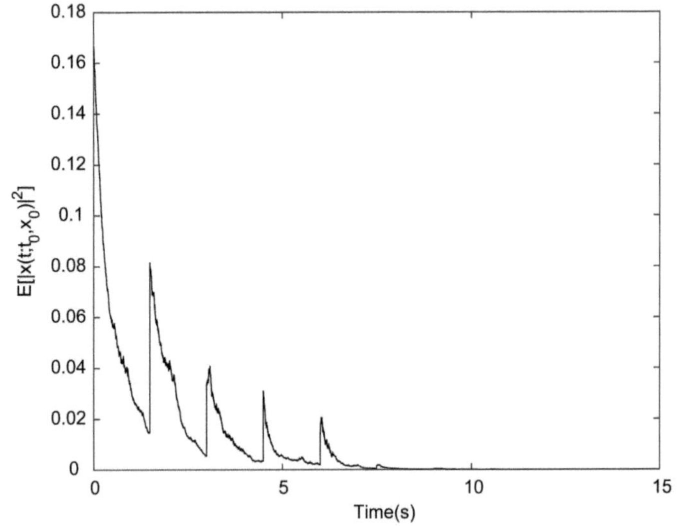

Fig. 4.2 Monte-Carlo simulation for $h = 1.5$

processes). We have considered 1000 realizations. The corresponding results are resumed by Fig. 4.2.

We now fix $h = 0.75$. For such value, the LMI (4.58) is unfeasible. By using the feedback gain (4.59), we performed again a Monte-Carlo simulation (with 1000 realizations), the results are shown in 4.3.

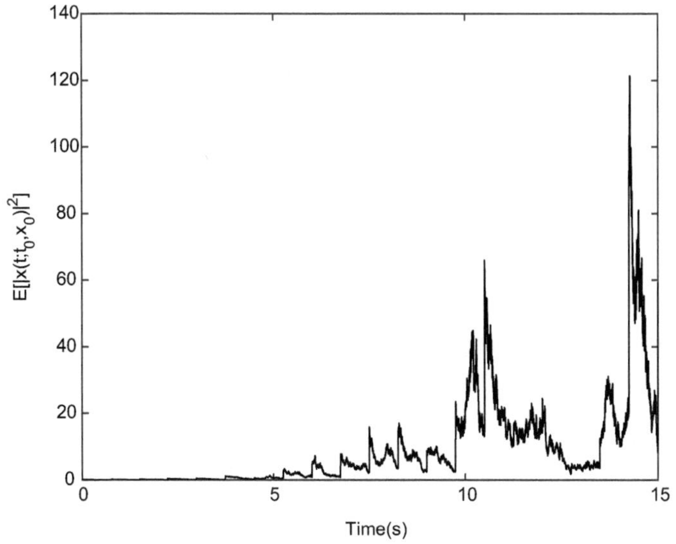

Fig. 4.3 Monte-Carlo simulation for $h = 0.75$

II. Further, we analyze the problem of stabilizability in mean square by impulses of the JLSS (4.10). According to the Definition 4.4 (a) and Remark 4.2 (b), the JLSS (4.10) is stabilizable in mean square by impulses if there exist the matrices $F_d(i) \in \mathbb{R}^{m_d \times n}$, $i \in \mathcal{N}$, with the property that the closed-loop JSLDE:

$$dx(t) = A_0(\eta(t))x(t)\mathrm{d}t + \sum_{l=1}^{r_1} A_l(\eta(t))x(t)\mathrm{d}w_l(t), \qquad kh < t \leq (k+1)h, \tag{4.60a}$$

$$\begin{aligned} x(kh^+) &= [A_{d0}(\eta(kh)) + B_{d0}(\eta(kh))F_d(\eta(kh))]x(kh) \\ &+ \sum_{l=1}^{r_2} w_{dl}(k)(A_{dl}(\eta(kh)) + B_{dl}(\eta(kh))F_d(\eta(kh)))x(kh), \quad k \in \mathbb{Z}_+ \end{aligned} \tag{4.60b}$$

is ESME. Invoking the equivalence $(i) \Leftrightarrow (vi)$ from Theorem 3.4 we deduce that the JSLDE (4.60) is ESMS if and only if the non-homogeneous forward GLJLDE:

$$\begin{aligned} \dot{X}_i(t) &= A_0(i)X_i(t) + X_i(t)A_0^\top(i) + \sum_{l=1}^{r_1} A_l(i)X_i(t)A_l^\top(i) + \sum_{j=1}^{N} q_{ji}X_j(t) + I_n, \\ &kh < t \leq (k+1)h, \end{aligned} \tag{4.61a}$$

$$X_i(kh^+) = \sum_{l=0}^{r_2}(A_{dl}(i) + B_{dl}(i)F_d(i))X_i(kh)(A_{dl}(i) + B_{dl}(i)F_d(i))^\top + I_n, \quad k \in \mathbb{Z} \tag{4.61b}$$

has a bounded solution $t \to \mathbf{X}(t) = (X_1(t),\ X_2(t), \ldots, X_N(t)) : \mathbb{R} \to \mathcal{S}_n^N$ which is a periodic function of period h and satisfies $X_i(t) > 0$, for all $(t, i) \in \mathbb{R} \times \mathcal{N}$.

Let $\mathbf{X} \to \mathcal{L}[\mathbf{X}] = (\mathcal{L}_1[\mathbf{X}],\ \mathcal{L}_2[\mathbf{X}], \ldots, \mathcal{L}_N[\mathbf{X}]) : \mathcal{S}_n^N \to \mathcal{S}_n^N$ be the linear operator defined as

$$\mathcal{L}_i[\mathbf{X}] \triangleq A_0(i)X_i + X_iA_0^\top(i) + \sum_{l=1}^{r_1} A_l(i)X_iA_l^\top(i) + \sum_{j=1}^{N} q_{ji}X_j, \tag{4.62}$$

$i \in \mathcal{N}$, for all $\mathbf{X} = (X_1,\ X_2, \ldots, X_N) \in \mathcal{S}_n^N$. With this notation the differential equation (4.61a) may be written in a compact form as

$$\dot{\mathbf{X}}(t) = \mathcal{L}[\mathbf{X}(t)] + \mathbb{I}^N, \quad kh < t \le (k+1)h \tag{4.63}$$

where we have denoted $\mathbb{I}^N = (I_N,\ I_N, \ldots, I_N) \in \mathcal{S}_n^N$. Now we can prove a result which provides necessary and sufficient conditions for the mean-square stabilizability by impulses of the JLSS of type (4.10).

Proposition 4.3 *Under the Assumption 1.2 the following are equivalent:*

(i) the JLSS (4.10) is mean-square stabilizable by impulses;

(ii) the forward non-homogeneous GLJLDE (4.61) has a unique bounded solution $t \to \mathbf{Z}(t) = (Z_1(t),\ Z_2(t), \ldots, Z_N(t)) : \mathbb{R} \to \mathcal{S}_n^N$ which is a periodic function of period h and satisfies $Z_i(t) > 0$, for all $(t, i) \in \mathbb{R} \times \mathcal{N}$;

(iii) there exist matrices $Z_i \in \mathcal{S}_n$, $K_i \in \mathbb{R}^{m_d \times n}$, $i \in \mathcal{N}$, solving the following system of LMIs:

$$\begin{pmatrix} -Z_i & \mathcal{W}_1(i, Z_1, Z_2, \ldots, Z_N, K_i) \\ \mathcal{W}_1^\top(i, Z_1, Z_2, \ldots, Z_N, K_i) & -\mathcal{W}_2(i, Z_1, \ldots, Z_N) \end{pmatrix} < 0 \tag{4.64}$$

$i \in \mathcal{N}$, where

$$\mathcal{W}_1(i, Z_1, Z_2, \ldots, Z_N, K_i) = (A_{d0}(i)e^{\mathcal{L}h}[\mathbf{Z}](i) + B_{d0}(i)K_i,$$
$$A_{d1}(i)e^{\mathcal{L}h}[\mathbf{Z}](i) + B_{d1}(i)K_i, \ldots, A_{dr_2}(i)e^{\mathcal{L}h}[\mathbf{Z}](i) + B_{dr_2}(i)K_i) \in \mathbb{R}^{n \times \hat{n}}$$

$$\mathcal{W}_2(i, Z_1, Z_2, \ldots, Z_N) = diag(e^{\mathcal{L}h}[\mathbf{Z}](i), e^{\mathcal{L}h}[\mathbf{Z}](i), \ldots) \in \mathcal{S}_{\hat{n}}, \quad \hat{n} := (r_2+1)n.$$

If (Z_i, K_i), $1 \le i \le N$ is a solution of the system of LMIs (4.64), then the feedback gain matrices $F_d(i)$ defined by

$$F_d(i) = K_i\left(e^{\mathcal{L}h}[\mathbf{Z}](i)\right)^{-1}, \quad 1 \le i \le N \tag{4.65}$$

achieve the exponential stability in mean square of the corresponding closed-loop JSLDE of type (4.60).

(iv) *There exist functions* $t \to Z(t,i) : [0,\ h] \to \mathcal{S}_n$ *and matrices* $V(i) \in \mathbb{R}^{m_d \times n}$, $i \in \mathcal{N}$, *satisfying the following system of Lyapunov-type linear matrix differential inequalities with jumps:*

$$-\dot{Z}(t,i) + \mathcal{L}_i[\mathbb{Z}(t)] \leq 0, \qquad 0 < t \leq h \tag{4.66a}$$

$$\begin{pmatrix} Z(0,i) - I_n & \hat{\mathcal{W}}_1(Z(h,i), V(i), i) \\ \hat{\mathcal{W}}_1^\top(Z(h,i), V(i), i) & \hat{\mathcal{W}}_2(Z(h,i)) \end{pmatrix} > 0 \tag{4.66b}$$

$i \in \mathcal{N}$, *where* $\mathcal{L}_i[\mathbb{Z}]$ *is defined in (4.62) and* $\hat{\mathcal{W}}_1(Z(h,i), V(i), i) \triangleq (A_{d0}(i)Z(h,i) + B_{d0}(i)V(i),\ \ldots,\ A_{dr_2}(i)Z(h,i) + B_{dr_2}(i)V(i)) \in \mathbb{R}^{n\times\hat{n}}$, $\hat{\mathcal{W}}_2(Z(h,i)) \triangleq \mathrm{diag}(Z(h,i)\ \ Z(h,i)\ \ \ldots\ \ Z(h,i)) \in \mathcal{S}_{\hat{n}}$ *with* $\hat{n} := (r_2+1)n$.

If $(Z(\cdot,i), V(i))$, $i \in \mathcal{N}$ *is a solution of (4.66), then the control law* $u_d(k) = V(\eta(kh))Z^{-1}(h,\eta(kh))x(kh)$, $k \in \mathbb{Z}_+$ *stabilizes (by impulses) the JLSS (4.1).*

Proof The equivalence $(i) \Leftrightarrow (ii)$ is obtained directly from the chain of implications $(i) \Rightarrow (v) \Rightarrow (vi) \Rightarrow (i)$ from Theorem 3.4 with $\tilde{H}(i) = \tilde{\Gamma}(i) = I_n$, $i \in \mathcal{N}$, applied in the case of the closed-loop JSLDE (4.60).

Let us prove now the implication $(ii) \Rightarrow (iii)$. Suppose that (ii) holds, let $t \to \mathbf{X}(t) = (X_1(t), X_2(t), \ldots, X_N(t)) : \mathbb{R} \to \mathcal{S}_n^N$ be the unique bounded positive definite and $h-$periodic solution of the non-homogeneous forward GLJLDE (4.61). From (4.63) we deduce that

$$X_i(0) = X_i(h) = e^{\mathcal{L}h}[\mathbf{X}(0^+)](i) + M_i(h), \quad i \in \mathcal{N}, \tag{4.67}$$

where $M_i(h) \in \mathcal{S}_n$ is the ith matrix component of $\int_0^h e^{\mathcal{L}(h-s)}[\mathbb{I}^N]ds \in \mathcal{S}_n^N$.

Substituting $X_i(0)$ in (4.61b) written for $k = 0$, we obtain:

$$X_i(0^+) = \sum_{l=0}^{r_2}(A_{dl}(i) + B_{dl}(i)F_d(i))e^{\mathcal{L}h}[\mathbf{X}(0^+)](i)(A_{dl}(i) + B_{dl}(i)F_d(i))^\top + H(i) \tag{4.68}$$

where we have denoted

$$H(i) := I_n + \sum_{l=1}^{r_2}(A_{dl}(i) + B_{dl}(i)F_d(i))M_i(h)(A_{dl}(i) + B_{dl}(i)F_d(i))^\top. \tag{4.69}$$

Further we consider the following splitting of the operator $\mathcal{L}$ defined in (4.62):

$$\mathcal{L}[\mathbf{X}] = \mathcal{L}_0[\mathbf{X}] + \mathcal{L}_1[\mathbf{X}] \tag{4.70}$$

with $\mathcal{L}_j[\mathbf{X}] = (\mathcal{L}_{j1}[\mathbf{X}], \mathcal{L}_{j2}[\mathbf{X}], \dots, \mathcal{L}_{jN}[\mathbf{X}]),\ j = 0, 1$ are described by

$$\mathcal{L}_{0i}[\mathbf{X}] = (A_0(i) + \frac{1}{2}q_{ii}I_n)X_i + X_i(A_0(i) + \frac{1}{2}q_{ii}I_n)^\top, \tag{4.71a}$$

$$\mathcal{L}_{1i}[\mathbf{X}] = \sum_{l=1}^{r_1} A_l(i)X_iA_l^\top(i) + \sum_{k=1,\ k\neq i}^{N} q_{ki}X_k \tag{4.71b}$$

for all $i \in \mathcal{N}$, $\mathbf{X} = (X_1, X_2, \dots, X_N) \in \mathcal{S}_n^N$. If we take into account that the decay rate q_{ki} are satisfying (1.22) we deduce that $\mathcal{L}_{1i}[\mathbf{X}] \geq 0$ for all $\mathbf{X} \in \mathcal{S}_n^{N+}$. Hence, (4.70) gives $\mathcal{L}_i[\mathbf{X}] \geq \mathcal{L}_{0i}[\mathbf{X}]$, for all $i \in \mathcal{N}$, $\mathbf{X} \in \mathcal{S}_n^{N+}$. Applying the time-invariant version of Theorem 2.2.5 from Drăgan et al. [1] we may infer that

$$e^{\mathcal{L}t}[\mathbf{X}](i) \geq e^{\mathcal{L}_0 t}[\mathbf{X}](i) = e^{(A_0(i)+\frac{1}{2}q_{ii}I_n)t}X_ie^{(A_0(i)+\frac{1}{2}q_{ii}I_n)^\top t} \tag{4.72}$$

for all $t \geq 0$, $i \in \mathcal{N}$, $\mathbf{X} = (X_1, X_2, \dots, X_N) \in \mathcal{S}_n^{N+}$. From (4.72) we deduce that

$$\int_0^h e^{\mathcal{L}(h-s)}[\mathbb{I}^N]ds \in \mathcal{S}_n^{N+}.$$

This allows us to conclude via (4.69) that $H(i) \geq I_n$, for all $i \in \mathcal{N}$. Thus we may deduce via (4.68)–(4.69) that $\mathbf{X}(0^+) = (X_1(0^+), X_2(0^+), \dots, X_N(0^+))$ solves the following system of LMIs:

$$\sum_{l=0}^{r_2}(A_{dl}(i) + B_{dl}(i)F_d(i))e^{\mathcal{L}h}[\mathbf{X}(0^+)](A_{dl}(i) + B_{dl}(i)F_d(i))^\top - X_i(0^+) < 0. \tag{4.73}$$

On the other hand, (4.61b) written for $k = 0$, gives $X_i(0^+) > 0$, $i \in \mathcal{N}$, because $X_i(0) > 0$, $1 \leq i \leq N$. In this case, (4.72) yields

$$e^{\mathcal{L}h}[\mathbf{X}(0^+)](i) > 0, \quad \text{for all } i \in \mathcal{N},$$

because $e^{(A_0(i)+\frac{1}{2}q_{ii}I_n)h}$ is an invertible matrix. Employing the Schur complement techniques we obtain that (4.73) is equivalent to

$$\begin{pmatrix} -X_i(0^+) & \mathcal{W}_1(i, X_1(0^+), \dots, X_N(0^+), \tilde{K}(i)) \\ \mathcal{W}_1^\top(i, X_1(0^+), \dots, X_N(0^+), \tilde{K}(i)) & -\mathcal{W}_2(i, X_1(0^+), \dots, X_N(0^+)) \end{pmatrix} < 0$$

where $\tilde{K}(i) := F_d(i)e^{\mathcal{L}h}[\mathbf{X}(0^+)](i)$. Hence, the pairs of matrices $(X_i(0^+), \tilde{K}(i))$, $i \in \mathcal{N}$ solve the system of LMIs (4.64). This means that (iii) holds if (ii) is true.

Let us prove now the implication $(iii) \Rightarrow (ii)$. If the pairs of matrices (Z_i, K_i), $i \in \mathcal{N}$ solve the system of LMIs (4.64) with $Z_i = Z_i^\top > 0$, $1 \leq i \leq N$, then the Schur complement techniques allows us to deduce that $\mathbf{Z} = (Z_1, Z_2, \dots, Z_N)$ satisfies

$$-Z_i + \sum_{l=0}^{r_2} (A_{dl}(i) + B_{dl}(i)F_d(i))e^{\mathscr{L}h}[\mathbf{Z}](i)(A_{dl}(i) + B_{dl}(i)F_d(i))^\top < 0, \ \ i \in \mathscr{N},$$

where $F_d(i)$ are defined as in (4.65). Let $\Gamma(i) \in \mathscr{S}_n$ be defined by

$$\Gamma(i) := Z_i - \sum_{l=1}^{r_2} (A_{dl}(i) + B_{dl}(i)F_d(i))e^{\mathscr{L}h}[\mathbf{Z}](i)(A_{dl}(i) + B_{dl}(i)F_d(i))^\top, \ \ i \in \mathscr{N}.$$

Therefore, $\mathbf{Z}$ solves the following linear equation on $\mathscr{S}_n^N$:

$$Z_i = \sum_{l=1}^{r_2} (A_{dl}(i) + B_{dl}(i)F_d(i))e^{\mathscr{L}h}[\mathbf{Z}](i)(A_{dl}(i) + B_{dl}(i)F_d(i))^\top + \Gamma(i), \tag{4.74}$$

with $\Gamma(i) > 0, \ 1 \leq i \leq N$. For each $i \in \mathscr{N}$, let $t \to X_i(t) : \mathbb{R} \to \mathscr{S}_n$ be defined by

$$X_i(t) := e^{\mathscr{L}(t-kh)}[\mathbf{Z}](i) \ \text{ if } \ t \in (kh, (k+1)h], \ \ k \in \mathbb{Z}. \tag{4.75}$$

One sees that

$$X_i(kh^+) := \lim_{\substack{t \to kh \\ t > kh}} X_i(t) = Z_i$$

and

$$X_i(kh) = e^{\mathscr{L}h}[\mathbf{Z}](i), \ \ \text{for all} \ \ i \in \mathscr{N}, \ \ k \in \mathbb{Z}.$$

Thus (4.74) becomes:

$$X_i(kh^+) = \sum_{l=1}^{r_2} (A_{dl}(i) + B_{dl}(i)F_d(i))X_i(kh)(A_{dl}(i) + B_{dl}(i)F_d(i))^\top + \Gamma(i), \tag{4.76}$$

for all $i \in \mathscr{N}$, $k \in \mathbb{Z}$. On the other hand (4.72) and (4.75) allows us to deduce that

$$X_i(t) \geq e^{(A_0(i)+\frac{1}{2}q_{ii}I_n)t} Z_i e^{(A_0(i)+\frac{1}{2}q_{ii}I_n)^\top t} > 0 \tag{4.77}$$

for all $kh < t \leq (k+1)h$, $k \in \mathbb{Z}$, $i \in \mathscr{N}$. Employing (4.75)–(4.77) we conclude that $\mathbf{X}(\cdot) = (X_1(\cdot), X_2(\cdot), \ldots, X_N(\cdot))$ is a positive definite and $h-$periodic solution of the following non-homogeneous forward GLJLDE on $\mathscr{S}_n^N$:

$$\dot{X}_i(t) = \mathscr{L}_i[\mathbf{X}(t)], \qquad kh < t \leq (k+1)h \tag{4.78a}$$

$$X_i(kh^+) = \sum_{l=0}^{r_2} (A_{dl}(i) + B_{dl}(i)F_d(i))X_i(kh)(A_{dl}(i) + B_{dl}(i)F_d(i))^\top + \Gamma(i),$$
$$i \in \mathscr{N}, \ \ k \in \mathbb{Z}. \tag{4.78b}$$

Invoking the implication $(vi) \Rightarrow (i)$ from Theorem 3.4 in the case of (4.78) we deduce that the JSLDE of type (4.60) corresponding to the feedback gains $F_d(i)$ defined via (4.65) is ESMS. Thus the proof is complete.

Remark 4.4 (a) If $\mathscr{L}[\cdot]$ is a linear operator introduced via (4.62), then $e^{\mathscr{L}h}[\mathbf{Z}] = (e^{\mathscr{L}h}[\mathbf{Z}](1), e^{\mathscr{L}h}[\mathbf{Z}](2), \ldots, e^{\mathscr{L}h}[\mathbf{Z}](N)) \in \mathscr{S}_n^N$ involved in (4.64) is defined by

$$e^{\mathscr{L}h}[\mathbf{Z}] = \sum_{k=0}^{\infty} \frac{h^k}{k!} \mathscr{L}^k[\mathbf{Z}] \tag{4.79}$$

where $\mathscr{L}^k[\mathbf{Z}] := \mathscr{L}[\mathscr{L}^{k-1}[\mathbf{Z}]]$, $k \geq 1$, with $\mathscr{L}^0[\mathbf{Z}] = \mathbf{Z}$. The ith component element of (4.79) is

$$e^{\mathscr{L}h}[\mathbf{Z}](i) = \sum_{k=0}^{\infty} \frac{h^k}{k!} \mathscr{L}^k[\mathbf{Z}](i) \tag{4.80}$$

and

$$\begin{aligned} \mathscr{L}^k[\mathbf{Z}](i) =& A_0(i)\mathscr{L}^{k-1}[\mathbf{Z}](i) + \mathscr{L}^{k-1}[\mathbf{Z}](i)A_0^\top(i) + \sum_{l=1}^{r_1} A_l(i)\mathscr{L}^{k-1}[\mathbf{Z}](i)A_l^\top(i) \\ &+ \sum_{j=1}^{N} q_{ji}\mathscr{L}^{k-1}[\mathbf{Z}](j), \quad k \geq 1, \end{aligned} \tag{4.81}$$

with $\mathscr{L}^0[\mathbf{Z}](i) = Z_i$, $i \in \mathscr{N}$.

(b) Any element $\mathbf{Z}$ of the vector space $\mathscr{S}_n^N$ may be represented as

$$\mathbf{Z} = \sum_{i=1}^{N} \sum_{\mu=1}^{n} \sum_{\nu=\mu}^{n} Z_{i\mu\nu} \boldsymbol{\Xi}^{i\mu\nu} \tag{4.82}$$

where $Z_{i\mu\nu} \in \mathbb{R}$ and $\boldsymbol{\Xi}^{i\mu\nu} \in \mathscr{S}_n^N$ being defined by $\boldsymbol{\Xi}^{i\mu\nu} = \left(\Xi_1^{i\mu\nu}, \Xi_2^{i\mu\nu}, \ldots, \Xi_N^{i\mu\nu}\right)$, with

$$\Xi_j^{i\mu\nu} = \begin{cases} 0, & \text{if } \ j \neq i \\ \Xi^{\mu\nu} & \text{if } \ j = i, \end{cases}$$

$\Xi^{\mu\nu}$ being the matrices described in (4.48). From (4.82) one sees that $\boldsymbol{\Xi}^{i\mu\nu}$, $1 \leq i \leq N$, $1 \leq \mu \leq \nu \leq n$ is a system of generators of the vector space $\mathscr{S}_n^N$. It can be proved that they are a basis of this vector space. Further, we may write:

$$e^{\mathscr{L}h}[\mathbf{Z}] = \sum_{i=1}^{N} \sum_{\mu=1}^{n} \sum_{\nu=\mu}^{n} Z_{i\mu\nu} \tilde{\boldsymbol{\Xi}}^{i\mu\nu} \tag{4.83a}$$

$$\tilde{\boldsymbol{\Xi}}^{i\mu\nu} := e^{\mathscr{L}h}[\boldsymbol{\Xi}^{i\mu\nu}], \quad 1 \leq i \leq N, \quad 1 \leq \mu \leq \nu \leq n. \tag{4.83b}$$

In conclusion, in order to solve the system of LMIs (4.64), we need first to compute $\tilde{\boldsymbol{\Xi}}^{i\mu\nu}$ via (4.83b).

To this end one uses a suitable truncation of the series (4.80) written successively for $\mathbf{Z}$ replaced by $\boldsymbol{\Xi}^{i\mu\nu}$, $1 \le i \le N$, $1 \le \mu \le \nu \le n$.

III. Let us now analyze the problem of stabilizability in mean square by state feedback of the JLSS (4.10). According to Definition 4.3 (a) and Remark 4.1 (b) the JLSS (4.10) is stabilizable in mean square by state feedback if for each $i \in \mathcal{N}$ there exists a piecewise continuous function which is periodic of period h, $t \to F_c(t,i) : \mathbb{R} \to \mathbb{R}^{m_c \times n}$ and a matrix $F_d(i) \in \mathbb{R}^{m_d \times n}$ with the property that the closed-loop JSLDE:

$$\begin{aligned} dx(t) &= (A_0(\eta(t)) + B_0(\eta(t))F_c(t,\eta(t)))x(t)dt \\ &\quad + \sum_{l=1}^{r_1}(A_l(\eta(t)) + B_l(\eta(t))F_c(t,\eta(t)))x(t)dw_l(t), \quad kh < t \le (k+1)h, \end{aligned} \tag{4.84a}$$

$$\begin{aligned} x(kh^+) &= [A_{d0}(\eta(kh)) + B_{d0}(\eta(kh))F_d(\eta(kh))]x(kh) \\ &\quad + \sum_{l=1}^{r_2} w_{dl}(k)(A_{dl}(\eta(kh)) + B_{dl}(\eta(kh))F_d(\eta(kh)))x(kh), \quad k \in \mathbb{Z}_+ \end{aligned} \tag{4.84b}$$

is ESMS. The JSLDE (4.84) is periodic in time in the sense of Definition 3.2 with $\theta_c = h$ and $\theta_d = 1$. Hence, $\frac{\theta_c}{\theta_d h} = 1$. In this case (4.9) yields $\tilde{\theta} = h$. It is worth mentioning that even if the JLSS (4.10) is time invariant, the corresponding closed-loop JSLDE (4.84) is in the periodic case. This means that criteria for stabilizability in mean square by state feedback may be obtained using criteria for exponential stability in mean square for a JSLDE with periodic coefficients in the sense of Definition 3.2. Thus we obtain:

Proposition 4.4 *Under the Assumption 1.2 the following are equivalent:*

(i) *the JLSS (4.10) is mean-square stabilizable by state feedback;*

(ii) *there exist the pairs of functions* $t \to (Z_i(t), K_i(t)) : [0,h] \to \mathcal{S}_n \times \mathbb{R}^{m_c \times n}$ *and the pairs of matrices* $(Z_{di}, K_{di}) \in \mathcal{S}_n \times \mathbb{R}^{m_d \times n}$, $i \in \mathcal{N}$, *with the properties:*

 (a) $Z_i(\cdot)$ *is differentiable on* $(0,h)$, *left continuous in* $t = h$ *and* $Z_i(t) > 0$, *for all* $t \in [0,h]$;

 (b) $K_i(\cdot)$ *is continuous on* $[0,h)$ *and satisfies* $K_i(0) = K_i(h)$;

 (c) $Z_{di} > 0$;

 (d) *the functions* $Z_i(\cdot)$, $K_i(\cdot)$ *and the matrices* Z_{di}, K_{di} *solve the following system of linear matrix differential inequalities (LMDIs):*

$$\begin{pmatrix} -\dot{Z}_i(t) + \Theta_0(i, \mathbf{Z}(t), K_i(t)) & \Theta_1(i, Z_i(t), K_i(t)) \\ \Theta_1^\top(i, Z_i(t), K_i(t)) & -\Theta_2(Z_i(t)) \end{pmatrix} < 0, \quad 0 \le t < h, \tag{4.85a}$$

$$\begin{pmatrix} -Z_{di} & \mathcal{W}_1(i, Z_i(h), K_{di}) \\ \mathcal{W}_1^\top(i, Z_i(h), K_{di}) & -\mathcal{W}_2(Z_i(h)) \end{pmatrix} < 0, \tag{4.85b}$$

$$Z_i(0^+) = Z_{di}, \quad i \in \mathcal{N}, \tag{4.85c}$$

where

$$\Theta_0(i, \mathbf{Z}(t), K_i(t)) := A_0(i)Z_i(t) + Z_i(t)A_0^\top(i) + B_0(i)K_i(t) + K_i^\top(t)B_0^\top(i) + \sum_{j=1}^{N} q_{ji}Z_j(t)$$

$$\Theta_1(i, Z_i(t), K_i(t)) := (A_1(i)Z_i(t) + B_1(i)K_i(t) \;\; A_2(i)Z_i(t) + B_2(i)K_i(t) \;\; \ldots \;\; A_{r_1}(i)Z_i(t) + B_{r_1}(i)K_i(t)) \in \mathbb{R}^{n\times\check{n}}$$

$$\Theta_2(Z_i(t)) := diag\{Z_i(t), Z_i(t), \ldots, Z_i(t)\} \in \mathcal{S}_{\check{n}} \;\; \check{n} := r_1 n$$

$$\mathcal{W}_1(i, Z_i(h), K_{di}) := (A_{d0}(i)Z_i(h) + B_{d0}(i)K_{di} \;\; A_{d1}(i)Z_i(h) + B_{d1}(i)K_{di} \;\; \ldots \; A_{dr_2}(i)Z_i(h) + B_{dr_2}(i)K_{di}) \in \mathbb{R}^{n\times\hat{n}}$$

$$\mathcal{W}_2(Z_i(h)) = diag\{Z_i(h), Z_i(h), \ldots, Z_i(h)\} \in \mathcal{S}_{\hat{n}} \;\; \hat{n} := (r_2 + 1)n.$$

If $(Z_i(\cdot), K_i(\cdot); Z_{di}, K_{di})$, $i \in \mathcal{N}$ *is a solution of the LMDI (4.85), then the pair of controls* $(u_c(t), u_d(k))$ *defined by*

$$u_c(t) = F_c(t, \eta(t))x(t) = K_{\eta(t)}(t - [\frac{t}{h}]h)Z_{\eta(t)}^{-1}(t - [\frac{t}{h}]h)x(t) \tag{4.86a}$$

$$u_d(k) = F_d(\eta(kh))x(kh) = K_d(\eta(kh))Z_{\eta(kh)}^{-1}(h)x(kh) \tag{4.86b}$$

$$k = \left[\frac{t}{h}\right] \tag{4.86c}$$

(where $\left[\frac{t}{h}\right]$ *denotes the integer part of the real number* $\frac{t}{h}$*) achieves the exponential stability in mean square of the corresponding closed-loop JSLDE (4.84).*

Proof Applying the equivalence $(i) \Leftrightarrow (vii)$ from Theorem 3.3 in the case of the closed-loop JSLDE (4.84) we deduce that this system is ESMS if and only if there exists a function $t \to \mathbf{Z}(t) = (Z_1(t), Z_2(t), \ldots, Z_N(t)) : \mathbb{R} \to \mathcal{S}_n^N$ with the properties:

(α) it is uniform positive on $\mathbb{R}$;
(β) it is a periodic function of period $\tilde{\theta} = h$;
(γ) it is a differentiable on each interval $(kh, (k+1)h)$ and is left continuous in each $t = kh$, $k \in \mathbb{Z}$;
(δ) it solves the following non-homogeneous forward differential inequality of Lyapunov type with jumps:

$$-\dot{Z}_i(t) + \mathscr{L}_{iF_c}(t)[\mathbf{Z}(t)] \le 0, \quad kh < t \le (k+1)h, \tag{4.87a}$$

$$\sum_{l=0}^{r_2}(A_{dl}(i)+B_{dl}(i)F_d(i))Z_i(kh)(A_{dl}+B_{dl}F_d(i))^\top - Z_i(kh^+) < 0, \, k \in \mathbb{Z}, \tag{4.87b}$$

$i \in \mathscr{N}$, where

$$\mathscr{L}_{iF_c}(t)[\mathbf{Z}(t)] = (A_0(i) + B_0(i)F_c(t,i))Z_i(t) + Z_i(t)(A_0(i) + B_0(i)F_c(t,i))^\top$$
$$+\sum_{l=1}^{r_1}(A_l(i) + B_l(i)F_c(t,i))Z_i(t)(A_l(i) + B_l(i)F_c(t,i))^\top + \sum_{j=1}^{N} q_{ji}Z_j(t).$$

Due to the periodicity property of the function $Z_i(\cdot)$, we may consider only the restriction of (4.87) to the interval $(0, h]$. Further, employing the Schur complement technique, we deduce that the restriction of (4.87) to $(0, h]$ is equivalent to (4.85). Thus the proof is complete.

Following a similar reasoning as in the proof of Proposition 4.4, we obtain:

Proposition 4.5 *Assume:*

(a) the Assumption 1.2 is fulfilled;
(b) the JLSS (4.1) is in the periodic case in the sense of the Definition 4.1.

Under these conditions the following are equivalent:

(i) the JLSS (4.1) is mean-square stabilizable by state feedback;
(ii) there exist functions $t \to (Z_i(t), K_i(t)) : [0, \tilde{\theta}] \to \mathscr{S}_n \times \mathbb{R}^{m_c \times n}$ and pairs of matrices $(Z_d(k,i), K_d(k,i)) \in \mathscr{S}_n \times \mathbb{R}^{m_d \times n}$, $0 \le k \le \tilde{\mathbf{k}}$, $i \in \mathscr{N}$, ($\tilde{\theta}$ and $\tilde{\mathbf{k}}$ being defined in (4.9)) having the properties:

(α) $Z_i(\cdot)$ are differentiable on each interval $(kh, (k+1)h)$, are left-continuous in each $t = kh$, $0 < k \le \tilde{\mathbf{k}}$ and $Z_i(t) > 0$, $t \in [0, \tilde{\theta}]$;
(β) $K_i(\cdot)$ are continuous on each interval $[kh, (k+1)h)$, $0 \le k \le \tilde{\mathbf{k}} - 1$ and satisfy $K_i(0) = K_i(\tilde{\theta})$;
(γ) $Z_d(k,i) > 0$, $(k,i) \in \{0, 1, \ldots, \tilde{\mathbf{k}} - 1\} \times \mathscr{N}$;
(δ) $Z_i(\cdot)$, $K_i(\cdot)$ and the matrices $Z_d(k,i)$, $K_d(t,i)$ solve the following LMDIs:

$$\begin{pmatrix} -\dot{Z}_i(t) + \tilde{\Theta}_0(t, i, \mathbf{Z}(t), K_i(t)) & \tilde{\Theta}_1(t, i, Z_i(t), K_i(t)) \\ \tilde{\Theta}_1^\top(t, i, Z_i(t), K_i(t)) & -\tilde{\Theta}_2(Z_i(t)) \end{pmatrix} < 0, \quad kh < t \le (k+1)h, \tag{4.88a}$$

$$\begin{pmatrix} -Z_d(k, i) & \tilde{\mathscr{W}}_1(k, i, Z_i(kh), K_d(k, i)) \\ \tilde{\mathscr{W}}_1^\top(k, i, Z_i(kh), K_d(k, i)) & -\tilde{\mathscr{W}}_2(Z_i(kh)) \end{pmatrix} < 0, \tag{4.88b}$$

$$Z_i(kh^+) = Z_d(k, i), \quad 0 \le k \le \tilde{\mathbf{k}} - 1, \quad i \in \mathscr{N}, \tag{4.88c}$$

where

$$\tilde{\Theta}_0(t, i, \mathbf{Z}(t), K_i(t)) := A_0(t, i)Z_i(t) + Z_i(t)A_0^\top(t, i) + B_0(t, i)K_i(t) + K_i^\top(t)B_0^\top(t, i) + \sum_{j=1}^{N} q_{ji}Z_j(t)$$

$$\tilde{\Theta}_1(t, i, Z_i(t), K_i(t)) := (A_1(t, i)Z_i(t) + B_1(t, i)K_i(t) \quad A_2(t, i)Z_i(t) + B_2(t, i)K_i(t) \ \ldots \ A_{r_1}(t, i)Z_i(t) + B_{r_1}(t, i)K_i(t)) \in \mathbb{R}^{n\times\check{n}} \tilde{\Theta}_2(Z_i(t))$$
$$:= diag\{Z_i(t), Z_i(t), \ldots, Z_i(t)\} \in \mathscr{S}_{\check{n}} \ \check{n}$$
$$= r_1 n \tilde{\mathscr{W}}_1(k, i, Z_i(kh), K_d(k, i))$$
$$:= (A_{d0}(kh, i)Z_i(kh) + B_{d0}(kh, i)K_d(k, i) \quad A_{d1}(kh, i)Z_i(kh) + B_{d1}(kh, i)K_d(k, i) \ldots A_{dr_2}(kh, i)Z_i(kh) + B_{dr_2}(kh, i)K_d(k, i)) \in \mathbb{R}^{n\times\hat{n}} \tilde{\mathscr{W}}_2(Z_i(kh))$$
$$= diag\{Z_i(kh), Z_i(kh), \ldots, Z_i(kh)\} \in \mathscr{S}_{\hat{n}}, \quad \hat{n} = (r_2 + 1)n.$$

Under the considered assumptions, if $(Z_i(\cdot), K_i(\cdot); Z_d(k, i), K_d(k, i))$, $0 \le k \le \tilde{\mathbf{k}}$, $i \in \mathscr{N}$ *is a solution of (4.85), then the pair of controls* $(u_c(t), u_d(k))$ *defined by*

$$u_c(t) = F_c(t, \eta(t))x(t) = K_{\eta(t)}(t - [\frac{t}{\tilde{\theta}}]\tilde{\theta})Z^{-1}_{\eta(t)}(t - [\frac{t}{\tilde{\theta}}]\tilde{\theta})x(t)$$

$$u_d(k) = K_d(k - \left[\frac{k}{\tilde{\mathbf{k}}}\right]\tilde{\mathbf{k}}, \eta(kh))Z^{-1}_{\eta(kh)}(kh - \left[\frac{k}{\tilde{\mathbf{k}}}\right]\tilde{\mathbf{k}}h)x(kh)$$

achieves the exponential stability in mean square of the corresponding closed-loop JSLDE (4.18).

4.3 Mean-Square Stabilizability of Linear Stochastic Sampled-Data Controlled Systems

Let us consider the linear stochastic controlled system:

$$\begin{aligned} dx(t) =&(A_0(t, \eta(t))x(t) + B_0(t, \eta(t))u(t))dt \\ &+ \sum_{l=1}^{r_1}(A_l(t, \eta(t))x(t) + B_l(t, \eta(t))u(t))dw_l(t), \quad t \in \mathbb{R}_+ \end{aligned} \tag{4.89}$$

where $x(t) \in \mathbb{R}^n$ is the vector of the state parameters and $u(t) \in \mathbb{R}^m$ is the vector of the control at instance time t. In (4.89), $\{w(t)\}_{t\geq 0}$ ($w(t) = (w_1(t)\ w_2(t)\ \dots\ w_{r_1}(t))^\top$) is an r_1-dimensional standard Wiener process and $\{\eta(t)\}_{t\geq 0}$ is a standard homogeneous Markov process with the set of the states $\mathcal{N} = \{1, 2, \dots, N\}$. The stochastic processes $\{w(t)\}_{t\geq 0}$ and $\{\eta(t)\}_{t\geq 0}$ are defined on a given probability space $(\Omega, \mathfrak{F}, \mathcal{P})$ and are assumed to be independent stochastic processes. According to Definition 4.1.1 in Drăgan et al. [1], the controlled system (4.89) is stochastic stabilizable if there exist continuous matrix-valued functions $t \to F(t, i) : \mathbb{R}_+ \to \mathbb{R}^{m\times n}$, $i \in \mathcal{N}$, with the property that the closed-loop system

$$\begin{aligned} dx(t) =&(A_0(t, \eta(t)) + B_0(t, \eta(t))F(t, \eta(t)))x(t)dt \\ &+ \sum_{l=1}^{r_1}(A_l(t, \eta(t)) + B_l(t, \eta(t))F(t, \eta(t)))x(t)dw_l(t) \end{aligned} \tag{4.90}$$

is exponentially stable in mean square with conditioning (ESMS-C). This means that its trajectories $x(t; t_0, x_0)$ satisfy

$$\mathbb{E}[|x(t; t_0, x_0)|^2 \mid \eta(t_0) = i] \leq \beta e^{-\alpha(t-t_0)}|x_0|^2, \tag{4.91}$$

for all $t \geq t_0 \geq 0$, $x_0 \in \mathbb{R}^n$, $i \in \mathcal{N}$. In (4.91), $\alpha > 0$, $\beta \geq 1$ are constants that do not depend upon $t_0 \in \mathbb{R}_+$, $x_0 \in \mathbb{R}^n$ and neither the initial probability distributions of the Markov process. The closed-loop system (4.90) is obtained by coupling the controlled system (4.89) with a control law in a state feedback form:

$$u(t) = F(t, \eta(t))x(t). \tag{4.92}$$

One sees that to implement the control law (4.92) it is necessary to know the values of the state parameters $x(t)$ at each time instant $t \geq 0$.

In this section, we consider the case when the states $x(t)$ of the system (4.89) are available for measurements only at some time instances τ_k, $k \in \mathbb{Z}_+$. In this case, the control takes constant values on the time interval between two time instances τ_k where the measurements of the states are possible, that is

$$u(t) = u(k), \quad \tau_k < t \leq \tau_{k+1}, \quad k \in \mathbb{Z}_+. \tag{4.93}$$

Substituting (4.93) in (4.89) yields:

$$\begin{aligned} dx(t) =& (A_0(t, \eta(t))x(t) + B_0(t, \eta(t))u(k))dt \\ &+ \sum_{l=1}^{r_1} (A_l(t, \eta(t))x(t) + B_l(t, \eta(t))u(k))dw_l(t), \quad \tau_k < t \leq \tau_{k+1}, \quad t \in \mathbb{R}_+. \end{aligned} \tag{4.94}$$

Proceeding as in the case of the system (1.31) (with $G_j(\cdot, i) = 0, i \in \mathcal{N}$) one shows that if $x(t)$ is a solution of (4.94) and if $\varphi(\cdot)$ is defined by $\varphi(t) = u(k), \quad \tau_k < t \leq \tau_{k+1}, \quad k \in \mathbb{Z}_+$, then $\mathsf{x}(t) \stackrel{\Delta}{=} (x^\top(t) \quad \varphi^\top(t))^\top$ solves the following ICLSS:

$$d\mathsf{x}(t) = \mathscr{A}_0(t, \eta(t))\mathsf{x}(t)dt + \sum_{l=1}^{r_1} \mathscr{A}_l(t, \eta(t))\mathsf{x}(t)dw_l(t), \quad \tau_k < t \leq \tau_{k+1}, \tag{4.95a}$$

$$\mathsf{x}(\tau_k^+) = \mathscr{A}_d\mathsf{x}(\tau_k) + \mathscr{B}_d u(k), \quad k \in \mathbb{Z}_+, \tag{4.95b}$$

where

$$\mathscr{A}_l(t, i) := \begin{pmatrix} A_l(t, i) & B_l(t, i) \\ 0 & 0 \end{pmatrix} \in \mathbb{R}^{(n+m)\times(n+m)}, 0 \leq l \leq r_1, \tag{4.96a}$$

$$\mathscr{A}_d = \begin{pmatrix} I_n & 0 \\ 0 & 0 \end{pmatrix} \in \mathbb{R}^{(n+m)\times(n+m)}, \tag{4.96b}$$

$$\mathscr{B}_d = \begin{pmatrix} 0 \\ I_m \end{pmatrix} \in \mathbb{R}^{(n+m)\times m}. \tag{4.96c}$$

Conversely, if $\mathsf{x}(t)$ is a solution of ICLSS (4.95)–(4.96) then $x(t)$ formed by first n scalar components of $\mathsf{x}(t)$ is a solution of (4.94).

If the ICLSS (4.95) is mean-square stabilizable by impulses in a state feedback form, then there exist bounded matrix-valued sequences $\{F_d(\tau_k, i)\}_{k\in\mathbb{Z}_+} \subset \mathbb{R}^{m\times(n+m)}$, $i \in \mathcal{N}$, with the property that the closed-loop system:

$$d\mathsf{x}(t) = \mathscr{A}_0(t, \eta(t))\mathsf{x}(t)dt + \sum_{l=1}^{r_1} \mathscr{A}_l(t, \eta(t))\mathsf{x}(t)dw_l(t), \quad \tau_k < t \leq \tau_{k+1}, \tag{4.97a}$$

$$\mathsf{x}(\tau_k^+) = (\mathscr{A}_d + \mathscr{B}_d F_d(\tau_k, \eta(\tau_k)))\mathsf{x}(\tau_k), \quad k \in \mathbb{Z}_+, \tag{4.97b}$$

is SESMS. Based on Definition 3.1 (a) and the Proposition 3.2 (i), we deduce that in this case there exist two constants $\beta \geq 1, \alpha > 0$ such that the trajectories of (4.97) satisfy

$$\mathbb{E}[|\mathsf{x}(t)|^2 \mid \eta(t_0) = i] \leq \beta e^{-\alpha(t-t_0)}\mathbb{E}[|\mathsf{x}(t_0)|^2 \mid \eta(t_0) = i] \tag{4.98}$$

for all $t \geq t_0 \geq 0$, $i \in \mathcal{N}$ and any initial probability distribution of the Markov process. If we take into account the structure of the coefficients of (4.97) given in (4.96) we deduce that

$$\begin{aligned} \dot{\varphi}(t) &= 0, \quad \tau_k < t \leq \tau_{k+1} \\ \varphi(\tau_k^+) &= F_k(\tau_k, \eta(\tau_k))\mathbf{x}(\tau_k) \end{aligned} \tag{4.99}$$

which leads to $\varphi(t) = F_d(\tau_k, \eta(\tau_k))\mathbf{x}(\tau_k), \quad \tau_k \leq t \leq \tau_{k+1}$.

Let $(F_{d1} \quad F_{d2}(\tau_k, i))$ be the partition of the matrix $F_d(\tau_k, i)$ such that $F_{d1}(\tau_k, i) \in \mathbb{R}^{m\times n}$ and $F_{d2}(\tau_k, i) \in \mathbb{R}^{m\times m}$. Thus, (4.99) becomes:

$$\varphi(t) = F_{d1}(\tau_k, \eta(\tau_k))x(\tau_k) + F_{d2}(\tau_k, \eta(\tau_k))\varphi(\tau_k), \quad \tau_k \leq t \leq \tau_{k+1}. \tag{4.100}$$

Further, (4.97a) and (4.100) give:

$$\begin{aligned} dx(t) &= [A_0(t, \eta(t))x(t) + B_0(t, \eta(t))(F_{d1}(\tau_k, \eta(\tau_k))x(\tau_k) + F_{d2}(\tau_k, \eta(\tau_k))\varphi(\tau_k))]dt \\ &+ \sum_{l=1}^{r_1}[A_l(t, \eta(t))x(t) + B_l(t, \eta(t))(F_{d1}(\tau_k, \eta(\tau_k))x(\tau_k) + F_{d2}(\tau_k, \eta(\tau_k))\varphi(\tau_k))]dw_l(t), \\ &\quad \tau_k < t \leq \tau_{k+1}, \qquad (4.101a) \\ \varphi(\tau_{k+1}) &= F_{d1}(\tau_k, \eta(\tau_k))x(\tau_k) + F_{d2}(\tau_k, \eta(\tau_k))\varphi(\tau_k), \quad k \geq k_0, \qquad (4.101b) \end{aligned}$$

where k_0 is such that $\tau_{k_0-1} < t_0 \leq \tau_{k_0}$. If we consider that $\mathbf{x}(t) = (x^\top(t) \quad \varphi^\top(t))^\top$ we deduce from (4.98) that

$$\mathbb{E}[|x(t; t_0, x_0)|^2 \mid \eta(t_0) = i] \leq \beta e^{-\alpha(t-t_0)}(|x_0|^2 + |\varphi_0|^2) \tag{4.102}$$

for all $t \geq t_0 \geq 0$, $x_0 \in \mathbb{R}^n$, $\varphi_0 \in \mathbb{R}^m$, $i \in \mathcal{N}$ and any initial probability distribution of the Markov process.

Now we are in position to introduce the following definition:

Definition 4.5 We say that the linear stochastic controlled system (4.89) is **mean-square stabilizable by sampled state measurements** if there exist sequences of matrices $\{F_{d1}(\tau_k, i)\}_{k\in\mathbb{Z}_+} \subset \mathbb{R}^{m\times n}$ and $\{F_{d2}(\tau_k, i)\}_{k\in\mathbb{Z}_+} \subset \mathbb{R}^{m\times m}$, $i \in \mathcal{N}$ with the property that the closed-loop system (4.101) is ESMS-C.

Remark 4.5 From (4.101b) one may obtain recursively that the vectors $\varphi(\tau_k)$ involved in the formula of the stabilizing control

$$u(t) = F_{d1}(\tau_k, \eta(\tau_k))x(\tau_k) + F_{d2}(\tau_k, \eta(\tau_k))\varphi(\tau_k), \quad \tau_k < t \leq \tau_{k+1} \tag{4.103}$$

depend upon the measured values $x(\tau_j)$, $k_0 \leq j \leq k-1$ of the states $x(\cdot)$ and the initial value $\varphi(t_0) = \varphi_0 \in \mathbb{R}^m$. If there are no other considerations that require a certain value of φ_0, in applications one may take $\varphi_0 = 0$.

From the previous developments, one sees that the mean-square stabilizability by impulses in a state feedback form of the ICLSS (4.95)–(4.96) is a sufficient condition for the mean-square stabilizability of the linear stochastic controlled system (4.89) by sampled state measurements. Hence, the result obtained in Proposition 4.3 specialized to the case of the time-invariant version of (4.95)–(4.96) provides the matrices $F_{d1} \in \mathbb{R}^{m\times n}$ and $F_{d2} \in \mathbb{R}^{m\times m}$, $i \in \mathcal{N}$, with the property that the control law

$$\begin{aligned}
&u(t) = F_{d1}(\eta(\tau_k))x(\tau_k) + F_{d2}(\eta(\tau_k))\varphi(\tau_k), \quad \tau_k < t \le \tau_{k+1}\\
&\varphi(\tau_{k+1}) = F_{d1}(\eta(\tau_k))x(\tau_k) + F_{d2}(\eta(\tau_k))\varphi(\tau_k), \quad k \ge k_0,\\
&\varphi(\tau_{k_0}) = \varphi_0
\end{aligned}$$

stabilizes the closed-loop system (4.97) with $A_l(t,i) = A_l(i)$, $B_l(t,i) = B_l(i)$, for all $(t,i) \in \mathbb{R}_+ \times \mathcal{N}$.

Example 4.3 The following example is adapted from Costa et al. (2013). This example addresses the control of coupled electrical machines. The authors in Costa et al. (2013) treated such a problem in the continuous-time case. In what follows, we will consider the sampled-data stabilization problem. A detailed description of the system under consideration can be found in Costa et al. (2013). The involved Markov process has N = 4 states. The system matrices describing its dynamics are given below:

$$A_0(1) = \begin{bmatrix} 0 & 1 & 0 & 0 \\ -\omega_1^2 - \varepsilon & 0 & -\varepsilon & 0 \\ 0 & 0 & 0 & 1 \\ -\varepsilon & 0 & -\omega_2^2 - \varepsilon & -2\alpha_2 \end{bmatrix}$$

$$A_0(2) = \begin{bmatrix} 0 & 1 & 0 & 0 \\ -\omega_1^2 - \varepsilon & 0 & \varepsilon & 0 \\ 0 & 0 & 0 & 1 \\ \varepsilon & 0 & -\omega_2^2 - \varepsilon & -2\alpha_2 \end{bmatrix}$$

$$A_0(3) = \begin{bmatrix} 0 & 1 & 0 & 0 \\ -\omega_1^2 + \varepsilon & 0 & -\varepsilon & 0 \\ 0 & 0 & 0 & 1 \\ -\varepsilon & 0 & -\omega_2^2 + \varepsilon & -2\alpha_2 \end{bmatrix}$$

$$A_0(4) = \begin{bmatrix} 0 & 1 & 0 & 0 \\ -\omega_1^2 + \varepsilon & 0 & \varepsilon & 0 \\ 0 & 0 & 0 & 1 \\ \varepsilon & 0 & -\omega_2^2 + \varepsilon & -2\alpha_2 \end{bmatrix}$$

where:

- $\varepsilon = 0.1$: tuning parameter for the amplitude of the adverse coupling mechanism;
- $\omega_1 = 1$: natural frequency of the first machine;
- $\omega_2 = 10$: natural frequency of the second machine;
- $\alpha_1 = 1$: damping coefficient of the first machine;
- $\alpha_2 = 1$: damping coefficient of the second machine;

The transition rate matrix is given by:

$$\Pi = \begin{bmatrix} -(\eta_1+\eta_2) & \eta_2 & \eta_1 & 0 \\ \eta_2 & -(\eta_1+\eta_2) & 0 & \eta_1 \\ \eta_1 & 0 & -(\eta_1+\eta_2) & \eta_2 \\ 0 & \eta_1 & \eta_2 & -(\eta_1+\eta_2) \end{bmatrix}$$

where $\eta_1 = 2.5$ and $\eta_2 = 0.5$.
For the first test, we consider the case without state/control-dependent noises and fix $h = 0.25$. This corresponds to the case addressed in Geromel and Gabriel [5]. The LMI (4.66) is found to be feasible, which is in adequacy with the results from Geromel and Gabriel [5]. The obtained stabilizing feedback gains are given as follows:

$$F_d(1) = \begin{bmatrix} -0.1201 & -0.9949 & 0.4499 & -0.0130 & 0.0006 \end{bmatrix}$$
$$F_d(2) = \begin{bmatrix} -0.1214 & -0.9953 & 0.4503 & -0.0150 & 0.0006 \end{bmatrix}$$
$$F_d(3) = \begin{bmatrix} -0.1471 & -1.0111 & 0.4507 & -0.0130 & 0.0007 \end{bmatrix}$$
$$F_d(4) = \begin{bmatrix} -0.1485 & -1.0115 & 0.4511 & -0.0150 & 0.0007 \end{bmatrix}.$$

Notice that the gains are almost the same which may indicate the existence of a mode independent solution. The same conclusion has been reported in Geromel and Gabriel [5].

The Fig. 4.4 shows the evolution of the closed-loop system states corresponding to a single realization of the Markovian process η_t. While Fig. 4.5 shows the sample path of the Markovian process η_t.

As a second scenario, we consider the case with state and control dependent noises. We have randomly generated the matrices $A_1(j)$ and fix $B_1(j) = 0.1 \times B_0(j)$, $j = 1, \ldots, N$. For $h = 0.1$, the LMI (4.66) is feasible. The evolution of the closed-loop system state corresponding to a single realization of the different stochastic processes is depicted in Fig. 4.6. Figure 4.7 shows the sample path of the Markovian process η_t. We have also performed a Monte-Carlo simulation (corresponding to different realizations of the stochastic processes). The corresponding results are resumed by Fig. 4.8.

In the case when the system (4.89) is time periodic, that is, there exists $\theta_c > 0$ such that $A_l(t+\theta_c, i) = A_l(t, i)$, $B_l(t+\theta_c, i) = B_l(t, i)$, for all $(t, i) \in \mathbb{R}_+ \times \mathcal{N}$, one may choose the sampling period $h > 0$ such that $\frac{\theta_c}{h}$ be a rational number. In this case, the ICLSS (4.95)–(4.96) is time periodic in the sense of the Definition 4.1.

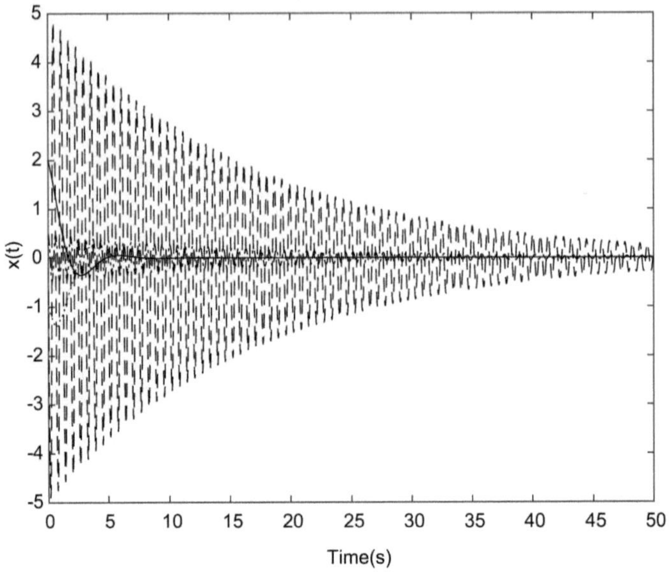

Fig. 4.4 Closed-loop states evolution

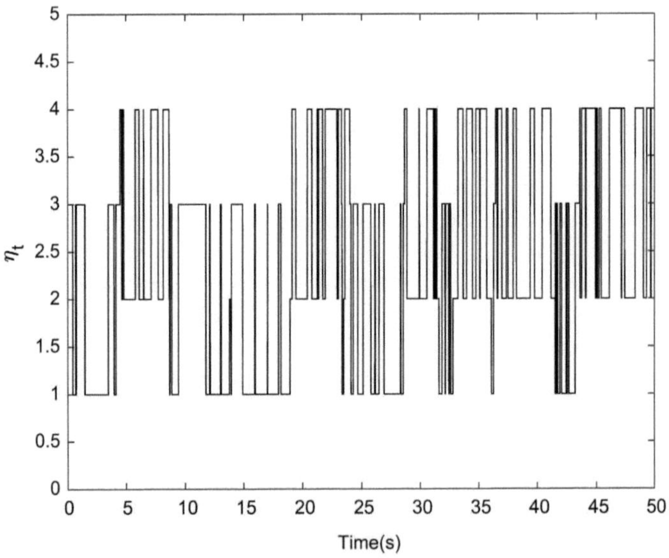

Fig. 4.5 Sample path of η_t

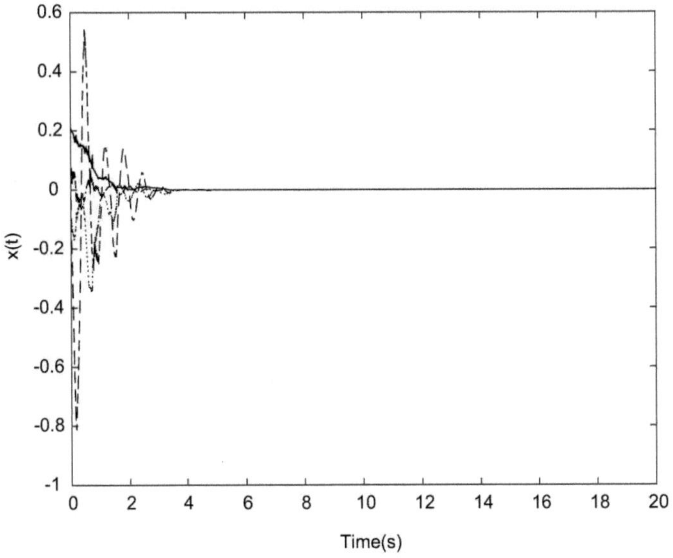

Fig. 4.6 Closed-loop states evolution

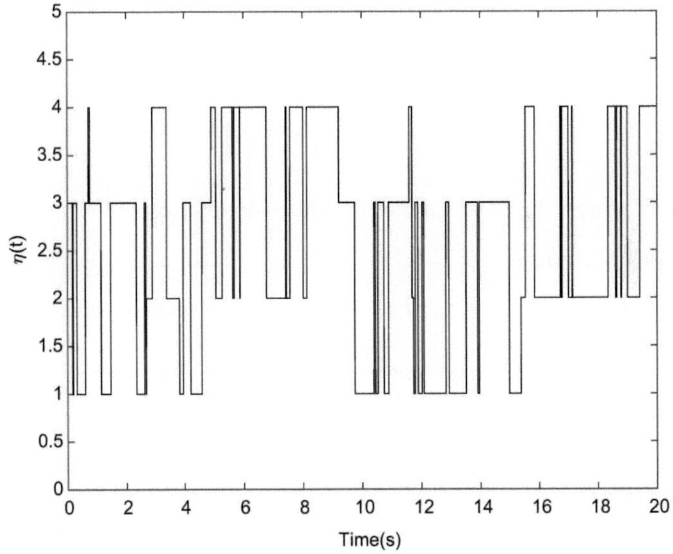

Fig. 4.7 Sample path of η_t

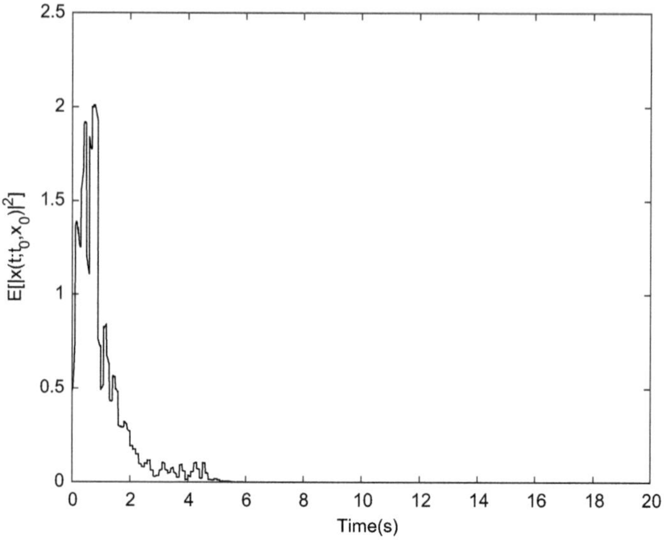

Fig. 4.8 Monte-Carlo simulation for $h = 0.1$

If $\tilde{p}$, $\tilde{q}$ are two coprime natural numbers such that $\frac{\theta_c}{h} = \frac{\tilde{p}}{\tilde{q}}$ we set:

$$\tilde{\theta} := \tilde{q}\theta_c \tag{4.104a}$$

$$\tilde{\mathbf{k}} := \tilde{p}. \tag{4.104b}$$

Employing Theorem 3.3 in the case of the closed-loop system (4.97) we have:

Proposition 4.6 *Assume:*

(a) $A_l(\cdot, i)$, $B_l(\cdot, i)$, $0 \le l \le r_1$, $i \in \mathcal{N}$ *are periodic matrix-valued functions of period* $\theta_c > 0$;

(b) $\tau_k = kh$, *with* $k > 0$ *is such that* $\frac{\theta_c}{h}$ *is a rational number.*

Under these conditions, the following are equivalent:

(i) the ICLSS (4.95)–(4.96) is mean-square stabilizable by impulses in a linear feedback form;

(ii) there exist functions $t \to Z_i(t) : [0, \tilde{\theta}] \to \mathcal{S}_{n+m}$ *and pairs of matrices* $(Z_d(k, i), K_d(k, i)) \in \mathcal{S}_{n+m} \times \mathbb{R}^{m\times(n+m)}$, *for all* $(k, i) \in \{0, 1, \ldots, \tilde{\mathbf{k}} - 1\} \times \mathcal{N}$ *having the properties:*

(α) $Z_i(\cdot)$ *are differentiable on each interval* $(kh, (k+1)h)$ *and are left continuous in each* $kh \in (0, \tilde{\theta}]$;

(β) $Z_i(t) \succ 0$, *for all* $t \in [0, \tilde{\theta}]$ *and* $Z_d(k, i) \succ 0$, *for all* $0 \le k \le \tilde{\mathbf{k}} - 1$, $i \in \mathcal{N}$;

(γ) $(Z_i(\cdot), (Z_d(k,i), K_d(k,i)))$, $0 \le k \le \tilde{\mathbf{k}} - 1$, $i \in \mathcal{N}$, *solve the following system of linear differential inequalities with jumps*

$$
\begin{aligned}
& -\dot{Z}_i(t) + \mathscr{A}_0(t,i)Z_i(t) + Z_i(t)\mathscr{A}_0^\top(t,i) + \sum_{l=1}^{r_1} \mathscr{A}_l(t,i)Z_i(t)\mathscr{A}_l^\top(t,i) \\
& + \sum_{j=1}^{N} q_{ji} Z_j(t) \preceq 0, \quad kh < t \le (k+1)h, \qquad (4.105a) \\
& Z_i(0) = Z_i(\tilde{\theta}h), \qquad (4.105b) \\
& Z_i(kh^+) = Z_d(k,i), \qquad (4.105c) \\
& \begin{pmatrix} -Z_d(k,i) & \mathscr{A}_d Z_i(kh) + \mathscr{B}_d K_d(k,i) \\ * & -Z_i(kh) \end{pmatrix} \prec 0, \ 0 \le k \le \tilde{\mathbf{k}} - 1. \qquad (4.105d)
\end{aligned}
$$

If $(Z_i(\cdot), (Z_d(k,i), K_d(k,i)))$, $0 \le k \le \tilde{\mathbf{k}} - 1$, $i \in \mathcal{N}$ *is a solution of (4.105), then the matrices*

$$
F_d(kh,i) := K_d(k,i)Z_i^{-1}(kh), \quad (k,i) \in \{0,1,\ldots,\tilde{\mathbf{k}}-1\} \times \mathcal{N} \qquad (4.106)
$$

have the property that the control law

$$
u(k) = F_d((k - [\frac{k}{\tilde{\mathbf{k}}}]\tilde{\mathbf{k}})h, \eta(kh))\mathbf{x}(kh)
$$

achieves the mean-square stability of the ICLSS (4.95)–(4.96).

Proof The implication $(i) \Rightarrow (ii)$ can be proved applying the implication $(i) \Rightarrow (v)$ from Theorem 3.3, in the case of the closed-loop system (4.97) together with the Schur complement techniques. Conversely, if (4.105) is feasible, one obtains that

$$
X_i(t) := Z_i(t - [\frac{t}{\tilde{\theta}}]\tilde{\theta}), \quad i \in \mathcal{N}
$$

solve a system of linear matrix differential inequalities with jumps of the form:

$$
\begin{aligned}
& \dot{X}_i(t) \succeq \mathscr{A}_0(t,i)X_i(t) + X_i(t)\mathscr{A}_0^\top(t,i) + \sum_{l=1}^{r_1} \mathscr{A}_l(t,i)X_i(t)\mathscr{A}_l^\top(t,i) \\
& \quad + \sum_{j=1}^{N} q_{ji} X_j(t), \quad kh < t \le (k+1)h, \qquad (4.107a) \\
& (\mathscr{A}_d + \mathscr{B}_d F_d(kh,i))X_i(kh)(\mathscr{A}_d + \mathscr{B}_d F_d(kh,i))^\top - X_i(kh^+) \prec 0, \qquad (4.107b)
\end{aligned}
$$

$F(kh, i)$ being computed via (4.106). Invoking the implication $(vii) \Rightarrow (i)$ from Theorem 3.3 in the case of (4.107) we may conclude that the closed-loop system (4.97) is ESMS. Thus the proof is complete.

4.4 Mean-Square Detectability of a Jump Linear Stochastic System

In this section we define the concept of mean-square detectability of a jump linear stochastic system and we provide methods which allow us to decide if a given jump linear stochastic system is detectable or not.

4.4.1 Model Description. Basic Definitions

Let us consider the uncontrolled jump linear stochastic system (JLSS) having the evolution described by

$$\mathrm{d}x(t) = A_0(t, \eta(t))x(t)\mathrm{d}t + \sum_{l=1}^{r_1} A_l(t, \eta(t))x(t)\mathrm{d}w_l(t), \quad \tau_k < t \leq \tau_{k+1}, \tag{4.108a}$$

$$x(\tau_k^+) = A_{d0}(\tau_k, \eta(\tau_k))x(\tau_k) + \sum_{l=1}^{r_2} w_{dl}(k)A_{dl}(\tau_k, \eta(\tau_k))x(\tau_k). \tag{4.108b}$$

Together with the outputs

$$\mathrm{d}x_c(t) = C_{c0}(t, \eta(t))x(t)\mathrm{d}t + \sum_{l=1}^{r_1} C_{cl}(t, \eta(t))x(t)\mathrm{d}w_l(t), \quad t \in \mathbb{R}_+ \tag{4.109a}$$

$$z_d(\tau_k) = C_{d0}(\tau_k, \eta(\tau_k))x(\tau_k) + \sum_{l=1}^{r_2} w_{dl}(k)C_{dl}(\tau_k, \eta(\tau_k))x(\tau_k), \quad k \in \mathbb{Z}_+ \tag{4.109b}$$

where $\{w(t)\}_{t\geq 0}$ $(w(t) = (w_1(t), \ldots, w_{r_1}(t))^\top)$, $\{w_d(k)\}_{k\in\mathbb{Z}_+}$ $(w_d(k) = (w_{d1}(k), w_{d2}(k), \ldots, w_{dr_2}(k))^\top)$ and $\{\eta(t)\}_{t\geq 0}$ are the stochastic processes introduced in Chap. 1 and satisfy the Assumption 1.2. In (4.108) and (4.109), $x(t) \in \mathbb{R}^n$ are the vectors of the state parameters and $z_c(t) \in \mathbb{R}^{n_{z_c}}$, $z_d(\tau_k) \in \mathbb{R}^{n_{z_d}}$ are the parameters of the outputs which evolve in continuous time and in discrete time, respectively. The JLSS (4.108)–(4.109) may be identified with a triple of the form

$$((\mathbf{C}_c(\cdot) \mid \mathbf{C}_d(\cdot)), (\mathbf{A}(\cdot) \mid \mathbf{A}_d(\cdot)), Q) \tag{4.110}$$

where $\mathbf{A}(\cdot)$, $\mathbf{A}_d(\cdot)$ are as in (4.4a), (4.5a), respectively, and

$$\mathbf{C}_c(\cdot) = (\mathbb{C}_{c0}(\cdot), \mathbb{C}_{c1}(\cdot), \ldots, \mathbb{C}_{cr_1}(\cdot)) \tag{4.111a}$$
$$\mathbf{C}_d(\cdot) = (\mathbb{C}_{d0}(\cdot), \mathbb{C}_{d1}(\cdot), \ldots, \mathbb{C}_{dr_2}(\cdot)) \tag{4.111b}$$

$$\mathbb{C}_{cl}(\cdot) = (C_{cl}(\cdot, 1), C_{cl}(\cdot, 2), \ldots, C_{cl}(\cdot, N)) \tag{4.112a}$$
$$\mathbb{C}_{dl}(\cdot) = (C_{dl}(\cdot, 1), C_{dl}(\cdot, 2), \ldots, C_{dl}(\cdot, N)), \tag{4.112b}$$

$C_{cl}(\cdot, i)$ and $C_{dl}(\cdot, i)$ being the matrix coefficients of (4.109a) and (4.109b), respectively. $Q \in \mathbb{R}^{N\times N}$ is the generator matrix of the Markov process $\eta(t)$, $t \geq 0$.

We will consider the following special cases of JLSS (4.108)–(4.109):

(i) **The JLSS without jump Markov perturbations.**
In this case Eq. (4.108) which are describing the state evolution takes the form:

$$dx(t) = A_0(t)x(t)dt + \sum_{l=1}^{r_1} A_l(t)x(t)dw_l(t), \quad \tau_k < t \leq \tau_{k+1}, \tag{4.113a}$$

$$x(\tau_k^+) = \left(A_{d0}(\tau_k) + \sum_{l=1}^{r_2} w_{dl}(k)A_{dl}(\tau_k)\right)x(\tau_k), \tag{4.113b}$$

$k \in \mathbb{Z}_+$, and Eqs. (4.109) which are describing the evolution of the outputs, become:

$$d_c(t) = C_{c0}(t)x(t)dt + \sum_{l=1}^{r_1} C_{cl}(t)x(t)dw_l(t), \quad t \in \mathbb{R}_+ \tag{4.114a}$$

$$z_d(\tau_k) = \left(C_{d0}(\tau_k) + \sum_{l=1}^{r_2} w_{dl}(k)C_{dl}(\tau_k)\right)x(\tau_k), \quad k \in \mathbb{Z}_+. \tag{4.114b}$$

(ii) **The JLSS subject to Markovian switching but without state multiplicative white noise perturbations.**
In this case, the equations which are describing the state evolution, are:

$$\dot{x}(t) = A_0(t, \eta(t))x(t), \quad \tau_k < t \leq \tau_{k+1}, \tag{4.115a}$$
$$x(\tau_k^+) = A_{d0}(\tau_k, \eta(\tau_k))x(\tau_k), \quad k \in \mathbb{Z}_+ \tag{4.115b}$$

while the evolution of the outputs $(z_c(t),\ z_d(t))$ is described by:

$$z_c(t) = C_{c0}(t, \eta(t))x(t), \quad t \in \mathbb{R}_+ \tag{4.116a}$$
$$z_d(\tau_k) = C_{d0}(\tau_k, \eta(\tau_k))x(\tau_k), \quad k \in \mathbb{Z}_+. \tag{4.116b}$$

The uncontrolled JLSS (4.113)–(4.114) may be identify to the pair

$$((\hat{\mathbf{C}}_c(\cdot) \mid \hat{\mathbf{C}}_d(\cdot)), (\hat{\mathbf{A}}(\cdot) \mid \hat{\mathbf{A}}_d(\cdot)))$$

where

$$\hat{\mathbf{C}}_c(\cdot) = (C_{c0}(\cdot), C_{c1}(\cdot), \ldots, C_{cr_1}(\cdot)),$$
$$\hat{\mathbf{C}}_d(\cdot) = (C_{d0}(\cdot), C_{d1}(\cdot), \ldots, C_{dr_2}(\cdot)),$$

$C_{cl}(\cdot)$, $C_{dl}(\cdot)$ being the matrix coefficients of (4.114).

$$\hat{\mathbf{A}}(\cdot) = (A_0(\cdot), A_1(\cdot), \ldots, A_{r_1}),$$
$$\hat{\mathbf{A}}_d(\cdot) = (A_{d0}(\cdot), A_{d1}(\cdot), \ldots, A_{dr_2}(\cdot)),$$

where $A_l(\cdot)$, $A_{dl}(\cdot)$ are the matrix coefficients of (4.113).

Now we define the concept of mean-square detectability of an uncontrolled JLSS.

Definition 4.6

(a) We say that the JLSS (4.108)–(4.109) is detectable in the mean-square sense if there exist matrix-valued functions $t \to H_c(t, i) : \mathbb{R}_+ \to \mathbb{R}^{n \times n_{zc}}$ which are continuous on each interval (τ_k, τ_{k+1}) and having finite lateral limits in any τ_k and the sequences $\{H_d(\tau_k, i)\}_{k \geq 0} \subset \mathbb{R}^{n \times n_{zd}}$, $i \in \mathcal{N}$ with the property that the following resulting JSLDE

$$dx(t) = (A_0(t, \eta(t)) + H_c(t, \eta(t))C_{c0}(t, \eta(t)))x(t)dt \tag{4.117a}$$
$$+ \sum_{l=1}^{r_1}(A_l(t, \eta(t)) + H_c(t, \eta(t))C_{cl}(t, \eta(t))x(t))dw_l(t), \ \tau_k < t \leq \tau_{k+1}$$
$$x(\tau_k^+) = (A_{d0}(\tau_k, \eta(\tau_k)) + H_d(\tau_k, \eta(\tau_k))C_{d0}(\tau_k, \eta(\tau_k)))x(\tau_k) \tag{4.117b}$$
$$+ \sum_{l=1}^{r_2} w_{dl}(k)(A_{dl}(\tau_k, \eta(\tau_k)) + H_d(\tau_k, \eta(\tau_k))C_{dl}(\tau_k, \eta(\tau_k)))x(\tau_k), \ k \in \mathbb{Z}_+$$

is SESMS.

(b) We say that the JLSS without Markovian switching (4.113)–(4.114) is detectable in the mean-square sense if there exist a matrix-valued function $t \to H_c(t) : \mathbb{R}_+ \to \mathbb{R}^{n \times n_{zc}}$ which is continuous on each interval (τ_k, τ_{k+1}) with finite lateral limits in any τ_k and a matrix-valued sequence $\{H_d(\tau_k)\}_{k \in \mathbb{Z}_+} \subset \mathbb{R}^{n \times n_{zd}}$ such that the resulting JSLDE

$$dx(t) = (A_0(t) + H_c(t)C_{c0}(t))x(t)dt + \sum_{l=1}^{r_1}(A_l(t) + H_c(t)C_{cl}(t))x(t)dw_l(t),$$

$$\tau_k < t \le \tau_{k_1} \tag{4.118a}$$

$$x(\tau_k^+) = (A_{d0}(\tau_k) + H_d(\tau_k)C_{d0}(\tau_k) + \sum_{l=1}^{r_2} w_{dl}(k)(A_{dl}(\tau_k) + H_d(\tau_k)C_{dl}(\tau_k)))x(\tau_k),$$

$$k \in \mathbb{Z}_+ \tag{4.118b}$$

is ESMS.

(c) We say that the JLSS with Markovian jumping (4.115)–(4.116) is detectable in mean square if there exist matrix-valued functions $t \to H_c(t,i) : \mathbb{R}_+ \to \mathbb{R}^{n \times n_{z_c}}$ continuous on each interval (τ_k, τ_{k+1}) with finite lateral limits in any τ_k and the matrix-valued sequences $\{H_d(\tau_k, i)\}_{k\in\mathbb{Z}_+} \subset \mathbb{R}^{n\times n_{z_d}}$ with the property that the resulting JSLDE

$$\dot{x}(t) = (A_0(t,\eta(t)) + H_c(t,\eta(t))C_{c0}(t,\eta(t)))x(t), \quad \tau_k < t \le \tau_{k_1} \tag{4.119a}$$

$$x(\tau_k^+) = (A_{d0}(\tau_k,\eta(\tau_k)) + H_d(\tau_k,\eta(\tau_k))C_{d0}(\tau_k,\eta(\tau_k)))x(\tau_k),\ k \in \mathbb{Z}_+ \tag{4.119b}$$

is SESMS.

(d) If there exist matrix-valued sequences $\{H_d(\tau_k,i)\}_{k\in\mathbb{Z}_+}$ $(\{H_d(\tau_k)\}_{k\in\mathbb{Z}_+})$ such that the resulting JSLDE (4.117) and (4.119) with $H_c(\cdot,i) \equiv 0,\ i \in \mathcal{N}$ is SESMS (the resulting JSLDE (4.118) with $H_c(\cdot) \equiv 0$ is ESMS), we say that the JLSSs (4.108)–(4.109), (4.115)–(4.116) and (4.113)–(4.114), respectively, are mean-square detectable by impulses.

The pairs of the form $\left(H_c(\cdot,i), \{H_d(\tau_k,i)\}_{k\in\mathbb{Z}_+}\right),\ i \in \mathcal{N}$ $\left(\left(H_c(\cdot), \{H_d(\tau_k)\}_{k\in\mathbb{Z}_+}\right),\right.$, respectively$\left.\right)$ for which the resulting JSLDEs (4.117), (4.119) are SESMS (the resulting JSLDE (4.118) is ESMS, respectively) will be named pairs of stabilizing gains.

Remark 4.6

(a) It can be proved that if the JLSS (4.108)–(4.109) is periodic in the sense of Definition 4.1, then it is mean-square detectable if and only if among the pairs of stabilizing gains there exist pairs of the form $(H_c(\cdot,i), \{H_d(kh,i)\}_{k\in\mathbb{Z}_+}),\ i \in \mathcal{N}$ in which $H_c(\cdot,i),\ i \in \mathcal{N}$ are periodic matrix-valued functions with period $\tilde{\theta}$ and $\{H_d(kh,i)\}_{k\in\mathbb{Z}_+},\ i \in \mathcal{N}$ are periodic sequences of period $\tilde{\mathbf{k}}$, $\tilde{\theta}$ and $\tilde{\mathbf{k}}$ being introduced in (4.8)–(4.9).

(b) Without loss of generality one may prove that if the JLSS (4.108)–(4.109) is time invariant in the sense of the Definition 4.2 it is mean-square detectable if and only if among the pairs of stabilizing gains there exist pairs of the form $(H_c(\cdot,i), \{H_d(kh,i)\}_{k\in\mathbb{Z}_+}),\ i \in \mathcal{N}$ in which, $H_c(\cdot,i)$ are matrix-valued functions continuous on each interval $(kh, (k+1)h)$, periodic with period h and $H_d(kh,i) = H_d(i) \in \mathbb{R}^{n\times n_{z_d}}$.

(c) The conclusions from (a) and (b) can be also applied in the case of the JLSSs (4.113)–(4.114) and (4.115)–(4.116).

4.4.2 Necessary and Sufficient Conditions for the Mean-Square Detectability of a JLSS

In this subsection we shall show how we may use the results from Sect. 3.4 to obtain necessary and sufficient conditions for mean-square detectability of the JLSS under consideration.

First, let us remark that the equivalence $(i) \Leftrightarrow (ii)$ from Theorem 3.2 applied in the case of a resulting system of the form (4.117), or the equivalence $(i) \Leftrightarrow (ii)$ from Corollary 3.2 applied in the case of the resulting system (4.118), allow us to obtain:

Proposition 4.7 *Assume:*

(a) the Assumptions 1.2 and 2.1 are fulfilled;
(b) $t \to (A_l(t,i), C_{cl}(t,i)) : \mathbb{R}_+ \to \mathbb{R}^{n\times n} \times \mathbb{R}^{n_{zc}\times n}$, $i \in \mathcal{N}$, $0 \le l \le r_1$ *are bounded and continuous matrix-valued functions;*
(c) $\{A_{dl}(\tau_k,i)\}_{k\in\mathbb{Z}_+} \subset \mathbb{R}^{n\times n}$, $\{C_{dl}(\tau_k,i)\}_{k\in\mathbb{Z}_+} \subset \mathbb{R}^{n_{zd}\times n}$, $i \in \mathcal{N}$, $0 \le l \le r_2$ *are bounded matrix-valued sequences.*

Under these conditions the following are equivalent:

(i) the JLSS (4.108)–(4.109) is mean-square detectable;
(ii) there exist matrix-valued functions $t \to H_c(t,i) : \mathbb{R}_+ \to \mathbb{R}^{n\times n_{zc}}$ *which are continuous on each interval* (τ_k, τ_{k+1}) *with finite lateral limits in each* τ_k *and bounded matrix-valued sequences* $\{H_d(\tau_k,i)\}_{k\in\mathbb{Z}_+} \subset \mathbb{R}^{n\times n_{zd}}$, $i \in \mathcal{N}$, *with the property that the following non-homogeneous backward GLJLDE:*

$$\begin{aligned} -\dot{Y}_i(t) &= (A_0(t,i) + H_c(t,i)C_{c0}(t,i))^\top Y_i(t) + Y_i(t)(A_0(t,i) + H_c(t,i)C_{c0}(t,i)) \\ &\quad + \sum_{l=1}^{r_1} (A_l(t,i) + H_c(t,i)C_{cl}(t,i))^\top Y_i(t)(A_l(t,i) + H_c(t,i)C_{cl}(t,i)) \\ &\quad + \sum_{j=1}^{N} q_{ij} Y_j(t) + I_n, \quad \tau_k < t < \tau_{k_1} \end{aligned} \tag{4.120a}$$

$$\begin{aligned} Y_i(\tau_k^-) &= \sum_{l=0}^{r_2} (A_{dl}(\tau_k,i) + H_d(\tau_k,i)C_{dl}(\tau_k,i))^\top Y_i(\tau_k)(A_{dl}(\tau_k,i) \\ &\quad + H_d(\tau_k,i)C_{dl}(\tau_k,i)) + I_n, \quad k \in \mathbb{Z}_+ \end{aligned} \tag{4.120b}$$

has a unique solution $t \to \tilde{\mathbf{Y}}(t) = (\tilde{Y}_1(t), \tilde{Y}_2(t), \ldots, \tilde{Y}_N(t)) : \mathbb{R}_+ \to \mathcal{S}_n^N$ *which is bounded and uniform positive.*
(iii) There exist functions $t \to (X_i(t), K_c(t,i)) : \mathbb{R}_+ \to \mathcal{S}_n \times \mathbb{R}^{n\times n_{zc}}$ *and the sequences* $\{Z_d(\tau_k,i)\}_{k\in\mathbb{Z}_+} \subset \mathcal{S}_n$, $\{K_d(\tau_k,i)\}_{k\in\mathbb{Z}_+} \subset \mathbb{R}^{n\times n_{zd}}$, $i \in \mathcal{N}$, *with properties:*

(α) *$X_i(\cdot)$ are differentiable on each interval (τ_k, τ_{k+1}) and right continuous in any τ_k, $k \in \mathbb{Z}_+$;*

(β) *$K_c(\cdot, i)$ are continuous on each interval (τ_k, τ_{k+1}) and have finite lateral limits in any $t = \tau_k$, $k \in \mathbb{Z}_+$;*

(γ) *there exist positive constants c_j, such that*

$$0 \prec c_1 I_n \preceq X_i(t) \preceq c_2 I_n, \quad \textit{for all} \quad t \in \mathbb{R}_+$$

$$0 \prec c_3 I_n \preceq Z_d(\tau_k, i) \preceq c_4 I_n$$

$$\|K_d(\tau_k, i)\| \leq c_5, \quad \textit{for all} \quad k \in \mathbb{Z}_+;$$

(δ) *$(X_i(\cdot), K_c(\cdot, i), \{Z_d(\tau_k, i), K_d(\tau_k, i)\}_{k \in \mathbb{Z}_+})$, $i \in \mathcal{N}$ solves the following system of linear matrix differential inequalities with jumps:*

$$\begin{pmatrix} -\dot{Z}_i(t) + \Psi_1(t, \mathbf{X}(t), K_c(t, i), i) & \Psi_2(t, X_i(t), K_c(t, i), i) \\ * & -\Psi_3(X_i(t)) \end{pmatrix} \preceq -\xi_1 I,$$

$$\tau_k \leq t < \tau_{k+1}, \tag{4.121a}$$

$$X_i(\tau_k^-) = Z_d(\tau_k, i), \tag{4.121b}$$

$$\begin{pmatrix} -Z_d(\tau_k, i) & \Upsilon_1(k, X_i(\tau_k), K_d(\tau_k, i), i) \\ * & -\Upsilon_2(X_i(\tau_k)) \end{pmatrix} \preceq -\xi_2 I, \quad k \in \mathbb{Z}_+, \tag{4.121c}$$

$i \in \mathcal{N}$, where

$$\Psi_1(t, \mathbf{X}(t), K_c(t, i), i) := A_0^\top(t, i) X_i(t) + X_i(t) A_0(t, i) + C_{c0}^\top(t, i) K_c^\top(t, i) + K_c(t, i) C_{c0}(t, i) + \sum_{j=1}^{N} q_{ij} X_j(t)$$

$$\Psi_2(t, X_i(t), K_c(t, i), i) := \operatorname*{row}_{l=1}^{r_1}\{(X_i(t) A_l(t, i) + K_c(t, i) C_{cl}(t, i))^\top\}$$

$$\Psi_3(X_i(t)) := \operatorname*{diag}_{l=1}^{r_1}\{X_i(t)\} \in \mathcal{S}_{n_1}, \quad n_1 = r_1 \cdot n$$

$$\Upsilon_1(k, X_i(\tau_k), K_d(\tau_k, i), i) := \operatorname*{row}_{l=0}^{r_2}\{(X_i(\tau_k) A_{dl}(\tau_k, i) + K_d(\tau_k, i) C_{dl}(\tau_k, i))^\top\}$$

$$\Upsilon_2(X_i(\tau_k)) := \operatorname*{diag}_{l=0}^{r_2}\{X_i(\tau_k)\} \in \mathcal{S}_{n_2}, \quad n_2 = (r_2 + 1)n.$$

If $(X_i(\cdot), K_c(\cdot, i), \{Z_d(\tau_k, i), K_d(\tau_k, i)\}_{k \in \mathbb{Z}_+})$, $i \in \mathcal{N}$ is a solution of (4.121) then $(H_c(\cdot, i), \{H_d(\tau_k, i)\}_{k \in \mathbb{Z}_+})$, $i \in \mathcal{N}$ defined by

$$H_c(t,i) := X_i^{-1}(t)K_c(t,i), \quad t \in \mathbb{R}_+$$

$$H_d(\tau_k,i) := X_i^{-1}(\tau_k)K_d(\tau_k,i), \quad k \in \mathbb{Z}_+$$

provide a pair of stabilizing gains for the JLSS (4.108)–(4.109).

Proof The equivalence $(i) \Leftrightarrow (ii)$ follows from the chain of implications $(i) \Rightarrow (ii) \Rightarrow (iii) \Rightarrow (i)$ from Theorem 3.2 applied in the case of the resulting system (4.117). The equivalence $(ii) \Leftrightarrow (iii)$ is obtained using Schur complement technique together with the solutions $X_i(t) = Y_i^{-1}(t)$, $K_c(t,i) = Y_i(t)H_c(t,i)$, $K_d(\tau_k,i) = Y_i(\tau_k)H_d(\tau_k,i)$.

Proposition 4.8 *Assume:*

(a) the Assumptions 1.1 and 2.1 are fulfilled;
(b) $t \to (A_l(t), C_{cl}(t)) : \mathbb{R}_+ \to \mathbb{R}^{n\times n} \times \mathbb{R}^{n_{zc}\times n}$, $0 \le l \le r_1$ *are continuous and bounded matrix-valued functions;*
(c) $\{A_{dl}(\tau_k)\}_{k\in\mathbb{Z}_+} \subset \mathbb{R}^{n\times n}$, $\{C_{dl}(\tau_k)\}_{k\in\mathbb{Z}_+} \subset \mathbb{R}^{n_{zd}\times n}$, $0 \le l \le r_2$ *are bounded matrix-valued sequences.*

Under these conditions the following are equivalent:

(i) the JLSS (4.113)–(4.114) is mean-square detectable;
(ii) there exist a function $t \to H_c(t) : \mathbb{R}_+ \to \mathbb{R}^{n\times n_{zc}}$ *continuous on each interval* (τ_k, τ_{k+1}) *with finite lateral limits in any* τ_k, $k \in \mathbb{Z}_+$ *and a bounded matrix-valued sequence* $\{H_d(\tau_k)_{k\in\mathbb{Z}_+} \subset \mathbb{R}^{n\times n_{zd}}$, *with the property that the following non-homogeneous backward GLJLDE:*

$$\begin{aligned}
-\dot{Y}(t) &= (A_0(t) + H_c(t)C_{c0}(t))^\top Y(t) + Y(t)(A_0(t) + H_c(t)C_{c0}(t)) \\
&\quad + \sum_{l=1}^{r_1}(A_l(t) + H_c(t)C_{cl}(t))^\top Y(t)(A_l(t) + H_c(t)C_{cl}(t)) + I_n, \\
&\quad \tau_k \le t < \tau_{k_1}
\end{aligned} \tag{4.122a}$$

$$\begin{aligned}
Y(\tau_k^-) &= \sum_{l=0}^{r_2}(A_{dl}(\tau_k) + H_d(\tau_k)C_{dl}(\tau_k))^\top Y(\tau_k)(A_{dl}(\tau_k) \\
&\quad + H_d(\tau_k)C_{dl}(\tau_k)) + I_n, \quad k \in \mathbb{Z}_+
\end{aligned} \tag{4.122b}$$

has a unique bounded solution $t \to \tilde{Y}(t) : \mathbb{R}_+ \to \mathcal{S}_n$ *which is uniform positive;*
(iii) there exist functions $t \to (Z(t), K_c(t)) : \mathbb{R}_+ \to \mathcal{S}_n \times \mathbb{R}^{n\times n_{zc}}$ *and matrices* $Z_d(\tau_k) \in \mathcal{S}_n$, $K_d(\tau_k) \in \mathbb{R}^{n\times n_{zd}}$, $k \in \mathbb{Z}_+$ *with the properties:*

(α) $Z(\cdot)$ *is differentiable on each interval* (τ_k, τ_{k+1}) *and right continuous in* $t = \tau_k$, $k \in \mathbb{Z}_+$;
(β) $K_c(\cdot)$ *is a continuous function on each interval* (τ_k, τ_{k+1}) *having finite lateral limits in* $t = \tau_k$, $k \in \mathbb{Z}_+$;

(γ) *there exist positive constants c_j, such that*

$$0 \prec c_1 I_n \preceq Z(t) \preceq c_2 I_n, \quad \textit{for all} \;\; t \in \mathbb{R}_+$$

$$0 \prec c_3 I_n \preceq Z_d(\tau_k) \preceq c_4 I_n$$

$$\|K_d(\tau_k)\| \leq c_5, \quad \textit{for all} \;\; k \in \mathbb{Z}_+;$$

(δ) $(Z(\cdot), K_c(\cdot), \{Z_d(\tau_k), K_d(\tau_k)\}_{k\in\mathbb{Z}_+})$ *solves the following linear matrix differential inequality with jumps*

$$\begin{pmatrix} -\dot{Z}(t) + \hat{\Psi}_1(t, Z(t)), K_c(t)) & \hat{\Psi}_2(t, Z(t), K_c(t)) \\ * & -\hat{\Psi}_3(Z(t)) \end{pmatrix} \preceq -\hat{\xi}_1 I,$$

$$\tau_k \leq t < \tau_{k+1}, \tag{4.123a}$$

$$Z(\tau_k^-) = Z_d(\tau_k), \quad k \in \mathbb{Z}_+ \tag{4.123b}$$

$$\begin{pmatrix} -Z_d(\tau_k) & \hat{\Upsilon}_1(k, Z(\tau_k), K_d(\tau_k)) \\ * & -\hat{\Upsilon}_2(Z(\tau_k)) \end{pmatrix} \preceq -\hat{\xi}_2 I, \quad k \in \mathbb{Z}_+ \tag{4.123c}$$

where we have denoted:

$$\hat{\Psi}_1(t, Z(t), K_c(t)) : = A_0^\top(t)Z(t) + Z(t)A_0(t) + C_{c0}^\top(t)K_c^\top(t) + K_c(t)C_{c0}(t)$$

$$\hat{\Psi}_2(t, Z(t), K_c(t)) := \operatorname*{row}_{l=1}^{r_1}\{(Z(t)A_l(t) + K_c(t)C_{cl}(t))^\top\}$$

$$\hat{\Psi}_3(Z(t)) := \operatorname*{diag}_{l=1}^{r_1}\{Z(t)\} \in \mathscr{S}_{n_1}, \quad n_1 = r_1 \cdot n$$

$$\hat{\Upsilon}_1(k, Z(\tau_k), K_d(\tau_k)) := \operatorname*{row}_{l=0}^{r_2}\{(Z(\tau_k)A_{dl}(\tau_k) + K_d(\tau_k)C_{dl}(\tau_k))^\top\}$$

$$\hat{\Upsilon}_2(Z(\tau_k)) := \operatorname*{diag}_{l=0}^{r_2}\{Z(\tau_k)\} \in \mathscr{S}_{n_2}, \quad n_2 = (r_2 + 1)n.$$

If $(Z(\cdot), K_c(\cdot), \{Z_d(\tau_k), K_d(\tau_k)\}_{k\in\mathbb{Z}_+})$ *is a solution of (4.123) then* $(H_c(\cdot),$
$\{H_d(\tau_k)\}_{k\in\mathbb{Z}_+})$, *defined by*

$$H_c(t) := Z^{-1}(t)K_c(t), \quad t \in \mathbb{R}_+$$

$$H_d(\tau_k) := Z^{-1}(\tau_k)K_d(\tau_k), \quad k \in \mathbb{Z}_+$$

is a pair of stabilizing gains for the JLSS (4.113)–(4.114).

Proof The equivalence $(i) \Leftrightarrow (ii)$ follows from the chain of implications $(i) \Rightarrow (ii) \Rightarrow (iii) \Rightarrow (i)$ from Corollary 3.2 used in the case of the resulting system (4.118). The equivalence $(ii) \Leftrightarrow (iii)$ may be proved employing the Schur complement technique to show that the solvability of (4.122) is equivalent to the solvability of (4.123). Details are omitted.

4.5 A Criterion for Mean-Square Exponential Stability

The equivalence $(i) \Leftrightarrow (iii)$ from Theorem 3.2 shows that a necessary and sufficient condition for the strong exponential stability in mean square of a JSLDE of type (3.1) consists of the existence of a bounded and continuous matrix-valued function $t \to (\tilde{H}_1(t), \tilde{H}_2(t), \ldots, \tilde{H}_N(t)) : \mathbb{R}_+ \to \mathcal{S}_n^{N+}$ and of a bounded and uniform positive sequence $\{\boldsymbol{\Gamma}_k\}_{k \in \mathbb{Z}_+} \subset Int\ \mathcal{S}_n^{N+}$ such that the corresponding non-homogeneous backward GLJLDE of type (3.32) has a bounded and uniform positive solution $t \to (\tilde{Y}_1(t), \tilde{Y}_2(t), \ldots, \tilde{Y}_N(t)) : \mathbb{R}_+ \to Int\ \mathcal{S}_n^{N+}$. In this section we show that under the detectability assumption a kind of mean-square stability of a jump stochastic linear differential equation of type (3.1) (or equivalently (4.108)) is achieved even in the case when the non-homogeneous backward GLJLDE (3.32) has the forcing term positive semidefinite instead of uniform positive definite and has a positive semidefinite global solution instead of a uniform positive one.

Involving the triple $((\mathbf{C}_c(\cdot) \mid \mathbf{C}_d(\cdot)), (\mathbf{A}(\cdot) \mid \mathbf{A}_d(\cdot)), Q)$ associated to the JLSS (4.108)–(4.109) we introduce the following non-homogeneous GLJLDE on the Hilbert space $\mathcal{S}_n^N$:

$$-\dot{Y}(t,i) = A_0^\top(t,i)Y(t,i) + Y(t,i)A_0(t,i) + \sum_{l=1}^{r_1} A_l^\top(t,i)Y(t,i)A_l(t,i)$$

$$+ \sum_{j=1}^{N} q_{ij}Y(t,j) + \sum_{l=0}^{r_1} C_{cl}^\top(t,i)C_{cl}(t,i), \quad \tau_k \leq t < \tau_{k+1} \tag{4.124a}$$

$$Y(\tau_k^-, i) = \sum_{l=0}^{r_2} A_{dl}^\top(\tau_k, i)Y(\tau_k, i)A_{dl}(\tau_k, i) + \sum_{l=0}^{r_2} C_{dl}^\top(\tau_k, i)C_{dl}(\tau_k, i), \quad k \in \mathbb{Z}_+. \tag{4.124b}$$

Employing the notation introduced in (2.114) and (2.116), respectively, we rewrite (4.124) as:

$$\dot{Y}(t,i) + \mathcal{L}_i^*(t)[\mathbf{Y}(t)] = -\sum_{l=0}^{r_1} C_{cl}^\top(t,i)C_{cl}(t,i), \quad \tau_k \leq t < \tau_{k+1} \tag{4.125a}$$

$$Y(\tau_k^-, i) = \mathcal{J}_i^*(\tau_k)[Y(\tau_k, i)] + \sum_{l=0}^{r_2} C_{dl}^\top(\tau_k, i)C_{dl}(\tau_k, i), \quad k \in \mathbb{Z}_+. \tag{4.125b}$$

Beside (4.124) we introduce the following GLJLDE on the space $\mathcal{S}_n$ associated to the JLSS (4.113)–(4.114):

$$-\dot{Y}(t) = A_0^\top(t)Y(t) + Y(t)A_0(t) + \sum_{l=1}^{r_1} A_l^\top(t)Y(t)A_l(t)$$
$$+ \sum_{l=0}^{r_1} C_{cl}^\top(t)C_{cl}(t), \quad \tau_k \le t < \tau_{k+1} \tag{4.126a}$$
$$Y(\tau_k^-) = \sum_{l=0}^{r_2} A_{dl}^\top(\tau_k)Y(\tau_k)A_{dl}(\tau_k) + \sum_{l=0}^{r_2} C_{dl}^\top(\tau_k)C_{dl}(\tau_k), \quad k \in \mathbb{Z}_+. \tag{4.126b}$$

The next result highlights the role of the property of mean-square detectability in obtaining a criterion on the stability in mean square of the JSLDE (4.108).

Theorem 4.1 *Assume:*

(a) the Assumptions 1.2, 2.1 and 4.1 are fulfilled;
(b) the triple $((\mathbf{C}_c(\cdot) \mid \mathbf{C}_d(\cdot)), (\mathbf{A}(\cdot) \mid \mathbf{A}_d(\cdot)), Q)$ is mean-square detectable.

Under these conditions the following hold:

(i) if the GLJLDE (4.124) has a solution $t \to (\tilde{Y}_1(t), \tilde{Y}_2(t), \ldots, \tilde{Y}_N(t)) : \mathbb{R}_+ \to \mathcal{S}_n^N$ such that

$$0 \prec \tilde{Y}_i(t) \preceq \mu I_n, \tag{4.127}$$

for all $t \in \mathbb{R}_+$, then the JSLDE (4.108) is ASMS;

(ii) if the JLSS (4.108)–(4.109) is time-periodic in the sense of Definition 4.1 then the following are equivalent:

(α) the JSLDE (4.108) is ESMS;
(β) the GLJLDE (4.124) has a solution $t \to (\tilde{Y}_1(t), \tilde{Y}_2(t), \ldots, \tilde{Y}_N(t)) : \mathbb{R}_+ \to \mathcal{S}_n^N$ which satisfies (4.127).

Proof Let $(t_0, x_0) \in \mathbb{R}_+ \times \mathbb{R}^n$ be arbitrary but fixed and let us denote $x(t) := x(t; t_0, x_0)$ the solution of (4.108) having the initial value $x(t_0) = x_0$. Let $k_0 \in \mathbb{Z}$ be such that $\tau_{k_0-1} < t_0 \le \tau_{k_0}$. We consider the function

$$v(t, x, i) := x^\top \tilde{Y}(t, i)x, \quad (t, x, i) \in \mathbb{R}_+ \times \mathbb{R}^n \times \mathcal{N},$$

$(\tilde{Y}(t, 1), \tilde{Y}(t, 2), \ldots, \tilde{Y}(t, N))$ being the bounded on $\mathbb{R}_+$ and positive semidefinite solution of (4.124).

Applying the Itô's formula (see for example, Theorem 1.11.4 from Drăgan et al. [1]) to the function $v(t, x, i)$ and to the solution $x(t)$ of (4.108) we get:

$$\mathbb{E}[v(b_k, x(b_k), \eta(b_k)) \mid \eta(t_0) = j] - \mathbb{E}[v(a_k, x(a_k), \eta(a_k)) \mid \eta(t_0) = j]$$
$$= \mathbb{E}\left[\int_{a_k}^{b_k} x^\top(t)\left(\dot{\tilde{Y}}(t, \eta(t)) + \mathscr{L}^*_{\eta(t)}(t)[\tilde{\mathbf{Y}}]\right)x(t)\mathrm{d}t \mid \eta(t_0) = j\right] \tag{4.128}$$

for all $\tau_k < a_k < b_k < \tau_{k+1}$, $k \in \mathbb{Z}_+$ such that $\tau_k \geq t_0$. Substituting (4.125a) in (4.128) we obtain

$$\sum_{l=0}^{r_1} \mathbb{E}\left[\int_{a_k}^{b_k} |C_{cl}(t, \eta(t))x(t)|^2 \mathrm{d}t \mid \eta(t_0) = j\right]$$
$$= \mathbb{E}[v(a_k, x(a_k), \eta(a_k)) \mid \eta(t_0) = j] - \mathbb{E}[v(b_k), x(b_k), \eta(b_k)) \mid \eta(t_0) = j].$$

Letting $a_k \searrow \tau_k$ and $b_k \nearrow \tau_{k+1}$ we further obtain via Lemma 3.1 that:

$$\sum_{l=0}^{r_1} \int_{\tau_k}^{\tau_{k+1}} \mathbb{E}[|C_{cl}(t, \eta(t))x(t)|^2 | \eta(t_0) = j]\mathrm{d}t = \mathbb{E}[x^\top(\tau_k^+)\tilde{Y}(\tau_k, \eta(\tau_k))x(\tau_k^+) | \eta(t_0) = j]$$
$$- \mathbb{E}[x^\top(\tau_{k+1})\tilde{Y}(\tau_{k+1}^-, \eta(\tau_{k+1}))x(\tau_{k+1} \mid \eta(t_0) = j]. \tag{4.129}$$

Since the $\sigma-$algebras $\sigma[w_{dl}(k)]$ are independent of the $\sigma-$algebra $\mathscr{H}_{\tau_k}$, and $x(\tau_k)$ is measurable with respect to $\mathscr{H}_{\tau_k}$, we may obtain successively that:

$$\mathbb{E}[x^\top(\tau_k^+)\tilde{Y}(\tau_k, \eta(\tau_k))x(\tau_k^+) \mid \eta(t_0) = j]$$
$$= \mathbb{E}\left[\mathbb{E}[x^\top(\tau_k)\left(A_{d0}(\tau_k, \eta(\tau_k)) + \sum_{l=1}^{r_2} w_{dl}(k)A_{dl}(\tau_k, \eta(\tau_k))\right)^\top \tilde{Y}(\tau_k, \eta(\tau_k))\right.$$
$$\left.\times \left(A_{d0}(\tau_k, \eta(\tau_k)) + \sum_{l=1}^{r_2} w_{dl}(k)A_{dl}(\tau_k, \eta(\tau_k))\right) x(\tau_k)|\mathscr{H}_{\tau_k}] \mid \eta(t_0) = j\right]$$
$$= \mathbb{E}\left[x^\top(\tau_k)\left(\sum_{l=0}^{r_2} A_{dl}^\top(\tau_k, \eta(\tau_k))\tilde{Y}(\tau_k, \eta(\tau_k))A_{dl}(\tau_k, \eta(\tau_k))\right)x(\tau_k) \mid \eta(t_0) = j\right].$$

Employing (4.124b) written for $\tilde{Y}(\tau_k, \eta(\tau_k))$ we may conclude that

$$\mathbb{E}[x^\top(\tau_k^+)\tilde{Y}(\tau_k, \eta(\tau_k))x(\tau_k^+) \mid \eta(t_0) = j] \tag{4.130}$$
$$= \mathbb{E}[x^\top(\tau_k)\tilde{Y}(\tau_k^-, \eta(\tau_k))x(\tau_k)|\eta(t_0) = j] - \sum_{l=0}^{r_2} \mathbb{E}[|C_{cl}(\tau_k, \eta(\tau_k))x(\tau_k)|^2 \, |\eta(t_0) = j].$$

Further, (4.129) and (4.130) yield

$$\sum_{l=0}^{r_1}\int_{\tau_k}^{\tau_{k+1}} \mathbb{E}[|C_{cl}(t,\eta(t))x(t)|^2 \mid \eta(t_0)=j]dt + \sum_{l=0}^{r_2}\mathbb{E}[|C_{dl}(\tau_k,\eta(\tau_k))x(\tau_k)|^2 \mid \eta(t_0)=j]$$
$$= \mathbb{E}[x^\top(\tau_k)\tilde{Y}(\tau_k^-,\eta(\tau_k))x(\tau_k) \mid \eta(t_0)=j]$$
$$- \mathbb{E}[x^\top(\tau_{k+1})\tilde{Y}(\tau_{k+1}^-,\eta(\tau_{k+1}))x(\tau_{k+1}) \mid \eta(t_0)=j] \tag{4.131}$$

for all $k \in \mathbb{Z}_+$ with $k \geq k_0$, where k_0 is such that $\tau_{k_0-1} < t_0 \leq \tau_{k_0}$, $j \in \mathcal{N}$.

If $t_0 \in (\tau_{k_0-1}, \tau_{k_0})$ then, applying Itô's formula to the function $v(t,x,i)$ and to the stochastic process $x(t)$ on the interval $[t_0, \tau_{k_0}]$ we obtain via (4.124a), written for $\tilde{Y}(t,i)$ that

$$\sum_{l=0}^{r_1}\int_{t_0}^{\tau_{k_0}} \mathbb{E}[(C_{cl}(t,\eta(t))x(t))|^2 \mid \eta(t_0)=j] = x_0^\top\tilde{Y}(t_0,j)x_0$$
$$- \mathbb{E}[x^\top(\tau_{k_0})\tilde{Y}(\tau_{k_0}^-,\eta(\tau_{k_0}))x(\tau_{k_0})x(\tau_{k_0}) \mid \eta(t_0)=j]. \tag{4.132}$$

From (4.131) we get:

$$\sum_{l=0}^{r_1}\int_{\tau_{k_0}}^{\tau_k} \mathbb{E}[|C_{cl}(t,\eta(t))x(t)|^2 \mid \eta(t_0)=j]dt$$
$$+\sum_{l=0}^{r_2}\sum_{\iota=k_0}^{k-1}\mathbb{E}[|C_{dl}(\tau_\iota,\eta(\tau_\iota))x(\tau_\iota)|^2 \mid \eta(t_0)=j]$$
$$= \mathbb{E}[x^\top(\tau_{k_0})\tilde{Y}(\tau_{k_0}^-,\eta(\tau_{k_0}))x(\tau_{k_0}) \mid \eta(t_0)=j]$$
$$- \mathbb{E}[x^\top(\tau_k)\tilde{Y}(\tau_k^-,\eta(\tau_k))x(\tau_k) \mid \eta(t_0)=j],$$

for all $k \geq k_0$, $j \in \mathcal{N}$. This equality together with (4.132) lead to:

$$\sum_{l=0}^{r_1}\int_{t_0}^{\tau_k} \mathbb{E}[|C_{cl}(t,\eta(t))x(t)|^2 \, |\eta(t_0)=j]dt + \sum_{l=0}^{r_2}\sum_{\iota=k_0}^{k-1}\mathbb{E}[|C_{dl}(\tau_\iota,\eta(\tau_\iota))x(\tau_\iota)|^2 \mid \eta(t_0)=j]$$
$$= x_0^\top\tilde{Y}(t_0,j)x_0 - \mathbb{E}[x^\top(\tau_k)\tilde{Y}(\tau_k^-,\eta(\tau_k))x(\tau_k) \mid \eta(t_0)=j], \tag{4.133}$$

for all $k > k_0$, $j \in \mathcal{N}$. On the other hand (4.124b) written for $\tilde{Y}(\cdot,j)$ together with (4.127) give:

$$0 \preceq \tilde{Y}(\tau_k^-,i) \preceq \hat{\mu}I_n, \quad \text{for all } k \in \mathbb{Z}_+, \tag{4.134}$$

$\hat{\mu} > 0$ being a constant not depending upon $k \in \mathbb{Z}_+$. Since the sequence $\{\tau_k\}_{k\in\mathbb{Z}_+}$ satisfies the Assumption 2.1 it follows that it is an increasing and unbounded sequence. Under these conditions (4.133) together with (4.127) and (4.134) allow us to conclude that

$$\sum_{l=0}^{r_1}\int_{t_0}^{\infty}\mathbb{E}[|C_{cl}(t,\eta(t))x(t)|^2 \mid \eta(t_0)=j]\mathrm{d}t$$
$$+\sum_{l=0}^{r_2}\sum_{k=k_0}^{\infty}\mathbb{E}[|C_{dl}(\tau_k,\eta(\tau_k))x(\tau_k)|^2 \mid \eta(t_0)=j] \leq \mu|x_0|^2, \tag{4.135}$$

for all $j \in \mathcal{N}$ and $(t_0, x_0) \in \mathbb{R}_+ \times \mathbb{R}^n$. Further, from the mean-square detectability of the triple $((\mathbf{C}_c(\cdot) \mid \mathbf{C}_d(\cdot)), (\mathbf{A}(\cdot) \mid \mathbf{A}_d(\cdot)), Q)$ we deduce that there exist functions $t \to H_c(t,i) : \mathbb{R}_+ \to \mathbb{R}^{n\times n_{zc}}$ continuous on each interval (τ_k, τ_{k+1}) with finite lateral limits on each $t = \tau_k$, $k \in \mathbb{Z}_+$ and bounded matrix-valued sequences $\{H_d(\tau_k, i)\}_{k\in\mathbb{Z}_+} \subset \mathbb{R}^{n\times n_{zd}}$, $i \in \mathcal{N}$ such that the resulting system of type (4.117) is SESMS. Let us rewrite the JSLDE (4.108) in the form:

$$\mathrm{d}x(t) = [(A_0(t,\eta(t)) + H_c(t,\eta(t))C_{c0}(t,\eta(t)))x(t) + f_0(t)]\mathrm{d}t \tag{4.136a}$$
$$+\sum_{l=1}^{r_1}[(A_l(t,\eta(t)) + H_c(t,\eta(t))C_{cl}(t,\eta(t)))x(t) + f_l(t)]\mathrm{d}w_l(t), \quad \tau_k < t \leq \tau_{k_1}$$
$$x(\tau_k^+) = (A_{d0}(\tau_k,\eta(\tau_k)) + H_d(\tau_k,\eta(\tau_k))C_{d0}(\tau_k,\eta(\tau_k)))x(\tau_k) + g_0(k) \tag{4.136b}$$
$$+\sum_{l=1}^{r_2} w_{dl}(k)[(A_{dl}(\tau_k,\eta(\tau_k)) + H_d(\tau_k,\eta(\tau_k))C_{dl}(\tau_k,\eta(\tau_k)))x(\tau_k) + g_l(k)],$$

$k \geq k_0$, where we have denoted

$$f_l(t) = -H_c(t,\eta(t))C_{cl}(t,\eta(t))x(t), \quad 0 \leq l \leq r_1 \tag{4.137a}$$
$$g_l(k) = -H_d(\tau_k,\eta(\tau_k))C_{dl}(\tau_k,\eta(\tau_k))x(\tau_k)), \quad 0 \leq l \leq r_2. \tag{4.137b}$$

From (4.135) and (4.137a) it follows that $f_l(\cdot) \in \mathcal{L}^2_{\mathcal{H}}([t_0,\infty); \mathbb{R}^n)$ and from (4.135) and (4.137b) we may conclude that the sequences $\{g_l(k)\}_{k\geq k_0}$, $0 \leq k \leq r_2$, lie in $\ell^2_{\mathcal{H}}\{k_0,\infty; \mathbb{R}^n\}$. Applying Theorem 3.5 (iii) in the case of the system (4.136) we deduce that

$$\lim_{t\to\infty}\mathbb{E}[|x(t)|^2 \mid \eta(t_0)=j] = 0,$$

for all $j \in \mathcal{N}$, $(t_0, x_0) \in \mathbb{R}_+ \times \mathbb{R}^n$. This means that the JSLDE (4.108) is ASMS according to Definition 3.1 (f). Thus the proof of part (i) is complete.

Now we prove (ii). We begin by the implication $(\alpha) \Rightarrow (\beta)$. Let us remark that if the JSLDE (4.108) is ESMS, then based on the implication $(iii) \Rightarrow (i)$ from Theorem 3.1 (4.108) is SESMS. According to Definition 3.1 (a) it follows that the GLJLDE of the form (2.117) associated to (4.108) is exponentially stable. Under these conditions, Theorem 2.13 applied in the case of the non-homogeneous backward GLJLDE (4.124) allows us to deduce that (4.124) has a solution $t \to (\tilde{Y}(t,1), \tilde{Y}(t,2), \ldots, \tilde{Y}(t,N))$ which is a periodic function of period $\tilde{\theta}$ and $\tilde{Y}(t,i) \succ 0$, for all $(t,i) \in \mathbb{R}_+ \times \mathcal{N}$. Hence, this solution satisfies a condition of type (4.127). Thus the validity of the implication $(\alpha) \Rightarrow (\beta)$ is confirmed.

To prove the opposite implication, let us remark that if (β) holds then, based on (i) of the statement, the JSLDE (4.108) is ASMS. Finally, the implication $(iv) \Rightarrow (iii)$ of Theorem 3.1 shows that (4.108) is ESMS. Thus the proof ends.

The next result provides a criterion for exponential stability in mean-square of the JSLDE (4.113).

Theorem 4.2 *Assume:*

(a) the Assumptions 1.2, 2.1 and 4.1 in the special case $\mathcal{N} = \{1\}$ *are fulfilled;*
(b) the pair $((\mathscr{C}_c(\cdot) \mid \mathscr{C}_d(\cdot)), (\mathscr{A}(\cdot) \mid \mathscr{A}_d(\cdot)))$ *is mean-square detectable.*

Under these conditions the following hold:

(i) if the non-homogeneous backward GLJLDE (4.126) has a solution $t \to \tilde{Y}(t) : \mathbb{R}_+ \to \mathscr{S}_n$ *such that*

$$0 \prec \tilde{Y}(t) \preceq \tilde{\mu} I_n, \tag{4.138}$$

for all $t \in \mathbb{R}_+$, $\tilde{\mu}$ *being a positive constant, then the JSLDE (4.113) is ASMS;*
(ii) if the uncontrolled JLSS (4.113)–(4.114) is time periodic in the sense of Definition 4.1 then the following are equivalent:

(α) the JSLDE (4.113) is ESMS;
(β) the GLJLDE (4.126) has a solution $t \to \tilde{Y}(t) : \mathbb{R}_+ \to \mathscr{S}_n$ *which satisfies a condition of the form (4.138).*

Proof The proof of this result follows similar lines as for the proof of Theorem 4.1. The details are omitted here.

4.6 Duality

It is well known that in the time-invariant deterministic framework (see e.g., Anderson and Moore [6]), a pair of matrices (A, B) is stabilizable if and only if its dual pair $(B^\top, A^\top)$ is detectable. Similarly, a pair of matrices (C, A) is detectable if and only if its dual pair $(A^\top, C^\top)$ is stabilizable.

In the present section we analyze the duality of the concept of mean-square stabilizability and mean-square detectability defined in Sects. 4.2 and 4.4, respectively, for jump linear stochastic systems (JLSSs).

4.6.1 The Case of a Jump Linear Stochastic System Free of Markovians Switching

Let us consider the time periodic JLSS having the state-space representation described by:

$$\mathrm{d}x(t) = (A_0(t)x(t) + B_0(t)u_c(t))\mathrm{d}t + \sum_{l=1}^{r_1}(A_l(t)x(t) + B_l(t)u_c(t))\mathrm{d}w_l(t),$$

$$kh < t \le (k+1)h \tag{4.139a}$$

$$y(kh^+) = A_{d0}(kh)x(kh) + B_{d0}(kh)u_d(k) + \sum_{p=1}^{r_2} w_{dp}(k)(A_{dp}(kh)x(kh)$$

$$+ B_{dp}(kh)u_d(k)), \quad k \in \mathbb{Z}_+ \tag{4.139b}$$

where $x(t) \in \mathbb{R}^n$, $u_c(t) \in \mathbb{R}^{m_c}$, $u_d(k) \in \mathbb{R}^{m_d}$. The stochastic processes $\{w(t)\}_{t\ge 0}$ $(w(t) = (w_1(t) \; w_2(t) \; \dots \; w_{r_1}(t))^\top)$ and $\{w_d(k)\}_{k\in\mathbb{Z}_+}$ $(w_d(k) = (w_{d1}(k) \; w_{d2}(k) \; \dots \; w_{dr_2}(k))^\top)$ involved in (4.139) satisfy the Assumption 1.1. Assume that $t \to (A_l(t), B_l(t)) : \mathbb{R} \to \mathbb{R}^{n\times n} \times \mathbb{R}^{n\times m_c}$, $0 \le l \le r_1$ are matrix-valued functions continuous on each interval $(kh, (k+1)h)$ having finite lateral limits in any $\tau_k = kh$, $k \in \mathbb{Z}$. Additionally, we assume that they are periodic functions of period $\theta_c > 0$.

In (4.139b), $\{A_{dp}(kh)\}_{k\in\mathbb{Z}} \subset \mathbb{R}^{n\times n}$, $\{B_{dp}(kh)\}_{k\in\mathbb{Z}} \subset \mathbb{R}^{n\times m_d}$ are matrix-valued sequences which are periodic of period $\theta_d \ge 1$. We assume that the numbers θ_c, θ_d, h satisfy condition (4.8). If the JLSS (4.139), or equivalently, the pair $((\mathscr{A}(\cdot) \mid \mathscr{A}_d(\cdot)), (\mathscr{B}(\cdot) \mid \mathscr{B}_d(\cdot)))$ is mean-square stabilizable by state feedback, then, according to Remark 4.1 there exist a piecewise continuous function $t \to F_c(t) : \mathbb{R}_+ \to \mathbb{R}^{m_c\times n}$ which is periodic with period $\widetilde{\theta}$ and a sequence $\{F_d(kh)\}_{k\in\mathbb{Z}_+} \subset \mathbb{R}^{m_d\times n}$ which is periodic with period $\widetilde{\mathbf{k}}$ ($\widetilde{\theta}$, $\widetilde{\mathbf{k}}$ being defined in (4.9)) such that the closed-loop system

$$\mathrm{d}x(t) = (A_0(t) + B_0(t)F_c(t))x(t)\mathrm{d}t + \sum_{l=1}^{r_1}(A_l(t) + B_l(t)F_c(t))x(t)\mathrm{d}w_l(t),$$

$$kh < t \le (k+1)h \tag{4.140a}$$

$$y(kh^+) = (A_{d0}(kh) + B_{d0}(kh)F_d(kh))x(kh) + \sum_{p=1}^{r_2} w_{dp}(k)(A_{dp}(kh)$$

$$+ B_{dp}(kh)F_d(kh))x(kh), \quad k \in \mathbb{Z}_+ \tag{4.140b}$$

is ESMS.

The chain of implications $(i) \Rightarrow (v) \Rightarrow (vi) \Rightarrow (i)$ from Corollary 3.3 together with the updates

$$A_l(t) \leftarrow A_l(t) + B_l(t)F_c(t), \quad 0 \le l \le r_1$$

$$A_{dp}(kh) \leftarrow A_{dp}(kh) + B_{dp}(kh)F_d(kh), \quad 0 \le p \le r_2$$

allows us to deduce that the closed-loop JSLDE (4.140) is ESMS if and only if the non-homogeneous forward GLJLDE:

$$\dot{X}(t) = (A_0(t) + B_0(t)F_c(t))X(t) + X(t)(A_0(t) + B_0(t)F_c(t))^\top \tag{4.141a}$$

$$+ \sum_{l=1}^{r_1} (A_l(t) + B_l(t)F_c(t))X(t)(A_l(t) + B_l(t)F_c(t))^\top + I_n, \quad kh < t \le (k+1)h$$

$$X(kh^+) = \sum_{p=0}^{r_2} (A_{dp}(kh) + B_{dp}(kh)F_d(kh))X(kh)(A_{dp}(kh) + B_{dp}(kh)F_d(kh))^\top + I_n \tag{4.141b}$$

$k \in \mathbb{Z}$, has a solution $t \to \widetilde{X}(t) : \mathbb{R} \to \mathscr{S}_n$ which is a periodic function of period $\widetilde{\theta}$ and satisfies

$$0 < \nu_1 I_n \le \widetilde{X}(t) \le \nu_2 I_n, \quad \text{for all } t \in \mathbb{R}. \tag{4.142}$$

We set $\widetilde{Y}(t) \triangleq \widetilde{X}(-t)$, $t \in \mathbb{R}$. By direct calculations involving (4.141) and (4.142) we obtain that $t \to \widetilde{Y}(t) : \mathbb{R} \to \mathscr{S}_n$ satisfies:

$$\dot{\widetilde{Y}}(t) + (A_0(-t) + B_0(-t)F_c(-t))\widetilde{Y}(t) + \widetilde{Y}(t)(A_0(-t) + B_0(-t)F_c(-t))^\top$$

$$+ \sum_{l=1}^{r_1} (A_l(-t) + B_l(-t)F_c(-t))\widetilde{Y}(t)(A_l(-t) + B_l(-t)F_c(-t))^\top + I_n = 0$$

$$-kh < -t \le -kh + h, \tag{4.143a}$$

$$\widetilde{Y}(kh^-) := \lim_{\substack{\varepsilon \to 0 \\ \varepsilon > 0}} \widetilde{Y}(kh - \varepsilon) = \lim_{\substack{\varepsilon \to 0 \\ \varepsilon > 0}} \widetilde{X}(-kh + \varepsilon) = \widetilde{X}(-kh^+)$$

$$= \sum_{p=0}^{r_2} (A_{dp}(-kh) + B_{dp}(-kh)F_d(-kh))\widetilde{Y}(kh)(A_{dp}(-kh) + B_{dp}(-kh)F_d(-kh))^\top$$

$$+ I_n, \quad k \in \mathbb{Z} \tag{4.143b}$$

$$0 < \nu_1 I_n \le \widetilde{Y}(t) \le \nu_2 I_n, \quad \text{for all } t \in \mathbb{R}. \tag{4.143c}$$

We set

$$\begin{aligned}
&\widehat{A}_l(t) := A_l^\top(-t), \quad \widehat{C}_{cl}(t) := B_l^\top(-t), \quad 0 \le l \le r_1,\\
&H_c(t) := F_c^\top(-t), \quad t \in \mathbb{R},\\
&\widehat{A}_{dp}(kh) := A_{dp}^\top(-kh), \quad \widehat{C}_{dp}(kh) := B_{dp}^\top(-kh), \quad 0 \le p \le r_2,\\
&H_d(kh) := F_d^\top(-kh), \quad k \in \mathbb{Z}.
\end{aligned} \tag{4.144}$$

Using this notation, (4.143) becomes:

$$-\dot{\widetilde{Y}}(t) = (\widehat{A}_0(t) + H_c(t)\widehat{C}_0(t))^\top \widetilde{Y}(t) + \widetilde{Y}(t)(\widehat{A}_0(t) + H_c(-t)\widehat{C}_0(t))$$
$$+ \sum_{l=1}^{r_1} (\widehat{A}_l(t) + H_c(t)\widehat{C}_l(t))^\top \widetilde{Y}(t)(\widehat{A}_l(t) + H_c(t)\widehat{C}_l(t)) + I_n$$
$$(k-1)h \le t < kh, \tag{4.145a}$$
$$\widetilde{Y}(kh^-) = \sum_{l=0}^{r_2} (\widehat{A}_{dp}(kh) + H_d(kh)\widehat{C}_{dp}(kh))^\top \widetilde{Y}(kh)(\widehat{A}_{dp}(kh) + H_d(-kh)\widehat{C}_{dp}(kh))$$
$$+ I_n, \quad k \in \mathbb{Z}, \tag{4.145b}$$
$$0 < \nu_1 I_n \le \widetilde{Y}(t) \le \nu_2 I_n, \quad \text{for all} \quad t \in \mathbb{R}. \tag{4.145c}$$

The equivalence $(ii) \Leftrightarrow (i)$ from Corollary 3.3 together with the updates

$$A_l(t) \leftarrow \widehat{A}_l(t) + H_c(t)\widehat{C}_{cl}(t), \quad 0 \le l \le r_1,$$
$$A_{dp}(kh) \leftarrow \widehat{A}_{dp}(kh) + H_d(kh)\widehat{C}_{dp}(kh), \quad 0 \le p \le r_2$$

guarantee that the feasibility of (4.145) is equivalent to the exponential stability in mean square of the following JSLDE:

$$\mathrm{d}x(t) = (\widehat{A}_0(t) + H_c(t)\widehat{C}_{c0}(t))x(t)\mathrm{d}t + \sum_{l=1}^{r_1} (\widehat{A}_l(t) + H_c(t)\widehat{C}_{cl}(t))$$
$$\cdot\, x(t)\mathrm{d}w_l(t), \quad kh < t \le (k+1)h, \tag{4.146a}$$
$$x(kh^+) = (\widehat{A}_{d0}(kh) + H_d(kh)\widehat{C}_{d0}(kh))x(kh) + \sum_{p=1}^{r_2} w_{dp}(k)$$
$$\cdot\, (\widehat{A}_{dp}(kh) + H_d(kh)\widehat{C}_{dp}(kh))x(kh), \quad k \in \mathbb{Z}_+. \tag{4.146b}$$

According to the Definition 4.6 (b), the exponential stability in mean square of the JSLDE (4.146) is equivalent to the mean-square detectability of the JLSS whose state evolution is described by:

$$\mathrm{d}x(t) = \widehat{A}_0(t)x(t)\mathrm{d}t + \sum_{l=1}^{r_1} \widehat{A}_l(t)x(t)\mathrm{d}w_l(t), \quad kh < t \le (k+1)h, \tag{4.147a}$$
$$x(kh^+) = \left(\widehat{A}_{d0}(kh) + \sum_{p=1}^{r_2} w_{dp}(k)\widehat{A}_{dp}(kh)\right) x(kh), \quad k \in \mathbb{Z}_+, \tag{4.147b}$$

and the outputs modeled by

$$\mathrm{d}_c(t) = \widehat{C}_{c0}(t)x(t)\mathrm{d}t + \sum_{l=1}^{r_1} \widehat{C}_{cl}(t)x(t)\mathrm{d}w_l(t), \quad t \in \mathbb{R}_+ \tag{4.148a}$$

$$z_d(kh) = \left(\widehat{C}_{d0}(kh) + \sum_{p=1}^{r_2} w_{dp}(k)\widehat{C}_{dp}(kh) \right) x(kh), \quad k \in \mathbb{Z}_+. \tag{4.148b}$$

Thus, we have proved:

Proposition 4.9 *Under the Assumption 1.1, the following are equivalent:*

(i) the JLSS (4.139) is mean-square stabilizable by state feedback;
(ii) the JLSS (4.147)–(4.148) whose coefficients are defined in (4.144), is mean-square detectable.

In the sequel, the JLSS (4.147)–(4.148) whose coefficients are computed by (4.144), will be named **the dual jump linear stochastic system** of the JLSS (4.139). If we set

$$\widehat{\mathscr{A}}(\cdot) := (A_0^\top(\cdot),\ A_1^\top(\cdot), \ldots, A_{r_1}^\top(\cdot)),$$

$$\widehat{\mathscr{A}_d}(\cdot) := (A_{d0}^\top(\cdot),\ A_{d1}^\top(\cdot), \ldots, A_{dr_2}^\top(\cdot)),$$

$$\widehat{\mathscr{C}_c}(\cdot) := (B_0^\top(\cdot),\ B_1^\top(\cdot), \ldots, B_{r_1}^\top(\cdot)),$$

$$\widehat{\mathscr{C}_d}(\cdot) := (B_{d0}^\top(\cdot),\ B_{d1}^\top(\cdot), \ldots, B_{dr_2}^\top(\cdot)),$$

we may identify the dual JLSS (4.147)–(4.148) with the pair $\big((\widehat{\mathscr{C}_c}(\cdot) \mid \widehat{\mathscr{C}_d}(\cdot)), (\widehat{\mathscr{A}}(\cdot) \mid \widehat{\mathscr{A}_d}(\cdot))\big)$. With this notation, Proposition 4.9 may be restated as:

Proposition 4.10 *Under the Assumption 1.1 the following are equivalent:*

(i) the pair $((\mathscr{A}(\cdot) \mid \mathscr{A}_d(\cdot)), (\mathscr{B}(\cdot) \mid \mathscr{B}_d(\cdot)))$ is mean-square stabilizable by state feedback;
(ii) the dual pair $\big((\widehat{\mathscr{C}_c}(\cdot) \mid \widehat{\mathscr{C}_d}(\cdot)), (\widehat{\mathscr{A}}(\cdot) \mid \widehat{\mathscr{A}_d}(\cdot))\big)$ is mean-square detectable.

Let us consider the time-invariant JLSS:

$$\mathrm{d}x(t) = (A_0x(t) + B_0u_c(t))\mathrm{d}t + \sum_{l=1}^{r_1}(A_lx(t) + B_lu_c(t))\mathrm{d}w_l(t),$$
$$kh < t \le (k+1)h, \tag{4.149a}$$

$$y(kh^+) = A_{d0}x(kh) + B_{d0}u_d(k) + \sum_{p=1}^{r_2} w_{dp}(k)(A_{dp}x(kh)$$
$$+ B_{dp}u_d(k)), \quad k \in \mathbb{Z}_+. \tag{4.149b}$$

The JLSS (4.149) may be identified with the following pair of sequences of matrices $((\mathscr{A} \mid \mathscr{A}_d), (\mathscr{B} \mid \mathscr{B}_d))$ where $\mathscr{A} = \{A_0, A_1, \ldots, A_{r_1}\}$, $\mathscr{A}_d = \{A_{d0}, A_{d1}, \ldots, A_{dr_2}\}$,

$\mathscr{B} = \{B_0, B_1, \ldots, B_{r_1}\}$, $\mathscr{B}_d = \{B_{d0}, B_{d1}, \ldots, B_{dr_2}\}$. Since the JLSS (4.149) is a special case of (4.139) it can be followed step by step the previous reasoning done in the case of (4.139) but invoking Corollary 3.4 instead of Corollary 3.3 to obtain:

Corollary 4.1 *Under the Assumption 1.1, the following are equivalent:*

(i) the pair $((\mathscr{A} \mid \mathscr{A}_d), (\mathscr{B} \mid \mathscr{B}_d))$ *is mean-square stabilizable by state feedback;*
(ii) the pair $((\widehat{\mathscr{C}_c} \mid \widehat{\mathscr{C}_d}), (\widehat{\mathscr{A}} \mid \widehat{\mathscr{A}_d}))$ *is mean-square detectable, where*

$$\begin{aligned} &\widehat{\mathscr{A}} = \{\widehat{A}_0, \widehat{A}_1, \ldots, \widehat{A}_{r_1}\}, \quad \widehat{A}_l := A_l^{\top}, \quad 0 \le l \le r_1, \\ &\widehat{\mathscr{A}_d} = \{\widehat{A}_{d0}, \widehat{A}_{d1}, \ldots, \widehat{A}_{dr_2}\}, \quad \widehat{A}_{dp} := A_{dp}^{\top}, \quad 0 \le l \le r_2, \\ &\widehat{\mathscr{C}_c} = \{\widehat{C}_{c0}, \widehat{C}_{c1}, \ldots, \widehat{C}_{cr_1}\}, \quad \widehat{C}_{cl} := B_l^{\top}, \quad 0 \le l \le r_1, \\ &\widehat{\mathscr{C}_d} = \{\widehat{C}_{d0}, \widehat{C}_{d1}, \ldots, \widehat{C}_{dr_2}\}, \quad \widehat{C}_{dp} := B_{dp}^{\top}, \quad 0 \le l \le r_2. \end{aligned} \tag{4.150}$$

The equivalence from Corollary 4.1 allows us to prove the following result which highlights the role of the property of the mean-square stabilizability by state feedback to derive a criterion for mean-square stabilizability.

Theorem 4.3 *Under the Assumption 1.1, if the JLSS (4.149) is mean-square stabilizable by state feedback, then the following are equivalent:*

(i) the jump stochastic linear differential equation

$$dx(t) = A_0 x(t)dt + \sum_{l=1}^{r_1} A_l x(t) dw_l(t), \quad kh < t \le (k+1)h \tag{4.151a}$$

$$x(kh^+) = (A_{d0} + \sum_{p=1}^{r_2} w_{dp}(k) A_{dp}) x(kh), \quad k \in \mathbb{Z}_+, \tag{4.151b}$$

is ESMS;

(ii) the forward generalized Lyapunov-type jump linear differential equation (GLJLDE):

$$\dot{X}(t) = A_0 X(t) + X(t) A_0^{\top} + \sum_{l=1}^{r_1} A_l X(t) A_l^{\top} + \sum_{l=0}^{r_1} B_l B_l^{\top} \; kh < t \le (k+1)h \tag{4.152a}$$

$$X(kh^+) = \sum_{p=0}^{r_2} A_{dp} X(kh) A_{dp}^{\top} + \sum_{p=0}^{r_2} B_{dp} B_{dp}^{\top}, \quad k \in \mathbb{Z} \tag{4.152b}$$

has a solution $t \to \widetilde{X}(t) : \mathbb{R} \to \mathscr{S}_n$ *having the properties* $0 \le \widetilde{X}(t) \le \tilde{c} I_n$ *and* $\widetilde{X}(t+h) = \widetilde{X}(t)$, *for all* $t \in \mathbb{R}$.

Proof The implication $(i) \Rightarrow (ii)$ can be proved specializing the results from Theorem 2.3, Theorem 2.4 and Theorem 2.9 to the case of the GLJLDE (4.152).
It remains to show that the implication $(ii) \Rightarrow (i)$ holds.

Setting $\widetilde{Y}(t) := \widetilde{X}(-t),\ \ t \in \mathbb{R}$, by direct calculation we obtain that $t \to \widetilde{Y}(t) : \mathbb{R} \to \mathcal{S}_n$ solves the following non-homogeneous backward GLJLDE:

$$-\dot{\widetilde{Y}}(t) = \widehat{A}_0^\top \widetilde{Y}(t) + \widetilde{Y}(t)\widehat{A}_0 + \sum_{l=1}^{r_1} \widehat{A}_l^\top \widetilde{Y}(t)\widehat{A}_l(t) + \sum_{l=0}^{r_1} \widehat{C}_{cl}^\top \widehat{C}_{cl},$$

$$kh \le t < (k+1)h, \tag{4.153a}$$

$$\widetilde{Y}(kh^-) = \sum_{p=0}^{r_2} \widehat{A}_{dp}^\top \widetilde{Y}(kh)\widehat{A}_{dp} + \sum_{p=0}^{r_2} \widehat{C}_{dp}^\top \widehat{C}_{dp}, \ \ k \in \mathbb{Z}, \tag{4.153b}$$

$$0 \le \widetilde{Y}(t) \le \tilde{c} I_n, \ \ \text{for all}\ \ t \in \mathbb{R}, \tag{4.153c}$$

where $\widehat{A}_l, \widehat{C}_{cl}, 0 \le l \le r_1, \widehat{A}_{dp}, \widehat{C}_{dp}, 0 \le p \le r_2$ were defined in (4.150).

The GLJLDE (4.153) is associated to the pair $((\widehat{\mathscr{C}_c \mid \mathscr{C}_d}), (\widehat{\mathscr{A} \mid \mathscr{A}_d}))$. On the other hand, from Corollary 4.1 we deduce that under the considered assumptions $((\widehat{\mathscr{C}_c \mid \mathscr{C}_d}), (\widehat{\mathscr{A} \mid \mathscr{A}_d}))$ is mean-square detectable. So, Theorem 4.2 applied in the case of the pair $((\widehat{\mathscr{C}_c \mid \mathscr{C}_d}), (\widehat{\mathscr{A} \mid \mathscr{A}_d}))$ and of the equation (4.153) allows us to conclude that the JSLDE:

$$\mathrm{d}x(t) = \widehat{A}_0 x(t)\mathrm{d}t + \sum_{l=1}^{r_1} \widehat{A}_l x(t)\mathrm{d}w_l(t), \ \ kh < t \le (k+1)h \tag{4.154a}$$

$$x(kh^+) = (\widehat{A}_{d0} + \sum_{p=1}^{r_2} w_{dp}(k)\widehat{A}_{dp})x(kh), \ \ k \in \mathbb{Z}_+, \tag{4.154b}$$

is ESMS. Further, the implication $(ii) \Rightarrow (i)$ from Theorem 3.1, Definition 3.1 (a) and (3.17) applied in the case of the JSLDE (4.154) give

$$\|\widehat{\mathbf{T}}(t, t_0)\| \le \widehat{\beta} e^{-\widehat{\alpha}(t-t_0)}, \tag{4.155}$$

for all $t \ge t_0,\ t, t_0 \in \mathbb{R}$, where $\widehat{\beta} \ge 1,\ \widehat{\alpha} > 0$ do not depend upon t, t_0. In (4.155), $\widehat{\mathbf{T}}(t, t_0)$ is the linear evolution operator on $\mathcal{S}_n$ defined by the following forward jump linear differential equation (FJLDE):

$$\dot{Z}(t) = \widehat{\mathscr{L}}[Z(t)], \ \ kh < t \le (k+1)h \tag{4.156a}$$

$$Z(kh^+) = \widehat{\mathscr{I}}[Z(kh)], \ \ k \in \mathbb{Z}, \tag{4.156b}$$

where

$$\widehat{\mathscr{L}}[Z] \triangleq \widehat{A}_0 Z + Z\widehat{A}_0^\top + \sum_{l=1}^{r_1} \widehat{A}_l Z \widehat{A}_l^\top, \tag{4.157a}$$

$$\widehat{\mathscr{I}}[Z] \triangleq \sum_{p=0}^{r_2} \widehat{A}_{dp} Z \widehat{A}_{dp}^\top, \tag{4.157b}$$

for all $Z \in \mathcal{S}_n$. From (4.155) we may deduce via the implication $(i) \Rightarrow (ii)$ from Corollary 2.2 applied in the case of the FJLDE (4.156), that the spectral radius of the linear operator $Z \to \widehat{\mathcal{L}_d}[Z] \triangleq e^{\widehat{\mathcal{L}}h}[\widehat{\mathcal{I}}[Z]] : \mathcal{S}_n \to \mathcal{S}_n$ satisfies:

$$\rho(\widehat{\mathcal{L}_d}) < 1. \tag{4.158}$$

Let us consider the linear operator $\breve{\mathcal{L}_d} : \mathcal{S}_n \to \mathcal{S}_n$ defined by

$$\breve{\mathcal{L}_d}[Z] \triangleq \widehat{\mathcal{I}}[e^{\widehat{\mathcal{L}}h}[Z]], \quad \text{for all} \quad Z \in \mathcal{S}_n.$$

One sees that a complex number $\lambda \neq 0$ is an eigenvalue of the linear operator $\breve{\mathcal{L}_d}$ if and only if, it is an eigenvalue of the operator $\widehat{\mathcal{L}_d}$. It means that (4.158) holds if and only if the spectral radius of the linear operator $\breve{\mathcal{L}_d}$ satisfies $\rho\breve{\mathcal{L}_d} < 1$ or, equivalently

$$\rho(\breve{\mathcal{L}_d}^*) < 1. \tag{4.159}$$

Combining (2.111), (2.114), (4.150) together with (4.157) we may infer that

$$\breve{\mathcal{L}_d}^* = e^{\mathcal{L}h}\mathcal{I},$$

where the operators $\mathcal{L}$ and $\mathcal{I}$ are associated to the JSLDE (4.151) by

$$\mathcal{L}[X] = A_0X + XA_0^\top + \sum_{l=1}^{r_1} A_lXA_l^\top$$

and

$$\mathcal{I}[X] = \sum_{p=0}^{r_2} A_{dp}XA_{dp}^\top.$$

Since (4.159) is equivalent to $\rho(e^{\mathcal{L}h}\mathcal{I}) < 1$ we may conclude via $(ii) \Rightarrow (i)$ from Corollary 2.2 that the FJLDE on $\mathcal{S}_n$:

$$\dot{X}(t) = \mathcal{L}[X(t)], \quad kh < t \leq (k+1)h \tag{4.160a}$$

$$X(kh^+) = \mathcal{I}[X(kh)], \quad k \in \mathbb{Z}, \tag{4.160b}$$

is exponentially stable. This means that if $\mathbf{T}(t, t_0)$ is the linear evolution operator on $\mathcal{S}_n$ defined by (4.160) we have

$$\|\mathbf{T}(t, t_0)\| \leq \beta e^{-\alpha(t-t_0)}, \tag{4.161}$$

for all $t \geq t_0,\ t, t_0 \in \mathbb{R}$. This conclusion follows combining (4.161) with (3.17) written for the solution of (4.151). Thus the proof ends.

With a slight modification of the reasonings made in the first part of this section, we may prove:

Proposition 4.11 *Under the Assumption 1.1, if the JLSS (4.113)–(4.114) is time periodic in the sense of Definition 4.1, the following are equivalent:*

1. *the JLSS (4.113)–(4.114) is mean-square detectable;*
2. *the corresponding dual JLSS:*

$$dx(t) = (\bar{A}_0(t)x(t) + \bar{B}_0(t)u_c(t))dt + \sum_{l=1}^{r_1}(\bar{A}_l(t)x(t) + \bar{B}_l(t)u_c(t))dw_l(t),$$
$$kh < t \leq (k+1)h$$
$$x(kh^+) = \bar{A}_{d0}(kh)x(kh) + \bar{B}_{d0}(kh)u_d(k) + \sum_{p=1}^{r_2} w_{dp}(k)(\bar{A}_{dp}(kh)x(kh) + \bar{B}_{dp}(kh)u_d(k)),$$

is mean-square stabilizable by state feedback, where

$$\bar{A}_l(t) := A_l^\top(-t),\ \bar{B}_l(t) := C_{cl}^\top(-t),\ 0 \leq l \leq r_1,\ t \in \mathbb{R},$$
$$\bar{A}_{dp}(kh) := A_{dp}^\top(-kh),\ \ \bar{B}_{dp}(kh) := C_{dp}^\top(-kh),\ 0 \leq p \leq r_2,\ k \in \mathbb{Z}.$$

Proof The proof is left to the readers.

4.6.2 The Case of a Jump Linear Stochastic System Subject to Markovians Switching

Let us consider the time periodic jump linear stochastic system described by:

$$\mathrm{d}x(t) = (A_0(t,\eta(t))x(t) + B_0(t,\eta(t))u_c(t))\mathrm{d}t + \sum_{l=1}^{r_1}(A_l(t,\eta(t))x(t) + B_l(t,\eta(t))u_c(t))\mathrm{d}w_l(t),\quad kh < t \leq (k+1)h \tag{4.162a}$$

$$x(kh^+) = A_{d0}(kh,\eta(kh))x(kh) + B_{d0}(kh,\eta(kh))u_d(k) + \sum_{p=1}^{r_2} w_{dp}(k)(A_{dp}(kh,\eta(kh))x(kh) + B_{dp}(kh,\eta(kh)))u_d(k),\quad k \in \mathbb{Z}_+ \tag{4.162b}$$

where $x(t) \in \mathbb{R}^n$, $u_c(t) \in \mathbb{R}^{m_c}$, $u_d(k) \in \mathbb{R}^{m_d}$ are the states and the control parameters, respectively. The stochastic processes $\{w(t)\}_{t\geq 0}$ $(w(t) = (w_1(t)\ w_2(t) \ldots w_{r_1}(t))^\top)$

$\{w_d(k)\}_{k\in\mathbb{Z}_+}$ $(w_d(k) = (w_{d1}(k)\ w_{d2}(k)\dots w_{dr_2}(k))^\top)$ and $\{\eta(t)\}_{t\geq 0}$ are satisfying Assumption 1.2. For each $i \in \mathcal{N}$, $t \to (A_l(t,i),\ B_l(t,i)) : \mathbb{R} \to \mathbb{R}^{n\times n} \times \mathbb{R}^{n\times m_c}$ are continuous functions on each interval $(kh, (k+1)h)$ having finite lateral limits in any $\tau_k = kh$, $k \in \mathbb{Z}$. Additionally, we assume that they are periodic functions of period $\theta_c > 0$. In (4.162b), $\{A_{dp}(kh,i)\}_{k\in\mathbb{Z}} \subset \mathbb{R}^{n\times n}$, $\{B_{dp}(kh,i)\}_{k\in\mathbb{Z}} \subset \mathbb{R}^{n\times m_d}$, $0 \leq p \leq r_2$, are periodic matrix-valued sequences which are periodic with period $\theta_d \geq 1$. We assume that the numbers θ_c, θ_d, h are satisfying condition (4.8). Under these conditions if the JLSS (4.162) is mean-square stabilizable by state feedback then according to the Remark 4.1, there exist piecewise continuous functions $t \to F_c(t,i) : \mathbb{R} \to \mathbb{R}^{m_c\times n}$ which are periodic with period $\widetilde{\theta}$ and matrix-valued sequences $\{F_d(kh,i)\}_{k\in\mathbb{Z}} \subset \mathbb{R}^{m_d\times n}$, $i \in \mathcal{N}$, which are periodic with period $\widetilde{\mathbf{k}}$ ($\widetilde{\theta}$, $\widetilde{\mathbf{k}}$ being defined as in (4.9)), with the property that the closed-loop jump stochastic linear differential equation (JSLDE):

$$dx(t) = (A_0(t,\eta(t)) + B_0(t,\eta(t))F_c(t,\eta(t)))x(t)dt \tag{4.163a}$$
$$+\sum_{l=1}^{r_1}(A_l(t,\eta(t)) + B_l(t,\eta(t))F_c(t,\eta(t)))dw_l(t),\quad kh < t \leq (k+1)h$$

$$x(kh^+) = (A_{d0}(kh,\eta(kh)) + B_{d0}(kh,\eta(kh))F_d(kh,\eta(kh)))x(kh) \tag{4.163b}$$
$$+\sum_{p=1}^{r_2} w_{dp}(k)(A_{dp}(kh,\eta(kh)) + B_{dp}(kh,\eta(kh))F_d(kh,\eta(kh)))x(kh),\quad k \in \mathbb{Z}_+$$

is ESMS. The chain of implications $(i) \Rightarrow (v) \Rightarrow (vi) \Rightarrow (i)$ together with the updates $A_l(t,i) \leftarrow A_l(t,i) + B_l(t,i)F_c(t,i)$, $0 \leq l \leq r_1$, $(t,i) \in \mathbb{R}\times\mathcal{N}$, $A_{dp}(kh,i) \leftarrow A_{dp}(kh,i) + B_{dp}(kh,i)F_d(kh,i)$, $0 \leq p \leq r_2$, $(k,i) \in \mathbb{Z}\times\mathcal{N}$ allow us to deduce that the JSLDE (4.163) is ESMS if and only if the non-homogeneous forward GLJLDE:

$$\dot{X}(t,i) = (A_0(t,i) + B_0(t,i)F_c(t,i))X(t,i) + X(t,i)(A_0(t,i) + B_0(t,i)F_c(t,i))^\top$$
$$+\sum_{l=1}^{r_1}(A_l(t,i) + B_l(t,i)F_c(t,i))X(t,i)(A_l(t,i) + B_l(t,i)F_c(t,i))^\top$$
$$+\sum_{j=1}^{N} q_{ji}X(t,j) + I_n,\quad kh < t \leq (k+1)h \tag{4.164a}$$

$$X(kh^+,i) = \sum_{p=0}^{r_2}(A_{dp}(kh,i) + B_{dp}(kh,i)F_d(kh,i))X(kh,i)$$
$$\cdot (A_{dp}(kh,i) + B_{dp}(kh,i)F_d(kh,i))^\top + I_n,\quad k \in \mathbb{Z} \tag{4.164b}$$

$i \in \mathcal{N}$ has a solution $t \to \widetilde{\mathbf{X}}(t) = (\widetilde{X}(t,1),\dots,\widetilde{X}(t,N)) : \mathbb{R} \to \mathcal{S}_n^N$ which is a periodic function of period $\widetilde{\theta}$ and satisfies

$$0 < \nu_1 I_n \leq \widetilde{X}(t,i) \leq \nu_2 I_n, \tag{4.165}$$

for all $(t,i) \in \mathbb{R} \times \mathcal{N}$ and some positive constants ν_1 and ν_2. We set

$$\widetilde{Y}(t,i) \triangleq \widetilde{X}(-t,i), \quad \text{for all } \ (t,i) \in \mathbb{R} \times \mathcal{N}.$$

By direct calculation involving (4.164)–(4.165) we obtain that $t \to \widetilde{\mathbf{Y}}(t) = (\widetilde{Y}(t,1), \dots, \widetilde{Y}(t,N)) : \mathbb{R} \to \mathcal{S}_n^N$ satisfies:

$$\begin{aligned}
&\dot{\widetilde{Y}}(t,i) + (A_0(-t,i) + B_0(-t,i)F_c(-t,i))\widetilde{Y}(t,i) + \widetilde{Y}(t,i)(A_0(-t,i) + B_0(-t,i)F_c(-t,i))^\top \\
&\quad + \sum_{l=1}^{r_1} (A_l(-t,i) + B_l(-t,i)F_c(-t,i))\widetilde{Y}(t,i)(A_l(-t,i) + B_l(-t,i)F_c(-t,i))^\top \\
&\quad + \sum_{j=1}^{N} q_{ji}\widetilde{Y}(t,j) + I_n = 0, \quad -(k+1)h < -t \le -kh
\end{aligned} \tag{4.166a}$$

$$\begin{aligned}
\widetilde{Y}(kh^-,i) &= \sum_{p=0}^{r_2} (A_{dp}(-kh,i) + B_{dp}(-kh,i)F_d(-kh,i))\widetilde{Y}(kh,i) \\
&\quad \cdot (A_{dp}(-kh,i) + B_{dp}(-kh,i)F_d(-kh,i))^\top + I_n, \quad k \in \mathbb{Z}
\end{aligned} \tag{4.166b}$$

$$0 < \nu_1 I_n \le \widetilde{Y}(t,i) \le \nu_2 I_n, \quad \text{for all } \ t \in \mathbb{R}, \ i \in \mathcal{N}. \tag{4.166c}$$

Setting

$$\begin{aligned}
&\widehat{A}_l(t,i) \triangleq A_l^\top(-t,i), \quad \widehat{C}_{cl}(t,i) \triangleq B_l^\top(-t,i), \quad 0 \le l \le r_1, \\
&\widehat{A}_{dp}(kh,i) \triangleq A_{dp}^\top(-kh,i), \quad \widehat{C}_{dp}(kh,i) \triangleq B_{dp}^\top(-kh,i), \quad k \in \mathbb{Z}, \quad 0 \le p \le r_2, \\
&H_c(t,i) \triangleq F_c^\top(-t,i), \quad t \in \mathbb{R}, \quad H_d(kh,i) \triangleq F_d^\top(-kh,i), \quad k \in \mathbb{Z},
\end{aligned} \tag{4.167}$$

$i \in \mathcal{N}$, we may rewrite (4.166) as:

$$\begin{aligned}
&-\dot{\widetilde{Y}}(t,i) = (\widehat{A}_0(t,i) + H_c(t,i)\widehat{C}_{c0}(t,i))^\top \widetilde{Y}(t,i) + \widetilde{Y}(t,i)(\widehat{A}_0(t,i) + H_c(t,i)\widehat{C}_{c0}(t,i)) \\
&\quad + \sum_{l=1}^{r_1} (\widehat{A}_l(t,i) + H_c(t,i)\widehat{C}_{cl}(t,i))^\top \widetilde{Y}(t,i)(\widehat{A}_l(t,i) + H_c(t,i)\widehat{C}_c(t,i)) \\
&\quad + \sum_{j=1}^{N} \widehat{q}_{ij}\widetilde{Y}(t,j) + I_n, \quad kh \le t < (k+1)h,
\end{aligned} \tag{4.168a}$$

$$\begin{aligned}
\widetilde{Y}(kh^-,i) &= \sum_{p=0}^{r_2} (\widehat{A}_{dp}(kh,i) + H_d(kh,i)\widehat{C}_{dp}(kh,i))^\top \widetilde{Y}(kh,i) \\
&\quad \cdot (\widehat{A}_{dp}(kh,i) + H_d(kh,i)\widehat{C}_{dp}(kh,i)) + I_n, \quad k \in \mathbb{Z}
\end{aligned} \tag{4.168b}$$

$$0 < \nu_1 I_n \le \widetilde{Y}(t,i) \le \nu_2 I_n, \quad \text{for all } \ t \in \mathbb{R}, \tag{4.168c}$$

$i \in \mathcal{N}$, where $\widehat{q}_{ij} := q_{ji}$ for all $i, j \in \mathcal{N}$. Let us assume that besides the condition (1.22) the components q_{ij} of the generator matrix Q satisfy the additional condition

$$\sum_{j=1}^{N} q_{ji} = 0, \quad 1 \le i \le N. \tag{4.169}$$

Under these conditions we may invoke the equivalence $(iii) \Leftrightarrow (i)$ from Theorem 3.3 together with the updates $A_l(t, i) \leftarrow \widehat{A}_l(t, i) + H_c(t, i)\widehat{C}_{cl}(t, i)$, $(t, i) \in \mathbb{R} \times \mathcal{N}$, $0 \le l \le r_1$ $\quad A_{dp}(kh, i) \leftarrow \widehat{A}_{dp}(kh, i) + H_d(kh, i)\widehat{C}_{dp}(kh, i)$, $(k, i) \in \mathbb{Z} \times \mathcal{N}$, $0 \le p \le r_2$, and $q_{ij} \leftarrow \hat{q}_{ij}$ to deduce that under (4.169), (4.168) is satisfied if and only if the following JSLDE:

$$\begin{aligned} dx(t) &= (\widehat{A}_0(t, \widehat{\eta}(t)) + H_c(t, \widehat{\eta}(t))\widehat{C}_{c0}(t, \widehat{\eta}(t)))x(t)dt \\ &\quad + \sum_{l=1}^{r_1} (\widehat{A}_l(t, \widehat{\eta}(t)) + H_c(t, \widehat{\eta}(t))\widehat{C}_{cl}(t, \widehat{\eta}(t)))x(t), \quad kh < t \le (k+1)h, \end{aligned} \tag{4.170a}$$

$$\begin{aligned} x(kh^+) &= (\widehat{A}_{d0}(kh, \widehat{\eta}(kh)) + H_d(kh, \widehat{\eta}(kh))\widehat{C}_{d0}(kh, \widehat{\eta}(kh)))x(kh) \\ &\quad + \sum_{p=1}^{r_2} w_{dp}(k)(\widehat{A}_{dp}(kh, \widehat{\eta}(kh)) + H_d(kh, \widehat{\eta}(kh))\widehat{C}_{dp}(kh, \widehat{\eta}(kh)))x(kh), \quad k \in \mathbb{Z}_+ \end{aligned} \tag{4.170b}$$

is ESMS, where $\{\widehat{\eta}(t)\}_{t \ge 0}$ is a standard homogeneous Markov process on the probability space $(\Omega, \mathfrak{F}, \mathscr{P})$ taking values in the finite set $\mathcal{N} = \{1, 2, \ldots, N\}$ having the transition semigroup $\widehat{P}(t) \stackrel{\Delta}{=} e^{Q^\top t}$, $t \ge 0$.

According to the Definition 4.6 (a) the ESMS of the JSLDE (4.170) is equivalent to the mean-square detectability of the JLSS whose state evolution is described by:

$$dx(t) = \widehat{A}_0(t, \widehat{\eta}(t))x(t)dt + \sum_{l=1}^{r_1} \widehat{A}_l(t, \widehat{\eta}(t))x(t)dw_l(t), \quad kh < t \le (k+1)h \tag{4.171a}$$

$$x(kh^+) = \widehat{A}_{d0}(kh, \widehat{\eta}(kh))x(kh) + \sum_{p=1}^{r_2} w_{dp}(k)\widehat{A}_{dp}(kh, \widehat{\eta}(kh))x(kh), k \in \mathbb{Z}_+ \tag{4.171b}$$

and the outputs evolution described by

$$d_c(t) = \widehat{C}_{c0}(t, \widehat{\eta}(t))x(t)dt + \sum_{l=1}^{r_1} \widehat{C}_{cl}(t, \widehat{\eta}(t))x(t)dw_l(t), \quad t \in \mathbb{R}_+ \tag{4.172a}$$

$$z_d(kh) = \widehat{C}_{d0}(kh, \widehat{\eta}(kh))x(kh) + \sum_{p=1}^{r_2} w_{dp}(k)\widehat{C}_{dp}(kh, \widehat{\eta}(kh))x(kh), k \in \mathbb{Z}_+. \tag{4.172b}$$

Thus we have proved:

Proposition 4.12 *Under the Assumption 1.2 if the entries of the generator matrix Q of the Markov process satisfy the additional conditions (4.169) then the following are equivalent:*

(i) the JLSS (4.162) is mean-square stabilizable by linear state feedback;
(ii) the JLSS (4.171)–(4.172) whose coefficients were defined in (4.167) is mean-square detectable.

Remark 4.7 The JLSS (4.171)–(4.172) whose coefficients were defined in (4.167) and they are affected by the standard homogeneous Markov process $\{\widehat{\eta}(t)\}_{t\geq 0}$ having the transition semigroup

$$\widehat{P}(t) = e^{Q^{\top}t}, \quad t \geq 0, \tag{4.173}$$

will be named **the dual jump linear stochastic system of the JLSS** (4.162). It is worth noticing that the dual JLSS (4.171)–(4.172) can be defined mathematically rigorously only if the generator matrix Q of the Markov process $\{\eta(t)\}_{t\geq 0}$ which perturb the coefficients of the JLSS (4.162) satisfies a condition of the form (4.169).

Let us consider the generalized Lyapunov-type jump linear differential equation (GLJLDE) on the Hilbert space $\mathscr{S}_n^N$:

$$\dot{\mathbf{Z}}(t) = \widehat{\mathscr{L}}_{\mathbf{H}_c}(t)[\mathbf{Z}(t)], \quad kh < t \leq (k+1)h \tag{4.174a}$$

$$\mathbf{Z}(kh^+) = \widehat{\mathscr{I}}_{\mathbf{H}_d}(kh)[\mathbf{Z}(kh)], \quad k \in \mathbb{Z}, \tag{4.174b}$$

where $\mathbf{Z}(t) = (Z(t,1),\ Z(t,2), \dots, Z(t,N)) \in \mathscr{S}_n^N$,

$$\widehat{\mathscr{L}}_{\mathbf{H}_c}(t)[\mathbf{Z}] = (\widehat{\mathscr{L}}_{\mathbf{H}_c}(t)[\mathbf{Z}](1),\ \widehat{\mathscr{L}}_{\mathbf{H}_c}(t)[\mathbf{Z}](2), \dots, \widehat{\mathscr{L}}_{\mathbf{H}_c}(t)[\mathbf{Z}](N))$$

$$\widehat{\mathscr{I}}_{\mathbf{H}_d}(kh)[\mathbf{Z}] = (\widehat{\mathscr{I}}_{\mathbf{H}_d}(kh)[\mathbf{Z}](1),\ \widehat{\mathscr{I}}_{\mathbf{H}_d}(kh)[\mathbf{Z}](2), \dots, \widehat{\mathscr{I}}_{\mathbf{H}_d}(kh)[\mathbf{Z}](N))$$

with

$$\widehat{\mathscr{L}}_{\mathbf{H}_c}(t)[\mathbf{Z}](i) := (\widehat{A}_0(t,i) + H_c(t,i)\widehat{C}_{c0}(t,i))Z(i) + Z(i)(\widehat{A}_0(t,i) + H_c(t,i)\widehat{C}_{c0}(t,i))^{\top}$$
$$+\sum_{l=1}^{r_1}(\widehat{A}_l(t,i)+Hc(t,i)\widehat{C}_{cl}(t,i))Z(i)(\widehat{A}_l(t,i)+H_c(t,i)\widehat{C}_{cl}(t,i))^{\top}+\sum_{j=1}^{N} q_{ij}Z(j) \tag{4.175a}$$

$$\widehat{\mathscr{I}}_{\mathbf{H}_d}(kh)[\mathbf{Z}](i) = \sum_{p=0}^{r_2}(\widehat{A}_{dp}(kh,i) + H_d(kh,i)\widehat{C}dp(kh,i))Z(i)$$
$$\cdot (\widehat{A}_{dp}(kh,i) + H_d(kh,i)\widehat{C}dp(kh,i))^{\top}. \tag{4.175b}$$

With this notation we may rewrite (4.168) in a compact form as:

$$-\dot{\widetilde{\mathbf{Y}}}(t) = \widehat{\mathscr{L}}^*_{\mathbf{H}_c}(t)[\widetilde{\mathbf{Y}}(t)] + \mathbf{I}_n, \quad kh < t \leq (k+1)h, \tag{4.176a}$$

$$\widetilde{\mathbf{Y}}(kh^-) = \widehat{\mathscr{I}}^*_{\mathbf{H}_d}(kh)[\widetilde{\mathbf{Y}}(kh)] + \mathbf{I}_n, \quad k \in \mathbb{Z}, \tag{4.176b}$$

$$0 \prec \nu_1 \mathbf{I}_n \preceq \widetilde{\mathbf{Y}}(t) \preceq \nu_2 \mathbf{I}_n, \quad t \in \mathbb{R}, \tag{4.176c}$$

where $\mathbf{I}_n := (I_n,\ I_n, \ldots, I_n) \in \mathscr{S}_n^N$. According to Remark 2.10 the GLJLDE (4.174) defines a positive evolution on the ordered Hilbert space $(\mathscr{S}_n^N, \mathscr{S}_n^{N+})$. Hence, the implication $(iii) \Rightarrow (i)$ from Theorem 2.10 applied in the case of non-homogeneous backward GLJLDE (4.176) allows us to conclude that the GLJLDE (4.174) is exponentially stable.

To provide an interpretation in the framework of systems theory of the exponential stability of the generalized Lyapunov-type jump linear differential equation (4.174), let us introduced a new concept of detectability. To this end, let us consider $\widehat{\mathscr{L}}_c : \mathscr{S}_n^N \to \mathscr{S}_n^N$ and $\widehat{\mathscr{L}}_d : \mathscr{S}_n^N \to \mathscr{S}_n^N$ defined as:

$$\widehat{\mathscr{L}}_c(t)[\mathbf{Z}] = (\widehat{\mathscr{L}}_c(t)[\mathbf{Z}](1), \widehat{\mathscr{L}}_c(t)[\mathbf{Z}](2), \ldots, \widehat{\mathscr{L}}_c(t)[\mathbf{Z}](N))$$

$$\widehat{\mathscr{L}}_d(kh)[\mathbf{Z}] = (\widehat{\mathscr{L}}_d(kh)[\mathbf{Z}](1), \widehat{\mathscr{L}}_d(kh)[\mathbf{Z}](2), \ldots, \widehat{\mathscr{L}}_d(kh)[\mathbf{Z}](N))$$

with

$$\widehat{\mathscr{L}}_c(t)[\mathbf{Z}](i) := \widehat{A}_0(t,i)Z(i) + Z(i)\widehat{A}_0^\top(t,i) + \sum_{l=1}^{r_1} \widehat{A}_l(t,i)Z(i)\widehat{A}_l^\top(t,i) + \sum_{j=1}^{N} q_{ij}Z(j) \tag{4.177a}$$

$$\widehat{\mathscr{L}}_d(kh)[\mathbf{Z}](i) = \sum_{p=0}^{r_2} \widehat{A}_{dp}(kh,i)Z(i)\widehat{A}_{dp}^\top(kh,i), \tag{4.177b}$$

$i \in \mathcal{N}$, for all $\mathbf{Z} = (Z(1), Z(2), \ldots, Z(N)) \in \mathscr{S}_n^N$. Further we denote:

$$\widehat{\mathbf{C}}_c(\cdot) \triangleq (\widehat{\mathbb{C}}_{c0}(\cdot), \widehat{\mathbb{C}}_{c1}(\cdot), \ldots, \widehat{\mathbb{C}}_{cr_1}(\cdot)), \qquad \widehat{\mathbf{C}}_d(\cdot) \triangleq (\widehat{\mathbb{C}}_{d0}(\cdot), \widehat{\mathbb{C}}_{d1}(\cdot), \ldots, \widehat{\mathbb{C}}_{dr_2}(\cdot))$$

where

$$\widehat{\mathbb{C}}_{cl}(\cdot) := (\widehat{C}_{cl}(\cdot, 1), \widehat{C}_{cl}(\cdot, 2), \ldots, \widehat{C}_{cl}(\cdot, N)) \tag{4.178a}$$

$$\widehat{\mathbb{C}}_{dp}(\cdot) := (\widehat{C}_{dp}(\cdot, 1), \widehat{C}_{dp}(\cdot, 2), \ldots, \widehat{C}_{dp}(\cdot, N)), \tag{4.178b}$$

$\widehat{C}_{cl}(\cdot, i)$, $0 \leq l \leq r_1$ and $\widehat{C}dp(\cdot, i)$, $0 \leq p \leq r_2$ being defined as in (4.167).

Definition 4.7 We say that the pair $((\widehat{\mathbf{C}}_c(\cdot) \mid \widehat{\mathbf{C}}_d(\cdot)), (\widehat{\mathscr{L}}_c(\cdot) \mid \widehat{\mathscr{L}}_d(\cdot)))$ is detectable if there exists a piecewise continuous and periodic with period $\tilde{\theta}$ function $t \to \mathbf{H}_c(t) = (H_c(t,1), H_c(t,2), \ldots, Hc(t,N)) : \mathbb{R} \to \mathscr{M}_{m_c n}^N$ and the periodic of period

$\tilde{\mathbf{k}}$ matrix-valued sequences $\{H_d(kh,i)\}_{k\in\mathbb{Z}} \subset \mathbb{R}^{m_d\times n}$, $i \in \mathcal{N}$, with the property that the corresponding GLJLDE of type (4.174)–(4.175) is exponentially stable.

Summarizing the previous reasonings we may state:

Proposition 4.13 *Assume that the Assumption 1.2 holds and the JLSS (4.162) is time periodic (in the sense of Definition 4.1). Under these conditions the following are equivalent:*

(i) the JLSS (4.162) is mean-square stabilizable by linear state feedback;
(ii) the corresponding triple $((\widehat{\mathbf{C}}_c(\cdot) \mid \widehat{\mathbf{C}}_d(\cdot)), (\widehat{\mathcal{L}}_c(\cdot) \mid \widehat{\mathcal{L}}_d(\cdot)), \{kh\}_{k\in\mathbb{Z}})$ *is mean-square detectable.*

Remark 4.8 The triple $((\widehat{\mathbf{C}}_c(\cdot) \mid \widehat{\mathbf{C}}_d(\cdot)), (\widehat{\mathcal{L}}_c(\cdot) \mid \widehat{\mathcal{L}}_d(\cdot)), \{kh\}_{k\in\mathbb{Z}})$ associated to the JLSS (4.162) via (4.177)–(4.178) will be named **dual triple** associated to the time periodic JLSS (4.162) without appealing to the condition (4.169).

In the last part of this section, let us make a short analyze between the duality of the property of mean-square detectability and the property of mean-square stabilizability in the case of a JLSS of the form (4.108)–(4.109).

Assume that the JLSS (4.108)–(4.109) is time periodic, its coefficients can be extended in a natural way by periodicity to the whole real axis. In this case, for each $i \in \mathcal{N}$ we define:

$$\begin{aligned} &\breve{A}_l(t,i) := A_l^\top(-t,i), \quad \breve{B}_l(t,i) := C_{cl}^\top(-t,i), \quad 0 \le l \le r_1, \quad t \in \mathbb{R}, \qquad (4.179)\\ &\breve{A}_{dp}(kh,i) : = A_{dp}^\top(-kh,i), \ \breve{B}_{dp}(kh,i) : = C_{dp}^\top(-kh,i), k \in \mathbb{Z}, \ 0 \le p \le r_2. \end{aligned}$$

With this notation we define the linear operators $\breve{\mathcal{L}}_c(t) : \mathcal{S}_n^N \to \mathcal{S}_n^N$ and $\breve{\mathcal{L}}_d(kh) : \mathcal{S}_n^N \to \mathcal{S}_n^N$ as:

$$\breve{\mathcal{L}}_c(t)[\mathbf{Z}] = (\breve{\mathcal{L}}_c(t)[\mathbf{Z}](1), \breve{\mathcal{L}}_c(t)[\mathbf{Z}](2), \ldots, \breve{\mathcal{L}}_c(t)[\mathbf{Z}](N))$$

$$\breve{\mathcal{L}}_d(kh)[\mathbf{Z}] = (\breve{\mathcal{L}}_d(kh)[\mathbf{Z}](1), \breve{\mathcal{L}}_d(kh)[\mathbf{Z}](2), \ldots, \breve{\mathcal{L}}_d(kh)[\mathbf{Z}](N))$$

where

$$\breve{\mathcal{L}}_c(t)[\mathbf{Z}](i) = \breve{A}_0(t,i)Z(i) + Z(i)\breve{A}_0^\top(t,i) + \sum_{l=1}^{r_1} \breve{A}_l(t,i)Z(i)\breve{A}_l^\top(t,i) + \sum_{j=1}^{N} q_{ij}Z(j) \qquad (4.180a)$$

$$\breve{\mathcal{L}}_d(kh)[\mathbf{Z}](i) = \sum_{p=0}^{r_2} \breve{A}_{dp}(kh,i)Z(i)\breve{A}_{dp}^\top(kh,i), \quad k \in \mathbb{Z}. \qquad (4.180b)$$

Also, we denote

$$\breve{\mathbf{B}}_c(\cdot)=(\breve{\mathbb{B}}_0(\cdot), \breve{\mathbb{B}}_1(\cdot), \ldots, \breve{\mathbb{B}}_{r_1}(\cdot)), \quad \breve{\mathbf{B}}_d(\cdot)=(\breve{\mathbb{B}}_{d0}(\cdot), \breve{\mathbb{B}}_{d1}(\cdot), \ldots, \breve{\mathbb{B}}_{dr_2}(\cdot))$$

where

$$\breve{\mathbb{B}}_l(\cdot) = (\breve{B}_l(\cdot, 1), \breve{B}_l(\cdot, 2), \ldots, \breve{B}_l(\cdot, N)) \tag{4.181a}$$

$$\breve{\mathbb{B}}_{dp}(\cdot) = (\breve{B}_{dp}(\cdot, 1), \breve{B}_{dp}(\cdot, 2), \ldots, \breve{C}_{dp}(\cdot, N)), \tag{4.181b}$$

$0 \leq l \leq r_1, 0 \leq p \leq r_2$, $\breve{B}_l(\cdot, i)$ and $\breve{B}_{dp}(\cdot, i)$ being defined in (4.179). The following definition extends to the operator case, the concept of mean-square stabilizability introduced in Definition 4.3.

Definition 4.8 We say that the triple $((\breve{\mathscr{L}}_c(\cdot) \mid \breve{\mathscr{L}}_d(\cdot)), (\breve{\mathbf{B}}_c(\cdot) \mid \breve{\mathbf{B}}_d(\cdot)), \{kh\}_{k\in\mathbb{Z}})$ is stabilizable by linear state feedback if there exist bounded and piecewise continuous periodic of period $\tilde{\theta}$ functions $t \to H_c(t, i) : \mathbb{R} \to \mathbb{R}^{m_c \times n}$, $i \in \mathcal{N}$ and matrix-valued sequences $\{H_d(kh, i)\}_{k\in\mathbb{Z}} \subset \mathbb{R}^{m_d \times n}$, $i \in \mathcal{N}$ periodic of period $\tilde{\mathbf{k}}$ ($\tilde{\theta}$ and $\tilde{\mathbf{k}}$ being introduced in (4.9)) with the property that the corresponding generalized Lyapunov-type jump linear differential equation

$$\begin{aligned}\dot{Z}(t, i) &= (\breve{A}_0(t, i) + \breve{B}_0(t, i)H_c(t, i))Z(t, i) + Z(t, i)(\breve{A}_0(t, i) + \breve{B}_0(t, i)H_c(t, i))^\top \\ &\quad + \sum_{l=1}^{r_1}(\breve{A}_l(t, i) + \breve{B}_l(t, i)H_c(t, i))Z(t, i)(\breve{A}_l(t, i) + \breve{B}_l(t, i)H_c(t, i))^\top \\ &\quad + \sum_{j=1}^{N} q_{ij} Z(t, j), \quad kh < t \leq (k+1)h\end{aligned} \tag{4.182a}$$

$$\begin{aligned}Z(kh^+, i) &= \sum_{p=0}^{r_2}(\breve{A}_{dp}(kh, i) + \breve{B}_{dp}(kh, i)H_d(kh, i))Z(kh, i) \\ &\quad \cdot (\breve{A}_{dp}(kh, i) + \breve{B}_{dp}(kh, i)H_d(kh, i))^\top,\end{aligned} \tag{4.182b}$$

$i \in \mathcal{N}$ is exponentially stable.

With a slight modification of the reasonings from the first part of this subsection, we can obtain:

Proposition 4.14 *Assume that the Assumption 1.2 holds and the JLSS (4.108)–(4.109) is time periodic. Under these conditions, the following equivalences are true:*

(a) if the components q_{ij} of the generator matrix Q of the Markov process are satisfying the additional conditions (4.169), then the following are equivalent:

(i) the JLSS (4.108)–(4.109) is mean-square detectable;
(ii) the dual jump linear stochastic system

$$\begin{aligned}dx(t) &= (\breve{A}_0(t, \widehat{\eta}(t))x(t) + \breve{B}_0(t, \widehat{\eta}(t))u_c(t))dt \\ &\quad + \sum_{l=1}^{r_1}(\breve{A}_l(t, \widehat{\eta}(t))x(t) + \breve{B}_l(t, \widehat{\eta}(t)))dw_l(t), \quad kh < t \leq (k+1)h,\end{aligned} \tag{4.183a}$$

$$x(kh^+) = (\breve{A}_{d0}(kh, \widehat{\eta}(kh))x(kh) + \breve{B}_{d0}(kh, \widehat{\eta}(kh))u_d(k)) \tag{4.183b}$$
$$+ \sum_{p=1}^{r_2} w_{dp}(k)(\breve{A}_{dp}(kh, \widehat{\eta}(kh))x(kh) + \breve{B}_{dp}(kh, \widehat{\eta}(kh))u_d(k)),\ k \in \mathbb{Z}_+,$$

is mean-square stabilizable by linear state feedback, where $\breve{A}_l(\cdot, i)$, $\breve{B}_l(\cdot, i)$, $0 \le l \le r_1$, $\breve{A}_{dp}(\cdot, i)$, $\breve{B}_{dp}(\cdot, i)$, $0 \le p \le r_2$ *are defined as in (4.179) and* $\{\widehat{\eta}(t)\}_{t\ge 0}$ *is a standard homogeneous Markov process independent of* $\{w(t)\}_{t\ge 0}$ *and* $\{w_d(k)\}_{k\in\mathbb{Z}_+}$ *defined on the probability space* $(\Omega, \mathfrak{F}, \mathscr{P})$ *taking values in the set* $\mathscr{N} = \{1, 2, \ldots, N\}$ *and having the transition semigroup* $\widehat{P}(t)$ *defined in (4.173).*

(b) The following are equivalent:
(j) the JLSS (4.108)–(4.109) is mean-square detectable;
(jj) the corresponding triple $((\breve{\mathscr{L}}_c(\cdot) \mid \breve{\mathscr{L}}_d(\cdot)), (\breve{\mathbf{B}}(\cdot) \mid \breve{\mathbf{B}}_d(\cdot)), \{kh\}_{k\in\mathbb{Z}})$ *is stabilizable by linear state feedback.*

Proof is left to the readers.

Remark 4.9

(a) the triple $((\breve{\mathscr{L}}_c(\cdot) \mid \breve{\mathscr{L}}_d(\cdot)), (\breve{\mathbf{B}}_c(\cdot) \mid \breve{\mathbf{B}}_d(\cdot)), \{kh\}_{k\in\mathbb{Z}})$ defined by (4.180)–(4.181) will be named **dual triple** associated to the system (4.108)–(4.109).
(b) If the condition (4.169) is fulfilled, then the stabilizability by state feedback of the triple $((\breve{\mathscr{L}}_c(\cdot) \mid \breve{\mathscr{L}}_d(\cdot)), (\breve{\mathbf{B}}_c(\cdot) \mid \breve{\mathbf{B}}_d(\cdot)), \{kh\}_{k\in\mathbb{Z}})$ is equivalent to the mean-square stabilizability by linear state feedback of the JLSS (4.183).
(c) The results stated in Propositions 4.11-4.14 may be adapted mutatis-mutandis to the case when the JLSS (4.108)–(4.109), JLSS (4.113)–(4.114) and JLSS (4.162), respectively, are time invariant, because they may be viewed as being time periodic with period $\theta_c = h$, $\theta_d = 1$, $\tilde{\theta} = h$ and $\tilde{\mathbf{k}} = 1$.

4.7 Notes and References

Systems structural properties play a central role in control theory. In this chapter, we addressed stabilizability and detectability properties.

Stochastic stabilizability for systems described by linear Itô differential equations has been introduced in Willems [7] where necessary and sufficient conditions for mean-square stabilizability are derived in terms of a solution to a Riccati-type algebraic equation. Ghaoui [8] and Rami and Zhou [9] obtained mean-square stabilizability necessary and sufficient conditions in terms of linear matrix inequalities (LMIs). Different from the aforementioned approaches, the authors in Zhang and Chen [10] proposed a spectral operator approach and obtained necessary and sufficient conditions for mean-square stabilizability which has been applied to characterize some solutions of stochastic algebraic Riccati equations. Zhang et al. [11]

further developed the spectral approach and introduced the notions of critical stabilizability and weak stabilizability. Recently, Hou et al. [12] introduced the notion of $\mathscr{D}(0, \alpha)$-stabilizability, $0 < \alpha \leq 1$, for which a necessary and sufficient condition is also proposed via an LMI-based approach, thanks to operator spectral theory.

For the class of MJLSs, Ji and Chizeck [13]; Ji et al. [14]; Feng et al. [15] used Lyapunov's second method in order to obtain necessary and sufficient conditions for second moment stabilizability of both continuous-time and discrete-time MJLSs. In Costa and Boukas [16], the authors introduced the notion of quadratic stabilizability of the class of continuous-time MJLSs with norm-bound uncertainties in the parameters and obtained a necessary and sufficient mean-square quadratic stabilizability condition via a LMI optimization problem. In Fragoso and Costa [17], necessary and sufficient mean-square stabilizability conditions are obtained using a LMI approach when only partial information on the mode parameter is available to the controller. The notion of uniform stabilizability has been defined and characterized in Lee and Dullerud [18, 19]. For stochastic systems that include both Markov jump perturbations as well as state/control dependent noise, one can refer to Drăgan et al. [1] and the references therein.

On the other hand, detectability notions arise naturally in control and filtering problems. They can be traced back to the late sixties Wohnam [20]; Kwakernaak and Sivan [21] and have been continuously studied over time as they play important roles in theory and practice. Then, detectability has been studied for different plants and problems, and consequently, it has been defined in different ways. The seminal paper Anderson and Moore [22] considers a definition for time-varying systems, which is adapted to several other contexts afterwards, including Markov jump systems Costa and Val [23]; Costa and do Val [24]; Shen et al. [25], Shen [26]. Sometimes the starting point is the stabilizability of a dual system Fragoso et al. [27]; Freiling and Hochhaus [28]. This characterization is known in literature as stochastic detectability. Note however that the stabilizability of the dual system does not give rise to a realizable observer equation. We also mention here that in the stochastic framework the property that all unstable modes produce some non-zero output is only a necessary but not sufficient condition for stabilizability of the dual system while in the deterministic case one gets equivalence between these two definitions. The discussion above shows that the definitions of detectability and observability in the stochastic framework offer some very specific challenges, when compared to there deterministic counterparts. Several authors devoted there efforts to address some of the above raised points. Different notions and tests are gathered and demonstrated equivalent in special scenarios. Along this line, Shen et al. [25]; Shen [26] in the context of linear systems both in discrete-time and continuous-time, with multiplicative noise and parameters affected by Markovian switching; Zhang and Tan [29]; Zhang et al. [30] compares a number of detectability notions, including some new ones, for time-varying stochastic systems. See also Tai [31]; Li et al. [32] and a fairly complete discussion in Zhang et al. [30] and references therein. We also cite here the pioneering works of Damm [33]; Zhang et al. [11] where the authors succeeded to prove the equivalence between the stochastic version of the PBH-test and the *natural* concept of detectability of the system in the sense that all unstable modes produce some non-zero output. This natural concept is the so-called *exact detectability* notion (Definition 3.1 from Zhang et al. [11]). A PBH criterion for exact observability of a stochastic linear system described by Itô differential equations was established in Theorem 4 from Zhang and Chen [10]. Such a criterion was then extended in Damm [33] and Zhang et al. [11] to the exact detectability of the same class of stochastic systems. Recently, in Drăgan et al. [34] the authors considered a large class of linear stochastic systems with time-varying coefficients simultaneously affected by state multiplicative white noise perturbations and Markovian switching, both in discrete and continuous time. They introduced the notion of exact detectability for this class of systems and a PBH test for it. The main tool used is the theory of positive operators.

Regarding structural properties of impulsive stochastic systems, and to the best of the authors knowledge, there are currently very few results about structural properties available in the literature. We can cite Shen et al. [35] where the problem of controllability of linear Itô-type impulsive stochastic systems in Hilbert spaces has been addressed. The authors introduced the concepts of null controllability and approximate null controllability for this class of systems. They obtained necessary and sufficient conditions for the null controllability and the approximate null controllability by using a backward stochastic differential equations viewpoint. In Chen et al. [36], the authors studied observability of the same class of linear impulsive stochastic systems. The concept of 2-pth moment observability is first introduced. Using the generalized $\mathscr{H}$-representation technique Zhang and Chen [37], the authors established a deterministic expanded system for the linear stochastic impulsive systems and then showed that the 2-pth moment observability of the original system is equivalent to the observability of the associated deterministic system. For the class of linear stochastic impulsive systems with Markovian jump parameters, and to the best of the authors knowledge, there are no results in the literature about structural properties.

References

1. Drăgan, V., Morozan, T., Stoica, A.M.: Mathematical Methods in Robust Control of Linear Stochastic Systems. Springer-Verlag, New York (2013)
2. Drăgan, V., Morozan, T., Stoica, A.M.: Mathematical Methods in Robust Control of Discrete-Time Linear Stochastic Systems. Springer-Verlag, New York (2010)
3. Ichikawa, A., Katayama, H.: Linear Time Varying Systems and Sampled-data Systems. Springer, London (2001)
4. Briat, C.: Stability analysis and stabilization of stochastic linear impulsive, switched and sampled-data systems under dwell-time constraints. Automatica **74**(1), 279–287 (2016)
5. Geromel, J.C., Gabriel, G.W.: Optimal h_2 state feedback sampled-data control design of Markov jump linear systems. Automatica **54**(1), 182–188 (2015)
6. Anderson, B., Moore, J.: Optimal Control: Linear Quadratic Methods. Prentice-Hall, Englewood Cliffs, NJ (1989)
7. Willems, J.: in Stochastic Problems in Dynamics. Pitman, London (1977)
8. Ghaoui, L.E.: State-feedback control of systems with multiplicative noise via linear matrix inequalities. Syst. Cont. Lett. **24**, 223–228 (1995)

9. Rami, M.A., Zhou, X.Y.: Linear matrix inequalities, riccati equations, and inde8nite stochastic linear quadratic control. IEEE Trans. Automat. Cont. **45**, 1131–1142 (2000)
10. Zhang, W., Chen, B.S.: On stabilizability and exact observability of stochastic systems with their applications. Automatica **40**, 87–94 (2004)
11. Zhang, W., Zhang, H., Chen, B.S.: Generalized lyapunov equation approach to state-dependendent stochastic stabilization/detectability criterion. IEEE Trans. Automat. Cont. **53**, 1630–1642 (2008)
12. Hou, T., Zhang, W., Chen, B.S.: Study on general stability and stabilizability of linear discrete-time stochastic systems. Asian J. Cont. **13**, 977–987 (2011)
13. Ji, Y., Chizeck, H.J.: Controllability, stabilizability, and jump linear quadratic control. IEEE Trans. Automat. Contr. **35**, 777–788 (1990)
14. Ji, Y., Chizeck, H.J., Feng, X., Loparo, K.A.: Stability and control of discrete-time jump linear systems. Cont. Theo. Adv. Technol. **7**, 247–270 (1991)
15. Feng, X., Loparo, K.A., Ji, Y., Chizeck, H.J.: Stochastic stability properties of jump linear systems. IEEE Trans. Automat. Contr. **37**, 38–53 (1992)
16. Costa, O.L.V., Boukas, E.K.: Necessary and sufficient condition for robust stability and stabilizability of continuous-time linear systems with markovian jumps. J. Optim. Theory Appl. **99**, 359–379 (1998)
17. Fragoso, M.D., Costa, O.L.V.: Mean square stabilizability of continuous-time linear systems with partial information on the markovian jumping parameters. Stoch. Anal. Appl. **22**, 99–111 (2004)
18. Lee, J.W., Dullerud, G.E.: Uniform stabilization of discrete-time switched and markovian jump linear systems. Automatica **42**, 205–218 (2006)
19. Lee, J.W., Dullerud, G.E.: A stability and contractiveness analysis of discrete-time markovian jump linear systems. Automatica **43**, 168–173 (2007)
20. Wohnam, W.M.: On a matrix Riccati equation of stochastic control. SIAM J. Control Optim. **6**, 312–326 (1968)
21. Kwakernaak, H., Sivan, R.: Linear Optimal Control Systmes. Wiley-Interscience, New York (1972)
22. Anderson, B.D.O., Moore, J.B.: Detectability and stabilizability of time-varying discrete-time linear systems. SIAM J. Control Optim. **19**, 20–32 (1981)
23. Costa, E., Val, J.D.: On the detectability and observability of discrete-time Markovian jump linear systems. Syst. Control Lett. **44**, 135–145 (2001)
24. Costa, E.F., do Val, J.B.R.: On the observability and detectability of continuous-time Markov jump linear systems. SIAM J. Cont. Optim. **41**, 1295–1314 (2002)
25. Shen, L., Sun, J., Wu, Q.: Observability and detectability of discrete-time stochastic systems with Markovian jump. Syst. Control Lett. **62**, 37–42 (2013)
26. Shen, L.: On the detectability and observability of continuous stochastic Markov jump linear systems. J. Math. Anal. Appl. **424**(2), 878–891 (2015). https://doi.org/10.1016/j.jmaa.2014.11.042. http://www.sciencedirect.com/science/article/pii/S0022247X14010841
27. Fragoso, M.D., Costa, O.L.V., de Souza, C.E.: A new approach to linearly perturbed Riccati equations in stochastic control. Appl. Math. Optim. **37**, 99–126 (1998)
28. Freiling, G., Hochhaus, A.: On a class of rational matrix differential equations arising in stochastic control. Linear Algebra Appl. **379**, 43–68 (2004)
29. Zhang, W., Tan, C.: On detectability and observability of discrete-time stochastic Markov jump systems with state-dependent noise. Asian J. Cont. **15**, 1366–1375 (2013)
30. Zhang, W., Zheng, W.X., Chen, B.S.: Detectability, observability and Lyapunov-type theorems of linear discrete time-varying stochastic systems with multiplicative noise. Int. J. Cont. **90**(11), 2490–2507 (2017). https://doi.org/10.1080/00207179.2016.1257152
31. Tai, H.: On detectability and observability of discrete-time stochastic Markova jump systems with state-dependent noise. Syst. Cont. Lett. **8**, 425–428 (1987)
32. Li, Z.Y., Wang, Y., Zhou, B., Duan, G.R.: On unified concepts of detectability and observability for continuous-time stochastic systems. Appl. Math. Comput. **217**(2), 521–536 (2010). https://doi.org/10.1016/j.amc.2010.05.086. http://www.sciencedirect.com/science/article/pii/S0096300310006557

33. Damm, T.: On detectability of stochastic systems. Automatica **43**(5), 928–933 (2007)
34. Drăgan, V., Costa, E.F., Popa, I.L., Aberkane, S.: Exact detectability of discrete-time and continuous-time linear stochastic systems with periodic coefficients: a unified approach. IEEE Trans. Autom. Cont. **67**, 5730–5745 (2022)
35. Shen, L., Sun, J., Wu, Q.: Controllability of linear impulsive stochastic systems in Hilbert spaces. Automatica **49**, 1026–1030 (2013)
36. Chen, X., Luo, S., Chen, W.H.: Moment observability and output feedback stabilisation for linear stochastic impulsive systems. Int. J. Syst. Sci. **54**, 1015–1032 (2023)
37. Zhang, W., Chen, B.S.: $\mathcal{H}$-representation and applications to generalized Lyapunov equations and linear stochastic systems. IEEE Trans. Autom. Cont. **57**, 3009–3022 (2012)

Chapter 5
Jump Matrix Riccati Differential Equations

Abstract This Chapter is devoted to the study of several Riccati-type differential equations, namely: • ***jump matrix Riccati differential equations of stochastic control/filtering***: such equations arise in connection with the solution of several optimal control/filtering problems for a large class of jump stochastic linear differential equation. • ***jump matrix linear differential equation with Riccati-type jumping operator***: such equations are related to some optimal control problems for a class of jump linear stochastic system controlled by impulses, *i.e.*, the control actions take places only at the jump time instances. The above equations can be viewed as particular cases of a larger class of jump matrix nonlinear differential equations. In order to obtain the most general results, we shall systematically study such a general mathematical object without referring to any optimal control/filtering problem. The obtained results will then be specialized in a second step in order to cope with the Riccati-type equations cited above. We begin by analyzing the problem of the global existence on a whole given interval $[t_0, T]$ of a solution with given terminal values of a backward jump matrix nonlinear differential equation. Then, we shall study the problem of the existence on an unbounded interval $[t_0, \infty)$ of some particular solutions such as the maximal solution, the minimal solution and the stabilizing solution, respectively. We provide conditions which guarantee the existence and the uniqueness of such global solutions. The proofs are mainly based on positivity properties of linear evolution operators defined by the generalized Lyapunov differential equations with jumps studied in Chap. 2. The last part of this chapter will be devoted to the numerical aspects of computing such global solutions for the considered jump Riccati-type differential equations. We shall provide Newton-like iterative algorithms and obtain the conditions that guarantee the global convergence of such numerical schemes.

5.1 A Class of Jump Matrix Nonlinear Differential Equations

5.1.1 Several Conventions of Notation

If $N \geq 1$ is a natural number $\mathscr{N} := \{1, 2, \ldots, N\}$. Beside the Hilbert space $\mathscr{S}_n^N$ introduced in Sect. 2.5.1, in the sequel, we shall use the following vector spaces

V. Drăgan et al., *Robust Control of Jump Linear Stochastic Systems*, Lecture Notes in Control and Information Sciences 497, https://doi.org/10.1007/978-3-031-84070-8_5

whose elements are finite sequences of matrices (which are not necessarily symmetric matrices). If m, n, N are fixed natural numbers, we denote

$$\mathscr{M}_{nm}^{N} \stackrel{\Delta}{=} \underbrace{\mathbb{R}^{n\times m} \times \mathbb{R}^{n\times m} \times \ldots \times \mathbb{R}^{n\times m}}_{N \text{ times}}.$$

So, the elements $\mathbf{X}$ of $\mathscr{M}_{nm}^{N}$ are finite sequences of matrices of the form $\mathbf{X} = (X(1), X(2), \ldots, X(N))$ where $X(j) \in \mathbb{R}^{n\times m}$, $1 \le j \le N$, are arbitrary matrices of size $n \times m$. If $m = n$, we shall write $\mathscr{M}_{n}^{N}$ instead of $\mathscr{M}_{nn}^{N}$. It is obvious that $\mathscr{S}_{n}^{N} \subset \mathscr{M}_{n}^{N}$. On the vector space $\mathscr{M}_{nm}^{N}$ we consider the inner product

$$\langle \mathbf{M}, \mathbf{S} \rangle = \sum_{i=1}^{N} Tr[S^{\top}(i)M(i)], \tag{5.1}$$

for all $\mathbf{M} = (M(1), M(2), \ldots, M(N))$, $\mathbf{S} = (S(1), S(2), \ldots, S(N)) \in \mathscr{M}_{nm}^{N}$. When $m = n$, (5.1) coincides to (2.111) if $\mathbf{M}, \mathbf{S} \in \mathscr{S}_{n}^{N} \subset \mathscr{M}_{n}^{N}$. Equipped with the inner product (5.1), $\mathscr{M}_{nm}^{N}$ becomes a real Hilbert space. By $|\cdot|_2$ we denote the norm induced on $\mathscr{M}_{nm}^{N}$ by the inner product (5.1), that is $|\mathbf{M}|_2 = (\langle \mathbf{M}, \mathbf{M} \rangle)^{1/2}$, for all $\mathbf{M} \in \mathscr{M}_{nm}^{N}$. To ease the understanding of the formulae from this chapter, we introduce the following conventions regarding some algebraic operations involving elements of the vector spaces of the form $\mathscr{M}_{nm}^{N}$.

(C1) If $\mathbf{B} = (B(1), B(2), \ldots, B(N)) \in \mathscr{M}_{nm}^{N}$, $\mathbf{X} = (X(1), X(2), \ldots, X(N)) \in \mathscr{M}_{mp}^{N}$ then $\mathbf{BX} = (B(1)X(1), B(2)X(2), \ldots, B(N)X(N)) \in \mathscr{M}_{np}^{N}$.

(C2) If $\mathbf{C} = (C(1), C(2), \ldots, C(N)) \in \mathscr{M}_{nm}^{N}$ then $\mathbf{C}^{\top} := (C^{\top}(1), C^{\top}(2), \ldots, C^{\top}(N)) \in \mathscr{M}_{mn}^{N}$.

(C3) If $\mathbf{A} = (A(1), A(2), \ldots, A(N)) \in \mathscr{M}_{n}^{N}$ is such that $\det A(i) \neq 0$, $1 \le i \le N$, then $\mathbf{A}^{-1} := (A^{-1}(1), A^{-1}(2), \ldots, A^{-1}(N)) \in \mathscr{M}_{n}^{N}$.

Throughout this chapter $\mathfrak{T} = \{\tau_k\}_{k\in\mathbb{Z}_+}$ is a sequence of real numbers such that: $0 = \tau_0 < \tau_1 < \ldots < \tau_k < \tau_{k+1} < \ldots$.

If $(\mathscr{X}, \|\cdot\|)$ is a normed space, then $\mathscr{PC}\{\mathscr{I}, \mathscr{X}; \mathfrak{T}\}$ denotes the set of all functions $f : \mathscr{I} \subset \mathbb{R} \to \mathscr{X}$ which are continuous on each interval $(\tau_k, \tau_{k+1}) \cap \mathscr{I}$ having finite lateral limits at the borders of each intervals. In this chapter we shall use the cases $\mathscr{X} \leftarrow \mathscr{M}_{mn}^{N}$ and $\mathscr{X} \leftarrow \mathscr{S}_{n}^{N}$.

If $\mathscr{X}$, $\mathscr{Y}$ are normed spaces then $\mathscr{B}(\mathscr{X}, \mathscr{Y})$ denotes the set of all linear and bounded operators $\mathscr{L} : \mathscr{X} \to \mathscr{Y}$. If $\mathscr{Y} = \mathscr{X}$ we write $\mathscr{B}(\mathscr{X})$ instead of $\mathscr{B}(\mathscr{X}, \mathscr{X})$.

5.1.2 A Class of Jump Matrix Nonlinear Differential Equations

On the Hilbert space $\mathcal{S}_n^N$ we consider the backward jump nonlinear differential equation:

$$\begin{aligned}
-\dot{\mathbf{X}}(t) &= \mathbf{A}^\top(t)\mathbf{X}(t) + \mathbf{X}(t)\mathbf{A}(t) + \boldsymbol{\Pi}_{c1}[\mathbf{X}(t)] \\
&\quad - (\mathbf{X}(t)\mathbf{B}(t) + \boldsymbol{\Pi}_{c2}(t)[\mathbf{X}(t)] + \mathbf{L}_c(t)) \\
&\quad \cdot (\boldsymbol{\Pi}_{c3}(t)[\mathbf{X}(t)] + \mathbf{R}_c(t))^{-1}(\mathbf{X}(t)\mathbf{B}(t) + \boldsymbol{\Pi}_{c2}(t)[\mathbf{X}(t)] + \mathbf{L}_c(t))^\top \\
&\quad + \mathbf{M}_c(t), \quad \tau_k \le t < \tau_{k+1} \qquad (5.2a)
\end{aligned}$$

$$\begin{aligned}
\mathbf{X}(\tau_k^-) &= \boldsymbol{\Pi}_{d1}(\tau_k)[\mathbf{X}(\tau_k)] - (\boldsymbol{\Pi}_{d2}(\tau_k)[\mathbf{X}(\tau_k)]+\mathbf{L}_d(\tau_k))(\boldsymbol{\Pi}_{d3}(\tau_k)[\mathbf{X}(\tau_k)]+\mathbf{R}_d(\tau_k))^{-1} \\
&\quad \cdot (\boldsymbol{\Pi}_{d2}(\tau_k)[\mathbf{X}(\tau_k)] + \mathbf{L}_d(\tau_k))^\top + \mathbf{M}_d(\tau_k), \quad k \in \mathbb{Z}_+, \qquad (5.2b)
\end{aligned}$$

with unknown function $\mathbf{X}(\cdot) = (X(\cdot,1), X(\cdot,2), \dots, X(\cdot,N))$ taking values in $\mathcal{S}_n^N$.

In (5.2), the sequence $\mathfrak{T} = \{\tau_k\}_{k\in\mathbb{Z}_+}$ of the impulse time instances is satisfying the conditions from the Assumption 2.1. If $\mathbf{X} \in \mathcal{S}_n^N$, then $\boldsymbol{\Pi}_{cj}(t)[\mathbf{X}] = (\Pi_{cj}(t)[\mathbf{X}](1), \Pi_{cj}(t)[\mathbf{X}](2), \dots, \Pi_{cj}(t)[\mathbf{X}](N))$ and $\boldsymbol{\Pi}_{dj}(t)[\mathbf{X}] = (\Pi_{dj}(t)[\mathbf{X}](1), \Pi_{dj}(t)[\mathbf{X}](2), \dots, \Pi_{dj}(t)[\mathbf{X}](N))$, $j = 1, 2, 3$, where $\Pi_{c1}(t)[\mathbf{X}](i)$, $\Pi_{d1}(\tau_k)[\mathbf{X}](i) \in \mathcal{S}_n$, $\Pi_{c2}(t)[\mathbf{X}](i) \in \mathbb{R}^{n\times m_c}$, $\Pi_{d2}(\tau_k)[\mathbf{X}](i) \in \mathbb{R}^{n\times m_d}$ and $\Pi_{c3}(t)[\mathbf{X}](i) \in \mathcal{S}_{m_c}$, $\Pi_{d3}(\tau_k)[\mathbf{X}](i) \in \mathcal{S}_{m_d}$, for each $i \in \mathcal{N}$ and for all $\mathbf{X} \in \mathcal{S}_n^N$.

This allows us to write the following partition of (5.2)

$$\begin{aligned}
-\dot{X}(t,i) &= A^\top(t,i)X(t,i) + X(t,i)A(t,i) + \Pi_{c1}(t)[\mathbf{X}(t)](i) \qquad (5.3a) \\
&\quad - (X(t,i)B(t,i) + \Pi_{c2}(t)[\mathbf{X}(t)](i) + L_c(t,i)) \\
&\quad (\Pi_{c3}(t)[\mathbf{X}(t)](i) + R_c(t,i))^{-1} \\
&\quad \cdot (X(t,i)B(t,i) + \Pi_{c2}(t)[\mathbf{X}(t)](i) + L_c(t,i))^\top + M_c(t,i), \quad \tau_k \le t < \tau_{k+1} \\
X(\tau_k^-, i) &= \Pi_{d1}(\tau_k)[\mathbf{X}(\tau_k)](i) - (\Pi_{d2}(\tau_k)[\mathbf{X}(\tau_k)](i) + L_d(\tau_k, i)) \\
&\quad \cdot (\Pi_{d3}(\tau_k)[\mathbf{X}(\tau_k)](i) + R_d(\tau_k, i))^{-1}(\Pi_{d2}(\tau_k)[\mathbf{X}(\tau_k)](i) + L_d(\tau_k, i))^\top \\
&\quad + M_d(\tau_k, i), \quad k \in \mathbb{Z}_+, \quad i \in \mathcal{N}. \qquad (5.3b)
\end{aligned}$$

For each $t \in \mathbb{R}_+$ and for each $\tau_k \in \mathfrak{T}$ we consider the operators $\mathbf{X} \to \boldsymbol{\Pi}_c(t)[\mathbf{X}] : \mathcal{S}_n^N \to \mathcal{S}_{n+m_c}^N$, $\mathbf{X} \to \boldsymbol{\Pi}_d(\tau_k)[\mathbf{X}] : \mathcal{S}_n \to \mathcal{S}_{n+m_d}^N$ defined by

$$\boldsymbol{\Pi}_c(t)[\mathbf{X}] = (\Pi_c(t)[\mathbf{X}](1), \Pi_c(t)[\mathbf{X}](2), \dots, \Pi_c(t)[\mathbf{X}](N)) \qquad (5.4a)$$

$$\Pi_c(t)[\mathbf{X}](i) \triangleq \begin{pmatrix} \Pi_{c1}(t)[\mathbf{X}](i) & \Pi_{c2}(t)[\mathbf{X}](i) \\ \Pi_{c2}(t)[\mathbf{X}](i)^\top & \Pi_{c3}(t)[\mathbf{X}](i) \end{pmatrix} \qquad (5.4b)$$

and

$$\boldsymbol{\Pi}_d(\tau_k)[\mathbf{X}] = (\Pi_d(\tau_k)[\mathbf{X}](1), \Pi_d(\tau_k)[\mathbf{X}](2), \ldots, \Pi_d(\tau_k)[\mathbf{X}](N)) \tag{5.5a}$$

$$\Pi_d(\tau_k)[\mathbf{X}](i) \stackrel{\Delta}{=} \begin{pmatrix} \Pi_{d1}(\tau_k)[\mathbf{X}](i) & \Pi_{d2}(\tau_k)[\mathbf{X}](i) \\ \Pi_{d2}(\tau_k)[\mathbf{X}](i)^\top & \Pi_{d3}(\tau_k)[\mathbf{X}](i) \end{pmatrix} \tag{5.5b}$$

for all $\mathbf{X} \in \mathcal{S}_n^N$ and $i \in \mathcal{N}$. The coefficients of (5.2) are satisfying the assumption:

Assumption 5.1 (a) $t \to \mathbf{A}(t) = (A(t,1), A(t,2), \ldots, A(t,N)) : \mathbb{R}_+ \to \mathcal{M}_n^N$
$t \to \mathbf{B}(t) = (B(t,1), B(t,2), \ldots, B(t,N)) : \mathbb{R}_+ \to \mathcal{M}_{nm_c}^N$
$t \to \mathbf{L}_c(t) = (L_c(t,1), L_c(t,2), \ldots, L_c(t,N)) : \mathbb{R}_+ \to \mathcal{M}_{nm_c}^N$
$t \to \mathbf{M}_c(t) = (M_c(t,1), M_c(t,2), \ldots, M_c(t,N)) : \mathbb{R}_+ \to \mathcal{S}_n^N$
$t \to \mathbf{R}_c(t) = (R_c(t,1), R_c(t,2), \ldots, R_c(t,N)) : \mathbb{R}_+ \to \mathcal{S}_{m_c}^N$
are bounded and continuous functions.

(b) $t \to \boldsymbol{\Pi}_{c1}(t)[\cdot] : \mathbb{R}_+ \to \mathcal{B}(\mathcal{S}_n^N)$, $\quad t \to \boldsymbol{\Pi}_{c2}(t)[\cdot] : \mathbb{R}_+ \to \mathcal{B}(\mathcal{S}_n^N, \mathcal{M}_{nm_c}^N)$
$t \to \boldsymbol{\Pi}_{c3}(t)[\cdot] : \mathbb{R}_+ \to \mathcal{B}(\mathcal{S}_n^N, \mathcal{S}_{m_c}^N)$
are bounded and continuous operator valued functions.

(c) $\{\mathbf{M}_d(\tau_k)\}_{k\in\mathbb{Z}_+} \subset \mathcal{S}_n^N$, $\{\mathbf{L}_d(\tau_k)\}_{k\in\mathbb{Z}_+} \subset \mathcal{M}_{nm_d}^N$, $\{\mathbb{R}_d(\tau_k)\}_{k\in\mathbb{Z}_+} \subset \mathcal{S}_{m_d}^N$ are bounded matrix-valued sequences.

(d) $\{\boldsymbol{\Pi}_{d1}(\tau_k)[\cdot]\}_{k\in\mathbb{Z}_+} \subset \mathcal{B}(\mathcal{S}_n^N)$, $\{\boldsymbol{\Pi}_{d2}(\tau_k)[\cdot]\}_{k\in\mathbb{Z}_+} \subset \mathcal{B}(\mathcal{S}_n^N, \mathcal{M}_{nm_d}^N)$, $\{\boldsymbol{\Pi}_{d3}(\tau_k)[\cdot]\}_{k\in\mathbb{Z}_+} \subset \mathcal{B}(\mathcal{S}_n^N, \mathcal{S}_{m_d}^N)$ are operator valued bounded sequences.

(e) For each $t \in \mathbb{R}_+$, $\tau_k \in \mathfrak{T}$, $\boldsymbol{\Pi}_c(t)[\cdot]$ and $\boldsymbol{\Pi}_d(\tau_k)[\cdot]$ are linear positive operators, that is $\boldsymbol{\Pi}_c(t)[\mathbf{X}] \in \mathcal{S}_{n+m_c}^{N+}$ and $\boldsymbol{\Pi}_d(\tau_k)[\mathbf{X}] \in \mathcal{S}_{n+m_d}^{N+}$ if $\mathbf{X} \in \mathcal{S}_n^{N+}$.

Definition 5.1 A solution of the equation (5.2) is a function $t \to \mathbf{X}(t) = (X(t,1), X(t,2), \ldots, X(t,N)) : \mathcal{I}_{\mathbf{X}} \subset \mathbb{R}_+ \to \mathcal{S}_n^N$ having the properties:

(a) $\mathbf{X}(\cdot)$ is differentiable in any $t \in \mathcal{I}_{\mathbf{X}}$, $t \neq \tau_k$ and it is right continuous in $t = \tau_k$ for all $k \in \mathbb{Z}_+$ for which $\tau_k \in \mathcal{I}_{\mathbf{X}}$;

(b) the matrices $\Pi_{c3}(t)[\mathbf{X}(t)](i) + R_c(t,i)$, $(t,i) \in \mathcal{I}_{\mathbf{X}} \times \mathcal{N}$ and $\Pi_{d3}(\tau_k)[\mathbf{X}(\tau_k)](i) + R_d(\tau_k,i)$, $(\tau_k,i) \in (\mathfrak{T} \cap \mathcal{I}_{\mathbf{X}}) \times \mathcal{N}$ are invertible;

(c) $\mathbf{X}(t)$ satisfies (5.2) for all $t \in \mathcal{I}_{\mathbf{X}}$.

Here and in the sequel $\mathcal{I}_{\mathbf{X}} \subset \mathbb{R}_+$ is the maximal interval on which $\mathbf{X}(\cdot)$ is satisfying the conditions $(a) - (c)$ from Definition 5.1.

Remark 5.1 (a) If in (5.3b) we have that

$$\begin{aligned} &\Pi_{d1}(\tau_k)[\mathbf{X}(\tau_k)](i) = X(\tau_k,i), \quad \Pi_{d2}(\tau_k)[\mathbf{X}(\tau_k)](i) = 0, \\ &M_d(\tau_k,i) = 0, \quad L_d(\tau_k,i) = 0, \end{aligned} \tag{5.6}$$

for all $(\tau_k,i) \in \mathfrak{T} \times \mathcal{N}$, then (5.3b) reduces to $X(\tau_k^-,i) = X(\tau_k,i)$, for all $(\tau_k, i) \in \mathfrak{T} \times \mathcal{N}$. In this case (5.3) becomes

$$\begin{aligned} -\dot{X}(t,i) &= A^\top(t,i)X(t,i) + X(t,i)A(t,i) + \Pi_{c1}(t)[\mathbf{X}(t)](i) + Mc(t,i) \\ &- (X(t,i)B(t,i) + \Pi_{c2}(t)[\mathbf{X}(t)](i) + L_c(t,i))(\Pi_{c3}(t)[\mathbf{X}(t)](i) + R_c(t,i))^{-1} \\ &\cdot (X(t,i)B(t,i) + \Pi_{c2}(t)[\mathbf{X}(t)](i) + L_c(t,i))^\top, \quad t \in \mathbb{R}_+, \quad i \in \mathcal{N}, \end{aligned} \tag{5.7}$$

which is just the nonlinear backward differential equation (5.1) from Drăgan et al. [1] widely investigated in the Chap. 5 of the aforementioned reference.

(b) If in (5.3a) we have that

$$\begin{aligned} &A(t,i)=0,\quad B(t,i)=0,\quad \Pi_{cj}(t)[\mathbf{X}(t)](i)=0,\quad j=1,2,\\ &M_c(t,i)=0,\quad L_c(t,i)=0,\quad \text{for all }\ (t,i)\in\mathbb{R}_+\times\mathcal{N}, \end{aligned}\tag{5.8}$$

then, (5.3a) gives $X(t,i)=X(\tau_k^-,i)$, for all $t\in[\tau_{k-1},\tau_k)$, for any $(\tau_k,i)\in\mathfrak{T}\times\mathcal{N}$. In this case (5.3) reduces to:

$$\begin{aligned} X(\tau_k^-,i)=&\Pi_{d1}(\tau_k)[\mathbf{X}(\tau_{k+1}^-)](i)+M_d(\tau_k,i)-(\Pi_{d2}(\tau_k)[\mathbf{X}(\tau_{k+1}^-)](i)+L_d(\tau_k,i))\\ &\cdot(\Pi_{d3}(\tau_k)[\mathbf{X}(\tau_{k+1}^-)](i)+R_d(\tau_k,i))^{-1}(\Pi_{d2}(\tau_k)[\mathbf{X}(\tau_{k+1}^-)](i)+L_d(\tau_k,i))^\top, \end{aligned}\tag{5.9}$$

$(\tau_k,i)\in\mathfrak{T}\times\mathcal{N}$. So, if the equalities (5.8) holds, then any solution $t\to\mathbf{X}(t):\mathcal{I}_\mathbf{X}\to\mathcal{S}_n^N$ of (5.3) reduces to $\mathbf{X}(t)=\mathbf{X}(\tau_{k+1}^-),\ \ t\in[\tau_k,\tau_{k+1}),\ \ k\in\mathbb{Z}_+$, for which $[\tau_k,\tau_{k+1})\subset\mathcal{I}_\mathbf{X}$. The discrete-time backward nonlinear equation (5.9) satisfied by the sequence of the terminal values $\mathbf{X}(\tau_{k+1}^-)$ on each interval $[\tau_k,\tau_{k+1})$ coincides with the discrete-time backward nonlinear equation (5.8) from Drăgan et al. [2] widely investigated in Chap. 5 from the aforementioned reference. The developments in the present chapter are done under the assumption that both (5.6) as well as (5.8) are not fully satisfied.

Let us consider the following jump matrix Riccati differential equation of stochastic control (SJMRDE):

$$\begin{aligned} \dot X(t,i)=&A_0^\top(t,i)X(t,i)+X(t,i)A_0(t,i)+\sum_{l=1}^{r_1}A_l^\top(t,i)X(t,i)A_l(t,i)\\ &+\sum_{j=1}^{N}q_{ij}X(t,j)-(X(t,i)B_0(t,i)+\sum_{l=1}^{r_1}A_l^\top(t,i)X(t,i)B_l(t,i)+L_c(t,i))\\ &\left(\sum_{l=1}^{r_1}B_l^\top(t,i)X(t,i)B_l(t,i)+R_c(t,i)\right)^{-1}(X(t,i)B_0(t,i)\\ &+\sum_{l=1}^{r_1}A_l^\top(t,i)X(t,i)B_l(t,i)+L_c(t,i))^\top+M_c(t,i),\quad \tau_k\le t<\tau_{k+1}, \end{aligned}\tag{5.10a}$$

$$\begin{aligned} X(\tau_k^-,i)=&\sum_{l=0}^{r_2}A_{dl}^\top(\tau_k,i)X(\tau_k,i)A_{dl}(\tau_k,i)\\ &-\left(\sum_{l=0}^{r_2}A_{dl}^\top(\tau_k,i)X(\tau_k,i)B_{dl}(\tau_k,i)+L_d(\tau_k,i)\right)\\ &\left(\sum_{l=0}^{r_2}B_{dl}^\top(\tau_k,i)X(\tau_k,i)B_{dl}(\tau_k,i)+R_d(\tau_k,i)\right)^{-1} \end{aligned}\tag{5.10b}$$

$$\left(\sum_{l=0}^{r_2} A_{dl}^{\top}(\tau_k,i)X(\tau_k,i)B_{dl}(\tau_k,i)+L_d(\tau_k,i)\right)^{\top}$$
$$+M_d(\tau_k,i),\quad k\in\mathbb{Z}_+,\quad i\in\mathcal{N}.$$

The solution of the SJMRDEs of type (5.10) are involved in the solution of some optimal control problems which require the minimization of a cost functional of the form:

$$\begin{aligned}J(t_0,\infty,x_0;u_c,u_d) &= \int_{t_0}^{\infty}\mathbb{E}\left[\begin{pmatrix}x(t)\\u_c(t)\end{pmatrix}^{\top}\begin{pmatrix}M_c(t,\eta(t)) & L_c(t,\eta(t))\\ L_c^{\top}(t,\eta(t)) & R_c(t,\eta(t))\end{pmatrix}\begin{pmatrix}x(t)\\u_c(t)\end{pmatrix}\right]\mathrm{d}t\\ &+\sum_{k=k_0}^{\infty}\mathbb{E}\left[\begin{pmatrix}x(\tau_k)\\u_d(k)\end{pmatrix}^{\top}\begin{pmatrix}M_d(\tau_k,\eta(\tau_k)) & L_d(\tau_k,\eta(\tau_k))\\ L_d^{\top}(\tau_k,\eta(\tau_k)) & R_d(\tau_k,\eta(\tau_k))\end{pmatrix}\begin{pmatrix}x(\tau_k)\\u_d(k)\end{pmatrix}\right]\end{aligned}\tag{5.11}$$

along of the trajectories of an impulsive controlled linear stochastic system of type (4.1) determined by some suitable sets of admissible controls.

Such optimal control problems will be analyzed in detail in the next chapter.

In (5.11), $k_0\in\mathbb{Z}_+$ is such that $\tau_{k_0-1}<t_0\leq\tau_{k_0}$. We assume that the coefficients of the SJMRDE (5.10) satisfy the next assumption.

Assumption 5.2

(a) For each $i\in\mathcal{N}$, $t\to(A_l(t,i),B_l(t,i)):\mathbb{R}_+\to\mathbb{R}^{n\times n}\times\mathbb{R}^{n\times m_c}$, $0\leq l\leq r_1$, $t\to(M_c(t,i),L_c(t,i),R_c(t,i)):\mathbb{R}_+\to\mathcal{S}_n\times\mathbb{R}^{n\times m_c}\times\mathcal{S}_{m_c}$ are bounded and continuous matrix-valued functions;

(b) $\{A_{dl}(\tau_k,i)\}_{k\in\mathbb{Z}_+}\subset\mathbb{R}^{n\times n}$, $\{B_{dl}(\tau_k,i)\}_{k\in\mathbb{Z}_+}\subset\mathbb{R}^{n\times m_d}$, $0\leq l\leq r_2$, $\{M_d(\tau_k,i)\}_{k\in\mathbb{Z}_+}\subset\mathcal{S}_n$, $\{L_d(\tau_k,i)\}_{k\in\mathbb{Z}_+}\subset\mathbb{R}^{n\times m_d}$, $\{R_d(\tau_k,i)\}_{k\in\mathbb{Z}_+}\subset\mathcal{S}_{m_d}$, $i\in\mathcal{N}$ are bounded matrix valued sequences;

(c) the scalars q_{ij}, $(i,j)\in\mathcal{N}\times\mathcal{N}$ are satisfying the conditions (1.22).

Setting

$$A(t,i):=A_0(t,i)+\frac{1}{2}q_{ii}I_n \qquad B(t,i):=B_0(t,i)\tag{5.12a}$$

$$\Pi_{c1}(t)[\mathbf{X}](i)=\sum_{l=1}^{r_1}A_l^{\top}(t,i)X(i)A_l(t,i)+\sum_{j=1,j\neq i}^{N}q_{ij}X(j)\tag{5.12b}$$

$$\Pi_{c2}(t)[\mathbf{X}](i)=\sum_{l=1}^{r_1}A_l^{\top}(t,i)X(i)B_l(t,i)\tag{5.12c}$$

$$\Pi_{c3}(t)[\mathbf{X}](i)=\sum_{l=1}^{r_1}B_l^{\top}(t,i)X(i)B_l(t,i)\tag{5.12d}$$

$$\Pi_{d1}(\tau_k)[\mathbf{X}](i) = \sum_{l=0}^{r_2} A_{dl}^\top(\tau_k, i)X(i)A_{dl}(\tau_k, i) \tag{5.12e}$$

$$\Pi_{d2}(\tau_k)[\mathbf{X}](i) = \sum_{l=0}^{r_2} A_{dl}^\top(\tau_k, i)X(i)B_{dl}(\tau_k, i) \tag{5.12f}$$

$$\Pi_{d3}(\tau_k)[\mathbf{X}](i) = \sum_{l=0}^{r_2} B_{dl}^\top(\tau_k, i)X(i)B_{dl}(\tau_k, i) \tag{5.12g}$$

$i \in \mathcal{N}$, $\mathbf{X} = (X(1), X(2), \ldots, X(N)) \in \mathcal{S}_n^N$, we obtain that if Assumption 5.2 is fulfilled then the SJMRDE (5.10) is a special case of (5.3) (or equivalently, a special case of (5.2)) satisfying Assumption 5.1. This entitles us to name the nonlinear equations of the form (5.2), for which Assumption 5.1 is fulfilled,"the generalized backward jump matrix Riccati differential equations" (GBJMRDE). Often we shall use the following more compact form of a GBJMRDE (5.2):

$$-\dot{\mathbf{X}}(t) = \mathcal{R}_c(t, \mathbf{X}(t)), \quad \tau_k \le t < \tau_{k+1} \tag{5.13a}$$

$$\mathbf{X}(\tau_k^-) = \mathcal{R}_d(\tau_k, \mathbf{X}(\tau_k)), \quad k \in \mathbb{Z}_+ \tag{5.13b}$$

where $(t, \mathbf{X}) \to \mathcal{R}_c(t, \mathbf{X}) : Dom\ \mathcal{R}_c \to \mathcal{S}_n^N$ and $(\tau_k, \mathbf{X}) \to \mathcal{R}_d(\tau_k, \mathbf{X}) : Dom\ \mathcal{R}_d \to \mathcal{S}_n^N$ are described by the right hand side of (5.2a) and the right hand side of (5.2b), respectively, while
$Dom\ \mathcal{R}_c = \{(t, \mathbf{X}) \in \mathbb{R}_+ \times \mathcal{S}_n^N \mid \det(\Pi_{c3}(t)[\mathbf{X}](i) + R_c(t, i)) \neq 0, \ \ 1 \le i \le N\}$ and $Dom\ \mathcal{R}_d = \{(\tau_k, \mathbf{X}) \in \mathfrak{T} \times \mathcal{S}_n^N \mid \det(\Pi_{d3}(\tau_k)[\mathbf{X}](i) + R_d(\tau, i)) \neq 0, \ \ 1 \le i \le N\}$.
The operator $\mathcal{R}_c$ will be named continuous-time generalized Riccati-type operator, while $\mathcal{R}_d$ will be named discrete-time generalized Riccati-type operator.
The next definition extends to this framework the concepts introduced in Definition 4.1 and Definition 4.2 in the case of a jump linear stochastic system.

Definition 5.2

(i) We say that the GBJMRDE (5.2) or equivalently the pair of generalized Riccati-type operators $(\mathcal{R}_c, \mathcal{R}_d)$ is in the period case if the following conditions are simultaneously satisfied:

(a) $\tau_k = kh$, for all $k \in \mathbb{Z}_+$, where $h > 0$ does not depend upon k;
(b) there exists a real number $\theta_c > 0$ with the property that $\mathbf{A}(t + \theta_c) = \mathbf{A}(t)$, $\mathbf{B}(t + \theta_c) = \mathbf{B}(t)$, $\mathbf{M}_c(t + \theta_c) = \mathbf{M}_c(t)$, $\mathbf{L}_c(t + \theta_c) = \mathbf{L}_c(t), \mathbf{R}_c(t + \theta_c) = \mathbf{R}_c(t), \boldsymbol{\Pi}_c(t + \theta_c)[\cdot] = \boldsymbol{\Pi}_c(t)[\cdot]$ for all $t \in \mathbb{R}_+$;
(c) there exists a natural number $\theta_d \ge 1$ with the properties: $\mathbf{M}_d((k + \theta_d)h) = \mathbf{M}_d(kh)$, $\mathbf{L}_d((k + \theta_d)h) = \mathbf{L}_d(kh), \mathbf{R}_d((k + \theta_d)h) = \mathbf{R}_d(kh)$, $\boldsymbol{\Pi}_d((k + \theta_d)h)[\cdot] = \boldsymbol{\Pi}_d(kh)[\cdot]$ for all $k \in \mathbb{Z}_+$;
(d) $\frac{\theta_c}{\theta_d h}$ is a rational number.

(ii) We say that the GBJMRDE (5.2), or equivalently the pair of the generalized Riccati-type operators $(\mathscr{R}_c, \mathscr{R}_d)$ is in the time-invariant case if the following conditions are simultaneously satisfied

(a) $\tau_k = kh$, for all $k \in \mathbb{Z}_+$, where $h > 0$ does not depend upon k;

(b) $\mathbf{A}(t) = \mathbf{A} \in \mathscr{M}_n^N$, $\mathbf{B}(t) = \mathbf{B} \in \mathscr{M}_{nm_c}^N$, $\mathbf{M}_c(t) = \mathbf{M} \in \mathscr{S}_n^N$, $\mathbf{L}_c(t) = \mathbf{L}_c \in \mathscr{M}_{nm_c}^N$, $\mathbf{R}_c(t) = \mathbf{R}_c \in \mathscr{S}_{m_c}^N$, $\boldsymbol{\Pi}_c(t)[\cdot] = \boldsymbol{\Pi}_c[\cdot] \in \mathscr{B}(\mathscr{S}_n^N, \mathscr{S}_{n+m_c}^N)$ for all $t \in \mathbb{R}_+$;

(c) $\mathbf{M}_d(kh) = \mathbf{M}_d \in \mathscr{S}_n^N$, $\mathbf{L}_d(kh) = \mathbf{L}_d \in \mathscr{M}_{nm_d}^N$, $\mathbf{R}_d(kh) = \mathbf{R}_d \in \mathscr{S}_{m_d}^N$, $\boldsymbol{\Pi}_d(kh)[\cdot] = \boldsymbol{\Pi}_d[\cdot] \in \mathscr{B}(\mathscr{S}_n^N, \mathscr{S}_{n+m_D}^N)$ for all $k \in \mathbb{Z}_+$.

If the property (d) from part (i) of the above definition holds, then, there exist natural coprime numbers $\tilde{p}$, $\tilde{q}$ with the property that $\frac{\theta_c}{\theta_d h} = \frac{\tilde{p}}{\tilde{q}}$.

We set:

$$\tilde{\theta} := \tilde{q}\theta_c = \tilde{p}\theta_d h \tag{5.14a}$$

$$\tilde{\mathbf{k}} := \tilde{p}\theta_d. \tag{5.14b}$$

Summing up, we can say that a GBJMRDE (5.2) as well as the pair of operators $(\mathscr{R}_c, \mathscr{R}_d)$ are defined by a quintuple of the form

$$\boldsymbol{\Sigma} = (\mathbf{A}, \mathbf{B}, (\boldsymbol{\Pi}_c \mid \boldsymbol{\Pi}_d), (\boldsymbol{\Theta}_c \mid \boldsymbol{\Theta}_d), \mathfrak{T}) \tag{5.15}$$

where

$$t \to (\mathbf{A}(t), \mathbf{B}(t)) : \mathbb{R}_+ \to \mathscr{M}_n^N \times \mathscr{M}_{nm_c}^N,$$

$$t \to \boldsymbol{\Pi}_c(t)[\cdot] : \mathbb{R}_+ \to \mathscr{B}(\mathscr{S}_n^N, \mathscr{S}_{n+m_c}^N),\ \boldsymbol{\Pi}_d(\tau_k)[\cdot] \in \mathscr{B}(\mathscr{S}_n^N, \mathscr{S}_{n+m_d}^N),\ \ k \in \mathbb{Z}_+,$$

$$t \to \boldsymbol{\Theta}_c(t) = (\Theta_c(t,1), \Theta_c(t,2), \ldots, \Theta_c(t,N)) : \mathbb{R}_+ \to \mathscr{S}_{n+m_c}^N$$

with $\Theta_c(t,i) = \begin{pmatrix} M_c(t,i) & L_c(t,i) \\ L_c^\top(t,i) & R_c(t,i) \end{pmatrix} \in \mathscr{S}_{n+m_c}$.

For each $\tau_k \in \mathfrak{T}$, $\boldsymbol{\Theta}_d(\tau_k) = (\Theta_d(\tau_k,1), \Theta_d(\tau_k,2), \ldots, \Theta_d(\tau_k,N)) \in \mathscr{S}_{n+m_d}^N$ with the partition $\Theta_d(\tau_k,i) = \begin{pmatrix} M_d(\tau_k,i) & L_d(\tau_k,i) \\ L_d^\top(\tau_k,i) & R_d(\tau_k,i) \end{pmatrix}$, $i \in \mathscr{N}$. $\mathfrak{T} = \{\tau_k\}_{k \in \mathbb{Z}_+}$ is an increasing sequence of real numbers such that $\tau_0 = 0$ and satisfies the Assumption 2.1.

Now we introduce two sets of functions associated to the quintuple $\boldsymbol{\Sigma}$. So, if $\mathscr{I} = [t_0, T] \subset \mathbb{R}_+$ or $\mathscr{I} = [t_0, \infty) \subset \mathbb{R}_+$ are given intervals, we denote $\Gamma^{\Sigma}[\mathscr{I}]$ the set of all bounded functions $t \to \mathbf{Z}(t) = (Z(t,1), \ldots, Z(t,N)) : \mathscr{I} \to \mathscr{S}_n^N$ continuously differentiable in any $t \neq \tau_k$, right continuous in any $\tau_k \in \mathscr{I}$ and satisfying the following system of linear matrix differential inequalities with jumps:

$$\mathfrak{D}_c^{\Sigma}(t)[\mathbf{Z}](i) := \begin{pmatrix} \dot{Z}(t,i) + \Gamma_1(t,\mathbf{Z}(t),i) + M_c(t,i) & \Gamma_2(t,\mathbf{Z}(t),i) + L_c(t,i) \\ \Gamma_2^{\top}(t,\mathbf{Z}(t),i) + L_c^{\top}(t,i) & \Pi_{c3}(t)[\mathbf{Z}(t)](i) + R_c(t,i) \end{pmatrix} \geq 0,$$

$$\tau_k \leq t < \tau_{k+1}, \tag{5.16a}$$

$$\mathfrak{D}_d^{\Sigma}(\tau_k)[\mathbf{Z}](i) := \tag{5.16b}$$

$$\begin{pmatrix} -Z(\tau_k^-,i) + \Pi_{d1}(\tau_k)[\mathbf{Z}(\tau_k)](i) + M_d(\tau_k,i) & \Pi_{d2}(\tau_k)[\mathbf{Z}(\tau_k)](i) + L_d(\tau_k,i) \\ (\Pi_{d2}(\tau_k)[\mathbf{Z}(\tau_k)](i))^{\top} + L_d^{\top}(\tau_k,i) & \Pi_{d3}(\tau_k)[\mathbf{Z}(\tau_k)](i) + R_d(\tau_k,i) \end{pmatrix} \geq 0$$

$$\Pi_{c3}(t)[\mathbf{Z}(t)](i) + R_c(t,i) \geq \gamma_c I_{m+c} > 0, \quad \text{for all } t \in \mathcal{I} \tag{5.16c}$$

$$\Pi_{d3}(\tau_k)[\mathbf{Z}(\tau_k)](i) + R_d(\tau_k,i) \geq \gamma_d I_{m_d} > 0, \quad \text{for all } \tau_k \in \mathfrak{T} \cap \mathcal{I} \tag{5.16d}$$

for $i \in \mathcal{N}$. In (5.16a) we denoted

$$\Gamma_1(t,\mathbf{Z}(t),i) \triangleq A^{\top}(t,i)Z(t,i) + Z(t,i)A(t,i) + \Pi_{c1}(t)[\mathbf{Z}(t)](i),$$
$$\Gamma_2(t,\mathbf{Z}(t),i) \triangleq Z(t,i)B(t,i) + \Pi_{c2}[\mathbf{Z}(t)](i).$$

Beside the set $\Gamma^{\Sigma}[\mathcal{I}]$ an important role will be played by its subset $\widetilde{\Gamma}^{\Sigma}[\mathcal{I}]$ which consists all bounded functions $t \to \mathbf{Z}(t) : \mathcal{I} \to \mathcal{S}_n^N$ which are continuous differentiable in any $t \neq \tau_k$, right continuous in any $\tau_k \in \mathfrak{T} \cap \mathcal{I}$ satisfying the linear matrix differential inequality (5.16a), (5.16c) and (5.16d) together with

$$\mathfrak{D}_d^{\Sigma}(\tau_k)[\mathbf{Z}](i) \geq \hat{\gamma} I_{n+m_d} > 0, \tag{5.17}$$

for all $k \in \mathbb{Z}_+$ for which $\tau_k \in \mathcal{I}$.

5.1.3 Some Equivalent Representation Formulae of the Generalized Riccati-Type Operators

Let $t \to \mathbf{W}_c(t) = (W_c(t,1), W_c(t,2), \ldots, W_c(t,N)) : \mathcal{I} \subset \mathbb{R}_+ \to \mathcal{M}_{m_c n}^N$. We set

$$\boldsymbol{\Pi}_{c\mathbf{W}_c}(t)[\mathbf{X}] = (\Pi_{c\mathbf{W}_c}(t)[\mathbf{X}](1), \Pi_{c\mathbf{W}_c}(t)[\mathbf{X}](2), \ldots, \Pi_{c\mathbf{W}_c}(t)[\mathbf{X}](N)) \tag{5.18a}$$

$$\Pi_{c\mathbf{W}_c}(t)[\mathbf{X}](i) := \begin{pmatrix} I_n \\ W_c(t,i) \end{pmatrix}^{\top} \Pi_c(t)[\mathbf{X}](i) \begin{pmatrix} I_n \\ W_c(t,i) \end{pmatrix} \tag{5.18b}$$

$i \in \mathcal{N}$, $\mathbf{X} \in \mathcal{S}_n^N$. Similarly we define:

$$\boldsymbol{\Theta}_{c\mathbf{W}_c}(t) = (\Theta_{c\mathbf{W}_c}(t,1), \Theta_{c\mathbf{W}_c}(t,2), \ldots, \Theta_{c\mathbf{W}_c}(t,N)) \tag{5.19a}$$

$$\Theta_{c\mathbf{W}_c}(t,i) := \begin{pmatrix} I_n \\ W_c(t,i) \end{pmatrix}^{\top} \Theta_c(t,i) \begin{pmatrix} I_n \\ W_c(t,i) \end{pmatrix}, \quad i \in \mathcal{N}. \tag{5.19b}$$

If $\mathbf{W}_d(\tau_k) = (W_d(\tau_k,1), W_d(\tau_k,2), \ldots, W_d(\tau_k,N)) \in \mathcal{M}_{m_d n}^N$, $k \in \mathbb{Z}_+$, we set

$$\boldsymbol{\Pi}_{d\mathbf{W}_d}(\tau_k)[\mathbf{X}] = (\Pi_{d\mathbf{W}_d}(\tau_k)[\mathbf{X}](1), \Pi_{d\mathbf{W}_d}(\tau_k)[\mathbf{X}](2), \ldots, \Pi_{d\mathbf{W}_d}(\tau_k)[\mathbf{X}](N)) \tag{5.20a}$$

$$\Pi_{d\mathbf{W}_d}(\tau_k)[\mathbf{X}](i) := \begin{pmatrix} I_n \\ W_d(\tau_k, i) \end{pmatrix}^\top \Pi_d(\tau_k)[\mathbf{X}](i) \begin{pmatrix} I_n \\ W_d(t, i) \end{pmatrix} \tag{5.20b}$$

for all $i \in \mathcal{N}$, $\mathbf{X} \in \mathcal{S}_n^N$. Similarly we define

$$\boldsymbol{\Theta}_{d\mathbf{W}_d}(\tau_k) = (\Theta_{d\mathbf{W}_d}(\tau_k, 1), \Theta_{d\mathbf{W}_d}(\tau_k, 2), \ldots, \Theta_{d\mathbf{W}_d}(\tau_k, N)) \tag{5.21a}$$

$$\Theta_{d\mathbf{W}_d}(\tau_k, i) := \begin{pmatrix} I_n \\ W_d(\tau_k, i) \end{pmatrix}^\top \Theta_d(\tau_k, i) \begin{pmatrix} I_n \\ W_d(\tau_k, i) \end{pmatrix}, \quad i \in \mathcal{N}. \tag{5.21b}$$

The next result will be repeatedly used in the proof of the main results of this chapter.

Lemma 5.1 *The following hold:*

(i) for any function $t \to \mathbf{W}_c(t) : \mathcal{I}_{\mathbf{W}_c} \subset \mathbb{R}_+ \to \mathcal{M}_{m_c n}^N$ *we have*

$$\begin{aligned} \mathcal{R}_c(t, \mathbf{X}) = {} & (\mathbf{A}(t) + \mathbf{B}(t)\mathbf{W}_c(t))^\top \mathbf{X} + \mathbf{X}(\mathbf{A}(t) + \mathbf{B}(t)\mathbf{W}_c(t)) \\ & + \boldsymbol{\Pi}_{c\mathbf{W}_c}(t)[\mathbf{X}] - (\mathbf{W}_c(t) - \mathbf{F}_c^{\mathbf{X}}(t))^\top (\boldsymbol{\Pi}_{c3}(t)[\mathbf{X}] + \mathbf{R}_c(t)) \\ & \cdot (\mathbf{W}_c(t) - \mathbf{F}_c^{\mathbf{X}}(t)) + \boldsymbol{\Theta}_{c\mathbf{W}_c}(t) \end{aligned} \tag{5.22}$$

for all $(t, \mathbf{X}) \in$ *Dom* $\mathcal{R}_c$ *for which* $t \in \mathcal{I}_{\mathbf{W}_c}$, *where*

$$\mathbf{F}_c^{\mathbf{X}}(t) = (F_c^{\mathbf{X}}(t, 1), F_c^{\mathbf{X}}(t, 2), \ldots, F_c^{\mathbf{X}}(t, N)) \tag{5.23a}$$

$$\begin{aligned} F_c^{\mathbf{X}}(t, i) \triangleq {} & -(\Pi_{c3}(t)[\mathbf{X}](i) + R_c(t, i))^{-1}(X(i)B(t, i) + \Pi_{c2}(t)[\mathbf{X}](i) \\ & + L_c(t, i))^\top, \quad 1 \le i \le N, \quad (t, \mathbf{X}) \in \textit{Dom } \mathcal{R}_c. \end{aligned} \tag{5.23b}$$

(ii) If $\{\mathbf{W}_d(\tau_k)\}_{k_0 \le k \le k_1} \subset \mathcal{M}_{m_d n}^N$ *then we have*

$$\begin{aligned} \mathcal{R}_d(\tau_k, \mathbf{X}) = {} & \boldsymbol{\Pi}_{d\mathbf{W}_d}(\tau_k)[\mathbf{X}] - (\mathbf{W}_d(\tau_k) - \mathbf{F}_d^{\mathbf{X}}(\tau_k))^\top (\boldsymbol{\Pi}_{d3}(\tau_k)[\mathbf{X}] + \mathbf{R}_d(\tau_k)) \\ & \cdot (\mathbf{W}_d(\tau_k) - \mathbf{F}_d^{\mathbf{X}}(\tau_k)) + \boldsymbol{\Theta}_{d\mathbf{W}_d}(\tau_k) \end{aligned} \tag{5.24}$$

for all $(\tau_k, \mathbf{X}) \in$ *Dom* $\mathcal{R}_d$, *with* $k \in \{k_0, k_0 + 1, \ldots, k_1\}$ *where we denoted*

$$\mathbf{F}_d^{\mathbf{X}}(\tau_k) = (F_d^{\mathbf{X}}(\tau_k, 1), F_d^{\mathbf{X}}(\tau_k, 2), \ldots, F_d^{\mathbf{X}}(\tau_k, N)) \tag{5.25a}$$

$$\begin{aligned} F_d^{\mathbf{X}}(\tau_k, i) \triangleq {} & -(\Pi_{d3}(\tau_k)[\mathbf{X}](i) + R_d(\tau_k, i))^{-1}(\Pi_{d2}(\tau_k)[\mathbf{X}](i) + L_d(\tau_k, i))^\top, \\ & 1 \le i \le N, \quad (\tau_k, \mathbf{X}) \in \textit{Dom } \mathcal{R}_d. \end{aligned} \tag{5.25b}$$

Proof The equality (5.22) is obtained applying Lemma 5.1.1 from Drăgan et al. [1] in the case of the operator $\mathscr{R}_c$, while the equality (5.24) can be obtained applying Lemma 5.1 from Drăgan et al. [2] in the case of the operator $\mathscr{R}_d$.

Let $\mathbf{X} \to \boldsymbol{\Pi}^*_{c\mathbf{W}_c}(t)[\mathbf{X}] = (\Pi^*_{c\mathbf{W}_c}(t)[\mathbf{X}](1), \Pi^*_{c\mathbf{W}_c}(t)[\mathbf{X}](2), \ldots, \Pi^*_{c\mathbf{W}_c}(t)[\mathbf{X}](N)) : \mathscr{S}_n^N \to \mathscr{S}_n^N$ be the adjoint with respect to the inner product (2.111) of the operator $\boldsymbol{\Pi}_{c\mathbf{W}_c}(t)[\cdot]$ defined by (5.18). We consider the linear operator $\mathscr{L}_{c\mathbf{W}_c}[\mathbf{X}] : \mathscr{S}_n^N \to \mathscr{S}_n^N$ defined by

$$\mathscr{L}_{c\mathbf{W}_c}(t)[\mathbf{X}]=(\mathbf{A}(t)+\mathbf{B}(t)\mathbf{W}_c(t))\mathbf{X}+\mathbf{X}(\mathbf{A}(t)+\mathbf{B}(t)\mathbf{W}_c(t))^\top+\boldsymbol{\Pi}^*_{c\mathbf{W}_c}(t)[\mathbf{X}]. \tag{5.26}$$

The adjoint $\mathscr{L}^*_{c\mathbf{W}_c}(t)[\cdot]$ with respect to the inner product (2.111) of the operator defined in (5.26) is given by

$$\mathscr{L}^*_{c\mathbf{W}_c}(t)[\mathbf{X}]=(\mathbf{A}(t)+\mathbf{B}(t)\mathbf{W}_c(t))^\top\mathbf{X}+\mathbf{X}(\mathbf{A}(t)+\mathbf{B}(t)\mathbf{W}_c(t))+\boldsymbol{\Pi}_{c\mathbf{W}_c}(t)[\mathbf{X}]. \tag{5.27}$$

$t \in \mathbb{R}_+$, $\mathbf{X} \in \mathscr{S}_n^N$. Combining (5.24) with (5.27) we obtain a new representation formula of the continuous-time generalized Riccati-type operator

$$\begin{aligned}\mathscr{R}_c(t, \mathbf{X}) &= \mathscr{L}^*_{c\mathbf{W}_c}(t)[\mathbf{X}] + \boldsymbol{\Theta}_{c\mathbf{W}_c}(t) - (\mathbf{W}_c(t) - \mathbf{F}_c^{\mathbf{X}}(t))^\top(\boldsymbol{\Pi}_{c3}(t)[\mathbf{X}] + \mathbf{R}_c(t)) \\ &\quad\cdot (\mathbf{W}_c(t) - \mathbf{F}_c^{\mathbf{X}}(t)), \quad (\forall) \quad (t, \mathbf{X}) \in \ Dom\ \mathscr{R}_c,\end{aligned} \tag{5.28}$$

for which $t \in \mathscr{I}$ and any function $t \to \mathbf{W}_c(t) : \mathscr{I} \subset \mathbb{R}_+ \to \mathscr{M}^N_{m_c n}$.

Lemma 5.2 *The following hold:*

(i) for any function $\mathbf{W}_c(\cdot)$ from $\mathscr{PC}\{\mathscr{I}, \mathscr{M}^N_{m_c n}; \mathfrak{T}\}$ we have:

$$\begin{aligned}&\mathscr{L}^*_{c\mathbf{W}_c}(t)[\mathbf{X}] + \boldsymbol{\Theta}_{c\mathbf{W}_c} = \mathscr{L}^*_{c\mathbf{F}_c^{\mathbf{X}}}(t)[\mathbf{X}] + \boldsymbol{\Theta}_c\mathbf{F}_c^{\mathbf{X}}(t) \\ &+ (\mathbf{F}_c^{\mathbf{X}}(t) - \mathbf{W}_c(t))^\top(\boldsymbol{\Pi}_{c3}(t)[\mathbf{X}] + \mathbf{R}_c(t))(\mathbf{F}_c^{\mathbf{X}}(t) - \mathbf{W}_c(t)),\end{aligned} \tag{5.29}$$

for all $(t, \mathbf{X}) \in\ Dom\ \mathscr{R}_c$ for which $t \in \mathscr{I}$. In (5.29), $\mathbf{F}_c^{\mathbf{X}}(t)$ is computed as in (5.23) and $\mathscr{L}_{c\mathbf{F}_c^{\mathbf{X}}}(t)[\cdot]$ is computed as in (5.27) with $\mathbf{W}_c(\cdot)$ replaced by $\mathbf{F}_c^{\mathbf{X}}(\cdot)$.

(ii)

$$\begin{aligned}&\boldsymbol{\Pi}_{d\mathbf{W}_d}(\tau_k)[\mathbf{X}] + \boldsymbol{\Theta}_{d\mathbf{W}_d}(\tau_k) = \boldsymbol{\Pi}_{d\mathbf{F}_d^{\mathbf{X}}}(\tau_k)[\mathbf{X}] + \boldsymbol{\Theta}_{d\mathbf{F}_d^{\mathbf{X}}}(\tau_k) \\ &+ (\mathbf{F}_d^{\mathbf{X}}(\tau_k) - \mathbf{W}_d(\tau_k))^\top(\boldsymbol{\Pi}_{d3}(\tau_k)[\mathbf{X}] + \mathbf{R}_d(\tau_k))(\mathbf{F}_d^{\mathbf{X}}(\tau_k) - \mathbf{W}_d(\tau_k)),\end{aligned} \tag{5.30}$$

for all $(\tau_k, \mathbf{X}) \in\ Dom\ \mathscr{R}_d$ for which $\tau_k \in \mathscr{I} \cap \mathfrak{T}$. In (5.30), $\mathbf{F}_d^{\mathbf{X}}(\tau_k)$ is computed as in (5.25) and $\boldsymbol{\Pi}_{d\mathbf{F}_d^{\mathbf{X}}}(\tau_k)[\cdot]$ is computed as in (5.20) for $\mathbf{W}_d(\tau_k)$ replaced by $\mathbf{F}_d^{\mathbf{X}}(\tau_k)$.

By direct calculation one obtains via (5.12) and (5.18) that in the case of the SJMRDE (5.10) the operator (5.26) takes the form:

$$\mathscr{L}_{c\mathbf{W}_c}(t)[\mathbf{X}] = (\mathscr{L}_{c\mathbf{W}_c}(t)[\mathbf{X}](1), \mathscr{L}_{c\mathbf{W}_c}(t)[\mathbf{X}](2)\ldots, \mathscr{L}_{c\mathbf{W}_c}(t)[\mathbf{X}](N)) \tag{5.31a}$$

$$\mathscr{L}_{c\mathbf{W}_c}(t)[\mathbf{X}](i) := (A_0(t,i) + B_0(t,i)W_c(t,i))X(i) + X(i)(A_0(t,i) + B_0(t,i)W_c(t,i))^\top$$
$$+\sum_{l=1}^{r_1}(A_l(t,i)+B_l(t,i)W_c(t,i))X(i)(A_l(t,i)+B_l(t,i)W_c(t,i))^\top+\sum_{j=1}^{N} q_{ji}X(j) \tag{5.31b}$$

for all $\mathbf{X} = (X(1), X(2), \ldots, X(N)) \in \mathscr{S}_n^N$. Also, by direct calculation, involving (5.12e)–(5.12g) and (5.20) one may show that in the special case of the SJMRDE (5.10) one gets that

$$\boldsymbol{\Pi}^*_{d\mathbf{W}_d}(\tau_k)[\mathbf{X}] = (\Pi^*_{d\mathbf{W}_d}(\tau_k)[\mathbf{X}](1), \Pi^*_{d\mathbf{W}_d}(\tau_k)[\mathbf{X}](2), \ldots, \Pi^*_{d\mathbf{W}_d}(\tau_k)[\mathbf{X}](N)) \tag{5.32a}$$

$$\Pi^*_{d\mathbf{W}_d}(\tau_k)[\mathbf{X}](i) := \sum_{l=0}^{r_2}(A_{dl}(\tau_k,i) + B_{dl}(\tau_k,i)W_d(\tau_k,i))X(i)$$
$$(A_{dl}(\tau_k,i) + B_{dl}(\tau_k,i)W_d(\tau_k,i))^\top, \tag{5.32b}$$

$i \in \mathscr{N}$, $\mathbf{X} = (X(1), X(2), \ldots, X(N)) \in \mathscr{S}_n^N$, $\boldsymbol{\Pi}^*_{d\mathbf{W}_d}(\tau_k)[\cdot]$ being the adjoint with respect to the inner product (2.111) of the operator $\boldsymbol{\Pi}_{d\mathbf{W}_d}(\tau_k)[\cdot]$ defined in (5.20).

5.1.4 A Class of Generalized Lyapunov-Type Jump Matrix Linear Differential Equations

Let us consider the following jump matrix linear differential equation on the Hilbert space $\mathscr{S}_n^N$:

$$\dot{\mathbf{X}}(t) = \mathscr{L}_{c\mathbf{W}_c}(t)[\mathbf{X}(t)], \quad \tau_k < t \le \tau_{k+1}, \tag{5.33a}$$
$$\mathbf{X}(\tau_k^+) = \boldsymbol{\Pi}^*_{d\mathbf{W}_d}(\tau_k)[\mathbf{X}(\tau_k)], \quad k_0(\mathscr{I}) \le k \le k_1(\mathscr{I}), \tag{5.33b}$$

where $k_0(\mathscr{I})$, $k_1(\mathscr{I})$ are integers defined by

$$k_0(\mathscr{I}) = \min\{k \in \mathbb{Z}_+ \mid \tau_k \in \mathscr{I}\} \tag{5.34a}$$
$$k_1(\mathscr{I}) = \max\{k \in \mathbb{Z}_+ \mid \tau_k \in \mathscr{I}\}. \tag{5.34b}$$

In the case when $\mathscr{I} \subset \mathbb{R}_+$ is a right unbounded interval, then $k_1(\mathscr{I}) = \infty$.
In (5.33a), $t \to \mathbf{W}_d(t) : \mathscr{I} \to \mathscr{M}^N_{m_c n}$ is a function from the class $\mathscr{PC}\{\mathscr{I}, \mathscr{M}^N_{m_c n}; \mathfrak{T}\}$, while in (5.33b), $\{\mathbf{W}_d(\tau_k)\}_{k_0(\mathscr{I})\le k\le k_1(\mathscr{I})} \subset \mathscr{M}^N_{m_d n}$ is a bounded sequence. From (5.31) and (5.32) one sees that in the particular case when the operators $\mathscr{L}_{c\mathbf{W}_c}(t)[\cdot]$ and $\boldsymbol{\Pi}_{d\mathbf{W}_d}(\tau_k)[\cdot]$ are constructed based on the coefficients of the SJMRDE (5.10), the

jump matrix linear differential equation (5.33) takes the form of a generalized Lyapunov jump linear differential equation of type (2.117). That is why, we shall name the equations of type (5.33) generalized Lyapunov jump matrix linear differential equations (GLJMLDEs).

The next result highlights an important property of the GLJMLDEs.

Lemma 5.3 *Under the Assumption 5.1, the GLJMLDE (5.33) generates a positive evolution on the ordered Hilbert space* $(\mathcal{S}_n^N, \mathcal{S}_n^{N+})$.

Proof Let us remark that (5.33) is a particular case of the FJLDE (2.2) if we take $(\mathcal{X}, \mathcal{X}^+) \leftarrow (\mathcal{S}_n^N, \mathcal{S}_n^{N+})$, $\mathcal{L}(t)[\cdot] \leftarrow \mathcal{L}_{c\mathbf{W}_c}(t)[\cdot]$ and $\mathcal{I}(\tau_k)[\cdot] \leftarrow \boldsymbol{\Pi}_{d\mathbf{W}_d}(\tau_k)[\cdot]$. According to Definition 2.5 (b) applied in the case of (5.33) we have to show that any solution $\mathbf{X}(t; t_0, \mathbf{X}_0)$ lies in $\mathcal{S}_n^{N+}$ for any $t \geq t_0$, $t, t_0 \in \mathcal{I}$, if $\mathbf{X}_0 \in \mathcal{S}_n^{N+}$. In order to confirm this behavior of the solutions of (5.33) we use the sufficiency part of Proposition 2.5 (i). To this end, we show that

$$\mathbf{T}_{\mathbf{W}_c\mathbf{W}_d}(t, t_0)\mathcal{S}_n^{N+} \subset \mathcal{S}_n^{N+}, \tag{5.35}$$

for all $t \geq t_0$, $t, t_0 \in \mathcal{I}$, where $\mathbf{T}_{\mathbf{W}_c\mathbf{W}_d}(t, t_0)$ is the linear evolution operator defined by (2.39) on $\mathcal{S}_n^N$. Specializing (2.5), (2.7) to the case of the linear evolution operator defined by (5.33) we remark that it is computed composing the linear evolution operator $T_{\mathbf{W}_c}(t, s)$ defined by the linear differential equation

$$\dot{\mathbf{Y}}(t) = \mathcal{L}_{c\mathbf{W}_c}(t)[\mathbf{Y}(t)], \tag{5.36}$$

$\tau_k \leq s \leq t \leq \tau_{k+1}$, by the operators $\boldsymbol{\Pi}_{d\mathbf{W}_d}(\tau_k)[\cdot]$, for all $t_0 \leq \tau_k \leq t$. Employing Corollary 2.2.6 in Drăgan et al. [1] in the case of the linear differential equation (5.36) when the operator $\mathcal{L}_{c\mathbf{W}_c}(t)[\cdot]$ is defined by (5.26), we deduce that

$$T_{\mathbf{W}_c}(t, s)\mathcal{S}_n^{N+} \subset \mathcal{S}_n^{N+}, \tag{5.37}$$

for all $t \geq s \geq 0$. On the other hand, (5.21) yields

$$\boldsymbol{\Pi}_{d\mathbf{W}_d}(\tau_k)[\mathbf{X}] \in \mathcal{S}_n^{N+} \text{ if } \mathbf{X} \in \mathcal{S}_n^{N+}. \tag{5.38}$$

Finally, (5.37), (5.38) allow us to conclude via (2.5)–(2.7) that (5.35) holds. Thus the proof is complete.

Let us consider the non-homogeneous backward GLJMLDE:

$$-\dot{\mathbf{Y}}(t) = \mathcal{L}^*_{c\mathbf{W}_c}(t)[\mathbf{Y}(t)] + \mathbf{M}_c(t), \quad \tau_k \leq t < \tau_{k+1} \tag{5.39a}$$

$$\mathbf{X}(\tau_k^-) = \boldsymbol{\Pi}_{d\mathbf{W}_d}(\tau_k)[\mathbf{Y}(\tau_k)] + \mathbf{M}_d(\tau_k), \quad k_0(\mathcal{I}) \leq k \leq k_1(\mathcal{I}). \tag{5.39b}$$

We assume that the following assumption is satisfied:

Assumption 5.3

(a) $\mathbf{W}_c(\cdot)$ lies in $\mathscr{PC}\{\mathscr{I}, \mathscr{M}^N_{m_cn}; \mathfrak{T}\}$, $\mathbf{M}_c(\cdot)$ lies in $\mathscr{PC}\{\mathscr{I}, \mathscr{S}^N_n; \mathfrak{T}\}$;
(b) for each $k_0(\mathscr{I}) \le k \le k_1(\mathscr{I})$, $\mathbf{W}_d(\tau_k) \in \mathscr{M}^N_{m_dn}$ while $\mathbf{M}_d(\tau_k) \in \mathscr{S}^N_n$.

If $T \in \mathscr{I}$ and $\mathbf{G} \in \mathscr{S}^N_n$ we denote $\mathbf{Y}(\cdot; T, \mathbf{G})$ the solution of (5.39) satisfying the terminal condition $\mathbf{Y}(T; T, \mathbf{G}) = \mathbf{G}$.

Concerning the solutions of the non-homogeneous backward GLJMLDE (5.39) we prove:

Proposition 5.1 *Under the Assumptions 5.1 and 5.3 the following hold:*

(i) the solution with given terminal values $\mathbf{Y}(\cdot; T, \mathbf{G})$ is defined on the whole interval $[0, T] \cap \mathscr{I}$ being right continuous in each $\tau_k \in [0, T] \cap \mathscr{I}$;
(ii) if $\mathbf{G} \succeq 0$, $\mathbf{M}_c(t) \succeq 0$, for all $t \in \mathscr{I}$, $\mathbf{M}_d(\tau_k) \succeq 0$, for all $k_0(\mathscr{I}) \le k \le k_1(\mathscr{I})$ then

$$\mathbf{Y}(t; T, \mathbf{G}) \succeq 0,$$

for all $t \in [0, T] \cap \mathscr{I}$, " $\succeq$ " being the order relation induced by the convex cone $\mathscr{S}^{N+}_n$.
(iii) Assume:

(a) $\mathscr{I} \subset \mathbb{R}_+$ is a right unbounded interval;
(b) the GLJMLDE (5.33) is exponentially stable;
(c) $\sup_{t\in\mathscr{I}}\{\|\mathbf{M}_c(t)\|_\infty\} < \infty$ and $\sup_{k\ge k_0(\mathscr{I})}\{\|\mathbf{M}_d(\tau_k)\|_\infty\} < \infty$. $\|\cdot\|_\infty$ being the norm on $\mathscr{S}^N_n$ defined in (2.112).

Under these conditions, the non-homogeneous backward GLJMLDE (5.39) has a unique bounded solution $\tilde{\mathbf{Y}}(\cdot) : \mathscr{I} \to \mathscr{S}^N_n$. Additionally, if $\mathbf{M}_c(t) \succeq 0$ for all $t \in \mathscr{I}$ and $\mathbf{M}_d(\tau_k) \succeq 0$ for all $k \ge k_0(\mathscr{I})$ then $\tilde{\mathbf{Y}}(t) \succeq 0$, for all $t \in \mathscr{I}$. If $\mathbf{M}_c(t) \succeq 0$ for all $t \in \mathscr{I}$, but $\{\mathbf{M}_d(\tau_k)\}_{k\ge k_0(\mathscr{I})}$ is a uniform positive sequence then the unique bounded on $\mathscr{I}$ solution of (5.39) is uniform positive.

Proof Let $T_{\mathbf{W}_c}(t, s)$ be the linear evolution operator on $\mathscr{S}^N_n$ defined by the linear differential equation (5.36). Let k_T be the natural number with the property that $T \in [\tau_{k_T}, \tau_{k_T+1}]$. For $T \in [\tau_{k-1}, \tau_k)$, $k \le k_T$ we have that

$$\mathbf{Y}(t; T, \mathbf{G}) = T^*_{\mathbf{W}_c}(\tau_k, t)[\mathbf{Y}(\tau_k^-; T, \mathbf{G})] + \int_t^{\tau_k} T^*_{\mathbf{W}_c}(s, t)[\mathbf{M}_c(s)]\mathrm{d}s. \tag{5.40}$$

If $T \in (\tau_{k_T}, \tau_{k_T+1})$ we have:

$$\mathbf{Y}(t; T, \mathbf{G}) = T^*_{\mathbf{W}_c}(T, t)[\mathbf{G}] + \int_t^{T} T^*_{\mathbf{W}_c}(s, t)[\mathbf{M}_c(s)]ds, \quad \tau_{k_T} \le t < T. \tag{5.41}$$

If $T = \tau_{k_T}$ we compute

$$\mathbf{Y}(\tau_{k_T}^-; T, \mathbf{G}) = \boldsymbol{\Pi}_{d\mathbf{W}_d}(\tau_{k_T})[\mathbf{G}]. \tag{5.42}$$

Employing (5.40)–(5.42) together with (5.39b) written for $k \le k_T$ we may conclude that $\mathbf{Y}(t; T, \mathbf{G})$ is well defined for each $t \in [0, T] \cap \mathscr{I}$ with possible discontinuities in $t = \tau_k$. This confirms the validity of the statement (i).

To show that (ii) holds, one takes into account that $T^*_{\mathbf{W}_c}(s,t)\mathcal{S}_n^{N+} \subset \mathcal{S}_n^{N+}$ if $\tau_k \ge s \ge t \ge \tau_{k-1}$. This allows us to infer, via (5.40)–(5.42) together with (5.39b) that $\mathbf{Y}(t; T, \mathbf{G}) \succeq 0$, for all $t \in [0, T] \cap \mathscr{I}$. The validity of the properties stated in (iii) is obtained applying Theorem 2.5 and Theorem 2.6 in the special case of the ordered space $(\mathcal{S}_n^N, \mathcal{S}_n^{N+})$ and of the GLJMLDEs (5.33) and (5.39). Thus the proof is complete.

5.1.5 Several Intermediate Facts

In this subsection we recall several known results which will be useful in the developments of this chapter. Let

$$S = \begin{pmatrix} S_{11} & S_{12} \\ S_{12}^\top & S_{22} \end{pmatrix} \in \mathcal{S}_{n+m}, \quad S_{11} \in \mathcal{S}_n.$$

If S_{22} is invertible, we define $S_{22}^\# := S_{11} - S_{12}S_{22}^{-1}S_{12}^\top$. $S_{22}^\#$ will be named Schur complement of the block S_{22}.

For the proof of the next result one may see Theorem 1 from Albert [3].

Lemma 5.4 ***[Schur complement.]*** *Let*

$$S = \begin{pmatrix} S_{11} & S_{12} \\ S_{12}^\top & S_{22} \end{pmatrix} \in \mathcal{S}_{n+m}$$

such that $S_{22} > 0$. Under this condition the following equivalences hold:

(i) $S \ge 0$ if and only if $S_{11} - S_{12}S_{22}^{-1}S_{12}^\top \ge 0$;
(ii) $S > 0$ if and only if $S_{11} - S_{12}S_{22}^{-1}S_{12}^\top > 0$.

Often we shall use the following version of the parameterized result stated in the previous lemma.

Lemma 5.5 ***[Parameterized Schur complement.]*** *Let $S(\cdot) : \mathscr{I} \to \mathcal{S}_{n+m}$ be a bounded function. Let*

$$\begin{pmatrix} S_{11}(t) & S_{12}(t) \\ S_{12}^\top(t) & S_{22}(t) \end{pmatrix}$$

be a partition of the matrix $S(t)$ *such that* $S_{11}(t) \in \mathscr{S}_n$, *for all* $t \in \mathscr{I}$. *Assume that there exists* $c_2 > 0$ *such that* $S_{22}(t) \geq c_2 I_m$, *for all* $t \in \mathscr{I}$. *Under these conditions the following equivalences hold:*

(i) $S(t) \geq 0$ *for all* $t \in \mathscr{I}$ *if and only if* $S_{11}(t) - S_{12}(t)S_{22}^{-1}(t)S_{12}^{\top}(t) \geq 0$, *for all* $t \in \mathscr{I}$;
(ii) *there exists* $c > 0$ *such that* $S(t) \geq cI_{n+m}$, *for all* $t \in \mathscr{I}$ *if and only if there exists* $c_1 > 0$ *such that* $S_{11}(t) - S_{12}(t)S_{22}^{-1}(t)S_{12}^{\top}(t) \geq c_1 I_n$, *for all* $t \in \mathscr{I}$.

Proof The proof can be done following step by step the proof of Lemma 5.1.2 from Drăgan et al. [1].

Remark 5.2 Lemma 5.5 allows us to deduce that the set $\Gamma^{\Sigma}[\mathscr{I}]$ contains all solutions $\mathbf{X}(\cdot) : \mathscr{I} \to \mathscr{S}_n^N$ of (5.2) which are satisfying the sign conditions:

$$\Pi_{c3}(t)[\mathbf{X}(t)](i) + R_c(t,i) \geq \gamma_c I_{m_c},$$

for all $(t,i) \in \mathscr{I} \times \mathscr{N}$ and $\quad \Pi_{d3}(\tau_k)[\mathbf{X}(\tau_k)](i) + R_d(\tau_k, i) \geq \gamma_d I_{m_d}$, for all $(\tau_k, i) \in (\mathfrak{T} \cap \mathscr{I}) \times \mathscr{N}$, γ_c, γ_d are positive constants.

Lemma 5.6 *Let*

$$S = \begin{pmatrix} S_{11} & S_{12} \\ S_{12}^{\top} & S_{22} \end{pmatrix} \quad \text{and} \quad \widetilde{S} = \begin{pmatrix} \widetilde{S}_{11} & \widetilde{S}_{12} \\ \widetilde{S}_{12}^{\top} & \widetilde{S}_{22} \end{pmatrix}$$

be the partition of two matrices from $\mathscr{S}_{n+m}$ *such that* S_{11} *and* $\widetilde{S}_{11}$ *are in* $\mathscr{S}_n$. *If* $S \geq \widetilde{S}$ *and* $\widetilde{S}_{22} > 0$ *then* $\quad S_{22}^{\#} \geq \widetilde{S}_{22}^{\#}$.

Proof follows the line of the proof of Corollary 2.6 from Freiling and Hochhaus [4].

5.2 A Comparison Theorem and Several Consequences

Let us consider two GBJMRDEs of type (5.2) written in a compact form as (5.13):

$$-\dot{\mathbf{X}}(t) = \widehat{\mathscr{R}}_c(t, \mathbf{X}(t)), \quad \tau_k \leq t < \tau_{k+1} \tag{5.43a}$$
$$\mathbf{X}(\tau_k^-) = \widehat{\mathscr{R}}_d(\tau_k, \mathbf{X}(\tau_k)), \quad k \in \mathbb{Z}_+ \tag{5.43b}$$

$$-\dot{\mathbf{Y}}(t) = \widetilde{\mathscr{R}}_c(t, \mathbf{Y}(t)), \quad \tau_k \leq t < \tau_{k+1} \tag{5.44a}$$
$$\mathbf{Y}(\tau_k^-) = \widetilde{\mathscr{R}}_d(\tau_k, \mathbf{Y}(\tau_k)), \quad k \in \mathbb{Z}_+. \tag{5.44b}$$

The pair of generalized Riccati-type operators $(\widehat{\mathscr{R}}_c, \widehat{\mathscr{R}}_d)$ is associated to the quintuple $\widehat{\boldsymbol{\Sigma}} = (\mathbf{A}, \mathbf{B}, (\boldsymbol{\Pi}_c \mid \boldsymbol{\Pi}_d), (\widehat{\boldsymbol{\Theta}}_c \mid \widehat{\boldsymbol{\Theta}}_d), \mathfrak{T})$ and the pair $(\widetilde{\mathscr{R}}_c, \widetilde{\mathscr{R}}_d)$ is associated to the quintuple $\widetilde{\boldsymbol{\Sigma}} = (\mathbf{A}, \mathbf{B}, (\boldsymbol{\Pi}_c \mid \boldsymbol{\Pi}_d), (\widetilde{\boldsymbol{\Theta}}_c \mid \widetilde{\boldsymbol{\Theta}}_d), \mathfrak{T})$, where the quadruple $(\mathbf{A}, \mathbf{B}, (\boldsymbol{\Pi}_c \mid \boldsymbol{\Pi}_d), \mathfrak{T})$ is described in connection with (5.15) and

$$\widehat{\boldsymbol{\Theta}}_c(t) = (\widehat{\Theta}_c(t,1), \widehat{\Theta}_c(t,2), \ldots, \widehat{\Theta}_c(t,N)) \in \mathscr{S}^N_{n+m_c},$$

$$\widehat{\boldsymbol{\Theta}}_d(\tau_k) = (\widehat{\Theta}_d(\tau_k,1), \widehat{\Theta}_d(\tau_k,2), \ldots, \widehat{\Theta}_d(\tau_k,N)) \in \mathscr{S}^N_{n+m_d}$$

with

$$\widehat{\Theta}_c(t,i) = \begin{pmatrix} \widehat{M}_c(t,i) & \widehat{L}_c(t,i) \\ \widehat{L}_c^\top(t,i) & \widehat{R}_c(t,i) \end{pmatrix},$$

$$\widehat{\Theta}_d(\tau_k,i) = \begin{pmatrix} \widehat{M}_d(\tau_k,i) & \widehat{L}_d(\tau_k,i) \\ \widehat{L}_d^\top(\tau_k,i) & \widehat{R}_d(\tau_k,i) \end{pmatrix},$$

where $\widehat{M}_c(t,i),\ \widehat{M}_d(\tau_k,i) \in \mathscr{S}_n$;

$$\widetilde{\boldsymbol{\Theta}}_c(t) = (\widetilde{\Theta}_c(t,1), \widetilde{\Theta}_c(t,2), \ldots, \widetilde{\Theta}_c(t,N)) \in \mathscr{S}^N_{n+m_c},$$

$$\widetilde{\boldsymbol{\Theta}}_d(\tau_k) = (\widetilde{\Theta}_d(\tau_k,1), \widetilde{\Theta}_d(\tau_k,2), \ldots, \widetilde{\Theta}_d(\tau_k,N)) \in \mathscr{S}^N_{n+m_d}$$

with

$$\widetilde{\Theta}_c(t,i) = \begin{pmatrix} \widetilde{M}_c(t,i) & \widetilde{L}_c(t,i) \\ \widetilde{L}_c^\top(t,i) & \widetilde{R}_c(t,i) \end{pmatrix},$$

$$\widetilde{\Theta}_d(\tau_k,i) = \begin{pmatrix} \widetilde{M}_d(\tau_k,i) & \widetilde{L}_d(\tau_k,i) \\ \widetilde{L}_d^\top(\tau_k,i) & \widetilde{R}_d(\tau_k,i) \end{pmatrix},$$

where $\widetilde{M}_c(t,i),\ \widetilde{M}_d(\tau_k,i) \in \mathscr{S}_n,\ i \in \mathscr{N}$.

Theorem 5.1 ***(Comparison theorem.)*** *Let $t \to \mathbf{X}(t) : \mathscr{I} \subset \mathbb{R}_+ \to \mathscr{S}_n^N$ and $t \to \mathbf{Y}(t) : \mathscr{I} \to \mathscr{S}_n^N$ be solutions of the GBJMRDE (5.43) and (5.44), respectively.*
Assume:

(a) $\widehat{\Theta}_c(t,i) \geq \widetilde{\Theta}_c(t,i)$, for all $(t,i) \in \mathscr{I} \times \mathscr{N}$ and $\widehat{\Theta}_d(\tau_k,i) \geq \widetilde{\Theta}_d(\tau_k,i)$ for all $i \in \mathscr{N}$ and for any $k \in \mathbb{Z}_+$ for which $\tau_k \in \mathscr{I}$;
(b)

$$\Pi_{c3}(t)[\mathbf{Y}(t)](i) + \widetilde{R}_c(t,i) > 0,$$

for all $(t,i) \in \mathscr{I} \times \mathscr{N}$,

$$\Pi_{d3}(\tau_k)[\mathbf{Y}(\tau_k)](i) + \widetilde{R}_d(\tau_k,i) > 0,$$

for all $i \in \mathscr{N}$, $k \in \mathbb{Z}_+$ such that $\tau_k \in \mathscr{I}$;
(c) there exists $T \in \mathscr{I}$ such that $X(T,i) \geq Y(T,i)$ for all $i \in \mathscr{N}$.

Under these conditions we have

$$X(t,i) \geq Y(t,i),$$

for all $(t, i) \in ([0, T] \cap \mathscr{I}) \times \mathscr{N}$.

Proof Let $\widehat{\mathbf{F}}_c(t) = (\widehat{F}_c(t, 1), \widehat{F}_c(t, 2), \ldots, \widehat{F}_c(t, N))$ and $\widetilde{\mathbf{F}}_c(t) = (\widetilde{F}_c(t, 1), \widetilde{F}_c(t, 2), \ldots, \widetilde{F}_c(t, N))$ computed as in (5.23) for $\mathbf{X} = \mathbf{X}(t)$ and $\mathbf{X} = \mathbf{Y}(t)$, respectively, for all $t \in \mathscr{I}$. Let also $\widehat{\mathbf{F}}_d(\tau_k) = (\widehat{F}_d(\tau_k, 1), \widehat{F}_d(\tau_k, 2), \ldots, \widehat{F}_d(\tau_k, N))$ and $\widetilde{\mathbf{F}}_d(\tau_k) = (\widetilde{F}_d(\tau_k, 1), \widetilde{F}_d(\tau_k, 2), \ldots, \widetilde{F}_d(\tau_k, N))$ be computed as in (5.25) for $\mathbf{X} = \mathbf{X}(\tau_k)$ and $\mathbf{X} = \mathbf{Y}(\tau_k)$, respectively, for all $k \in \mathbb{Z}_+$ for which $\tau_k \in \mathscr{I}$.

Employing (5.28) with $\mathbf{W}_c(t)$ replaced by $\widehat{\mathbf{F}}_c(t)$ and (5.24) with $\mathbf{W}_d(\tau_k)$ replaced by $\widehat{\mathbf{F}}_d(\tau_k)$ we rewrite (5.43) and (5.44) satisfied by $\mathbf{X}(t)$ and $\mathbf{Y}(t)$, respectively, as:

$$-\dot{\mathbf{X}}(t) = \mathscr{L}^*_{c\hat{\mathbf{F}}_c}(t)[\mathbf{X}(t)] + \widehat{\boldsymbol{\Theta}}_{c\hat{\mathbf{F}}_c}(t), \quad \tau_k \le t < \tau_{k+1} \tag{5.45a}$$

$$\mathbf{X}(\tau_k^-) = \boldsymbol{\Pi}_{d\hat{\mathbf{F}}_d}(\tau_k)[\mathbf{X}(\tau_k)] + \widehat{\boldsymbol{\Theta}}_{d\hat{\mathbf{F}}_d}(\tau_k), \quad k \in \mathbb{Z}_+, \ \tau_k \in \mathscr{I}, \tag{5.45b}$$

$$\begin{aligned} -\dot{\mathbf{Y}}(t) &= \mathscr{L}^*_{c\hat{\mathbf{F}}_c}(t)[\mathbf{Y}(t)] + \widetilde{\boldsymbol{\Theta}}_{c\hat{\mathbf{F}}_c}(t) - (\widehat{\mathbf{F}}_c(t) - \widetilde{\mathbf{F}}_c(t))^\top \\ &\quad \cdot (\boldsymbol{\Pi}_{c3}(t)[\mathbf{Y}(t)] + \widetilde{\mathbf{R}}_c(t))(\widehat{\mathbf{F}}_c(t) - \widetilde{\mathbf{F}}_c(t)), \quad \tau_k \le t < \tau_{k+1} \qquad (5.46a) \\ \mathbf{Y}(\tau_k^-) &= \boldsymbol{\Pi}_{d\hat{\mathbf{F}}_d}(\tau_k)[\mathbf{Y}(\tau_k)] + \widetilde{\boldsymbol{\Theta}}_{d\hat{\mathbf{F}}_d}(\tau_k) - (\widehat{\mathbf{F}}_d(\tau_k) - \widetilde{\mathbf{F}}_d(\tau_k))^\top \\ &\quad \cdot (\boldsymbol{\Pi}_{d3}(\tau_k)[\mathbf{Y}(\tau_k)] + \widetilde{\mathbf{R}}_d(\tau_k))(\widehat{\mathbf{F}}_d(\tau_k) - \widetilde{\mathbf{F}}_d(\tau_k)), \quad k \in \mathbb{Z}_+, \tau_k \in \mathscr{I}. \qquad (5.46b) \end{aligned}$$

Let, $\mathbf{Z}(t) = (Z(t, 1), Z(t, 2), \ldots, Z(t, N))$ be defined by $Z(t, i) := X(t, i) - Y(t, i)$, for all $(t, i) \in \mathscr{I} \times \mathscr{N}$. Subtracting (5.46) from (5.45) we obtain that $t \to \mathbf{Z}(t) : \mathscr{I} \to \mathscr{S}_n^N$ solves the following backward GLJMLDE of the form (5.39):

$$-\dot{\mathbf{Z}}(t) = \mathscr{L}^*_{c\hat{\mathbf{F}}_c}(t)[\mathbf{Z}(t)] + \boldsymbol{\Delta}_c(t), \quad \tau_k \le t < \tau_{k+1} \tag{5.47a}$$

$$\mathbf{Z}(\tau_k^-) = \boldsymbol{\Pi}_{d\hat{\mathbf{F}}_d}(\tau_k)[\mathbf{Z}(\tau_k)] + \boldsymbol{\Delta}_d(\tau_k), \quad k_0(\mathscr{I}) \le k \le k_1(\mathscr{I}), \tag{5.47b}$$

where we have denoted $\boldsymbol{\Delta}_c(t) = (\Delta_c(t, 1), \Delta_c(t, 2), \ldots, \Delta_c(t, N))$ with

$$\begin{aligned} \Delta_c(t, i) &:= \begin{pmatrix} I_n \\ \widehat{F}_c(t, i) \end{pmatrix}^\top (\widehat{\Theta}_c(t, i) - \widetilde{\Theta}_c(t, i)) \begin{pmatrix} I_n \\ \widehat{F}_c(t, i) \end{pmatrix} \\ &\quad + (\widehat{F}_c(t, i) - \widetilde{F}_c(t, i))^\top (\Pi_{c3}(t)[\mathbf{Y}(t)](i) + \widetilde{R}_c(t, i))(\widehat{F}_c(t, i) - \widetilde{F}_c(t, i)), \end{aligned}$$

for all $(t, i) \in \mathscr{I} \times \mathscr{N}$ and $\boldsymbol{\Delta}_d(\tau_k) = (\Delta_d(\tau_k, 1), \Delta_d(\tau_k, 2), \ldots, \Delta_d(\tau_k, N))$ with

$$\begin{aligned} \Delta_d(\tau_k, i) &:= \begin{pmatrix} I_n \\ \widehat{F}_d(\tau_k, i) \end{pmatrix}^\top (\widehat{\Theta}_d(\tau_k, i) - \widetilde{\Theta}_d(\tau_k, i)) \begin{pmatrix} I_n \\ \widehat{F}_d(\tau_k, i) \end{pmatrix} \\ &+ (\widehat{F}_d(\tau_k, i) - \widetilde{F}_d(\tau_k, i))^\top (\Pi_{d3}(\tau_k)[\mathbf{Y}(\tau_k)](i) + \widetilde{R}_d(\tau_k, i))(\widehat{F}_d(\tau_k, i) - \widetilde{F}_d(\tau_k, i)), \end{aligned}$$

for all $i \in \mathscr{N}$, $k_0(\mathscr{I}) \le k \le k_1(\mathscr{I})$, with $k_0(\mathscr{I})$, $k_1(\mathscr{I})$ defined in (5.34).

Based on the assumptions (a) and (b) from the statement, we may infer that

$$\boldsymbol{\Delta}_c(t) \succeq 0, \qquad t \in \mathscr{I}$$

and

$$\boldsymbol{\Delta}_d(\tau_k) \succeq 0, \qquad \forall \ k_0(\mathscr{I}) \leq k \leq k_1(\mathscr{I}).$$

In the same time, the assumption (c) allows us to deduce that $\mathbf{Z}(T) \succeq 0$. Applying Proposition 5.1 (ii) in the case of the non-homogeneous backward GLJMLDE (5.47), we may conclude that $\mathbf{Z}(t) \succeq 0$, for all $t \in [0, T] \cap \mathscr{I}$ which is equivalent to $X(t, i) \geq Y(t, i)$, for all $(t, i) \in ([0, T] \cap \mathscr{I}) \times \mathscr{N}$. Thus the proof is complete.

The next result provides a necessary and sufficient condition which guarantee the existence on the whole interval $[0, T]$ of the solution with given terminal values of the generalized jump matrix Riccati differential equation (5.2). Let $\boldsymbol{\Sigma} = (\mathbf{A}, \mathbf{B}, (\boldsymbol{\Pi}_c \mid \boldsymbol{\Pi}_d), (\boldsymbol{\Theta}_c \mid \boldsymbol{\Theta}_d), \mathfrak{T})$ be the quintuple defining the GBJMRDE (5.2) and the pair of generalized Riccati-type operators $(\mathscr{R}_c, \mathscr{R}_d)$. Let

$$\widetilde{\mathscr{D}}(\mathscr{R}_c, \mathscr{R}_d) \stackrel{\Delta}{=} \{(T, \mathbf{X}) \in (0, \infty) \times \mathscr{S}_n^N \mid \exists\, \mathbf{Z}(\cdot) \in \Gamma^{\Sigma}([0, T]) \text{ such that } \mathbf{X} \succeq \mathbf{Z}(T)\}.$$

Theorem 5.2 *Assume that the Assumption 5.1 holds. Let* $\mathbf{X}(\cdot; T, \mathbf{G})$ *be the solution of (5.2) satisfying the terminal condition* $\mathbf{X}(T; T, \mathbf{G}) = \mathbf{G}$.

Under these conditions the following are equivalent:

(i) $(T, \mathbf{G}) \in \widetilde{\mathscr{D}}(\mathscr{R}_c, \mathscr{R}_d)$,

(ii) $\mathbf{X}(\cdot; T, \mathbf{G})$ *is defined on the whole interval* $[0, T]$ *and satisfies the sign conditions:*

$$\Pi_{c3}(t)[\mathbf{X}(t; T, \mathbf{G})](i) + R_c(t, i) \geq \gamma_c I_{m_c}, \ \text{for all} \ (t, i) \in [0, T] \times \mathscr{N} \tag{5.48a}$$

$$\Pi_{d3}(\tau_k)[\mathbf{X}(\tau_k; T, \mathbf{G})](i) + R_d(\tau_k, i) \geq \gamma_d I_{m_d} \ \text{for all} \ (\tau_k, i) \in (\mathfrak{T} \cap [0, T]) \times \mathscr{N}, \tag{5.48b}$$

γ_c, γ_d *being positive constants which could depend upon the solution* $\mathbf{X}(\cdot; T, \mathbf{G})$.

Proof First we prove the implication $(i) \Rightarrow (ii)$. Let $k_T \in \mathbb{Z}_+$ be such that $T \in [\tau_{k_T}, \tau_{k_T+1})$. We consider the case $T \in (\tau_{k_T}, \tau_{k_T+1})$. According to a general result from the theory of nonlinear differential equations, the solution $\mathbf{X}(\cdot; T, \mathbf{G})$ of the differential equation

$$\dot{\mathbf{X}}(t) = -\mathscr{R}_c(t, \mathbf{X}(t))$$

is defined in a neighborhood of T. Let $\varepsilon_0 > 0$ be such that $(T - \varepsilon_0, T] \subset (\tau_{k_T}, T]$ is the maximal interval where the solution $\mathbf{X}(\cdot; T, \mathbf{G})$ is defined.

We show that $\mathbf{X}(t; T, \mathbf{G})$ is defined also in $t = T - \varepsilon_0$ and $[T - \varepsilon_0, T] = [\tau_{k_T}, T]$. Since $(T, \mathbf{G}) \in \widetilde{\mathscr{D}}(\mathscr{R}_c, \mathscr{R}_d)$ we deduce that there exists $\mathbf{Z}(\cdot) \in \Gamma^{\Sigma}([0, T])$ such that

$$\mathbf{G} \succeq \mathbf{Z}(T).$$

Applying Lemma 5.5 in the case of (5.16a) together with (5.16c) we deduce that $\mathbf{Z}(\cdot)$ is satisfying:

$$\dot{\mathbf{Z}}(t) + \mathscr{R}_c(t, \mathbf{Z}(t)) \succeq 0, \quad \tau_k \le t < \tau_{k+1}, \quad 0 \le k \le k_T \tag{5.49}$$

with the convention that when $k = k_T$ (5.49) holds for $t \in [\tau_{k_T}, T]$. Let

$$\widetilde{\mathbf{M}}_c(t) = (\widetilde{M}_c(t, 1), \widetilde{M}_c(t, 2), \dots, \widetilde{M}_c(t, N)) \in \mathscr{S}_n^N$$

be defined by

$$\widetilde{\mathbf{M}}_c(t) \stackrel{\Delta}{=} \dot{\mathbf{Z}}(t) + \mathscr{R}_c(t, \mathbf{Z}(t)), \quad t \in [\tau_k, \tau_{k+1}), \quad 0 \le k \le k_T.$$

From (5.49) it follows that $\widetilde{\mathbf{M}}_c(t) \succeq 0$, for all $t \in [0, T]$. Similarly, (5.16b) together with (5.16d) lead to

$$-\mathbf{Z}(\tau_k^-) + \mathscr{R}_d(\tau_k, \mathbf{Z}(\tau_k)) \succeq 0, \quad 0 \le k \le k_T. \tag{5.50}$$

Let

$$\widetilde{\mathbf{M}}_d(\tau_k) \stackrel{\Delta}{=} -\mathbf{Z}(\tau_k) + \mathscr{R}_d(\tau_k, \mathbf{Z}(\tau_k)), \quad 0 \le k \le k_T.$$

From (5.50) we deduce that $\widetilde{\mathbf{M}}_d(\tau_k) \succeq 0, \quad 0 \le k \le k_T$. Summing up, we may infer that $\mathbf{Z}(\cdot)$ solves the following GBJMRDE:

$$-\dot{\mathbf{Z}}(t) = \widetilde{\mathscr{R}}_c(t, \mathbf{Z}(t)), \quad \tau_k \le t < \tau_{k+1} \tag{5.51a}$$

$$\mathbf{Z}(\tau_k^-) = \widetilde{\mathscr{R}}_d(\tau_k, \mathbf{Z}(\tau_k)), \quad 0 \le k \le k_T \tag{5.51b}$$

where

$$\widetilde{\mathscr{R}}_c(t, \mathbf{Z}) := \mathscr{R}_c(t, \mathbf{Z}) - \widetilde{\mathbf{M}}_c(t), \quad \text{for all } \ (t, \mathbf{Z}) \in \ Dom\ \mathscr{R}_c$$

and

$$\widetilde{\mathscr{R}}_d(\tau_k, \mathbf{Z}) := \mathscr{R}_d(\tau_k, \mathbf{Z}) - \widetilde{\mathbf{M}}_d(\tau_k), \quad \text{for all } \ (\tau_k, \mathbf{Z}) \in \ Dom\ \mathscr{R}_d.$$

The pair of the generalized Riccati-type operators $(\widetilde{\mathscr{R}}_c, \widetilde{\mathscr{R}}_d)$ is associated to the quintuple $\widetilde{\boldsymbol{\Sigma}} = (\mathbf{A}, \mathbf{B}, (\boldsymbol{\Pi}_c \mid \boldsymbol{\Pi}_d), (\widetilde{\boldsymbol{\Theta}}_c \mid \widetilde{\boldsymbol{\Theta}}_d), \mathfrak{T})$, where

$$\widetilde{\boldsymbol{\Theta}}_c(t) = (\widetilde{\Theta}_c(t, 1), \widetilde{\Theta}_c(t, 2), \dots, \widetilde{\Theta}_c(t, N))$$

with

$$\widetilde{\Theta}_c(t,i) = \begin{pmatrix} M_c(t,i) - \widetilde{M}_c(t,i) & L_c(t,i) \\ L_c^\top(t,i) & R_c(t,i) \end{pmatrix}, \quad i \in \mathcal{N}$$

and

$$\widetilde{\boldsymbol{\Theta}}_d(\tau_k) = (\widetilde{\Theta}_d(\tau_k,1), \widetilde{\Theta}_d(\tau_k,2), \ldots, \widetilde{\Theta}_d(\tau_k,N))$$

with

$$\widetilde{\Theta}_d(\tau_k,i) = \begin{pmatrix} M_d(\tau_k,i) - \widetilde{M}_d(\tau_k,i) & L_d(\tau_k,i) \\ L_d^\top(\tau_k,i) & R_d(\tau_k,i) \end{pmatrix}, \quad i \in \mathcal{N}.$$

Hence,

$$\Theta_c(t,i) - \widetilde{\Theta}_c(t,i) = \begin{pmatrix} \widetilde{M}_c(t,i) & 0 \\ 0 & 0 \end{pmatrix} \geq 0, \tag{5.52a}$$

$$\Theta_d(\tau_k,i) - \widetilde{\Theta}_d(\tau_k,i) = \begin{pmatrix} \widetilde{M}_d(\tau_k,i) & 0 \\ 0 & 0 \end{pmatrix} \geq 0, \quad i \in \mathcal{N}. \tag{5.52b}$$

Employing, (5.16c), (5.16d) and (5.52) together with $\mathbf{G} \succeq \mathbf{Z}(T)$ we may conclude that the assumptions of the Theorem 5.1 are satisfied in the case of the pair of equations (5.2) and (5.51). Hence,

$$\mathbf{X}(t;T,\mathbf{G}) \succeq \mathbf{Z}(t) \quad \text{for all} \quad t \in [T-\varepsilon_0, T]. \tag{5.53}$$

Combining (5.16c), (5.16d) and (5.53) we may infer that (5.48) are satisfied for all $(t,i) \in [T-\varepsilon_0, T] \times \mathcal{N}$ and for all $(\tau_k, i) \in (\mathfrak{T} \cap (T-\varepsilon_0, T]) \times \mathcal{N}$.

Let us consider now the non-homogeneous backward GLJMLDE:

$$-\dot{\mathbf{Y}}(t) = \mathbf{A}^\top(t)\mathbf{Y}(t) + \mathbf{Y}(t)\mathbf{A}(t) + \boldsymbol{\Pi}_{c1}(t)[\mathbf{Y}(t)] + \mathbf{M}_c(t), \quad \tau_k \leq t < \tau_{k+1} \tag{5.54a}$$

$$\mathbf{Y}(\tau_k^-) = \boldsymbol{\Pi}_{d1}(\tau_k)[\mathbf{Y}(\tau_k)] + \mathbf{M}_d(\tau_k), \quad 0 \leq k \leq k_T. \tag{5.54b}$$

One sees that (5.54) is the particular case of (5.39) with $\mathbf{W}_c(t) \equiv 0$ and $\mathbf{W}_d(\tau_k) \equiv 0$. If $\mathbf{Y}(\cdot;T,\mathbf{G})$ is the solution of (5.54) satisfying $\mathbf{Y}(T;T,\mathbf{G}) = \mathbf{G}$ we may deduce via Proposition 5.1 (i) that this solution is defined on the whole interval $[0,T]$ and it is bounded on this interval.

Let $t \to \boldsymbol{\Upsilon}(t) : (T-\varepsilon_0, T] \to \mathcal{S}_n^N$ be defined by

$$\boldsymbol{\Upsilon}(t) := \mathbf{Y}(t;T,\mathbf{G}) - \mathbf{X}(t;T,\mathbf{G}).$$

Subtracting (5.2) (satisfied by $\mathbf{X}(\cdot;T,\mathbf{G})$) from (5.54) (satisfied by $\mathbf{Y}(\cdot;T,\mathbf{G})$) we obtain that $\boldsymbol{\Upsilon}(\cdot)$ solves the following non-homogeneous backward GLJMLDE with given terminal value:

$$\dot{\boldsymbol{\Upsilon}}(t) + \mathbf{A}^\top(t)\boldsymbol{\Upsilon}(t) + \boldsymbol{\Upsilon}(t)\mathbf{A}(t) + \boldsymbol{\Pi}_{c1}(t)[\boldsymbol{\Upsilon}(t)] + \widetilde{\boldsymbol{\Delta}}_c(t) = 0, \quad \tau_k \le t < \tau_{k+1} \tag{5.55a}$$

$$\boldsymbol{\Upsilon}(\tau_k^-) = \boldsymbol{\Pi}_{d1}(\tau_k)[\boldsymbol{\Upsilon}(\tau_k)] + \widetilde{\boldsymbol{\Delta}}(\tau_k), \quad k \in \mathbb{Z}_+ \text{ for which } \tau_k \in (T - \varepsilon_0, T], \tag{5.55b}$$

$$\boldsymbol{\Upsilon}(T) = 0, \tag{5.55c}$$

where

$$\widetilde{\boldsymbol{\Delta}}_c(t) = (\widetilde{\Delta}_c(t,1), \widetilde{\Delta}_c(t,2), \ldots, \widetilde{\Delta}_c(t,N))$$

with

$$\begin{aligned}\widetilde{\Delta}_c(t,i) \stackrel{\Delta}{=} &(\Pi_{c2}(t)[\mathbf{X}(t;T,\mathbf{G})](i) + L_c(t,i))(\Pi_{c3}(t)[\mathbf{X}(t;T,\mathbf{G})](i) + R_c(t,i))^{-1} \\ &\cdot (\Pi_{c2}(t)[\mathbf{X}(t;T,\mathbf{G})](i) + L_c(t,i))^\top\end{aligned}$$

and

$$\widetilde{\boldsymbol{\Delta}}_d(\tau_k) = (\widetilde{\Delta}_d(\tau_k,1), \widetilde{\Delta}_d(\tau_k,2), \ldots, \widetilde{\Delta}_d(\tau_k,N))$$

with

$$\begin{aligned}\widetilde{\Delta}_d(\tau_k,i) \stackrel{\Delta}{=} &(\Pi_{d2}(\tau_k)[\mathbf{X}(\tau_k;T,\mathbf{G})](i) \\ &+ L_d(\tau_k,i))(\Pi_{d3}(\tau_k)[\mathbf{X}(\tau_k;T,\mathbf{G})](i) + R_d(\tau_k,i))^{-1} \\ &\cdot (\Pi_{d2}(\tau_k)[\mathbf{X}(\tau_k;T,\mathbf{G})](i) + L_d(\tau_k,i))^\top,\end{aligned}$$

for all $i \in \mathcal{N}$. It is clear that $\widetilde{\Delta}_c(t,i) \ge 0$, for all $(t,i) \in (T-\varepsilon_0, T] \times \mathcal{N}$ and $\widetilde{\Delta}_d(\tau_k,i) \ge 0$, for all $i \in \mathcal{N}$ and any $k \in \mathbb{Z}_+$ for which $\tau_k \in (T-\varepsilon_0, T]$.
Thus, Proposition 5.1 (ii) applied in the case of the problem with given terminal values (5.55) allows us to deduce that

$$\mathbf{Y}(t;T,\mathbf{G}) \succeq \mathbf{X}(t;T,\mathbf{G}), \tag{5.56}$$

for all $t \in (T-\varepsilon_0, T]$. Employing a classical result from the theory of differential equations we may conclude via (5.53) and (5.56) that $\mathbf{X}(\cdot;T,\mathbf{G})$ is defined on the whole interval $[\tau_{k_T}, T]$ and (5.48) and (5.53) together with (5.56) hold for any $t \in [\tau_{k_T}, T]$. Hence, $\mathbf{X}(\tau_{k_T}^-;T,\mathbf{G})$ can be computed by

$$\mathbf{X}(\tau_{k_T}^-;T,\mathbf{G}) = \mathscr{R}_d(\tau_{k_T}, \mathbf{X}(\tau_{k_T};T,\mathbf{G})). \tag{5.57}$$

Further, (5.53) written for $t = \tau_{k_T}$ together with the Assumption 5.1 (e) yields

$$\begin{aligned}&\begin{pmatrix} \Pi_{d1}(\tau_{k_T})[\mathbf{X}(\tau_{k_T};T,\mathbf{G})](i) + M_d(\tau_{k_T},i) & \Pi_{d2}(\tau_{k_T})[\mathbf{X}(\tau_{k_T};T,\mathbf{G})](i) + L_d(\tau_{k_T},i) \\ (\Pi_{d2}(\tau_{k_T})[\mathbf{X}(\tau_{k_T};T,\mathbf{G})](i) + L_d(\tau_{k_T},i))^\top & \Pi_{d3}(\tau_{k_T})[\mathbf{X}(\tau_{k_T};T,\mathbf{G})](i) + R_d(\tau_{k_T},i) \end{pmatrix} \\ &\ge \begin{pmatrix} \Pi_{d1}(\tau_{k_T})[\mathbf{Z}(\tau_{k_T})](i) + M_d(\tau_{k_T},i) & \Pi_{d2}(\tau_{k_T})[\mathbf{Z}(\tau_{k_T})](i) + L_d(\tau_{k_T},i) \\ (\Pi_{d2}(\tau_{k_T})[\mathbf{Z}(\tau_{k_T})](i) + L_d(\tau_{k_T},i))^\top & \Pi_{d3}(\tau_{k_T})[\mathbf{Z}(\tau_{k_T})](i) + R_d(\tau_{k_T},i) \end{pmatrix}, \quad i \in \mathcal{N}.\end{aligned}$$

Applying Lemma 5.6 in the case of previous matrix inequalities we deduce that for each $i \in \mathcal{N}$ we have that

$$\mathscr{R}_d(\tau_{k_T}; \mathbf{X}(\tau_{k_T}; T, \mathbf{G}))(i) \geq \mathscr{R}_d(\tau_{k_T}; \mathbf{Z}(\tau_{k_T}))(i),$$

or equivalently

$$\mathscr{R}_d(\tau_{k_T}; \mathbf{X}(\tau_{k_T}; T, \mathbf{G})) \succeq \mathscr{R}_d(\tau_{k_T}; \mathbf{Z}(\tau_{k_T})). \tag{5.58}$$

Employing (5.50), (5.57) and (5.58) we may conclude that

$$\mathbf{X}(\tau_{k_T}^-; T, \mathbf{G}) \succeq \mathbf{Z}(\tau_{k_T}^-). \tag{5.59}$$

On the other hand, (5.2b) and (5.54b) written for k replaced by k_T yield

$$\mathbf{Y}(\tau_{k_T}^-; T, \mathbf{G}) \succeq \mathbf{X}(\tau_{k_T}; T, \mathbf{G}). \tag{5.60}$$

In the case when $T = \tau_T$, $\mathbf{X}(T^-; T, \mathbf{G})$ can be computed via (5.57) because

$$\mathbf{G} \succeq \mathbf{Z}(T) = \mathbf{Z}(\tau_{k_T})$$

which guarantees the invertability of the matrices $\Pi_{d3}(\tau_{k_T})[\mathbf{G}](i) + R_d(\tau_{k_T}, i)$, $i \in \mathcal{N}$. One checks then (5.59) and (5.60) still hold in the case $\tau_{k_T} = T$, too.

Letting $t \to \tau_{k_T}$, $t < k_T$, in (5.16c) we get

$$\Pi_{c3}(\tau_{k_T})[\mathbf{Z}(\tau_{k_T}^-)](i) + R_c(\tau_{k_T}, i) \geq \gamma I_{m_c} > 0.$$

This inequality together with (5.59) yields

$$\Pi_{c3}(\tau_{k_T})[\mathbf{X}(\tau_{k_T}^-; T, \mathbf{G})](i) + R_c(\tau_{k_T}, i) \geq \gamma I_{m_c}.$$

Hence, $(\tau_{k_T}, \mathbf{X}(\tau_{k_T}^-; T, \mathbf{G}))$ lies in $Dom\ \mathscr{R}_c$.

Taking $\mathbf{G}_1 \stackrel{\Delta}{=} \mathbf{X}(\tau_{k_T}^-; T, \mathbf{G})$ as initial condition at the instant time τ_{k_T}, we deduce that there exists an interval $(\tau_{k_T} - \varepsilon_1, \tau_{k_T}] \subset [\tau_{k_T-1}, \tau_{k_T}]$ where the unique solution $\mathbf{X}(\cdot; \tau_{k_T}, \mathbf{G}_1)$ of the differential equation (5.2a) is defined and satisfies the terminal condition $\mathbf{X}(\tau_{k_T}; \tau_{k_T}, \mathbf{G}_1) = \mathbf{G}_1$. Applying again Theorem 5.1 one shows that

$$\mathbf{X}(t; \tau_{k_T}, \mathbf{G}_1) \succeq \mathbf{Z}(t),$$

for all $t \in (\tau_{k_T} - \varepsilon_1, \tau_{k_T}]$. Also, we have that

$$\mathbf{Y}(t; T, \mathbf{G}) = \mathbf{Y}(t; \tau_{k_T}, \mathbf{Y}(\tau_{k_T}^-; T, \mathbf{G})) \succeq \mathbf{X}(t; \tau_{k_T}, \mathbf{G}_1),$$

for all $t \in (\tau_{k_T} - \varepsilon_1, \tau_{k_T}]$. This allows us to conclude that $\mathbf{X}(\cdot; \tau_{k_T}, \mathbf{G}_1)$ is defined on the whole interval $[\tau_{k_T-1}; \tau_{k_T}]$. If we take into account that $\mathbf{X}(\cdot; \tau_{k_T}, \mathbf{G}_1)$ is the restriction of $\mathbf{X}(\cdot; T, \mathbf{G})$ to the interval $[\tau_{k_T-1}; \tau_{k_T}]$ we may deduce that the unique solution $\mathbf{X}(\cdot; T, \mathbf{G})$ of (5.2) with the terminal condition $\mathbf{X}(T; T, \mathbf{G}) = \mathbf{G}$ is defined on the whole interval $[\tau_{k_T-1}; T]$ and (5.53) together with (5.56) hold for any $t \in [\tau_{k_T-1}; T]$.

Let us assume now that for a $0 \leq k \leq \tau_{k_T-1}$ the solution $\mathbf{X}(\cdot; T, \mathbf{G})$ of (5.2) is defined on the interval $[\tau_k, T]$ and (5.53) together with (5.56) hold for any $t \in [\tau_k, T]$. Employing (5.16c) and (5.16d) together with (5.53) we may deduce that (5.48) are true on the whole interval $[\tau_k, T]$. Particularly, (5.48b) guarantees the invertability of the matrices $\Pi_{d3}(\tau_k)[\mathbf{X}(\tau_k; T, \mathbf{G})](i) + R_d(\tau_k, i)$. This means that $(\tau_k, \mathbf{X}(\tau_k; T, \mathbf{G}))$ lies in *Dom* $\mathcal{R}_d$. So, we may compute $\mathbf{X}(\tau_k^-; T, \mathbf{G})$ by

$$\mathbf{X}(\tau_k^-; T, \mathbf{G}) = \mathcal{R}_d(\tau_k, \mathbf{X}(\tau_k; T, \mathbf{G})). \tag{5.61}$$

Proceeding as in the case $k = k_T$ we may conclude via Lemma 5.6, that

$$\mathcal{R}_d(\tau_k, \mathbf{X}(\tau_k; T, \mathbf{G})) \succeq \mathcal{R}_d(\tau_k, \mathbf{Z}(\tau_k)). \tag{5.62}$$

Combining (5.50), (5.61) and (5.62) we may infer that

$$\mathbf{X}(\tau_k^-; T, \mathbf{G}) \succeq \mathbf{Z}(\tau_k^-). \tag{5.63}$$

Also, (5.2b) and (5.54b) together with (5.53) yield

$$\mathbf{Y}(\tau_k^-; T, \mathbf{G}) \succeq \mathbf{X}(\tau_k^-; T, \mathbf{G}). \tag{5.64}$$

Further, letting $t \to \tau_k$, $t < \tau_k$ in (5.16c) we get, via (5.63), that

$$\Pi_{c3}(\tau_k)[\mathbf{X}(\tau_k^-; T, \mathbf{G})](i) + R_c(\tau_k, i) \geq \gamma I_{m_c} > 0,$$

for all $i \in \mathcal{N}$. Hence, $(\tau_k, \mathbf{X}(\tau_k^-; T, \mathbf{G})) \in$ *Dom* $\mathcal{R}_c$. Setting $\mathbf{G}_k = \mathbf{X}(\tau_k^-; T, \mathbf{G})$ and applying a classical result from the theory of the differential equation, we deduce that the solution $\mathbf{X}(\cdot; \tau_k, \mathbf{G}_k)$ of the differential equation (5.2a) satisfying the terminal condition $\mathbf{X}(\tau_k; \tau_k, \mathbf{G}_k) = \mathbf{G}_k$ is defined in a neighborhood of τ_k. Let $(\tau_k - \varepsilon_k, \tau_k]$ be the maximal interval when $\mathbf{X}(t; \tau_k, \mathbf{G}_k)$ is defined for $t \in [\tau_{k-1}, \tau_k] \cap (\tau_k - \varepsilon_k, \tau_k]$. Applying again Theorem 5.1 in the case of the pair of GBJMRDEs (5.2) and (5.51) we obtain via (5.63) that

$$\mathbf{X}(t; \tau_k, \mathbf{G}_k) \succeq \mathbf{Z}(t), \tag{5.65}$$

for all $t \in (\tau_k - \varepsilon_k, \tau_k]$, which is equivalent to

$$\mathbf{X}(t; T, \mathbf{G}) \succeq \mathbf{Z}(t), \quad t \in (\tau_k - \varepsilon_k, T].$$

This means that (5.48) hold on the interval $(\tau_k - \varepsilon_k, T]$, then from (5.2a), (5.54a) and (5.64) together with (5.48a) we deduce that

$$\mathbf{Y}(t;T,\mathbf{G}) = \mathbf{Y}(t;\tau_k,\mathbf{Y}(\tau_k^-;T,\mathbf{G})) \succeq \mathbf{X}(t;\tau_k,\mathbf{G}_k), \quad t \in (\tau_k - \varepsilon_k, \tau_k]. \tag{5.66}$$

Employing (5.65) and (5.66) we may conclude that $\mathbf{X}(\cdot;\tau_k,\mathbf{G}_k)$ is defined on the whole interval $[\tau_{k-1},\tau_k]$. So, $\mathbf{X}(t;T,\mathbf{G})$ is defined for any $t \in [\tau_{k-1},T]$ and additionally (5.53) and (5.56) hold on the whole interval $[\tau_{k-1},T]$. Further, (5.16c) and (5.16d) together with (5.53) allow us to deduce that (5.48) holds for any $t \in [\tau_{k-1},T]$. Thus we may conclude that the solution $\mathbf{X}(\cdot;T,\mathbf{G})$ of (5.2) can be extended to the whole interval $[0,T] = [\tau_0,T]$ such that the inequalities (5.48), (5.53) and (5.56) still hold on this interval.

The implication $(ii) \Rightarrow (i)$ is obvious because according to Remark 5.2, $\mathbf{Z}(\cdot) = \mathbf{X}(\cdot;T,\mathbf{G})$ lie in Γ^{Σ}. Thus the proof is complete.

Corollary 5.1 *Let us assume that for an interval* $[0,T]$ *the function* $\mathbf{Z}(t) \equiv 0$ *lies in* $\Gamma^{\Sigma}(0,T)$, $\sum$ *being the quintuple which defines the GBJMRDE (5.2). Under these conditions, for any* $\mathbf{G} = (G(1), G(2), \ldots, G(N)) \in \mathcal{S}_n^{N+}$, *the solution* $\mathbf{X}(\cdot;T,\mathbf{G})$ *of (5.2) with the terminal condition* $\mathbf{X}(T;T,\mathbf{G}) = \mathbf{G}$ *is defined on the whole interval* $[0,T]$ *and satisfies* $0 \preceq \mathbf{X}(t;T,\mathbf{G}) \preceq \mathbf{Y}(t;T,\mathbf{G})$, *for all* $t \in [0,T]$, $\mathbf{Y}(\cdot;T,\mathbf{G})$ *being the solution of (5.54).*

5.3 The Maximal Solution of a Generalized Backward Jump Matrix Riccati-Type Differential Equation

Unlike the previous section, when we treated the existence of solutions with given terminal values of a GBJMRDE (5.2), in the present section as well as the next sections, we investigate the problem of the existence of some solutions of (5.2) which are defined on the whole semiaxis $\mathbb{R}_+$, they being identified by a property which ensure their unicity without prior knowledge of a boundary value. Particularly, in the present section we study the problem of the existence of a solution of (5.2) which has a maximality property with respect to the set $\Gamma^{\Sigma}[\mathbb{R}_+]$. We recall that $\Gamma^{\Sigma}[\mathbb{R}_+]$ is the set of functions which satisfy the conditions (5.16) when $\mathcal{I} = \mathbb{R}_+$.

Definition 5.3 A solution $t \to \mathbf{X}_{max}(t) = (X_{max}(t,1), X_{max}(t,2), \ldots, X_{max}(t,N)) : \mathbb{R}_+ \to \mathcal{S}_n^N$ of the GBJMRDE (5.2) is named maximal solution with respect to the set $\Gamma^{\Sigma}[\mathbb{R}_+]$, or equivalently, maximal solution of (5.2) (for shortness), if

$$\mathbf{X}_{max}(t) \succeq \mathbf{Z}(t) \ \text{ for all } \ t \in \mathbb{R}_+ \tag{5.67}$$

for any $\mathbf{Z}(\cdot) \in \Gamma^{\Sigma}[\mathbb{R}_+]$.

Remark 5.3 Under the Assumption 5.1 (e), we may infer via (5.16c), (5.16d) and (5.67) that the maximal solution $\mathbf{X}_{max}(\cdot)$, if any, is satisfying the sign conditions:

$$\Pi_{c3}(t)[\mathbf{X}_{max}(t)](i) + R_c(t,i) \geq \gamma_c I_{m_c} \ \text{ for all } \ t \in \mathbb{R}_+, \tag{5.68a}$$

$$\Pi_{d3}(\tau_k)[\mathbf{X}_{max}(\tau_k)](i) + R_d(\tau_k,i) \geq \gamma_d I_{m_d} \ \text{ for all } \ \tau_k \in \mathfrak{T}, \tag{5.68b}$$

$i \in \mathcal{N}$, where $\gamma_c > 0$, $\gamma_d > 0$ are constants which do not depend upon $t \in \mathbb{R}_+$, $k \in \mathbb{Z}_+$. Hence, under the Assumption 5.1 (*e*), the maximal solution of (5.2), if any, lies in $\Gamma^{\Sigma}[\mathbb{R}_+]$.

In the developments from this chapter, when we invoke a quadruple of the form $(\mathbf{A}, \mathbf{B}, (\boldsymbol{\Pi}_c \mid \boldsymbol{\Pi}_d), \mathfrak{T})$ we refer to a quadruple where $\mathbf{A}(\cdot)$, $\mathbf{B}(\cdot)$ are matrix-valued functions, $\boldsymbol{\Pi}_c(\cdot)$ is an operator valued function, while $\{\boldsymbol{\Pi}_d(\tau_k)\}_{k\in\mathbb{Z}_+}$ is an operator valued sequence and $\mathfrak{T}$ is the sequence of non-negative real numbers, which were described in connection with (5.15).

Definition 5.4 We say that the quadruple $(\mathbf{A}, \mathbf{B}, (\boldsymbol{\Pi}_c \mid \boldsymbol{\Pi}_d), \mathfrak{T})$ is stabilizable if there exist a bounded function $\mathbf{F}_c(\cdot) = (F_c(\cdot, 1), F_c(\cdot, 2), \ldots, F_c(\cdot, N)) \in \mathcal{PC}\{\mathbb{R}_+, \mathcal{M}^N_{m_c n}; \mathfrak{T}\}$ and a bounded sequence $\{\mathbf{F}_d(\tau_k)\}_{k\in\mathbb{Z}_+} \subset \mathcal{M}^N_{m_d n}$, $\mathbf{F}_d(\tau_k) = (F_d(\tau_k, 1), F_d(\tau_k, 2), \ldots, F_d(\tau_k, N))$ with the property that the following jump matrix linear differential equation (JMLDE) on the Hilbert space $\mathcal{S}^N_n$

$$\dot{\mathbf{X}}(t) = \mathcal{L}_{c\mathbf{F}_c}(t)[\mathbf{X}(t)], \quad \tau_k < t \le \tau_{k+1} \tag{5.69a}$$

$$\mathbf{X}(\tau_k^+) = \boldsymbol{\Pi}^*_{d\mathbf{F}_d}(\tau_k)[\mathbf{Z}(\tau_k)], \quad k \in \mathbb{Z}_+, \tag{5.69b}$$

is exponentially stable.

In (5.69a), $\mathbf{X} \to \mathcal{L}_{c\mathbf{F}_c}(t)[\mathbf{X}] : \mathcal{S}^N_n \to \mathcal{S}^N_n$ is described by

$$\mathcal{L}_{c\mathbf{F}_c}(t)[\mathbf{X}] = (\mathbf{A}(t) + \mathbf{B}(t)\mathbf{F}_c(t))\mathbf{X} + \mathbf{X}(\mathbf{A}(t) + \mathbf{B}(t)\mathbf{F}_c(t))^\top + \boldsymbol{\Pi}^*_{c\mathbf{F}_c}(t)[\mathbf{X}] \tag{5.70}$$

where $\boldsymbol{\Pi}^*_{c\mathbf{F}_c}(t)[\cdot]$ is the adjoint of the operator $\mathbf{X} \to \boldsymbol{\Pi}_{c\mathbf{F}_c}(t)[\mathbf{X}] : \mathcal{S}^N_n \to \mathcal{S}^N_n$ described as in (5.18) for $\mathbf{W}_c(\cdot)$ replaced by $\mathbf{F}_c(\cdot)$.

In (5.69b), $\boldsymbol{\Pi}^*_{d\mathbf{F}_d}(\tau_k)[\cdot]$ is the adjoint of the operator $\mathbf{X} \to \boldsymbol{\Pi}_{d\mathbf{F}_d}(\tau_k)[\mathbf{X}] : \mathcal{S}^N_n \to \mathcal{S}^N_n$ defined by (5.20) with $\mathbf{W}_d(\tau_k)$ replaced by $\mathbf{F}_d(\tau_k)$.

Remark 5.4 In the particular case when the elements of the quadruple $(\mathbf{A}, \mathbf{B}, (\boldsymbol{\Pi}_c \mid \boldsymbol{\Pi}_d), \mathfrak{T})$ are defined by (5.12) using the matrix coefficients of the jump linear stochastic system (4.1), the concept of stabilizability introduced in Definition 5.4 reduces to the stabilizability in mean square by state feedback introduced in the Definition 4.3.

The next result provides a necessary and sufficient condition for the existence of maximal and bounded solution of the equation (5.2) satisfying the sign conditions (5.68).

Theorem 5.3 *Under the Assumption 5.1, if the quadruple* $(\mathbf{A}, \mathbf{B}, (\boldsymbol{\Pi}_c \mid \boldsymbol{\Pi}_d), \mathfrak{T})$ *is stabilizable then the following are equivalent:*

(i) the GBJMRDE (5.2) has a unique maximal and bounded solution $t \to \mathbf{X}_{max}(t) : \mathbb{R}_+ \to \mathcal{S}^N_n$ *which is satisfying the sign condition (5.68);*
(ii) the set $\Gamma^{\Sigma}[\mathbb{R}_+]$ *is not empty.*

Proof The validity of the implication $(i) \Rightarrow (ii)$ is obvious, because Remark 5.2 and Remark 5.3 allows us to conclude that the maximal solution $\mathbf{X}_{max}(\cdot)$, if any, lie in $\Gamma^{\Sigma}[\mathbb{R}_+]$. It remains to prove the validity of the implication $(ii) \Rightarrow (i)$. If (ii) holds, there exists a function $t \to \mathbf{Z}(t) : \mathbb{R}_+ \to \mathscr{S}_n^N$ satisfying (5.16). Applying Lemma 5.5 in the case of (5.16a) together with (5.16c) written for the function $\mathbf{Z}(\cdot)$, we obtain:

$$\dot{\mathbf{Z}}(t) + \mathscr{R}_c(t, \mathbf{Z}(t)) \succeq 0, \quad \tau_k \leq t < \tau_{k+1}, \quad k \in \mathbb{Z}_+. \tag{5.71}$$

Similarly, Lemma 5.5 applied in the case of (5.16b) together with (5.16d) yield

$$-\mathbf{Z}(\tau_k^-) + \mathscr{R}_d(\tau_k, \mathbf{Z}(\tau_k)) \succeq 0, \quad k \in \mathbb{Z}_+. \tag{5.72}$$

Let $\widehat{\mathbf{M}}_c(t) = \big(\widehat{M}_c(t,1), \widehat{M}_c(t,2), \ldots, \widehat{M}_c(t,N)\big)$ and $\widehat{\mathbf{M}}_d(\tau_k) = (\widehat{M}_d(\tau_k,1), \widehat{M}_d(\tau_k,2), \ldots, \widehat{M}_d(\tau_k,N))$ be defined by

$$\widehat{\mathbf{M}}_c(t) := \dot{\mathbf{Z}}(t) + \mathscr{R}_c(t, \mathbf{Z}(t)), \quad t \in \mathbb{R}_+$$

and

$$\widehat{\mathbf{M}}_d(\tau_k) := -\mathbf{Z}(\tau_k^-) + \mathscr{R}_d(\tau_k, \mathbf{Z}(\tau_k)), \quad k \in \mathbb{Z}_+.$$

So, from (5.71) and (5.72) we may infer that $\mathbf{Z}(\cdot)$ is a solution of the following generalized backward jump matrix Riccati-type differential equation

$$-\dot{\mathbf{Z}}(t) = \mathscr{R}_c(t, \mathbf{Z}(t)) - \widehat{\mathbf{M}}_c(t), \quad \tau_k \leq t < \tau_{k+1} \tag{5.73a}$$

$$\mathbf{Z}(\tau_k^-) = \mathscr{R}_d(\tau_k, \mathbf{Z}(\tau_k)) - \widehat{\mathbf{M}}_d(\tau_k), \quad k \in \mathbb{Z}_+. \tag{5.73b}$$

Also, (5.71) and (5.72) guarantee that

$$\widehat{\mathbf{M}}_c(t) \succeq 0 \ \text{ for all } \ t \in \mathbb{R}_+, \tag{5.74a}$$

$$\widehat{\mathbf{M}}_d(\tau_k) \succeq 0 \ \text{ for all } \ t \in \mathbb{Z}_+, \tag{5.74b}$$

hold. If the quadruple $(\mathbf{A}, \mathbf{B}, (\boldsymbol{\Pi}_c \mid \boldsymbol{\Pi}_d), \mathfrak{T})$ is stabilizable there exists a bounded function $\mathbf{F}_c(\cdot) \in \mathscr{PC}\{\mathbb{R}_+, \mathscr{M}_{m_c n}^N; \mathfrak{T}\}$ and a bounded sequence $\{\mathbf{F}_d(\tau_k)\}_{k \in \mathbb{Z}_+} \subset \mathscr{M}_{m_d n}^N$ such that the jump matrix linear differential equation (5.69) to be exponentially stable.

Let $\mathbf{X}^{(0)}(\cdot) = (X^{(0)}(\cdot,1), X^{(0)}(\cdot,2), \ldots, X^{(0)}(\cdot,N))$ be the unique bounded on $\mathbb{R}_+$ solution of the following non-homogeneous backward jump matrix linear differential equation:

$$-\dot{\mathbf{X}}(t) = \mathscr{L}^*_{c\mathbf{F}_c}(t)[\mathbf{X}(t)] + \boldsymbol{\Theta}_{c\mathbf{F}_c}(t), \quad \tau_k \leq t < \tau_{k+1}, \tag{5.75a}$$

$$\mathbf{X}(\tau_k^-) = \boldsymbol{\Pi}_{d\mathbf{F}_d}(\tau_k)[\mathbf{X}(\tau_k)] + \boldsymbol{\Theta}_{d\mathbf{F}_d}(\tau_k) + \varepsilon^2 \mathbf{I}_n, \quad k \in \mathbb{Z}_+, \tag{5.75b}$$

where $\mathbf{I}_n = (I_n \ I_n \ \ldots \ I_n) \in \mathscr{S}_n^N$ and $\boldsymbol{\Theta}_{c\mathbf{F}_c}(t)$ and $\boldsymbol{\Theta}_{d\mathbf{F}_d}(\tau_k)$ are computed as in (5.19) and (5.21), respectively, with $\mathbf{W}_c(t)$ replaced by $\mathbf{F}_c(t)$, and $\mathbf{W}_d(\tau_k)$, replaced by $\mathbf{F}_d(\tau_k)$, respectively. Invoking Proposition 5.1 (iii), we deduce that the equation (5.75) has a unique solution $t \to \mathbf{X}^{(0)}(t) = (X^{(0)}(t,1), X^{(0)}(t,2), \ldots, X^{(0)}(t,N)) : \mathbb{R}_+ \to \mathscr{S}_n^N$ with the property that

$$\sup_{t\in\mathbb{R}_+} \|\mathbf{X}^{(0)}(t)\|_\infty < \infty$$

$\|\cdot\|_\infty$ being the norm on $\mathscr{S}_n^N$ introduced by (2.112). We show that

$$\mathbf{X}^{(0)}(t) \succeq \mathbf{Z}(t),$$

for all $t \in \mathbb{R}_+$, for any $\mathbf{Z}(\cdot) \in \Gamma^\Sigma[\mathbb{R}_+]$. To this end, we apply Lemma 5.1 together with (5.28) taking $\mathbf{W}_c(t) = \mathbf{F}_c(t)$ and $\mathbf{W}_d(\tau_k) = \mathbf{F}_d(\tau_k)$ to rewrite the GBJMRDE (5.73) in the form

$$\begin{aligned} -\dot{\mathbf{Z}}(t) &= \mathscr{L}^*_{c\mathbf{F}_c}(t)[\mathbf{Z}(t)] + \boldsymbol{\theta}_{c\mathbf{F}_c}(t) - \widehat{\mathbf{M}}_c(t) && (5.76a)\\ &\quad - (\mathbf{F}_c(t) - \mathbf{F}_c^{\mathbf{Z}}(t))^\top(\boldsymbol{\Pi}_{c3}(t)[\mathbf{Z}(t)] + \mathbf{R}_c(t))(\mathbf{F}_c(t) - \mathbf{F}_c^{\mathbf{Z}}(t)), \quad \tau_k \le t < \tau_{k+1} \\ \mathbf{Z}(\tau_k^-) &= \boldsymbol{\Pi}_{d\mathbf{F}_d}(\tau_k)[\mathbf{Z}(\tau_k)] + \boldsymbol{\theta}_{d\mathbf{F}_d}(\tau_k) && (5.76b)\\ &\quad - (\mathbf{F}_d(\tau_k) - \mathbf{F}_d^{\mathbf{Z}}(\tau_k))^\top(\boldsymbol{\Pi}_{d3}(\tau_k)[\mathbf{Z}(\tau_k)] + \mathbf{R}_d(\tau_k))(\mathbf{F}_d(\tau_k) - \mathbf{F}_d^{\mathbf{Z}}(\tau_k)), \quad k \in \mathbb{Z}_+, \end{aligned}$$

where $\mathbf{F}_c^{\mathbf{Z}}(t)$ and $\mathbf{F}_d^{\mathbf{Z}}(\tau_k)$, respectively, are computed as in (5.23) and (5.25), respectively, with $\mathbf{X}$ replaced by $\mathbf{Z}(t)$ and $\mathbf{Z}(\tau_k)$, respectively.

Subtracting (5.76) from (5.75) we obtain that

$$\mathbf{Y}^{(0)}(\cdot) := \mathbf{X}^{(0)}(\cdot) - \mathbf{Z}(\cdot)$$

is a bounded solution of the following non-homogeneous backward JMLDE:

$$-\dot{\mathbf{Y}}^{(0)}(t) = \mathscr{L}^*_{c\mathbf{F}_c}(t)[\mathbf{Y}^{(0)}(t)] + \mathbf{H}_c^{(0)}(t), \quad \tau_k \le t < \tau_{k+1}, \tag{5.77a}$$

$$\mathbf{Y}^{(0)}(\tau_k^-) = \boldsymbol{\Pi}_{d\mathbf{F}_d}(\tau_k)[\mathbf{Y}^{(0)}(\tau_k)] + \mathbf{H}_d^{(0)}(\tau_k), \quad k \in \mathbb{Z}_+, \tag{5.77b}$$

where

$$\mathbf{H}_c^{(0)}(t) := \widehat{\mathbf{M}}_c(t) + (\mathbf{F}_c(t) - \mathbf{F}_c^{\mathbf{Z}}(t))^\top(\boldsymbol{\Pi}_{c3}(t)[\mathbf{Z}(t)] + \mathbf{R}_c(t))(\mathbf{F}_c(t) - \mathbf{F}_c^{\mathbf{Z}}(t)), \quad t \in \mathbb{R}_+$$

and

$$\begin{aligned} \mathbf{H}_d^{(0)}(\tau_k) :=& \varepsilon^2\mathbf{I}_n + \widehat{\mathbf{M}}_d(\tau_k) + (\mathbf{F}_d(\tau_k) - \mathbf{F}_d^{\mathbf{Z}}(\tau_k))^\top(\boldsymbol{\Pi}_{d3}(\tau_k)[\mathbf{Z}(\tau_k)] + \mathbf{R}_d(\tau_k)) \\ &\cdot (\mathbf{F}_d(\tau_k) - \mathbf{F}_d^{\mathbf{Z}}(\tau_k)), \quad k \in \mathbb{Z}_+. \end{aligned}$$

Employing (5.16c) and (5.16d) together with (5.74) we may deduce that

$$0 \preceq \mathbf{H}_c^{(0)}(t) \preceq \nu_c^{(0)}\mathbf{I}_n, \quad \text{for all } t \in \mathbb{R}_+$$

and

$$\varepsilon^2 \mathbf{I}_n \preceq \mathbf{H}_d^{(0)}(\tau_k) \preceq \nu_d^{(0)} \mathbf{I}_n, \quad \text{for all } \ t \in \mathbb{Z}_+.$$

Based on Proposition 5.1 (iii) we may conclude that (5.77) has a unique bounded on $\mathbb{R}_+$ solution and it satisfies $\mathbf{Y}^{(0)}(t) \succeq c^{(0)}\mathbf{I}_n, \quad$ for all $\ t \in \mathbb{R}_+$, or, equivalently,

$$\mathbf{X}^{(0)}(t) - \mathbf{Z}(t) \succeq c^{(0)}\mathbf{I}_n, \tag{5.78}$$

for all $t \in \mathbb{R}_+$ and for any $\mathbf{Z}(\cdot) \in \Gamma^{\Sigma}[\mathbb{R}_+]$, $c^{(0)}$ being a positive constant.

Combining (5.16c), (5.16d) and (5.78) we obtain:

$$\Pi_{c3}(t)[\mathbf{X}^{(0)}(t)](i) + R_c(t,i) \geq \gamma_c I_{m_c}, \quad \text{for all } \ (t,i) \in \mathbb{R}_+ \times \mathcal{N}, \tag{5.79}$$

$$\Pi_{d3}(\tau_k)[\mathbf{X}^{(0)}(\tau_k)](i) + R_d(\tau_k,i) \geq \gamma_d I_{m_d}, \quad \text{for all } \ (\tau_k,i) \in \mathfrak{T} \times \mathcal{N}, \tag{5.80}$$

γ_c, γ_d being positive constants. This allows us to introduce the gain matrices

$$\mathbf{F}_c^{(0)}(t) = (F_c^{(0)}(t,1), F_c^{(0)}(t,2), \ldots, F_c^{(0)}(t,N)) \tag{5.81a}$$

$$\begin{aligned} F_c^{(0)}(t,i) &:= -(\Pi_{c3}(t)[\mathbf{X}^{(0)}(t)](i) + R_c(t,i))^{-1} \\ &\cdot (X^{(0)}(t,i)B(t,i) + \Pi_{c2}(t)[\mathbf{X}^{(0)}(t)](i) + L_c(t,i))^{\top}, \quad (t,i) \in \mathbb{R}_+ \times \mathcal{N}, \end{aligned} \tag{5.81b}$$

$$\mathbf{F}_d^{(0)}(\tau_k) = (F_d^{(0)}(\tau_k,1), F_d^{(0)}(\tau_k,2), \ldots, F_d^{(0)}(\tau_k,N)) \tag{5.82a}$$

$$\begin{aligned} F_d^{(0)}(\tau_k,i) &:= -(\Pi_{d3}(\tau_k)[\mathbf{X}^{(0)}(\tau_k)](i) + R_d(\tau_k,i))^{-1} \\ &\cdot (\Pi_{d2}(\tau_k)[\mathbf{X}^{(0)}(\tau_k)](i) + L_d(\tau_k,i))^{\top}, \quad k \in \mathbb{Z}_+. \end{aligned} \tag{5.82b}$$

We show that the jump matrix linear differential equation

$$\dot{\mathbf{X}}(t) = \mathcal{L}_{c\mathbf{F}_c^{(0)}}(t)[\mathbf{X}(t)], \quad \tau_k < t \leq \tau_{k+1}, \tag{5.83a}$$

$$\mathbf{X}(\tau_k^+) = \boldsymbol{\Pi}^*_{d\mathbf{F}_d^{(0)}}(\tau_k)[\mathbf{X}(\tau_k)], \quad k \in \mathbb{Z}_+, \tag{5.83b}$$

is exponentially stable, where $\mathcal{L}_{c\mathbf{F}_c^{(0)}}(t)[\cdot]$ is the linear operator defined by (5.70) for $\mathbf{F}_c(t)$ replaced by $\mathbf{F}_c^{(0)}(t)$ and $\boldsymbol{\Pi}^*_{d\mathbf{F}_d^{(0)}}(\tau_k)[\cdot]$ is the adjoint of the linear operator introduced via (5.20) taking $\mathbf{W}_d(\tau_k) = \mathbf{F}_d^{(0)}(\tau_k)$.

So, applying Lemma 5.2 we rewrite (5.75) as:

$$\begin{aligned} -\dot{\mathbf{X}}^{(0)}(t) &= \mathcal{L}^*_{c\mathbf{F}_c^{(0)}}(t)[\mathbf{X}^{(0)}(t)] + \boldsymbol{\Theta}_{c\mathbf{F}_c^{(0)}}(t) + (\mathbf{F}_c^{(0)}(t) - \mathbf{F}_c(t))^{\top} \\ &\cdot (\boldsymbol{\Pi}_{c3}(t)[\mathbf{X}^{(0)}(t)] + \mathbf{R}_c(t))(\mathbf{F}_c^{(0}(t) - \mathbf{F}_c(t)), \quad \tau_k \leq t < \tau_{k+1}, \end{aligned} \tag{5.84a}$$

$$\begin{aligned} \mathbf{X}^{(0)}(\tau_k^-) &= \boldsymbol{\Pi}_{d\mathbf{F}_d^{(0)}}(\tau_k)[\mathbf{X}^{(0)}(\tau_k)] + \boldsymbol{\Theta}_{d\mathbf{F}_d^{(0)}}(\tau_k) + (\mathbf{F}_d^{(0)}(\tau_k) - \mathbf{F}_d(\tau_k))^{\top} \\ &\cdot (\boldsymbol{\Pi}_{d3}(\tau_k)[\mathbf{X}^{(0)}(\tau_k)] + \mathbf{R}_d(\tau_k))(\mathbf{F}_d^{(0}(\tau_k) - \mathbf{F}_d(\tau_k)) + \varepsilon^2\mathbf{I}_n, \quad k \in \mathbb{Z}_+. \end{aligned} \tag{5.84b}$$

On the other hand, Lemma 5.1 together with (5.28) applied in the case of (5.73) taking $\mathbf{W}_c(t) = \mathbf{F}_c^{(0)}(t)$ and $\mathbf{W}_d(\tau_k) = \mathbf{F}_d^{(0)}(\tau_k)$ yield

$$-\dot{\mathbf{Z}}(t) = \mathscr{L}^*_{c\mathbf{F}_c^{(0)}}(t)[\mathbf{Z}(t)] + \boldsymbol{\Theta}_{c\mathbf{F}_c^{(0)}}(t) - (\mathbf{F}_c^{(0)}(t) - \mathbf{F}_c(t))^\top \tag{5.85a}$$
$$\cdot (\boldsymbol{\Pi}_{c3}(t)[\mathbf{Z}(t)] + \mathbf{R}_c(t))(\mathbf{F}_c^{(0}(t) - \mathbf{F}_c^{\mathbf{Z}}(t)) - \widehat{\mathbf{M}}_c(t), \quad \tau_k \le t < \tau_{k+1},$$
$$\mathbf{Z}(\tau_k^-) = \boldsymbol{\Pi}_{d\mathbf{F}_d^{(0)}}(\tau_k)[\mathbf{Z}(\tau_k)] + \boldsymbol{\Theta}_{d\mathbf{F}_d^{(0)}}(\tau_k) - (\mathbf{F}_d^{(0)}(\tau_k) - \mathbf{F}_d^{\mathbf{Z}}(\tau_k))^\top \tag{5.85b}$$
$$\cdot (\boldsymbol{\Pi}_{d3}(\tau_k)[\mathbf{Z}(\tau_k)] + \mathbf{R}_d(\tau_k))(\mathbf{F}_d^{(0}(\tau_k) - \mathbf{F}_d^{\mathbf{Z}}(\tau_k)) - \widehat{\mathbf{M}}_d(\tau_k), \quad k \in \mathbb{Z}_+.$$

Subtracting (5.85) from (5.84), we obtain via (5.78) that $\mathbf{Y}^{(0)}(\cdot) = \mathbf{X}^{(0)}(\cdot) - \mathbf{Z}(\cdot)$ is a uniformly positive and bounded on $\mathbb{R}_+$ solution of the following non-homogeneous backward JMLDE

$$-\dot{\mathbf{Y}}^{(0)}(t) = \mathscr{L}^*_{c\mathbf{F}_c^{(0)}}(t)[\mathbf{Y}^{(0)}(t)] + \widehat{\mathbf{H}}_c^{(0)}(t), \quad \tau_k \le t < \tau_{k+1}, \tag{5.86a}$$
$$\mathbf{Y}^{(0)}(\tau_k^-) = \boldsymbol{\Pi}_{d\mathbf{F}_d^{(0)}}(\tau_k)[\mathbf{Y}^{(0)}(\tau_k)] + \widehat{\mathbf{H}}_d^{(0)}(\tau_k), \quad k \in \mathbb{Z}_+, \tag{5.86b}$$

where

$$\widehat{\mathbf{H}}_c^{(0)}(t) = \widehat{\mathbf{M}}_c(t) + (\mathbf{F}_c^{(0)}(t) - \mathbf{F}_c(t))^\top(\boldsymbol{\Pi}_{c3}(t)[\mathbf{X}^{(0)}(t)] + \mathbf{R}_c(t))(\mathbf{F}_c^{(0)}(t) - \mathbf{F}_c(t))$$
$$+ (\mathbf{F}_c^{(0)}(t) - \mathbf{F}_c^{\mathbf{Z}}(t))^\top(\boldsymbol{\Pi}_{c3}(t)[\mathbf{Z}(t)] + \mathbf{R}_c(t))(\mathbf{F}_c^{(0)}(t) - \mathbf{F}_c^{\mathbf{Z}}(t))$$

$$\widehat{\mathbf{H}}_d^{(0)}(\tau_k) = \varepsilon^2\mathbf{I}_n + \widehat{\mathbf{M}}_d(\tau_k) + (\mathbf{F}_d^{(0)}(\tau_k) - \mathbf{F}_d(\tau_k))^\top(\boldsymbol{\Pi}_{d3}(\tau_k)[\mathbf{X}^{(0)}(\tau_k)] + \mathbf{R}_d(\tau_k))$$
$$\cdot (\mathbf{F}_d^{(0)}(\tau_k) - \mathbf{F}_d(\tau_k)) + (\mathbf{F}_d^{(0)}(\tau_k) - \mathbf{F}_d^{\mathbf{Z}}(\tau_k))^\top(\boldsymbol{\Pi}_{d3}(\tau_k)[\mathbf{Z}(\tau_k)] + \mathbf{R}_d(\tau_k))$$
$$\cdot (\mathbf{F}_d^{(0)}(\tau_k) - \mathbf{F}_d^{\mathbf{Z}}(\tau_k)).$$

Employing (5.16c), (5.16d), (5.74), (5.79) and (5.80) we may infer that

$$\widehat{\mathbf{H}}_c^{(0)}(t) \succeq 0, \quad \text{for all } \ t \in \mathbb{R}_+$$

and

$$\mathbf{H}_d^{(0)}(\tau_k) \succeq \varepsilon^2\mathbf{I}_m \gg 0, \quad \text{for all } \tau_k \in \mathfrak{T}.$$

Taking into account that based on Lemma 5.3 the jump matrix linear differential equation (5.83) generates a positive evolution on the ordered Hilbert space $(\mathscr{S}_n^N, \mathscr{S}_n^{N+})$, we may invoke the implication $(iii) \Rightarrow (i)$ from Theorem 2.7 to obtain via (5.86) that (5.83) is exponentially stable.

Starting from $\mathbf{X}^{(0)}(\cdot)$, $\mathbf{F}_c^{(0)}(\cdot)$, $\mathbf{F}_d^{(0)}(\cdot)$ we construct the sequences $\{\mathbf{X}^{(p)}(\cdot)\}_{p\ge1}$, $\{\mathbf{F}_c^{(p)}(\cdot)\}_{p\ge1}$, $\{\mathbf{F}_d^{(p)}(\cdot)\}_{p\ge1}$ as: $\mathbf{X}^{(p)}(\cdot)$ is the unique bounded on $\mathbb{R}_+$ solution of the following non-homogeneous backward jump matrix differential equation (BJMLDE):

$$-\dot{\mathbf{X}}^{(p)}(t) = \mathscr{L}^*_{c\mathbf{F}_c^{(p-1)}}(t)[\mathbf{X}^{(p)}(t)] + \boldsymbol{\Theta}_{c\mathbf{F}_c^{(p-1)}}(t), \quad \tau_k \leq t < \tau_{k+1}, \tag{5.87a}$$

$$\mathbf{X}^{(p)}(\tau_k^-) = \boldsymbol{\Pi}_{d\mathbf{F}_d^{(p-1)}}(\tau_k)[\mathbf{X}^{(p)}(\tau_k)] + \boldsymbol{\Theta}_{d\mathbf{F}_d^{(p-1)}}(\tau_k) + \frac{\varepsilon^2}{p+1}\mathbf{I}_n, \quad k \in \mathbb{Z}_+, \tag{5.87b}$$

$$\mathbf{F}_c^{(p)}(\cdot) = (F_c^{(p)}(\cdot, 1), F_c^{(p)}(\cdot, 2), \ldots, F_c^{(p)}(\cdot, N)) \tag{5.88a}$$

$$\begin{aligned} F_c^{(p)}(t,i) := &-(\Pi_{c3}(t)[\mathbf{X}^{(p)}(t)](i) + R_c(t,i))^{-1}(X^{(p)}(t,i)B(t,i) \\ &+ \Pi_{c2}(t)[\mathbf{X}^{(p)}(t)](i) + L_c(t,i))^\top, \quad \text{for all } (t,i) \in \mathbb{R}_+ \times \mathcal{N} \end{aligned} \tag{5.88b}$$

$$\mathbf{F}_d^{(p)}(\tau_k) = (F_d^{(p)}(\tau_k, 1), F_d^{(p)}(\tau_k, 2), \ldots, F_d^{(p)}(\tau_k, N)) \tag{5.89a}$$

$$\begin{aligned} F_d^{(p)}(\tau_k, i) := &-(\Pi_{d3}(\tau_k)[\mathbf{X}^{(p)}(\tau_k)](i) + R_d(\tau_k, i))^{-1} \\ &\cdot (\Pi_{d2}(\tau_k)[\mathbf{X}^{(p)}(\tau_k)](i) + L_d(\tau_k, i))^\top, \quad \text{for all } (\tau_k, i) \in \mathfrak{T} \times \mathcal{N}. \end{aligned} \tag{5.89b}$$

In (5.87a) the operator $\mathscr{L}_{c\mathbf{F}_c^{(p-1)}}(t)[\cdot]$ is described by (5.18) with $\mathbf{W}_c(\cdot)$ replaced by $\mathbf{F}_c^{(p-1)}(\cdot)$ and $\boldsymbol{\Theta}_{c\mathbf{F}_c^{(p-1)}}(t)$ is computed as in (5.19) with $\mathbf{W}_c(\cdot)$ replaced by $\mathbf{F}_c^{(p-1)}(\cdot)$. In (5.87b), $\boldsymbol{\Pi}_{d\mathbf{F}_d^{(p-1)}}(\tau_k)[\cdot]$ and $\boldsymbol{\Theta}_{d\mathbf{F}_d^{(p-1)}}(\tau_k)$ are computed as in (5.20) and (5.21), respectively, with $\mathbf{W}_d(\tau_k)$ replaced by $\mathbf{F}_d^{(p-1)}(\tau_k)$.

We show by induction with respect to p that the following items hold:

(a_p) for any $\mathbf{Z}(\cdot)$ from $\Gamma^\Sigma[\mathbb{R}_+]$ we have

$$\mathbf{X}^{(p)}(t) - \mathbf{Z}(t) \succeq c^{(p)}\mathbf{I}_n,$$

for all $t \in \mathbb{R}_+$, where $c^{(p)}$ is a positive constant not depending upon $\mathbf{Z}(\cdot)$;

(b_p) the jump matrix linear differential equation

$$\dot{\mathbf{X}}(t) = \mathscr{L}_{c\mathbf{F}_c^{(p)}}(t)[\mathbf{X}(t)], \quad \tau_k < t \leq \tau_{k+1}, \tag{5.90a}$$

$$\mathbf{X}(\tau_k^+) = \boldsymbol{\Pi}^*_{d\mathbf{F}_d^{(p)}}(\tau_k)[\mathbf{X}(\tau_k)], \quad k \in \mathbb{Z}_+, \tag{5.90b}$$

is exponentially stable;

(c_p)

$$\mathbf{X}^{(p-1)}(t) \succeq \mathbf{X}^{(p)}(t), \quad \text{for all } t \in \mathbb{R}_+.$$

First, we show that $(a_p) - (c_p)$ hold for $p = 1$. Since the JMLDE (5.83) is exponentially stable, it follows that the BJMLDE (5.87) written for $p = 1$ has a unique bounded on $\mathbb{R}_+$ solution $\mathbf{X}^{(1)}(\cdot) = (X^{(1)}(\cdot, 1), X^{(1)}(\cdot, 2), \ldots, X^{(1)}(\cdot, N))$. Subtracting (5.85) written for an arbitrary $\mathbf{Z}(\cdot) \in \Gamma^\Sigma[\mathbb{R}_+]$ from (5.87) written for $p = 1$, we obtain

$$\mathbf{Y}^{(1)}(\cdot) \triangleq \mathbf{X}^{(1)}(\cdot) - \mathbf{Z}(\cdot)$$

is a bounded on $\mathbb{R}_+$ solution of the following non-homogeneous BJMLDE:

$$-\dot{\mathbf{Y}}^{(1)}(t) = \mathscr{L}^*_{c\mathbf{F}_c^{(0)}}(t)[\mathbf{Y}^{(1)}(t)] + \mathbf{H}_c^{(1)}(t), \quad \tau_k \leq t < \tau_{k+1} \tag{5.91a}$$

$$\mathbf{Y}^{(1)}(\tau_k^-) = \boldsymbol{\Pi}_{d\mathbf{F}_d^{(0)}}(\tau_k)[\mathbf{Y}^{(1)}(\tau_k)] + \mathbf{H}_d^{(1)}(\tau_k), \quad k \in \mathbb{Z}_+, \tag{5.91b}$$

where

$$\mathbf{H}_c^{(1)}(t) := \widehat{\mathbf{M}}_c(t) + (\mathbf{F}_c^{(0)}(t) - \mathbf{F}_c^{\mathbf{Z}}(t))^\top (\boldsymbol{\Pi}_{c3}(t)[\mathbf{Z}(t)] + \mathbf{R}_c(t))(\mathbf{F}_c^{(0)}(t) - \mathbf{F}_c^{\mathbf{Z}}(t)),$$

$t \in \mathbb{R}_+$ and

$$\mathbf{H}_d^{(1)}(\tau_k) := \frac{\varepsilon^2}{2}\mathbf{I}_n + \widehat{\mathbf{M}}_d(\tau_k) + (\mathbf{F}_d^{(0)}(\tau_k) - \mathbf{F}_d^{\mathbf{Z}}(\tau_k))^\top (\boldsymbol{\Pi}_{d3}(\tau_k)[\mathbf{Z}(\tau_k)] + \mathbf{R}_d(\tau_k)) \cdot (\mathbf{F}_d^{(0)}(\tau_k) - \mathbf{F}_d^{\mathbf{Z}}(\tau_k)), \quad k \in \mathbb{Z}_+.$$

Employing (5.16c), (5.16d), (5.74), (5.81) and (5.82) we may conclude that there exist positive constants $\nu_c^{(1)}$, $\nu_d^{(1)}$ with the property that

$$0 \preceq \mathbf{H}_c^{(1)}(t) \preceq \nu_c^{(1)}\mathbf{I}_n, \quad \text{for all } t \in \mathbb{R}_+$$

$$\frac{\varepsilon^2}{2}\mathbf{I}_n \preceq \mathbf{H}_d^{(1)}(\tau_k) \preceq \nu_d^{(1)}\mathbf{I}_n, \quad \text{for all } t \in \mathbb{Z}_+.$$

Invoking Proposition 5.1 (iii) we may infer that (5.91) has a unique bounded on $\mathbb{R}_+$ solution which is uniform positive. Thus we obtained that

$$\mathbf{X}^{(1)}(t) - \mathbf{Z}(t) \succeq c^{(1)}\mathbf{I}_n, \quad \text{for all } t \in \mathbb{R}_+ \tag{5.92}$$

and any $\mathbf{Z}(\cdot) \in \Gamma^\Sigma[\mathbb{R}_+]$, ($c^{(1)}$ being a positive constant). Hence, (a_p) holds when $p = 1$. Further, (5.16c) and (5.16d) together with (5.92) leads to:

$$\Pi_{c3}(t)[\mathbf{X}^{(1)}(t)](i) + R_c(t,i) \geq \gamma_c I_{m_c}, \quad \text{for all } (t,i) \in \mathbb{R}_+ \times \mathscr{N} \tag{5.93a}$$

$$\Pi_{d3}(\tau_k)[\mathbf{X}^{(1)}(\tau_k)](i) + R_d(\tau_k,i) \geq \gamma_d I_{m_d}, \quad \text{for all } (\tau_k,i) \in \mathfrak{T} \times \mathscr{N}. \tag{5.93b}$$

This allows us to deduce that the gain matrices $\mathbf{F}_c^{(1)}(t)$, $t \in \mathbb{R}_+$ and $\mathbf{F}_d^{(1)}(\tau_k)$, $k \in \mathbb{Z}_+$ can be computed by (5.88) and (5.89), respectively, written for $p = 1$. We show now that the JMLDE (5.90) written for $p = 1$ is exponentially stable. To this end, we shall apply Lemma 5.2 in the case of (5.87) written for $p = 1$ to obtain:

$$-\dot{\mathbf{X}}^{(1)}(t) = \mathscr{L}^*_{c\mathbf{F}_c^{(1)}}(t)[\mathbf{X}^{(1)}(t)] + \boldsymbol{\Theta}_{c\mathbf{F}_c^{(1)}}(t) + (\mathbf{F}_c^{(1)}(t) - \mathbf{F}_c^{(0)}(t))^\top \cdot (\boldsymbol{\Pi}_{c3}(t)[\mathbf{X}^{(1)}(t)] + \mathbf{R}_c(t))(\mathbf{F}_c^{(1)}(t) - \mathbf{F}_c^{(0)}(t)), \quad \tau_k \leq t < \tau_{k+1} \tag{5.94a}$$

$$\mathbf{X}^{(1)}(\tau_k^-) = \boldsymbol{\Pi}_{d\mathbf{F}_d^{(1)}}(\tau_k)[\mathbf{X}^{(1)}(\tau_k)] + \boldsymbol{\Theta}_{d\mathbf{F}_d^{(1)}}(\tau_k) + (\mathbf{F}_d^{(1)}(\tau_k) - \mathbf{F}_d^{(0)}(\tau_k))^\top$$

$$\cdot (\boldsymbol{\Pi}_{d3}(\tau_k)[\mathbf{X}^{(1)}(\tau_k)] + \mathbf{R}_d(\tau_k))(\mathbf{F}_d^{(1)}(\tau_k) - \mathbf{F}_d^{(0)}(\tau_k)) + \frac{\varepsilon^2}{2}\mathbf{I}_n, \quad k \in \mathbb{Z}_+. \tag{5.94b}$$

On the other hand, Lemma 5.1 together with (5.28) applied in the case of (5.73) taking $\mathbf{W}_c(t) = \mathbf{F}_c^{(1)}(t)$ and $\mathbf{W}_d(\tau_k) = \mathbf{F}_d^{(1)}(\tau_k)$ give:

$$\begin{aligned}-\dot{\mathbf{Z}}(t) &= \mathscr{L}^*_{c\mathbf{F}_c^{(1)}}(t)[\mathbf{Z}(t)] + \boldsymbol{\Theta}_{c\mathbf{F}_c^{(1)}}(t) - (\mathbf{F}_c^{(1)}(t) - \mathbf{F}_c^{\mathbf{Z}}(t))^\top \\ &\quad \cdot (\boldsymbol{\Pi}_{c3}(t)[\mathbf{Z}(t)] + \mathbf{R}_c(t))(\mathbf{F}_c^{(1)}(t) - \mathbf{F}_c^{\mathbf{Z}}(t)) - \widehat{\mathbf{M}}_c(t), \quad \tau_k \le t < \tau_{k+1}\end{aligned} \tag{5.95a}$$

$$\begin{aligned}\mathbf{Z}(\tau_k^-) &= \boldsymbol{\Pi}_{d\mathbf{F}_d^{(1)}}(\tau_k)[\mathbf{Z}(\tau_k)] + \boldsymbol{\Theta}_{d\mathbf{F}_d^{(1)}}(\tau_k) - (\mathbf{F}_d^{(1)}(\tau_k) - \mathbf{F}_d^{\mathbf{Z}}(\tau_k))^\top \\ &\quad \cdot (\boldsymbol{\Pi}_{d3}(\tau_k)[\mathbf{Z}(\tau_k)] + \mathbf{R}_d(\tau_k))(\mathbf{F}_d^{(1)}(\tau_k) - \mathbf{F}_d^{\mathbf{Z}}(\tau_k)) - \widehat{\mathbf{M}}_d(\tau_k), \quad k \in \mathbb{Z}_+.\end{aligned} \tag{5.95b}$$

Subtracting (5.95) from (5.94), we obtain via (5.92) that

$$\mathbf{Y}^{(1)}(\cdot) = \mathbf{X}^{(1)}(\cdot) - \mathbf{Z}(\cdot)$$

is a uniform positive and bounded on $\mathbb{R}_+$ solution of the following non-homogeneous BJMLDE:

$$-\dot{\mathbf{Y}}^{(1)}(t) = \mathscr{L}^*_{c\mathbf{F}_c^{(1)}}(t)[\mathbf{Y}^{(1)}(t)] + \widehat{\mathbf{H}}_c^{(1)}(t), \quad \tau_k \le t < \tau_{k+1} \tag{5.96a}$$

$$\mathbf{Y}^{(1)}(\tau_k^-) = \boldsymbol{\Pi}_{d\mathbf{F}_d^{(1)}}(\tau_k)[\mathbf{Y}^{(1)}(\tau_k)] + \widehat{\mathbf{H}}_d^{(1)}(\tau_k), \quad k \in \mathbb{Z}_+ \tag{5.96b}$$

where

$$\begin{aligned}\widehat{\mathbf{H}}_c^{(1)}(t) &:= \widehat{\mathbf{M}}_c(t) + (\mathbf{F}_c^{(1)}(t) - \mathbf{F}_c^{(0)}(t))^\top(\boldsymbol{\Pi}_{c3}(t)[\mathbf{X}^{(1)}(t)] + \mathbf{R}_c(t)) \\ &\quad (\mathbf{F}_c^{(1)}(t) - \mathbf{F}_c^{(0)}(t)) \\ &\quad + (\mathbf{F}_c^{(1)}(t) - \mathbf{F}_c^{\mathbf{Z}}(t))^\top(\boldsymbol{\Pi}_{c3}(t)[\mathbf{Z}(t)] + \mathbf{R}_c(t))(\mathbf{F}_c^{(1)}(t) - \mathbf{F}_c^{\mathbf{Z}}(t))\end{aligned}$$

$$\begin{aligned}\widehat{\mathbf{H}}_d^{(1)}(\tau_k) &:= \frac{\varepsilon^2}{2}\mathbf{I}_n + \widehat{\mathbf{M}}_d(\tau_k) + (\mathbf{F}_d^{(1)}(\tau_k) - \mathbf{F}_d^{(0)}(\tau_k))^\top(\boldsymbol{\Pi}_{d3}(\tau_k)[\mathbf{X}^{(1)}(\tau_k)] + \mathbf{R}_d(\tau_k)) \\ &\quad \cdot (\mathbf{F}_d^{(1)}(\tau_k) - \mathbf{F}_d^{(0)}(\tau_k)) + (\mathbf{F}_d^{(1)}(\tau_k) - \mathbf{F}_d^{\mathbf{Z}}(\tau_k))^\top(\boldsymbol{\Pi}_{d3}(\tau_k)[\mathbf{Z}(\tau_k)] + \mathbf{R}_d(\tau_k)) \\ &\quad \cdot (\mathbf{F}_d^{(1)}(\tau_k) - \mathbf{F}_d^{\mathbf{Z}}(\tau_k)).\end{aligned}$$

Employing (5.16c), (5.16d), (5.74) and (5.93) we deduce that there exist positive constants $\widehat{\nu}_c^{(1)}$, $\widehat{\nu}_d^{(1)}$ with the property that

$$0 \preceq \widehat{\mathbf{H}}_c^{(1)}(t) \preceq \widehat{\nu}_c^{(1)}\mathbf{I}_n, \quad \text{for all } t \in \mathbb{R}_+$$

and

$$\frac{\varepsilon^2}{2}\mathbf{I}_n \preceq \widehat{\mathbf{H}}_d^{(1)}(\tau_k) \preceq \widehat{\nu}_d^{(1)}\mathbf{I}_n, \quad \text{for all } \ \tau_k \in \mathfrak{T}.$$

On the other hand, Lemma 5.3 allows us to infer that the jump matrix linear differential equation (5.90) define a positive evolution on the ordered Hilbert space $(\mathscr{S}_n^N, \mathscr{S}_n^{N+})$. Hence, the implication $(iii) \Rightarrow (i)$ from Theorem 2.7 applied in the case of the equation (5.90) written for $p = 1$ allows us to conclude via (5.92) and (5.96) that the item (b_p) holds when $p = 1$.

Further, subtracting (5.87) written for $p = 1$ from (5.84) we obtain that

$$\boldsymbol{\Delta}^{(1)}(\cdot) := \mathbf{X}^{(0)}(\cdot) - \mathbf{X}^{(1)}(\cdot)$$

is a bounded on $\mathbb{R}$ solution of the following non-homogeneous BJMLDE:

$$-\dot{\boldsymbol{\Delta}}^{(1)}(t) = \mathscr{L}^*_{c\mathbf{F}_c^{(0)}}(t)[\boldsymbol{\Delta}^{(1)}(t)] + \widetilde{\mathbf{H}}_c^{(1)}(t), \quad \tau_k \le t < \tau_{k+1} \tag{5.97a}$$

$$\boldsymbol{\Delta}^{(1)}(\tau_k^-) = \boldsymbol{\Pi}_{d\mathbf{F}_d^{(0)}}(\tau_k)[\boldsymbol{\Delta}^{(1)}(\tau_k)] + \widetilde{\mathbf{H}}_d^{(1)}(\tau_k), \quad k \in \mathbb{Z}_+ \tag{5.97b}$$

where

$$\widetilde{\mathbf{H}}_c^{(1)}(t) := (\mathbf{F}_c^{(0)}(t) - \mathbf{F}_c(t))^\top(\boldsymbol{\Pi}_{c3}(t)[\mathbf{X}^{(0)}(t)] + \mathbf{R}_c(t))(\mathbf{F}_c^{(0)}(t) - \mathbf{F}_c(t)), \quad t \in \mathbb{R}_+$$

and

$$\begin{aligned}\widetilde{\mathbf{H}}_d^{(1)}(\tau_k) &:= \frac{\varepsilon^2}{2}\mathbf{I}_n + (\mathbf{F}_d^{(0)}(\tau_k) - \mathbf{F}_d(\tau_k))^\top(\boldsymbol{\Pi}_{d3}(\tau_k)[\mathbf{X}^{(0)}(\tau_k)] + \mathbf{R}_d(\tau_k)) \\ &\quad \cdot (\mathbf{F}_d^{(0)}(\tau_k) - \mathbf{F}_d(\tau_k)), \quad k \in \mathbb{Z}_+.\end{aligned}$$

Employing again (5.79) and (5.80), we deduce that

$$\widetilde{\mathbf{H}}_c^{(1)}(t) \succeq 0, \quad \text{for all } \ t \in \mathbb{R}_+$$

and

$$\widetilde{\mathbf{H}}_d^{(1)}(\tau_k) \succeq \frac{\varepsilon^2}{2}\mathbf{I}_n \gg 0, \quad \text{for all } \ k \in \mathbb{Z}_+.$$

Since (5.83) is exponentially stable, we may conclude via Proposition 5.1 (iii) that the unique bounded solution of (5.97) is uniform positive on $\mathbb{R}_+$. This confirms the validity of item (c_p) for $p = 1$.

Take now $q > 1$. Let us assume that the items $(a_p) - (c_p)$ hold for each $1 \le p < q - 1$. We show that they still hold when $p = q$. If (b_p) holds when $p = q - 1$ we deduce via Proposition 5.1 (iii) that the non-homogeneous BJMLDE (5.87) written for p replaced by q has a unique bounded on $\mathbb{R}_+$ solution $\mathbf{X}^{(q)}(\cdot)$. In order to check that the item (a_p) holds when $p = q$, we apply Lemma 5.1 together with (5.28) taking $\mathbf{W}_c(t) = \mathbf{F}_c^{(q-1)}(t)$ and $\mathbf{W}_d(\tau_k) = \mathbf{F}_d^{(q-1)}(\tau_k)$ to rewrite (5.73) satisfied by an arbitrary $\mathbf{Z}(\cdot) \in \Gamma^{\Sigma}[\mathbb{R}_+]$ as:

$$\begin{aligned}
-\dot{\mathbf{Z}}(t) = &\mathscr{L}^*_{c\mathbf{F}_c^{(q-1)}}(t)[\mathbf{Z}(t)] + \boldsymbol{\Theta}_{c\mathbf{F}_c^{(q-1)}}(t) - (\mathbf{F}_c^{(q-1)}(t) - \mathbf{F}_c^{\mathbf{Z}}(t))^\top \qquad (5.98a)\\
&\cdot (\boldsymbol{\Pi}_{c3}(t)[\mathbf{Z}(t)] + \mathbf{R}_c(t))(\mathbf{F}_c^{(q-1)}(t) - \mathbf{F}_c^{\mathbf{Z}}(t)) - \widehat{\mathbf{M}}_c(t), \quad \tau_k \le t < \tau_{k+1}\\
\mathbf{Z}(\tau_k^-) = &\boldsymbol{\Pi}_{d\mathbf{F}_d^{(q-1)}}(\tau_k)[\mathbf{Z}(\tau_k)] + \boldsymbol{\Theta}_{d\mathbf{F}_d^{(q-1)}}(\tau_k) - (\mathbf{F}_d^{(q-1)}(\tau_k) - \mathbf{F}_d^{\mathbf{Z}}(\tau_k))^\top \qquad (5.98b)\\
&\cdot (\boldsymbol{\Pi}_{d3}(\tau_k)[\mathbf{Z}(\tau_k)] + \mathbf{R}_d(\tau_k))(\mathbf{F}_d^{(q-1)}(\tau_k) - \mathbf{F}_d^{\mathbf{Z}}(\tau_k)) - \widehat{\mathbf{M}}_d(\tau_k), \quad k \in \mathbb{Z}_+.
\end{aligned}$$

Subtracting (5.98) from (5.87) written for p replaced by q we deduce that

$$\mathbf{Y}^{(q)}(\cdot) := \mathbf{X}^{(q)}(\cdot) - \mathbf{Z}(\cdot)$$

is a bounded on $\mathbb{R}_+$ solution of the following BJMLDE:

$$-\dot{\mathbf{Y}}^{(q)}(t) = \mathscr{L}^*_{c\mathbf{F}_c^{(q-1)}}(t)[\mathbf{Y}^{(q)}(t)] + \mathbf{H}_c^{(q)}(t), \quad \tau_k \le t < \tau_{k+1} \qquad (5.99a)$$

$$\mathbf{Y}^{(q)}(\tau_k^-) = \boldsymbol{\Pi}_{d\mathbf{F}_d^{(q-1)}}(\tau_k)[\mathbf{Y}^{(q)}(\tau_k)] + \mathbf{H}_d^{(q)}(\tau_k), \quad k \in \mathbb{Z}_+ \qquad (5.99b)$$

where

$$\begin{aligned}
\mathbf{H}_c^{(q)}(t) :=& \widehat{\mathbf{M}}_c(t) + (\mathbf{F}_c^{(q-1)}(t) - \mathbf{F}_c^{\mathbf{Z}}(t))^\top (\boldsymbol{\Pi}_{c3}(t)[\mathbf{Z}(t)] + \mathbf{R}_c(t))\\
&\cdot (\mathbf{F}_c^{(q-1)}(t) - \mathbf{F}_c^{\mathbf{Z}}(t)), \quad t \in \mathbb{R}_+
\end{aligned}$$

$$\begin{aligned}
\mathbf{H}_d^{(q)}(\tau_k) :=& \frac{\varepsilon}{q+1}\mathbf{I}_n + \widehat{\mathbf{M}}_d(\tau_k) + (\mathbf{F}_d^{(q-1)}(\tau_k) - \mathbf{F}_d^{\mathbf{Z}}(\tau_k))^\top (\boldsymbol{\Pi}_{d3}(\tau_k)[\mathbf{Z}(\tau_k)] + \mathbf{R}_d(\tau_k))\\
&(\mathbf{F}_d^{(q-1)}(\tau_k) - \mathbf{F}_d^{\mathbf{Z}}(\tau_k)), \quad k \in \mathbb{Z}_+.
\end{aligned}$$

Invoking again (5.16c), (5.16d) and (5.74), we may conclude that there exist positive constants $\nu_c^{(q)}$ and $\nu_d^{(q)}$ such that

$$0 \preceq \mathbf{H}_c^{(q)}(t) \preceq \nu_c^{(q)}\mathbf{I}_n,$$

for all $t \in \mathbb{R}_+$ and

$$\frac{\varepsilon^2}{q+1}\mathbf{I}_n \preceq \mathbf{H}_d^{(q)}(\tau_k) \preceq \nu_d^{(q)}\mathbf{I}_n,$$

for all $k \in \mathbb{Z}_+$. So, we may conclude via Proposition 5.1 (iii) that the unique bounded on $\mathbb{R}_+$ solution of (5.99) is uniform positive. Hence,

$$\mathbf{X}^{(q)}(t) - \mathbf{Z}(t) \succeq c^{(q)}\mathbf{I}_n, \quad \text{for all } t \in \mathbb{R}_+, \qquad (5.100)$$

$c^{(q)}$ being a positive constant. Thus, we have shown that (a_p) holds for $p = q$. Combining (5.16c), (5.16d) with (5.100) we obtain that

$$\Pi_{c3}(t)[\mathbf{X}^{(q)}(t)](i) + R_c(t,i) \geq \gamma_c I_{m_c}, \quad \text{for all} \quad (t,i) \in \mathbb{R}_+ \times \mathcal{N} \tag{5.101a}$$

$$\Pi_{d3}(\tau_k)[\mathbf{X}^{(q)}(\tau_k)](i) + R_d(\tau_k,i) \geq \gamma_d I_{m_d}, \quad \text{for all} \quad (\tau_k,i) \in \mathfrak{T} \times \mathcal{N}. \tag{5.101b}$$

Hence, the gain matrices $\mathbf{F}_c^{(q)}(t)$ and $\mathbf{F}_d^{(q)}(\tau_k)$ can be computed via (5.88) and (5.89), respectively, written for $p = q$.

Let us show that (b_p) is true, when $p = q$. To this end, we apply Lemma 5.2 in the case of the equation (5.87) (written for $p = q$) to obtain:

$$\begin{aligned} -\dot{\mathbf{X}}^{(q)}(t) &= \mathcal{L}^*_{c\mathbf{F}_c^{(q)}}(t)[\mathbf{X}^{(q)}(t)] + \boldsymbol{\Theta}_{c\mathbf{F}_c^{(q)}}(t) + (\mathbf{F}_c^{(q)}(t) - \mathbf{F}_c^{(q-1)}(t))^\top \\ &\quad \cdot (\boldsymbol{\Pi}_{c3}(t)[\mathbf{X}^{(q)}(t)] + \mathbf{R}_c(t))(\mathbf{F}_c^{(q)}(t) - \mathbf{F}_c^{(q-1)}(t)), \quad \tau_k \leq t < \tau_{k+1}, \end{aligned} \tag{5.102a}$$

$$\begin{aligned} \mathbf{X}^{(q)}(\tau_k^-) &= \boldsymbol{\Pi}_{d\mathbf{F}_d^{(q)}}(\tau_k)[\mathbf{X}^{(q)}(\tau_k)] + \boldsymbol{\Theta}_{d\mathbf{F}_d^{(q)}}(\tau_k) + (\mathbf{F}_d^{(q)}(\tau_k) - \mathbf{F}_d^{(q-1)}(\tau_k))^\top \\ &\quad \cdot (\boldsymbol{\Pi}_{d3}(\tau_k)[\mathbf{X}^{(q)}(\tau_k)] + \mathbf{R}_d(\tau_k))(\mathbf{F}_d^{(q)}(\tau_k) - \mathbf{F}_d^{(q-1)}(\tau_k)) \frac{\varepsilon^2}{q+1} \mathbf{I}_n, \quad k \in \mathbb{Z}_+. \end{aligned} \tag{5.102b}$$

Further, Lemma 5.1 together with (5.28) applied to the equation (5.73) taking $\mathbf{F}_c^{(q)}(t)$ instead of $\mathbf{W}_c(t)$, and $\mathbf{F}_d^{(q)}(\tau_k)$ instead of $\mathbf{W}_d(\tau_k)$ give

$$\begin{aligned} -\dot{\mathbf{Z}}(t) &= \mathcal{L}^*_{c\mathbf{F}_c^{(q)}}(t)[\mathbf{Z}(t)] + \boldsymbol{\Theta}_{c\mathbf{F}_c^{(q)}}(t) - (\mathbf{F}_c^{(q)}(t) - \mathbf{F}_c^{\mathbf{Z}}(t))^\top \\ &\quad \cdot (\boldsymbol{\Pi}_{c3}(t)[\mathbf{Z}(t)] + \mathbf{R}_c(t))(\mathbf{F}_c^{(q)}(t) - \mathbf{F}_c^{\mathbf{Z}}(t)) - \widehat{\mathbf{M}}_c(t), \quad \tau_k \leq t < \tau_{k+1} \end{aligned} \tag{5.103a}$$

$$\begin{aligned} \mathbf{Z}(\tau_k^-) &= \boldsymbol{\Pi}_{d\mathbf{F}_d^{(q)}}(\tau_k)[\mathbf{Z}(\tau_k)] + \boldsymbol{\Theta}_{d\mathbf{F}_d}(\tau_k) - (\mathbf{F}_d^{(q)}(\tau_k) - \mathbf{F}_d^{\mathbf{Z}}(\tau_k))^\top \\ &\quad \cdot (\boldsymbol{\Pi}_{d3}(\tau_k)[\mathbf{Z}(\tau_k)] + \mathbf{R}_d(\tau_k))(\mathbf{F}_d^{(q)}(\tau_k) - \mathbf{F}_d^{\mathbf{Z}}(\tau_k)) - \widehat{\mathbf{M}}_d(\tau_k), \quad k \in \mathbb{Z}_+. \end{aligned} \tag{5.103b}$$

Subtracting (5.103) from (5.102) we obtain via (5.100) that

$$\mathbf{Y}^{(q)}(\cdot) := \mathbf{X}^{(q)}(\cdot) - \mathbf{Z}(\cdot)$$

is a bounded and uniform positive solution of the following BJMLDE:

$$-\dot{\mathbf{Y}}^{(q)}(t) = \mathcal{L}^*_{c\mathbf{F}_c^{(q)}}(t)[\mathbf{Y}^{(q)}(t)] + \widehat{\mathbf{H}}_c^{(q)}(t), \quad \tau_k \leq t < \tau_{k+1} \tag{5.104a}$$

$$\mathbf{Y}^{(q)}(\tau_k^-) = \boldsymbol{\Pi}_{d\mathbf{F}_d^{(q)}}(\tau_k)[\mathbf{Y}^{(q)}(\tau_k)] + \widehat{\mathbf{H}}_d^{(q)}(\tau_k), \quad k \in \mathbb{Z}_+ \tag{5.104b}$$

where

$$\widehat{\mathbf{H}}_c^{(q)}(t) := \widehat{\mathbf{M}}_c(t) + (\mathbf{F}_c^{(q)}(t) - \mathbf{F}_c^{(q-1)}(t))^\top(\boldsymbol{\Pi}_{c3}(t)[\mathbf{X}^{(q)}(t)] + \mathbf{R}_c(t))$$
$$\cdot (\mathbf{F}_c^{(q)}(t) - \mathbf{F}_c^{(q-1)}(t)) + (\mathbf{F}_c^{(q)}(t) - \mathbf{F}_c^{\mathbf{Z}}(t))^\top(\boldsymbol{\Pi}_{c3}(t)[\mathbf{Z}(t)] + \mathbf{R}_c(t))$$
$$\cdot (\mathbf{F}_c^{(q)}(t) - \mathbf{F}_c^{\mathbf{Z}}(t))$$

and

$$\mathbf{H}_d^{(q)}(\tau_k) := \frac{\varepsilon^2}{q+1}\mathbf{I}_n + \widehat{\mathbf{M}}_d(\tau_k) + (\mathbf{F}_d^{(q)}(\tau_k) - \mathbf{F}_d^{(q-1)}(\tau_k))^\top(\boldsymbol{\Pi}_{d3}(\tau_k)[\mathbf{X}^{(q)}(\tau_k)] + \mathbf{R}_d(\tau_k))$$
$$(\mathbf{F}_d^{(q)}(\tau_k) - \mathbf{F}_d^{(q-1)}(\tau_k)) + +(\mathbf{F}_d^{(q)}(\tau_k) - \mathbf{F}_d^{\mathbf{Z}}(\tau_k))^\top(\boldsymbol{\Pi}_{d3}(\tau_k)[\mathbf{Z}(\tau_k)] + \mathbf{R}_d(\tau_k))$$
$$(\mathbf{F}_d^{(q)}(\tau_k) - \mathbf{F}_d^{\mathbf{Z}}(\tau_k)).$$

Employing (5.16c), (5.16d) and (5.74) together with (5.101) we may infer that

$$0 \preceq \widehat{\mathbf{H}}_c^{(q)}(t) \preceq \widehat{\nu}_c^{(q)}\mathbf{I}_n,$$

for all $t \in \mathbb{R}_+$ and

$$\frac{\varepsilon^2}{q+1}\mathbf{I}_n \preceq \widehat{\mathbf{H}}_d^{(q)}(\tau_k) \preceq \widehat{\nu}_d^{(q)}\mathbf{I}_n, \quad \text{for all} \quad k \in \mathbb{Z}_+.$$

The implication $(iii) \Rightarrow (i)$ from Theorem 2.7 applied in the case of the non-homogeneous backward jump matrix linear differential equation (5.104) allows us to conclude that the jump matrix linear differential equation (5.90) (written for $p = q$) is exponentially stable. Thus, we have shown that the item (b_p) holds for $p = q$. To check the validity of item (c_p) for $p = q$, let us rewrite (5.87) when $p = q - 1$, as:

$$-\dot{\mathbf{X}}^{(q-1)}(t) = \mathscr{L}^*_{c\mathbf{F}_c^{(q-1)}}(t)[\mathbf{X}^{(q-1)}(t)] + \boldsymbol{\Theta}_{c\mathbf{F}_c^{(q-1)}}(t) + (\mathbf{F}_c^{(q-1)}(t) - \mathbf{F}_c^{(q-2)}(t))^\top$$
$$\cdot (\boldsymbol{\Pi}_{c3}(t)[\mathbf{X}^{(q-1)}(t)] + \mathbf{R}_c(t))(\mathbf{F}_c^{(q-1)}(t) - \mathbf{F}_c^{(q-2)}(t)),$$
$$\tau_k \le t < \tau_{k+1}, \tag{5.105a}$$
$$\mathbf{X}^{(q-1)}(\tau_k^-) = \boldsymbol{\Pi}_{d\mathbf{F}_d^{(q-1)}}(\tau_k)[\mathbf{X}^{(q-1)}(\tau_k)] + \boldsymbol{\Theta}_{d\mathbf{F}_d^{(q-1)}}(\tau_k) + (\mathbf{F}_d^{(q-1)}(\tau_k) - \mathbf{F}_d^{(q-2)}(\tau_k))^\top$$
$$\cdot(\boldsymbol{\Pi}_{d3}(\tau_k)[\mathbf{X}^{(q-1)}(\tau_k)] + \mathbf{R}_d(\tau_k))(\mathbf{F}_d^{(q-1)}(\tau_k) - \mathbf{F}_d^{(q-2)}(\tau_k))$$
$$+ \frac{\varepsilon^2}{q}\mathbf{I}_n, k \in \mathbb{Z}_+. \tag{5.105b}$$

Subtracting (5.87) written for $p = q$ from (5.105) we deduce that

$$\boldsymbol{\Delta}^{(q)}(\cdot) := \mathbf{X}^{(q-1)}(\cdot) - \mathbf{X}^{(q)}(\cdot)$$

is a bounded solution of the following non-homogeneous BJMLDE:

$$-\dot{\boldsymbol{\Delta}}^{(q)}(t) = \mathscr{L}^*_{c\mathbf{F}_c^{(q-1)}}(t)[\boldsymbol{\Delta}^{(q)}(t)] + \widetilde{\mathbf{H}}_c^{(q)}(t), \quad \tau_k \le t < \tau_{k+1} \tag{5.106a}$$

$$\boldsymbol{\Delta}^{(q)}(\tau_k^-) = \boldsymbol{\Pi}_{d\mathbf{F}_d^{(q-1)}}(\tau_k)[\boldsymbol{\Delta}^{(q)}(\tau_k)] + \widetilde{\mathbf{H}}_d^{(q)}(\tau_k), \quad k \in \mathbb{Z}_+ \tag{5.106b}$$

where

$$\widetilde{\mathbf{H}}_c^{(q)}(t) := (\mathbf{F}_c^{(q-1)}(t) - \mathbf{F}_c^{(q-2)}(t))^\top (\boldsymbol{\Pi}_{c3}(t)[\mathbf{X}^{(q-1)}(t)] + \mathbf{R}_c(t))$$
$$\cdot (\mathbf{F}_c^{(q-1)}(t) - \mathbf{F}_c^{(q-2)}(t)),$$

$$\widetilde{\mathbf{H}}_d^{(q)}(\tau_k) := \frac{\varepsilon^2}{q(q+1)}\mathbf{I}_n + (\mathbf{F}_d^{(q-1)}(\tau_k) - \mathbf{F}_d^{(q-2)}(\tau_k))^\top (\boldsymbol{\Pi}_{d3}(\tau_k)[\mathbf{Z}(\tau_k)] + \mathbf{R}_d(\tau_k))$$
$$(\mathbf{F}_d^{(q-1)}(\tau_k) - \mathbf{F}_d^{\mathbf{Z}}(\tau_k)), \quad k \in \mathbb{Z}_+.$$

Since (a_p) holds for $p = q-1$, we may deduce via (5.16c) and (5.16d), that (5.105) are true for q replaced by $q-1$, too. Hence, $0 \preceq \widetilde{\mathbf{H}}_c^{(q)}(t) \preceq \widetilde{\nu}_c^{(q)}\mathbf{I}_n$, for all $t \in \mathbb{R}_+$ and

$$\frac{\varepsilon^2}{q(q+1)}\mathbf{I}_n \preceq \widetilde{\mathbf{H}}_d^{(q)}(\tau_k) \preceq \widetilde{\nu}_d^{(q)}\mathbf{I}_n, \quad \text{for all} \quad k \in \mathbb{Z}_+.$$

Now we may conclude via Proposition 5.1 (iii) that the unique bounded on $\mathbb{R}_+$ solution of (5.106) is uniform positive. This yields

$$\mathbf{X}^{(q-1)}(t) \succeq \mathbf{X}^{(q)}(t),$$

for all $t \in \mathbb{R}_+$ which confirms the validity of item (c_p) for $p = q$. Thus, the iterative procedure (5.87)–(5.89) may be continued and the properties described by (a_p) – (c_p) are still valid for any $p = 0, 1, \ldots$.

So, we have obtained the decreasing matrix valued sequences

$$X^{(0)}(t,i) \ge X^{(1)}(t,i) \ge \ldots \ge X^{(p)}(t,i) \ge X^{(p+1)}(t,i) \ge \ldots \ge Z(t,i), \tag{5.107}$$

for all $(t,i) \in \mathbb{R}_+ \times \mathscr{N}$ and for any $\mathbf{Z}(\cdot) = (Z(\cdot,1), Z(\cdot,2), \ldots, Z(\cdot,N)) \in \Gamma^{\Sigma}[\mathbb{R}_+]$. From (5.107), we deduce that for each $(t,i) \in \mathbb{R}_+ \times \mathscr{N}$ the matrix-valued sequence $\{X^{(p)}(t,i)\}_{p\ge1}$ is convergent. Let

$$\widetilde{X}(t,i) := \lim_{p\to\infty} X^{(p)}(t,i), \quad \text{for all} \quad (t,i) \in \mathbb{R}_+ \times \mathscr{N}. \tag{5.108}$$

We show that $\widetilde{\mathbf{X}}(t) = (\widetilde{X}(t,1), \widetilde{X}(t,2), \ldots, \widetilde{X}(t,N)), \quad t \in \mathbb{R}_+$ is just the maximal and bounded solution of (5.2). From (5.107) together with the monotonicity property of the euclidean norm of the matrices in $\mathscr{S}_n^{N+}$, we deduce that

$$\sup_{t\in\mathbb{R}_+} \|\mathbf{X}^{(p)}(t)\|_\infty \le \widetilde{c}, \quad \text{for all} \quad p \ge 1 \tag{5.109a}$$

$$\sup_{t \in \mathbb{R}_+} \|\widetilde{\mathbf{X}}(t)\|_\infty \leq \widetilde{c}, \tag{5.109b}$$

where

$$\widetilde{c} = \sup_{t \in \mathbb{R}_+} \|\mathbf{X}^{(0)}(t) - \mathbf{Z}(t)\|_\infty + \sup_{t \in \mathbb{R}_+} \|\mathbf{Z}(t)\|_\infty,$$

$\|\cdot\|_\infty$ being the norm on the space $\mathscr{S}_n^N$ introduced by (2.112). From (5.88), (5.89), (5.101) and (5.108), (5.109) we may infer that

$$\lim_{p \to \infty} (F_c^{(p)}(t,i) - F_c^{(p-1)}(t,i)) = 0, \quad \text{for all } \ (t,i) \in \mathbb{R}_+ \times \mathcal{N} \tag{5.110a}$$

$$\lim_{p \to \infty} (F_d^{(p)}(\tau_k,i) - F_d^{(p-1)}(\tau_k,i)) = 0, \quad \text{for all } \ (\tau_k,i) \in \mathfrak{T} \times \mathcal{N}. \tag{5.110b}$$

Further, employing Lemma 5.2 for $\mathbf{W}_c(t) \leftarrow \mathbf{F}_c^{(p-1)}(t)$, $\mathbf{W}_d(\tau_k) \leftarrow \mathbf{F}_d^{(p-1)}(\tau_k)$ combined with Lemma 5.1 written for $\mathbf{W}_c(t) \leftarrow \mathbf{F}_c^{(p)}(t)$, $\mathbf{W}_d(\tau_k) \leftarrow \mathbf{F}_d^{(p)}(\tau_k)$ we obtain the following equivalent version of (5.87):

$$\begin{aligned} -\dot{X}^{(p)}(t,i) &= \mathscr{R}_c(t, \mathbf{X}^{(p)}(t))(i) + (F_c^{(p)}(t,i) - F_c^{(p-1)}(t,i))^\top \\ &\quad \cdot (\Pi_{c3}(t)[\mathbf{X}^{(p)}(t)](i) + R_c(t,i))(F_c^{(p)}(t,i) - F_c^{(p-1)}(t,i)), \\ &\quad \tau_k \leq t < \tau_{k+1} \qquad (5.111a) \\ X^{(p)}(\tau_k^-,i) &= \mathscr{R}_d(\tau_k, \mathbf{X}^{(p)}(\tau_k))(i) + (F_d^{(p)}(\tau_k,i) - F_d^{(p-1)}(\tau_k,i))^\top \\ &\quad \cdot (\Pi_{d3}(\tau_k)[\mathbf{X}^{(p)}(\tau_k)](i) + R_d(\tau_k,i))(F_d^{(p)}(\tau_k,i) - F_d^{(p-1)}(\tau_k,i)), \\ &\quad k \in \mathbb{Z}_+. \qquad (5.111b) \end{aligned}$$

Bearing in mind that the generalized Riccati operator $\mathbf{X} \to \mathscr{R}_d(\tau_k, \mathbf{X})(i) : \mathscr{S}_n^N \to \mathscr{S}_n$ is described by the right hand side of (5.3b), we obtain (via (5.108) (5.109a) and (5.110b)) that:

$$\begin{aligned} &\lim_{p \to \infty} [\mathscr{R}_d(\tau_k, \mathbf{X}^{(p)}(\tau_k))(i) + (F_d^{(p)}(\tau_k,i) - F_d^{(p-1)}(\tau_k,i))^\top \\ &\cdot (\Pi_{d3}(\tau_k)[\mathbf{X}^{(p)}(\tau_k)](i) + R_d(\tau_k,i))(F_d^{(p)}(\tau_k,i) - F_d^{(p-1)}(\tau_k,i))] \\ &= \mathscr{R}_d(\tau_k, \widetilde{\mathbf{X}}(\tau_k))(i), \end{aligned}$$

for all $(\tau_k, i) \in \mathfrak{T} \times \mathcal{N}$. Taking the limit for $p \to \infty$ in both sides of (5.111b), one gets

$$\lim_{p \to \infty} X^{(p)}(\tau_k, i) = \mathscr{R}_d(\tau_k, \widetilde{\mathbf{X}}(\tau_k))(i) \stackrel{\Delta}{=} S(\tau_k, i), \tag{5.112}$$

for all $(\tau_k, i) \in \mathfrak{T} \times \mathcal{N}$. Integrating (5.111a) on an arbitrary interval $[\tau_{k-1}, \tau_k]$ we obtain

$$X^{(p)}(t,i) - X(\tau_k^-, i) = \int_t^{\tau_k} \Psi_i(s, \mathbf{X}^{(p)}(s))ds \tag{5.113}$$

where we have denoted

$$\Psi_i(s, \mathbf{X}^{(p)}(s)) := \mathscr{R}_c(s, \mathbf{X}^{(p)}(s))(i) + (F_c^{(p)}(s,i) - F_c^{(p-1)}(s,i))^\top$$
$$\cdot (\Pi_{c3}(s)[\mathbf{X}^{(p)}(s)](i) + R_c(s,i))(F_c^{(p)}(s,i) - F_c^{(p-1)}(s,i)),$$

for all $\tau_{k-1} \le t \le s \le \tau_k$, $i \in \mathcal{N}$. Further, (5.108), (5.109a) and (5.110a) allow us to deduce that

$$\lim_{p\to\infty} \Psi_i(s, \mathbf{X}^{(p)}(s)) = \mathscr{R}_c(s, \widetilde{\mathbf{X}}(s))(i), \tag{5.114}$$

for all $\tau_{k-1} \le t \le s \le \tau_k$, $k \in \mathbb{Z}_+$, $i \in \mathcal{N}$. By a standard reason based on Lebesgue's theorem of dominant convergence, one shows that

$$\lim_{p\to\infty} \int_t^{\tau_k} \Psi_i(s, \mathbf{X}^{(p)}(s))ds = \int_t^{\tau_k} \mathscr{R}_c(s, \widetilde{\mathbf{X}}(s))ds.$$

Taking the limit for $p \to \infty$ in both sides of (5.113) we obtain (via (5.112)) that

$$\widetilde{X}(t,i) - S(\tau_k, i) = \int_t^{\tau_k} \mathscr{R}_c(s, \widetilde{\mathbf{X}}(s))(i)ds, \tag{5.115}$$

for all $\tau_{k-1} \le t \le \tau_k$, $k \in \mathbb{Z}_+$, $i \in \mathcal{N}$. Letting $t \to \tau_k$, $t < \tau_k$, in (5.115) we have

$$\widetilde{X}(\tau_k^-, i) = \lim_{\substack{t\to\tau_k \\ t<\tau_k}} \widetilde{X}(t,i) = S(\tau_k, i).$$

Thus (5.112) leads to

$$\widetilde{X}(\tau_k^-, i) = \mathscr{R}_d(\tau_k, \widehat{\mathbf{X}}(\tau_k))(i),$$

for all $k \in \mathbb{Z}_+$, $i \in \mathcal{N}$, or, in a compact form

$$\widetilde{\mathbf{X}}(\tau_k^-) = \mathscr{R}_d(\tau_k, \widetilde{\mathbf{X}}(\tau_k)), \quad k \in \mathbb{Z}_+. \tag{5.116}$$

On the other hand $s \to \mathscr{R}_c(s, \widetilde{\mathbf{X}}(s))(i)$ is a continuous function. So, the right hand side of (5.115) is a differentiable function with respect to t. Differentiating in both sides of (5.115), we obtain

$$-\dot{\widetilde{X}}(t,i) = \mathscr{R}_c(t, \widetilde{\mathbf{X}}(t))(i), \quad \tau_k \le t < \tau_{k+1}, \quad i \in \mathcal{N},$$

or, in a compact form

$$-\dot{\widetilde{\mathbf{X}}}(t) = \mathscr{R}_c(t, \widetilde{\mathbf{X}}(t)), \quad \tau_k \leq t < \tau_{k+1}, \quad k \in \mathbb{Z}_+. \tag{5.117}$$

From (5.116) and (5.117) together with (5.109b) we may conclude that $t \to \widetilde{\mathbf{X}}(t) : \mathbb{R}_+ \to \mathscr{S}_n^N$ is a bounded solution of (5.13). On the other hand, (5.107) and (5.108) give

$$\widetilde{\mathbf{X}}(t) \succeq \mathbf{Z}(t),$$

for all $t \in \mathbb{R}_+$ and for an arbitrary $\mathbf{Z}(\cdot) \in \Gamma^{\Sigma}[\mathbb{R}_+]$. We obtain that $\widetilde{\mathbf{X}}(\cdot)$ is a maximal solution of (5.13), or equivalently of (5.2). The maximality property gives uniqueness to the solution $\widetilde{\mathbf{X}}(\cdot)$. Thus the proof is complete.

Corollary 5.2 *For the quintuple Σ associated to the GBJMRDE (5.2), via (5.15), assume:*

(a) the quadruple $(\mathbf{A}, \mathbf{B}, (\boldsymbol{\Pi}_c \mid \boldsymbol{\Pi}_d), \mathfrak{T})$ is stabilizable;
(b) $\Gamma^{\Sigma}[\mathbb{R}_+]$ is nonempty.

Then the following hold:

(i) if the GBJMRDE (5.2) is in the periodic case in the sense of Definition 5.2 (i), then the unique maximal and bounded solution of (5.2) is a periodic function of period $\tilde{\theta}$, where $\tilde{\theta}$ is introduced in (5.14a);
(ii) if the GBJMRDE (5.2) is time invariant in the sense of the Definition 5.2 (ii), then the unique maximal and bounded solution of (5.2) is a periodic function of period h.

Proof (i) If the GBJMRDE (5.2) is in the periodic case, then, according to Corollary 5.4 (a) we may infer that the quadruple $(\mathbf{A}, \mathbf{B}, (\boldsymbol{\Pi}_c \mid \boldsymbol{\Pi}_d), \mathfrak{T})$ is stabilizable if and only if there exists a pair of stabilizing feedback gains of the form $(\widetilde{\mathbf{F}}_c(\cdot), \{\widetilde{\mathbf{F}}_d(kh)\}_{k\in\mathbb{Z}_+})$ with $\widetilde{\mathbf{F}}_c(\cdot) \in \mathscr{PC}\{\mathbb{R}_+, \mathscr{M}^N_{m_c n}; \mathfrak{T}\}$ being a periodic function of period $\tilde{\theta}$ and $\{\widetilde{\mathbf{F}}_d(kh)\}_{k\in\mathbb{Z}_+} \subset \mathscr{M}^N_{m_d n}$ is a periodic matrix valued sequence of period $\tilde{\mathbf{k}}$ with $\tilde{\theta}$ and $\tilde{\mathbf{k}}$ computed as in (5.14).

Applying Theorem 2.13 (ii) in the case of the non-homogeneous backward jump matrix linear differential equation (5.75) when $\mathbf{F}_c(\cdot)$ is replaced by $\widetilde{\mathbf{F}}_c(\cdot)$ and $\mathbf{F}_d(kh)$ are replaced by $\widetilde{\mathbf{F}}_d(kh)$, $k \in \mathbb{Z}_+$, we may infer that the unique bounded on $\mathbb{R}_+$ solution $\mathbf{X}^{(0)}(\cdot)$ is a periodic function of period $\tilde{\theta}$ and $\{\mathbf{X}^{(0)}(kh)\}_{k\in\mathbb{Z}_+}$ is a periodic sequence of period $\tilde{\mathbf{k}}$. Proceeding inductively for $p = 1, 2, \ldots$ one may check that in this case Theorem 2.13 (ii) guarantees that the unique bounded on $\mathbb{R}_+$ solution $\mathbf{X}^{(p)}(\cdot)$ of the non-homogeneous backward jump matrix linear differential equation (5.87) is a periodic function of period $\tilde{\theta}$ and $\mathbf{F}_c^{(p)}(\cdot)$ is a periodic function of period $\tilde{\theta}$, while $\{\mathbf{F}_d(kh)\}_{k\in\mathbb{Z}_+}$ computed in (5.89) is a periodic sequence of period $\tilde{\mathbf{k}}$. So, (5.108) yields

$$\widetilde{X}(t, i) = \lim_{p\to\infty} X^{(p)}(t, i) = \lim_{p\to\infty} X^{(p)}(t + \tilde{\theta}, i) = \widetilde{X}(t + \tilde{\theta}, i),$$

for all $(t, i) \in \mathbb{R}_+ \times \mathcal{N}$. This shows that the function $\widetilde{\mathbf{X}}(\cdot) = (\widetilde{X}(\cdot, 1), \widetilde{X}(\cdot, 2), \ldots, \widetilde{X}(\cdot, N))$ is a periodic function of period $\tilde{\theta}$. Bearing in mind that $\widetilde{\mathbf{X}}(\cdot)$ computed via (5.108) is just the unique maximal and bounded solution $\mathbf{X}_{max}(\cdot)$ of (5.2), we may conclude that $\mathbf{X}_{max}(\cdot)$ is a periodic function of period $\tilde{\theta}$. Thus (i) is proved.

Part (ii) can be viewed as a special case of (i), because in the time-invariant case, we may take $\theta_c = h$, $\theta_d = 1$. These lied to $\tilde{\theta} = h$. This complete the proof.

Remark 5.5 If the GBJMRDE (5.2) is in the periodic case, then its coefficients can be extended by periodicity to the whole real axis $\mathbb{R}$. The Assumption 5.1 remains true for any $t \in \mathbb{R}$ and $k \in \mathbb{Z}$. The sequence $\mathfrak{T} = \{\tau_k \mid k \in \mathbb{Z}_+\}$ is replaced by $\mathfrak{T} = \{kh \mid k \in \mathbb{Z}\}$. The unique bounded on $\mathbb{R}_+$ solution $\mathbf{X}^{(p)}(\cdot)$ of (5.75) and (5.87), respectively, may be extended in a natural way by periodicity to the whole real axis. In this case (5.108) takes place for any $(t, i) \in \mathbb{R} \times \mathcal{N}$. This means that in the periodic case $\mathbf{X}_{max}(\cdot)$ provided by Theorem 5.3 is the unique maximal and bounded on $\mathbb{R}$ solution of (5.2).

Let $\boldsymbol{\Theta}_c^{(p)}(t) = (\Theta_c^{(p)}(t, 1), \Theta_c^{(p)}(t, 2), \ldots, \Theta_c^{(p)}(t, N))$ be such that

$$\Theta_c^{(p)}(t, i) = \begin{pmatrix} M_c^{(p)}(t, i) & L_c^{(p)}(t, i) \\ (L_c^{(p)}(t, i))^\top & R_c^{(p)}(t, i) \end{pmatrix} \in \mathcal{S}_{n+m_c}$$

with $M_c^{(p)}(t, i) \in \mathcal{S}_n$ and $\boldsymbol{\Theta}_d^{(p)}(\tau_k) = (\Theta_d^{(p)}(\tau_k, 1), \Theta_d^{(p)}(\tau_k, 2), \ldots, \Theta_d^{(p)}(\tau_k, N))$ be such that

$$\Theta_d^{(p)}(\tau_k, i) = \begin{pmatrix} M_d^{(p)}(\tau_k, i) & L_d^{(p)}(\tau_k, i) \\ (L_d^{(p)}(\tau_k, i))^\top & R_d^{(p)}(\tau_k, i) \end{pmatrix} \in \mathcal{S}_{n+m_d}$$

with $M_d^{(p)}(\tau_k, i) \in \mathcal{S}_n$, for all $(\tau_k, i) \in \mathfrak{T} \times \mathcal{N}$.
Beside the quintuple Σ described by (5.15) we consider the quintuples $\Sigma^{(p)}$, $p \in \mathbb{Z}_+$, where $\Sigma^{(p)} = (\mathbf{A}, \mathbf{B}, (\boldsymbol{\Pi}_c \mid \boldsymbol{\Pi}_d), (\boldsymbol{\Theta}_c^{(p)} \mid \boldsymbol{\Theta}_d^{(p)}), \mathfrak{T})$ where $\mathbf{A}$, $\mathbf{B}$, $\boldsymbol{\Pi}_c$, $\boldsymbol{\Pi}_d$, $\mathfrak{T}$ are as in (5.15) and $\boldsymbol{\Theta}_c^{(p)}$, $\boldsymbol{\Theta}_d^{(p)}$ are as above.

The operator valued functions $\mathfrak{D}_c^{\Sigma^{(p)}}(t)[\cdot]$ and $\mathfrak{D}_d^{\Sigma^{(p)}}(\tau_k)[\cdot]$ are as in (5.16a) and (5.16b), respectively, with $M_c(t, i)$, $L_c(t, i)$, $R_c(t, i)$ replaced by $M_c^{(p)}(t, i)$, $L_c^{(p)}(t, i)$, $R_c^{(p)}(t, i)$ and $M_d(\tau_k, i)$, $L_d(\tau_k, i)$, $R_d(\tau_k, i)$ replaced by $M_d^{(p)}(\tau_k, i)$, $L_d^{(p)}(\tau_k, i)$, $R_d^{(p)}(\tau_k, i)$, respectively.

For each $p \in \mathbb{Z}_+$ the quintuple $\Sigma^{(p)}$ defines the following GBJMRDE of the form (5.2):

$$\begin{aligned} -\dot{\mathbf{X}}(t) &= \mathbf{A}^\top(t)\mathbf{X}(t) + \mathbf{X}(t)\mathbf{A}(t) + \boldsymbol{\Pi}_{c1}[\mathbf{X}(t)] - (\mathbf{X}(t)\mathbf{B}(t) + \boldsymbol{\Pi}_{c2}(t)[\mathbf{X}(t)] + \mathbf{L}_c^{(p)}(t)) \\ &\quad \cdot (\boldsymbol{\Pi}_{c3}(t)[\mathbf{X}(t)] + \mathbf{R}_c^{(p)}(t))^{-1}(\mathbf{X}(t)\mathbf{B}(t) + \boldsymbol{\Pi}_{c2}(t)[\mathbf{X}(t)] + \mathbf{L}_c^{(p)}(t))^\top \\ &\quad + \mathbf{M}_c^{(p)}(t), \quad \tau_k \le t < \tau_{k+1} \end{aligned} \tag{5.118a}$$

$$\begin{aligned} \mathbf{X}(\tau_k^-) &= \boldsymbol{\Pi}_{d1}(\tau_k)[\mathbf{X}(\tau_k)] - (\boldsymbol{\Pi}_{d2}(\tau_k)[\mathbf{X}(\tau_k)] + \mathbf{L}_d^{(p)}(\tau_k))(\boldsymbol{\Pi}_{d3}(\tau_k)[\mathbf{X}(\tau_k)] + \mathbf{R}_d^{(p)}(\tau_k))^{-1} \\ &\quad \cdot (\boldsymbol{\Pi}_{d2}(\tau_k)[\mathbf{X}(\tau_k)] + \mathbf{L}_d^{(p)}(\tau_k))^\top + \mathbf{M}_d^{(p)}(\tau_k), \quad k \in \mathbb{Z}_+. \end{aligned} \tag{5.118b}$$

The next result may be viewed as a comparison theorem between the maximal solution $\mathbf{X}_{max}^{(p)}(\cdot)$ of (5.118) when $p \in \mathbb{Z}_+$.

Theorem 5.4 *Assume:*

(a) the Assumption 5.1 is fulfilled;
(b) the quadruple $(\mathbf{A}, \mathbf{B}, (\boldsymbol{\Pi}_c \mid \boldsymbol{\Pi}_d), \mathfrak{T})$ *is stabilizable and the set* $\Gamma^{\Sigma}[\mathbb{R}_+]$ *is not empty;*
(c)

$$\Theta_c^{(p)}(t,i) \geq \Theta_c^{(p+1)}(t,i) \geq \Theta_c(t,i), \quad \textit{for all} \ (t,i) \in \mathbb{R}_+ \times \mathcal{N}; \tag{5.119a}$$

$$\Theta_d^{(p)}(\tau_k,i) \geq \Theta_d^{(p+1)}(\tau_k,i) \geq \Theta_d(\tau_k,i), \quad \textit{for all} \ (\tau_k,i) \in \mathfrak{T} \times \mathcal{N}; \tag{5.119b}$$

(d) for each $p \in \mathbb{Z}_+$ *and* $i \in \mathcal{N}$, $(M_c^{(p)}(\cdot,i), L_c^{(p)}(\cdot,i), R_c^{(p)}(\cdot,i)) : \mathbb{R}_+ \to \mathcal{S}_n \times \mathbb{R}^{n \times m_c} \times \mathcal{S}_{m_c}$ *are bounded and continuous matrix-valued functions and* $\{M_d^{(p)}(\tau_k,i), L_d^{(p)}(\tau_k,i), R_d^{(p)}(\tau_k,i)\}_{k \in \mathbb{Z}_+} \subset \mathcal{S}_n \times \mathbb{R}^{n \times m_d} \times \mathcal{S}_{m_d}$ *are bounded sequences.*

Under these conditions the following hold:

(i)

$$\Gamma^{\Sigma}[\mathbb{R}_+] \subset \Gamma^{\Sigma^{(p)}}[\mathbb{R}_+] \subset \Gamma^{\Sigma^{(p-1)}}[\mathbb{R}_+], \quad \textit{for all} \ p \geq 1;$$

(ii) if $\mathbf{X}_{max}^{(p)}(\cdot)$ *is the maximal and bounded on* $\mathbb{R}_+$ *solution of (5.118) satisfying the sign conditions*

$$R_c^{(p)}(t,i) + \Pi_{c3}(t)[\mathbf{X}_{max}^{(p)}(t)](i) \geq \gamma_c^{(p)} I_{m_c} > 0, \ (\forall) \ (t,i) \in \mathbb{R}_+ \times \mathcal{N}; \tag{5.120a}$$

$$R_d^{(p)}(\tau_k,i) + \Pi_{d3}(\tau_k)[\mathbf{X}_{max}^{(p)}(\tau_k)](i) \geq \gamma_d^{(p)} I_{m_d} > 0, \ (\forall) \ (\tau_k,i) \in \mathfrak{T} \times \mathcal{N} \tag{5.120b}$$

and if $\mathbf{X}_{max}(\cdot)$ *is the maximal and bounded on* $\mathbb{R}_+$ *solution of the equation (5.2) satisfying the sign conditions (5.68), then*

$$\mathbf{X}_{max}^{(p)}(t) \succeq \mathbf{X}_{max}^{(p+1)}(t) \succeq \mathbf{X}_{max}(t), \quad \textit{for all} \ t \in \mathbb{R}_+, \ p \geq 0; \tag{5.121}$$

(iii) if

$$\lim_{p\to\infty} \Theta_c^{(p)}(t,i) = \Theta_c(t,i), \quad \textit{for all} \ (t,i) \in \mathbb{R}_+ \times \mathcal{N}; \tag{5.122a}$$

$$\lim_{p\to\infty} \Theta_d^{(p)}(\tau_k,i) = \Theta_c(t,i), \quad \textit{for all} \ (\tau_k,i) \in \mathfrak{T} \times \mathcal{N} \tag{5.122b}$$

then

$$\lim_{p\to\infty} X_{max}^{(p)}(t,i) = X_{max}(t,i), \quad \textit{for all} \ (t,i) \in \mathbb{R}_+ \times \mathcal{N}.$$

Proof (*i*) Let $(\mathfrak{D}_c^{\Sigma}(t)[\cdot](i), \mathfrak{D}_d^{\Sigma}(\tau_k)[\cdot](i))$ and $(\mathfrak{D}_c^{\Sigma^{(p)}}(t)[\cdot](i), \mathfrak{D}_d^{\Sigma^{(p)}}(\tau_k)[\cdot](i))$ be the pairs of generalized dissipation matrices defined as in (5.16a) and (5.16b), respectively, for the quintuple Σ and the quintuple $\Sigma^{(p)}$, respectively.

If $\widehat{\mathbf{Z}}(\cdot) = (\widehat{Z}(\cdot, 1), \widehat{Z}(\cdot, 2), \ldots, \widehat{Z}(\cdot, N))$ is a differentiable function in any $t \neq \tau_k$, we have:

$$\mathfrak{D}_c^{\Sigma^{(p)}}(t)[\widehat{\mathbf{Z}}(t)](i) = (\mathfrak{D}_c^{\Sigma}(t)[\widehat{\mathbf{Z}}(t)](i) + \Theta_c^{(p)}(t, i) - \Theta_c(t, i) \tag{5.123a}$$

$$\mathfrak{D}_d^{\Sigma^{(p)}}(\tau_k)[\widehat{\mathbf{Z}}(\tau_k)](i) = (\mathfrak{D}_c^{\Sigma}(\tau_k)[\widehat{\mathbf{Z}}(\tau_k)](i) + \Theta_d^{(p)}(\tau_k, i) - \Theta_d(\tau_k, i). \tag{5.123b}$$

Let $\mathbf{Z}(\cdot) \in \Gamma^{\Sigma}[\mathbb{R}_+]$ be arbitrary. This means that $t \to \mathbf{Z}(t) : \mathbb{R}_+ \to \mathcal{S}_n^N$ is a bounded function which is differentiable in any $t \neq \tau_k$, right continuous in $t = \tau_k$, $k \in \mathbb{Z}_+$ and satisfies (5.16a)–(5.16d). Employing (5.16a)–(5.16b) together with (5.119) we obtain from (5.123) written for $\widehat{\mathbf{Z}}(\cdot)$ replaced by $\mathbf{Z}(\cdot)$ that:

$$\mathfrak{D}_c^{\Sigma^{(p)}}(t)[\mathbf{Z}(t)](i) \geq 0, \quad \text{for all} \quad (t, i) \in \mathbb{R}_+ \times \mathcal{N} \tag{5.124a}$$

$$\mathfrak{D}_d^{\Sigma^{(p)}}(\tau_k)[\mathbf{Z}(\tau_k)](i) \geq 0, \quad \text{for all} \quad (\tau_k, i) \in \mathfrak{T} \times \mathcal{N}. \tag{5.124b}$$

Further, the inequalities (5.16c), (5.16d) and (5.119) give:

$$\Pi_{c3}(t)[\mathbf{Z}(t)](i) + R_c^{(p)}(t, i) \geq \gamma_c I_{m_c} > 0, \quad \text{for all} \quad (t, i) \in \mathbb{R}_+ \times \mathcal{N} \tag{5.125a}$$

$$\Pi_{d3}(\tau_k)[\mathbf{Z}(\tau_k)](i) + Rd^{(p)}(\tau_k, i) \geq \gamma_d I_{m_d} > 0, \quad \text{for all} \quad (\tau_k, i) \in \mathfrak{T} \times \mathcal{N}. \tag{5.125b}$$

From (5.124) and (5.125) we deduce that $\mathbf{Z}(\cdot) \in \Gamma^{\Sigma^{(p)}}[\mathbb{R}_+]$, which confirm the validity of the inclusion

$$\Gamma^{\Sigma}[\mathbb{R}_+] \subset \Gamma^{\Sigma^{(p)}}[\mathbb{R}_+], \quad \text{for any} \quad p \in \mathbb{Z}_+.$$

The inclusion

$$\Gamma^{\Sigma^{(p)}}[\mathbb{R}_+] \subset \Gamma^{\Sigma^{(p-1)}}[\mathbb{R}_+]$$

may be proved in the same way. Thus (*i*) is true.

Since the quadruple $(\mathbf{A}, \mathbf{B}, (\boldsymbol{\Pi}_c \mid \boldsymbol{\Pi}_d), \mathfrak{T})$ is stabilizable and the set $\Gamma^{\Sigma^{(p)}}[\mathbb{R}_+]$ is not empty, we may apply Theorem 5.3 to obtain that for each $p = 0, 1, \ldots$ the GBJMRDE (5.118) has a maximal and bounded on $\mathbb{R}_+$ solution $\mathbf{X}_{max}^{(p)}(\cdot)$ satisfying a sign condition of the form (5.120).

Let $\mathbf{X}_{max}(\cdot)$ be the maximal and bounded on $\mathbb{R}_+$ solution of (5.2) which is satisfying the sign condition (5.68). Since $\mathbf{X}_{max}(\cdot) \in \Gamma^{\Sigma}[\mathbb{R}_+] \subset \Gamma^{\Sigma^{(p+1)}}[\mathbb{R}_+]$ we may infer that

$$\mathbf{X}_{max}(t) \preceq \mathbf{X}_{max}^{(p+1)}(t),$$

for all $t \in \mathbb{R}_+$, $p \in \mathbb{Z}_+$. Further $\mathbf{X}_{max}^{(p+1)}(\cdot) \in \Gamma^{\Sigma^{(p+1)}}[\mathbb{R}_+] \subset \Gamma^{\Sigma^{(p)}}[\mathbb{R}_+]$, which leads to

$$\mathbf{X}_{max}^{(p+1)}(t) \preceq \mathbf{X}_{max}^{(p)}(t),$$

for all $t \in \mathbb{R}_+$, for any $p \in \mathbb{Z}_+$. This confirms the validity of (ii).

To prove (iii), let us remark that (5.121) guarantees the convergence of the sequences $\{X_{max}^{(p)}(t, i)\}_{p \geq 0}$, $(t, i) \in \mathbb{R}_+ \times \mathcal{N}$. Let

$$\widetilde{Y}(t, i) \stackrel{\Delta}{=} \lim_{p \to \infty} X_{max}^{(p)}(t, i), \quad \text{for all} \quad (t, i) \in \mathbb{R}_+ \times \mathcal{N}.$$

Employing (5.122) together with the Lebesgue's theorem of dominant convergence, one shows that $\widetilde{\mathbf{Y}}(t) = (\widetilde{Y}(t, 1), \widetilde{Y}(t, 2), \ldots, \widetilde{Y}(t, N))$, $t \in \mathbb{R}_+$ is a solution of (5.2). From (5.121) one obtains that $\widetilde{\mathbf{Y}}(t) \succeq \mathbf{X}_{max}(t)$, $\quad t \in \mathbb{R}_+$. Bearing in mind that $\mathbf{X}max(\cdot)$ is the maximal solution of (5.2) we may conclude that

$$\widetilde{\mathbf{Y}}(t) = \mathbf{X}_{max}(t)$$

which ends the proof.

5.4 The Stabilizing Solution of a Generalized Backward Jump Matrix Riccati Differential Equation

5.4.1 The General Case

Among the solutions defined on the whole real semiaxis $\mathbb{R}_+$ of a GBJMRDE (5.2), a privileged place is occupied by the **stabilizing solution.** This kind of solution plays a central role in the solving of some different optimal control problems on infinite time horizon.

Definition 5.5 A solution $t \to \mathbf{X}_s(t) = (X_s(t, 1), X_s(t, 2), \ldots, X_s(t, N)) : \mathbb{R}_+ \to \mathcal{S}_n^N$ of (5.2) is named **stabilizing solution** if the following jump matrix linear differential equation on $\mathcal{S}_n^N$

$$\dot{\mathbf{Y}}(t) = \mathcal{L}_{c\mathbf{F}_{c,s}}(t)[\mathbf{Y}(t)], \quad \tau_k < t \leq \tau_{k+1} \tag{5.126a}$$

$$\mathbf{Y}(\tau_k^+) = \boldsymbol{\Pi}^*_{d\mathbf{F}_{d,s}}(\tau_k)[\mathbf{Y}(\tau_k)], \quad k \in \mathbb{Z}_+ \tag{5.126b}$$

is exponentially stable, where $\mathbf{F}_{c,s}(t) = (F_{c,s}(t, 1), F_{c,s}(t, 2), \ldots, F_{c,s}(t, N))$

$$\begin{aligned} F_{c,s}(t, i) :=& -(\Pi_{c3}(t)[\mathbf{X}_s(t)](i) + R_c(t, i))^{-1} \\ &\cdot (X_s(t, i)B^\top(t, i) + \Pi_{c2}(t)[\mathbf{X}_s(t)](i) + L_c(t, i))^\top, \quad (t, i) \in \mathbb{R}_+ \times \mathcal{N} \end{aligned} \tag{5.127}$$

and $\mathbf{F}_{d,s}(\tau_k) = (F_{d,s}(\tau_k, 1), F_{d,s}(\tau_k, 2), \ldots, F_{d,s}(\tau_k, N))$

$$F_{d,s}(\tau_k, i) := -\left(\Pi_{d3}(\tau_k)[\mathbf{X}_s(\tau_k)](i) + R_d(\tau_k, i)\right)^{-1} \cdot \left(\Pi_{d2}(\tau_k)[\mathbf{X}_s(\tau_k)](i) + L_d(\tau_k, i)\right)^{\top}, \quad k \in \mathscr{Z}_+. \tag{5.128}$$

In (5.126a), the linear operator $\mathscr{L}_{c\mathbf{F}_{c,s}}(t)[\cdot]$ is defined as in (5.26), but for $\mathbf{W}_c(\cdot) \leftarrow \mathbf{F}_{c,s}(\cdot)$, while in (5.126b), $\boldsymbol{\Pi}^*_{d\mathbf{F}_{d,s}}(\tau_k)[\cdot]$ is the adjoint with respect to the inner product (2.111) of the linear operator described by (5.20) in which $\mathbf{W}_d(\tau_k)$ is replaced by $\mathbf{F}_{d,s}(\tau_k)$. To make the choice of the jump matrix linear differential equation (5.126) for the characterization of the concept of stabilizing solution of a GBJMRDE (5.2) more intuitive, let us consider the particular case of this equation given in (5.10). According to (5.31) and (5.32), the jump matrix linear differential equation (5.126) becomes

$$\begin{aligned}\dot{Y}(t, i) &= (A_0(t, i) + B_0(t, i)F_{c,s}(t, i))Y(t, i) + Y(t, i)(A_0(t, i) + B_0(t, i)F_{c,s}(t, i))^{\top} \\ &\quad + \sum_{l=1}^{r_1}(A_l(t, i) + B_l(t, i)F_{c,s}(t, i))Y(t, i)(A_l(t, i) + B_l(t, i)F_{c,s}(t, i))^{\top} \\ &\quad + \sum_{j=1}^{N} q_{ji}Y(t, j), \quad \tau_k < t \le \tau_{k+1}\end{aligned} \tag{5.129a}$$

$$\begin{aligned}Y(\tau_k^+, i) &= \sum_{p=0}^{r_2}(A_{dp}(\tau_k, i) + B_{dp}(\tau_k, i)F_{d,s}(\tau_k, i))Y(\tau_k, i) \\ &\quad \cdot (A_{dp}(\tau_k, i) + B_{dp}(\tau_k, i)F_{d,s}(\tau_k, i))^{\top}, \quad k \in \mathbb{Z}_+,\end{aligned} \tag{5.129b}$$

$i \in \mathscr{N}$, having unknown function $\mathbf{Y}(t) = (Y(t, 1), Y(t, 2), \ldots, Y(t, N))$. In this particular case, the gain matrices $F_{c,s}(t, i)$, $F_{d,s}(\tau_k, i)$ introduced by (5.127) and (5.128), take the form:

$$\begin{aligned}F_{c,s}(t, i) &= -\left(\sum_{l=1}^{r_1} B_l^{\top}(t, i)X_s(t, i)B_l(t, i) + R_c(t, i)\right)^{-1} \\ &\quad \cdot \left(B_0^{\top}(t, i)X_s(t, i) + \sum_{l=1}^{r_1} B_l^{\top}(t, i)X_s(t, i)A_l(t, i) + L_c^{\top}(t, i)\right)\end{aligned} \tag{5.130a}$$

$$\begin{aligned}F_{d,s}(\tau_k, i) &= -\left(\sum_{p=0}^{r_2} B_{dp}^{\top}(\tau_k, i)X_s(\tau_k, i)B_{dp}(\tau_k, i) + R_d(\tau_k, i)\right)^{-1} \\ &\quad \cdot \left(\sum_{p=0}^{r_2} B_{dp}^{\top}(\tau_k, i)X_s(\tau_k, i)A_{dp}(\tau_k, i) + L_d^{\top}(\tau_k, i)\right), \quad (\tau_k, i) \in \mathfrak{T} \times \mathscr{N}.\end{aligned} \tag{5.130b}$$

Specializing Definition 5.5 to the case of a SJMRDE (5.10) we may infer that a solution $t \to \mathbf{X}_s(t) = (X_s(t, 1), X_s(t, 2), \dots, X_s(t, N))$ of (5.10) is a stabilizing solution if the corresponding jump matrix linear differential equation (5.129) is exponentially stable. Further, (5.129) is a generalized Lyapunov-type jump linear differential equation (GLJLDE) of type (2.117) associated to a jump stochastic linear differential equation (JSLDE) of the form

$$\begin{aligned} dx(t) &= -(A_0(t, \eta(t)) + B_0(t, \eta(t))F_{c,s}(t, \eta(t)))x(t)dt \\ &\quad + \sum_{l=1}^{r_1} (A_l(t, \eta(t)) + B_l(t, \eta(t))F_{c,s}(t, \eta(t)))x(t)dw_l(t), \\ &\qquad \tau_k < t \le \tau_{k+1} \end{aligned} \tag{5.131a}$$

$$\begin{aligned} x(\tau_k^+) &= (A_{d0}(\tau_k, \eta(\tau_k)) + B_{d0}(\tau_k, \eta(\tau_k))F_{d,s}(\tau_k, \eta(\tau_k)))x(\tau_k) \\ &\quad + \sum_{p=1}^{r_2} w_{dp}(k)(A_{dp}(\tau_k, \eta(\tau_k)) + B_{dp}(\tau_k, \eta(\tau_k))F_{d,s}(\tau_k, \eta(\tau_k)))x(\tau_k), \\ &\qquad k \in \mathbb{Z}_+, \end{aligned} \tag{5.131b}$$

where $\{w_l(t)\}_{t \ge 0}$, $1 \le l \le r_1$, $\{w_{dp}(k)\}_{k \in \mathbb{Z}_+}$, $1 \le p \le r_2$, $\{\eta(t)\}_{t \ge 0}$ are stochastic processes which are satisfying the Assumption 1.2.

Based on the Definition 3.1 (*a*) applied in the case of the JSLDE (5.131) we deduce that $\mathbf{X}_s(\cdot)$ is a stabilizing solution of the SJMRDE (5.10) in the sense of the Definition 5.5 if the JSLDE (5.131) is strong exponentially stable in mean square (SESMS). The JSLDE (5.131) is the closed-loop system obtained by coupling the controls in a state feedback form

$$c_c(t) = F_{c,s}(t, \eta(t))x(t) \tag{5.132a}$$

$$u_d(k) = F_{d,s}(\tau_k, \eta(\tau_k))x(\tau_k) \tag{5.132b}$$

to the controlled system (4.1).

In the developments from the next chapters in which the stochastic jump matrix Riccati differential equations (SJMRDEs) of the form (5.10) are involved, we shall use the following characterizations, equivalent to that from Definition 5.5 of the stabilizing solution.

Definition 5.6 A solution $t \to \mathbf{X}_s(t) : \mathbb{R}_+ \to \mathcal{S}_n^N$ of the SJMRDE (5.10) is stabilizing solution if the corresponding stochastic jump linear differential equation (5.131) is strong exponentially stable in mean square (SESMS), where the gain matrices $F_{c,s}(t, i)$, $F_{d,s}(\tau_k, i)$ are computed as in (5.130) based on the solution $\mathbf{X}_s(\cdot)$ under consideration.

Note 5.1 According to Theorem 3.1 in the case where the JSLDE (5.131) has periodic coefficients, the property of SESMS from the Definition 5.6 may be replaced by exponentially stable in mean square (ESMS) of the closed-loop system (5.131).

In the derivation of the main result from this section, we prefer the characterization given in Definition 5.5 for the concept of stabilizing solution due to the procedural advantages.

The next result establishes the solution between the maximal solution of a GBJMRDE of the form (5.2) and its stabilizing solution.

Proposition 5.2 *Under the Assumption 5.1, if the set $\Gamma^{\Sigma}[\mathbb{R}_+]$ is nonempty then the stabilizing and bounded on $\mathbb{R}_+$ solution of (5.2) coincides to the maximal and bounded on $\mathbb{R}_+$ solution of this GBJMRDE.*

Proof Let $\mathbf{Z}(\cdot) \in \Gamma^{\Sigma}[\mathbb{R}_+]$ be arbitrary. As we shown in the proof of Theorem 5.3, $\mathbf{Z}(\cdot)$ solves a GBJMRDE of the form (5.73) for which (5.74) holds. Applying Lemma 5.1 together with (5.28) taking $\mathbf{W}c(t) = \mathbf{F}_{c,s}(t)$ and $\mathbf{W}_d(\tau_k) = \mathbf{F}_{d,s}(\tau_k)$, respectively, we rewrite (5.73) as:

$$
\begin{aligned}
-\dot{\mathbf{Z}}(t) &= \mathscr{L}^*_{c\mathbf{F}_{c,s}}(t)[\mathbf{Z}(t)] + \boldsymbol{\Theta}_{c\mathbf{F}_{c,s}}(t) - (\mathbf{F}_{c,s}(t) - \mathbf{F}_c^{\mathbf{Z}}(t))^\top && (5.133a)\\
&\quad \cdot ((\boldsymbol{\Pi})c3(t)[\mathbf{Z}(t)] + \mathbf{R}_c(t))(\mathbf{F}_{c,s}(t) - \mathbf{F}^{\mathbf{Z}}(t)) - \widehat{\mathbf{M}}_c(t), \quad \tau_k \le t < \tau_{k+1} \\
\mathbf{Z}(\tau_k^-) &= \boldsymbol{\Pi}_{d\mathbf{F}_{d,s}}(\tau_k)[\mathbf{Z}(\tau_k)] + \boldsymbol{\Theta}_{d\mathbf{F}_{d,s}}(\tau_k) - (\mathbf{F}_{d,s}(\tau_k) - \mathbf{F}_d^{\mathbf{Z}}(\tau_k))^\top && (5.133b)\\
&\quad \cdot (\boldsymbol{\Pi}_{d3}(\tau_k)[\mathbf{Z}(\tau_k)] + \mathbf{R}_d(\tau_k))(\mathbf{F}_{d,s}(\tau_k) - \mathbf{F}_d^{\mathbf{Z}}(\tau_k)) - \widehat{\mathbf{M}}_d(\tau_k), \quad k \in \mathbb{Z}_+.
\end{aligned}
$$

On the other hand, Lemma 5.1 together with (5.28) applied for the equation (5.2) satisfied by its stabilizing solution $\mathbf{X}_s(\cdot)$ if any, give

$$
\begin{aligned}
-\dot{\mathbf{X}}_s(t) &= \mathscr{L}^*_{c\mathbf{F}_{c,s}}(t)[\mathbf{X}_s(t)] + \boldsymbol{\Theta}_{c\mathbf{F}_{c,s}}(t), \quad \tau_k \le t < \tau_{k+1} && (5.134a)\\
\mathbf{X}_s(\tau_k^-) &= \boldsymbol{\Pi}_{d\mathbf{F}_{d,s}}(\tau_k)[\mathbf{X}_s(\tau_k)] + \boldsymbol{\Theta}_{d\mathbf{F}_{d,s}}(\tau_k), \quad k \in \mathbb{Z}_+. && (5.134b)
\end{aligned}
$$

Subtracting (5.133) from (5.134) we obtain that $\mathbf{Y}(\cdot) := \mathbf{X}_s(\cdot) - \mathbf{Z}(\cdot)$ is a bounded on $\mathbb{R}_+$ solution of the following non-homogeneous backward JMLDE

$$
\begin{aligned}
-\dot{\mathbf{Y}}(t) &= \mathscr{L}^*_{c\mathbf{F}_{c,s}}(t)[\mathbf{Y}(t)] + \mathbf{H}_c(t), \quad \tau_k \le t < \tau_{k+1} && (5.135a)\\
\mathbf{Y}(\tau_k^-) &= \boldsymbol{\Pi}_{d\mathbf{F}_{d,s}}(\tau_k)[\mathbf{Y}(\tau_k)] + \mathbf{H}_d(\tau_k), \quad k \in \mathbb{Z}_+ && (5.135b)
\end{aligned}
$$

where $\mathbf{H}_c(t) := \widehat{M}_c(t) + (\mathbf{F}_{c,s}(t) - \mathbf{F}_c^{\mathbf{Z}}(t))^\top(\boldsymbol{\Pi}_{c3}(t)[\mathbf{Z}(t)] + \mathbf{R}_c(t))(\mathbf{F}_{c,s}(t) - \mathbf{F}_c^{\mathbf{Z}}(t))$ and $\mathbf{H}_d(\tau_k) := \widehat{M}_d(\tau_k) + (\mathbf{F}_{d,s}(\tau_k) - \mathbf{F}_d^{\mathbf{Z}}(\tau_k))^\top(\boldsymbol{\Pi}_{d3}(\tau_k)[\mathbf{Z}(\tau_k)] + \mathbf{R}_d(\tau_k)) \cdot (\mathbf{F}_{d,s}(\tau_k) - \mathbf{F}_d^{\mathbf{Z}}(\tau_k))$. Employing (5.16c), (5.16d) and (5.74) we deduce that there exist positive constants ν_c, ν_d such that

$$
0 \preceq \mathbf{H}_c(t) \preceq \nu_c \mathbf{I}_n, \quad \text{for all } t \in \mathbb{R}_+
$$

$$
0 \preceq \mathbf{H}_d(\tau_k) \preceq \nu_d \mathbf{I}_n, \quad \text{for all } \tau_k \in \mathfrak{T}.
$$

Based on Proposition 5.1 (iii) we may conclude that (5.135) has a unique bounded on $\mathbb{R}_+$ solution and that solution lies in $\mathscr{S}_n^{N+}$. Thus we obtained that

$$\mathbf{X}_s(t) \succeq \mathbf{Z}(t), \quad \text{for all} \quad t \in \mathbb{R}_+ \tag{5.136}$$

and for arbitrary $\mathbf{Z}(\cdot) \in \Gamma^{\Sigma}[\mathbb{R}_+]$. Particularly, $\mathbf{X}_s(t) \succeq \mathbf{X}_{max}(t), \quad \text{for all} \quad t \in \mathbb{R}_+$. On the other hand (5.16c), (5.16d) and (5.136) give

$$\Pi_{c3}(t)[\mathbf{X}_s(t)](i) + R_c(t,i) \geq \gamma_c I_{m_c}, \quad \text{for all} \quad (t,i) \in \mathbb{R}_+ \times \mathcal{N} \tag{5.137a}$$

$$\Pi_{d3}(\tau_k)[\mathbf{X}_s(\tau_k)](i) + R_d(\tau_k,i) \geq \gamma_d I_{m_d}, \quad \text{for all} (\tau_k,i) \in \mathfrak{T} \times \mathcal{N}, \tag{5.137b}$$

where γ_c, γ_d are positive constants. These allow us to conclude via Remark 5.3, that $\mathbf{X}_s(\cdot) \in \Gamma^{\Sigma}[\mathbb{R}_+]$. Hence, $\mathbf{X}_{max}(t) \succeq \mathbf{X}_s(t), \quad \text{for all} \quad t \in \mathbb{R}_+$ which shows that $\mathbf{X}_s(t) = \mathbf{X}_{max}(t), \quad \text{for all} \quad t \in \mathbb{R}_+$. This complete the proof.

Corollary 5.3 *Under the Assumption 5.1, if the set $\Gamma^{\Sigma}[\mathbb{R}_+]$ is not empty, the GBJMRDE (5.2) may have at most one bounded on $\mathbb{R}_+$ and stabilizing solution.*

Proof The proof results from Proposition 5.2 together with the unicity of the maximal and bounded on $\mathbb{R}_+$ solution.

Theorem 5.5 *Under the Assumption 5.1, if the set $\Gamma^{\Sigma}[\mathbb{R}_+]$ is not empty, the following hold:*

(i) if the GBJMRDE (5.2) is in the periodic case in the sense of the Definition 5.2 (i), then its bounded on $\mathbb{R}_+$ and stabilizing solution, if any, is a periodic function of period $\tilde{\theta}$, where $\tilde{\theta}$ is introduced via (5.14a);
(ii) if the GBJMRDE (5.2) is time invariant in the sense of Definition 5.2 (ii), then its bounded on $\mathbb{R}_+$ and stabilizing solution, if any, is a periodic function of period h.

Proof If (5.2), or equivalently (5.13), is time periodic, in the sense of the Definition 5.2 (i), we have

$$\mathscr{R}_c(t+\tilde{\theta}, \mathbf{X}) = \mathscr{R}_c(t, \mathbf{X}), \quad \text{for all} \quad (t, \mathbf{X}) \in \; Dom\ \mathscr{R}_c$$

and

$$\mathscr{R}_d(kh+\tilde{\mathbf{k}}h, \mathbf{X}) = \mathscr{R}_d(kh, \mathbf{X}), \quad \text{for all} \quad (kh, \mathbf{X}) \in \; Dom\ \mathscr{R}_d,$$

$\tilde{\mathbf{k}}$ being introduced in (5.14b). If $\mathbf{X}_s(\cdot) = (X_s(\cdot,1), X_s(\cdot,2), \ldots, X_s(\cdot,N))$ is its bounded and stabilizing solution we set

$$\widetilde{X}(t,i) := X_s(t+\tilde{\theta}, i) \quad \text{for all} \quad (t,i) \in \mathbb{R}_+ \times \mathcal{N}.$$

Employing the periodicity properties of the generalized Riccati-type operators $\mathscr{R}_c$, $\mathscr{R}_d$ (from above) one may check that $t \to \widetilde{\mathbf{X}}(t) = (\widetilde{X}(t,1), \widetilde{X}(t,2), \ldots, \widetilde{X}(t,N)) : \mathbb{R}_+ \to \mathscr{S}_n^N$ is a solution of (5.13), hence for (5.2) too.
Let us show that $\widetilde{\mathbf{X}}(\cdot)$ is also a stabilizing solution of (5.2). To this end, we have to show that the generalized Lyapunov-type jump matrix linear differential equation on $\mathscr{S}_n^N$:

$$\dot{\mathbf{Y}}(t) = \mathscr{L}_{c\widetilde{\mathbf{F}}_c}(t)[\mathbf{Y}(t)], \quad kh < t \le (k+1)h \tag{5.138a}$$

$$\mathbf{Y}(kh^+) = \boldsymbol{\Pi}^*_{d\widetilde{\mathbf{F}}_d}(kh)[\mathbf{Y}(kh)], \quad k \in \mathbb{Z}_+, \tag{5.138b}$$

is exponentially stable, where $\widetilde{\mathbf{F}}_c(t) = (F_c(t, 1), F_c(t, 2), \ldots, F_c(t, N))$ is computed as in (5.23) with $\mathbf{X}$ replaced by $\widetilde{\mathbf{X}}(t)$ and $\widetilde{\mathbf{F}}_d(kh) = (F_d(kh, 1), F_d(kh, 2), \ldots, F_d(kh, N))$ are computed as in (5.25) taking $\mathbf{X}$ replaced by $\widetilde{\mathbf{X}}(kh)$ and τ_k replaced by kh, $k \in \mathbb{Z}_+$. Let $\widetilde{\mathbf{Y}}(t; t_0, \mathbf{H}_0)$, $t \ge t_0 \ge 0$ be the solution of (5.138) satisfying the initial condition

$$\widetilde{\mathbf{Y}}(t_0; t_0, \mathbf{H}_0) = \mathbf{H}_0,$$

with $\mathbf{H}_0 \in \mathscr{S}_n^N$. Let $\widetilde{\mathbf{T}}(t, t_0)$ be the linear evolution operator on $\mathscr{S}_n^N$ defined by the linear differential equation (5.138). This means that $\widetilde{\mathbf{T}}(t, t_0)\mathbf{H}_0 \stackrel{\Delta}{=} \widetilde{\mathbf{Y}}(t; t_0, \mathbf{H}_0)$, for all $t \ge t_0 \ge 0$, $\mathbf{H}_0 \in \mathscr{S}_n^N$. From the periodicity property of the coefficients of (5.2) and from the definition of $\widetilde{\mathbf{X}}(\cdot)$ one obtains the following equalities:

$$\mathscr{L}_{c\widetilde{\mathbf{F}}_c}(t)[\mathbf{Y}] = \mathscr{L}_{c\mathbf{F}_{c,s}}(t + \tilde{\theta})[\mathbf{Y}], \quad \text{for all } (t, \mathbf{Y}) \in \mathbb{R}_+ \times \mathscr{S}_n^N$$

$$\boldsymbol{\Pi}_{d\widetilde{\mathbf{F}}_d}(kh)[\mathbf{Y}] = \boldsymbol{\Pi}_{d\mathbf{F}_{d,s}}(kh + \tilde{\mathbf{k}}h)[\mathbf{Y}], \quad \text{for all } (k, \mathbf{Y}) \in \mathbb{Z}_+ \times \mathscr{S}_n^N,$$

where $\mathbf{F}_{c,s}(t)$, $\mathbf{F}_{d,s}(kh)$ are associated to the stabilizing solution $\mathbf{X}_s(\cdot)$ via (5.127) and (5.128), with $\tau_k = kh$, respectively. These allows us to rewrite (5.138) satisfied by the solution $\widetilde{\mathbf{Y}}(\cdot; t_0, \mathbf{H}_0)$ as:

$$\begin{aligned}
&\dot{\widetilde{\mathbf{Y}}}(t; t_0, \mathbf{H}_0) = \mathscr{L}_{c\mathbf{F}_{c,s}}(t + \tilde{\theta})[\widetilde{\mathbf{Y}}(t; t_0, \mathbf{H}_0)], \quad kh < t \le (k+1)h \\
&\widetilde{\mathbf{Y}}(kh^+; t_0, \mathbf{H}_0) = \boldsymbol{\Pi}^*_{d\mathbf{F}_{d,s}}(kh + \tilde{\mathbf{k}}h)[\widetilde{\mathbf{Y}}(kh; t_0, \mathbf{H}_0)], \quad k \in \mathbb{Z}_+, \\
&\widetilde{\mathbf{Y}}(t_0; t_0, \mathbf{H}_0) = \mathbf{H}_0.
\end{aligned}$$

From the unicity of the solution of the given initial values problem we may infer that

$$\widetilde{\mathbf{Y}}(t; t_0, \mathbf{H}_0) = \mathbf{S}(t + \tilde{\theta}; t_0 + \tilde{\theta}, \mathbf{H}_0), \tag{5.139}$$

for all $t \ge t_0 \ge 0$, $\mathbf{H}_0 \in \mathscr{S}_n^N$, where $t \to \mathbf{S}(t; \tau, \mathbf{H}_0)$ is the solution of the following problem with given initial values:

$$\dot{\mathbf{S}}(t) = \mathscr{L}_{c\widetilde{\mathbf{F}}_{c,s}}(t)[\mathbf{S}(t)], \quad kh < t \le (k+1)h \tag{5.140a}$$

$$\mathbf{S}(kh^+) = \boldsymbol{\Pi}^*_{d\widetilde{\mathbf{F}}_{d,s}}(kh)[\mathbf{S}(kh)], \quad k \in \mathbb{Z}_+, \tag{5.140b}$$

$$\mathbf{S}(\tau) = \mathbf{H}_0. \tag{5.140c}$$

Let $\mathbb{T}(t; t_0)$, $t \ge t_0 \ge 0$, be the linear evolution operator generated by the generalized Lyapunov-type jump matrix linear differential equation (5.140).
Since (5.140) is associated to the stabilizing solution $\mathbf{X}_s(\cdot)$ we deduce that there exist $\beta \ge 1$, $\alpha > 0$, such that

$$\|\mathbb{T}(t;t_0)\| \le \beta e^{-\alpha(t-t_0)}, \quad \text{for all } t \ge t_0 \ge 0.$$

From (5.139) we obtain: $\mathbf{T}(t,t_0)\mathbf{H}_0 = \mathbb{T}(t+\tilde{\theta}; t_0+\tilde{\theta})\mathbf{H}_0$, for all $t \ge t_0 \ge 0$, $\mathbf{H}_0 \in \mathcal{S}_n^N$. This allows us to deduce that

$$\|\widetilde{\mathbf{Y}}(t;t_0,\mathbf{H}_0)\| = \|\mathbb{T}(t+\tilde{\theta}; t_0+\tilde{\theta})\mathbf{H}_0\| \le \beta e^{-\alpha(t-t_0)}\|\mathbf{H}_0\|,$$

for all $t \ge t_0 \ge 0$, $\mathbf{H}_0 \in \mathcal{S}_n^N$. This confirms the exponential stability property of (5.138). So, we may conclude that $\widetilde{\mathbf{X}}(\cdot)$ is a stabilizing and bounded on $\mathbb{R}_+$ solution of (5.2), because (5.138) is just the generalized jump matrix linear differential equation of type (5.126) associated to the solution $\widetilde{\mathbf{X}}(\cdot)$. On the other hand, the GBJMRDE (5.2) may has at most one bounded and stabilizing solution because the set $\Gamma^{\Sigma}[\mathbb{R}_+]$ is not empty. Hence $\mathbf{X}_s(t) = \widetilde{\mathbf{X}}(t)$, for all $t \in \mathbb{R}_+$ or equivalently

$$\mathbf{X}_s(t) = \mathbf{X}_s(t+\tilde{\theta}), \quad \text{for all } t \in \mathbb{R}_+$$

which confirms that the bounded and stabilizing solution of (5.2) is a periodic function of period $\tilde{\theta}$. This ends the proof of part (i).

The validity of the statement (ii) is obtained from (i) because, if the GBJMRDE (5.2) is time invariant, it can be viewed time periodic if we take $\theta_c = h$ and $\theta_d = 1$. In this case $\tilde{\theta} = h$. Thus the proof is complete.

Remark 5.6 As in the case of the maximal and bounded solution $\mathbf{X}_{max}(\cdot)$, if the GBJMRDE (5.2) is time periodic or time invariant, we may consider that its bounded and stabilizing solution, if any, is defined on the whole real axis.

The next result provides a set of necessary and sufficient conditions for the existence of the bounded on $\mathbb{R}_+$ and stabilizing solution of the GBJMRDE (5.2) which is satisfying some sign conditions.

Theorem 5.6 *Under the Assumption 5.1 the following are equivalent:*

(i) the quadruple $(\mathbf{A}, \mathbf{B}, (\boldsymbol{\Pi}_c \mid \boldsymbol{\Pi}_d), \mathfrak{T})$ is stabilizable and the set $\widetilde{\Gamma}^{\Sigma}[\mathbb{R}_+]$ is not empty;

(ii) the GBJMRDE (5.2) has a bounded and stabilizing solution $t \to \mathbf{X}_s(t)$: $\mathbb{R}+ \to \mathcal{S}_n^N$ satisfying the sign conditions (5.137).

Proof If (i) holds, then according to the implication $(ii) \Rightarrow (i)$ from Theorem 5.3 we obtain that the GBJMRDE (5.2) has a maximal and bounded solution $t \to \widetilde{\mathbf{X}}(t)$: $\mathbb{R}_+ \to \mathcal{S}_n^N$ which satisfies the sign condition (5.68). We show that if the set $\widetilde{\Gamma}^{\Sigma}[\mathbb{R}_+]$ is not empty the maximal and bounded solution $\widetilde{\mathbf{X}}(\cdot)$ is a stabilizing solution too.

Let $\widetilde{\mathbf{F}}_c(t) = (\widetilde{F}_c(t,1), \widetilde{F}_c(t,2), \ldots, \widetilde{F}_c(t,N))$ and $\widetilde{\mathbf{F}}_d(\tau_k) = (\widetilde{F}_d(\tau_k,1), \widetilde{F}_d(\tau_k, 2), \ldots, \widetilde{F}_d(\tau_k,N))$ be the feedback gains associated to the solution $\widetilde{\mathbf{X}}(\cdot)$. This means that for each $i \in \mathcal{N}$, $\widetilde{F}_c(t,i)$ are computed as in (5.23) for $\mathbf{X}$ replaced by $\widetilde{\mathbf{X}}(t)$, $t \in \mathbb{R}_+$ and $\widetilde{F}_d(\tau_k,i)$ are computed as in (5.25) for $\mathbf{X}$ replaced by $\widetilde{\mathbf{X}}(\tau_k)$, $\tau_k \in \mathfrak{T}$.

Applying Lemma 5.1 together with (5.28) in the case of the equation (5.13) written for its solution $\widetilde{\mathbf{X}}(\cdot)$ we obtain:

$$-\dot{\widetilde{\mathbf{X}}}(t) = \mathscr{L}^*_{c\widetilde{\mathbf{F}}_c}(t)[\widetilde{\mathbf{X}}(t)] + \boldsymbol{\Theta}_{c\widetilde{\mathbf{F}}_c}(t), \quad \tau_k \le t < \tau_{k+1}, \tag{5.141a}$$

$$\widetilde{\mathbf{X}}(\tau_k^-) = \boldsymbol{\Pi}_{d\widetilde{\mathbf{F}}_d}(\tau_k)[\widetilde{\mathbf{X}}(\tau_k)] + \boldsymbol{\Theta}_{d\widetilde{\mathbf{F}}_d}(\tau_k), \quad k \in \mathbb{Z}_+. \tag{5.141b}$$

Let $\mathbf{Z}(\cdot) \in \widetilde{\Gamma}^{\Sigma}[\mathbb{R}_+]$ be arbitrary. Applying Lemma 5.5 (i) in the case of (5.16a) and (5.16c), and Lemma 5.5 (ii) in the case of (5.17) satisfied by $\mathbf{Z}(\cdot)$, we obtain that this function solves a jump differential equation of the form

$$-\dot{\mathbf{Z}}(t) = \mathscr{R}_c(t, \mathbf{Z}(t)) - \widehat{\mathbf{M}}_c(t), \quad \tau_k \le t < \tau_{k+1} \tag{5.142a}$$

$$\mathbf{Z}(\tau_k^-) = \mathscr{R}_d(\tau_k, \mathbf{Z}(\tau_k)) - \widehat{\mathbf{M}}_d(\tau_k), \quad k \in \mathbb{Z}_+, \tag{5.142b}$$

where $\widehat{\mathbf{M}}_c(t) = (\widehat{M}_c(t,1), \widehat{M}_c(t,2), \ldots, \widehat{M}_c(t,N)) \in \mathscr{S}_n^N$ and $\widehat{\mathbf{M}}_d(\tau_k) = (\widehat{M}_d(\tau_k,1), \widehat{M}_d(\tau_k,2), \ldots, \widehat{M}_d(\tau_k,N)) \in \mathscr{S}_n^N$ are such that

$$\widehat{M}_c(t,i) \ge 0, \quad \text{for all } (t,i) \in \mathbb{R}_+ \times \mathcal{N} \tag{5.143a}$$

$$\widehat{M}_d(\tau_k,i) \ge \widehat{\gamma} I_n > 0, \quad \text{for all } (\tau_k,i) \in \mathfrak{T} \times \mathcal{N}. \tag{5.143b}$$

Employing Lemma 5.1 together with (5.28) we rewrite (5.142) as:

$$\begin{aligned} -\dot{\mathbf{Z}}(t) = {} & \mathscr{L}^*_{c\widetilde{\mathbf{F}}_c}(t)[\mathbf{Z}(t)] + \boldsymbol{\Theta}_{c\widetilde{\mathbf{F}}_c}(t) - \widehat{\mathbf{M}}_c(t) - (\widetilde{\mathbf{F}}_c(t) - \mathbf{F}_c^{\mathbf{Z}}(t))^\top \\ & \cdot (\boldsymbol{\Pi}_{c3}(t)[\mathbf{Z}(t)] + \mathbf{R}_c(t))(\widetilde{\mathbf{F}}_c(t) - \mathbf{F}_c^{\mathbf{Z}}(t)), \quad \tau_k \le t < \tau_{k+1}, \end{aligned} \tag{5.144a}$$

$$\begin{aligned} \mathbf{Z}(\tau_k^-) = {} & \boldsymbol{\Pi}_{d\widetilde{\mathbf{F}}_d}(\tau_k)[\mathbf{Z}(\tau_k)] + \boldsymbol{\Theta}_{d\widetilde{\mathbf{F}}_d}(\tau_k) - \widehat{\mathbf{M}}_d(\tau_k) - (\widetilde{\mathbf{F}}_d(\tau_k) - \mathbf{F}_d^{\mathbf{Z}}(\tau_k))^\top \\ & \cdot (\boldsymbol{\Pi}_{d3}(\tau_k)[\mathbf{Z}(\tau_k)] + \mathbf{R}_d(\tau_k))(\widetilde{\mathbf{F}}_d(\tau_k) - \mathbf{F}_d^{\mathbf{Z}}(\tau_k)), \quad k \in \mathbb{Z}_+. \end{aligned} \tag{5.144b}$$

Subtracting (5.144) from (5.141) and bearing in mind that $\widetilde{\mathbf{X}}(t) \succeq \mathbf{Z}(t)$, for all $t \in \mathbb{R}_+$ we obtain that $\mathbf{Y}(\cdot) := \widetilde{\mathbf{X}}(\cdot) - \mathbf{Z}(\cdot)$ is a bounded and non-negative solution of the following non-homogeneous backward jump linear differential equation BJMLDE on the ordered Hilbert space $(\mathscr{S}_n^N, \mathscr{S}_n^{N+})$:

$$-\dot{\mathbf{Y}}(t) = \mathscr{L}^*_{c\widetilde{\mathbf{F}}_c}(t)[\mathbf{Y}(t)] + \mathbf{H}_c(t), \quad \tau_k \le t < \tau_{k+1} \tag{5.145a}$$

$$\mathbf{Y}(\tau_k^-) = \boldsymbol{\Pi}_{d\widetilde{\mathbf{F}}_d}(\tau_k)[\mathbf{Y}(\tau_k)] + \mathbf{H}_d(\tau_k), \quad k \in \mathbb{Z}_+, \tag{5.145b}$$

where

$$\mathbf{H}_c(t) := \widehat{\mathbf{M}}_c(t) + (\widetilde{\mathbf{F}}_c(t) - \mathbf{F}_c^{\mathbf{Z}}(t))^\top (\boldsymbol{\Pi}_{c3}(t)[\mathbf{Z}(t)] + \mathbf{R}_c(t))(\widetilde{\mathbf{F}}_c(t) - \mathbf{F}_c^{\mathbf{Z}}(t)),$$

for all $t \in \mathbb{R}_+$ and

$$\begin{aligned} \mathbf{H}_d(\tau_k) := {} & \widehat{\mathbf{M}}_d(\tau_k) + (\widetilde{\mathbf{F}}_d(\tau_k) - \mathbf{F}_d^{\mathbf{Z}}(\tau_k))^\top (\boldsymbol{\Pi}_{d3}(\tau_k)[\mathbf{Z}(\tau_k)] + \mathbf{R}_d(\tau_k)) \\ & (\widetilde{\mathbf{F}}_d(\tau_k) - \mathbf{F}_d^{\mathbf{Z}}(\tau_k)), \end{aligned}$$

for all $k \in \mathbb{Z}_+$. Employing (5.16c), (5.16d) and (5.143) we may infer that

$$0 \preceq \mathbf{H}_c(t) \preceq \nu_c \mathbf{I}_n, \quad \text{for all } \ t \in \mathbb{R}_+$$

and

$$0 \prec \check{\gamma}\mathbf{I}_n \preceq \mathbf{H}_d(\tau_k) \preceq \nu_d \mathbf{I}_n, \quad \text{for all } \ k \in \mathbb{Z}_+.$$

The implication $(iii) \Rightarrow (i)$ from Theorem 2.7 applied in the case of BJLDE (5.145) allows us to conclude that the FJLDE on ordered space $(\mathcal{S}_n^N, \mathcal{S}_n^{N+})$:

$$\dot{\mathbf{Y}}(t) = \mathcal{L}_{c\widetilde{\mathbf{F}}_c}(t)[\mathbf{Y}(t)], \quad \tau_k \le t < \tau_{k+1} \tag{5.146a}$$

$$\mathbf{Y}(\tau_k^-) = \boldsymbol{\Pi}^*_{d\widetilde{\mathbf{F}}_d}(\tau_k)[\mathbf{Y}(\tau_k)], \quad k \in \mathbb{Z}_+, \tag{5.146b}$$

is exponentially stable. This means that $\widetilde{\mathbf{X}}(\cdot)$ is a stabilizing solution of the GBJM-RDE (5.2), because (5.146) coincides to (5.126) when $\mathbf{F}_{c,s}(t) \leftarrow \widetilde{\mathbf{F}}_c(t),\ t \in \mathbb{R}_+$ and $\mathbf{F}_{d,s}(\tau_k) \leftarrow \widetilde{\mathbf{F}}_d(\tau_k),\ k \in \mathbb{Z}_+$. Thus, the validity of the implication $(i) \Rightarrow (ii)$ is confirmed.

Now we prove the implication $(ii) \Rightarrow (i)$. Let $\mathbf{X}_s(\cdot)$ be the bounded on $\mathbb{R}_+$ and stabilizing solution of (5.2). This means that the corresponding generalized Lyapunov-type jump matrix linear differential equation (5.126) is exponentially stable.

On the other hand, (5.126) coincides to the jump matrix linear differential equation (5.69) when $\mathbf{F}_c(t) \leftarrow \widetilde{\mathbf{F}}_{c,s}(t),\ t \in \mathbb{R}_+$ and $\mathbf{F}_d(\tau_k) \leftarrow \widetilde{\mathbf{F}}_{d,s}(\tau_k),\ k \in \mathbb{Z}_+$. Hence, the quadruple $(\mathbf{A}, \mathbf{B}, (\boldsymbol{\Pi}_c \mid \boldsymbol{\Pi}_d), \mathfrak{T})$ is stabilizable. It remains to show that the set $\widetilde{\Gamma}^{\Sigma}[\mathbb{R}_+]$ is not empty if (ii) holds. To this end we rewrite (5.2) as:

$$-\dot{\mathbf{X}}(t) = \mathcal{L}^*_{c\mathbf{F}_{c,s}}(t)[\mathbf{X}(t)] + \boldsymbol{\Theta}_{c\mathbf{F}_{c,s}}(t) - \boldsymbol{\Upsilon}_{c\mathbf{F}_{c,s}}(t)[\mathbf{X}(t)], \quad \tau_k \le t < \tau_{k+1} \tag{5.147a}$$

$$\mathbf{X}(\tau_k^-) = \boldsymbol{\Pi}_{d\mathbf{F}_{d,s}}(\tau_k)[\mathbf{X}(\tau_k)] + \boldsymbol{\Theta}_{d\mathbf{F}_{d,s}}(\tau_k) - \boldsymbol{\Upsilon}_{d\mathbf{F}_{d,s}}(\tau_k)[\mathbf{X}(\tau_k)], \quad k \in \mathbb{Z}_+, \tag{5.147b}$$

where $\mathbf{F}_{c,s}(t),\ t \in \mathbb{R}_+$ and $\mathbf{F}_{d,s}(\tau_k),\ k \in \mathbb{Z}_+$ are associated to the stabilizing solution $\mathbf{X}_s(\cdot)$ via (5.127) and (5.128), respectively, while

$$\boldsymbol{\Upsilon}_{c\mathbf{F}_{c,s}}(t)[\mathbf{X}] = (\Upsilon_{c\mathbf{F}_{c,s}}(t)[\mathbf{X}](1), \Upsilon_{c\mathbf{F}_{c,s}}(t)[\mathbf{X}](2), \ldots, \Upsilon_{c\mathbf{F}_{c,s}}(t)[\mathbf{X}](N))$$

and

$$\boldsymbol{\Upsilon}_{d\mathbf{F}_{d,s}}(\tau_k)[\mathbf{X}] = (\Upsilon_{d\mathbf{F}_{d,s}}(\tau_k)[\mathbf{X}](1), \Upsilon_{d\mathbf{F}_{d,s}}(\tau_k)[\mathbf{X}](2), \ldots, \Upsilon_{d\mathbf{F}_{d,s}}(\tau_k)[\mathbf{X}](N))$$

with

$$\Upsilon_{c\mathbf{F}_{c,s}}(t)[\mathbf{X}](i) := (\mathcal{P}_{c\mathbf{F}_{c,s}}(t)[\mathbf{X}](i))^\top (\Pi_{c3}(t)[\mathbf{X}](i) + R_c(t,i))^{-1} \mathcal{P}_{c\mathbf{F}_{c,s}}(t)[\mathbf{X}](i) \tag{5.148a}$$

$$\begin{aligned}\mathcal{P}_{c\mathbf{F}_{c,s}}(t)[\mathbf{X}](i) := (&X(i)B(t,i) + \Pi_{c2}(t)[\mathbf{X}](i) + F_{c,s}^\top(t,i)\Pi_{c3}(t)[\mathbf{X}](i) \\ &+ L_c(t,i) + F_{c,s}^\top(t,i)R_c(t,i))^\top, \quad \text{for all } \ t \in \mathbb{R}_+\end{aligned} \tag{5.148b}$$

$$\begin{aligned}\Upsilon_{d\mathbf{F}_{d,s}}(\tau_k)[\mathbf{X}](i) &:= (\mathscr{P}_{d\mathbf{F}_{d,s}}(\tau_k)[\mathbf{X}](i))^\top(\Pi_{d3}(\tau_k)[\mathbf{X}](i)+R_d(\tau_k,i))^{-1}\\ &\quad\cdot \mathscr{P}_{d\mathbf{F}_{d,s}}(\tau_k)[\mathbf{X}](i) \qquad (5.149a)\\ \mathscr{P}_{d\mathbf{F}_{d,s}}(\tau_k)[\mathbf{X}](i) &:= (\Pi_{d2}(\tau_k)[\mathbf{X}](i)+F_{d,s}^\top(\tau_k,i)\Pi_{d3}(\tau_k)[\mathbf{X}](i)\\ &\quad + L_d(\tau_k,i)+F_{d,s}^\top(\tau_k,i)R_d(\tau_k,i))^\top, \quad \text{for all } \tau_k\in\mathfrak{T}, \qquad (5.149b)\end{aligned}$$

for any $\mathbf{X}=(X(1),X(2),\ldots,X(N))\in\mathscr{S}_n^N$, $i\in\mathscr{N}$.
From (5.127), (5.148) and (5.128), (5.149), respectively, one sees that

$$\mathscr{P}_{c\mathbf{F}_{c,s}}(t)[\mathbf{X}_s(t)](i)=0, \quad \text{for all } t\in\mathbb{R}_+ \qquad (5.150a)$$

$$\mathscr{P}_{d\mathbf{F}_{d,s}}(\tau_k)[\mathbf{X}_s(\tau_k)](i)=0, \quad \text{for all } k\in\mathbb{Z}_+, \qquad (5.150b)$$

$i\in\mathscr{N}$. We denote $\mathbb{F}_s\triangleq(\mathbf{F}_{c,s},\mathbf{F}_{d,s})$ the pair of the stabilizing feedback gains associated to the stabilizing solution $\mathbf{X}_s$ (via (5.127)–(5.128), respectively).

Let $\mathbf{T}_{\mathbb{F}_s}(t,t_0)$, $t\ge t_0\ge 0$ be the linear evolution operator on $\mathscr{S}_n^N$ defined by the forward jump linear differential equation (5.126). According to the Remark 2.2 (c) we may infer that $\mathbf{X}_c(\cdot)$ is the stabilizing solution of the GBJMRDE (5.2) if and only if

$$\|\mathbf{T}_{\mathbb{F}_s}(t,t_0)\|\le\beta e^{-\alpha(t-t_0)},$$

for all $t\ge t_0\ge 0$, where $\beta\ge 1$, $\alpha>0$, are not depending upon t and t_0.

Employing a representation formula as (2.87) in the case of forward jump linear differential equation (5.147) written for $\mathbf{X}(\cdot)\leftarrow\mathbf{X}_s(\cdot)$, we obtain via (5.150):

$$\mathbf{X}_s(t)=\sum_{l=k}^{\infty}\mathbf{T}_{\mathbb{F}_s}^*(\tau_l^-,t)[\boldsymbol{\Theta}_{d\mathbf{F}_{d,s}}(\tau_l)]+\int_t^{\infty}\mathbf{T}_{\mathbb{F}_s}(\sigma,t)[\boldsymbol{\Theta}_{c\mathbf{F}_{c,s}}(\sigma)]d\sigma, \qquad (5.151)$$

if $\tau_{k-1}<t\le\tau_k$, $k\in\mathbb{Z}_+$, $k\ge 1$.

Let $\mathscr{PC}_b(\mathbb{R}_+,\mathscr{S}_n^N;\mathfrak{T})$ be the Banach space of bounded functions $\mathbf{Z}(\cdot):\mathbb{R}_+\to\mathscr{S}_n^N$ continuous on each interval $[\tau_k,\tau_{k+1})$ with finite left limits in each τ_k, $k\in\mathbb{Z}$. It is obvious that $\mathbf{X}_s(\cdot)$ lies in $\mathscr{PC}_b(\mathbb{R}_+,\mathscr{S}_n^N;\mathfrak{T})$ and satisfies

$$\boldsymbol{\Pi}_{c3}(t)[\mathbf{X}_s(t)]+\mathbf{R}_c(t)\gg 0, \quad t\in\mathbb{R}_+$$

and

$$\boldsymbol{\Pi}_{d3}(\tau_k)[\mathbf{X}_s(\tau_k)]+\mathbf{R}_d(\tau_k)\gg 0, \quad \tau_k\in\mathfrak{T}.$$

Hence, there exists an open set $\mathscr{U}\subset\mathscr{PC}_b(\mathbb{R}_+,\mathscr{S}_n^N;\mathfrak{T})$ such that $\mathbf{X}_s(\cdot)\in\mathscr{U}$ and

$$\boldsymbol{\Pi}_{c3}(t)[\mathbf{Z}(t)]+\mathbf{R}_c(t)\gg 0, \quad t\in\mathbb{R}_+$$

$$\boldsymbol{\Pi}_{d3}(\tau_k)[\mathbf{Z}(\tau_k)]+\mathbf{R}_d(\tau_k)\gg 0, \quad \tau_k\in\mathfrak{T}$$

for any $\mathbf{Z}(\cdot) \in \mathscr{U}$. We obtain the function $\boldsymbol{\Psi} : \mathscr{U} \times \mathbb{R} \to \mathscr{PC}_b(\mathbb{R}_+, \mathscr{S}_n^N; \mathfrak{T})$ defined as

$$\boldsymbol{\Psi}(\mathbf{X}(\cdot), \delta)(t) \triangleq \sum_{l=k}^{\infty} \mathbf{T}^*_{\mathbb{F}_s}(\tau_l^-, t)[\boldsymbol{\Theta}_{d\mathbf{F}_{d,s}}(\tau_l) - \boldsymbol{\Upsilon}_{d\mathbf{F}_{d,s}}(\tau_l)[\mathbf{X}(\tau_l)] + \delta \mathbf{I}_n]$$
$$+ \int_t^{\infty} \mathbf{T}^*_{\mathbb{F}_s}(\sigma, t)[\boldsymbol{\Theta}_{c\mathbf{F}_{c,s}}(\sigma) - \boldsymbol{\Upsilon}_{c\mathbf{F}_{c,s}}(\sigma)[\mathbf{X}(\sigma)] + \delta \mathbf{I}_n] d\sigma - \mathbf{X}(t)$$
$$\tau_{k-1} < t \le \tau_k, \quad k \in \mathbb{Z}_+, \quad k \ge 1. \tag{5.152}$$

We shall apply the implicit function theorem in the case of the equation:

$$\boldsymbol{\Psi}(\mathbf{X}(\cdot), \delta) = 0 \tag{5.153}$$

to deduce the existence of a smooth function $\mathbf{X}_\delta(\cdot) \in \mathscr{U}$ which satisfies

$$\mathbf{X}_\delta(t) \triangleq \sum_{l=k}^{\infty} \mathbf{T}^*_{\mathbb{F}_s}(\tau_l^-, t)[\boldsymbol{\Theta}_{d\mathbf{F}_{d,s}}(\tau_l) - \boldsymbol{\Upsilon}_{d\mathbf{F}_{d,s}}(\tau_l)[\mathbf{X}_\delta(\tau_l)] + \delta \mathbf{I}_n]$$
$$+ \int_t^{\infty} \mathbf{T}^*_{\mathbb{F}_s}(\sigma, t)[\boldsymbol{\Theta}_{c\mathbf{F}_{c,s}}(\sigma) - \boldsymbol{\Upsilon}_{c\mathbf{F}_{c,s}}(\sigma)[\mathbf{X}_\delta(\sigma)] + \delta \mathbf{I}_n] d\sigma$$
$$\tau_{k-1} < t \le \tau_k, \quad k \in \mathbb{Z}_+, \quad k \ge 1, \tag{5.154}$$

for $|\delta|$ sufficiently small. Comparing (5.151) to (5.154) we may deduce that $(\mathbf{X}_s(\cdot), 0)$ is a solution of (5.153).

We show that $\mathscr{D}_1\boldsymbol{\Psi}(\mathbf{X}_s(\cdot), 0) : \mathscr{PC}_b(\mathbb{R}_+, \mathscr{S}_n^N; \mathfrak{T}) \to \mathscr{PC}_b(\mathbb{R}_+, \mathscr{S}_n^N; \mathfrak{T})$ is an isomorphism, $\mathscr{D}_1\boldsymbol{\Psi}(\cdot, \cdot)$ being the derivative of $\boldsymbol{\Psi}(\cdot, \cdot)$ with respect to its first argument. Since

$$\mathscr{D}_1\boldsymbol{\Psi}(\mathbf{X}_s(\cdot), 0)\boldsymbol{\Delta}(\cdot) = \lim_{\varepsilon \to 0} \frac{1}{\varepsilon}(\boldsymbol{\Psi}(\mathbf{X}_s(\cdot) + \varepsilon \boldsymbol{\Delta}(\cdot), 0) - \boldsymbol{\Psi}(\mathbf{X}_s(\cdot), 0))$$

we obtain via (5.148)–(5.150) that $\mathscr{D}_1\boldsymbol{\Psi}(\mathbf{X}_s(\cdot), 0)\boldsymbol{\Delta}(\cdot) = -\boldsymbol{\Delta}(\cdot)$, for all $\boldsymbol{\Delta}(\cdot) \in \mathscr{PC}_b(\mathbb{R}_+, \mathscr{S}_n^N; \mathfrak{T})$. Hence $\mathscr{D}_1\boldsymbol{\Psi}(\mathbf{X}_s(\cdot), 0) = -\mathbb{I}_{\mathscr{PC}_b}$, $\mathbb{I}_{\mathscr{PC}_b}$ being the identity operator on the Banach space $\mathscr{PC}_b(\mathbb{R}_+, \mathscr{S}_n^N; \mathfrak{T})$ which is an isomorphism. Also, one sees that $\mathscr{D}_1\boldsymbol{\Psi}(\mathbf{X}(\cdot), \delta)$ is continuous in $(\mathbf{X}(\cdot), \delta) = (\mathbf{X}_s(\cdot), 0)$. Applying the implicit functions theorem (see for example Schwartz [5]) we deduce the existence of a $\widetilde{\delta} > 0$ and a smooth function $\delta \to \mathbf{X}_\delta(\cdot) : (-\widetilde{\delta}, \widetilde{\delta}) \to \mathscr{U}$ satisfying

$$\boldsymbol{\Psi}(\mathbf{X}_\delta(\cdot), \delta) = 0, \quad \text{for all } \delta \in (-\widetilde{\delta}, \widetilde{\delta}).$$

This means that $\mathbf{X}_\delta(\cdot)$ satisfies (5.154) if $\delta \in (-\widetilde{\delta}, \widetilde{\delta})$. From (5.154) one sees that $\mathbf{X}_\delta(\cdot)$ is differentiable in any $t \neq \tau_k$, $k \in \mathbb{Z}_+$. This allows us to obtain from (5.154) that $\mathbf{X}_\delta(\cdot)$ satisfies

$$-\dot{\mathbf{X}}_\delta(t) = \mathscr{R}_c(t, \mathbf{X}_\delta(t)) + \delta \mathbf{I}_n, \quad \tau_k \leq t < \tau_{k+1} \tag{5.155a}$$

$$\mathbf{X}_\delta(\tau_k^-) = \mathscr{R}_d(\tau_k, \mathbf{X}_\delta(\tau_k)) + \delta \mathbf{I}_n, \quad k \in \mathbb{Z}_+. \tag{5.155b}$$

On the other hand from $\mathbf{X}_\delta(\cdot) \in \mathscr{U}$ we deduce that there exist positive constants γ_c, γ_d which may depend upon δ, such that

$$\boldsymbol{\Pi}_{c3}(t)[\mathbf{X}_\delta(t)] + \mathbf{R}_c(t) \succeq \gamma_c \mathbf{I}_n, \quad \text{for all } \ t \in \mathbb{R}_+ \tag{5.156a}$$

$$\boldsymbol{\Pi}_{d3}(\tau_k)[\mathbf{X}_\delta(\tau_k)] + \mathbf{R}_d(\tau_k) \succeq \gamma_d \mathbf{I}_n, \quad \text{for all } \ t \in \mathbb{Z}_+. \tag{5.156b}$$

Finally, (5.155)–(5.156) together with Lemma 5.5 allow us to conclude that $\mathbf{X}_\delta(\cdot) \in \widetilde{\Gamma}^{\Sigma}[\mathbb{R}_+]$ if $\delta \in (-\widetilde{\delta}, 0)$. Thus the proof ends.

The next definition is a particularization of the Definition 5.2 to the case of the quadruple $(\mathbf{A}, \mathbf{B}, (\boldsymbol{\Pi}_c \mid \boldsymbol{\Pi}_d), \mathfrak{T})$ in which $\mathbf{A}(\cdot)$, $\mathbf{B}(\cdot)$, $\boldsymbol{\Pi}_c(\cdot)$, $\{\boldsymbol{\Pi}_d(\tau_k)\}_{k \in \mathbb{Z}_+}$ are as in (5.15) and $\mathfrak{T} = \{\tau_k\}_{k \in \mathbb{Z}_+} \subset \mathbb{R}_+$ is a sequence satisfying conditions from 2.1.

Definition 5.7 We say that the quadruple $(\mathbf{A}, \mathbf{B}, (\boldsymbol{\Pi}_c \mid \boldsymbol{\Pi}_d), \mathfrak{T})$ is:

(i) time periodic, if its components $\mathbf{A}(\cdot)$, $\mathbf{B}(\cdot)$, $\boldsymbol{\Pi}_c(\cdot)$, $\boldsymbol{\Pi}_d(\cdot)$, $\mathfrak{T}$ are satisfying $(a) - (d)$ from Definition 5.2 (i).

(ii) time invariant if its components $\mathbf{A}(\cdot)$, $\mathbf{B}(\cdot)$, $\boldsymbol{\Pi}_c(\cdot)$, $\boldsymbol{\Pi}_d(\cdot)$, $\mathfrak{T}$ are satisfying $(a) - (c)$ from Definition 5.2 (ii).

The following special form of the GBJMRDE (5.2) will be involved in the proof of the next result:

$$\begin{aligned} -\dot{\mathbf{X}}(t) &= \mathbf{A}^\top(t)\mathbf{X}(t) + \mathbf{X}(t)\mathbf{A}(t) + \boldsymbol{\Pi}_{c1}[\mathbf{X}(t)] - (\mathbf{X}(t)\mathbf{B}(t) + \boldsymbol{\Pi}_{c2}(t)[\mathbf{X}(t)]) \\ &\quad (\mathbf{I}_{m_c} + \boldsymbol{\Pi}_{c3}(t)[\mathbf{X}(t)])^{-1}(\mathbf{X}(t)\mathbf{B}(t) + \boldsymbol{\Pi}_{c2}(t)[\mathbf{X}(t)])^\top, \quad kh \leq t < (k+1)h \end{aligned} \tag{5.157a}$$

$$\begin{aligned} \mathbf{X}(kh^-) &= \boldsymbol{\Pi}_{d1}(kh)[\mathbf{X}(kh)] - (\boldsymbol{\Pi}_{d2}(kh)[\mathbf{X}(kh)](\mathbf{I}_{m_d} + \boldsymbol{\Pi}_{d3}(kh)[\mathbf{X}(kh)])^{-1} \\ &\quad \cdot (\boldsymbol{\Pi}_{d2}(kh)[\mathbf{X}(kh)])^\top + \mathbf{I}_n, \quad k \in \mathbb{Z}_+, \end{aligned} \tag{5.157b}$$

where $h > 0$ is given. The GBJMRDE (5.157) is defined by the quintuple $\Sigma_0 = (\mathbf{A}, \mathbf{B}, (\boldsymbol{\Pi}_c \mid \boldsymbol{\Pi}_d), (\boldsymbol{\Theta}_{c0} \mid \boldsymbol{\Theta}_{d0}), \mathfrak{T}_0)$ where $\mathbf{A}(\cdot)$, $\mathbf{B}(\cdot)$, $\boldsymbol{\Pi}_c(\cdot)$ are those from (5.2), while $\boldsymbol{\Theta}_{c0}(t) = (\Theta_{c0}(t, 1), \Theta_{c0}(t, 2), \ldots, \Theta_{c0}(t, N))$ and $\boldsymbol{\Theta}_{d0}(kh) = (\Theta_{d0}(kh, 1), \Theta_{d0}(kh, 2),$
$\ldots, \Theta_{d0}(kh, N))$ with

$$\Theta_{c0}(t, i) = \begin{pmatrix} 0 & 0 \\ 0 & I_{m_c} \end{pmatrix}, \quad \text{for all } \ (t, i) \in \mathbb{R}_+ \times \mathscr{N}$$

and

$$\Theta_{d0}(kh, i) = \begin{pmatrix} I_n & 0 \\ 0 & I_{m_d} \end{pmatrix}, \quad \text{for all } (k, i) \in \mathbb{Z}_+ \times \mathcal{N}, \; \mathfrak{T}_0 = \{kh\}_{k \in \mathbb{Z}_+}.$$

The next result shows that without loss of generality in the case of a time-periodic quadruple, we may restrict the definition of stabilizability to pairs of time periodic stabilizing feedback gains.

Corollary 5.4 *Under the Assumption 5.1, the following equivalences hold:*

(i) if the quadruple $(\mathbf{A}, \mathbf{B}, (\boldsymbol{\Pi}_c \mid \boldsymbol{\Pi}_d), \mathfrak{T})$ *is time periodic then the following are equivalent:*

(a) the quadruple $(\mathbf{A}, \mathbf{B}, (\boldsymbol{\Pi}_c \mid \boldsymbol{\Pi}_d), \mathfrak{T})$ *is stabilizable (in the sense of Definition 5.4);*

(b) there exist pairs of the form $(\mathbf{F}_c(\cdot), \{\mathbf{F}_d(kh)\}_{k \in \mathbb{Z}_+})$ *where* $\mathbf{F}_c(\cdot) \in \mathcal{PC}(\mathbb{R}_+, \mathcal{M}^N_{m_c n}; \mathfrak{T})$ *is a periodic function of period* $\tilde{\theta}$ *and* $\{\mathbf{F}_d(kh)\}_{k \in \mathbb{Z}_+} \subset \mathcal{M}^N_{m_d n}$ *is a matrix valued sequence periodic of period* $\tilde{\mathbf{k}}$ *having the property that the corresponding generalized jump matrix linear differential equation (5.69) is exponentially stable (*$\tilde{\theta}$ *and* $\tilde{\mathbf{k}}$ *being introduced in (5.14));*

(ii) if the quadruple $(\mathbf{A}, \mathbf{B}, (\boldsymbol{\Pi}_c \mid \boldsymbol{\Pi}_d), \mathfrak{T})$ *is time invariant then the following are equivalent:*

(c) the quadruple $(\mathbf{A}, \mathbf{B}, (\boldsymbol{\Pi}_c \mid \boldsymbol{\Pi}_d), \mathfrak{T})$ *is stabilizable (in the sense of Definition 5.4);*

(d) there exist pairs of the form $(\mathbf{F}_c(\cdot), \mathbf{F}_d) \in \mathcal{PC}(\mathbb{R}_+, \mathcal{M}^N_{m_c n}; \mathfrak{T}) \times \mathcal{M}^N_{m_d n}$ *with* $\mathbf{F}_c(\cdot)$ *a periodic function of period* h *such that the corresponding generalized jump matrix linear differential equation of the form (5.69) be exponentially stable.*

Proof The implications $(b) \Rightarrow (a)$ and $(d) \Rightarrow (c)$ are obvious. To prove the implications $(a) \Rightarrow (b)$ and $(c) \Rightarrow (d)$ we first apply the implication $(i) \Rightarrow (ii)$ from Theorem 5.6 in the case of GBJMRDE (5.157) to obtain the existence of the stabilizing and bounded on $\mathbb{R}_+$ solution $\widetilde{\mathbf{X}}_s(\cdot)$ of this equation. To this end, we take into account that both in the time-periodic case as well as in the time-invariant case, the set $\widetilde{\Gamma}^{\Sigma_0}[\mathbb{R}_+]$ is not empty, because $\mathbf{Z}(t) = 0$, for all $t \in \mathbb{R}_+$ lies in $\widetilde{\Gamma}^{\Sigma_0}[\mathbb{R}_+]$. Further, Theorem 5.5 enables us to deduce that in the time-periodic case $\widetilde{\mathbf{X}}_s(\cdot)$ is a periodic function of period $\tilde{\theta}$ and in the time-invariant case $\widetilde{\mathbf{X}}_s(\cdot)$ is a periodic function of period h. The conclusion is obtained taking $\mathbf{F}_c(t) = \widetilde{\mathbf{F}}_{c,s}(t)$, $t \in \mathbb{R}_+$, $\mathbf{F}_d(kh) = \widetilde{\mathbf{F}}_{d,s}(kh)$, for all $k \in \mathbb{Z}_+$, $\widetilde{\mathbf{F}}_{c,s}(\cdot)$, $\widetilde{\mathbf{F}}_{d,s}(\cdot)$ being associated to the stabilizing solution $\widetilde{\mathbf{X}}_s(\cdot)$ via (5.127) and (5.128), respectively. Thus the proof ends.

Corollary 5.5 *Under the Assumption 5.1 if the quadruple* $(\mathbf{A}, \mathbf{B}, (\boldsymbol{\Pi}_c \mid \boldsymbol{\Pi}_d), \mathfrak{T})$ *is stabilizable, then the following equivalences hold:*

(i) if the GBJMRDE (5.2) is in the time periodic case, the following are equivalent:

(a) the set $\widetilde{\Gamma}^{\Sigma}[\mathbb{R}_+]$ *is not empty;*
(b) there exist periodic functions of period $\tilde{\theta}$, $\mathbf{Z}(\cdot) \in \widetilde{\Gamma}^{\Sigma}[\mathbb{R}_+]$;

(ii) if the GBJMRDE (5.2) is time invariant, then the following are equivalent:

(c) the set $\widetilde{\Gamma}^{\Sigma}[\mathbb{R}_+]$ *is not empty;*
(d) there exist periodic functions of period h, $\mathbf{Z}(\cdot) \in \widetilde{\Gamma}^{\Sigma}[\mathbb{R}_+]$.

Proof The implications $(b) \Rightarrow (a)$ and $(d) \Rightarrow (c)$ are obvious. To prove the implications $(a) \Rightarrow (b)$ and $(c) \Rightarrow (d)$, respectively, let us remark that under the considered assumption we know from the proof of implication $(ii) \Rightarrow (i)$ in Theorem 5.6 that there exist a a real number $\widetilde{\delta} > 0$ and a smooth function $\delta \rightarrow \mathbf{X}_\delta(\cdot) : (-\widetilde{\delta}, 0) \rightarrow \mathcal{PC}_b(\mathbb{R}_+, \mathcal{S}_n^N; \mathfrak{T})$ such that for each $-\widetilde{\delta} < \delta < 0$, $\mathbf{X}_\delta(\cdot)$ is differentiable on each interval (τ_k, τ_{k+1}) and satisfies

$$-\dot{\mathbf{X}}_\delta(t) = \mathcal{R}_c(t, \mathbf{X}_\delta(t)) + \delta\mathbf{I}_n, \quad \tau_k \leq t < \tau_{k+1} \tag{5.158a}$$
$$\mathbf{X}_\delta(\tau_k^-) = \mathcal{R}_d(\tau_k, \mathbf{X}_\delta(\tau_k)) + \delta\mathbf{I}_n, \quad k \in \mathbb{Z}_+. \tag{5.158b}$$
$$\boldsymbol{\Pi}_{c3}(t)[\mathbf{X}_\delta(t)] + \mathbf{R}_c(t) \succeq 0, \quad t \in \mathbb{R}_+ \tag{5.158c}$$
$$\boldsymbol{\Pi}_{d3}(\tau_k)[\mathbf{X}_\delta(\tau_k)] + \mathbf{R}_d(\tau_k) \succeq 0, \quad t \in \mathbb{Z}_+. \tag{5.158d}$$

We choose $-\widetilde{\delta} < \delta < \delta_1 < 0$ fixed. We consider the quintuple $\Sigma_{\delta_1} = (\mathbf{A}, \mathbf{B}, (\boldsymbol{\Pi}_c \mid \boldsymbol{\Pi}_d), (\boldsymbol{\Theta}_{c\delta_1} \mid \boldsymbol{\Theta}_{d\delta_1}), \mathfrak{T})$ where $\mathbf{A}, \mathbf{B}, \boldsymbol{\Pi}_c, \boldsymbol{\Pi}_d, \mathfrak{T}$ are as in (5.2), while $\boldsymbol{\Theta}_{c\delta_1}(t) = (\Theta_{c\delta_1}(t, 1), \Theta_{c\delta_1}(t, 2), \ldots, \Theta_{c\delta_1}(t, N))$ and $\boldsymbol{\Theta}_{d\delta_1}(\tau_k) = (\Theta_{d\delta_1}(\tau_k, 1), \Theta_{d\delta_1}(\tau_k, 2), \ldots, \Theta_{d\delta_1}(\tau_k, N))$ with

$$\Theta_{c\delta_1}(t, i) = \begin{pmatrix} M_c(t, i) + \delta_1 I_n & L_c(t, i) \\ L_c^\top(t, i) & R_c(t, i) \end{pmatrix}, \quad \text{for all } (t, i) \in \mathbb{R}_+ \times \mathcal{N}$$

and

$$\Theta_{d\delta_1}(\tau_k, i) = \begin{pmatrix} M_d(\tau_k, i) + \delta_1 I_n & L_d(\tau_k, i) \\ L_d^\top(\tau_k, i) & R_d(\tau_k, i) \end{pmatrix}, \quad \text{for all } (\tau_k, i) \in \mathfrak{T} \times \mathcal{N}.$$

Employing (5.158) together with Lemma 5.5 one may check that

$$\mathbf{X}_\delta(\cdot) \in \widetilde{\Gamma}^{\Sigma_{\delta_1}}[\mathbb{R}_+].$$

Applying implication $(i) \Rightarrow (ii)$ from Theorem 5.6 in the case of GBJMRDE of the form (5.2) but associated to quintuple Σ_{δ_1} we may infer that the equation has a stabilizing and bounded on $\mathbb{R}_+$ solution $\mathbf{X}_{\delta_1,s}(\cdot)$ satisfying the sign conditions of the form (5.137). This means that $\mathbf{X}_{\delta_1,s}(\cdot)$ verifies a system of equations and inequalities of the form (5.158), but with δ replaced by δ_1. This allows us to conclude that $\mathbf{X}_{\delta_1,s}(\cdot) \in \widetilde{\Gamma}^{\Sigma}[\mathbb{R}_+]$. Additionally from Theorem 5.5 applied for the equation of type (5.2) associated to the quintuple Σ_{δ_1}, we deduce that $\mathbf{X}_{\delta_1}(\cdot)$ is a periodic function of period $\tilde{\theta}$ if the GBJMRDE (5.2) is time periodic. Furthermore $\mathbf{X}_{\delta_1,s}(\cdot)$ is a periodic function of period h if the GBJMRDE (5.2) is time invariant. Thus the proof is complete.

Remark 5.7 The results proved in Corollary 5.4 and Corollary 5.5 show that in the case when the GBJMRDE (5.2) is time periodic or time invariant, respectively, to test if the conditions from (i) of Theorem 5.6 are true or not, it is sufficient to look for pairs of stabilizing feedback gains for the quadruple $(\mathbf{A}, \mathbf{B}, (\boldsymbol{\Pi}_c \mid \boldsymbol{\Pi}_d), \mathfrak{T})$ which are time periodic and to check if there exist periodic functions $\mathbf{Z}(\cdot)$ in $\widetilde{\Gamma}^{\Sigma}[\mathbb{R}_+]$.

5.4.2 Stochastic Jump Matrix Riccati Differential Equations

The results derived in the previous sections in the general case of a backward jump matrix Riccati-type differential equation of type (5.2) allow us to obtain conditions which guarantee the existence of some globally defined solutions of a stochastic jump matrix Riccati differential equation (SJMRDE) (5.10), as well as of other special forms of this SJMRDE which will be involved in the solution of the optimal control problems analyzed in the next chapters.

Employing (5.12) together with (5.16a)–(5.16d) we obtain that in the special case of the SJMRDE (5.10), the set $\Gamma^{\Sigma}[\mathbb{R}_+]$ consists of all bounded functions $t \to \mathbf{Z}(t) = (Z(t,1), Z(t,2), \ldots, Z(t,N)) : \mathbb{R}_+ \to \mathcal{S}_n^N$ continuously differentiable on each interval (τ_k, τ_{k+1}), right continuous in any $t = \tau_k$, $k \in \mathbb{Z}_+$, which are satisfying the following system of linear matrix differential inequalities with jumps:

$$\widetilde{\mathfrak{D}}_c^{\Sigma}(t)[\mathbf{Z}(t)](i) \geq 0, \quad t \in [\tau_k, \tau_{k+1}), \tag{5.159a}$$

$$\widetilde{\mathfrak{D}}_d^{\Sigma}(\tau_k)[\mathbf{Z}(\tau_k)](i) \geq 0, \quad k \in \mathbb{Z}_+, \tag{5.159b}$$

$$\sum_{l=1}^{r_1} B_l^{\top}(t,i)Z(t,i)B_l(t,i) + R_c(t,i) \geq \gamma_c I_{m_c} > 0, \quad (t,i) \in \mathbb{R}_+ \times \mathcal{N}, \tag{5.159c}$$

$$\sum_{p=0}^{r_2} B_{dp}^{\top}(\tau_k,i)Z(\tau_k,i)B_{dp}(\tau_k,i) + R_d(\tau_k,i) \geq \gamma_d I_{m_d} > 0, \quad \text{for all } \tau_k \in \mathfrak{T}, \tag{5.159d}$$

$i \in \mathcal{N}$ where

$$\widetilde{\mathfrak{D}}_c^{\Sigma}(t)[\mathbf{Z}(t)](i) \triangleq \begin{pmatrix} \mathbb{D}_{c11}(t)[\mathbf{Z}(t)](i) & \mathbb{D}_{c12}(t)[\mathbf{Z}(t)](i) \\ \mathbb{D}_{c12}^{\top}[\mathbf{Z}(t)](i) & \mathbb{D}_{c22}(t)[\mathbf{Z}(t)](i) \end{pmatrix}$$

$$\begin{aligned} \mathbb{D}_{c11}(t)[\mathbf{Z}(t)](i) \triangleq\ & \dot{Z}(t,i) + A_0^{\top}(t,i)Z(t,i) + Z(t,i)A_0(t,i) \\ & + \sum_{l=1}^{r_1} A_l^{\top}(t,i)Z(t,i)A_l(t,i) + \sum_{j=1}^{N} q_{ij}Z(t,j) + M_c(t,i) \end{aligned}$$

$$\mathbb{D}_{c12}(t)[\mathbf{Z}(t)](i) \stackrel{\Delta}{=} Z(t,i)B_0(t,i) + \sum_{l=1}^{r_1} A_l^\top(t,i)Z(t,i)B_l(t,i) + L_c(t,i) \tag{5.160}$$

$$\mathbb{D}_{c22}(t)[\mathbf{Z}(t)](i) \stackrel{\Delta}{=} \sum_{l=1}^{r_1} B_l^\top(t,i)Z(t,i)B_l(t,i) + R_c(t,i), \quad (t,i) \in \mathbb{R}_+ \times \mathcal{N},$$

and

$$\widetilde{\mathfrak{D}}_d^\Sigma(\tau_k)[\mathbf{Z}(\tau_k)](i) \stackrel{\Delta}{=} \begin{pmatrix} \mathbb{D}_{d11}(\tau_k)[\mathbf{Z}(\tau_k)](i) & \mathbb{D}_{d12}(\tau_k)[\mathbf{Z}(\tau_k)](i) \\ \mathbb{D}_{d12}^\top[\mathbf{Z}(\tau_k)](i) & \mathbb{D}_{d22}(\tau_k)[\mathbf{Z}(\tau_k)](i) \end{pmatrix}$$

$$\mathbb{D}_{d11}(\tau_k)[\mathbf{Z}(\tau_k)](i) \stackrel{\Delta}{=} \sum_{p=0}^{r_2} A_{dp}^\top(\tau_k,i)Z(\tau_k,i)A_{dp}(\tau_k,i) + M_d(\tau_k,i) - Z(\tau_k^-,i)$$

$$\mathbb{D}_{d12}(t)[\mathbf{Z}(\tau_k)](i) \stackrel{\Delta}{=} \sum_{p=0}^{r_2} A_{dp}^\top(\tau_k,i)Z(\tau_k,i)B_{dp}(\tau_k,i) + L_d(\tau_k,i)$$

$$\mathbb{D}_{d22}(\tau_k)[\mathbf{Z}(\tau_k)](i) \stackrel{\Delta}{=} \sum_{p=0}^{r_2} B_{dp}^\top(\tau_k,i)Z(\tau_k,i)B_{dp}(\tau_k,i) + R_d(\tau_k,i),$$
$$(\tau_k,i) \in \mathfrak{T} \times \mathcal{N}. \tag{5.161}$$

The subset $\widetilde{\Gamma}^\Sigma[\mathbb{R}_+]$ of Γ^Σ consists of all functions $\mathbf{Z}(\cdot)$ from $\Gamma^\Sigma[\mathbb{R}_+]$ which are satisfying (5.159a) and (5.159c) together with stronger constrains

$$\widetilde{\mathfrak{D}}_d^\Sigma(\tau_k)[\mathbf{Z}(\tau_k)](i) \geq \gamma I_{n+m_d} > 0, \quad \text{for all } k \in \mathbb{Z}_+, \quad i \in \mathcal{N}. \tag{5.162}$$

To facilitate the flow reading, let us recall the JLSS (4.1):

$$\begin{aligned} dx(t) &= (A_0(t,\eta(t))x(t) + B_0(t,\eta(t))u_c(t))dt \\ &\quad + \sum_{l=1}^{r_1}(A_l(t,\eta(t))x(t) + B_l(t,\eta(t))u_c(t))dw_l(t), \qquad \tau_k < t \leq \tau_{k+1} \end{aligned} \tag{5.163a}$$

$$\begin{aligned} x(\tau_k^+) &= A_{d0}(\tau_k,\eta(\tau_k))x(\tau_k) + B_{d0}(\tau_k,\eta(\tau_k))u_d(k) \\ &\quad + \sum_{l=1}^{r_2} w_{dl}(k)\left(A_{dl}(\tau_k,\eta(\tau_k))x(\tau_k) + B_{dl}(\tau_k,\eta(\tau_k))u_d(k)\right), \quad k \in \mathbb{Z}_+. \end{aligned} \tag{5.163b}$$

Bearing in mind the Remark 5.4 we obtain immediately from Theorem 5.3:

Corollary 5.6 *Under the Assumptions 1.2, 5.2 if the JLSS (5.163) is mean-square stabilizable by linear state feedback the following are equivalent:*

(*i*) *the SJMRDE (5.10) has a maximal and bounded solution* $t \to \mathbf{X}_{max}(t) = (X_{max}(t, 1), X_{max}(t, 2), \dots, X_{max}(t, N)) : \mathbb{R}_+ \to \mathcal{S}_n^N$ *satisfying the sign conditions:*

$$\sum_{l=1}^{r_1} B_l^\top(t,i)X_{max}(t,i)B_l(t,i)+R_c(t,i) \geq \gamma_c I_{m_c} > 0, \quad (t,i) \in \mathbb{R}_+ \times \mathcal{N} \tag{5.164a}$$

$$\sum_{p=0}^{r_2} B_{dp}^\top(\tau_k,i)X_{max}(\tau_k,i)B_{dp}(\tau_k,i)+R_d(\tau_k,i) \geq \gamma_d I_{m_d} > 0, \quad (\tau_k,i) \in \mathfrak{T} \times \mathcal{N}; \tag{5.164b}$$

(*ii*) *the set* $\Gamma^\Sigma[\mathbb{R}_+]$ *associated to the SJMRDE (5.10) is not empty.*

In the case of the stochastic jump matrix Riccati differential equation (5.10), the concept of **stabilizing solution** was introduced in Definition 5.6. A necessary and sufficient condition for the existence of the stabilizing and bounded on $\mathbb{R}_+$ solution of the SJMRDE (5.10), is obtained directly particularizing the result proved in Theorem 5.6. Thus, we get:

Corollary 5.7 *Under the Assumptions 1.2, 5.2 the following are equivalent:*

(*i*) *the SJMRDE (5.10) has a unique bounded and stabilizing solution* $t \to \mathbf{X}_s(t) = (X_s(t, 1), X_s(t, 2), \dots, X_s(t, N)) : \mathbb{R}_+ \to \mathcal{S}_n^N$ *satisfying the sign conditions:*

$$\sum_{l=1}^{r_1} B_l^\top(t,i)X_s(t,i)B_l(t,i) + R_c(t,i) \geq \nu_c I_{m_c} > 0, \quad (t,i) \in \mathbb{R}_+ \times \mathcal{N} \tag{5.165a}$$

$$\sum_{p=0}^{r_2} B_{dp}^\top(\tau_k,i)X_s(\tau_k,i)B_{dp}(\tau_k,i) + R_d(\tau_k,i) \geq \nu_d I_{m_d} > 0, \quad (\tau_k,i) \in \mathfrak{T} \times \mathcal{N}; \tag{5.165b}$$

(*ii*) *the JLSS (5.163) is mean-square stabilizable by linear state feedback and the set* $\widetilde{\Gamma}^\Sigma[\mathbb{R}_+]$ *associated to the SJMRDE (5.10) is not empty.*

The following definition is an adaptation of the Definition 5.2 to the case of the SJMRDE (5.10) .

Definition 5.8 We say that the SJMRDE (5.10) is:

(i) with time periodic coefficients if the following conditions are simultaneously fulfilled:

(a) $\tau_k = kh$, for all $k \in \mathbb{Z}_+$ where $h > 0$ does not depend upon k;

(b) there exists a real number $\theta_c > 0$ with the property that $A_l(t+\theta_c, i) = A_l(t,i)$, $B_l(t+\theta_c, i) = B_l(t,i)$, $0 \le l \le r_1$, $M_c(t+\theta_c, i) = M_c(t,i)$, $L_c(t+\theta_c, i) = L_c(t,i)$, $R_c(t+\theta_c, i) = R_c(t,i)$, for all $(t,i) \in \mathbb{R}_+ \times \mathcal{N}$;

(c) there exists a natural number $\theta_d \ge 1$ with the properties: $A_{dp}((k+\theta_d)h, i) = A_{dp}(kh,i)$, $B_{dp}((k+\theta_d)h, i) = B_{dp}(kh,i)$, $0 \le p \le r_2$, $M_d((k+\theta_d)h, i) = M_d(kh,i)$, $L_d((k+\theta_d)h, i) = L_d(kh,i)$, $R_d((k+\theta_d)h, i) = R_d(kh,i)$, for all $k \in \mathbb{Z}_+$, $i \in \mathcal{N}$;

(d) $\frac{\theta_c}{\theta_d h}$ is a rational number;

(ii) time invariant if the following conditions are simultaneously fulfilled:

(α) $\tau_k = kh$, for all $k \in \mathbb{Z}_+$, $h > 0$ does not depend upon k;

(β) $A_l(t,i) = A_l(i) \in \mathbb{R}^{n\times n}$, $B_l(t,i) = B_l(i) \in \mathbb{R}^{n\times m_c}$, $0 \le l \le r_1$, $M_c(t,i) = M_c(i) \in \mathcal{S}_n$, $L_c(t,i) = L_c(i) \in \mathbb{R}^{n\times m_c}$, $R_c(t,i) = R_c(i) \in \mathcal{S}_{m_c}$, for all $(t,i) \in \mathbb{R}_+ \times \mathcal{N}$;

(γ) $A_{dp}(kh,i) = A_{dp}(i) \in \mathbb{R}^{n\times n}$, $B_{dp}(kh,i) = B_{dp}(i) \in \mathbb{R}^{n\times m_d}$, $0 \le p \le r_2$, $M_d(kh,i) = M_d(i) \in \mathcal{S}_n$, $L_d(kh,i) = L_d(i) \in \mathbb{R}^{n\times m_d}$, $R_d(kh,i) = R_d(i) \in \mathcal{S}_{m_d}$, for all $(k,i) \in \mathbb{Z}_+ \times \mathcal{N}$.

If the numbers θ_c, θ_d and h are satisfying condition (d) from the first part of this definition, there exist coprime natural numbers $\tilde{p}$, $\tilde{q}$ such that

$$\frac{\theta_c}{\theta_d h} = \frac{\tilde{p}}{\tilde{q}}.$$

In this case we define

$$\tilde{\theta} := \tilde{q}\theta_c = \tilde{p}\theta_d h \tag{5.166a}$$

$$\tilde{\mathbf{k}} := \tilde{p}\theta_d. \tag{5.166b}$$

The result proved in Theorem 5.5 specialized to the case of the SJMRDE (5.10) becomes:

Corollary 5.8 *Under the Assumption 5.2, if the set $\Gamma^{\Sigma}[\mathbb{R}_+]$ associated to the equation (5.10) is not empty, then the following hold:*

(i) if the SJMRDE (5.10) is with time periodic coefficients, then its unique bounded and stabilizing solution, if any, $t \to \mathbf{X}_s(t) : \mathbb{R}_+ \to \mathcal{S}_n^N$ is a $\tilde{\theta}$ periodic function;

(ii) if the SJMRDE (5.10) is time invariant, then its unique bounded and stabilizing solution $t \to \mathbf{X}_s(t) : \mathbb{R}_+ \to \mathcal{S}_n^N$, if any, is a h–periodic function.

The statement of Corollary 5.5 adapted to the case of the SJMRDE (5.10) becomes:

Corollary 5.9 *Under the Assumption 5.2 if the JLSS (5.163) is mean-square stabilizable by a linear state feedback the following equivalences hold:*

(i) if the SJMRDE (5.10) is with time periodic coefficients, then the following are equivalent:

(a) the set $\widetilde{\Gamma}^{\Sigma}[\mathbb{R}_+]$ associated to this equation is not empty;

(b) there exist periodic functions of period $\tilde{\theta}$, $t \to \mathbf{Z}(t) : [0, \tilde{\theta}] \to \mathcal{S}_n^N$ differentiable for $t \neq kh$, right continuous in any $t = kh$, $0 \leq k < \tilde{\mathbf{k}}$ which are satisfying the system of linear matrix differential inequalities with jumps (5.159a) with $\tau_k = kh$, (5.159c) together with

$$\mathfrak{D}_d^{\Sigma}(kh)[\mathbf{Z}(kh)](i) > 0, \tag{5.167}$$

$0 \leq k \leq \tilde{\mathbf{k}} - 1$, where $\tilde{\theta}$, $\tilde{\mathbf{k}}$ are those from (5.166);

(ii) if the SJMRDE (5.10) is time invariant, then the following are equivalent:

(c) the set $\widetilde{\Gamma}^{\Sigma}[\mathbb{R}_+]$ associated to SJMRDE (5.10) is not empty;

(d) there exist functions $t \to \mathbf{Z}(t) : [0, h] \to \mathcal{S}_n^N$ differentiable on the interval $(0, h)$, right continuous in $t = 0$, $\mathbf{Z}(h) = \mathbf{Z}(0)$ and satisfies (5.159a) and (5.159c) for $t \in [0, h]$ together with (5.167) written for k replaced by 1.

5.4.3 The Case of a Backward Jump Matrix Linear Differential Equation with Riccati-Type Jumping Operator

In this subsection, the following kind of stochastic jump matrix Riccati differential equations on the space $\mathcal{S}_n^N$ is considered:

$$-\dot{X}(t,i) = A_0^{\top}(t,i)X(t,i) + X(t,i)A_0(t,i) + \sum_{l=1}^{r_1} A_l^{\top}(t,i)X(t,i)A_l(t,i) + \sum_{j=1}^{N} q_{ij}X(t,j) + M_c(t,i), \quad \tau_k \leq t < \tau_{k+1} \tag{5.168a}$$

$$X(\tau_k^-, i) = \sum_{p=0}^{r_2} A_{dp}^{\top}(\tau_k,i)X(\tau_k,i)A_{dp}(\tau_k,i) - \left(\sum_{p=0}^{r_2} A_{dp}^{\top}(\tau_k,i)X(\tau_k,i)B_{dp}(\tau_k,i) + L_d(\tau_k,i)\right) \cdot \left(\sum_{p=0}^{r_2} B_{dp}^{\top}(\tau_k,i)X(\tau_k,i)B_{dp}(\tau_k,i) + R_d(\tau_k,i)\right)^{-1} \tag{5.168b}$$

$$\left(\sum_{p=0}^{r_2} B_{dp}^{\top}(\tau_k, i)X(\tau_k, i)A_{dp}(\tau_k, i) + L_d^{\top}(\tau_k, i)\right) + M_d(\tau_k, i), \quad k \in \mathbb{Z}_+.$$

One sees that (5.168) is a special case of (5.10) for $B_l(t, i) = 0,\ 0 \leq l \leq r_1$, $L_c(t, i) = 0,\ R_c(t, i) = 0,\ (t, i) \in \mathbb{R}_+ \times \mathcal{N}$. That is why it will be named backward jump matrix linear differential equation (BJMLDE) with Riccati-type jumping operator.

The BJMLDE (5.168) is involved in solution of an optimal control problem described by a jump linear stochastic system controlled by impulses, of the form:

$$dx(t) = A_0(t, \eta(t))x(t)dt + \sum_{l=1}^{r_1} A_l(t, \eta(t))x(t)dw_l(t), \quad \tau_k < t \leq \tau_{k+1} \tag{5.169a}$$

$$\begin{aligned} x(\tau_k^+) &= A_{d0}(\tau_k, \eta(\tau_k))x(\tau_k) + B_{d0}(\tau_k, \eta(\tau_k))u_d(k) \\ &+ \sum_{p=1}^{r_2} w_{dp}(k)\left(A_{dp}(\tau_k, \eta(\tau_k))x(\tau_k) + B_{dp}(\tau_k, \eta(\tau_k))u_d(k)\right), \quad k \in \mathbb{Z}_+, \end{aligned} \tag{5.169b}$$

and the performance criterion of the form

$$\begin{aligned} J(t_0, \infty; u_d) &= \mathbb{E}\left[\int_{t_0}^{\infty} x_{u_d}^{\top}(t)M_c(t, \eta(t))x_{u_d}(t)dt\right] \\ &+ \sum_{k=k_0}^{\infty} \mathbb{E}[x_{u_d}^{\top}(\tau_k)M_d(\tau_k, \eta(\tau_k))x_{u_d}(\tau_k) \\ &+ 2x_{u_d}^{\top}(\tau_k)L_d(\tau_k, \eta(\tau_k))u_d(k) + u_d^{\top}(k)R_d(\tau_k, \eta(\tau_k))u_d(k)]. \end{aligned} \tag{5.170}$$

In (5.170), $x_{u_d}(\cdot)$ is the solution of (5.169) determined by the input $u_d = \{u_d(k)\}_{k \geq k_0}$ and satisfying $x(t_0) = x_0$. This kind of optimal control problem will be analyzed in the next chapter.

Definition 5.9 A solution $t \to \mathbf{X}_s(t) = (X_s(t, 1), X_s(t, 2), \ldots, X_s(t, N)) : \mathbb{R}_+ \to \mathcal{S}_n^N$ of (5.168) is named stabilizing solution if the matrices

$$\sum_{p=0}^{r_2} B_{dp}^{\top}(\tau_k, i)X_s(\tau_k, i)B_{dp}(\tau_k, i) + R_d(\tau_k, i), \quad (k, i) \in \mathbb{Z}_+ \times \mathcal{N},$$

are invertible and the closed-loop JSLDE:

$$\mathrm{d}x(t) = (A_0(t,\eta(t))x(t)\mathrm{d}t + \sum_{l=1}^{r_1}(A_l(t,\eta(t))x(t)\mathrm{d}w_l(t), \quad \tau_k < t \le \tau_{k+1} \tag{5.171a}$$

$$\begin{aligned} x(\tau_k^+) &= (A_{d0}(\tau_k,\eta(\tau_k)) + B_{d0}(\tau_k,\eta(\tau_k))F_s(\tau_k,\eta(\tau_k)))x(\tau_k) \\ &\quad + \sum_{p=1}^{r_2} w_{dp}(k)\left(A_{dp}(\tau_k,\eta(\tau_k)) + B_{dp}(\tau_k,\eta(\tau_k))F_s(\tau_k,\eta(\tau_k))\right)x(\tau_k), \\ &\qquad k \in \mathbb{Z}_+, \end{aligned} \tag{5.171b}$$

is SESMS, where

$$\begin{aligned} F_s(\tau_k,i) &:= -\left(\sum_{p=0}^{r_2} B_{dp}^\top(\tau_k,i)X_s(\tau_k,i)B_{dp}(\tau_k,i) + R_d(\tau_k,i)\right)^{-1} \\ &\quad \cdot \left(\sum_{p=0}^{r_2} B_{dp}^\top(\tau_k,i)X_s(\tau_k,i)A_{dp}(\tau_k,i) + L_d^\top(\tau_k,i)\right), \quad (\tau_k,i) \in \mathfrak{T}\times\mathcal{N}. \end{aligned} \tag{5.172}$$

The set $\widetilde{\Gamma}^\Sigma[\mathbb{R}_+]$ associated to the BJMLDE (5.168) consists of all bounded functions $t \to \mathbb{Z}(t) = (Z(t,1), Z(t,2), \ldots, Z(t,N)) : \mathbb{R}_+ \to \mathcal{S}_n^N$ continuously differentiable on each interval (τ_k, τ_{k+1}), right continuous in $t = \tau_k$, $k \in \mathbb{Z}_+$, which are satisfying the linear matrix differential inequalities with jumps:

$$\begin{aligned} &\dot{Z}(t,i) + A_0^\top(t,i)Z(t,i) + Z(t,i)A_0(t,i) + \sum_{l=1}^{r_1} A_l^\top(t,i)Z(t,i)A_l(t,i) \\ &+ \sum_{j=1}^{N} q_{ij}Z(t,j) + M_c(t,i) \ge 0, \quad t \in [\tau_k,\tau_{k+1}), \end{aligned} \tag{5.173a}$$

$$\mathfrak{D}_d^\Sigma(\tau_k)[\mathbf{Z}(\tau_k)](i) \ge \gamma_d I_{n+m_d} > 0, \quad k \in \mathbb{Z}_+, \tag{5.173b}$$

where $\mathfrak{D}_d^\Sigma(\tau_k)[\mathbf{Z}(\tau_k)](i)$ is computed as in (5.161). Bearing in mind that the BJMLDE (5.168) is a particular case of (5.10), the results stated in Corollary 5.6 - Corollary 5.9 can be easily adapted to the case of the maximal solution and of the stabilizing solution of (5.168). To this end, we have to take into account that the set $\widetilde{\Gamma}^\Sigma[\mathbb{R}_+]$ associated to (5.168) consists of all functions $\mathbf{Z}(\cdot)$ which are satisfying (5.173), and the property of mean-square stabilizability of the JLSS (5.163) is replaced by property of mean-square stabilizability by impulses in a state feedback form of the JLSS (5.169). The details are left to the readers.

5.5 The Minimal Positive Semidefinite Solution of a Stochastic Jump Matrix Riccati Differential Equation

Besides the maximal solution and the stabilizing solution of a SJMRDE (5.10), an important role in the solution of some optimal control problems is played by another globally defined on $\mathbb{R}_+$ solution of this equation, namely the minimal positive semidefinite solution.

Definition 5.10 A solution $t \to \mathbf{X}_{min}(t) = (X_{min}(t, 1), X_{min}(t, 2), \ldots, X_{min}(t, N)) : \mathbb{R}_+ \to \mathcal{S}_n^N$ of the SJMRDE (5.10) is **the minimal solution** in the set of positive semidefinite solutions of this equation, if

$$0 \preceq \mathbf{X}_{min}(t) \preceq \widehat{\mathbf{X}}(t), \quad t \in \mathbb{R}_+$$

for any solution of (5.10), $t \to \widehat{\mathbf{X}}(t) : \mathbb{R}_+ \to \mathcal{S}_n^{N+}$.

The development from this section are done under the following assumption:

Assumption 5.4 The set $\Gamma^{\Sigma}[\mathbb{R}_+]$ associated to the SJMRDE (5.10) contains the function $\mathbf{Z}(\cdot) = (Z(\cdot, 1), Z(\cdot, 2), \ldots, Z(\cdot, N))$ with $Z(t, i) = 0$, for all $(t, i) \in \mathbb{R}_+ \times \mathcal{N}$.

Remark 5.8 Employing (5.159)–(5.161) we may infer that the Assumption 5.4 is equivalent to the following constrains on the weights matrices from (5.11):

$$\begin{pmatrix} M_c(t, i) & L_c(t, i) \\ L_c^\top(t, i) & R_c(t, i) \end{pmatrix} \geq 0 \tag{5.174a}$$

$$R_c(t, i) \geq \gamma_c I_{m_c}, \quad \text{for all } (t, i) \in \mathbb{R}_+ \times \mathcal{N}, \tag{5.174b}$$

$$\begin{pmatrix} M_d(\tau_k, i) & L_d(\tau_k, i) \\ L_d^\top(\tau_k, i) & R_d(\tau_k, i) \end{pmatrix} \geq 0 \tag{5.175a}$$

$$R_d(\tau_k, i) \geq \gamma_d I_{m_d}, \quad \text{for all } (\tau_k, i) \in \mathfrak{T} \times \mathcal{N}, \tag{5.175b}$$

γ_c, γ_d being positive constants.

The following theorem provides a set of conditions which guarantee the existence of the minimal positive semidefinite solution of a SJMRDE.

Theorem 5.7 *Assume:*

(a) the Assumptions 1.2, 5.2 and 5.4 hold;
(b) the JLSS (5.163) is mean-square stabilizable by linear state feedback.

Under these conditions the SJMRDE (5.10) has a unique minimal positive semidefinite solution $t \to \mathbf{X}_{min}(t) : \mathbb{R}_+ \to \mathcal{S}_n^{N+}$.

Proof For each $T > 0$ we consider $\mathbf{X}_T(\cdot) = (X_T(\cdot, 1), X_T(\cdot, 2), \ldots, X_T(\cdot, N))$ the solution of the equation (5.10) with the terminal condition $X_T(T, i) = 0$, $1 \leq i \leq N$. Since $\mathbf{Z}(\cdot) = (0, 0, \ldots, 0) \in \Gamma^{\Sigma}[\mathbb{R}_+] \subset \Gamma^{\Sigma}[0, T]$, we deduce via Corollary 5.1 applied in the case of the SJMRDE (5.10) (viewed as a particular case of (5.2)) that $\mathbf{X}_T(t) = (X_T(t, 1), X_T(t, 2), \ldots, X_T(t, N))$ is defined on the whole interval $[0, T]$ and satisfies

$$X_T(t, i) \geq 0, \quad \text{for all} \quad (t, i) \in [0, T] \times \mathcal{N}. \tag{5.176}$$

Further, according to (b) from the statement there exist bounded and piecewise continuous functions with finite lateral limits $t \to \widetilde{F}_c(t, i) : \mathbb{R}_+ \to \mathbb{R}^{m_c \times n}$ and bounded sequences $\{\widetilde{F}_d(\tau_k, i)\}_{k \in \mathbb{Z}_+} \subset \mathbb{R}^{m_d \times n}$, $i \in \mathcal{N}$, with the property that the closed-loop system

$$\begin{aligned} \mathrm{d}x(t) &= (A_0(t, \eta(t)) + B_0(t, \eta(t))\widetilde{F}_c(t, \eta(t)))x(t)\mathrm{d}t \\ &\quad + \sum_{l=1}^{r_1} (A_l(t, \eta(t)) + B_l(t, \eta(t))\widetilde{F}_c(t, \eta(t)))x(t)\mathrm{d}w_l(t), \quad \tau_k < t \leq \tau_{k+1} \end{aligned} \tag{5.177a}$$

$$\begin{aligned} x(\tau_k^+) &= (A_{d0}(\tau_k, \eta(\tau_k)) + B_{d0}(\tau_k, \eta(\tau_k))\widetilde{F}_d(\tau_k, \eta(\tau_k)))x(\tau_k) \\ &\quad + \sum_{p=1}^{r_2} w_{dp}(k) \left(A_{dp}(\tau_k, \eta(\tau_k)) + B_{dp}(\tau_k, \eta(\tau_k))\widetilde{F}_d(\tau_k, \eta(\tau_k))\right) x(\tau_k), \\ k &\in \mathbb{Z}_+, \end{aligned} \tag{5.177b}$$

is strong exponentially stable in mean square (SESMS). Employing Definition 3.1 (a) together with parts (i) and (iv) from Theorem 2.13 in the case of the closed-loop system (5.177) we may infer that the non-homogeneous backward GLJLDE:

$$\begin{aligned} -\dot{Y}(t, i) &= (A_0(t, i) + B_0(t, i)\widetilde{F}_c(t, i))^\top Y(t, i) + Y(t, i)(A_0(t, i) + B_0(t, i)\widetilde{F}_c(t, i)) \\ &\quad + \sum_{l=1}^{r_1} (A_l(t, i) + B_l(t, i)\widetilde{F}_c(t, i))^\top Y(t, i)(A_l(t, i) + B_l(t, i)\widetilde{F}_c(t, i)) \\ &\quad + \sum_{j=1}^{N} q_{ij} Y(t, j) + \Theta_{c\widetilde{\mathbf{F}}_c}(t, i), \quad \tau_k \leq t < \tau_{k+1} \end{aligned} \tag{5.178a}$$

$$\begin{aligned} Y(\tau_k^-, i) &= \sum_{p=0}^{r_2} (A_{dp}(\tau_k, i) + B_{dp}(\tau_k, i)\widetilde{F}_d(\tau_k, i))^\top Y(\tau_k, i) \\ &\quad \cdot (A_{dp}(\tau_k, i) + B_{dp}(\tau_k, i)\widetilde{F}_d(\tau_k, i)) + \Theta_{d\widetilde{\mathbf{F}}_d}(\tau_k, i), \quad k \in \mathbb{Z}_+ \end{aligned} \tag{5.178b}$$

has a unique solution $t \to \widetilde{\mathbf{Y}}(t) = (\widetilde{Y}(t, 1), \widetilde{Y}(t, 2), \ldots, \widetilde{Y}(t, N)) : \mathbb{R}_+ \to \mathcal{S}_n^N$ with the property that

$$0 \leq \widetilde{Y}(t, i) \leq \tilde{c} I_n, \quad \text{for all} \quad (t, i) \in \mathbb{R}_+ \times \mathcal{N}, \tag{5.179}$$

when

$$\Theta_{c\widetilde{\mathbf{F}}_c}(t,i) \triangleq \begin{pmatrix} I_n \\ \widetilde{F}_c(t,i) \end{pmatrix}^\top \begin{pmatrix} M_c(t,i) & L_c(t,i) \\ L_c^\top(t,i) & R_c(t,i) \end{pmatrix} \begin{pmatrix} I_n \\ \widetilde{F}_c(t,i) \end{pmatrix}$$

and

$$\Theta_{d\widetilde{\mathbf{F}}_d}(\tau_k,i) \triangleq \begin{pmatrix} I_n \\ \widetilde{F}_d(\tau_k,i) \end{pmatrix}^\top \begin{pmatrix} M_d(\tau_k,i) & L_d(\tau_k,i) \\ L_d^\top(\tau_k,i) & R_d(\tau_k,i) \end{pmatrix} \begin{pmatrix} I_n \\ \widetilde{F}_d(\tau_k,i) \end{pmatrix}.$$

On the other hand, Lemma 5.1 and (5.28)–(5.32) written for $W_c(t,i)$ and $W_d(\tau_k,i)$ replaced by $\widetilde{F}_c(t,i)$ and $\widetilde{F}_d(\tau_k,i)$, respectively, allow us to rewrite (5.10) satisfied by $\mathbf{X}_T(\cdot)$ in the form

$$\begin{aligned} -\dot{X}_T(t,i) &= (A_0(t,i)+B_0(t,i)\widetilde{F}_c(t,i))^\top X_T(t,i) \\ &\quad + X_T(t,i)(A_0(t,i)+B_0(t,i)\widetilde{F}_c(t,i)) \\ &\quad + \sum_{l=1}^{r_1}(A_l(t,i)+B_l(t,i)\widetilde{F}_c(t,i))^\top X_T(t,i)(A_l(t,i)+B_l(t,i)\widetilde{F}_c(t,i)) \\ &\quad + \sum_{j=1}^{N} q_{ij}X_T(t,j) + \Theta_{c\widetilde{\mathbf{F}}_c}(t,i) - \mathscr{H}(t,i), \quad \tau_k \le t < \tau_{k+1} \end{aligned} \tag{5.180a}$$

$$\begin{aligned} X_T(\tau_k^-,i) &= \sum_{p=0}^{r_2}(A_{dp}(\tau_k,i)+B_{dp}(\tau_k,i)\widetilde{F}_d(\tau_k,i))^\top X_T(\tau_k,i) \\ &\quad \cdot (A_{dp}(\tau_k,i)+B_{dp}(\tau_k,i)\widetilde{F}_d(\tau_k,i)) + \Theta_{d\widetilde{\mathbf{F}}_d}(\tau_k,i) - \mathscr{H}_d(\tau_k,i), \quad k\in\mathbb{Z}_+ \end{aligned} \tag{5.180b}$$

where we denoted

$$\begin{aligned} \mathscr{H}_c(t,i) &\triangleq (\widetilde{F}_c(t,i)-F_c^{\mathbf{X}}(t,i))^\top \left(\sum_{l=1}^{r_1} B_l^\top(t,i)X_T(t,i)B_l(t,i)+R_c(t,i)\right) \\ &\quad \cdot (\widetilde{F}_c(t,i)-F_c^{\mathbf{X}}(t,i)), \quad (t,i)\in[0,T]\times\mathcal{N}, \end{aligned}$$

and

$$\begin{aligned} \mathscr{H}_d(\tau_k,i) &\triangleq (\widetilde{F}_d(\tau_k,i)-F_d^{\mathbf{X}}(\tau_k,i))^\top \\ &\quad \left(\sum_{p=0}^{r_2} B_{dp}^\top(\tau_k,i)X_T(\tau_k,i)B_{dp}(\tau_k,i)+R_d(\tau_k,i)\right) \\ &\quad \cdot (\widetilde{F}_d(\tau_k,i)-F_d^{\mathbf{X}}(\tau_k,i)), \quad (\tau_k,i)\in(\mathfrak{T}\cap[0,T])\times\mathcal{N}, \end{aligned}$$

$F_c^{\mathbf{X}}(t,i)$ and $F_d^{\mathbf{X}}(\tau_k,i)$ being computed as in (5.23) and (5.25), respectively, with $\mathbf{X}$ replaced by $\mathbf{X}_T(t)$ and $\mathbf{X}$ replaced by $\mathbf{X}_T(\tau_k)$, respectively.

Invoking (5.174b), (5.175b) and (5.178b), we may conclude that

$$\mathscr{H}_c(t,i) \geq 0, \quad \text{for all } (t,i) \in [0,T] \times \mathscr{N} \tag{5.181a}$$

$$\mathscr{H}_d(\tau_k,i) \geq 0, \quad \text{for all } (\tau_k,i) \in (\mathfrak{T} \cap [0,T]) \times \mathscr{N}. \tag{5.181b}$$

Subtracting (5.180) from (5.178) we obtain that $\boldsymbol{\Delta}_T(\cdot) = (\Delta(\cdot,1), \Delta(\cdot,2), \ldots, \Delta_T(\cdot,N))$, $\Delta_T(t,i) := \widetilde{Y}(t,i) - X_T(t,i)$, $(t,i) \in [0,T] \times \mathscr{N}$ solves the following terminal value problem (TVP):

$$\begin{aligned}
-\dot{\Delta}_T(t,i) &= (A_0(t,i) + B_0(t,i)\widetilde{F}_c(t,i))^\top \Delta_T(t,i) \\
&\quad + \Delta_T(t,i)(A_0(t,i) + B_0(t,i)\widetilde{F}_c(t,i)) \\
&\quad + \sum_{l=1}^{r_1} (A_l(t,i) + B_l(t,i)\widetilde{F}_c(t,i))^\top \Delta_T(t,i)(A_l(t,i) + B_l(t,i)\widetilde{F}_c(t,i)) \\
&\quad + \sum_{j=1}^{N} q_{ij} \mathrm{d}_\mathrm{T}(\mathrm{t},\mathrm{j}) + \mathscr{H}_c(\mathrm{t},\mathrm{i}), \quad \tau_k \leq \mathrm{t} < \tau_{k+1}
\end{aligned} \tag{5.182a}$$

$$\begin{aligned}
\Delta_T(\tau_k^-,i) &= \sum_{p=0}^{r_2} (A_{dp}(\tau_k,i) + B_{dp}(\tau_k,i)\widetilde{F}_d(\tau_k,i))^\top \Delta_T(\tau_k,i) \\
&\quad \cdot (A_{dp}(\tau_k,i) + B_{dp}(\tau_k,i)\widetilde{F}_d(\tau_k,i)) + \mathscr{H}_d(\tau_k,i), \\
&\qquad (\tau_k,i) \in (\mathfrak{T} \cap [0,T]) \times \mathscr{N}
\end{aligned} \tag{5.182b}$$

$$\Delta_T(T,i) = \widetilde{Y}(T,i) \geq 0, \quad i \in \mathscr{N}. \tag{5.182c}$$

According to Proposition 5.1 (ii) applied in the case of TVP (5.182) we may conclude via (5.181) and (5.182c) that $\Delta_T(t,i) \geq 0$, for all $(t,i) \in [0,T] \times \mathscr{N}$. Thus, we have obtained that

$$0 \leq X_T(t,i) \leq \widetilde{Y}(t,i) \leq \tilde{c} I_n, \quad \text{for all } (t,i) \in [0,T] \times \mathscr{N}, \tag{5.183}$$

and for $T > 0$, $\tilde{c}$ does not depend on T.

Let $0 < T_1 < T_2$ be arbitrary. Employing (5.174b), (5.175b) and (5.183) written for $X_{T_1}(t,i)$ we may infer that

$$\sum_{l=1}^{r_1} B_l^\top(t,i) X_{T_1}(t,i) B_l(t,i) + R_c(t,i) \geq \gamma_c I_{m_c}, \quad t \in [0,T_1]$$

and

$$\sum_{p=0}^{r_2} B_{dp}^\top(\tau_k,i) X_{T_1}(\tau_k,i) B_{dp}(\tau_k,i) + R_d(\tau_k,i) \geq \gamma_d I_{m_d}, \quad \tau_k \in \mathfrak{T} \cap [0,T_1], \quad i \in \mathscr{N}.$$

Since $X_{T_2}(T_1, i) \geq 0 = X_{T_1}(T_1, i)$ we may conclude via Theorem 5.1 together with (5.183) that

$$0 \leq X_{T_1}(t, i) \leq X_{T_2}(t, i) \leq \tilde{c} I_n,$$

for all $(t, i) \in [0, T_1] \times \mathcal{N}$, $0 < T_1 < T_2$. Hence the sequences $\{X_T(t, i)\}_{T \geq 0}$ are convergent. Let $X_{min}(t, i)$, $(t, i) \in \mathbb{R}_+ \times \mathcal{N}$, be defined by

$$X_{min}(t, i) := \lim_{T \to \infty} X_T(t, i). \tag{5.184}$$

Proceeding as in the proof of Theorem 5.3 one may show via Lebesgue's theorem of dominate convergence that $t \to \mathbf{X}_{min}(t) = (X_{min}(t, 1), X_{min}(t, 2), \ldots, X_{min}(t, N)) : \mathbb{R}_+ \to \mathcal{S}_n^N$ is a solution of the SJMRDE (5.10). From (5.183) and (5.184) we obtain that $0 \leq X_{min}(t, i) \leq \tilde{c} I_n$, for all $(t, i) \in \mathbb{R}_+ \times \mathcal{N}$.
Let $t \to \widehat{\mathbf{X}}(t) = (\widehat{X}(t, 1), \widehat{X}(t, 2), \ldots, \widehat{X}(t, N)) : \mathbb{R}_+ \to \mathcal{S}_n^{N+}$ be an arbitrary bounded and positive semidefinite solution of (5.10). For each $T > 0$ we have

$$\widehat{X}(T, i) \geq 0 = X_T(T, i), \quad \text{for all } i \in \mathcal{N}.$$

Invoking again Theorem 5.1 we may infer that

$$0 \leq X_T(t, i) \leq \widehat{X}(t, i), \quad \text{for all } (t, i) \in [0, T] \times \mathcal{N}.$$

Taking the limit for $T \to \infty$ we obtain that $0 \leq X_{min}(t, i) \leq \widehat{X}(t, i)$, for all $(t, i) \in \mathbb{R}_+ \times \mathcal{N}$. Thus we have shown that $X_{min}(\cdot)$ defined by (5.184) is just the minimal positive semidefinite and bounded on $\mathbb{R}_+$ solution of the SJMRDE (5.10).

Let us assume by contrary that (5.10) has at least two minimal positive semidefinite and bounded solutions $\mathbf{X}_{min}^{(j)}(\cdot)$, $j = 1, 2$. Then, from the minimality property of $\mathbf{X}_{min}^{(1)}(\cdot)$ we deduce that

$$\mathbf{X}_{min}^{(1)}(t) \preceq \mathbf{X}_{min}^{(2)}(t), \quad \text{for all } t \in \mathbb{R}_+,$$

while the minimality property of the solution $\mathbf{X}_{min}^{(2)}(\cdot)$ yields

$$\mathbf{X}_{min}^{(2)}(t) \preceq \mathbf{X}_{min}^{(1)}(t), \quad \text{for all } t \in \mathbb{R}_+.$$

Hence $\mathbf{X}_{min}^{(1)}(t) = \mathbf{X}_{min}^{(2)}(t)$, for all $t \in \mathbb{R}_+$. Thus the proof is complete.

Using the notation introduced in the proof of the previous theorem, we obtain:

Proposition 5.3 *Assume that the assumptions of Theorem 5.7 hold. Under these conditions, if the SJMRDE (5.10) has time periodic coefficients (in the sense of Definition 5.8 (i)) then its minimal positive semidefinite solution $\mathbf{X}_{min}(\cdot)$ is a periodic function of period $\tilde{\theta}$ with $\tilde{\theta}$ defined in (5.166a). Moreover, if the SJMRDE (5.10) is time invariant (in the sense of Definition 5.8 (ii)) then, its minimal positive semidefinite solution $\mathbf{X}_{min}(\cdot)$ is a periodic function with the period equal to h.*

Proof Let $t \to \mathbf{X}_{T+\tilde{\theta}}(t+\tilde{\theta}) : [0, T] \to \mathcal{S}_n^N$. By direct calculations involving the periodicity property of the coefficients of (5.10) one may check that $\mathbf{X}_{T+\tilde{\theta}}(\cdot + \tilde{\theta})$ is the solution of (5.10) which is satisfying the terminal condition

$$\mathbf{X}_{T+\tilde{\theta}}(T+\tilde{\theta}) = 0 = \mathbf{X}_T(T).$$

From the unicity of the solution of the SJMRDE (5.10) with given terminal condition we deduce that $\mathbf{X}_{T+\tilde{\theta}}(t+\tilde{\theta}) = \mathbf{X}_T(t), \quad \text{for all} \quad t \in [0, T]$. Taking the limit for $T \to \infty$ in the last equality, we obtain via (5.184) that

$$X_{min}(t+\tilde{\theta}, i) = X_{min}(t, i), \quad \text{for all} \quad (t, i) \in \mathbb{R}_+ \times \mathcal{N},$$

which confirms the periodicity property of the minimal solution $\mathbf{X}_{min}(\cdot)$ of (5.10). To end the proof, let us remark that if (5.10) is time invariant, it may be viewed as a time periodic equation in the sense of Definition 5.8 (i) with $\theta_c = h$ and $\theta_d = 1$. In this case (5.166) holds for $\tilde{\theta} = h$ and $\tilde{\mathbf{k}} = 1$. Thus the proof is complete.

Combining Theorem 5.3 and Corollary 5.2 on one hand and Theorem 5.7 and Proposition 5.3 on the other hand, we obtain:

Corollary 5.10 *Under the assumptions from Theorem 5.7, the SJMRDE (5.10) has two solutions* $t \to \mathbf{X}_{max}(t) : \mathbb{R}_+ \to \mathcal{S}_n^{N+}$ *and* $t \to \mathbf{X}_{min}(t) : \mathbb{R}_+ \to \mathcal{S}_n^{N+}$ *which are satisfying*

$$0 \preceq \mathbf{X}_{min}(t) \preceq \widetilde{\mathbf{X}}(t) \preceq \mathbf{X}_{max}(t) \preceq \nu \mathbf{I}_n, \quad t \in \mathbb{R}_+, \tag{5.185}$$

for any positive semidefinite and bounded solution $t \to \widetilde{\mathbf{X}}(t) : \mathbb{R}_+ \to \mathcal{S}_n^{N+}$ *of (5.10),* $\nu > 0$ *and* $\mathbf{I}_n = (I_n, I_n, \ldots, I_n) \in \mathcal{S}_n^N$. *If the SJMRDE (5.10) is time periodic (in the sense of the Definition 5.8 (i)) then both* $\mathbf{X}_{max}(\cdot)$ *and* $\mathbf{X}_{min}(\cdot)$ *are periodic functions of period* $\tilde{\theta}$ *with* $\tilde{\theta}$ *computed as in (5.166a). If the SJMRDE (5.10) is in the time-invariant case (in the sense of Definition 5.8 (ii)) then both* $\mathbf{X}_{max}(\cdot)$ *and* $\mathbf{X}_{min}(\cdot)$ *are periodic functions of period equal to the length h of the dwell time.*

Now we assume that the SJMRDE (5.10) has periodic coefficients in the sense of Definition 5.8. Based on the matrix coefficients of (5.10) we introduce the following auxiliary jump linear stochastic system having the state evolution described by

$$dx(t) = \widehat{A}_0(t, \eta(t))x(t)dt + \sum_{l=1}^{r_1} \widehat{A}_l(t, \eta(t))x(t)dw_l(t), \quad kh < t \leq (k+1)h \tag{5.186a}$$

$$x(kh^+) = \widehat{A}_{d0}(kh, \eta(kh))x(kh) + \sum_{p=1}^{r_2} w_{dp}(k)\widehat{A}_{dp}(kh, \eta(kh))x(kh), \quad k \in \mathbb{Z}_+ \tag{5.186b}$$

and the outputs described by

$$z_c(t) = \widehat{C}_c(t, \eta(t))x(t), \quad t \in \mathbb{R}_+ \tag{5.187a}$$

$$z_d(kh) = \widehat{C}_d(kh, \eta(kh))x(kh), \ k \in \mathbb{Z}_+ \tag{5.187b}$$

where

$$\widehat{A}_l(t,i) := A_l(t,i) - B_l(t,i)R_c^{-1}(t,i)L_c^\top(t,i), \quad 0 \le l \le r_1, \quad t \in \mathbb{R}_+ \tag{5.188a}$$

$$\widehat{A}_{dp}(kh,i) := A_{dp}(kh,i) - B_{dp}(kh,i)R_d^{-1}(kh,i)L_d^\top(kh,i), \ 0 \le p \le r_2, \ k \in \mathbb{Z}_+ \tag{5.188b}$$

$$\widehat{C}_c(t,i) := (M_c(t,i) - L_c(t,i)R_c^{-1}(t,i)L_c^\top(t,i))^{1/2}, \quad t \in \mathbb{R}_+ \tag{5.188c}$$

$$\widehat{C}_d(kh,i) := (M_d(kh,i) - L_d(kh,i)R_d^{-1}(kh,i)L_d^\top(kh,i))^{1/2}, \quad k \in \mathbb{Z}_+, \tag{5.188d}$$

$i \in \mathcal{N}$. The formulae (5.188) are well defined if (5.174) and (5.175) hold. According to Definition 4.6 and Remark 4.6, the JLSS (5.186)–(5.187) is mean-square detectable if and only if there exist $\tilde{\theta}$ periodic and piecewise continuous functions with finite lateral limits $t \to H_c(t,i) : \mathbb{R}_+ \to \mathbb{R}^{m_c \times n}$ and $\tilde{\mathbf{k}}$ periodic sequences $\{H_d(kh,i)\}_{k \in \mathbb{Z}_+} \subset \mathbb{R}^{m_d \times n}$, $i \in \mathcal{N}$, ($\tilde{\theta}$ and $\tilde{\mathbf{k}}$ being introduced in (5.166)) such that the resulting JSLDE:

$$\begin{aligned} \mathrm{d}x(t) = {} & (\widehat{A}_0(t,\eta(t)) + H_c(t,\eta(t))\widehat{C}_c(t,\eta(t)))x(t)\mathrm{d}t \\ & + \sum_{l=1}^{r_1} \widehat{A}_l(t,\eta(t))x(t)\mathrm{d}w_l(t), \quad kh < t \le (k+1)h \end{aligned} \tag{5.189a}$$

$$\begin{aligned} x(kh^+) = {} & (\widehat{A}_{d0}(kh,\eta(kh)) + H_d(kh,\eta(kh))\widehat{C}_d(kh,\eta(kh)))x(kh) \\ & + \sum_{p=1}^{r_2} w_{dp}(k)\widehat{A}_{dp}(kh,\eta(kh))x(kh), \ k \in \mathbb{Z}_+ \end{aligned} \tag{5.189b}$$

is ESMS. Now, we are in position to prove:

Theorem 5.8 *Assume:*

(a) *the Assumptions 1.2, 5.2 and 5.4 are fulfilled;*
(b) *the SJMRDE (5.10) has time periodic coefficients (in the sense of the Definition 5.8);*
(c) *the auxiliary JLSS (5.186)–(5.187) is mean-square detectable.*

Under these conditions, any positive semidefinite and $\tilde{\theta}$ periodic solution of (5.10) is a stabilizing solution of this equation.

Proof Let $\widetilde{\mathbf{X}}(\cdot) = (\widetilde{X}(\cdot,1), \widetilde{X}(\cdot,2), \ldots, \widetilde{X}(\cdot,N))$ be a positive semidefinite and $\tilde{\theta}$ periodic solution of (5.10). We have to show that the closed-loop JSLDE:

$$\mathrm{d}x(t) = (A_0(t,\eta(t)) + B_0(t,\eta(t))\widetilde{F}_c(t,\eta(t)))x(t)\mathrm{d}t$$

$$+\sum_{l=1}^{r_1}(A_l(t,\eta(t))+B_l(t,\eta(t))\widetilde{F}_c(t,\eta(t)))x(t)\mathrm{d}w_l(t),$$
$$kh < t \le (k+1)h \tag{5.190a}$$
$$x(kh^+) = (A_{d0}(kh,\eta(kh))+B_{d0}(kh,\eta(kh))\widetilde{F}_d(kh,\eta(kh)))x(kh)$$
$$+\sum_{p=1}^{r_2} w_{dp}(k)\left(A_{dp}(kh,\eta(kh))+B_{dp}(kh,\eta(kh))\widetilde{F}_d(kh,\eta(kh))\right)x(kh),$$
$$k\in\mathbb{Z}_+, \tag{5.190b}$$

is ESMS, where

$$\widetilde{F}_c(t,i) := -\left(\sum_{l=1}^{r_1} B_l^\top(t,i)\widetilde{X}(t,i)B_l(t,i)+R_c(t,i)\right)^{-1}$$
$$\left(B_0^\top(t,i)\widetilde{X}(t,i)+\sum_{l=1}^{r_1}B_l^\top(t,i)\widetilde{X}(t,i)A_l(t,i)+L_c^\top(t,i)\right),$$
$$(t,i)\in\mathbb{R}_+\times\mathcal{N},$$

$$\widetilde{F}_d(kh,i) := -\left(\sum_{p=0}^{r_2} B_{dp}^\top(kh,i)\widetilde{X}(kh,i)B_{dp}(kh,i)+R_d(kh,i)\right)^{-1}$$
$$\left(\sum_{p=0}^{r_2} B_{dp}^\top(kh,i)\widetilde{X}(kh,i)A_{dp}(kh,i)+L_d^\top(kh,i)\right),\quad (k,i)\in\mathbb{Z}_+\times\mathcal{N}.$$

By direct calculation we rewrite Eq. (5.10) where $X(t,i)$ is replaced by $\widetilde{X}(t,i)$, $(t,i)\in\mathbb{R}_+\times\mathcal{N}$, as:

$$-\dot{\widetilde{X}}(t,i) = \mathcal{L}^*_{c\widetilde{\mathbf{F}}_c}(t)[\widetilde{\mathbf{X}}(t)](i)+\Theta_{c\widetilde{\mathbf{F}}_c}(t,i),\quad kh\le t<(k+1)h \tag{5.191a}$$
$$\widetilde{X}(kh^-,i) = \mathcal{L}^*_{d\widetilde{\mathbf{F}}_d}(kh)[\widetilde{\mathbf{X}}(kh)](i)+\Theta_{d\widetilde{\mathbf{F}}_d}(kh,i),\quad k\in\mathbb{Z}_+, \tag{5.191b}$$

where

$$\mathcal{L}^*_{c\widetilde{\mathbf{F}}_c}(t)[\widetilde{\mathbf{X}}(t)](i) := (A_0(t,i)+B_0(t,i)\widetilde{F}_c(t,i))^\top\widetilde{X}(t,i)+\widetilde{X}(t,i)(A_0(t,i)+B_0(t,i)\widetilde{F}_c(t,i))$$
$$+\sum_{l=1}^{r_1}(A_l(t,i)+B_l(t,i)\widetilde{F}_c(t,i))^\top\widetilde{X}(t,i)(A_l(t,i)+B_l(t,i)\widetilde{F}_c(t,i))$$
$$+\sum_{j=1}^{N}q_{ij}\widetilde{X}(t,j), \tag{5.192a}$$

$$\Theta_{c\widetilde{\mathbf{F}}_c}(t,i) := \begin{pmatrix} I_n \\ \widetilde{F}_c(t,i) \end{pmatrix}^\top \begin{pmatrix} M_c(t,i) & L_c(t,i) \\ L_c^\top(t,i) & R_c(t,i) \end{pmatrix} \begin{pmatrix} I_n \\ \widetilde{F}_c(t,i) \end{pmatrix} \tag{5.192b}$$

$$\mathscr{L}^*_{d\widetilde{\mathbf{F}}_d}(kh)[\widetilde{\mathbf{X}}(kh)](i) := \sum_{p=0}^{r_2} (A_{dp}(kh,i) + B_{dp}(kh,i)\widetilde{F}_d(kh,i))^\top$$
$$\cdot \widetilde{X}(kh,i)(A_{dp}(kh,i) + B_{dp}(kh,i)\widetilde{F}_d(kh,i)) \tag{5.192c}$$

$$\Theta_{d\widetilde{\mathbf{F}}_d}(kh,i) := \begin{pmatrix} I_n \\ \widetilde{F}_d(kh,i) \end{pmatrix}^\top \begin{pmatrix} M_d(kh,i) & L_d(kh,i) \\ L_d^\top(kh,i) & R_d(kh,i) \end{pmatrix} \begin{pmatrix} I_n \\ \widetilde{F}_d(kh,i) \end{pmatrix}. \tag{5.192d}$$

Bearing in mind (5.174) and (5.175) we may rewrite (5.192b) and (5.192d), respectively, as:

$$\begin{aligned} \Theta_{c\widetilde{\mathbf{F}}_c}(t,i) &= (M_c(t,i) - L_c(t,i)R_c^{-1}(t,i)L_c^\top(t,i)) \\ &\quad + (\widetilde{F}_c(t,i) + R_c^{-1}(t,i)L_c^\top(t,i))^\top R_c(t,i)(\widetilde{F}_c(t,i) + R_c^{-1}(t,i)L_c^\top(t,i)) \\ &= \sum_{l=0}^{r_1} \mathbb{C}_{cl}^\top(t,i)\mathbb{C}_{cl}(t,i) \end{aligned} \tag{5.193a}$$

$$\begin{aligned} \Theta_{d\widetilde{\mathbf{F}}_d}(kh,i) &= (M_d(kh,i) - L_d(kh,i)R_d^{-1}(kh,i)L_d^\top(kh,i)) \\ &\quad + (\widetilde{F}_d(kh,i) + R_d^{-1}(kh,i)L_c^\top(kh,i))^\top R_d(kh,i)(\widetilde{F}_d(kh,i) \\ &\quad + R_d^{-1}(kh,i)L_d^\top(kh,i)) \\ &= \sum_{p=0}^{r_2} \mathbb{C}_{dp}^\top(kh,i)\mathbb{C}_{dp}(kh,i) \end{aligned} \tag{5.193b}$$

where

$$\mathbb{C}_{cl}(t,i) = \begin{pmatrix} \mathscr{C}_{cl}^0(t,i) \\ \mathscr{C}_{cl}^1(t,i) \\ \vdots \\ \mathscr{C}_{cl}^{r_1}(t,i) \end{pmatrix} \qquad \mathbb{C}_{dp}(kh,i) = \begin{pmatrix} \mathscr{C}_{dp}^0(kh,i) \\ \mathscr{C}_{dp}^1(kh,i) \\ \vdots \\ \mathscr{C}_{dp}^{r_2}(kh,i) \end{pmatrix}$$

with, for $l = 0$ we have

$$\mathscr{C}_{c0}^0(t,i) = \begin{pmatrix} \widehat{C}_c(t,i) \\ (r_1+1)^{-1/2} R_c^{1/2}(t,i)(\widetilde{F}_c(t,i) + R_c^{-1}(t,i)L_c^\top(t,i)) \end{pmatrix} \in \mathbb{R}^{(n+m_c)\times n}$$

$\mathscr{C}_{c0}^j(t,i) = 0 \in \mathbb{R}^{m_c\times n}$, $1 \le j \le r_1$, while for $1 \le l \le r_1$ $\mathscr{C}_{cl}^0(t,i) = 0 \in \mathbb{R}^{(n+m_c)\times n}$, $\mathscr{C}_{cl}^j(t,i) = 0 \in \mathbb{R}^{m_c\times n}$ if $j \ne l$ and

$$\mathscr{C}_{cl}^j(t,i) = \left((r_1+1)^{-1/2} R_c^{1/2}(t,i)(\widetilde{F}_c(t,i) + R_c^{-1}(t,i)L_c^\top(t,i)\right) \ \text{ if } \ j = l.$$

Similarly, for $p = 0$,

$$\mathscr{C}_{d0}^{0}(kh,i)=\begin{pmatrix}\widehat{C}_d(kh,i)\\(r_2+1)^{-1/2}R_d^{1/2}(kh,i)(\widetilde{F}_d(kh,i)+R_d^{-1}(kh,i)L_d^{\top}(kh,i))\end{pmatrix}\in\mathbb{R}^{(n+m_d)\times n}$$

$$\mathscr{C}_{d0}^{j}(kh,i)=0\in\mathbb{R}^{m_d\times n},\quad 1\le j\le r_2,\quad \text{while for }\ 1\le p\le r_2,$$

$$\mathscr{C}_{dp}^{0}(kh,i)=0\in\mathbb{R}^{(n+m_d)\times n},\quad \mathscr{C}_{dp}^{j}(kh,i)=0\in\mathbb{R}^{m_d\times n}\ \text{ if }\ j\neq p$$

and

$$\mathscr{C}_{dp}^{j}(kh,i)=\left((r_2+1)^{-1/2}R_d^{1/2}(kh,i)(\widetilde{F}_d(kh,i)+R_d^{-1}(kh,i)L_d^{\top}(kh,i)\right)$$
$$\text{if }\ j=p.$$

Plugging (5.193) in (5.191) we obtain that $\widetilde{\mathbf{X}}(\cdot)$ solves the non-homogeneous backward generalized Lyapunov jump linear differential equation:

$$\dot{\widetilde{X}}(t,i)+\mathscr{L}^{*}_{c\widetilde{\mathbf{F}}_c}(t)[\widetilde{\mathbf{X}}(t)](i)+\sum_{l=1}^{r_1}\mathbb{C}_{cl}^{\top}(t,i)\mathbb{C}_{cl}(t,i)=0,\quad kh\le t<(k+1)h \tag{5.194a}$$

$$X(kh^{-},i)=\mathscr{L}^{*}_{d\widetilde{\mathbf{F}}_d}(kh)[\mathbf{X}(kh)](i)+\sum_{p=0}^{r_2}\mathbb{C}_{dp}^{\top}(kh,i)\mathbb{C}_{dp}(kh,i),\quad k\in\mathbb{Z}_+. \tag{5.194b}$$

Bearing in mind (5.192) and (5.193) we may infer that (5.194) is a non-homogeneous backward GLJLDE of the form (4.124) associated to the JLSS having the states evolution described by (5.190) and the outputs described by:

$$d\mathbf{z}_c(t)=\mathbb{C}_{c0}(t,\eta(t))x(t)dt+\sum_{l=1}^{r_1}\mathbb{C}_{cl}(t,\eta(t))x(t)dw_l(t),\quad t\in\mathbb{R}_+ \tag{5.195a}$$

$$\mathbf{z}_d(kh)=\mathbb{C}_{d0}(kh,\eta(kh))x(kh)+\sum_{p=1}^{r_2}w_{dp}(k)\mathbb{C}_{dp}(kh,\eta(kh))x(kh),\ k\in\mathbb{Z}_+. \tag{5.195b}$$

So, to show that (5.190) is ESMS we apply Theorem 4.1 in the case of the JLSS (5.190) with the outputs (5.195) together with the non-homogeneous backward GLJLDE (5.194). We have to show that the JLSS (5.190) together with the outputs (5.195) is mean-square detectable. To this end we chose

$$\widetilde{H}_c(t,i)=(\widetilde{H}_c^{0}(t,i)\ \ \widetilde{H}_c^{1}(t,i)\ldots\widetilde{H}_c^{r_1}(t,i)),$$

$$\widetilde{H}_d(kh,i)=(\widetilde{H}_d^{0}(kh,i)\ \ \widetilde{H}_d^{1}(kh,i)\ldots\widetilde{H}_d^{r_2}(kh,i)),$$

defined by

$$\widetilde{H}_c^0(t,i) = (H_c(t,i) \quad -(r_1+1)^{1/2}B_0(t,i)R_c^{-1/2}(t,i)) \in \mathbb{R}^{n\times(n+m_c)}$$

and

$$\widetilde{H}_c^l(t,i) = -(r_1+1)^{1/2}B_l(t,i)R_c^{-1/2}(t,i), \quad 1 \le l \le r_1,$$

while

$$\widetilde{H}_d^0(kh,i) = (H_d(kh,i) \quad -(r_2+1)^{1/2}B_{d0}(kh,i)R_d^{-1/2}(kh,i)) \in \mathbb{R}^{n\times(n+m_d)}$$

and

$$\widetilde{H}_d^p(kh,i) = -(r_2+1)^{1/2}B_{dp}(kh,i)R_d^{-1/2}(kh,i), \quad 1 \le p \le r_2.$$

By direct calculation involving the explicit formulae for $\widetilde{H}_c^l(t,i)$ and $\mathbb{C}_{cl}(t,i)$, $0 \le l \le r_1$, $\widetilde{H}_d^p(kh,i)$ and $\mathbb{C}_{dp}(kh,i)$, $0 \le p \le r_2$, detailed before, we obtain that the resulting system:

$$\begin{aligned} dx(t) &= (A_0(t,\eta(t)) + B_0(t,\eta(t))\widetilde{F}_c(t,\eta(t)) + \widetilde{H}_c(t,\eta(t))\mathbb{C}_{c0}(t,\eta(t)))x(t)dt \\ &\quad + \sum_{l=1}^{r_1}(A_l(t,\eta(t)) + B_l(t,\eta(t))\widetilde{F}_c(t,\eta(t)) + \widetilde{H}_c(t,\eta(t))\mathbb{C}_{cl}(t,\eta(t))) \\ &\quad x(t)dw_l(t), \; kh < t \le (k+1)h \end{aligned} \tag{5.196a}$$

$$\begin{aligned} x(kh^+) &= (A_{d0}(kh,\eta(kh)) + B_{d0}(kh,\eta(kh))\widetilde{F}_d(kh,\eta(kh)) \\ &\quad + \widetilde{H}_d(kh,\eta(kh))\mathbb{C}_{d0}(kh,\eta(kh)))x(kh) \\ &\quad + \sum_{p=1}^{r_2} w_{dp}(k)(A_{dp}(kh,\eta(kh)) + B_{dl}(kh,\eta(kh))\widetilde{F}_d(kh,\eta(kh)) \\ &\quad + \widetilde{H}_d(kh,\eta(kh))\mathbb{C}_{dp}(kh,\eta(kh)))x(kh), \quad k \in \mathbb{Z}_+, \end{aligned} \tag{5.196b}$$

coincides with (5.189) and hence it is ESMS. Thus, we have shown that the JLSS having the states evolution described by (5.190) and the outputs evolution described by (5.195) is mean-square detectable. So, we may apply Theorem 4.1 to conclude that the closed-loop JSLDE (5.190) is ESMS. Thus the proof ends.

A first consequence of the result proved in the previous theorem is:

Corollary 5.11 *Under the assumption of Theorem 5.8 the SJMRDE (5.10) may have at most one $\tilde{\theta}$ periodic and positive semidefinite solution.*

Proof According to Theorem 5.8, each positive semidefinite $\tilde{\theta}$ periodic solution of (5.10) is a stabilizing solution of this equation. Then, Corollary 5.3 applied in the case of the equation (5.10) which is a special case of (5.2) allows us to conclude that the stabilizing and bounded solution is unique.

The next result extends to this framework a well-known result from the deterministic framework according to which under the stabilizing assumption, the detectability property of a controlled linear system is a sufficient condition for the existence of a bounded and stabilizing solution of a matrix Riccati differential equation.

Theorem 5.9 *Assume:*

(a) the Assumptions 1.2, 5.2 and 5.4 hold;
(b) the SJMRDE (5.10) has time periodic coefficients (in the sense of the Definition 5.8);
(c) the controlled jump linear stochastic system (5.188) is mean-square stabilizable by linear state feedback;
(d) the auxiliary jump linear stochastic system (5.186)–(5.187) is mean-square detectable.

Under these conditions, the stochastic jump matrix Riccati differential equation (5.10) has a unique positive semidefinite solution which is $\tilde{\theta}-$periodic function. Moreover, it is just the stabilizing solution of (5.10).

Proof Invoking Corollary 5.10 we may infer that under the assumptions $(a) - (c)$ from the statement, the SJMRDE (5.10) has two positive semidefinite solutions $\mathbf{X}_{max}(\cdot)$ and $\mathbf{X}_{min}(\cdot)$ which are $\tilde{\theta}-$periodic functions.

If in addition, the assumption (d) holds, both the maximal and minimal positive semidefinite solutions, $\mathbf{X}_{max}(\cdot)$, and $\mathbf{X}_{min}(\cdot)$ of (5.10) are stabilizing solutions for (5.10). From the uniqueness of the bounded and stabilizing solution of a SJMRDE, it follows that under the assumption (d) we have

$$\mathbf{X}_{max}(t) = \mathbf{X}_{min}(t), \quad \text{for all} \quad t \in \mathbb{R}_+.$$

Thus the proof is complete.

Remark 5.9 Employing (5.185) we may conclude that under the assumptions of Theorem 5.9, the stochastic jump matrix Riccati differential equation may have only one bounded and positive semidefinite solution.

5.5.1 The Minimal Positive Semidefinite Solution of a Backward Jump Matrix Linear Differential Equation with Riccati-Type Jumping Operator

Bearing in mind (5.172) we may infer that in the case when the SJMRDE (5.10) reduces to a BJMLDE with Riccati-type jumping operator of the form (5.168), the Assumption 5.4 holds if and only if the condition

$$M_l(t, i) \geq 0, \quad \text{for all} \quad (t, i) \in \mathbb{R}_+ \times \mathcal{N} \tag{5.197}$$

is satisfied together with (5.175).

The Definition 5.10 for the minimal and positive semidefinite solution of (5.10) is also applicable in the case of the BJMLDE of the form (5.168).

In the case of the equation (5.168) the Theorem 5.7 takes the form:

Theorem 5.10 *Assume:*

(a) the Assumptions 1.2, 5.2 hold;
(b) the weights matrices from (5.170) satisfy the sign conditions (5.175) and (5.197);
(c) the ICLSS (5.169) is mean-square stabilizable by impulses in a linear state feedback form.

Under these conditions the BJMLDE (5.168) has a unique minimal positive semidefinite solution $t \to \mathbf{X}_{min}(t) : \mathbb{R}_+ \to \mathcal{S}_n^{N+}$.

In the special case of a BJMLDE with Riccati-type jumping operator, the results from Proposition 5.3 and Corollary 5.10 stated from the SJMRDE (5.10) become:

Corollary 5.12 *Under the assumptions of Theorem 5.10 the following hold:*

(i) if the coefficients of the BJMLDE (5.168) are time-periodic (in the sense of Definition 5.8 (i)) then the unique minimal and positive semidefinite solution $\mathbf{X}_{min}(\cdot)$ of this equation is a periodic function of period $\tilde{\theta}$ where $\tilde{\theta}$ is defined in (5.166a);
(ii) if the BJMLDE (5.168) is time invariant (in the sense of Definition 5.8 (ii)) then its unique minimal and positive semidefinite solution $\widetilde{\mathbf{X}}_{min}(\cdot)$ is a periodic function of period equal to the length h of the dwell time.

Corollary 5.13 *Under the assumptions from Theorem 5.10 the BJMLDE with Riccati-type jumping operator (5.168) has two solutions* $t \to \mathbf{X}_{max}(t) : \mathbb{R}_+ \to \mathcal{S}_n^{N+}$ *and* $t \to \mathbf{X}_{min}(t) : \mathbb{R}_+ \to \mathcal{S}_n^{N+}$ *with the property that*

$$0 \preceq \mathbf{X}_{min}(t) \preceq \widetilde{\mathbf{X}}(t) \preceq \mathbf{X}_{max}(t) \preceq \nu \mathbf{I}_n, \quad \text{for all } \ t \in \mathbb{R}_+,$$

for any bounded and positive semidefinite solutions $t \to \widetilde{\mathbf{X}}(t) : \mathbb{R}_+ \to \mathcal{S}_n^{N+}$ *of (5.168),* $\nu > 0$ *being a constant and* $\mathbf{I}_n = (I_n \ I_n \ \dots \ I_n) \in \mathcal{S}_n^{N+}$*. If the BJMLDE (5.168) is time periodic (in the sense of the Definition 5.8 (i)) then, both $\mathbf{X}_{max}(\cdot)$, as well as $\mathbf{X}_{min}(\cdot)$ are $\tilde{\theta}$ periodic functions with $\tilde{\theta}$ computed as in (5.166a).*

If the BJMLDE (5.168) is time invariant (in the sense of Definition 5.8 (ii)) then both $\mathbf{X}_{max}(\cdot)$, as well as $\mathbf{X}_{min}(\cdot)$ are periodic functions with the period equal to the length h of the dwell time.

Further, we assume that the BJMLDE (5.168) is time periodic (in the sense of the Definition 5.8). Employing the matrix coefficients of (5.168) we construct the following auxiliary JLSS having the evolution of the states described by

$$dx(t) = A_0(t, \eta(t))x(t)dt + \sum_{l=1}^{r_1} A_l(t, \eta(t))x(t)dw_l(t), \quad kh < t \leq (k+1)h \tag{5.198a}$$

$$x(kh^+) = \widehat{A}_{d0}(kh, \eta(kh)) + \sum_{p=1}^{r_2} w_{dp}(k)\widehat{A}_{dp}(kh, \eta(kh))x(kh), \quad k \in \mathbb{Z}_+, \tag{5.198b}$$

and the outputs:

$$z_d(kh) = \widehat{C}_d(kh, \eta(kh))x(kh), \quad k \in \mathbb{Z}_+ \tag{5.199}$$

where $\widehat{A}_{dp}(kh, i)$, $0 \le p \le r_2$, $k \in \mathbb{Z}_+$, $i \in \mathscr{N}$ are defined as in (5.188b) and $\widehat{C}_d(kh, i), k \in \mathbb{Z}_+, i \in \mathscr{N}$ are defined as in (5.188d). We recall that the JLSS (5.198)–(5.199) is mean-square detectable by impulses if there exist matrix-valued sequences $\{H_d(kh, i)\}_{k \in \mathbb{Z}_+}$ which are periodic of period $\tilde{\mathbf{k}}$ (with $\tilde{\mathbf{k}}$ as in (5.166b)) such that the resulting JSLDE:

$$\mathrm{d}x(t) = A_0(t, \eta(t))x(t)\mathrm{d}t + \sum_{l=1}^{r_1} A_l(t, \eta(t))x(t)\mathrm{d}w_l(t), \quad kh < t \le (k+1)h \tag{5.200a}$$

$$x(kh^+) = (\widehat{A}_{d0}(kh, \eta(kh)) + H_d(kh, \eta(kh))\widehat{C}_d(kh, \eta(kh)))x(kh) + \sum_{p=1}^{r_2} w_{dp}(k)\widehat{A}_{dp}(kh, \eta(kh))x(kh), \quad k \in \mathbb{Z}_+, \tag{5.200b}$$

is ESMS.

In the particular case of the BJMLDE (5.168), Theorem 5.8 becomes:

Corollary 5.14 *Assume:*

(a) the Assumptions 1.2, 5.2 hold;
(b) the weights matrices from (5.170) are satisfying the sign conditions (5.175) and (5.197);
(c) the BJMLDE (5.168) has time periodic coefficients (in the sense of Definition 5.8);
(d) the auxiliary JLSS (5.198) and (5.199) is mean-square detectable by impulses.

Under these conditions any positive semidefinite and $\tilde{\theta}$ periodic solution of (5.168) is a stabilizing solution of this equation.

Finally, summarizing the results stated in Corollary 5.13 and Corollary 5.14 we obtain the following version of Theorem 5.9 adapted to the particular case of the BJMLDE (5.168).

Corollary 5.15 *Assume:*

(a) the Assumptions 1.2, 5.2 hold;
(b) the weights matrices from (5.170) are satisfying the sign conditions (5.175) and (5.197);
(c) the BJMLDE (5.168) has periodic coefficients (in the sense of Definition 5.8);

(d) the ICLSS (5.169) is mean-square stabilizable by impulses in form of linear state feedback;
(e) the auxiliary JLSS (5.198)–(5.199) is mean-square detectable by impulses.

Under these conditions the BJMLDE with Riccati-type jumping operator (5.168) has a unique positive semidefinite solution $t \to \widetilde{\mathbf{X}}(t) : \mathbb{R} \to \mathcal{S}_n^{N+}$*. Moreover* $\widetilde{\mathbf{X}}(\cdot)$ *is a* $\tilde{\theta}$*−periodic function and it coincides to the unique bounded on* $\mathbb{R}$ *and stabilizing solution of the BJMLDE (5.168).*

From Theorem 5.9 and Corollary 5.15, respectively, one sees that under a detectability assumption of some auxiliary JLSSs adequately chosen, the SJM-RDE (5.10) and BJMLDE (5.168), respectively, may have at most a single positive semidefinite solution.

The next example shows that in the absence of the assumption (*e*) from Corollary 5.15 a BJMLDE of the form (5.168) may has at least two positive semidefinite solutions.

Example 5.1 Consider a special case of a time-invariant BJMLDE with Riccati-type jumping operator with $\mathcal{N} = 1$, $q_{11} = 0$, $n = 2$, $m_d = 1$,

$$-\dot{X}(t) = A_0^\top X(t) + X(t)A_0 + M_c, \quad kh \le t < (k+1)h \tag{5.201a}$$

$$\begin{aligned} X(kh^+) &= A_{d0}^\top X(kh)A_{d0} + A_{d1}^\top X(kh)A_{d1} + M_d \\ &\quad - (A_{d0}^\top X(kh)B_{d0} + A_{d1}^\top X(kh)B_{d1} + L_d) \\ &\quad (R_d + B_{d0}^\top X(kh)B_{d0} + B_{d1}^\top X(kh)B_{d1})^{-1} \\ &\quad \cdot (B_{d0}^\top X(kh)A_{d0} + B_{d1}^\top X(kh)A_{d1} + L_d^\top), \quad k \in \mathbb{Z}, \end{aligned} \tag{5.201b}$$

where

$$A_0 = \begin{pmatrix} a & a-b \\ 0 & b \end{pmatrix}, \quad a, b \in \mathbb{R}, \quad M_c = \begin{pmatrix} 0 & 0 \\ 0 & 0 \end{pmatrix},$$

$$A_{d0} = \begin{pmatrix} \frac{4}{5}\mathrm{e}^{-ah} & \frac{8}{5}(\mathrm{e}^{-ah} - \mathrm{e}^{-bh}) \\ 0 & \frac{8}{5}\mathrm{e}^{-ah} \end{pmatrix}, \quad A_{d1} = \frac{3}{5}\begin{pmatrix} \mathrm{e}^{-ah} & \mathrm{e}^{-ah} - \mathrm{e}^{-bh} \\ 0 & \mathrm{e}^{-bh} \end{pmatrix}$$

$$B_{d0} = \begin{pmatrix} 3\mathrm{e}^{-ah} - \mathrm{e}^{-bh} \\ \mathrm{e}^{-bh} \end{pmatrix}, \quad B_{d1} = \begin{pmatrix} 0 \\ 0 \end{pmatrix} \quad M_d = \begin{pmatrix} \frac{16}{25} & 0 \\ 0 & 0 \end{pmatrix} \quad L_d = \begin{pmatrix} 0 \\ 0 \end{pmatrix},$$

$R_d = 1$, $h > 0$ be fixed. In this particular case, the auxiliary JLSS (5.198) and (5.199) becomes

$$\mathrm{d}x(t) = A_0 x(t)\mathrm{d}t, \quad kh < t \le (k+1)h \tag{5.202a}$$

$$x(kh^+) = (A_{d0} + w_d(k)A_{d1})x(kh), \tag{5.202b}$$

$$z_d(kh) = \widehat{C}_d x(kh), \quad k \in \mathbb{Z}_+, \tag{5.202c}$$

with

$$\widehat{C}_d = \begin{pmatrix} \frac{4}{5} & 0 \\ 0 & 0 \end{pmatrix}$$

and $\{w_d(k)\}_{k \in \mathbb{Z}_+}$ is a sequence of independent random variables with zero mean and variance 1. We first show that the JLSS (5.202) is not mean-square detectable by impulses.

If JLSS (5.202) would be mean-square detectable by impulses, then it would exists $H_d \in \mathbb{R}^{2\times 2}$ such that the resulting JSLDE:

$$dx(t) = A_0 x(t)dt, \quad kh < t \leq (k+1)h \tag{5.203a}$$

$$x(kh^+) = (A_{d0} + H_d\widehat{C}_d)x(kh) + w_d(k)A_{d1}x(kh), \quad k \in \mathbb{Z}_+ \tag{5.203b}$$

is ESMS. According to the equivalence $(i) \Leftrightarrow (iii)$ from Corollary 3.4, we may infer that the JSLDE (5.203) is ESMS if and only if would exists a positive definite matrix $\Gamma \in \mathcal{S}_2^1$ with the property that the non-homogeneous backward GLJLDE:

$$-\dot{Y}(t) = A_0^\top Y(t) + Y(t)A_0, \quad kh \leq t < (k+1)h \tag{5.204a}$$

$$\begin{aligned} Y(kh^-) = {} & (A_{d0} + H_d\widehat{C}_d)^\top Y(kh)(A_{d0} + H_d\widehat{C}_d) \\ & + A_{d1}^\top Y(kh)A_{d1} + \Gamma, \quad k \in \mathbb{Z} \end{aligned} \tag{5.204b}$$

has a unique solution $t \to \widetilde{Y}(t) : \mathbb{R} \to \mathcal{S}_2^1$ which is a periodic function with period h and satisfies

$$\widetilde{Y}(t) > 0, \quad \text{for all } \ t \in \mathbb{R}.$$

From (5.204b) together with the periodicity property of $\widetilde{Y}(\cdot)$ we deduce that

$$\widetilde{Y}((k+1)h^-) = \widetilde{Y}(kh^-) \triangleq Y, \quad \text{for all } \ k \in \mathbb{Z}.$$

Invoking again (5.204b), we deduce that

$$Y \geq \Gamma > 0.$$

From (5.204a) we obtain

$$Y(kh) = e^{A_0^\top h} Y e^{A_0 h}.$$

Substituting this equality in (5.204b) we get, via (5.201b), that Y solves the linear equation:

$$Y = (\widehat{A}_{d0} + \widehat{H}_d\widehat{C}_d)^\top Y(\widehat{A}_{d0} + \widehat{H}_d\widehat{C}_d) + \frac{9}{25}Y + \Gamma \tag{5.205}$$

where

$$\widehat{A}_{d0} = \begin{pmatrix} \frac{4}{5} & 0 \\ 0 & \frac{8}{5} \end{pmatrix}, \quad \widehat{H}_d = e^{A_0 h} H_d.$$

Setting

$$Y = \begin{pmatrix} y_1 & y_2 \\ y_2 & y_3 \end{pmatrix}, \quad \Gamma = \begin{pmatrix} \gamma_1 & \gamma_2 \\ \gamma_2 & \gamma_3, \end{pmatrix}$$

it follows that $y_3 > 0$ and $\gamma_3 > 0$. Partitioning (5.205) we obtain that its 22-component leads to

$$\frac{16}{25} y_3 + \gamma_3 = 0$$

which is in contradiction with $y_3 > 0$ and $\gamma_3 > 0$. Hence, the auxiliary JLSS (5.202) is not mean-square detectable by impulses.

Let $\widetilde{X}(t) : \mathbb{R} \to \mathscr{S}_2^1$ and $t \to \widetilde{\widetilde{X}}(t) : \mathbb{R} \to \mathscr{S}_2^1$ be defined by

$$\widetilde{X}(t) = \mathrm{e}^{A_0^\top((k+1)h-t)} \widetilde{\mathbb{X}} \mathrm{e}^{A_0((k+1)h-t)}, \quad kh \le t < (k+1)h, \quad k \in \mathbb{Z} \tag{5.206a}$$

$$\widetilde{\widetilde{X}}(t) = \mathrm{e}^{A_0^\top((k+1)h-t)} \widetilde{\widetilde{\mathbb{X}}} \mathrm{e}^{A_0((k+1)h-t)}, \quad kh \le t < (k+1)h, \quad k \in \mathbb{Z} \tag{5.206b}$$

where

$$\widetilde{\mathbb{X}} = \begin{pmatrix} \frac{1}{2}(7+\sqrt{2}) & -3(3+\sqrt{2}) \\ -3(3+\sqrt{2}) & 3(11+6\sqrt{2}) \end{pmatrix} \tag{5.207a}$$

$$\widetilde{\widetilde{\mathbb{X}}} = \begin{pmatrix} \frac{1}{2}(1+\sqrt{2}) & 0 \\ 0 & 0 \end{pmatrix}. \tag{5.207b}$$

By direct calculation one may check that $\widetilde{X}(\cdot)$ and $\widetilde{\widetilde{X}}(\cdot)$ defined by (5.206) are positive semidefinite solutions of BJMLDE (5.201). Moreover, $\widetilde{X}(\cdot)$ is the stabilizing solution of (5.201). To show this, one verify that the closed-loop JSLDE:

$$\mathrm{d}x(t) = A_0 x(t)\mathrm{d}t, \quad kh < t \le (k+1)h \tag{5.208a}$$

$$x(kh^+) = (A_{d0} + B_{d0}\widetilde{F}(kh))x(kh) + w_d(k)A_{d1}x(kh), \quad k \in \mathbb{Z}_+ \tag{5.208b}$$

is ESMS, where

$$\widetilde{F}(kh) := -(1 + B_{d0}^\top \widetilde{X}(kh) B_{d0})^{-1} B_{d0}^\top \widetilde{X}(kh) A_{d0}.$$

From (5.206a) we get

$$\widetilde{X}(kh) = \mathrm{e}^{A_0^\top h} \widetilde{\mathbb{X}} \mathrm{e}^{A_0 h}.$$

Thus,

$$\widetilde{F}(kh) = -(1 + \widehat{B}_{d0}^\top \widetilde{\widetilde{\mathbb{X}}} \widehat{B}_{d0})^{-1} \widetilde{B}_{d0}^\top \widetilde{\mathbb{X}} \widehat{A}_{d0} \triangleq \widetilde{F} \tag{5.209}$$

with

$$\widehat{B}_{d0} = e^{A_0 h} B_{d0} = \begin{pmatrix} 2 \\ 1 \end{pmatrix}, \quad \widehat{A}_{d0} = e^{A_0 h} A_{d0} = \begin{pmatrix} \frac{4}{5} & 0 \\ 0 & \frac{8}{5} \end{pmatrix}.$$

Taking into account (5.209) we deduce that (5.208) is a time-invariant JSLDE. Invoking again the equivalence $(i) \Leftrightarrow (iii)$ from Corollary 3.4, we may infer that the JSLDE (5.208) is ESMS if and only if there exists $\Gamma \in \mathcal{S}_2^1$, $\Gamma > 0$ such that the non-homogeneous backward GLJLDE:

$$-\dot{Y}(t) = A_0^\top Y(t) + Y(t) A_0, \quad kh \le t < (k+1)h \tag{5.210a}$$

$$\begin{aligned} Y(kh^-) &= (A_{d0} + B_{d0}\widetilde{F})^\top Y(kh)(A_{d0} + B_{d0}\widetilde{F}) \\ &\quad + A_{d1}^\top Y(kh) A_{d1} + \Gamma, \quad k \in \mathbb{Z} \end{aligned} \tag{5.210b}$$

has a unique solution $t \to \widehat{Y}(t) : \mathbb{R} \to \mathcal{S}_2^1$ which is a periodic function of period h and

$$\widehat{Y}(t) > 0, \quad \text{for all } t \in \mathbb{R}.$$

From the periodicity property of the function $\widehat{Y}(\cdot)$ we deduce via (5.210b) that

$$\widehat{Y}((k+1)h^-) = \widehat{Y}(kh^-) \stackrel{\Delta}{=} \widehat{Y}, \quad \text{for all } k \in \mathbb{Z}.$$

From (5.210a) we obtain

$$\widehat{Y}(kh) = e^{A_0^\top h} \widehat{Y} e^{A_0 h}.$$

Plugging this equality in (5.210b) we obtain that $\widehat{Y}$ solves the linear equation:

$$\widehat{Y} = (\widehat{A}_{d0} + \widehat{B}_{d0}\widetilde{F})^\top \widehat{Y} (\widehat{A}_{d0} + \widehat{B}_{d0}\widetilde{F}) + \frac{9}{25} Y + \Gamma. \tag{5.211}$$

To show that (5.210) has a $h-$periodic and positive definite solution it is sufficient to show that (5.211) has a positive definite solution $\widehat{Y}$. To this end we rewrite (5.211) in an equivalent form:

$$\widehat{Y} = \mathscr{A}^\top \widehat{Y} \mathscr{A} + \Gamma \tag{5.212}$$

where

$$\mathscr{A} = \frac{5}{4}(\widehat{A}_{d0} + \widehat{B}_{d0}\widetilde{F}).$$

By direct calculation one sees that the eigenvalues of the matrix $\mathscr{A}$ are located in the disk $|\lambda| < 1$. Hence, (5.212) has a unique positive definite solution if $\Gamma > 0$. If $\widehat{Y}$ is the unique positive definite solution of (5.212) we set

$$\widehat{Y}(t) = e^{A_0^\top((k+1)h - t)} \widehat{Y} e^{A_0((k+1)h - t)}, \quad \text{for } t \in [kh, (k+1)h), \quad k \in \mathbb{Z}.$$

Thus, $\widehat{Y}$ defined above, is the h−periodic and positive definite solution of (5.210). This guarantee that the JSLDE (5.208) is ESMS.

5.6 Stochastic Jump Matrix Riccati Differential Equations of Filtering

Consider the forward jump matrix Riccati-type differential equation:

$$\begin{aligned}\dot{Y}(t,i) &= A_0(i)Y(t,i) + Y(t,i)A_0^\top(i) + \sum_{l=1}^{r_1} A_l(i)Y(t,i)A_l^\top(i)\\ &\quad + \sum_{j=1}^{N} q_{ji}Y(t,j) - \left(Y(t,i)C_{c0}^\top(i) + \sum_{l=1}^{r_1} A_l(i)Y(t,i)C_{cl}^\top(i) + L_c(i)\right) \qquad (5.213a)\\ &\quad \cdot \left(\sum_{l=1}^{r_1} C_{cl}(i)Y(t,i)C_{cl}^\top(i) + R_c(i)\right)^{-1}\\ &\quad \left(C_{c0}(i)Y(t,i) + \sum_{l=1}^{r_1} C_{cl}(i)Y(t,i)A_l^\top(i) + L_c^\top(i)\right)\\ &\quad + M_c(i), \quad kh < t \le (k+1)h\\ Y(kh^+,i) &= \sum_{p=0}^{r_2} A_{dp}(i)Y(kh,i)A_{dp}^\top(i)\\ &\quad - \left(\sum_{p=0}^{r_2} A_{dp}(i)Y(kh,i)C_{dp}^\top(i) + L_d(i)\right)\\ &\quad \cdot \left(\sum_{p=0}^{r_2} C_{dp}(i)Y(kh,i)C_{dp}^\top(i) + R_d(i)\right)^{-1}\\ &\quad \left(\sum_{p=0}^{r_2} C_{dp}(i)Y(kh,i)A_{dp}^\top(i) + L_d^\top(i)\right)\\ &\quad + M_d(i), \quad k \in \mathbb{Z}, \qquad (5.213b)\end{aligned}$$

$i \in \mathcal{N}$, with the unknown function $\mathbf{Y}(t) = (Y(t,1), Y(t,2), \dots, Y(t,N)) \in \mathcal{S}_n^N$. In (5.213), $A_l(i) \in \mathbb{R}^{n\times n}$, $C_{cl}(i) \in \mathbb{R}^{n_{zc}\times n}$, $0 \le l \le r_1$, $A_{dp}(i) \in \mathbb{R}^{n\times n}$, $C_{dp}(i) \in \mathbb{R}^{n_{zd}\times n}$, $0 \le p \le r_2$, $M_c(i) \in \mathcal{S}_n$, $R_c(i) \in \mathcal{S}_{n_{zc}}$, $L_c(i) \in \mathbb{R}^{n\times n_{zc}}$, $M_d(i) \in \mathcal{S}_n$, $R_d(i) \in \mathcal{S}_{n_{zd}}$, $L_d(i) \in \mathbb{R}^{n\times n_{zd}}$, $i \in \mathcal{N}$ are known matrices and $q_{ij} \in \mathbb{R}$ are satisfying conditions (1.22).

Beside the FJMRDE (5.213) we consider the JLSS with state evolution described by

$$\mathrm{d}x(t) = A_0(\eta(t))x(t)\mathrm{d}t + \sum_{l=1}^{r_1} A_l(\eta(t))x(t)\mathrm{d}w_l(t), \quad kh < t \le (k+1)h \tag{5.214a}$$

$$x(kh^+) = A_{d0}(\eta(kh))x(kh) + \sum_{p=1}^{r_2} w_{dp}(k)A_{dp}(\eta(kh))x(kh), \quad k \in \mathbb{Z}_+ \tag{5.214b}$$

and the outputs

$$dz_c(t) = C_{c0}(\eta(t))x(t)\mathrm{d}t + \sum_{l=1}^{r_1} C_{cl}(\eta(t))x(t)\mathrm{d}w_l(t), \quad t \in \mathbb{R}_+ \tag{5.215a}$$

$$z_d(kh) = C_{d0}(\eta(kh))x(kh) + \sum_{p=1}^{r_2} w_{dp}(k)C_{dp}(\eta(kh))x(kh), \quad k \in \mathbb{Z}_+ \tag{5.215b}$$

where the stochastic processes $\{w(t)\}_{t\ge 0}$ $(w(t) = (w_1(t), w_2(t), dots, w_{r_1}(t))^\top)$, $\{w_d(k)\}_{k\in\mathbb{Z}_+}$ $(w_d(k) = (d_{d1}(k), \dots, d_{dr_2}(k))^\top)$ and $\{\eta(t)\}_{t\ge 0}$ satisfy the Assumption 1.2.

Definition 5.11 A solution $t \to \mathbf{Y}_s(t) = (Y_s(t,1), \dots, Y_s(t,N)) : \mathbb{R} \to \mathcal{S}_n^N$ of (5.213) is stabilizing solution if the matrices $\sum_{l=1}^{r_1} C_{cl}(i)Y_s(t,i)C_{cl}^\top(i) + R_c(i), \quad t \in \mathbb{R}$ and $\sum_{p=0}^{r_2} C_{dp}(i)Y_s(kh,i)C_{dp}^\top(i) + R_d(i), \quad k \in \mathbb{Z}, \quad i \in \mathcal{N}$ are invertible and the closed-loop system:

$$\begin{aligned}\mathrm{d}x(t) &= (A_0(\eta(t)) + K_{c,s}(t,\eta(t))C_{c0}(\eta(t)))x(t)\mathrm{d}t \\ &\quad + \sum_{l=1}^{r_1}(A_l(\eta(t)) + K_{c,s}(t,\eta(t))C_{cl}(\eta(t)))x(t)\mathrm{d}w_l(t), \quad kh < t \le (k+1)h\end{aligned} \tag{5.216a}$$

$$\begin{aligned}x(kh^+) &= (A_{d0}(\eta(kh)) + K_{d,s}(kh,\eta(kh))C_{d0}(\eta(kh)))x(kh) \\ &\quad + \sum_{p=1}^{r_2} w_{dp}(k)(A_{dp}(\eta(kh)) + K_{d,s}(kh,\eta(kh))C_{dp}(\eta(kh)))x(kh), \quad k \in \mathbb{Z}_+,\end{aligned} \tag{5.216b}$$

is ESMS, where

$$\begin{aligned}K_{c,s}(t,i) &\triangleq -\left(Y_s(t,i)C_{c0}^\top(i) + \sum_{l=1}^{r_1} A_l(i)Y_s(t,i)C_{cl}^\top(i) + L_c(i)\right) \\ &\quad \cdot \left(\sum_{l=1}^{r_1} C_{cl}(i)Y_s(t,i)C_{cl}^\top(i) + R_c(i)\right)^{-1}\end{aligned} \tag{5.217a}$$

$$K_{d,s}(kh,i) \stackrel{\Delta}{=} -\left(\sum_{p=0}^{r_2} C_{dp}(i)Y_s(kh,i)C_{dp}^\top(i) + L_d(i)\right)$$
$$\cdot\left(\sum_{p=0}^{r_2} C_{dp}(i)Y_s(kh,i)C_{dp}^\top(i) + R_d(i)\right)^{-1}, \quad i \in \mathscr{N}. \tag{5.217b}$$

To derive a set of conditions that guarantee the existence and the unicity of the stabilizing and bounded solution of the FJMRDE (5.213), we shall use a duality procedure similar to that from Section 4.6 through the FJMRDE (5.213) is transformed into a GBJMRDE of type (5.2) to which we apply the developments from Sect. 5.4.

If $t \to \mathbf{Y}(t) = (Y(t,1),\ldots,Y(t,N)) : \mathscr{I}_\mathbf{Y} \subset \mathbb{R} \to \mathscr{S}_n^N$ is a solution of (5.213), we define $\mathbf{X}(t) = (X(t,1),\ldots,X(t,N))$ by $X(t,i) \stackrel{\Delta}{=} Y(-t,i),\ \ t \in \mathscr{I}_\mathbf{X},\ \ i \in \mathscr{N}$. By direct calculation we get that $t \to \mathbf{X}(t) : \mathscr{I}_\mathbf{X} \subset \mathbb{R} \to \mathscr{S}_n^N$ solves

$$\begin{aligned} -\dot{X}(t,i) &= \breve{A}^\top(i)X(t,i) + X(t,i)\breve{A}(i) + \breve{\Pi}_{c1}[\mathbf{X}(t)](i) \\ &\quad - (X(t,i)\breve{B}(i) + \breve{\Pi}_{c2}[\mathbf{X}(t)](i) + L_c(i))(\breve{\Pi}_{c3}[\mathbf{X}(t)](i) + R_c(i))^{-1} \\ &\quad \cdot (X(t,i)\breve{B}(i) + \breve{\Pi}_{c2}[\mathbf{X}(t)](i) + L_c(i))^\top + M_c(i), \\ &\quad\quad kh \le t < (k+1)h \end{aligned} \tag{5.218a}$$

$$\begin{aligned} X(kh^-,i) &= \breve{\Pi}_{d1}[\mathbf{X}(kh)](i) - (\breve{\Pi}_{d2}[\mathbf{X}(kh)](i) + L_d(i)) \\ &\quad \cdot (\breve{\Pi}_{d3}[\mathbf{X}(kh)](i) + R_d(i))^{-1}(\breve{\Pi}_{d2}[\mathbf{X}(kh)](i) + L_d(i))^\top + M_d(i), \\ &\quad\quad k \in \mathbb{Z} \end{aligned} \tag{5.218b}$$

where

$$\breve{A}(i) \stackrel{\Delta}{=} A_0^\top(i) + \frac{1}{2}q_{ii}I_n, \qquad \breve{B}(i) \stackrel{\Delta}{=} C_{c0}^\top(i) \tag{5.219a}$$

$$\breve{\Pi}_{c1}[\mathbf{X}](i) \stackrel{\Delta}{=} \sum_{l=1}^{r_1} A_l(i)X(i)A_l^\top(i) + \sum_{\substack{j=1\\ j\neq i}}^{N} q_{ji}X(j) \tag{5.219b}$$

$$\breve{\Pi}_{c2}[\mathbf{X}](i) \stackrel{\Delta}{=} \sum_{l=1}^{r_1} A_l(i)X(i)C_{cl}^\top(i) \tag{5.219c}$$

$$\breve{\Pi}_{c3}[\mathbf{X}](i) \stackrel{\Delta}{=} \sum_{l=1}^{r_1} C_{cl}(i)X(i)C_{cl}^\top(i) \tag{5.219d}$$

$$\breve{\Pi}_{d1}[\mathbf{X}](i) \stackrel{\Delta}{=} \sum_{p=0}^{r_2} A_{dp}(i)X(i)A_{dp}^\top(i) \tag{5.219e}$$

$$\breve{\Pi}_{d2}[\mathbf{X}](i) \stackrel{\Delta}{=} \sum_{p=0}^{r_2} A_{dp}(i)X(i)C_{dp}^\top(i) \tag{5.219f}$$

$$\breve{\Pi}_{d3}[\mathbf{X}](i) \triangleq \sum_{p=0}^{r_2} C_{dp}(i)X(i)C_{dp}^{\top}(i), \tag{5.219g}$$

for all $\mathbf{X} = (X(1), \ldots, X(N)) \in \mathcal{S}_n^N$. Further, we denote

$$\breve{\Pi}_c[\mathbf{X}](i) := \begin{pmatrix} \breve{\Pi}_{c1}[\mathbf{X}](i) & \breve{\Pi}_{c2}[\mathbf{X}](i) \\ (\breve{\Pi}_{c2}[\mathbf{X}](i))^{\top} & \breve{\Pi}_{c3}[\mathbf{X}](i) \end{pmatrix}$$

$$\breve{\Pi}_d[\mathbf{X}](i) := \begin{pmatrix} \breve{\Pi}_{d1}[\mathbf{X}](i) & \breve{\Pi}_{d2}[\mathbf{X}](i) \\ (\breve{\Pi}_{d2}[\mathbf{X}](i))^{\top} & \breve{\Pi}_{d3}[\mathbf{X}](i) \end{pmatrix}$$

$$\breve{\boldsymbol{\Pi}}_c[\mathbf{X}] = (\breve{\Pi}_c[\mathbf{X}](1), \ldots, \breve{\Pi}_c[\mathbf{X}](N)), \tag{5.220a}$$

$$\breve{\boldsymbol{\Pi}}_d[\mathbf{X}] = (\breve{\Pi}_d[\mathbf{X}](1), \ldots, \breve{\Pi}_d[\mathbf{X}](N)), \tag{5.220b}$$

for all $\mathbf{X} \in \mathcal{S}_n^N$, $\breve{\Theta}_c = (\breve{\Theta}_c(1), \breve{\Theta}_c(2), \ldots, \breve{\Theta}_c(N))$, $\breve{\Theta}_d = (\breve{\Theta}_d(1), \breve{\Theta}_d(2), \ldots, \breve{\Theta}_d(N))$,

$$\breve{\Theta}_c(i) = \begin{pmatrix} M_c(i) & L_c(i) \\ L_c^{\top}(i) & R_c(i) \end{pmatrix} \tag{5.221a}$$

$$\breve{\Theta}_d(i) = \begin{pmatrix} M_d(i) & L_d(i) \\ L_d^{\top}(i) & R_d(i) \end{pmatrix} \tag{5.221b}$$

$$\breve{\mathbf{A}} = (\breve{A}(1), \breve{A}(2), \ldots, \breve{A}(N)) \tag{5.222a}$$

$$\breve{\mathbf{B}} = (\breve{B}(1), \breve{B}(2), \ldots, \breve{B}(N)). \tag{5.222b}$$

In this way we may identify the GBJMRDE (5.218) with the quintuple

$$\breve{\Sigma} = (\breve{\mathbf{A}}, \breve{\mathbf{B}}, (\breve{\boldsymbol{\Pi}}_c \mid \breve{\boldsymbol{\Pi}}_d), (\breve{\boldsymbol{\Theta}}_c \mid \breve{\boldsymbol{\Theta}}_d), \mathfrak{T}), \tag{5.223}$$

where $\mathfrak{T} = \{kh\}_{k \in \mathbb{Z}}$.

First we highlight the following duality relation:

Proposition 5.4 *Under the Assumption 1.2 the following are equivalent:*

(i) the JLSS (5.214)–(5.215) is mean-square detectable;
(ii) the quadruple $(\breve{\mathbf{A}}, \breve{\mathbf{B}}, (\breve{\boldsymbol{\Pi}}_c \mid \breve{\boldsymbol{\Pi}}_d), \mathfrak{T})$ is stabilizable in the sense of Definition 5.4.

Proof According to Definition 4.6 and Remark 4.6 (b) if the JLSS (5.214)–(5.215) is mean-square detectable then there exists a bounded function

$$\mathbf{H}_c(\cdot) = (H_c(\cdot, 1), H_c(\cdot, 2), \ldots, Hc(\cdot, N)) \in \mathscr{PC}(\mathbb{R}, \mathscr{M}^N_{nn_{z_c}}; \mathfrak{T})$$

periodic with period h and

$$\mathbf{H}_d = (H_d(1), H_d(2), \ldots, H_d(N)) \in \mathscr{M}^N_{nn_{z_d}}$$

with the property that the resulting JSLDE

$$\begin{aligned} dx(t) &= (A_0(\eta(t)) + H_c(t, \eta(t))C_{c0}(\eta(t)))x(t)dt \qquad (5.224a)\\ &\quad + \sum_{l=1}^{r_1}(A_l(\eta(t)) + H_c(t, \eta(t))C_{c,l}(\eta(t)))x(t)dw_l(t), \quad kh < t \le (k+1)h \\ x(kh^+) &= (A_{d0}(\eta(kh)) + H_d(\eta(kh))C_{d0}(\eta(kh)))x(kh) \qquad (5.224b)\\ &\quad + \sum_{p=1}^{r_2} w_{dp}(k)(A_{dp}(\eta(kh)) + H_d(\eta(kh))C_{dp}(\eta(kh)))x(kh), \quad k \in \mathbb{Z}_+ \end{aligned}$$

is ESMS. Applying Theorem 3.3 with updates $A_l(t, i) \leftarrow A_l(i) + H_c(t, i)C_{cl}(i)$, $0 \le l \le r_1$ and $A_{dp}(kh, i) \leftarrow A_{dp}(i) + H_d(i)C_{dp}(i)$, $0 \le p \le r_2$, $\theta_d = \tilde{\theta} = h$, $\theta_d = \tilde{\mathbf{k}} = 1$ we may infer that the JSLDE (5.224) is ESMS if and only if the non-homogeneous backward GLJLDE:

$$\begin{aligned} -\dot{Y}(t, i) &= (A_0(i) + H_c(t, i)C_{c0}(i))^\top Y(t, i) + Y(t, i)(A_0(i) + H_c(t, i)C_{c0}(i)) \\ &\quad + \sum_{l=1}^{r_1}(A_l(i) + H_c(t, i)C_{cl}(i))^\top Y(t, i)(A_l(i) + H_c(i)C_{cl}(i)) \\ &\quad + \sum_{j=1}^{N} q_{ij}Y(t, j) + I_n, \quad kh \le t < (k+1)h, \qquad (5.225a) \\ Y(kh^-, i) &= \sum_{p=0}^{r_2}(A_{dp}(i) + H_d(i)C_{dp}(i))^\top Y(kh, i)(A_{dp}(i) + H_d(i)C_{dp}(i)) \\ &\quad + I_n, \quad k \in \mathbb{Z}, \qquad (5.225b) \end{aligned}$$

$i \in \mathcal{N}$, has a solution $t \to \widetilde{\mathbf{Y}} = (\widetilde{Y}(t, 1), \ldots, \widetilde{Y}(t, N)) : \mathbb{R} \to \mathscr{S}^N_n$ satisfying

$$0 < \nu_1 I_n \le \widetilde{Y}(t, i) \le \nu_2 I_n, \qquad (5.226)$$

for all $(t, i) \in \mathbb{R} \times \mathcal{N}$. Let $\mathbf{X}(\cdot) = (X(\cdot, 1), X(\cdot, 2), \ldots, X(\cdot, N))$ be defined by

$$X(t, i) \triangleq \widetilde{Y}(-t, i), \quad t \in \mathbb{R}.$$

By direct calculation we obtain from (5.225) and (5.226) that $t \to \mathbf{X}(t) : \mathbb{R} \to \mathscr{S}^N_n$ solves the following non-homogeneous forward GLJLDE:

$$\dot{X}(t,i) = (A_0(i) + H_c(-t,i)C_{c0}(i))^\top X(t,i) + X(t,i)$$
$$(A_0(i) + H_c(-t,i)C_{c0}(i))$$
$$+ \sum_{l=1}^{r_1}(A_l(i) + H_c(-t,i)C_{cl}(i))^\top X(t,i)(A_l(i) + H_c(-t,i)C_{cl}(i))$$
$$+ \sum_{j=1}^{N} q_{ij}X(t,j) + I_n, \quad kh \leq t < (k+1)h, \tag{5.227a}$$
$$X(kh^+,i) = \sum_{p=0}^{r_2}(A_{dp}(i) + H_d(i)C_{dp}(i))^\top X(kh,i)(A_{dp}(i) + H_d(i)C_{dp}(i))$$
$$+ I_n, \quad k \in \mathbb{Z}, \tag{5.227b}$$
$$0 < \nu_1 I_n \leq X(t,i) \leq \nu_2 I_n, \quad (t,i) \in \mathbb{R} \times \mathcal{N}. \tag{5.227c}$$

Since (5.227) is defined on the ordered Hilbert space $(\mathcal{S}_n^N, \mathcal{S}_n^{N+})$, we may conclude via Theorem 2.10. that the forward generalized Lyapunov-type jump linear differential equation:

$$\dot{X}(t,i) = (A_0(i) + \breve{B}(i)F_c(t,i))^\top X(t,i) + X(t,i)(A_0^\top(i) + \breve{B}(i)F_c(t,i))$$
$$+ \sum_{l=1}^{r_1}(A_l^\top(i) + C_{cl}^\top(i)F_c(t,i))X(t,i)(A_l^\top(i) + C_{cl}^\top(i)F_c(t,i))^\top$$
$$+ \sum_{j=1}^{N} q_{ij}X(t,j), \quad kh \leq t < (k+1)h, \tag{5.228a}$$
$$X(kh^+,i) = \sum_{p=0}^{r_2}(A_{dp}^\top(i) + C_{dp}^\top(i)F_d(i))X(kh,i)(A_{dp}^\top(i) + C_{dp}^\top(i)F_d(i))^\top,$$
$$k \in \mathbb{Z}, \tag{5.228b}$$

is exponentially stable, where $F_c(t,i) := H_c^\top(-t,i)$, $t \in \mathbb{R}$, $F_d(i) = H_d^\top(i)$, $i \in \mathcal{N}$. Bearing in mind the notation introduced by (5.219), (5.220) and (5.222), we rewrite (5.228) into a compact form:

$$\dot{\mathbf{X}}(t) = (\breve{\mathbf{A}} + \breve{\mathbf{B}}\mathbf{F}_c(t))\mathbf{X}(t) + \mathbf{X}(t)(\breve{\mathbf{A}} + \breve{\mathbf{B}}\mathbf{F}_c(t))^\top$$
$$+ \breve{\boldsymbol{\Pi}}_{c\mathbf{F}_c}^*[\mathbf{X}(t)], \quad kh \leq t < (k+1)h, \tag{5.229a}$$
$$\mathbf{X}(kh^+) = \breve{\boldsymbol{\Pi}}_{d\mathbf{F}_d}^*[\mathbf{X}(kh)], \quad k \in \mathbb{Z}, \tag{5.229b}$$

where $\breve{\boldsymbol{\Pi}}_{c\mathbf{F}_c}^*[\cdot]$ is the adjoint of the operator $\mathbf{X} \to \breve{\boldsymbol{\Pi}}_{c\mathbf{F}_c}[\mathbf{X}] : \mathcal{S}_n^N \to \mathcal{S}_n^N$ computed as in (5.18), but with $\boldsymbol{\Pi}_c(t)[\cdot]$ replaced by $\breve{\boldsymbol{\Pi}}_c[\cdot]$ defined as in (5.220a) and $W_c(\cdot,i)$ replaced by $F_c(\cdot,i)$. In (5.229b), $\breve{\boldsymbol{\Pi}}_{d\mathbf{F}_d}^*[\cdot]$ is the adjoint of the operator $\mathbf{X} \to \breve{\boldsymbol{\Pi}}_{d\mathbf{F}_d}[\mathbf{X}] : \mathcal{S}_n^N \to \mathcal{S}_n^N$ computed as in (5.20), but with $\boldsymbol{\Pi}_d(\tau_k)[\cdot]$ replaced by the operator $\breve{\boldsymbol{\Pi}}_d[\cdot]$ defined in (5.220b) and $W_d(i)$ replaced by $F_d(i)$. Since the linear differential equation (5.229) is a particular case of (5.69) we may conclude that the quadruple $(\breve{\mathbf{A}}, \breve{\mathbf{B}}, (\breve{\boldsymbol{\Pi}}_c \mid \breve{\boldsymbol{\Pi}}_d), \mathfrak{T})$ is stabilizable, because (5.229) is exponentially stable. Thus, the validity of the implication $(i) \Rightarrow (ii)$ is confirmed.

To show that the implication $(ii) \Rightarrow (i)$ holds, let us observe that if the quadruple $(\breve{\mathbf{A}}, \breve{\mathbf{B}}, (\breve{\boldsymbol{\Pi}}_c \mid \breve{\boldsymbol{\Pi}}_d), \mathfrak{T})$ is stabilizable, then, according to Corollary 5.4 there exist a pair of feedback gains $(\mathbf{F}_c(\cdot), \mathbf{F}_d \in \mathscr{PC}(\mathbb{R}, \mathscr{M}^N_{n_{z_c}n}, \mathfrak{T}) \times \mathscr{M}^N_{n_{z_d}n}$, $\mathbf{F}_c(\cdot)$ being a periodic function with period h, with the property that the corresponding linear differential equation of the form (5.229) is exponentially stable. Further, invoking $(i) \Rightarrow (ii)$ of Theorem 2.10 combined by the implication $(vi) \Rightarrow (i)$ from Theorem 3.3 one can show that the JSLDE of the form (5.224) with $H_c(t, i) = F_c^\top(-t, i)$, $t \in \mathbb{R}$ and $H_d(i) = F_d^\top(i)$, $i \in \mathscr{N}$ is ESMS. Hence, the JLSS (5.224)–(5.225) is mean-square detectable. Thus the proof ends.

Following step by step the line of the proof of Proposition 5.4 one obtains:

Proposition 5.5 *Under the Assumption 1.2 a function $t \to \mathbf{Y}_c(t) = (Y_s(t, 1), \ldots, Y_s(t, N)) : \mathbb{R} \to \mathscr{S}_n^N$ is a bounded and stabilizing solution of the FJMRDE (5.213) if and only if the function $t \to \mathbf{X}_s(t) = (X_s(t, 1), \ldots, X_s(t, N)) : \mathbb{R} \to \mathscr{S}_n^N$ defined by*

$$X_s(t, i) := Y_s(-t, i),$$

for all $(t, i) \in \mathbb{R} \times \mathscr{N}$ is a bounded and stabilizing solution of the corresponding GBJMRDE of the form (5.218).

Proof The proof is left to the readers.

In the case of the quintuple $\breve{\Sigma}$ defined in (5.223), beside the set of functions $\Gamma^{\breve{\Sigma}}[\mathbb{R}]$ which consist of a class of functions $\mathbf{S}(\cdot) : \mathbb{R} \to \mathscr{S}_n^N$ which are satisfying conditions of the form (5.16) adapted to the components of the quintuple $\breve{\Sigma}$ which correspond to the backward evolution, we also introduce the set $\Gamma_F^{\breve{\Sigma}}[\mathbb{R}]$ (corresponding to the forward evolution) consisting of all bounded functions $t \to \mathbf{Z}(t) = (Z(t, 1), Z(t, 2), \ldots, Z(t, N)) : \mathbb{R} \to \mathscr{S}_n^N$, continuously differentiable on each interval $(kh, (k+1)h]$, left continuous in $t = kh$, $k \in \mathbb{Z}$, and satisfying the linear matrix differential inequalities with jumps:

$$\begin{pmatrix} -\dot{Z}(t, i) + \Upsilon[\mathbf{Z}(t)](i) + M_c(i) & Z(t, i)C_{c0}^\top(i) + L_c(i) \\ C_{c0}(i)Z(t, i) + L_c^\top(i) & R_c(i) \end{pmatrix}$$
$$+ \sum_{l=1}^{r_1} \begin{pmatrix} A_l(i) \\ C_{cl}(i) \end{pmatrix} Z(t, i) \begin{pmatrix} A_l(i) \\ C_{cl}(i) \end{pmatrix}^\top \geq 0 \text{ for all } t \in (kh(k+1)h] \tag{5.230a}$$

$$\begin{pmatrix} -Z(kh^+, i) + M_d(i) & L_d(i) \\ L_d^\top(i) & R_d(i) \end{pmatrix} + \sum_{p=0}^{r_2} \begin{pmatrix} A_{dp}(i) \\ C_{dp}(i) \end{pmatrix} Z(kh, i) \begin{pmatrix} A_{dp}(i) \\ C_{dp}(i) \end{pmatrix}^\top \geq 0, \quad k \in \mathbb{Z}, \tag{5.230b}$$

$$\sum_{l=1}^{r_1} C_{cl}(i)Z(t,i)C_{cl}^\top(i) + R_c(i) \geq \gamma_c I_{n_{zc}} > 0, \quad \text{for all } t \in \mathbb{R} \tag{5.230c}$$

$$\sum_{p=0}^{r_2} C_{dp}(i)Z(kh,i)C_{dp}^\top(i) + R_d(i) \geq \gamma_d I_{n_{zd}} > 0, \quad \text{for all } k \in \mathbb{Z} \tag{5.230d}$$

where

$$\Upsilon[\mathbf{Z}(t)](i) \stackrel{\Delta}{=} A_0(i)Z(t,i) + Z(t,i)A_0^\top(i) + \sum_{j=1}^{N} q_{ji}Z(t,j), \quad i \in \mathcal{N}.$$

In the developments from this section, an important role will be played by the subset $\widetilde{\Gamma}_F^{\check{\Sigma}}[\mathbb{R}]$ of $\Gamma_F^{\check{\Sigma}}[\mathbb{R}]$ consisting of all functions $\mathbf{Z}(\cdot)$ from $\Gamma_F^{\check{\Sigma}}[\mathbb{R}]$ which (instead of (5.230b)) are satisfying the stronger constraint:

$$\begin{pmatrix} -Z(kh^+,i) + M_d(i) & L_d(i) \\ L_d^\top(i) & R_d(i) \end{pmatrix} + \sum_{p=0}^{r_2} \begin{pmatrix} A_{dp}(i) \\ C_{dp}(i) \end{pmatrix} Z(kh,i) \begin{pmatrix} A_{dp}(i) \\ C_{dp}(i) \end{pmatrix}^\top \geq \hat{\gamma} I_{n+n_{zd}} > 0,$$
$$\text{for all } k \in \mathbb{Z}, \ i \in \mathcal{N}. \tag{5.231}$$

Remark 5.10 For each $\mathbf{Z}(\cdot) \in \Gamma_F^{\check{\Sigma}}[\mathbb{R}]$ we set $\mathbf{Z}^\#(t) \stackrel{\Delta}{=} \mathbf{Z}(-t), \quad t \in \mathbb{R}$. The following equivalences hold:

(a) $\mathbf{Z}(\cdot) \in \Gamma_F^{\check{\Sigma}}[\mathbb{R}]$ if and only if $\mathbf{Z}^\#(\cdot)$ lies in $\Gamma^{\check{\Sigma}}[\mathbb{R}]$;
(b) $\mathbf{Z}(\cdot) \in \widetilde{\Gamma}_F^{\check{\Sigma}}[\mathbb{R}]$ if and only if $\mathbf{Z}^\#(\cdot)$ lies in $\widetilde{\Gamma}^{\check{\Sigma}}[\mathbb{R}]$.

Combining Corollary 5.5, Proposition 5.4 and Remark 5.10 (*b*) one obtains:

Corollary 5.16 *Under the Assumption 1.2, if the JLSS (5.214)–(5.215) is mean-square detectable, then the following are equivalent:*

(a) the set $\widetilde{\Gamma}_F^{\check{\Sigma}}[\mathbb{R}]$ *is not empty;*
(b) there exist periodic functions of period h, $\mathbf{Z}(\cdot)$ *which lie in* $\widetilde{\Gamma}_F^{\check{\Sigma}}[\mathbb{R}]$.

From Corollary 5.3, Proposition 5.5 and Remark 5.10 we obtain:

Corollary 5.17 *Under the Assumption 1.2, if the set* $\Gamma_F^{\check{\Sigma}}[\mathbb{R}]$ *is not empty, then the FJMRDE (5.213) has at most one bounded on* $\mathbb{R}$ *and stabilizing solution.*

Further, Theorem 5.5 together with Proposition 5.5 and Remark 5.10 give:

Corollary 5.18 *Under the Assumption 1.2 if the set* $\Gamma_F^{\check{\Sigma}}[\mathbb{R}]$ *is not empty, the unique bounded on* $\mathbb{R}$ *and stabilizing solution of the FJMRDE (5.213) if any, is a periodic function with period* h *and it is left continuous in each* kh, $k \in \mathbb{Z}$.

Employing Theorem 5.6 together with Proposition 5.4, Proposition 5.5 and Remark 5.10 (b) we obtain the following set of conditions that guarantee the existence of a stabilizing and bounded solution of the FJMRDE (5.213).

Theorem 5.11 *Under the Assumption 1.2 the following are equivalent:*

(a) the JLSS (5.214)–(5.215) is mean-square detectable and the set $\widetilde{\Gamma}_F^{\check{\Sigma}}[\mathbb{R}]$ *is not empty;*
(b) the FJMRDE (5.213) has a bounded on $\mathbb{R}$ *and stabilizing solution* $t \to \mathbf{Y}_s(t) = (Y_s(t,1), \dots, Y_s(t,N)) : \mathbb{R} \to \mathcal{S}_n^N$ *which is satisfying the following sign conditions:*

$$\sum_{l=1}^{r_1} C_{cl}(i)Y_s(t,i)C_{cl}^{\top}(i) + R_c(i) \geq \nu_c I_{n_{zc}} > 0, \;\; (\forall) \;\; t \in \mathbb{R}, \tag{5.232a}$$

$$\sum_{p=0}^{r_2} C_{dp}(i)Y_s(kh,i)C_{dp}^{\top}(i) + R_d(i) \geq \nu_d I_{n_{zd}} > 0, \;\; (\forall) \; k \in \mathbb{Z}, \; i \in \mathcal{N}. \tag{5.232b}$$

Proof The proof is left up to the reader.

Remark 5.11

(a) According to Corollary 5.16 to check if the set $\widetilde{\Gamma}_F^{\check{\Sigma}}[\mathbb{R}]$ is not empty, is sufficient to test the feasibility of the DLMI with jumps (5.230a), (5.230d) and (5.231) on the interval $[0, h]$.
(b) Invoking Corollary 5.18 we may infer that if $\mathbf{Y}_s(\cdot)$ is a unique bounded and stabilizing solution of the FJMRDE (5.213) then

$$Y_s(kh,i) = Y_s(0,i), \quad \text{for all} \;\; (k,i) \in \mathbb{Z} \times \mathcal{N}.$$

Hence, the gain matrices $K_{d,s}(kh,i)$ introduced by (5.217b) are not depending upon k, that is

$$K_{d,s}(kh,i) = K_{d,s}(i), \quad \text{for all} \;\; (k,i) \in \mathbb{Z} \times \mathcal{N},$$

where for each $i \in \mathcal{N}$

$$K_{d,s}(i) := -\Big(\sum_{p=0}^{r_2} A_{dp}(i)Y_s(0,i)C_{dp}^{\top}(i) + L_d(i)\Big)\Big(\sum_{p=0}^{r_2} C_{dp}(i)Y_s(0,i)C_{dp}^{\top}(i) + R_d(i)\Big)^{-1}.$$

5.6.1 A Class of Forward Jump Matrix Linear Differential Equations with Riccati-Type Jumping Operators.

In the developments from the next chapters the following special form of a FJMRDE (5.213) will be involved:

$$\dot{Y}(t,i) = A_0(i)Y(t,i) + Y(t,i)A_0^\top(i) + \sum_{l=1}^{r_1} A_l(i)Y(t,i)A_l^\top(i)$$

$$+ \sum_{j=1}^{N} q_{ji}Y(t,j) + M_c(i), \quad kh < t \leq (k+1)h \tag{5.233a}$$

$$Y(kh^+,i) = \sum_{p=0}^{r_2} A_{dp}(i)Y(kh,i)A_{dp}^\top(i) - \left(\sum_{p=0}^{r_2} A_{dp}(i)Y(kh,i)C_{dp}^\top(i) + L_d(i)\right)$$

$$\cdot \left(\sum_{p=0}^{r_2} C_{dp}(i)Y(kh,i)C_{dp}^\top(i) + R_d(i)\right)^{-1} \left(\sum_{p=0}^{r_2} C_{dp}(i)Y(kh,i)A_{dp}^\top(i) + L_d^\top(i)\right)$$

$$+ M_d(i), \quad k \in \mathbb{Z}, \quad i \in \mathcal{N}. \tag{5.233b}$$

The accompanying JLSS (5.214)–(5.215) reduces to:

$$\mathrm{d}x(t) = A_0(\eta(t))x(t)\mathrm{d}t + \sum_{l=1}^{r_1} A_l(\eta(t))x(t)\mathrm{d}w_l(t), \quad kh < t \leq (k+1)h \tag{5.234a}$$

$$x(kh^+) = (A_{d0}(\eta(kh)) + \sum_{p=1}^{r_2} w_{dp}(k)A_{dp}(\eta(kh)))x(kh) \tag{5.234b}$$

$$z_d(kh) = (C_{d0}(\eta(kh)) + \sum_{p=1}^{r_2} w_{dp}(k)C_{dp}(\eta(kh)))x(kh), \quad k \in \mathbb{Z}_+. \tag{5.234c}$$

A solution $t \to \widetilde{\mathbf{Y}}_s(t) = (\widetilde{Y}(t,1), \ldots, \widetilde{Y}_s(t,N)) : \mathbb{R} \to \mathcal{S}_n^N$ of (5.233) is named stabilizing solution if the matrices $\sum_{p=0}^{r_2} C_{dp}(i)\widetilde{Y}_s(kh,i)C_{dp}^\top(i) + R_d(i), \quad (k,i) \in \mathbb{Z} \times \mathcal{N}$ are invertible and the corresponding closed-loop system

$$\mathrm{d}x(t) = A_0(\eta(t))x(t)\mathrm{d}t + \sum_{l=1}^{r_1} A_l(\eta(t))x(t)\mathrm{d}w_l(t), \quad kh < t \leq (k+1)h \tag{5.235a}$$

$$x(kh^+) = \left(A_{d0}(\eta(kh)) + K_{d,s}(kh,\eta(kh))C_{d0}(\eta(kh))\right)x(kh)$$

$$+ \sum_{p=1}^{r_2} w_{dp}(k)\left(A_{dp}(\eta(kh)) + K_{d,s}(kh,\eta(kh))C_{dp}(\eta(kh))\right)x(kh), \quad k \in \mathbb{Z}_+, \tag{5.235b}$$

is ESMS, where $K_{d,s}(kh, i)$ are computed as in (5.217b) with $Y_s(kh, i)$ replaced by $\widetilde{Y}_s(kh, i)$. In the case of the FJMLDE (5.233) the set of the functions $\boldsymbol{\Gamma}_F^{\check{\Sigma}}[\mathbb{R}]$ consists of all bounded functions $t \to \mathbf{Z}(t) = (Z(t, 1), \ldots, Z(t, N)) : \mathbb{R} \to \mathcal{S}_n^N$ which are continuously differentiable on each interval $(kh, (k+1)h]$, left continuous in $t = kh$, $k \in \mathbb{Z}$, satisfying the following DLMI with jumps:

$$\dot{Z}(t, i) \leq A_0(i)Z(t, i) + Z(t, i)A_0^\top(i) + \sum_{l=1}^{r_1} A_l(i)Z(t, i)A_l^\top(i) + \sum_{j=1}^{N} q_{ji}Z(t, j), \quad \text{for all } t \in (kh, (k+1)h] \tag{5.236a}$$

$$\begin{pmatrix} -Z(kh^+, i) + M_d(i) & L_d(i) \\ L_d^\top(i) & R_d(i) \end{pmatrix} + \sum_{p=0}^{r_2} \begin{pmatrix} A_{dp}(i) \\ C_{dp}(i) \end{pmatrix} Z(kh, i) \begin{pmatrix} A_{dp}(i) \\ C_{dp}(i) \end{pmatrix}^\top \geq 0 \tag{5.236b}$$

$$\sum_{p=0}^{r_2} C_{dp}(i)Z(kh, i)C_{dp}^\top(i) + R_d(i) \geq \gamma_d I_{n_{z_d}}, \quad k \in \mathbb{Z}, \quad i \in \mathcal{N}. \tag{5.236c}$$

The subset $\widetilde{\boldsymbol{\Gamma}}_F^{\check{\Sigma}}[\mathbb{R}] \subset \boldsymbol{\Gamma}_F^{\check{\Sigma}}[\mathbb{R}]$ consists of all functions $\mathbf{Z}(\cdot) \in \boldsymbol{\Gamma}_F^{\check{\Sigma}}[\mathbb{R}]$ which are satisfying the DLMI (5.236a) with jumping operator of the form (5.231). The results stated in Corollary 5.17, Corollary 5.18 may be written mutatis-mutandis to the bounded and stabilizing solution of the FJMLDE (5.233). Finally, we provide the following criterion for the existence of the bounded and stabilizing solution of (5.233).

Theorem 5.12 *Under the Assumption 1.2 the following are equivalent:*

(i) the JLSS (5.234) is mean-square detectable by impulses and the set $\widetilde{\boldsymbol{\Gamma}}_F^{\check{\Sigma}}[\mathbb{R}]$ *is not empty;*
(ii) the FJMLDE with Riccati-type jumping operator (5.233) has a unique bounded and stabilizing solution $t \to \widetilde{\mathbf{Y}}_s(t) = (\widetilde{Y}_s(t, 1), \widetilde{Y}_s(t, 2), \ldots, \widetilde{Y}_s(t, N)) : \mathbb{R} \to \mathcal{S}_n^N$ *satisfying the sign conditions*

$$\sum_{p=0}^{r_2} C_{dp}(i)\widetilde{Y}_s(kh, i)C_{dp}^\top(i) + R_d(i) \geq \gamma_d I_{n_{z_d}} > 0, \quad k \in \mathbb{Z}, \quad i \in \mathcal{N}.$$

Proof The proof is left to the reader.

5.7 An Iterative Procedure for Numerical Computation of the Stabilizing Solution

5.7.1 *Numerical Computation of the Bounded and Stabilizing Solution of a Stochastic Jump Matrix Riccati Differential Equation*

In this section, we propose a method for a numerical computation of the bounded on $\mathbb{R}$ and stabilizing solution of the SJMRDE (5.10). If $\mathbf{F}_c(\cdot) = (F_c(\cdot, 1), \ldots, F_c(\cdot, N))$ is a function from $\mathscr{PC}(\mathbb{R}_+, \mathscr{M}^N_{m_c n}, \mathfrak{T})$ and $\{\mathbf{F}_d(\tau_k)\}_{k\in\mathbb{Z}_+} \subset \mathscr{M}^N_{m_d n}$ we define the linear operators $\mathbf{X} \to \mathscr{L}_{c,\mathbf{F}_c}(t)[\mathbf{X}] : \mathscr{S}^N_n \to \mathscr{S}^N_n$ and $\mathbf{X} \to \mathscr{L}_{d,\mathbf{F}_d}(t)[\mathbf{X}] : \mathscr{S}^N_n \to \mathscr{S}^N_n$ as $\mathscr{L}_{c,\mathbf{F}_c}[\mathbf{X}] = (\mathscr{L}_{c,\mathbf{F}_c}(t)[\mathbf{X}](1), \mathscr{L}_{c,\mathbf{F}_c}(t)[\mathbf{X}](2), \ldots, \mathscr{L}_{c,\mathbf{F}_c}(t)[\mathbf{X}](N))$ and

$$\mathscr{L}_{d,\mathbf{F}_d}[\mathbf{X}] = (\mathscr{L}_{d,\mathbf{F}_d}(\tau_k)[\mathbf{X}](1), \mathscr{L}_{d,\mathbf{F}_d}(\tau_k)[\mathbf{X}](2), \ldots, \mathscr{L}_{d,\mathbf{F}_d}(\tau_k)[\mathbf{X}](N))$$

with

$$\begin{aligned}\mathscr{L}_{c,\mathbf{F}_c}[\mathbf{X}](i) &\triangleq (A_0(t,i) + B_0(t,i)F_c(t,i))X(i) + X(i)(A_0(t,i) + B_0(t,i)F_c(t,i))^\top \\ &\quad + \sum_{l=1}^{r_1}(A_l(t,i) + B_l(t,i)F_c(t,i))X(i)(A_l(t,i) + B_l(t,i)F_c(t,i))^\top \\ &\quad + \sum_{j=1}^{N} q_{ji}X(j), \quad t \in \mathbb{R}_+ \end{aligned} \tag{5.237a}$$

$$\begin{aligned}\mathscr{L}_{d,\mathbf{F}_d}[\mathbf{X}](i) &\triangleq \sum_{p=0}^{r_2}(A_{dp}(\tau_k,i) + B_{dp}(\tau_k,i)F_d(\tau_k,i))X(i) \\ &\quad \cdot (A_{dp}(\tau_k,i) + B_{dp}(\tau_k,i)F_d(\tau_k,i))^\top, \quad \tau_k \in \mathfrak{T}, \end{aligned} \tag{5.237b}$$

for all $\mathbf{X} = (X(1), X(2), \ldots, X(N)) \in \mathscr{S}^N_n$. The adjoint operators with respect to the inner product (5.1) are described by $\mathscr{L}^*_{c,\mathbf{F}_c}[\mathbf{X}] = (\mathscr{L}^*_{c,\mathbf{F}_c}(t)[\mathbf{X}](1), \mathscr{L}_{c,\mathbf{F}_c}(t)^*[\mathbf{X}](2), \ldots, \mathscr{L}^*_{c,\mathbf{F}_c}(t)[\mathbf{X}](N))$ and

$$\mathscr{L}^*_{d,\mathbf{F}_d}[\mathbf{X}] = (\mathscr{L}^*_{d,\mathbf{F}_d}(\tau_k)[\mathbf{X}](1), \mathscr{L}^*_{d,\mathbf{F}_d}(\tau_k)[\mathbf{X}](2), \ldots, \mathscr{L}^*_{d,\mathbf{F}_d}(\tau_k)[\mathbf{X}](N))$$

where

$$\begin{aligned}\mathscr{L}^*_{c,\mathbf{F}_c}[\mathbf{X}](i) &\triangleq (A_0(t,i) + B_0(t,i)F_c(t,i))^\top X(i) \\ &\quad + X(i)(A_0(t,i) + B_0(t,i)F_c(t,i)) \\ &\quad + \sum_{l=1}^{r_1}(A_l(t,i) + B_l(t,i)F_c(t,i))^\top X(i)(A_l(t,i) + B_l(t,i)F_c(t,i))\end{aligned}$$

$$+\sum_{j=1}^{N} q_{ij} X(j), \quad t \in \mathbb{R}_+ \tag{5.238a}$$

$$\mathscr{L}^*_{d,\mathbf{F}_d}[\mathbf{X}](i) \triangleq \sum_{p=0}^{r_2} (A_{dp}(\tau_k, i) + B_{dp}(\tau_k, i) F_d(\tau_k, i))^\top X(i)$$
$$\cdot (A_{dp}(\tau_k, i) + B_{dp}(\tau_k, i) F_d(\tau_k, i)), \quad \tau_k \in \mathfrak{T}, \tag{5.238b}$$

$i \in \mathcal{N}$, for all $\mathbf{X} \in \mathscr{S}_n^N$.

The following kinds of notation will be used also:

$$\boldsymbol{\Theta}_{c,\mathbf{F}_c}(t) = (\Theta_{c,\mathbf{F}_c}(t, 1), \Theta_{c,\mathbf{F}_c}(t, 2), \ldots, \Theta_{c,\mathbf{F}_c}(t, N))$$

$$\boldsymbol{\Theta}_{d,\mathbf{F}_d}(\tau_k) = (\Theta_{d,\mathbf{F}_d}(\tau_k, 1), \Theta_{d,\mathbf{F}_d}(\tau_k, 2), \ldots, \Theta_{d,\mathbf{F}_d}(\tau_k, N))$$

where for each $i \in \mathcal{N}$ we define

$$\Theta_{c,\mathbf{F}_d}(t, i) \triangleq \begin{pmatrix} I_n \\ F_c(t, i) \end{pmatrix}^\top \begin{pmatrix} M_c(t, i) & L_c(t, i) \\ L_c^\top(t, i) & R_c(t, i) \end{pmatrix} \begin{pmatrix} I_n \\ F_c(t, i) \end{pmatrix}, \quad t \in \mathbb{R}_+ \tag{5.239a}$$

$$\Theta_{d,\mathbf{F}_d}(\tau_k, i) \triangleq \begin{pmatrix} I_n \\ F_d(\tau_k, i) \end{pmatrix}^\top \begin{pmatrix} M_d(\tau_k, i) & L_d(\tau_k, i) \\ L_d^\top(\tau_k, i) & R_d(\tau_k, i) \end{pmatrix} \begin{pmatrix} I_n \\ F_d(\tau_k, i) \end{pmatrix}, \quad \tau_k \in \mathfrak{T}. \tag{5.239b}$$

According to Definition 4.3 the JLSS (5.163) is mean-square stabilizable by linear state feedback if there exist a pair of feedback gains $(\mathbf{F}_c(\cdot), \{\mathbf{F}_d(\tau_k)\}_{k\in\mathbb{Z}_+})$, $\mathbf{F}_c(\cdot) \in \mathscr{PC}(\mathbb{R}_+, \mathscr{M}^N_{m_c n}, \mathfrak{T})$, and $\{\mathbf{F}_d(\tau_k)\}_{k\in\mathbb{Z}} \subset \mathscr{M}^N_{m_d n}$ with the property that the closed-loop JSLDE:

$$\mathrm{d}x(t) = (A_0(t, \eta(t)) + B_0(t, \eta(t)) F_c(t, \eta(t)))x(t)\mathrm{d}t$$
$$+ \sum_{l=1}^{r_1} (A_l(t, \eta(t)) + B_l(t, \eta(t)) F_c(t, \eta(t)))x(t)\mathrm{d}w_l(t), \quad \tau_k < t \le \tau_{k+1} \tag{5.240a}$$

$$x(\tau_k^+) = (A_{d0}(\tau_k, \eta(\tau_k)) + B_{d0}(\tau_k, \eta(\tau_k)) F_d(\tau_k, \eta(\tau_k)))x(\tau_k)$$
$$+ \sum_{p=1}^{r_2} w_{dp}(k)(A_{dp}(\tau_k, \eta(\tau_k)) + B_{dp}(\tau_k, \eta(\tau_k)) F_d(\tau_k, \eta(\tau_k)))x(\tau_k),$$
$$k \in \mathbb{Z}_+ \tag{5.240b}$$

is strongly exponentially stable in mean square (SESMS). From Definition 3.1 (*a*) we may infer that the JSLDE (5.240) is SESMS if and only if the corresponding generalized Lyapunov-type jump linear differential equation (GLJLDE) on the Hilbert space $\mathscr{S}_n^N$:

$$\dot{\mathbf{X}}(t) = \mathscr{L}_{c,\mathbf{F}_c}(t)[\mathbf{X}(t)], \quad \tau_k < t \leq \tau_{k+1} \tag{5.241a}$$

$$\mathbf{X}(\tau_k^+) = \mathscr{L}_{d,\mathbf{F}_d}(\tau_k)[\mathbf{X}(\tau_k)], \quad k \in \mathbb{Z}_+ \tag{5.241b}$$

is exponentially stable.

Now, we are in position to prove:

Lemma 5.7 *Assume:*

(a) the Assumptions 1.2, 2.1 and 5.2 are fulfilled;
(b) the JLSS (5.163) is mean-square stabilizable by linear state feedback;
(c) the set $\Gamma^{\Sigma}[\mathbb{R}_+]$ associated to the SJMRDE (5.10) is not empty.

Consider $\mathbf{F}_c(\cdot) \in \mathscr{PC}(\mathbb{R}_+, \mathscr{M}^N_{m_c n}, \mathfrak{T})$ and $\{\mathbf{F}_d(\tau_k)\}_{k \in \mathbb{Z}_+}$ a pair of stabilizing feedback gains for the JLSS (5.163). Let $\mathbf{X}^{(0)}(\cdot) = (X^{(0)}(\cdot, 1), X^{(0)}(\cdot, 2), \ldots, X^{(0)}(\cdot, N)) \in \mathscr{PC}(\mathbb{R}_+, \mathscr{S}^N_n, \mathfrak{T})$ be a bounded solution of the following jump linear matrix differential inequality (JLMDI) on the Hilbert space $\mathscr{S}^N_n$:

$$\dot{X}^{(0)}(t,i) = \mathscr{L}^*_{c,\mathbf{F}_c}(t)[\mathbf{X}^{(0)}(t)](i) + \Theta_{c,\mathbf{F}_c}(t,i) \leq 0, \quad \tau_k \leq t < \tau_{k+1} \tag{5.242a}$$

$$X^{(0)}(\tau_k^-, i) \geq \mathscr{L}^*_{d,\mathbf{F}_d}(\tau_k)[\mathbf{X}^{(0)}(\tau_k)](i) + \Theta_{d,\mathbf{F}_d}(\tau_k, i) + \varepsilon I_n, \quad k \in \mathbb{Z}_+ \tag{5.242b}$$

$i \in \mathscr{N}$, where $\varepsilon > 0$ is fixed. Under the considered assumptions any solution $\mathbf{X}^{(0)}(\cdot)$ of (5.242) is satisfying:

$$X^{(0)}(t,i) - Z(t,i) \geq \delta I_n, \tag{5.243}$$

for all $(t,i) \in \mathbb{R}_+ \times \mathscr{N}$, for an arbitrary $\mathbf{Z}(\cdot) = (Z(\cdot, 1), Z(\cdot, 2), \ldots, Z(\cdot, N)) \in \Gamma^{\Sigma}[\mathbb{R}_+]$ where $\delta > 0$ is a constant not depending upon $\mathbf{Z}(\cdot)$.

Proof Let $\mathbf{Z}(\cdot) \in \Gamma^{\Sigma}[\mathbb{R}_+]$ be arbitrary but fixed. Proceeding as in the proof of Theorem 5.3, we obtain that $\mathbf{Z}(\cdot)$ solves a backward jump matrix Riccati differential equation of the form (5.73)–(5.74) in which $\mathscr{R}_c(t, \mathbf{Z}(t))$ and $\mathscr{R}_d(\tau_k, \mathbf{Z}(\tau_k))$, respectively, are described by the right hand side of (5.10a) and the right hand side of (5.10b), respectively, with $X(\cdot, i)$ replaced by $Z(\cdot, i)$, $i \in \mathscr{N}$. Further, applying Lemma 5.1 taking $W_c(\cdot, i) = F_c(\cdot, i)$ and $W_d(\cdot, i) = F_d(\cdot, i)$, respectively, in the case of the differential equation of the form (5.73) satisfied by $\mathbf{Z}(\cdot)$, we obtain via (5.238)–(5.239) the following GLJLDE:

$$\begin{aligned} -\dot{Z}(t,i) = {} & \mathscr{L}^*_{c,\mathbf{F}_c}(t)[\mathbf{Z}(t)](i) + \Theta_{c,\mathbf{F}_c}(t,i) - \widehat{M}_c(t,i) \\ & - (F_c(t,i) - F^{\mathbf{Z}}_c(t,i))^{\top} \left(\sum_{l=1}^{r_1} B_l^{\top}(t,i) Z(t,i) B_l(t,i) + R_c(t,i)) \right) \\ & \cdot (F_c(t,i) - F^{\mathbf{Z}}_c(t,i)), \quad \tau_k \leq t < \tau_{k+1} \end{aligned} \tag{5.244a}$$

$$\begin{aligned} Z(\tau_k^-, i) = {} & \mathscr{L}^*_{d,\mathbf{F}_d}(\tau_k)[\mathbf{Z}(\tau_k)](i) + \Theta_{d,\mathbf{F}_d}(\tau_k, i) - \widehat{M}_d(\tau_k, i) \\ & - (F_d(\tau_k, i) - F^{\mathbf{Z}}_d(\tau_k, i))^{\top} \end{aligned}$$

$$\left(\sum_{p=0}^{r_2} B_{dp}^\top(\tau_k,i)Z(\tau_k,i)B_{dp}(\tau_k,i)+R_d(\tau_k,i))\right)$$
$$\cdot (F_d(\tau_k,i)-F_d^{\mathbf{Z}}(\tau_k,i)),\quad k\in\mathbb{Z}_+, \tag{5.244b}$$

where $\widehat{M}_c(t,i)\geq 0,\quad$ for all $\ (t,i)\in\mathbb{R}_+\times\mathcal{N}$, and $\widehat{M}_d(\tau_k,i)\geq 0,\quad$ for all $\ (\tau_k,i)\in\mathfrak{T}\times\mathcal{N}$, and

$$F_c^{\mathbf{Z}}(t,i)\stackrel{\Delta}{=}-\left(\sum_{l=1}^{r_1} B_l^\top(t,i)Z(t,i)B_l(t,i)+R_c(t,i)\right)^{-1} \tag{5.245a}$$
$$\cdot\left(B_0^\top(t,i)Z(t,i)+\sum_{l=1}^{r_1} B_l^\top(t,i)Z(t,i)A_l(t,i)+L_c(t,i)\right),\quad t\in\mathbb{R}_+,$$
$$F_d^{\mathbf{Z}}(\tau_k,i)\stackrel{\Delta}{=}-\left(\sum_{p=0}^{r_2} B_{dp}^\top(\tau_k,i)Z(\tau_k,i)B_{dp}(\tau_k,i)+R_d(\tau_k,i)\right)^{-1} \tag{5.245b}$$
$$\cdot\left(\sum_{p=0}^{r_2} B_{dp}^\top(\tau_k,i)Z(\tau_k,i)A_{dp}(\tau_k,i)+L_d^\top(\tau_k,i)\right),\quad \tau_k\in\mathfrak{T},$$

$i\in\mathcal{N}$. On the other hand, we set

$$\check{M}_c(t,i)\stackrel{\Delta}{=}-\dot{X}^{(0)}(t,i)-\mathcal{L}_{c,\mathbf{F}_c}^*(t)[\mathbf{X}^{(0)}(t)](i)-\Theta_{c,\mathbf{F}_c}(t,i),$$
$$\text{for all}\ \ (t,i)\in\mathbb{R}_+\times\mathcal{N}$$

and $\check{M}_d(\tau_k,i)\stackrel{\Delta}{=}X^{(0)}(\tau_k^-,i)-\mathcal{L}_{d,\mathbf{F}_d}^*(\tau_k)[\mathbf{X}^{(0)}(\tau_k)](i)-\Theta_{d,\mathbf{F}_d}(\tau_k,i)-\varepsilon I_n$, for all $(\tau_k,i)\in\mathfrak{T}\times\mathcal{N}$. From (5.242), it follows that $\check{M}_c(t,i)\geq 0,\quad$ for all $\ (t,i)\in\mathbb{R}_+\times\mathcal{N}$, $\check{M}_d(\tau_k,i)\geq 0,\quad$ for all $\ (\tau_k,i)\in\mathfrak{T}\times\mathcal{N}$ and $t\to\mathbf{X}^{(0)}(t)$ solves the backward jump linear differential equation:

$$-\dot{X}^{(0)}(t,i)=\mathcal{L}_{c,\mathbf{F}_c}^*(t)[\mathbf{X}^{(0)}(t)](i)+\Theta_{c,\mathbf{F}_c}(t,i) \tag{5.246a}$$
$$+\check{M}_c(t,i),\quad \tau_k\leq t<\tau_{k+1}$$
$$X^{(0)}(\tau_k^-,i)=\mathcal{L}_{d,\mathbf{F}_d}^*(\tau_k)[\mathbf{X}^{(0}(\tau_k)](i)+\Theta_{d,\mathbf{F}_d}(\tau_k,i)+\varepsilon I_n \tag{5.246b}$$
$$+\check{M}_d(\tau_k,i),\quad k\in\mathbb{Z}_+.$$

Subtracting (5.244) from (5.246) we obtain that $\boldsymbol{\Delta}^{(0)}(\cdot):=\mathbf{X}^{(0)}(\cdot)-\mathbf{Z}(\cdot)$ is a bounded on $\mathbb{R}_+$ solution on the following non-homogeneous backward linear differential equation with jumps:

$$-\dot{\Delta}^{(0)}(t,i)=\mathcal{L}_{c,\mathbf{F}_c}^*(t)[\boldsymbol{\Delta}^{(0)}(t)](i)+\mathfrak{M}_c^{(0)}(t,i),\quad \tau_k\leq t<\tau_{k+1} \tag{5.247a}$$

$$\Delta^{(0)}(\tau_k^-, i) = \mathcal{L}^*_{d,\mathbf{F}_d}(\tau_k)[\boldsymbol{\Delta}^{(0}(\tau_k)](i) + \mathfrak{M}_d^{(0)}(\tau_k, i), \quad k \in \mathbb{Z}_+, \tag{5.247b}$$

$i \in \mathcal{N}$, where

$$\mathfrak{M}_c^{(0)}(t,i) := \widehat{M}_c(t,i) + \check{M}_c(t,i) + (F_c(t,i) - F_c^{\mathbf{Z}}(t,i))^\top \cdot \left(\sum_{l=1}^{r_1} B_l^\top(t,i) Z(t,i) B_l(t,i) + R_c(t,i)\right)(F_c(t,i) - F_c^{\mathbf{Z}}(t,i)),$$
$$(t,i) \in \mathbb{R}_+ \times \mathcal{N}$$

$$\mathfrak{M}_d^{(0)}(\tau_k,i) := \varepsilon I_n + \widehat{M}_d(\tau_k,i) + \check{M}_d(\tau_k,i) + (F_d(\tau_k,i) - F_d^{\mathbf{Z}}(\tau_k,i))^\top \left(\sum_{p=0}^{r_2} B_{dp}^\top(\tau_k,i) Z(\tau_k,i) B_{dp}(\tau_k,i) + R_d(\tau_k,i)\right)(F_d(\tau_k,i) - F_d^{\mathbf{Z}}(\tau_k,i)), \ \tau_k \in \mathfrak{T}.$$

Since $\mathbf{Z}(\cdot)$ lies in the set $\Gamma^{\Sigma}[\mathbb{R}_+]$ associated to (5.10), we obtain via (5.159c) and (5.159d) that

$$\mathfrak{M}_c^{(0)}(t,i) \geq 0, \quad \text{for all} \quad (t,i) \in \mathbb{R}_+ \times \mathcal{N}$$

and

$$\mathfrak{M}_d^{(0)}(\tau_k,i) \geq \varepsilon I_n, \quad \text{for all} \quad (\tau_k,i) \in \mathfrak{T} \times \mathcal{N}.$$

Bearing in mind that the jump linear differential equation on $\mathcal{S}_n^N$ (5.241) is exponentially stable, we may conclude via Proposition 5.1 (iii) that the non-homogeneous linear differential equation with jumps (5.247) has a unique bounded on $\mathbb{R}_+$ solution and additionally that solution is uniform positive on $\mathbb{R}_+$. Thus the proof is complete.

Remark 5.12

(a) It is obvious that any solution $t \to \mathbf{X}(t) : \mathbb{R}_+ \to \mathcal{S}_n^N$ of the non-homogeneous backward GLJLDE:

$$-\dot{\mathbf{X}}(t) = \mathcal{L}^*_{c,\mathbf{F}_c}(t)[\mathbf{X}(t)] + \boldsymbol{\Theta}_{c,\mathbf{F}_c}(t) + \mathbf{I}_n, \quad \tau_k \leq t < \tau_{k+1} \tag{5.248a}$$
$$\mathbf{X}(\tau_k^-) = \mathcal{L}_{d,\mathbf{F}_d}(\tau_k)[\mathbf{X}(\tau_k)] + \boldsymbol{\Theta}_{d,\mathbf{F}_d}(\tau_k) + 2\varepsilon \mathbf{I}_n, \quad k \in \mathbb{Z}_+ \tag{5.248b}$$

$\varepsilon > 0$, $\mathbf{I}_n = (I_n, I_n, \ldots, I_n)$ is a solution of (5.242).

(b) If the SJMRDE (5.10) has a time periodic coefficients in the sense of Definition 5.8 (i), then, the BLMDI (5.242) has solutions $\mathbf{X}^{(0)}(\cdot)$ which are periodic functions of period $\tilde{\theta}$ with $\tilde{\theta}$ from (5.166a). Indeed, if the coefficients of (5.10) satisfy the conditions from Definition 5.8 (i) then based on Remark 4.1 (a) and Theorem 2.13 (ii), we may infer that the non-homogeneous backward GLJLDE (5.248) has a solution $t \to \mathbf{X}^{(0)}(t) : \mathbb{R} \to \mathcal{S}_n^N$ which is a periodic function with period $\tilde{\theta}$. From part (a) we may conclude that $\mathbf{X}^{(0)}(\cdot)$ is a $\tilde{\theta}-$periodic solution of the JLMDI (5.242).

(c) If the SJMRDE (5.10) is time invariant in the sense of Definition 5.8 (ii), then the JLMDI (5.242) has the solutions $\mathbf{X}^{(0)}(\cdot)$ which are periodic functions with the period h of the dwell time.
Indeed, if the coefficients of (5.10) satisfy the constraints from Definition 5.8 (ii) then, according to Remark 4.1 (b) and Theorem 2.13 (iii) we deduce that the non-homogeneous backward GLJLDE (5.248) has a solution $\mathbf{X}^{(0)}(\cdot)$ which is a periodic function of period h. This solution of (5.248) is also a $h-$periodic solution of (5.242).

(d) According to (b) and (c) from above, if the coefficients of the SJMRDE (5.10) satisfy the conditions from Definition 5.8 (i) or from Definition 5.8 (ii), respectively, for the numerical computation of a solution $\mathbf{X}^{(0)}(\cdot)$ of the JLMDI (5.242) it is sufficient to use a discretization method on one of the bounded interval $[0, \tilde{\theta}]$ or $[0, h]$, respectively.

Bearing in mind that a function $\mathbf{Z}(\cdot)$ from $\Gamma^{\Sigma}[\mathbb{R}_+]$ satisfies (5.159c) and (5.159d) we deduce from (5.243) that an arbitrary bounded on $\mathbb{R}_+$ solution $\mathbf{X}^{(0)}(\cdot)$ of the JLMDI (5.242) is satisfying the following sign conditions:

$$\sum_{l=1}^{r_1} B_l^\top(t,i)X^{(0)}(t,i)B_l(t,i) + R_c(t,i) \geq \nu_c I_{m_c} > 0, \ (\forall) \ (t,i) \in \mathbb{R}_+ \times \mathcal{N} \tag{5.249a}$$

$$\sum_{p=0}^{r_2} B_{dp}^\top(\tau_k,i)X^{(0)}(\tau_k,i)B_{dp}(\tau_k,i) + R_d(\tau_k,i) \geq \nu_d I_{m_d} > 0, (\tau_k,i) \in \mathfrak{T} \times \mathcal{N}. \tag{5.249b}$$

Under these conditions we may construct the gain functions

$$\mathbf{F}_c^{(0)}(\cdot) = (F_c^{(0)}(\cdot,1), F_c^{(0)}(\cdot,2), \ldots, F_c^{(0)}(\cdot,N)) \in \mathcal{PC}(\mathbb{R}_+, \mathcal{M}_{m_c n}^N, \mathfrak{T})$$

and the gain matrices

$$\mathbf{F}_d^{(0)}(\tau_k) = (F_d^{(0)}(\tau_k,1), F_d^{(0)}(\tau_k,2), \ldots, F_d^{(0)}(\tau_k,N)) \in \mathcal{M}_{m_d n}^N$$

by

$$\begin{aligned} F_c^{(0)}(t,i) &\triangleq \left(\sum_{l=1}^{r_1} B_l^\top(t,i)X^{(0)}(t,i)B_l(t,i) + R_c(t,i)\right)^{-1} \\ &\left(B_0^\top(t,i)X^{(0)}(t,i) + \sum_{l=1}^{r_1} B_l^\top(t,i)X^{(0)}(t,i)A_l(t,i) + L_c^\top(t,i)\right), \\ &(t,i) \in \mathbb{R}_+ \times \mathcal{N} \end{aligned} \tag{5.250a}$$

$$F_d^{(0)}(\tau_k,i) \triangleq \left(\sum_{p=0}^{r_2} B_{dp}^\top(\tau_k,i)X^{(0)}(\tau_k,i)B_{dp}(\tau_k,i)+R_d(\tau_k,i)\right)^{-1}$$
$$\left(\sum_{p=0}^{r_2} B_{dp}^\top(\tau_k,i)X^{(0)}(\tau_k,i)A_{dp}(\tau_k,i)+L_d^\top(\tau_k,i)\right),\quad (\tau_k,i)\in\mathfrak{T}\times\mathcal{N}. \tag{5.250b}$$

Lemma 5.8 *Under the Assumptions 1.2, 2.1 and 5.2 the pair of gains* $\left(\mathbf{F}_c^{(0)}(\cdot), \{\mathbf{F}_d^{(0)}(\tau_k)\}_{k\in\mathbb{Z}_+}\right)$ *defined by (5.250) stabilizes the JLSS (5.163).*

Proof We have to show that the forward jump linear differential equation (FJLDE) of the form (5.241) with $\mathbf{F}_c$, $\mathbf{F}_d$ replaced by $\mathbf{F}_c^{(0)}$, $\mathbf{F}_d^{(0)}$, respectively, is exponentially stable. Applying Lemma 5.2 we rewrite the backward jump linear differential equation (5.246) as:

$$-\dot{X}(t,i) = \mathcal{L}^*_{c,\mathbf{F}_c^{(0)}}(t)[\mathbf{X}^{(0)}(t)](i)+\Theta_{c,\mathbf{F}_c^{(0)}}(t,i)+(F_c^{(0)}(t,i)-F_c(t,i))^\top$$
$$\left(\sum_{l=1}^{r_1} B_l^\top(t,i)X^{(0)}(t,i)B_l(t,i)+R_c(t,i)\right)(F_c^{(0)}(t,i)-F_c(t,i))$$
$$+\check{M}_c(t,i),\quad \tau_k\le t<\tau_{k+1} \tag{5.251a}$$
$$X^{(0)}(\tau_k^-,i) = \mathcal{L}^*_{d,\mathbf{F}_d^{(0)}}(\tau_k)[\mathbf{X}^{(0}(\tau_k)](i)+\Theta_{d,\mathbf{F}_d^{(0)}}(\tau_k,i)+(F_d^{(0)}(\tau_k,i)-F_d(\tau_k,i))^\top$$
$$\left(\sum_{p=0}^{r_2} B_{dp}^\top(\tau_k,i)X^{(0)}(\tau_k,i)B_{dp}(\tau_k,i)+R_d(\tau_k,i)\right)$$
$$(F_d^{(0)}(\tau_k,i)-F_d(\tau_k,i))$$
$$+\check{M}_d(\tau_k,i)+\varepsilon I_n,\quad k\in\mathbb{Z}_+. \tag{5.251b}$$

Let $\mathbf{Z}(\cdot)=(Z(\cdot,1),Z(\cdot,2),\dots,Z(\cdot,N))\in\Gamma^\Sigma[\mathbb{R}_+]$ be arbitrary but fixed. Involving (5.159)–(5.161) one obtains that $\mathbf{Z}(\cdot)$ solves the following non-homogeneous backward linear differential equation with jump:

$$-\dot{Z}(t,i) = \mathcal{L}^*_{c,\mathbf{F}_c^{(0)}}(t)[\mathbf{Z}(t)](i)+\Theta_{c,\mathbf{F}_c^{(0)}}(t,i)-(F_c^{(0)}(t,i)-F_c^{\mathbf{Z}}(t,i))^\top$$
$$\left(\sum_{l=1}^{r_1} B_l^\top(t,i)Z(t,i)B_l(t,i)+R_c(t,i)\right)(F_c^{(0)}(t,i)-F_c^{\mathbf{Z}}(t,i))$$
$$-\widehat{M}_c(t,i),\quad \tau_k\le t<\tau_{k+1} \tag{5.252a}$$
$$Z(\tau_k^-,i) = \mathcal{L}^*_{d,\mathbf{F}_d^{(0)}}(\tau_k)[\mathbf{Z}(\tau_k)](i)+\Theta_{d,\mathbf{F}_d^{(0)}}(\tau_k,i)-(F_d^{(0)}(\tau_k,i)-F_d^{\mathbf{Z}}(\tau_k,i))^\top$$

$$\left(\sum_{p=0}^{r_2} B_{dp}^\top(\tau_k,i)Z(\tau_k,i)B_{dp}(\tau_k,i) + R_d(\tau_k,i))(F_d^{(0)}(\tau_k,i) - F_d^{\mathbf{Z}}(\tau_k,i)\right)$$
$$- \widehat{M}_d(\tau_k,i), \quad k \in \mathbb{Z}_+, \tag{5.252b}$$

where $\widehat{M}_c(t,i) \geq 0$ for all $t \geq 0$ and $\widehat{M}_d(\tau_k,i) \geq 0$ for all $\tau_k \in \mathfrak{T}$, $i \in \mathcal{N}$. Combining (5.251) and (5.252) we may infer via (5.243) that $\boldsymbol{\Delta}^{(0)}(\cdot) := \mathbf{X}^{(0)}(\cdot) - \mathbf{Z}(\cdot)$ is a bounded and uniform positive solution of the following non-homogeneous backward jump linear differential equation:

$$-\dot{\Delta}^{(0)}(t,i) = \mathcal{L}^*_{c,\mathbf{F}_c^{(0)}}(t)[\boldsymbol{\Delta}^{(0)}(t)](i) + \mathfrak{M}_c^{(0)}(t,i), \quad \tau_k \leq t < \tau_{k+1} \tag{5.253a}$$

$$\Delta^{(0)}(\tau_k^-,i) = \mathcal{L}^*_{d,\mathbf{F}_d^{(0)}}(\tau_k)[\boldsymbol{\Delta}^{(0}(\tau_k)](i) + \mathfrak{M}_d^{(0)}(\tau_k,i), \quad k \in \mathbb{Z}_+, \tag{5.253b}$$

$i \in \mathcal{N}$, where

$$\begin{aligned}\mathfrak{M}_c^{(0)}(t,i) &\triangleq (F_c^{(0)}(t,i) - F_c(t,i))^\top \left(\sum_{l=1}^{r_1} B_l^\top(t,i)X^{(0)}(t,i)B_l(t,i) + R_c(t,i)\right)\\ &\cdot (F_c^{(0)}(t,i) - F_c(t,i)) + (F_c^{(0)}(t,i) - F_c^{\mathbf{Z}}(t,i))^\top\\ &\left(\sum_{l=1}^{r_1} B_l^\top(t,i)Z(t,i)B_l(t,i) + R_c(t,i)\right)\\ &\cdot (F_c^{(0)}(t,i) - F^{\mathbf{Z}}(t,i)) + \check{M}_c(t,i) + \widehat{M}_c(t,i) \quad \text{for all} \quad (t,i) \in \mathbb{R}_+ \times \mathcal{N}\end{aligned}$$

$$\begin{aligned}\mathfrak{M}_d^{(0)}(\tau_k,i) &\triangleq (F_d^{(0)}(\tau_k,i) - F_d(\tau_k,i))^\top\\ &\left(\sum_{p=0}^{r_2} B_{dp}^\top(\tau_k,i)X^{(0)}(\tau_k,i)B_{dp}(\tau_k,i) + R_d(\tau_k,i)\right)\\ &\cdot (F_d^{(0)}(\tau_k,i) - F_d(\tau_k,i)) + (F_d^{(0)}(\tau_k,i) - F_d^{\mathbf{Z}}(\tau_k,i))^\top\\ &\cdot \left(\sum_{p=0}^{r_2} B_{dp}^\top(\tau_k,i)Z(\tau_k,i)B_{dp}(\tau_k,i) + R_d(\tau_k,i))(F_d^{(0)}(\tau_k,i) - F^{\mathbf{Z}}(\tau_k,i)\right)\\ &+ \check{M}_d(\tau_k,i) + \widehat{M}_d(\tau_k,i) \quad \text{for all} \quad (\tau_k,i) \in \mathfrak{T} \times \mathcal{N}.\end{aligned}$$

Based on (5.159c), (5.159d) together with (5.249) we may deduce that

$$\mathfrak{M}_c^{(0)}(t,i) \geq 0, \quad \text{for all} \quad (t,i) \in \mathbb{R}_+ \times \mathcal{N}$$

and $$\mathfrak{M}_d^{(0)}(\tau_k,i) \geq \varepsilon I_n > 0, \quad \text{for all} \quad (\tau_k,i) \in \mathfrak{T} \times \mathcal{N}.$$

The implication $(iii) \Rightarrow (i)$ from Theorem 2.7 in the case of the non-homogeneous backward jump linear differential equation (5.253) we may conclude that the FJMLDE:

$$\begin{aligned}\dot{\mathbf{X}}(t) &= \mathscr{L}_{c,\mathbf{F}_c^{(0)}}(t)[\mathbf{X}(t)], \quad \tau_k < t \le \tau_{k+1}\\ \mathbf{X}(\tau_k^+) &= \mathscr{L}_{d,\mathbf{F}_d^{(0)}}(\tau_k)[\mathbf{X}(\tau_k)], \quad k \in \mathbb{Z}_+\end{aligned}$$

is exponentially stable. Thus the proof ends.

Remark 5.13 We rewrite the equation (5.253) satisfied by $\boldsymbol{\Delta}^{(0)}(\cdot) \triangleq \mathbf{X}^{(0)}(\cdot) - \mathbf{Z}(\cdot)$ in the form:

$$\begin{aligned}-\dot{\Delta}^{(0)}(t,i) &= (\widetilde{A}_0(t,i) + B_0(t,i)F_c^{(0)}(t,i))^\top \Delta^{(0)}(t,i) + \Delta^{(0)}(t,i)\\ &\quad \cdot (\widetilde{A}_0(t,i) + B_0(t,i)F_c^{(0)}(t,i)) + \widetilde{\mathfrak{M}}_c^{(0)}(t,i), \quad \tau_k \le t < \tau_{k+1}\end{aligned} \tag{5.254a}$$

$$\begin{aligned}\Delta^{(0)}(\tau_k^-,i) &= (A_{d0}(\tau_k,i) + B_{d0}(\tau_k,i)F_d^{(0)}(\tau_k,i))^\top \Delta^{(0)}(\tau_k,i)\\ &\quad \cdot (A_{d0}(\tau_k,i) + B_{d0}(\tau_k,i)F_d^{(0)}(\tau_k,i)) + \widetilde{\mathfrak{M}}_d^{(0)}(\tau_k,i), \quad k \in \mathbb{Z}_+\end{aligned} \tag{5.254b}$$

where $\quad \widetilde{A}_0(t,i) := A_0(t,i) + \frac{1}{2}q_{ii}I_n \quad$ and

$$\begin{aligned}\widetilde{\mathfrak{M}}_c^{(0)}(t,i) :=& \sum_{l=1}^{r_1}(A_l(t,i) + B_l(t,i)F_c^{(0)}(t,i))^\top \Delta^{(0)}(t,i)\\ &(A_0(t,i) + B_0(t,i)F_c^{(0)}(t,i))\\ &+ \sum_{\substack{j=1\\ j\neq i}}^{N} q_{ij}\Delta^{(0)}(t,j) + \widehat{\mathfrak{M}}_c^{(0)}(t,i)\\ \widetilde{\mathfrak{M}}_d^{(0)}(\tau_k,i) :=& \sum_{p=1}^{r_2}(A_{dp}(\tau_k,i) + B_{dp}(\tau_k,i)F_d^{(0)}(\tau_k,i))^\top \Delta^{(0)}(\tau_k,i)(A_{dp}(\tau_k,i)\\ &+ B_{dp}(\tau_k,i)F_d^{(0)}(\tau_k,i)) + \widehat{\mathfrak{M}}_d^{(0)}(\tau_k,i).\end{aligned}$$

Since (5.243) holds we may infer that

$$\widetilde{\mathfrak{M}}_c^{(0)}(t,i) \ge \widehat{\mathfrak{M}}_c^{(0)}(t,i) \ge 0, \quad \text{for all } \ (t,i) \in \mathbb{R}_+ \times \mathscr{N}$$

and

$$\widetilde{\mathfrak{M}}_d^{(0)}(\tau_k,i) \ge \widehat{\mathfrak{M}}_d^{(0)}(\tau_k,i) \ge \varepsilon I_n > 0, \quad \text{for all } \ (\tau_k,i) \in \mathfrak{T} \times \mathscr{N}.$$

The implication $(iii) \Rightarrow (i)$ from Theorem 2.7 applied in the case of the non-homogeneous backward jump linear differential equation (5.254) allows us to deduce that for each $i \in \mathscr{N}$ the forward jump linear differential equation (FJLDE) on $\mathscr{S}_n^N$:

$$\dot{X}_i(t) = (\widetilde{A}_0(t,i) + B_0(t,i)F_c^{(0)}(t,i))X_i(t) \qquad (5.255a)$$
$$+ X_i(t)(\widetilde{A}_0(t,i) + B_0(t,i)F_c^{(0)}(t,i))^\top, \quad \tau_k < t \leq \tau_{k+1}$$
$$X_i(\tau_k^+) = (A_{d0}(\tau_k,i) + B_{d0}(\tau_k,i)F_d^{(0)}(\tau_k,i)) \qquad (5.255b)$$
$$\cdot X_i(\tau_k)(A_{d0}(\tau_k,i) + B_{d0}(\tau_k,i)F_d^{(0)}(\tau_k,i))^\top, \quad k \in \mathbb{Z}_+$$

is exponentially stable.

Lemma 5.9 *Under the Assumptions 2.1, 5.2, the following are equivalent:*

(i) the FJLDE on $\mathcal{S}_n^1$ (5.255) is exponentially stable;
(ii) the FJLDE on $\mathbb{R}^n$:

$$\dot{x}(t) = (\widetilde{A}_0(t,i) + B_0(t,i)F_c^{(0)}(t,i))x(t), \quad \tau_k < t \leq \tau_{k+1} \qquad (5.256a)$$
$$x(\tau_k^+) = (A_{d0}(\tau_k,i) + B_{d0}(\tau_k,i)F_d^{(0)}(\tau_k,i))x(\tau_k), \quad k \in \mathbb{Z}_+ \qquad (5.256b)$$

is exponentially stable.

Proof (hint). One shows that for each $i \in \mathcal{N}$, the exponential stability of the FJLDE (5.255) is equivalent to the exponential stability of the discrete-time Lyapunov equation

$$\mathbb{X}_i(k+1) = \mathbb{A}_i(k)\mathbb{X}_i(k)\mathbb{A}_i^\top(k) \qquad (5.257)$$

where

$$\mathbb{A}_i(k) := \Phi_i(\tau_{k+1},\tau_k)(A_{d0}(\tau_k,i) + B_{d0}(\tau_k,i)F_d^{(0)}(\tau_k,i)),$$

$\Phi_i(\tau_{k+1},\tau_k)$ being the value at $t = \tau_{k+1}$ of the matrix solution of the initial given value problem:

$$\dot{\Phi}_i(t,\tau_k) = (\widetilde{A}_0(t,i) + B_0(t,i)F_c^{(0)}(t,i))\Phi_i(t,\tau_k)$$
$$\Phi_i(\tau_k,\tau_k) = I_n.$$

Further, one shows that the exponential stability of the equation (5.257) is equivalent to the exponential stability of the discrete-time linear equation (DTLE) on $\mathbb{R}^n$:

$$\xi_i(k+1) = \mathbb{A}_i(k)\xi_i(k), \quad k \in \mathbb{Z}_+. \qquad (5.258)$$

One sees that (5.258) is just the accompanying DTLE of the form (2.9)–(2.10) of the FJLDE (5.256). The conclusion is obtained applying the equivalence (i) from Theorem 2.1 applied in the case of the pair formed by the FJLDE (5.256) and its accompanying DTLE (5.258). The details are left to the reader.

Starting from $X^{(0)}(t,i)$, $F_c^{(0)}(t,i)$, $(t,i) \in \mathbb{R}_+ \times \mathcal{N}$ and $F_d^{(0)}(\tau_k,i)$, $(\tau_k,i) \in \mathfrak{T} \times \mathcal{N}$, we construct iterative the sequences $\{X^{(q)}(t,i)\}_{q\geq 0}$, $\{F_c^{(q)}(t,i)\}_{q\geq 0}$, $\{F_d^{(q)}(\tau_k,i)\}_{q\geq 0}$ as $t \to X^{(q+1)}(t,i) : \mathbb{R}_+ \to \mathcal{S}_n^1$ is the unique bounded solution of the following backward jump matrix Lyapunov equation:

$$-\dot{X}^{(q+1)}(t,i) = (\widetilde{A}_0(t,i) + B_0(t,i)F_c^{(q)}(t,i))^\top X^{(q+1)}(t,i) \tag{5.259a}$$
$$+ X^{(q+1)}(t,i)(\widetilde{A}_0(t,i) + B_0(t,i)F_c^{(q)}(t,i)) + \varXi_c^{(q+1)}(t,i),$$
$$\tau_k \le t < \tau_{k+1}$$

$$X^{(q+1)}(\tau_k^-,i) = (A_{d0}(\tau_k,i) + B_{d0}(\tau_k,i)F_d^{(q)}(\tau_k,i))^\top X^{(q+1)}(\tau_k,i) \tag{5.259b}$$
$$\cdot (A_{d0}(\tau_k,i) + B_{d0}(\tau_k,i)F_d^{(q)}(\tau_k,i)) + \varXi_d^{(q+1)}(\tau_k,i), \quad k \in \mathbb{Z}_+$$

where

$$\varXi_c^{(q+1)}(t,i) := \Theta_{c,\mathbf{F}_c^{(q)}}(t,i) + \sum_{l=1}^{r_1}(A_l(t,i) + B_l(t,i)F_c^{(q)}(t,i))^\top X^{(q)}(t,i)$$
$$\cdot (A_l(t,i) + B_l(t,i)F_c^{(q)}(t,i)) + \sum_{\substack{j=1\\ j\neq i}}^{N} q_{ij}X^{(q)}(t,j), \quad (t,i) \in \mathbb{R}_+ \times \mathcal{N} \tag{5.260a}$$

$$\varXi_d^{(q+1)}(\tau_k,i) := \frac{\varepsilon}{q+2} + \Theta_{d,\mathbf{F}_d^{(q)}}(\tau_k,i) + \sum_{p=1}^{r_2}(A_{dp}(\tau_k,i) + B_{dp}(\tau_k,i)F_d^{(q)}(\tau_k,i))^\top$$
$$\cdot X^{(q)}(\tau_k,i)(A_{dp}(\tau_k,i) + B_{dp}(\tau_k,i)F_d^{(q)}(\tau_k,i)) \tag{5.260b}$$
$$\widetilde{A}_0(t,i) \stackrel{\Delta}{=} A_0(t,i) + \frac{1}{2}q_{ii}I_n, \tag{5.260c}$$

$i \in \mathcal{N}$, $\Theta_{c,\mathbf{F}_c^{(q)}}(t,i)$ and $\Theta_{d,\mathbf{F}_d^{(q)}}(\tau_k,i)$, respectively, are computed as in (5.239) with $F_c(t,i)$ and $F_d(\tau_k,i)$, respectively, replaced by $F_c^{(q)}(t,i)$ and $F_d^{(q)}(\tau_k,i)$, respectively. In (5.260b), $\varepsilon > 0$ is a fixed parameter. The gain matrices $F_c^{(q+1)}(t,i)$ and $F_d^{(q+1)}(\tau_k,i)$ are defined by:

$$F_c^{(q+1)}(t,i) \stackrel{\Delta}{=} -(R_c(t,i) + \sum_{l=1}^{r_1} B_l(t,i)X^{(q)}(t,i)B_l(t,i))^{-1}$$
$$(B_0^\top(t,i)X^{(p)}(t,i) + \sum_{l=1}^{r_1} B_l^\top(t,i)X^{(q)}(t,i)A_l(t,i) + L_c^\top(t,i)), \quad t \ge 0 \tag{5.261a}$$

$$F_d^{(q+1)}(\tau_k,i) \stackrel{\Delta}{=} -(R_d(\tau_k,i) + B_{d0}^\top(\tau_k,i)X^{(q+1)}(\tau_k,i)B_{d0}(\tau_k,i)$$
$$+ \sum_{p=1}^{r_2} B_{dp}^\top(\tau_k,i)X^{(q)}(\tau_k,i)B_{dp}(\tau_k,i))^{-1}$$
$$(B_{d0}^\top(\tau_k,i)X^{(q+1)}(\tau_k,i)A_{d0}(\tau_k,i)$$
$$+ \sum_{p=1}^{r_2} B_{dp}^\top(\tau_k,i)X^{(q)}(\tau_k,i)A_{dp}(\tau_k,i) + L_d^\top(\tau_k,i)), \quad \tau_k \in \mathfrak{T}, \tag{5.261b}$$

$i \in \mathcal{N}$. Further, we check inductively with respect to $q = 0, 1, \ldots$, that the following items hold:

(a_q)
$$X^{(q)}(t,i) - Z(t,i) \geq \gamma_q I_n, \quad \text{for all} \quad (t,i) \in \mathbb{R}_+ \times \mathcal{N},$$

where $\mathbf{Z}(\cdot) = (Z(\cdot,1), Z(\cdot,2), \ldots, Z(\cdot,N))$ is an arbitrary function from the set $\Gamma^{\Sigma}[\mathbb{R}_+]$, $\gamma_q > 0$ is a constant;

(b_q) for each $i \in \mathcal{N}$, the FJLDE:

$$\dot{X}_i(t) = (\widetilde{A}_0(t,i) + B_0(t,i)F_c^{(q)}(t,i))X_i(t) \tag{5.262a}$$
$$+ X_i(t)(\widetilde{A}_0(t,i) + B_0(t,i)F_c^{(q)}(t,i))^\top, \quad \tau_k < t \leq \tau_{k+1}$$
$$X_i(\tau_k^-) = (A_{d0}(\tau_k,i) + B_{d0}(\tau_k,i)F_d^{(q)}(\tau_k,i)) \tag{5.262b}$$
$$\cdot X_i(\tau_k)(A_{d0}(\tau_k,i) + B_{d0}(\tau_k,i)F_d^{(q)}(\tau_k,i))^\top, \quad k \in \mathbb{Z}_+$$

is exponentially stable;

(c_q)
$$X^{(q)}(t,i) \geq X^{(q+1)}(t,i), \quad \text{for all} \quad (t,i) \in \mathbb{R}_+ \times \mathcal{N}.$$

The validity of the items (a_0), (b_0) is confirmed by Lemma 5.7 and Lemma 5.8 together with Remark 5.13, respectively. If (b_0) holds, it follows that for each $i \in \mathcal{N}$, $X^{(1)}(t,i)$ is well defined as a unique bounded on $\mathbb{R}_+$ solution of the following non-homogeneous backward jump linear differential equation (BJLDE):

$$-\dot{X}^{(1)}(t,i) = (\widetilde{A}_0(t,i) + B_0(t,i)F_c^{(0)}(t,i))^\top X^{(1)}(t,i) \tag{5.263a}$$
$$+ X^{(1)}(t,i)(\widetilde{A}_0(t,i) + B_0(t,i)F_c^{(0)}(t,i)) + \Xi_c^{(1)}(t,i), \quad \tau_k \leq t < \tau_{k+1}$$
$$X^{(1)}(\tau_k^-,i) = (A_{d0}(\tau_k,i) + B_{d0}(\tau_k,i)F_d^{(0)}(\tau_k,i)) \tag{5.263b}$$
$$\cdot X^{(1)}(\tau_k,i)(A_{d0}(\tau_k,i) + B_{d0}(\tau_k,i)F_d^{(0)}(\tau_k,i)) + \Xi_d^{(1)}(\tau_k,i), \quad k \in \mathbb{Z}_+$$

where $\Xi_c^{(1)}(t,i)$ and $\Xi_d^{(1)}(\tau_k,i)$ are defined by (5.260a) and (5.260b) for $q = 0$.

Subtracting (5.263) from (5.251) we obtain that for each $i \in \mathcal{N}$ the function $t \to \Upsilon^{(0)}(t,i) \stackrel{\Delta}{=} X^{(0)}(t,i) - X^{(1)}(t,i) : \mathbb{R}_+ \to \mathcal{S}_n^1$ is a bounded solution of the following non-homogeneous BJLDE:

$$-\dot{\Upsilon}^{(0)}(t,i) = (\widetilde{A}_0(t,i) + B_0(t,i)F_c^{(0)}(t,i))^\top \Upsilon^{(0)}(t,i) \tag{5.264a}$$
$$+ \Upsilon^{(0)}(t,i)(\widetilde{A}_0(t,i) + B_0(t,i)F_c^{(0)}(t,i)) + \Psi_c^{(0)}(t,i), \quad \tau_k \leq t < \tau_{k+1}$$
$$\Upsilon^{(0)}(\tau_k^-,i) = (A_{d0}(\tau_k,i) + B_{d0}(\tau_k,i)F_d^{(0)}(\tau_k,i))\Upsilon^{(0)}(\tau_k,i) \tag{5.264b}$$
$$\cdot (A_{d0}(\tau_k,i) + B_{d0}(\tau_k,i)F_d^{(0)}(\tau_k,i)) + \Psi_d^{(0)}(\tau_k,i), \quad k \in \mathbb{Z}_+,$$

$i \in \mathcal{N}$, where

$$\Psi_c^{(0)}(t,i) \triangleq (F_c^{(0)}(t,i) - F_c(t,i))^\top (\sum_{l=1}^{r_1} B_l^\top(t,i)X^{(0)}(t,i)B_l(t,i) + R_c(t,i))$$
$$\cdot (F_c^{(0)}(t,i) - F_c(t,i)) + \check{M}_c(t,i) \text{ for all } (t,i) \in \mathbb{R}_+ \times \mathcal{N}$$

$$\Psi_d^{(0)}(\tau_k,i) \triangleq (F_d^{(0)}(\tau_k,i) - F_d(\tau_k,i))^\top$$
$$\left(\sum_{p=0}^{r_2} B_{dp}^\top(\tau_k,i)X^{(0)}(\tau_k,i)B_{dp}(\tau_k,i) + R_d(\tau_k,i)\right)$$
$$\cdot (F_d^{(0)}(\tau_k,i) - F_d(\tau_k,i)) + \check{M}_d(\tau_k,i) + \frac{\varepsilon}{2}I_n, \text{ for all } (\tau_k,i) \in \mathfrak{T} \times \mathcal{N}.$$

Bearing in mind the definitions of $\check{M}_c(t,i)$ and $\check{M}_d(\tau_k,i)$ together with (5.249) we deduce that $\Psi_c^{(0)}(t,i) \geq 0$, for all $(t,i) \in \mathbb{R}_+ \times \mathcal{N}$ and $\Psi_d^{(0)}(\tau_k,i) \geq \frac{\varepsilon}{2}I_n > 0$, for all $(\tau_k,i) \in \mathfrak{T} \times \mathcal{N}$. Since (b_0) holds, we deduce via part (i) and (v) of Theorem 2.13 applied in the case of the non-homogeneous BJLDE (5.264), that this equation has a unique bounded on $\mathbb{R}_+$ solution and memorize that this solution is uniform positive. Hence, $X^{(0)}(t,i) - X^{(1)}(t,i) \geq c_0 I_n$, for all $(t,i) \in \mathbb{R}_+ \times \mathcal{N}$, c_0 being a constant. Thus we have shown that item (c_q) holds when $q = 0$. Let us assume by induction that for an integer $q \geq 1$, (a_q), (b_q), (c_q) are fulfilled if (a_ι), (b_ι), (c_ι) hold for any $0 \leq \iota \leq q-1$.

Let $\mathbf{Z}(\cdot) = (Z(\cdot,1), Z(\cdot,2), \ldots, Z(\cdot,N)) \in \Gamma^\Sigma[\mathbb{R}_+]$ be arbitrary but fixed. By direct calculation involving (5.159)–(5.161) one shown that $\mathbf{Z}(\cdot)$ is a bounded on $\mathbb{R}_+$ solution of the following BJLDE:

$$\begin{aligned}
-\dot{Z}(t,i) &= (\widetilde{A}_0(t,i) + B_0(t,i)F_c^{(q-1)}(t,i))^\top Z(t,i) \\
&+ Z(t,i)(\widetilde{A}_0(t,i) + B_0(t,i)F_c^{(q-1)}(t,i)) \\
&+ \sum_{l=1}^{r_1}(A_l(t,i) + B_l(t,i)F_c^{(q-1)}(t,i))^\top Z(t,i)(A_l(t,i) \\
&+ B_l(t,i)F_c^{(q-1)}(t,i)) \\
&+ \sum_{\substack{j=1\\ j\neq i}}^{N} q_{ij}Z(t,j) \\
&- (F_c^{(q-1)} - F_c^{\mathbf{Z}}(t,i))^\top \left(\sum_{l=1}^{r_1} B_l^\top(t,i)Z(t,i)B_l(t,i) + R_c(t,i)\right) \\
&(F_c^{(q-1)}(t,i) - F_c^{\mathbf{Z}}(t,i)) \\
&- \widehat{M}_c(t,i) + \Theta_{c,\mathbf{F}_c^{(q-1)}}(t,i), \quad \tau_k \leq t < \tau_{k+1}
\end{aligned} \tag{5.265a}$$

$$\begin{aligned}
X(\tau_k^-,i) &= \sum_{p=0}^{r_2}(A_{dp}(\tau_k,i) + B_{dp}(\tau_k,i)F_d^{(p-1)}(\tau_k,i))^\top Z(\tau_k,i)(A_{dp}(\tau_k,i) \\
&+ B_{dp}(\tau_k,i)F_d^{(q-1)}(\tau_k,i)) + \Theta_{d,\mathbf{F}_d^{(q-1)}}(\tau_k,i)
\end{aligned}$$

$$-(F_d^{(p-1)}(\tau_k,i)-F_d^{\mathbf{Z}}(\tau_k,i))^\top$$
$$\cdot\left(\sum_{p=0}^{r_2} B_{dp}^\top(\tau_k,i)Z(\tau_k,i)B_{dp}(\tau_k,i)+R_d(\tau_k,i)\right)$$
$$\cdot(F_d^{(p-1)}(\tau_k,i)-F_d^{\mathbf{Z}}(\tau_k,i))-\widehat{M}_d(\tau_k,i),\ \ k\in\mathbb{Z}_+ \tag{5.265b}$$

where

$$\widehat{M}_c(t,i)\geq 0 \ \text{ for all } \ (t,i)\in\mathbb{R}_+\times\mathcal{N},$$
$$\widehat{M}_d(\tau_k,i)\geq 0 \ \text{ for all } \ (\tau_k,i)\in\mathfrak{T}\times\mathcal{N},$$

$F_c^{\mathbf{Z}}(t,i)$ and $F_d^{\mathbf{Z}}(\tau_k,i)$ were defined in (5.245). Subtracting (5.265) from (5.259) (written for q replaced by $q-1$), we obtain that for each $i\in\mathcal{N}$:

$$\Delta^{(q)}(\cdot,i)\stackrel{\Delta}{=}X^{(q)}(\cdot,i)-Z(\cdot,i)$$

is a bounded on $\mathbb{R}_+$ solution of the following non-homogeneous BJLDE

$$\begin{aligned}-\dot{\Delta}(t,i)&=(\widetilde{A}_0(t,i)+B_0(t,i)F_c^{(q-1)}(t,i))^\top\Delta(q)(t,i)\\&\quad+\Delta^{(q)}(t,i)(\widetilde{A}_0(t,i)+B_0(t,i)F_c^{(q-1)}(t,i))+\mathbb{M}_c^{(q)}(t,i),\\&\quad\tau_k\leq t<\tau_{k+1}\end{aligned} \tag{5.266a}$$
$$\begin{aligned}\Delta^{(q)}(\tau_k^-,i)&=(A_{d0}(\tau_k,i)+B_{d0}(\tau_k,i)F_d^{(q-1)}(\tau_k,i))^\top\\&\quad\cdot\Delta^{(q)}(\tau_k,i)(A_{d0}(\tau_k,i)+B_{d0}(\tau_k,i)F_d^{(q-1)}(\tau_k,i))+\mathbb{M}_d^{(q)}(\tau_k,i),\\&\quad k\in\mathbb{Z}_+\end{aligned} \tag{5.266b}$$

with

$$\begin{aligned}\mathbb{M}_c^{(q)}(t,i)&:=\sum_{l=1}^{r_1}(A_l(t,i)+B_l(t,i)F_c^{(q-1)}(t,i))^\top(X^{(q-1)}(t,i)-Z(t,i))\\&\quad\cdot(A_l(t,i)+B_l(t,i)F_c^{(q-1)}(t,i))+\sum_{\substack{j=1\\j\neq i}}q_{ij}(X^{(q-1)}(t,j)-Z(t,j))\\&\quad+(F_c^{(q-1)}(t,i)-F_c^{\mathbb{Z}}(t,i))\left(\sum_{l=1}^{r_1}B_l^\top(t,i)Z(t,i)B_l(t,i)+R_c(t,i)\right)\\&\quad\cdot(F_c^{(q-1)}(t,i)-F_c^{\mathbf{Z}}(t,i))+\widehat{M}_c(t,i),\end{aligned}$$

$$\begin{aligned}\mathbb{M}_d^{(q)}(\tau_k,i)&:=\sum_{p=1}^{r_2}(A_{dp}(\tau_k,i)+B_{dp}(\tau_k,i)F_d^{(q-1)}(\tau_k,i))^\top(X^{(q-1)}(\tau_k,i)-Z(\tau_k,i)\\&\quad\cdot(A_{dp}(\tau_k,i)+B_{dp}(\tau_k,i)F_d^{(q-1)}(\tau_k,i))+(F_d^{(q-1)}(\tau_k,i)-F_d^{\mathbf{Z}}(\tau_k,i))^\top\end{aligned}$$

$$\cdot\left(\sum_{p=0}^{r_2} B_{dp}^{\top}(\tau_k,i)Z(\tau_k,i)B_{dp}(\tau_k,i)+R_d(\tau_k,i)\right)$$
$$(F_d^{(q-1)}(\tau_k,i)-F_d^{\mathbf{Z}}(\tau_k,i))$$
$$+\widehat{M}_d(\tau_k,i)+\frac{\varepsilon}{q+1}I_n.$$

Bearing in mind that (a_{q-1}) is fulfilled and that $\mathbf{Z}(\cdot)$ is satisfying the constraints (5.159c) and (5.159d) we deduce that

$$\mathbb{M}_c^{(q)}(t,i)\geq 0 \ \text{ for all } \ (t,i)\in\mathbb{R}_+\times\mathcal{N},$$

$$\mathbb{M}_d^{(q)}(\tau_k,i)\geq\frac{\varepsilon}{q+1}I_n \ \text{ for all } \ (\tau_k,i)\in\mathfrak{T}\times\mathcal{N}.$$

Since (b_{q-1}) holds too, we may conclude via (i) and (v) from Theorem 2.13 applied in the case of (5.266) that this equation has a unique bounded on $\mathbb{R}_+$ solution and, moreover, that solution is uniform positive. Hence, there exists $\nu_2>0$ such that

$$X^{(q)}(t,i)-Z(t,i)\geq\nu_q I_n, \ \text{ for all } \ (t,i)\in\mathbb{R}_+\times\mathcal{N}.$$

Thus we have shown that (a_q) holds. Before proving that (b_q) is true, let us show that $F_c^{(q)}(t,i)$ and $F_d^{(q)}(\tau_k,i)$ are well defined by (5.261) with q replaced by $q-1$. Indeed, since (a_ι) holds for any $0\leq\iota\leq q$ we may infer that

$$\sum_{l=1}^{r_1} B_l^{\top}(t,i)X^{(q-1)}(t,i)B_l(t,i)+R_c(t,i)\geq\sum_{l=1}^{r_1} B_l^{\top}(t,i)Z(t,i)B_l(t,i)+R_c(t,i)$$
$$\geq\gamma_c I_{m_c}, \ \text{ for all } \ (t,i)\in\mathbb{R}_+\times\mathcal{N}$$

and

$$B_{d0}^{\top}(\tau_k,i)X^{(q)}(\tau_k,i)B_{d0}(\tau_k,i)+\sum_{p=1}^{r_2} B_{dp}^{\top}(\tau_k,i)X^{(q+1)}(\tau_k,i)B_{dp}(\tau_k,i)+R_d(\tau_k,i)$$

$$\geq\sum_{p=0}^{r_2} B_{dp}^{\top}(\tau_k,i)Z(\tau_k,i)B_{dp}(\tau_k,i)+R_d(\tau_k,i)\geq\gamma_d I_{m_d}, \ \text{ for all } \ (\tau_k,i)\in\mathfrak{T}\times\mathcal{N}.$$

Hence, the gain matrices $F_c^{(q)}(t,i)$, $F_d^{(q)}(\tau_k,i)$ are well defined by (5.261) written for q replaced by $q-1$.

Let us now consider the backward jump linear differential equation:

$$-\dot{X}^{(q)}(t,i)=(\widetilde{A}_0(t,i)+B_0(t,i)F_c^{(q-1)}(t,i))^{\top}X^{(q)}(t,i)$$
$$+X^{(q)}(t,i)(\widetilde{A}_0(t,i)+B_0(t,i)F_c^{(q-1)}(t,i))+\Xi_c^{(q)}(t,i),$$

$$\tau_k \le t < \tau_{k+1} \tag{5.267a}$$

$$\begin{aligned} X^{(q)}(\tau_k^-, i) &= (A_{d0}(\tau_k, i) + B_{d0}(\tau_k, i)F_d^{(q-1)}(\tau_k, i)) \\ &\cdot X^{(q)}(\tau_k, i)(A_{d0}(\tau_k, i) + B_{d0}(\tau_k, i)F_d^{(q-1)}(\tau_k, i)) + \Xi_d^{(q)}(\tau_k, i), \\ &k \in \mathbb{Z}_+ \end{aligned} \tag{5.267b}$$

obtained from (5.259) for q replaced by $q-1$. By direct calculation we rewrite (5.267) as:

$$\begin{aligned} -\dot{X}^{(q)}(t, i) &= (\widetilde{A}_0(t, i) + B_0(t, i)F_c^{(q)}(t, i))^\top X^{(q)}(t, i) \\ &+ X^{(q)}(t, i)(\widetilde{A}_0(t, i) + B_0(t, i)F_c^{(q)}(t, i)) \\ &+ \sum_{l=1}^{r_1}(A_l(t, i) + B_l(t, i)F_c^{(q)}(t, i))^\top \\ &\cdot X^{(q-1)}(t, i)(A_l(t, i) + B_l(t, i)F_c^{(q)}(t, i)) + \sum_{\substack{j=1 \\ j\neq i}}^{N} q_{ij}X^{(q-1)}(t, j) \\ &+ \theta_{c,\mathbf{F}_c^{(q)}}(t, i) - (F_c^{(q)}(t, i) - F_c^{(q-1)}(t, i))^\top \\ &\cdot \left(\sum_{l=1}^{r_1} B_l^\top(t, i)X^{(q-1)}(t, i)B_l(t, i) + R_c(t, i)\right) \\ &(F_c^{(q)}(t, i) - F_c^{(q-1)}(t, i)) \\ &\qquad \tau_k \le t < \tau_{k+1} \end{aligned} \tag{5.268a}$$

$$\begin{aligned} X(\tau_k^-, i) &= (A_{d0}(\tau_k, i) + B_{d0}(\tau_k, i)F_d(\tau_k, i))^\top X^{(q)}(\tau_k, i) \\ &\cdot (A_{d0}(\tau_k, i) + B_{d0}(\tau_k, i)F_d(\tau_k, i)) \\ &+ \sum_{p=1}^{r_2}(A_{dp}(\tau_k, i) + B_{dp}(\tau_k, i)F_d^{(q)}(\tau_k, i))^\top \\ &\cdot X^{(q-1)}(\tau_k, i)(A_{dp}(\tau_k, i) + B_{dp}(\tau_k, i)F_d^{(q)}(\tau_k, i)) \\ &+ \theta_{d,\mathbf{F}_d^{(q)}}(\tau_k, i) + (F_d^{(q)}(\tau_k, i) - F_d^{(q-1)}(\tau_k, i))^\top \\ &\cdot \Big(B_{d0}^\top(\tau_k, i)X^{(q)}(\tau_k, i)B_{d0}(\tau_k, i) \\ &+ \sum_{p=1}^{r_2} B_{dp}^\top(\tau_k, i)X^{(q-1)}(\tau_k, i)B_{dp}(\tau_k, i) + R_d(\tau_k, i)\Big) \\ &\cdot (F_d^{(q)}(\tau_k, i) - F_d^{(q-1)}(\tau_k, i)) + \frac{\varepsilon}{q+1}I_n, \quad k \in \mathbb{Z}_+. \end{aligned} \tag{5.268b}$$

Let $\mathbf{Z}(\cdot) = (Z(\cdot, 1), Z(\cdot, 2), \ldots, Z(\cdot, N)) \in \Gamma^{\Sigma}[\mathbb{R}_+]$ be arbitrary but fixed. By direct calculation involving (5.159)–(5.161) we obtain that $\mathbf{Z}(\cdot)$ solves the following backward jump differential equation:

$$-\dot{Z}(t,i) = (\widetilde{A}_0(t,i) + B_0(t,i)F_c^{(q)}(t,i))^\top Z(t,i)$$
$$+ Z(t,i)(\widetilde{A}_0(t,i) + B_0(t,i)F_c^{(q)}(t,i)) + \sum_{l=1}^{r_1}(A_l(t,i) + B_l(t,i)F_c^{(q)}(t,i))^\top$$
$$\cdot Z(t,i)(A_l(t,i) + B_l(t,i)F_c^{(q)}(t,i)) + \sum_{\substack{j=1\\ j\neq i}}^{N} q_{ij}Z(t,j)$$
$$+ \Theta_{c,\mathbf{F}_c^{(q)}}(t,i) - (F_c^{(q)}(t,i) - F_c^{\mathbf{Z}}(t,i))^\top$$
$$\cdot \left(\sum_{l=1}^{r_1} B_l^\top(t,i)Z(t,i)B_l(t,i) + R_c(t,i)\right)(F_c^{(q)}(t,i) - F_c^{\mathbf{Z}}(t,i))$$
$$- \widehat{M}_c(t,i), \quad \tau_k \leq t < \tau_{k+1} \tag{5.269a}$$

$$X(\tau_k^-,i) = \sum_{p=0}^{r_2}(A_{dp}(\tau_k,i) + B_{dp}(\tau_k,i)F_d^{(q)}(\tau_k,i))^\top$$
$$\cdot Z(\tau_k,i)(A_{dp}(\tau_k,i) + B_{dp}(\tau_k,i)F_d^{(q)}(\tau_k,i))$$
$$+ \Theta_{d,\mathbf{F}_d^{(q)}}(\tau_k,i) - (F_d^{(q)}(\tau_k,i) - F_d^{\mathbf{Z}}(\tau_k,i))^\top$$
$$\cdot \left(\sum_{p=0}^{r_2} B_{dp}^\top(\tau_k,i)Z(\tau_k,i)B_{dp}(\tau_k,i) + R_d(\tau_k,i)\right)$$
$$\cdot (F_d^{(q)}(\tau_k,i) - F_d^{\mathbf{Z}}(\tau_k,i)) - \widehat{M}_d(\tau_k,i), \quad k \in \mathbb{Z}_+, \tag{5.269b}$$

where $\widehat{M}_c(t,i) \geq 0$, for all $(t,i) \in \mathbb{R}_+ \times \mathcal{N}$, and $\widehat{M}_d(\tau_k,i) \geq 0$, for all $(\tau_k,i) \in \mathfrak{T} \times \mathcal{N}$. Taking into account that the items (a_ι) hold for $0 \leq \iota \leq q$ we deduce that for each $i \in \mathcal{N}$, $t \to \Delta^{(q)}(t,i) \stackrel{\Delta}{=} X^{(q)}(t,i) - Z(t,i) : \mathbb{R}_+ \to \mathcal{S}_n^1$ is a bounded and uniform positive on $\mathbb{R}_+$ solution of the following backward jump linear differential equation:

$$-\dot{\Delta}^{(q)}(t,i) = (\widetilde{A}_0(t,i) + B_0(t,i)F_c^{(q)}(t,i))^\top \Delta(q)(t,i)$$
$$+ \Delta^{(q)}(t,i)(\widetilde{A}_0(t,i) + B_0(t,i)F_c^{(q)}(t,i)) + \mathfrak{M}_c^{(q)}(t,i), \quad \tau_k \leq t < \tau_{k+1} \tag{5.270a}$$

$$\Delta^{(q)}(\tau_k^-,i) = (A_{d0}(\tau_k,i) + B_{d0}(\tau_k,i)F_d^{(q)}(\tau_k,i))^\top$$
$$\cdot \Delta^{(q)}(\tau_k,i)(A_{d0}(\tau_k,i) + B_{d0}(\tau_k,i)F_d^{(q)}(\tau_k,i)) + \mathfrak{M}_d^{(q)}(\tau_k,i), \quad k \in \mathbb{Z}_+ \tag{5.270b}$$

where

$$\mathfrak{M}_c^{(q)}(t,i) := \sum_{l=1}^{r_1}(A_l(t,i) + B_l(t,i)F_c^{(q)}(t,i))^\top (X^{(q-1)}(t,i) - Z(t,i))$$

$$
\begin{aligned}
&\cdot (A_l(t,i) + B_l(t,i)F_c^{(q)}(t,i)) + \sum_{\substack{j=1 \\ j\neq i}} q_{ij}(X^{(q-1)}(t,j) - Z(t,j)) \\
&+ (F_c^{(q)}(t,i) - F_c^{(q-1)}(t,i))^\top \left(\sum_{l=1}^{r_1} B_l^\top(t,i) X^{(q-1)}(t,i) B_l(t,i) + R_c(t,i) \right) \\
&\cdot (F_c^{(q)}(t,i) - F_c^{(q-1)}(t,i)) + (F_c^{(q)} - F_c^{\mathbf{Z}})^\top \left(\sum_{l=1}^{r_1} B_l^\top(t,i) Z(t,i) B_l(t,i) + R_c(t,i) \right) \\
&\cdot (F_c^{(q)} - F_c^{\mathbf{Z}}(t,i)) + \widehat{M}_c(t,i),
\end{aligned}
$$

$$
\begin{aligned}
\mathfrak{M}_d^{(q)}(\tau_k,i) := &\sum_{p=1}^{r_2} (A_{dp}(\tau_k,i) + B_{dp}(\tau_k,i)F_d^{(q)}(\tau_k,i))^\top (X^{(q-1)}(\tau_k,i) - Z(\tau_k,i) \\
&\cdot (A_{dp}(\tau_k,i) + B_{dp}(\tau_k,i)F_d^{(q)}(\tau_k,i)) + (F_d^{(q)}(\tau_k,i) - F_d^{(q-1)}(\tau_k,i))^\top \\
&\cdot (B_{d0}^\top(\tau_k,i) X^{(q)}(\tau_k,i) B_{d0}(\tau_k,i) \\
&+ \sum_{p=1}^{r_2} B_{dp}^\top(\tau_k,i) X^{(q-1)}(\tau_k,i) B_{dp}(\tau_k,i) + R_d(\tau_k,i)) \\
&\cdot (F_d^{(q)}(\tau_k,i) - F_d^{(q-1)}(\tau_k,i)) + (F_d^{(q)}(\tau_k,i) - F_d^{\mathbf{Z}}(\tau_k,i))^\top \\
&\cdot \left(\sum_{p=0}^{r_2} B_{dp}^{\rightarrow}(\tau_k,i) Z(\tau_k,i) B_{dp}(\tau_k,i) + R_d(\tau_k,i) \right) \\
&(F_d^{(q)}(\tau_k,i) - F_d^{\mathbf{Z}}(\tau_k,i)) \\
&+ \widehat{M}_d(\tau_k,i) + \frac{\varepsilon}{q+1} I_n.
\end{aligned}
$$

Since (a_{q-1}) holds and $Z(\cdot,i)$ satisfies (5.159c) and (5.159d) we may deduce that $\mathfrak{M}_c^{(q)}(t,i) \geq 0$, for all $(t,i) \in \mathbb{R}_+ \times \mathcal{N}$, and $\mathfrak{M}_d^{(q)}(\tau_k,i) \geq \frac{\varepsilon}{q+1} I_n$, for all $(\tau_k,i) \in \mathfrak{T} \times \mathcal{N}$. The implication $(iii) \Rightarrow (i)$ from Theorem 2.7 applied in the case of the non-homogeneous BJLDE (5.270) allows us to conclude that the FJLDE (5.262) is exponentially stable which confirms the validity of (b_2). Further, subtracting (5.259) from (5.268) we obtain that for each $i \in \mathcal{N}$ the function $t \to \Upsilon^{(q)}(t,i) := X^{(q)} - X^{(q+1)}(t,i) : \mathbb{R}_+ \to \mathcal{S}_n^1$ is a bounded solution of the following non-homogeneous BJMLDE:

$$
\begin{aligned}
-\dot{\Upsilon}^{(q)}(t,i) &= (\widetilde{A}_0(t,i) + B_0(t,i)F_c^{(q)}(t,i))^\top \Upsilon^{(q)}(t,i) \\
&\quad + \Upsilon^{(q)}(t,i)(\widetilde{A}_0(t,i) + B_0(t,i)F_c^{(q)}(t,i)) + \Psi_c^{(q)}(t,i), \\
&\quad \tau_k \leq t < \tau_{k+1} \qquad (5.271a) \\
\Upsilon^{(q)}(\tau_k^-,i) &= (A_{d0}(\tau_k,i) + B_{d0}(\tau_k,i)F_d^{(q)}(\tau_k,i))^\top
\end{aligned}
$$

$$\cdot \Upsilon^{(q)}(\tau_k, i)(A_{d0}(\tau_k, i) + B_{d0}(\tau_k, i)F_d^{(q)}(\tau_k, i)) + \Psi_d^{(q)}(\tau_k, i),\ k \in \mathbb{Z}_+, \tag{5.271b}$$

$i \in \mathcal{N}$, where

$$\Psi_c^{(q)}(t,i) \overset{\Delta}{=} (F_c^{(q)}(t,i) - F_c^{(q-1)}(t,i))^\top \left(\sum_{l=1}^{r_1} B_l^\top(t,i) X^{(q-1)}(t,i) B_l(t,i) + R_c(t,i) \right)$$
$$\cdot (F_c^{(q)}(t,i) - F_c^{(q-1)}(t,i))$$
$$+ \sum_{l=1}^{r_1} (A_l(t,i) + B_l(t,i)F_c^{(q)}(t,i))^\top (X^{(q-1)}(t,i) - X^{(q)}(t,i))$$
$$\cdot (A_l(t,i) + B_l(t,i)F_c^{(q)}(t,i)) + \sum_{\substack{j=1 \\ j \neq i}} q_{ij}(X^{(q-1)} - X^{(q)}(t,i)),$$

for all $(t,i) \in \mathbb{R}_+ \times \mathcal{N}$

$$\Psi_d^{(q)}(\tau_k, i) \overset{\Delta}{=} (F_d^{(q)}(\tau_k, i) - F_d^{(q-1)}(\tau_k, i))^\top (B_{d0}^\top(\tau_k, i) X^{(q)}(\tau_k, i) B_{d0}(\tau_k, i)$$
$$+ \sum_{p=1}^{r_2} B_{dp}^\top(\tau_k, i) X^{(q-1)}(\tau_k, i) B_{dp}(\tau_k, i) + R_d(\tau_k, i))$$
$$\cdot (F_d^{(q)}(\tau_k, i) - F_d^{(q-1)}(\tau_k, i))$$
$$+ \sum_{p=1}^{r_2} (A_{dp}(\tau_k, i) + B_{dp}(\tau_k, i)F_d^{(q)}(\tau_k, i))^\top$$
$$\cdot (X^{(q-1)}(\tau_k, i) - X^{(q)}(\tau_k, i))(A_{dp}(\tau_k, i) + B_{dp}(\tau_k, i)F_d^{(q)}(\tau_k, i))$$
$$+ \frac{\varepsilon}{(q+1)(q+2)} I_n, \text{ for all } (\tau_k, i) \in \mathfrak{T} \times \mathcal{N}.$$

One sees that $\Psi_c^{(q)}(t,i) \geq 0$, for all $(t,i) \in \mathbb{R}_+ \times \mathcal{N}$ and

$$\Psi_d^{(q)}(\tau_k, i) \geq \frac{\varepsilon}{(q+1)(q+2)} I_n > 0, \quad \text{for all } (\tau_k, i) \in \mathfrak{T} \times \mathcal{N},$$

because (c_{q-1}) holds according to the induction assumption. Further, the exponential stability of (5.262) allows us to conclude that the non-homogeneous BJMLDE (5.271) has a unique bounded on $\mathbb{R}_+$ solution and moreover that solution is uniform positive definite on $\mathbb{R}_+$. Hence, there exist a constant $\gamma_q > 0$ such that

$$X^{(q)}(t,i) - X^{(q+1)}(t,i) \geq \gamma_q I_n, \quad \text{for all } (t,i) \in \mathbb{R}_+ \times \mathcal{N},$$

which confirms the validity of (c_q).

Now, we are in position to prove:

Theorem 5.13 *Assume:*

(a) the Assumptions 1.2, 2.1 and 5.2 hold;
(b) the JLSS (5.163) is mean-square stabilizable by linear state feedback;
(c) the set $\Gamma^{\Sigma}[\mathbb{R}_+]$ associated to the SJMRDE (5.10) is not empty.

Under these conditions, for each $(t, i) \in \mathbb{R}_+ \times \mathcal{N}$, the sequence $\{X^{(q)}(t, i)\}_{q \geq 0}$ defined by (5.259)–(5.260) is convergent. If

$$\widetilde{X}(t, i) \stackrel{\Delta}{=} \lim_{q \to \infty} X^{(q)}(t, i), \quad \textit{for all} \ (t, i) \in \mathbb{R}_+ \times \mathcal{N} \tag{5.272}$$

then $\widetilde{\mathbf{X}}(\cdot) = (\widetilde{X}(\cdot, 1), \widetilde{X}(\cdot, 2), \dots, \widetilde{X}(\cdot, N))$ coincides to the maximal solution of the SJMRDE (5.10). If the set $\widetilde{\Gamma}^{\Sigma}[\mathbb{R}_+]$ is nonempty, then $\widetilde{\mathbf{X}}(\cdot)$ defined by (5.272) is just the bounded and stabilizing solution of (5.10).

Proof Since (a_q), (c_q) hold for any $q = 0, 1, \dots,$ we deduce that the sequences $\{X^{(q)}(t, i)\}_{q \geq 0}$ are convergent. Proceeding as in the proof of Theorem 5.3 one shows that $\widetilde{\mathbf{X}}(\cdot)$ defined by (5.272) is a bounded solution of the SJMRDE (5.10). Moreover, from (a_q), one gets that $\widetilde{\mathbf{X}}(t) \succeq \mathbf{Z}(t)$, for all $t \in \mathbb{R}_+$ and any $\mathbf{Z}(\cdot) \in \Gamma^{\Sigma}[\mathbb{R}_+]$. Bearing in mind the Definition 5.3 we may conclude that $\widetilde{\mathbf{X}}(\cdot)$ defined in (5.272) is the maximal solution of (5.10). Further, if the set $\widetilde{\Gamma}^{\Sigma}[\mathbb{R}_+]$ is nonempty then, according to Theorem 5.6 we deduce that $\widetilde{\mathbf{X}}(\cdot)$ is just the bounded and stabilizing solution of (5.10). Thus the proof ends.

5.7.2 The Numerical Computation of the Bounded and Stabilizing Solution of a Backward Jump Matrix Linear Differential Equation with Riccati-Type Jumping Operator

In this subsection we focus our attention on the numerical computation of the bounded and stabilizing solution of a BJMLDE of Riccati-type jumping operator (5.168). In this case, the role of JLSS (5.163) is played by the impulsive controlled linear stochastic system (ICLSS) (5.169). Based on the Definition 4.4 the ICLSS (5.169) is mean-square stabilizable by impulses in a linear state feedback form if there exist sequences of gain matrices $\{F_d(\tau_k, i)\}_{k \in \mathbb{Z}_+} \subset \mathbb{R}^{m_d \times n}$, $i \in \mathcal{N}$, with the property that the corresponding closed-loop JSLDE:

$$dx(t) = A_0(t, \eta(t))x(t)dt + \sum_{l=1}^{r_1} A_l(t, \eta(t))x(t)dw_l(t), \quad \tau_k < t \leq \tau_{k+1} \tag{5.273a}$$

$$\begin{aligned} x(\tau_k^+) &= (A_{d0}(\tau_k, \eta(\tau_k)) + B_{d0}(\tau_k, \eta(\tau_k))F_d(\tau_k, \eta(\tau_k)))x(\tau_k) \\ &+ \sum_{p=1}^{r_2} w_{dp}(k)(A_{dp}(\tau_k, \eta(\tau_k)) + B_{dp}(\tau_k, \eta(\tau_k))F_d(\tau_k, \eta(\tau_k)))x(\tau_k), \quad k \in \mathbb{Z}_+, \end{aligned} \tag{5.273b}$$

is strongly exponentially stable in mean square (SESMS). From the Definition 3.1 (a) it follows that the JSLDE (5.273) is SESMS if and only if the forward jump linear differential equation (FJLDE) on $\mathcal{S}_n^N$:

$$\begin{aligned} \dot{X}(t, i) &= A_0^\top(t, i)X(t, i) + X(t, i)A_0(t, i) + \sum_{l=1}^{r_1} A_l^\top(t, i)X(t, i)A_l(t, i) \\ &+ \sum_{j=1}^{N} q_{ji}X(t, j), \quad \tau_k < t \leq \tau_{k+1} \end{aligned} \tag{5.274a}$$

$$\begin{aligned} X(\tau_k^+) &= \sum_{p=0}^{r_2}(A_{dp}(\tau_k, i) + B_{dp}(\tau_k, i)F_d(\tau_k, i))X(\tau_k, i) \\ &\cdot (A_{dp}(\tau_k, i) + B_{dp}(\tau_k, i)F_d(\tau_k, i))^\top, \quad k \in \mathbb{Z}_+, \end{aligned} \tag{5.274b}$$

is exponentially stable.

Following step by step the proof of Lemma 5.7 one gets:

Lemma 5.10 *Assume that:*

(a) the Assumptions 1.2, 2.1 and 5.2 hold;
(b) the ICLSS (5.273) is mean-square stabilizable by impulses in a linear state feedback form;
(c) the set $\Gamma^\Sigma[\mathbb{R}_+]$ associated to the equation (5.168) is nonempty.

Let $\mathbf{F}_d(\tau_k) = (F_d(\tau_k, 1), F_d(\tau_k, 2), \ldots, F_d(\tau_k, N)) \in \mathcal{M}_{m_d n}^N$, $k \in \mathbb{Z}_+$ be such that the corresponding closed-loop JSLDE (5.274) is exponentially stable.
Let $\mathbf{X}^{(0)}(\cdot) = (X^{(0)}(\cdot, 1), X^{(0)}(\cdot, 2), \ldots, X^{(0)}(\cdot, N)) \in \mathcal{PC}(\mathbb{R}_+, \mathcal{S}_n^N; \mathfrak{T})$ be a bounded solution of the following jump linear matrix differential inequality (JLMDI) on the space $\mathcal{S}_n^N$:

$$\begin{aligned} &\dot{X}^{(0)}(t, i) + A_0^\top(t, i)X^{(0)}(t, i) + X^{(0)}(t, i)A_0(t, i) \\ &+ \sum_{l=1}^{r_1} A_l^\top(t, i)X^{(0)}(t, i)A_l(t, i) \\ &+ \sum_{j=1}^{N} q_{ij}X^{(0)}(t, j) + M_c(t, i) \leq 0, \quad \tau_k \leq t < \tau_{k+1} \end{aligned} \tag{5.275a}$$

$$X^{(0)}(\tau_k^-, i) = \sum_{p=0}^{r_2} (A_{dp}(\tau_k, i) + B_{dp}(\tau_k, i)F_d(\tau_k, i))^\top X^{(0)}(\tau_k, i)$$
$$\cdot (A_{dp}(\tau_k, i) + B_{dp}(\tau_k, i)F_d(\tau_k, i)) + \Theta_{d,\mathbf{F}_d}(\tau_k, i) + \varepsilon I_n, \quad k \in \mathbb{Z}_+ \tag{5.275b}$$

$i \in \mathcal{N}$, where $\varepsilon > 0$ is a fixed parameter and $F_d(\tau_k, i)$ are such that the FJLDE (5.274) is exponentially stable. Under the considered assumptions, the solution $\mathbf{X}^{(0)}(\cdot)$ of (5.275) is satisfying:

$$X^{(0)}(t, i) - Z(t, i) \geq \delta I_n, \quad \textit{for all} \quad (t, i) \in \mathbb{R}_+ \times \mathcal{N}, \tag{5.276}$$

where $\mathbf{Z}(\cdot) = (Z(\cdot, 1), Z(\cdot, 2), \ldots, Z(\cdot, N))$ is an arbitrary function from $\Gamma^\Sigma[\mathbb{R}_+]$, $\delta > 0$ being a constant not depending upon $\mathbf{Z}(\cdot)$.

We define

$$F_d^{(0)}(\tau_k, i) \triangleq -\left(\sum_{p=0}^{r_2} B_{dp}^\top(\tau_k, i) X^{(0)}(\tau_k, i) B_{dp}(\tau_k, i) + R_d(\tau_k, i) \right)^{-1} \left(\sum_{p=0}^{r_2} B_{dp}^\top(\tau_k, i) X^{(0)}(\tau_k, i) A_{dp}(\tau_k, i) + L_d^\top(\tau_k, i) \right) \tag{5.277}$$

$i \in \mathcal{N}$, $\mathbf{X}^{(0)}(\cdot)$ being a bounded solution of (5.275).

Starting from a bounded solution $\mathbf{X}^{(0)}(\cdot) \in \mathcal{PC}(\mathbb{R}_+, \mathcal{S}_n^N; \mathfrak{T})$ of (5.275) and the gain matrices $F_d^{(0)}(\tau_k, i)$ defined via (5.277) we construct iteratively the sequences: $\{X^{(q)}(t, i)\}_{q \geq 0}$, $\{F_d^{(q)}(\tau_k, i)\}_{q \geq o}$ in the following way: for each $q \geq 0$, $t \to X^{(q+1)}(t, i) : \mathbb{R}_+ \to \mathcal{S}_n^1$ is the unique bounded solution of the following BJLDE:

$$-\dot{X}^{(q+1)}(t, i) = \widetilde{A}_0(t, i) X^{(q+1)}(t, i) + X^{(q+1)}(t, i) \widetilde{A}_0(t, i)$$
$$+ \sum_{l=1}^{r_1} A_l^\top(t, i) X^{(q)}(t, i) A_l(t, i)$$
$$+ \sum_{\substack{j=1 \\ j \neq i}}^{N} q_{ij} X^{(q)}(t, j) + M_c(t, i) \quad \tau_k \leq t < \tau_{k+1} \tag{5.278a}$$

$$X^{(q+1)}(\tau_k^-, i) = (A_{d0}(\tau_k, i) + B_{d0}(\tau_k, i) F_d^{(q)}(\tau_k, i))^\top X^{(q+1)}(\tau_k, i)$$
$$\cdot (A_{d0}(\tau_k, i) + B_{d0}(\tau_k, i) F_d^{(q)}(\tau_k, i))$$
$$+ \sum_{p=1}^{r_2} (A_{dp}(\tau_k, i) + B_{dp}(\tau_k, i) F_d^{(q)}(\tau_k, i))^\top$$

$$\cdot X^{(q)}(\tau_k,i)(A_{dp}(\tau_k,i)+B_{dp}(\tau_k,i)F_d^{(q)}(\tau_k,i))$$
$$+M_d(\tau_k,i)+L_d(\tau_k,i)F_d^{(q)}(\tau_k,i)+(F_d^{(q)}(\tau_k,i))^\top L_d^\top(\tau_k,i)$$
$$+(F_d^{(q)}(\tau_k,i))^\top R_d(\tau_k,i)F_d^{(q)}(\tau_k,i))+\frac{\varepsilon}{q+1}I_n,\quad k\in\mathbb{Z}_+. \tag{5.278b}$$

$$F_d^{(q+1)}(\tau_k,i):=-\Big(B_{d0}^\top(\tau_k,i)X^{(q+1)}(\tau_k,i)B_{d0}(\tau_k,i)$$
$$+\sum_{p=1}^{r_2}B_{dp}^\top(\tau_k,i)X^{(q)}(\tau_k,i)B_{dp}(\tau_k,i)+R_d(\tau_k,i)\Big)^{-1}$$
$$\cdot(B_{d0}^\top(\tau_k,i)X^{(q+1)}(\tau_k,i)A_{d0}(\tau_k,i)$$
$$+\sum_{p=1}^{r_2}B_{dp}^\top(\tau_k,i)X^{(q)}(\tau_k,i)A_{dp}(\tau_k,i)+L_d^\top(\tau_k,i)) \tag{5.278c}$$

$i\in\mathcal{N}$, $\widetilde{A}_0(t,i)=A_0(t,i)+\frac{1}{2}q_{ii}I_n$.

Following the line of the reasonings from the previous subsection, one proves:

Theorem 5.14 *Assume that the assumptions of Lemma 5.10 are fulfilled. Under these conditions the sequences $\{X^{(q)}(t,i)\}_{q\geq 0}$, $(t,i)\in\mathbb{R}_+\times\mathcal{N}$ defined via (5.278) starting from an arbitrary bounded solution $\mathbf{X}^{(0)}(\cdot)$ from $\mathcal{PC}(\mathbb{R}_+,\mathcal{S}_n^N;\mathfrak{T})$ of (5.275) have the properties:*

$$X^{(q)}(t,i)\geq X^{(q+1)}(t,i)\geq Z(t,i),\quad \textit{for all}\quad (t,i)\in\mathbb{R}_+\times\mathcal{N}$$

$q=0,1,\ldots$, where $\mathbf{Z}(\cdot)=(Z(\cdot,1),Z(\cdot,2),\ldots,Z(\cdot,N))$ is an arbitrary function from $\Gamma^\Sigma[\mathbb{R}_+]$. If

$$\widetilde{X}(t,i)\triangleq\lim_{q\to\infty}X^{(q)}(t,i),\quad (t,i)\in\mathbb{R}_+\times\mathcal{N} \tag{5.279}$$

then $\widetilde{\mathbf{X}}(\cdot)=(\widetilde{X}(\cdot,1),\widetilde{X}(\cdot,2),\ldots,\widetilde{X}(\cdot,N))$ is the maximal solution of the BJMLDE (5.168). Further, if the set $\widetilde{\Gamma}^\Sigma[\mathbb{R}_+]$ associated to (5.168) is nonempty, then $\widetilde{\mathbf{X}}(\cdot)$ defined in (5.279) is just the bounded and stabilizing solution of the BJMLDE (5.168).

Proof Is similar to the proof of Theorem 5.13. The details are left to the reader.

Now we focus on the invariant version of a BJMLDE (5.168). In this case the JLMDI (5.173) becomes:

$$\dot{X}^{(0)}(t,i)+A_0^\top(i)X^{(0)}(t,i)+X^{(0)}(t,i)A_0(i)+\sum_{l=1}^{r_1}A_l^\top(i)X^{(0)}(t,i)A_l(i)$$
$$+\sum_{j=1}^{N}q_{ij}X^{(0)}(t,j)+M_c(i)\leq 0,\quad kh\leq t<(k+1)h, \tag{5.280a}$$

$$X^{(0)}(kh^-, i) \geq \sum_{p=0}^{r_2} (A_{dp}(i) + B_{dp}(i)F_d(i))^\top X^{(0)}(kh, i)(A_{dp}(i) + B_{dp}(i)F_d(i))$$

$$+ M_d(i) + L_d(i)F_d(i) + F_d^\top(i)L_d^\top(i) + F_d^\top(i)R_d(i)F_d(i) + \varepsilon I_n, \quad k \in \mathbb{Z}_+, \tag{5.280b}$$

$F_d(i) \in \mathbb{R}^{m_d \times n}$, $i \in \mathcal{N}$, being such that the time-invariant version of (5.274) is exponentially stable.

The Remark 5.12 is applicable in the case of JLMDI (5.280), too, because this is a particular case of the JLMDI (5.242). So, according to Remark 5.12 (a) and (c) we deduce that (5.280) has necessarily solutions $\mathbf{X}^{(0)}(\cdot) \in \mathcal{PC}(\mathbb{R}_+, \mathcal{S}_n^N; \mathfrak{T})$ ($\mathfrak{T} = \{kh\}_{k \in \mathbb{Z}_+}$) which are h-periodic functions. Hence, for numerical computations of such a solution, one may use a discretization procedure on the interval $[0, h]$. As a consequence of the periodicity property of the solution $\mathbf{X}^{(0)}(\cdot)$ it follows that $F_d^{(0)}(kh, i) = F_d^{(0)}(0, i) := F_d^{(0)}(i)$, for all $k \in \mathbb{Z}_+$ and $i \in \mathcal{N}$.

The time-invariant version of (5.278) is:

$$\begin{aligned} -\dot{X}^{(q+1)}(t, i) &= \widetilde{A}_0^\top(i)X^{(q+1)}(t, i) + X^{(q+1)}(t, i)\widetilde{A}_0(i) \\ &+ \sum_{l=1}^{r_1} A_l^\top(i)X^{(q)}(t, i)A_l(i) \\ &+ \sum_{\substack{j=1 \\ j \neq i}}^{N} q_{ij}X^{(0)}(t, j) + M_c(i) \quad \tau_k \leq t < \tau_{k+1} \end{aligned} \tag{5.281a}$$

$$\begin{aligned} X^{(q+1)}(\tau_k^-, i) &= (A_{d0}(i) + B_{d0}(i)F_d^{(q)}(kh, i))^\top X^{(q+1)}(kh, i)(A_{d0}(i) \\ &+ B_{d0}(i)F_d^{(q)}(kh, i)) \\ &\sum_{p=1}^{r_2} (A_{dp}(i) + B_{dp}(i)F_d^{(q)}(kh, i))^\top X^{(q)}(kh, i)(A_{dp}(i) \\ &+ B_{dp}(i)F_d^{(q)}(kh, i)) + M_d(i) + L_d(i)F_d^{(q)}(kh, i) \\ &+ (F_d^{(q)}(kh, i))^\top L_d^\top(i) \\ &+ (F_d^{(q)}(kh, i))^\top R_d(i)F_d^{(q)}(kh, i)) + \frac{\varepsilon}{q+2}I_n, \quad k \in \mathbb{Z}_+, \end{aligned} \tag{5.281b}$$

$$\begin{aligned} F_d^{(q+1)}(kh,i) &:= -(B_{d0}^\top(i)X^{(q+1)}(kh, i)B_{d0}(kh, i) \\ &+ \sum_{p=1}^{r_2} B_{dp}^\top(i)X^{(q)}(kh, i)B_{dp}(i) + R_d(i))^{-1} \\ &\cdot (B_{d0}^\top(i)X^{(q+1)}(kh, i)A_{d0}(i) \\ &+ \sum_{p=1}^{r_2} B_{dp}^\top(i)X^{(q)}(kh, i)A_{dp}(i) + L_d^\top(i)), \; k \in \mathbb{Z}_+. \end{aligned} \tag{5.281c}$$

One sees inductively for $q = 0, 1, \ldots,$ that for each $i \in \mathcal{N}$, the unique bounded solution $X^{(q+1)}(\cdot, i)$ of (5.281) is a periodic function of period h. Particularly

$$F_d^{(q+1)}(kh, i) = F_d^{(q+1)}(0, i) := F_d^{(q+1)}(i).$$

So, (5.281) may be rewritten as:

$$-\dot{X}^{(q+1)}(t, i) = \widetilde{A}_0^\top(i)X^{(q+1)}(t, i) + X^{(q+1)}(t, i)\widetilde{A}_0(i) + \widetilde{\Xi}_c^{(q+1)}(t, i), \quad kh \le t < (k+1)h \tag{5.282a}$$

$$X^{(q+1)}(\tau_k^-, i) = (A_{d0}(i) + B_{d0}(i)F_d^{(q)}(i))^\top X^{(q+1)}(kh, i) \cdot (A_{d0}(i) + B_{d0}(i)F_d^{(q)}(i)) + \widetilde{\Xi}_d^{(q+1)}(i), \tag{5.282b}$$

$$\begin{aligned} F_d^{(q+1)}(i) := &-(B_{d0}^\top(i)X^{(q+1)}(0, i)B_{d0}(i) \\ &+ \sum_{p=1}^{r_2} B_{dp}^\top(i)X^{(q)}(0, i)B_{dp}(i) + R_d(i))^{-1} \\ &\cdot (B_{d0}^\top(i)X^{(q+1)}(0, i)A_{d0}(i) \\ &+ \sum_{p=1}^{r_2} B_{dp}^\top(i)X^{(q)}(0, i)A_{dp}(i) + L_d^\top(i)), i \in \mathcal{N}, \end{aligned} \tag{5.282c}$$

where

$$\widetilde{\Xi}_c^{(q+1)}(t, i) = \sum_{l=1}^{r_1} A_l^\top(i)X^{(q)}(t, i)A_l(i) + \sum_{\substack{j=1 \\ j \ne i}}^{N} q_{ij}X^{(q)}(t, j) + M_c(i), \tag{5.283a}$$

$$\begin{aligned} \widetilde{\Xi}_d^{(q+1)}(i) = &\sum_{p=1}^{r_2}(A_{dp}(i) + B_{dp}(i)F_{dp}^{(q)}(i))^\top X^{(q)}(0, i)(A_{dp}(i) + B_{dp}(i)F_{dp}^{(q)}(i)) \\ &+ M_d(i) + L_d(i)F_d^{(q)}(i) + (F_d^{(q)}(i))^\top L_d^\top(i) + (F_d^{(q)}(i))^\top R_d(i)F_d^{(q)}(i) + \frac{\varepsilon}{q+2}I_n, \end{aligned} \tag{5.283b}$$

$i \in \mathcal{N}$. If $t \in [0, h)$ we obtain from (5.282a):

$$X^{(q+1)}(t, i) = e^{(\widetilde{A}_0(i))^\top (h-t)} X^{(q+1)}(h^-, i)e^{\widetilde{A}_0(i)(h-t)} + \int_t^h e^{\widetilde{A}_0^\top(i)(s-t)} \widetilde{\Xi}_c^{(q+1)}(s, i)e^{\widetilde{A}_0(i)(s-t)} ds. \tag{5.284}$$

Substituting this equality written for $t = 0$ in (5.282b) (written for $k = 0$) and taking into account that $X^{(q+1)}(0^-, i) = X^{(q+1)}(h^-, i)$ we obtain the following Stein equation which allows us to compute

$$X^{(q+1)}(h^-,i) = (\mathscr{A}_d^{(q+1)}(i))^\top X^{(q+1)}(h^-,i)\mathscr{A}_d^{(q+1)}(i) + \overset{\approx}{\Xi}{}_d^{(q+1)}(i), \quad i \in \mathscr{N}, \tag{5.285}$$

where

$$\mathscr{A}_d^{(q+1)}(i) := e^{\widetilde{A}_0(i)h}(A_{d0}(i) + B_{d0}(i)F_d^{(q)}(i))$$

and

$$\begin{aligned}\overset{\approx}{\Xi}{}_d^{(q+1)}(i) :=& (A_{d0}(i) + B_{d0}(i)F_d^{(q)}(i))^\top \int_0^h e^{\widetilde{A}_0^\top(i)s}\,\widetilde{\Xi}_c^{(q+1)}(s,i)e^{\widetilde{A}_0(i)s} \\ &\cdot (A_{d0}(i) + B_{d0}(i)F_d^{(q)}(i)) + \widetilde{\Xi}_d^{(q+1)}(i).\end{aligned}$$

The value $X^{(q+1)}(h^-,i)$ obtained as a solution of (5.285) is substituted in (5.284) to get the values of $X^{(q+1)}(t,i)$ for $t \in [0,h)$. The iterative process stops when

$$\max_{i\in\mathscr{N}} \lambda_{max}[X^{(q)}(0,i) - X^{(q+1)}(0,i)] < toll.$$

5.7.3 The Numerical Computation of the Stabilizing Solution of a Forward Jump Matrix Linear Differential Equation with Riccati-Type Jumping Operator

Now we focus our attention on an approximation method of the stabilizing solution of the following forward jump matrix linear differential equation (FJMLDE) with Riccati-type jumping operator:

$$\dot{Y}(t,i) = A_0(i)Y(t,i) + Y(t,i)A_0^\top(i) + \sum_{j=1}^N q_{ji}Y(t,j) + \widetilde{M}_c(i), \quad kh < t \le (k+1)h \tag{5.286a}$$

$$\begin{aligned}Y(kh^+,i) =& A_{d0}(i)Y(kh,i)A_{do}^\top(i) - (A_{d0}(i)Y(kh,i)C_{d0}^\top(i) + \widetilde{L}_d(i)) \\ &\cdot (C_{d0}(i)Y(kh,i)C_{d0}^\top(i) + \widetilde{R}_d(i))^{-1}(A_{d0}(i)Y(kh,i)C_{d0}^\top(i) + \widetilde{L}_d(i))^\top \\ &+ \widetilde{M}_d(i), \quad k \in \mathbb{Z}.\end{aligned} \tag{5.286b}$$

The FJMLDE (5.286) is a particular case of (5.233). Now, the accompanying JLSS (5.234) reduces to:

$$\dot{x}(t) = A_0(\eta(t))x(t), \quad kh < t \le (k+1)h \tag{5.287a}$$

$$x(kh^+) = A_{d0}(\eta(kh))x(kh), \tag{5.287b}$$

$$z_d(kh) = C_{d0}(\eta(kh))x(kh), \quad k \in \mathbb{Z}_+. \tag{5.287c}$$

The JLSS (5.287) is mean-square detectable by impulses if and only if there exist matrices $K_d(i) \in \mathbb{R}^{n \times n_{z_d}}$, $i \in \mathcal{N}$ such that the forward jump linear differential equation (FJLDE) on $\mathcal{S}_n^N$:

$$\dot{X}(t,i) = A_0(i)X(t,i) + X(t,i)A_0^\top(i) + \sum_{j=1}^{N} q_{ji}X(t,j), \quad kh < t \le (k+1)h \tag{5.288a}$$

$$X(kh^+,i) = (A_{d0}(i) + K_d(i)C_{d0}(i))X(kh,i)(A_{d0}(i) + K_d(i)C_{d0}(i))^\top, k \in \mathbb{Z}, \tag{5.288b}$$

is exponentially stable.

Let $t \to \mathbf{Y}^{(0)}(t) = (Y^{(0)}(t,1), Y^{(0)}(t,2), \ldots, Y^{(0)}(t,N)) : \mathbb{R} \to \mathcal{S}_n^N$ be a periodic solution of period h, of the following forward jump linear matrix differential inequality (FJLMDI):

$$\begin{aligned}\dot{Y}^{(0)}(t,i) &\ge A_0(i)Y^{(0)}(t,i) + Y^{(0)}(t,i)A_0^\top(i) \\ &+ \sum_{j=1}^{N} q_{ji}Y^{(0)}(t,j) + \widetilde{M}_c(i), \quad kh < t \le (k+1)h\end{aligned} \tag{5.289a}$$

$$\begin{aligned}Y^{(0)}(kh^+,i) &= (A_{d0}(i) + K_d(i)C_{d0}(i))Y^{(0)}(kh,i)(A_{d0}(i) + K_d(i)C_{d0}(i))^\top + \widetilde{M}_d(i) \\ &+ K_d(i)\widetilde{L}_d^\top(i) + \widetilde{L}_d(i)K_d^\top(i) + K_d(i)\widetilde{R}_d(i)K_d^\top(i) + \varepsilon I_n, \quad k \in \mathbb{Z},\end{aligned} \tag{5.289b}$$

where $\varepsilon > 0$ is a given fixed parameter, $K_d(i)$, $i \in \mathcal{N}$ are chosen such that the FJLDE (5.288) is exponentially stable. Applying Theorem 2.14 in the case of the forward jump Lyapunov differential equation

$$\begin{aligned}\dot{Y}(t,i) &= A_0(i)Y(t,i) + Y(t,i)A_0^\top(i) + \sum_{j=1}^{N} q_{ji}Y(t,j) + \widetilde{M}_c(i) + \delta^2 I_n, \\ &kh < t \le (k+1)h\end{aligned} \tag{5.290a}$$

$$\begin{aligned}Y(kh^+,i) &= (A_{d0}(i) + K_d(i)C_{d0}(i))Y(kh,i)(A_{d0}(i) + K_d(i)C_{d0}(i))^\top + \widetilde{M}_d(i) \\ &+ K_d(i)\widetilde{L}_d^\top(i) + \widetilde{L}_d(i)K_d^\top(i) + K_d(i)\widetilde{R}_d(i)K_d^\top(i) \\ &+ (\varepsilon + \rho^2)I_n, \quad k \in \mathbb{Z},\end{aligned} \tag{5.290b}$$

has a unique h-periodic solution $\widetilde{\mathbf{Y}}(\cdot)$. It is obvious that $\widetilde{\mathbf{Y}}(\cdot)$ is a h-periodic solution of the FJLMDI (5.289) too.

The reader can check that if the set $\boldsymbol{\Gamma}_F^{\check{\Sigma}}[\mathbb{R}]$ associated to (5.286) is not empty, then the solution $\mathbf{Y}^{(0)}(\cdot)$ of (5.289) satisfies the sign condition

$$C_{d0}(i)Y^{(0)}(kh,i)C_{d0}^\top(i) + \widetilde{R}_d(i) > 0, \quad \text{for all } k \in \mathbb{Z}.$$

In this case we define

$$K_d^{(0)} \triangleq -\left(A_{d0}(i)Y^{(0)}(0,i)C_{d0}^{\top}(i)+\widetilde{L}_d(i)\right) \cdot \left(C_{d0}(i)Y^{(0)}(0,i)C_{d0}^{\top}(i)+\widetilde{R}_d(i)\right)^{-1}, \quad i \in \mathcal{N}. \tag{5.291}$$

Taking $(Y^{(0)}, K^{(0)}(i))$, $i \in \mathcal{N}$ as first term, we construct iteratively the sequences $\{Y^{(q)}(t,i)\}_{q\geq 0}$ and $\{K_d^{(q)}(i)\}_{q\geq 0}$, $i \in \mathcal{N}$ in the following way: $Y^{(q+1)}(\cdot,i)$ is the unique periodic solution of period h of the uncoupled forward jump Lyapunov differential equation:

$$\dot{Y}^{(q+1)}(t,i) = \widetilde{A}_0(i)Y^{(q+1)}(t,i) + Y^{(q+1)}(t,i)\widetilde{A}_0^{\top}(i) + M_c^{(q+1)}(t,i), \quad kh < t \leq (k+1)h \tag{5.292a}$$

$$Y^{(q+1)}(kh^+,i) = (A_{d0}(i)+K_d^{(q)}(i)C_{d0}(i))Y^{(q+1)}(kh,i)(A_{d0}(i)+K_d^{(q)}(i)C_{d0}(i))^{\top} + M_d^{(q+1)}(i), \quad k \in \mathbb{Z}, \tag{5.292b}$$

$i \in \mathcal{N}$, where

$$\widetilde{A}_0(i) := A_0(i) + \frac{1}{2}q_{ii}I_n; \tag{5.293a}$$

$$M_c^{(q+1)}(t,i) := \widetilde{M}_c(i) + \sum_{\substack{j=1 \\ j\neq i}}^{N} q_{ji}Y^{(q)}(t,i), \quad (t,i) \in \mathbb{R}\times\mathcal{N}, \tag{5.293b}$$

$$M_d^{(q+1)}(i) = \widetilde{M}_d(i) + K_d^{(q)}(i)\widetilde{L}_d^{\top}(i) + \widetilde{L}_d(i)(K_d^{(q)}(i))^{\top} + K_d^{(q)}\widetilde{R}_d(i)(K_d^{(q)}(i))^{\top} + \frac{\varepsilon}{q+2}I_n, \quad i \in \mathcal{N}, \tag{5.293c}$$

$$K_d^{(q+1)}(i) \coloneqq -(A_{d0}(i)Y^{(q)}(0,i)C_{d0}^{\top}(i)+\widetilde{L}_d(i))(C_{d0}(i)Y^{(q)}(0,i)C_{d0}^{\top}(i)+\widetilde{R}_d(i))^{-1}, \tag{5.294}$$

$i \in \mathcal{N}$. One shows by induction with respect to $q = 0, 1, \ldots,$ that the FJLDE:

$$\dot{Y}(t,i) = \widetilde{A}_0(i)Y(t,i) + Y(t,i)\widetilde{A}_0^{\top}(i), \quad kh < t \leq (k+1)h$$
$$Y(kh^+,i) = (A_{d0}(i)+K_d^{(q)}(i)C_{d0}(i))Y(kh,i)(A_{d0}(i)+K_d^{(q)}(i)C_{d0}(i))^{\top}, \quad k \in \mathbb{Z},$$

is exponentially stable. This guarantee the existence of the $h-$periodic solution of (5.292). Further one shows that

$$Y^{(q)}(t,i) \geq Y^{(q+1)}(t,i) \geq Z(t,i), \quad (t,i) \in \mathbb{R}\times\mathcal{N},$$

where $\mathbf{Z}(\cdot) = (Z(\cdot,1), Z(\cdot,2), \dots, Z(\cdot,N)) \in \boldsymbol{\Gamma}_F^{\check{\Sigma}}[\mathbb{R}]$.
One shows that $\tilde{\tilde{Y}}(\cdot) = (\tilde{\tilde{Y}}(\cdot,1), \tilde{\tilde{Y}}(\cdot,2), \dots, \tilde{\tilde{Y}}(\cdot,N))$ is the maximal solution of (5.286) if

$$\tilde{\tilde{Y}}(t,i) := \lim_{q\to\infty} Y^{(q)}(t,i), \quad (t,i) \in \mathbb{R} \times \mathcal{N}.$$

If the set $\boldsymbol{\Gamma}_F^{\check{\Sigma}}[\mathbb{R}]$ is nonempty, then $\tilde{\tilde{\mathbf{Y}}}(\cdot)$ is just the stabilizing solution of (5.286).
To compute the solution $Y^{(q+1)}(\cdot,i)$ of (5.292) on the interval $[0,h]$ we write

$$\begin{aligned} Y^{(q+1)}(t,i) &= \mathrm{e}^{\widetilde{A}_0(i)t} Y^{(q+1)}(0^+,i)\mathrm{e}^{\widetilde{A}_0^\top(i)t} \\ &+ \int_0^t \mathrm{e}^{\widetilde{A}_0(i)(t-s)} M_c^{(q+1)}(s,i)\mathrm{e}^{\widetilde{A}_0^\top(i)(t-s)} ds, \quad t\in[0,h]. \end{aligned} \tag{5.295}$$

Substituting this equality written for $t = h$ in (5.292b) (written for $k = 1$) and taking into account that

$$Y^{(q+1)}(h^+,i) = Y^{(q+1)}(0^+,i), \quad i \in \mathcal{N}$$

we obtain the Stein equation:

$$Y^{(q+1)}(0^+,i) = \mathbb{A}_d^{(q+1)}(i) Y^{(q+1)}(0^+,i)(\mathbb{A}_d^{(q+1)}(i))^\top + \mathbb{M}_d^{(q+1)}(i) \tag{5.296}$$

where $\mathbb{A}_d^{(q+1)}(i) := (A_{d0}(i) + K_d^{(q)}(i)C_{d0}(i))\mathrm{e}^{\widetilde{A}_0(i)h}$ and

$$\begin{aligned} \mathbb{M}_d^{(q+1)}(i) :=& M_d^{(q+1)}(i) + (A_{d0} + K_d^{(q)}(i)C_{d0}(i)) \\ & \int_0^h \mathrm{e}^{\widetilde{A}_0(i)(h-s)} M_c^{(q+1)}(s,i)\mathrm{e}^{\widetilde{A}_0^\top(i)(h-s)} ds \\ & \cdot (A_{d0} + K_d^{(q)}(i)C_{d0}(i))^\top. \end{aligned}$$

The value of $Y^{(q+1)}(0^+,i)$ obtained as solution of (5.296) is plugged in (5.295) to obtain $Y^{(q+1)}(t,i)$ for $t \in [0,h]$. The value of $Y^{(q+1)}(0,i)$ involved in (5.294) is obtained from $Y^{(q+1)}(0,i) = Y^{(q+1)}(h,i)$ due to the periodicity property of $Y^{(q+1)}(\cdot,i)$. The algorithm stops when

$$\max_{i\in\mathcal{N}} \lambda_{max}[Y^{(q)}(h,i) - Y^{(q+1)}(h,i)] < toll.$$

5.8 Notes and References

Readers who are interested in the study of the Riccati equations (from both a theoretical and a numerical point of view) of stochastic control of linear stochastic systems subject to both multiplicative noise as well as Markovian switching perturbations can refer to Drăgan et al. [1] and the references therein.

To the best of the authors knowledge, problems of existence of different types of solutions of jump matrix Riccati differential equations related to different control/filtering problems of linear impulsive stochastic systems have not been addressed yet in the literature. Hence, this chapter can be viewed as the first rigorous study of such mathematical objects.

References

1. Drăgan, V., Morozan, T., Stoica, A.M.: Mathematical Methods in Robust Control of Linear Stochastic Systems. Springer, New York (2013)
2. Drăgan, V., Morozan, T., Stoica, A.M.: Mathematical Methods in Robust Control of Discrete-Time Linear Stochastic Systems. Springer, New York (2010)
3. Albert, A.: Conditions for positive and nonnegative definiteness in terms of pseudoinverses. SIAM J. Appl. Math. **17**, 434–440 (1969)
4. Freiling, G., Hochhaus, A.: On a class of rational matrix differential equations arising in stochastic control. Linear Algebra Appl. **379**, 43–68 (2004). https://doi.org/10.1016/S0024-3795(02)00651-1. https://www.sciencedirect.com/science/article/pii/S0024379502006511,specialIssueontheTenthILASConference(Auburn,2002)
5. Schwartz, L.: Analyse Mathématique, vol. I. Hermann, Paris, II (1967)

Chapter 6
The Linear Quadratic Optimal Control Problems for Jump Linear Stochastic Systems

Abstract As we have seen in Chap. 4, there exists a degree of freedom in choosing a stabilizing control law in a state feedback form for a controlled jump linear stochastic system. This freedom may be used in order to improve some suitable performance specifications. A natural performance criterion is described by a trade off between the cost of the deviation of an output from a required value and the *amount of energy* spent by the controller. Roughly speaking, this idea leads to a class of quadratic cost function of the form

$$J(u) = \int_0^\infty \mathbb{E}[|z(t) - z_d(t)|^2]\mathrm{d}t + \gamma^2 \int_0^\infty \mathbb{E}[|u(t)|^2]\mathrm{d}t.$$

The term $\int_0^\infty \mathbb{E}[|z(t) - z_d(t)|^2]\mathrm{d}t$ measures the cost of the deviation of the output $z(\cdot)$ from a desired value $z_d(\cdot)$ and $\int_0^\infty \mathbb{E}[|u(t)|^2]\mathrm{d}t$ measures the control effort. The coefficient γ^2 reflects the trade off between the two terms. In the present chapter we will address the problem of the design of a control law, in a state feedback form, which minimizes a wide class of quadratic cost functions. The minimization of the cost function is done along the state trajectories of a control jump linear stochastic system. Both the finite time horizon case as well as the infinite time horizon case, are considered. Before solving such optimal control problems we first give a fundamental result in Sect. 6.1 where we show that the considered quadratic cost functions can be expressed equivalently in terms of a solution of an adequately defined stochastic jump matrix Riccati differential equation (See Lemma 6.1). This equivalent version of the cost functions will be helpful in the derivation of the explicit formulae of the optimal controls. In Sect. 6.2, the finite time horizon case is addressed. We show that the feedback gain corresponding to the optimal controller is constructed based on the solution of an adequately defined stochastic jump matrix Riccati differential equation. Such a solution must satisfy a terminal value condition as well as the sign conditions on the quadratic terms of the aforementioned Riccati differential equation (see Theorem 6.1). Sections 6.3 and 6.4 are devoted to the infinite time horizon case. Two different cases will be analyzed:

(a) **The case of free end point of the optimal state trajectories:** the main result of this part is resumed in Theorem 6.2 where we show that the feedback gain

V. Drăgan et al., *Robust Control of Jump Linear Stochastic Systems*, Lecture Notes in Control and Information Sciences 497, https://doi.org/10.1007/978-3-031-84070-8_6

corresponding to the optimal controller is constructed based on the unique minimal positive semidefinite and bounded on $\mathbb{R}_+$ solution of an adequately defined stochastic jump matrix Riccati differential equation.

(b) **The case of zero end value of the optimal state trajectory:** in addition to the stabilization constraint imposed in this case, we also consider the general case of sign indefiniteness of the weighting matrices leading to a so called indefinite linear quadratic control problem. Such a constraint induces the issue of well posedness of the optimal control problem under consideration that has to be addressed before solving the control problem itself. Theorem 6.3 solves this issue by giving a set of conditions under which the control problem is well posed. The value function in this case is characterized via the unique maximal and bounded on $\mathbb{R}_+$ solution of an adequately defined stochastic jump matrix Riccati differential equation. Using this result, we then solve the optimal control problem by introducing the notion of *attainability* (see Definition 6.2). This is resumed in Theorem 6.4 where we show that the feedback gain corresponding to the optimal controller is constructed based on the unique maximal and bounded on $\mathbb{R}_+$ solution of a well defined Riccati differential equation.

We finally apply the obtained results to the case of sampled-data linear stochastic systems (Sect. 6.5).

6.1 Preliminaries

Let $\mathfrak{T} = \{\tau_k\}_{k\in\mathbb{Z}_+} \subset \mathbb{R}_+$ be the partition of the semiaxis $\mathbb{R}_+$ satisfying the conditions from the Assumption 2.1. Let t_0, T be real numbers such that

$$\tau_{\kappa_0-1} < t_0 \leq \tau_{\kappa_0} \leq \ldots \leq \tau_{\kappa_1-1} < T \leq \tau_{\kappa_1}.$$

For each $j \in \mathcal{N} = \{1, 2, \ldots, N\}$ we consider the function

$$(u_c, u_d) \to \widehat{J}_j(t_0, T, x_0; u_c, u_d) : \mathcal{L}^2_{\mathcal{H}}([t_0, T]; \mathbb{R}^{m_c}) \times \ell^2_{\mathcal{H}}\{\kappa_0, \kappa_1 - 1; \mathbb{R}^{m_d}\} \to \mathbb{R}$$

be defined by

$$\begin{aligned} \widehat{J}_j(t_0, T, x_0; u_c, u_d) = &\int_{t_0}^{T} \mathbb{E}[\mathcal{W}_c(t, x_{u_c,u_d}(t), u_c(t), \eta(t)) \mid \eta(t_0) = j]\mathrm{d}t \\ &+ \sum_{k=\kappa_0}^{\kappa_1-1} \mathbb{E}[\mathcal{W}_d(\tau_k, x_{u_c,u_d}(\tau_k), u_d(k), \eta(\tau_k)) \mid \eta(t_0) = j] \end{aligned} \tag{6.1}$$

where $x_{u_c,u_d}(\cdot) = x(\cdot; t_0, x_0, u_c, u_d)$ is the solution of the following problem with given initial value (IVP)

$$\begin{aligned}dx(t) &= (A_0(t,\eta(t))x(t) + B_0(t,\eta(t))u_c(t))dt \\ &+ \sum_{l=1}^{r_1}(A_l(t,\eta(t))x(t) + B_l(t,\eta(t))u_c(t))dw_l(t), \quad \tau_k \le t < \tau_{k+1}\end{aligned} \tag{6.2a}$$

$$\begin{aligned}x(\tau_k^+) &= A_{d0}(\tau_k,\eta(\tau_k))x(\tau_k) + B_{d0}(\tau_k,\eta(\tau_k))u_d(k) \\ &+ \sum_{p=1}^{r_2} w_{dp}(k)(A_{dp}(\tau_k,\eta(\tau_k))x(\tau_k) + B_{dp}(\tau_k,\eta(\tau_k))u_d(k)), \quad \kappa_0 \le k \le \kappa_1 - 1\end{aligned} \tag{6.2b}$$

$$x_{u_c,u_d}(t_0) = x_0 \tag{6.2c}$$

$x_0 \in \mathbb{R}^n$ being fixed. In (6.1) we have denoted:

$$\mathscr{W}_c(t, x_{u_c,u_d}(t), u_c(t), \eta(t)) = \begin{pmatrix} x_{u_c,u_d}(t) \\ u_c(t) \end{pmatrix}^\top \begin{pmatrix} M_c(t,\eta(t)) & L_c(t,\eta(t)) \\ L_c^\top(t,\eta(t)) & R_c(t,\eta(t)) \end{pmatrix} \begin{pmatrix} x_{u_c,u_d}(t) \\ u_c(t) \end{pmatrix} \tag{6.3a}$$

$$\begin{aligned}\mathscr{W}_d(\tau_k, x_{u_c,u_d}(\tau_k), u_d(k), \eta(\tau_k)) &= \begin{pmatrix} x_{u_c,u_d}(\tau_k) \\ u_d(k) \end{pmatrix}^\top \begin{pmatrix} M_d(\tau_k,\eta(\tau_k)) & L_d(\tau_k,\eta(\tau_k)) \\ L_d^\top(\tau_k,\eta(\tau_k)) & R_d(\tau_k,\eta(\tau_k)) \end{pmatrix} \\ &\cdot \begin{pmatrix} x_{u_c,u_d}(\tau_k) \\ u_d(k) \end{pmatrix}.\end{aligned} \tag{6.3b}$$

In (6.1) and (6.2), $\{w(t)\}_{t\ge 0}$ ($w(t) = (w_1(t)\ w_2(t)\ \ldots w_{r_1}(t))^\top$), $\{w_d(k)\}_{k\ge 0}$ ($w_d(k) = (w_{d1}(k)\ w_{d2}(k)\ \ldots\ w_{dr_2}(k))^\top$) and $\{\eta(t)\}_{t\ge 0}$ are stochastic processes which are satisfying the Assumption 1.2. The matrix coefficients from (6.2) and (6.3) satisfy the conditions from the Assumption 5.2. Based on Proposition 3.4 we may infer that $x_{u_c,u_d}(\cdot) \in \mathscr{L}^2_{\mathscr{H}}([t_0,T];\mathbb{R}^n)$ if $u_c(\cdot) \in \mathscr{L}^2_{\mathscr{H}}([t_0,T];\mathbb{R}^{m_c})$ and $\{u_d(k)\}_{\kappa_0\le k\le\kappa_1-1}$ is in $\ell^2_{\mathscr{H}}\{\kappa_0;\kappa_1-1;\mathbb{R}^{m_d}\}$. So, $J_j(t_0,T,x_0;u_c,u_d)$ is well defined by (6.1).

The equality from the following lemma will be widely involved in the developments from this chapter.

Lemma 6.1 *Assume that the Assumptions 1.2, 2.1 and 5.2 hold. Let* $\mathbf{X}(\cdot) = (X(\cdot,1), X(\cdot,2),\ldots,X(\cdot,N)) : \mathscr{I}^{\mathbf{X}} \subset \mathbb{R}_+ \to \mathscr{S}_n^N$ *be a solution of the SJMRDE of the form 5.10 constructed based on the matrix coefficients from (6.1)–(6.3). Assume that* $[t_0,T] \subset \mathscr{I}^{\mathbf{X}}$. *Under these conditions, the following equality holds:*

$$\begin{aligned}\widehat{J}_j(t_0,T,x_0;u_c,u_d) &= x_0^\top X(t_0^-,j)x_0 - \mathbb{E}[x_{u_c,u_d}^\top(T)X(T,\eta(T))x_{u_c,u_d}(T) \mid \eta(t_0)=j] \\ &+ \int_{t_0}^{T} \mathbb{E}[(u_c(t) - F_c^{\mathbf{X}}(t,\eta(t))x_{u_c,u_d}(t))^\top \mathbb{R}_c(t,X(t,\eta(t)),\eta(t)) \\ &\cdot (u_c(t) - F_c^{\mathbf{X}}(t,\eta(t))x_{u_c,u_d}(t)]dt + \sum_{k=\kappa_0}^{\kappa_1-1}\mathbb{E}[(u_d(k) - F_d^{\mathbf{X}}(\tau_k,\eta(\tau_k))x_{u_c,u_d}(\tau_k))^\top \\ &\cdot \mathbb{R}_d(\tau_k,X(\tau_k,\eta(\tau_k)),\eta(\tau_k))(u_d(k) - F_d^{\mathbf{X}}(\tau_k,\eta(\tau_k))x_{u_c,u_d}(\tau_k)) \mid \eta(t_0)=j]\end{aligned} \tag{6.4}$$

for all $u_c(\cdot) \in \mathscr{L}^2_{\mathbf{H}}([t_0, T]; \mathbb{R}^{m_c})$ *and* $\{u_d(k)\}_{\kappa_0 \le k \le \kappa_1 - 1} \in \ell^2_{\mathscr{H}}\{\kappa_0, \kappa_1 - 1; \mathbb{R}^{m_d}\}$ *where for each* $i \in \mathscr{N}$, *we have denoted*

$$\mathbb{R}_c(t, X(t,i), i) := \sum_{l=1}^{r_1} B_l^\top(t,i)X(t,i)B_l(t,i) + R_c(t,i), \quad t \in \mathscr{I}^{\mathbf{X}} \tag{6.5a}$$

$$\mathbb{R}_d(\tau_k, X(\tau_k, i), i) := \sum_{p=0}^{r_2} B_{dp}^\top(\tau_k, i)X(\tau_k, i)B_{dp}(\tau_k, i) + R_d(\tau_k, i), \; \tau_k \in \mathfrak{T} \cap \mathscr{I}^{\mathbf{X}} \tag{6.5b}$$

and

$$F_c^{\mathbf{X}}(t,i) := -\mathbb{R}_c^{-1}(t, X(t,i), i)(B_0^\top(t,i)X(t,i) + \sum_{l=1}^{r_1} B_l^\top(t,i)X(t,i)A_l(t,i) + L_c^\top(t,i)), \quad t \in \mathscr{I}^{\mathbf{X}} \tag{6.6a}$$

$$F_d^{\mathbf{X}}(\tau_k, i) := -\mathbb{R}_d^{-1}(\tau_k, X(\tau_k, i), i)(\sum_{p=0}^{r_2} B_{dp}^\top(\tau_k, i)X(\tau_k, i)A_{dp}(\tau_k, i) + L_d^\top(\tau_k, i)), \quad \tau_k \in \mathfrak{T} \cap \mathscr{I}^{\mathbf{X}}. \tag{6.6b}$$

If $t_0 \in (\tau_{\kappa_0 - 1}, \tau_{\kappa_0})$ *then* $X(t_0^-, j)$ *is replaced by* $X(t_0, j)$ *and if* $T = \tau_{\kappa_1}$ *then* $X(T, \eta(T))$ *is replaced by* $X(T^-, \eta(T))$ *in (6.4).*

Proof Let us consider intervals of the form $[a_k, b_k]$, $\kappa_0 - 1 \le k \le \kappa_1 - 1$ found in one of the following cases:

(α) $[a_k, b_k] \subset (\tau_k, \tau_{k+1})$ if $\kappa_0 \le k \le \kappa_1 - 2$
(β) $[a_{\kappa_1 - 1}, b_{\kappa_1 - 1}] \subset (\tau_{\kappa_1 - 1}, T)$
(γ) $[a_{\kappa_0 - 1}, b_{\kappa_0 - 1}] \subset (t_0, \tau_{\kappa_0})$ if $t_0 \in (\tau_{\kappa_0 - 1}, \tau_{\kappa_0})$.

Let us consider first the case $k \in \{\kappa_0, \kappa_0 + 1, \ldots, \kappa_1 - 2\}$. Applying Itô's formula on each interval $[a_k, b_k]$ to the function $v(t, x, i) = x^\top X(t,i)x$ and to the stochastic process $x_{u_c,u_d}(\cdot)$ which is satisfying (6.2a) and performing some algebraic calculations involving the differential Eq. (5.10a) satisfied by $X(\cdot, i)$, letting $a_k \searrow \tau_k$ and $b_k \nearrow \tau_{k+1}$ we obtain (via Lemma 3.1) that:

$$\begin{aligned}
&\int_{\tau_k}^{\tau_{k+1}} \mathbb{E}[\mathscr{W}_c(t, x_{u_c,u_d}(t), u_c(t), \eta(t)) \mid \eta(t_0) = j]dt \\
&= \mathbb{E}[x_{u_c,u_d}^\top(\tau_k^+)X(\tau_k, \eta(\tau_k))x_{u_c,u_d}(\tau_k^+) \mid \eta(t_0) = j] \\
&\quad - \mathbb{E}[x_{u_c,u_d}^\top(\tau_{k+1})X(\tau_{k+1}^-, \eta(\tau_{k+1}))x_{u_c,u_d}(\tau_{k+1}) \mid \eta(t_0) = j]
\end{aligned} \tag{6.7}$$

$$+\int_{\tau_k}^{\tau_{k+1}} \mathbb{E}[(u_c(t)-F_c^{\mathbf{X}}(t,\eta(t))x_{u_c,u_d}(t))^{\top}\mathbb{R}_c(t,X(t,\eta(t)),\eta(t))$$
$$\cdot(u_c(t)-F_c^{\mathbf{X}}(t,\eta(t))x_{u_c,u_d}(t)) \mid \eta(t_0)=j]\mathrm{d}t.$$

Bearing in mind the Assumption 1.2, the definition of the $\sigma-$algebras $\mathscr{H}_t$, $t\geq 0$ introduced in Sect. 1.2, we may infer that $\sigma-$algebra $\sigma[w_{dp}(k)]$, $0\leq p\leq r_2$ are independent of the $\sigma-$algebra $\mathscr{H}_{\tau_k}$. This observation together with the fact that the random variables $x_{u_c,u_d}(\tau_k)$, $u_d(k)$ and $\eta(\tau_k)$ are $\mathscr{H}_{\tau_k}-$measurable, allow us to deduce (via (6.2b)) that

$$\begin{aligned}
&\mathbb{E}[x_{u_c,u_d}^{\top}(\tau_k^+)X(\tau_k,\eta(\tau_k))x_{u_c,u_d}(\tau_k^+) \mid \eta(t_0)=j]\\
&=\mathbb{E}[\mathbb{E}[x_{u_c,u_d}^{\top}(\tau_k^+)X(\tau_k,\eta(\tau_k))x_{u_c,u_d}(\tau_k^+) \mid \mathscr{H}_{\tau_k}] \mid \eta(t_0)=j]\\
&=\sum_{p=0}^{r_2}\mathbb{E}[(A_{dp}(\tau_k,\eta(\tau_k))x_{u_c,u_d}(\tau_k)+B_{dp}(\tau_k,\eta(\tau_k))u_d(k))^{\top}X(\tau_k,\eta(\tau_k))\\
&\cdot(A_{dp}(\tau_k,\eta(\tau_k))x_{u_c,u_d}(\tau_k)+B_{dp}(\tau_k,\eta(\tau_k))u_d(k)) \mid \eta(t_0)=j].
\end{aligned}$$

Further, after several algebraic calculations involving (5.10b) satisfied by $X(\tau_k,i)$ together with (6.3b), one gets:

$$\begin{aligned}
&\mathbb{E}[x_{u_c,u_d}^{\top}(\tau_k^+)X(\tau_k,\eta(\tau_k))x_{u_c,u_d}(\tau_k^+) \mid \eta(t_0)=j]\\
&\ =\mathbb{E}[x_{u_c,u_d}^{\top}(\tau_k)X(\tau_k^-,\eta(\tau_k))x_{u_c,u_d}(\tau_k) \mid \eta(t_0)=j]\\
&\ \ +\mathbb{E}[(u_d(k)-F_d^{\mathbf{X}}(\tau_k,\eta(\tau_k))x_{u_c,u_d}(\tau_k))^{\top}\mathbb{R}_d(\tau_k,X(\tau_k,\eta(\tau_k)),\eta(\tau_k))\\
&\ \ \cdot(u_d(k)-F_d^{\mathbf{X}}(\tau_k,\eta(\tau_k))x_{u_c,u_d}(\tau_k)) \mid \eta(t_0)=j]\\
&\ \ -\mathbb{E}[\mathscr{W}_d(\tau_k,x_{u_c,u_d}(\tau_k),u_d(k),\eta(\tau_k)) \mid \eta(t_0)=j].
\end{aligned} \tag{6.8}$$

Substituting (6.8) in (6.7) and summing for $k=\kappa_0$ up to $k=\kappa_1-2$ we obtain:

$$\begin{aligned}
&\int_{\tau_{\kappa_0}}^{\tau_{\kappa_1-1}}\mathbb{E}[\mathscr{W}_c(t,x_{u_c,u_d}(t),u_c(t),\eta(t)) \mid \eta(t_0)=j]\mathrm{d}t\\
&\ +\sum_{k=\kappa_0}^{\kappa_1-2}\mathbb{E}[\mathscr{W}_d(\tau_k,x_{u_c,u_d}(\tau_k),u_d(k),\eta(\tau_k)) \mid \eta(t_0)=j]\\
&=\mathbb{E}[x_{u_c,u_d}^{\top}(\tau_{\kappa_0})X(\tau_{\kappa_0}^-,\eta(\tau_{\kappa_0}))x_{u_c,u_d}(\tau_{\kappa_0}) \mid \eta(t_0)=j]\\
&\ -\mathbb{E}[x_{u_c,u_d}^{\top}(\tau_{\kappa_1-1})X(\tau_{\kappa_1-1}^-,\eta(\tau_{\kappa_1-1}))x_{u_c,u_d}(\tau_{\kappa_1-1}) \mid \eta(t_0)=j]\\
&\ +\int_{\tau_{\kappa_0}}^{\tau_{\kappa_1-1}}\mathbb{E}[(u_c(t)-F_c^{\mathbf{X}}(t,\eta(t))x_{u_c,u_d}(t))^{\top}\mathbb{R}_c(t,X(t,\eta(t)),\eta(t))
\end{aligned} \tag{6.9}$$

$$\cdot (u_c(t) - F_c^{\mathbf{X}}(t, \eta(t))x_{u_c,u_d}(t)) \mid \eta(t_0) = j]\mathrm{d}t$$
$$+ \sum_{k=\kappa_0}^{\kappa_1 - 2} \mathbb{E}[(u_d(k) - F_d^{\mathbf{X}}(\tau_k, \eta(\tau_k))x_{u_c,u_d}(\tau_k))^\top \mathbb{R}_d(\tau_k, X(\tau_k, \eta(\tau_k), \eta(\tau_k))$$
$$\cdot (u_d(k) - F_d^{\mathbf{X}}(\tau_k, \eta(\tau_k))x_{u_c,u_d}(\tau_k)) \mid \eta(t_0) = j].$$

Applying again Itô's formula to the function $v(t, x, i) = x^\top X(t, i)x$ and to the stochastic process $x_{u_c,u_d}(\cdot)$ on the interval $[a_{\kappa_1-1}, b_{\kappa_1-1}] \subset (\tau_{\kappa_1-1}, T)$ and letting $a_{\kappa_1-1} \searrow \tau_{\kappa_1-1}$ and $b_{\kappa_1-1} \nearrow T$ we obtain via Lemma 3.1 that:

$$\begin{aligned}
&\int_{\tau_{\kappa_1-1}}^{T} \mathbb{E}[\mathscr{W}_c(t, x_{u_c,u_d}(t), u_c(t), \eta(t)) \mid \eta(t_0) = j]\mathrm{d}t \\
&= \mathbb{E}[x_{u_c,u_d}^\top(\tau_{\kappa_1-1}^+)X(\tau_{\kappa_1-1}, \eta(\tau_{\kappa_1-1}))x_{u_c,u_d}(\tau_{\kappa_1-1}^+) \mid \eta(t_0) = j] \\
&\quad - \mathbb{E}[x_{u_c,u_d}^\top(T)X(T^-, \eta(T))x_{u_c,u_d}(T) \mid \eta(t_0) = j] \\
&\quad + \int_{\tau_{\kappa_1-1}}^{T} \mathbb{E}[(u_c(t) - F_c^{\mathbf{X}}(t, \eta(t))x_{u_c,u_d}(t))^\top \mathbb{R}_c(t, X(t, \eta(t)), \eta(t)) \\
&\quad \cdot (u_c(t) - F_c^{\mathbf{X}}(t, \eta(t))x_{u_c,u_d}(t)) \mid \eta(t_0) = j]\mathrm{d}t.
\end{aligned} \tag{6.10}$$

Proceeding as in the case $k \leq \kappa_1 - 2$ we obtain that

$$\begin{aligned}
&\mathbb{E}[x_{u_c,u_d}^\top(\tau_{\kappa_1-1}^+)X(\tau_{\kappa_1-1}, \eta(\tau_{\kappa_1-1}))x_{u_c,u_d}(\tau_{\kappa_1-1}^+) \mid \eta(t_0) = j] \\
&= \mathbb{E}[x_{u_c,u_d}^\top(\tau_{\kappa_1-1})X(\tau_{\kappa_1-1}^-, \eta(\tau_{\kappa_1-1}))x_{u_c,u_d}(\tau_{\kappa_1-1}) \mid \eta(t_0) = j] \\
&\quad + \mathbb{E}[(u_d(\kappa_1 - 1) - F_d^{\mathbf{X}}(\tau_{\kappa_1-1}, \eta(\tau_{\kappa_1-1}))x_{u_c,u_d}(\tau_{\kappa_1-1}))^\top \\
&\quad \cdot \mathbb{R}_d(\tau_{\kappa_1-1}, X(\tau_{\kappa_1-1}, \eta(\tau_{\kappa_1-1})), \eta(\tau_{\kappa_1-1})) \\
&\quad \cdot (u_d(\kappa_1 - 1) - F_d^{\mathbf{X}}(\tau_{\kappa_1-1}, \eta(\tau_{\kappa_1-1}))x_{u_c,u_d}(\tau_{\kappa_1-1})) \mid \eta(t_0) = j] \\
&\quad - \mathbb{E}[\mathscr{W}_d(\tau_{\kappa_1-1}, x_{u_c,u_d}(\tau_{\kappa_1-1}), u_d(\kappa_1 - 1), \eta(\tau_{\kappa_1-1})) \mid \eta(t_0) = j].
\end{aligned}$$

Substituting this equality in (6.10) and then adding the obtained equality to (6.9) one gets:

$$\begin{aligned}
&\int_{\tau_{\kappa_0}}^{T} \mathbb{E}[\mathscr{W}_c(t, x_{u_c,u_d}(t), u_c(t), \eta(t)) \mid \eta(t_0) = j]\mathrm{d}t \\
&\quad + \sum_{k=\kappa_0}^{\kappa_1-1} \mathbb{E}[\mathscr{W}_d(\tau_k, x_{u_c,u_d}(\tau_k), u_d(k), \eta(\tau_k)) \mid \eta(t_0) = j] \\
&= \mathbb{E}[x_{u_c,u_d}^\top(\tau_{\kappa_0})X(\tau_{\kappa_0}^-, \eta(\tau_{\kappa_0}))x_{u_c,u_d}(\tau_{\kappa_0}) \mid \eta(t_0) = j]
\end{aligned}$$

$$- \mathbb{E}[x_{u_c,u_d}^\top(T)X(T^-,\eta(T))x_{u_c,u_d}(T) \mid \eta(t_0)=j] \tag{6.11}$$
$$+ \int_{\tau_{\kappa_0}}^{T} \mathbb{E}[(u_c(t) - F_c^{\mathbf{X}}(t,\eta(t))x_{u_c,u_d}(t))^\top \mathbb{R}_c(t, X(t,\eta(t)),\eta(t))$$
$$\cdot (u_c(t) - F_c^{\mathbf{X}}(t,\eta(t))x_{u_c,u_d}(t)) \mid \eta(t_0)=j]\mathrm{d}t$$
$$+ \sum_{k=\kappa_0}^{\kappa_1-1} \mathbb{E}[(u_d(k) - F_d^{\mathbf{X}}(\tau_k,\eta(\tau_k))x_{u_c,u_d}(\tau_k))^\top \mathbb{R}_d(\tau_k, X(\tau_k,\eta(\tau_k)),\eta(\tau_k))$$
$$\cdot (u_d(k) - F_d^{\mathbf{X}}(\tau_k,\eta(\tau_k))x_{u_c,u_d}(\tau_k)) \mid \eta(t_0)=j].$$

Let us remark that if $t_0 = \tau_{\kappa_0}$, then (6.11) is just (6.4). If t_0 is such that $\tau_{\kappa_0-1} < t_0 < \tau_{\kappa_0}$ then, applying Itô's formula to the function $v(t,x,i) = x^\top X(t,i)x$ and to the stochastic process $x_{u_c,u_d}(\cdot)$ on the interval $[a_{\kappa_0-1}, b_{\kappa_0-1}] \subset (t_0, \tau_{\kappa_0})$ and letting $a_{\kappa_0-1} \searrow t_0$ and $b_{\kappa_0-1} \nearrow \tau_{\kappa_0}$ we obtain based on Lemma 3.1 that

$$\int_{t_0}^{\tau_{\kappa_0}} \mathbb{E}[\mathscr{W}_c(t, x_{u_c,u_d}(t), u_c(t), \eta(t)) \mid \eta(t_0)=j]\mathrm{d}t$$
$$= x_0^\top X(t_0,j)x_0 - \mathbb{E}[x_{u_c,u_d}^\top(\tau_{\kappa_0})X(\tau_{\kappa_0}^-,\eta(\tau_{\kappa_0}))x_{u_c,u_d}(\tau_{\kappa_0}) \mid \eta(t_0)=j]$$
$$+ \int_{t_0}^{\tau_{\kappa_0}} \mathbb{E}[(u_c(t) - F_c^{\mathbf{X}}(t,\eta(t))x_{u_c,u_d}(t))^\top \mathbb{R}_c(t, X(t,\eta(t)),\eta(t))$$
$$\cdot (u_c(t) - F_c^{\mathbf{X}}(t,\eta(t))x_{u_c,u_d}(t)) \mid \eta(t_0)=j]\mathrm{d}t.$$

Adding this equality to (6.11) we obtain (6.4). Thus the proof is complete.

Let us consider now quadratic functionals of the form

$$\breve{J}_j(t_0,T,x_0;u_d) = \int_{t_0}^{T} \mathbb{E}[x_{u_d}^\top(t)M_c(t,\eta(t))x_{u_d}(t) \mid \eta(t_0)=j]\mathrm{d}t \tag{6.12}$$
$$+ \sum_{k=\kappa_0}^{\kappa_1-1} \mathbb{E}[\mathscr{W}_d(\tau_k, x_{u_d}(\tau_k), u_d(k), \eta(\tau_k)) \mid \eta(t_0)=j]$$

where $x_{u_d}(\cdot) = x(\cdot; t_0, x_0, u_d)$ is the solution of the following IVP:

$$\mathrm{d}x(t) = A_0(t,\eta(t))x(t)\mathrm{d}t + \sum_{l=1}^{r_1} A_l(t,\eta(t))x(t)\mathrm{d}w_l(t), \quad \tau_k < t \le \tau_{k+1} \tag{6.13a}$$

$$x(\tau_k^+) = A_{d0}(\tau_k,\eta(\tau_k))x(\tau_k) + B_{d0}(\tau_k,\eta(\tau_k))u_d(k) \tag{6.13b}$$

$$+\sum_{p=1}^{r_2} w_{dp}(k)(A_{dp}(\tau_k, \eta(\tau_k))x(\tau_k) + B_{dp}(\tau_k, \eta(\tau_k))u_d(k)),$$
$$k \in \{\kappa_0, \kappa_0 + 1, \ldots, \kappa_1 - 1\}$$
$$x_{u_d}(t_0) = x_0, \tag{6.13c}$$

where $x_0 \in \mathbb{R}^n$ is fixed. In (6.12), $\mathscr{W}_d(\tau_k, x_{u_d}(\tau_k), u_d(k), \eta(\tau_k))$ is as in (6.3b) but with $x_{u_c u_d}(\tau_k)$ replaced by $x_{u_d}(\tau_k)$. The pair formed by the functional (6.12) and IVP (6.13), is a particular case of the pair formed by the functional (6.1) and the IVP (6.2). The analogous of Lemma 6.1 is:

Lemma 6.2 *Assume that the Assumptions 1.2, 2.1, 5.2 are fulfilled. Let* $\mathbf{X}(\cdot) = (X(\cdot, 1), X(\cdot, 2), \ldots, X(\cdot, N)) : \mathscr{I}^{\mathbf{X}} \subset \mathbb{R}_+ \to \mathscr{S}_n^N$ *be a solution of the BJMLDE with Riccati-type jumping operator of the form (5.168) defined based on the matrix coefficients of (6.12) and (6.13).*

If the interval $[t_0, T] \subset \mathscr{I}^{\mathbf{X}}$ *then the following equality holds:*

$$\check{J}_j(t_0, T, x_0; u_d) = x_0^\top X(t_0^-, j)x_0 - \mathbb{E}[x_{u_d}^\top(T)X(T, \eta(T))x_{u_d}(T) \mid \eta(t_0) = j]$$
$$+\sum_{k=\kappa_0}^{\kappa_1-1} \mathbb{E}[(u_d(k) - F_d^{\mathbf{X}}(\tau_k, \eta(\tau_k))x_{u_d}(\tau_k))^\top \tag{6.14}$$
$$\cdot\, \mathbb{R}_d(\tau_k, X(\tau_k, \eta(\tau_k)), \eta(\tau_k))(u_d(k) - F_d^{\mathbf{X}}(\tau_k, \eta(\tau_k))x_{u_d}(\tau_k)) \mid \eta(t_0) = j]$$

for all $\{u_d(k)\}_{\kappa_0 \le k \le \kappa_1 - 1} \in \ell^2_{\mathscr{H}}\{\kappa_0, \kappa_1 - 1; \mathbb{R}^{m_d}\}$ *with* $\mathbb{R}_d(\tau_k, X(\tau_k, i), i)$ *and* $F_d^{\mathbf{X}}$ *defined as in (6.5b) and (6.6b), respectively, based on the matrix coefficients of (6.12) and (6.13). If* $t_0 \in (\tau_{\kappa_0 - 1}, \tau_{\kappa_0})$ *then,* $X(t_0^-, j)$ *is replaced by* $X(t_0, j)$ *and if* $T = \tau_{\kappa_1}$ *then* $X(T, \eta(T))$ *is replaced in (6.14) by* $X(T^-, \eta(T))$*.*

Proof Is obtained from (6.4) because (6.12) and (6.13) are special cases of (6.1) and (6.2), respectively, when $L_c(t, i) = 0$, $R_c(t, i) = 0$, $B_l(t, i) = 0$, $0 \le l \le r_1$, $(t, i) \in \mathbb{R}_+ \times \mathscr{N}$. In this case, the BJMLDE with Riccati-type jumping operator (5.168) is a particular case of the SJMRDE (5.10).

6.2 The Linear Quadratic Optimal Control Problem for a Jump Linear Stochastic System. The Finite Horizon Time Case

Let $[t_0, T] \subset \mathbb{R}_+$ be a bounded interval and $\mathfrak{T} \subset \mathbb{R}_+$ be a partition of the semiaxis $\mathbb{R}_+$ satisfying conditions from 2.1. If $\tau \ge 0$ is a given real number we denote

$$\kappa(\tau) \triangleq \inf\{\kappa \in \mathbb{Z}_+ \mid \tau \le \tau_\kappa\}. \tag{6.15}$$

Let us consider the quadratic functional

$$(u_c, u_d) \to J(t_0, T, x_0; u_c, u_d) : \mathcal{L}^2_{\mathcal{H}}([t_0, T], \mathbb{R}^{m_c}) \times \ell^2_{\mathcal{H}}\{\kappa(t_0), \kappa(T) - 1; \mathbb{R}^{m_d}\} \to \mathbb{R}$$

defined by

$$\begin{aligned} J(t_0, T, x_0; u_c, u_d) =& \mathbb{E}[x^\top_{u_c,u_d}(T) G(\eta(T)) x_{u_c,u_d}(T)] \\ &+ \int_{t_0}^{T} \mathbb{E}[\mathcal{W}_c(t, x_{u_c,u_d}(t), u_c(t), \eta(t))] \mathrm{d}t \\ &+ \sum_{k=\kappa(t_0)}^{\kappa(T)-1} \mathbb{E}[\mathcal{W}_d(\tau_k, x_{u_c,u_d}(\tau_k), u_d(k), \eta(\tau_k))] \end{aligned} \tag{6.16}$$

where $x_{u_c,u_d}(\cdot) = x(\cdot, t_0, x_0, u_c, u_d)$ is the solution of the IVP (6.2) determined by the input (u_c, u_d). According to Proposition 3.4 we deduce that $x_{u_c,u_d}(\cdot) \in \mathcal{L}^2_{\mathcal{H}}([t_0, T], \mathbb{R}^n)$ if $u_c(\cdot) \in \mathcal{L}^2_{\mathcal{H}}([t_0, T], \mathbb{R}^{m_c})$ and $u_d(k)$ is $\mathcal{H}_{\tau_k}$−measurable with

$$\mathbb{E}[|u_d(k)|^2] < \infty, \quad \text{for all } \kappa(t_0) \le k \le \kappa(T) - 1.$$

This allows us to consider the family of admissible controls $\mathcal{U}_{\text{ad}}$ consisting of all stochastic processes $(u_c(\cdot), \mathbf{u}_d)$ where $u_c(\cdot) \in \mathcal{L}^2_{\mathcal{H}}([t_0, T]; \mathbb{R}^{m_c})$ and $\mathbf{u}_d = \{u_d(k)\}_{\kappa(t_0) \le k \le \kappa(T)-1} \in \ell^2_{\mathcal{H}}\{\kappa(t_0), \kappa(T) - 1; \mathbb{R}^{m_d}\}$.

The optimal control problem which we are solving in this section, ask for finding a control $(\bar{u}_c(\cdot), \bar{\mathbf{u}}_d) \in \mathcal{U}_{\text{ad}}$ which satisfies the optimality condition:

$$J(t_0, T, x_0; \bar{u}_c(\cdot), \bar{\mathbf{u}}_d) = \inf\{J(t_0, T, x_0; u_c(\cdot), \mathbf{u}_d) \mid (u_c(\cdot), \mathbf{u}_d) \in \mathcal{U}_{\text{ad}}\}. \tag{6.17}$$

The controls $(\bar{u}_c(\cdot), \bar{\mathbf{u}}_d)$ which are satisfying (6.17) will be named optimal controls for the optimal control problem described by the quadratic functional (6.16) (often called **performance criterion or cost functional**). , the control system (6.2) and the set of admissible controls $\mathcal{U}_{\text{ad}}$.

In this section we provide conditions which guarantee the existence of an optimal control in a state feedback form. Let $\bar{\mathbf{X}}(\cdot) = (\bar{X}(\cdot, 1), \bar{X}(\cdot, 2), \ldots, \bar{X}(\cdot, N))$ be the solution with the terminal condition $\bar{X}(T, i) = G(i),\ \ i \in \mathcal{N}$ of the SJMRDE of type (5.10) constructed involving the matrix coefficients of the controlled system (6.2) and the weights matrices of the performance criterion (6.16).

The next theorem provides the control law in a state feedback form which is satisfying the optimality condition (6.17).

Theorem 6.1 *Assume:*

(a) the Assumptions 1.2 and 2.1 hold;

(b) the matrix coefficients of the controlled system (6.2) and the weights matrices from (6.16) satisfy the conditions from the Assumption 5.2;

(c) the solution $\bar{\mathbf{X}}(\cdot)$ of the SJMRDE of type (5.10) which is satisfying the terminal condition $X(T,i) = G(i),\quad i \in \mathcal{N}$, is defined on the whole interval $[t_0, T]$ and is satisfying the sign conditions:

$$\sum_{l=1}^{r_1} B_l^\top(t,i)\bar{X}(t,i)B_l(t,i) + R_c(t,i) \geq \gamma_c I_{m_c} > 0, \quad (\forall)\ \ (t,i) \in [t_0, T] \times \mathcal{N} \tag{6.18}$$

$$\sum_{p=0}^{r_2} B_{dp}^\top(\tau_k,i)\bar{X}(\tau_k,i)B_{dp}(\tau_k,i) + R_d(\tau_k,i) \geq \gamma_d I_{m_d} > 0, \quad (\forall)\ \ (\tau_k,i) \in (\mathfrak{T} \cap [t_0,T]) \times \mathcal{N}. \tag{6.19}$$

Let us consider the control law:

$$\bar{u}_c(t) = \bar{F}_c(t,\eta(t))\bar{x}(t), \quad t \in [t_0, T] \tag{6.20a}$$

$$\bar{u}_d(k) = \bar{F}_d(\tau_k,\eta(\tau_k))\bar{x}(\tau_k), \quad \kappa(t_0) \leq k \leq \kappa(T) - 1 \tag{6.20b}$$

where $\bar{F}_c(t,i)$ and $\bar{F}_d(\tau_k,i)$ are computed as in (6.6) but with $X(\cdot,i)$ replaced by $\bar{X}(\cdot,i)$, $i \in \mathcal{N}$ and $\bar{x}(\cdot)$ is the solution of the following IVP:

$$dx(t) = (A_0(t,\eta(t)) + B_0(t,\eta(t))\bar{F}_c(t,\eta(t)))x(t)dt + \sum_{l=1}^{r_1}(A_l(t,\eta(t)) + B_l(t,\eta(t))\bar{F}_c(t,\eta(t)))x(t)dw_l(t) \quad \tau_k < t \leq \tau_{k+1} \tag{6.21a}$$

$$x(\tau_k^+) = (A_{d0}(\tau_k,\eta(\tau_k)) + B_{d0}(\tau_k,\eta(\tau_k))\bar{F}_d(\tau_k,\eta(\tau_k)))x(\tau_k) + \sum_{p=1}^{r_2} w_{dp}(k)(A_{dp}(\tau_k,\eta(\tau_k)) + B_{dp}(\tau_k,\eta(\tau_k))\bar{F}_d(\tau_k,\eta(\tau_k)))x(\tau_k),$$
$$\kappa(t_0) \leq k \leq \kappa(T) - 1 \tag{6.21b}$$

$$\bar{x}(t_0) = x_0 \tag{6.21c}$$

obtained when (6.20) are plugged in (6.2). Under the considered Assumptions, the control law $(\bar{u}_c(\cdot), \bar{\mathbf{u}}_d)$ defined by (6.20)–(6.21) is satisfying the optimality condition (6.17) and it is the unique control law with this property. The minimal value of the cost (6.16) is:

$$J(t_0, T, x_0; \bar{u}_c(\cdot), \bar{\mathbf{u}}_d) = \begin{cases} \mathbb{E}[x_0^\top \bar{X}(t_0^-, \eta(t_0))x_0] & \text{if } t_0 = \tau_{\kappa(t_0)}, \\ \mathbb{E}[x_0^\top \bar{X}(t_0, \eta(t_0))x_0] & \text{if } t_0 < \tau_{\kappa(t_0)}. \end{cases}$$

Proof Applying Lemma 6.1 with the update $\mathbf{X}(\cdot) \leftarrow \bar{\mathbf{X}}(\cdot)$ we rewrite (6.16) as:

$$J(t_0, T, x_0; u_c, \mathbf{u}_d) = \mathbb{E}[x_0^\top \bar{X}(t_0^-, \eta(t_0))x_0] + \sum_{j=1}^{N} \pi_j(t_0) \left\{ \int_{t_0}^{T} \mathbb{E}[(u_c(t) - \bar{F}_c(t, \eta(t))x_{u_c,u_d}(t))^\top \right.$$
$$\mathbb{R}_c(t, \bar{X}(t, \eta(t)), \eta(t))(u_c(t) - \bar{F}_c(t, \eta(t))x_{u_c,u_d}(t)) \mid \eta(t_0) = j]\mathrm{d}t$$
$$+ \sum_{k=\kappa_0}^{\kappa(T)-1} \mathbb{E}[(u_d(k) - \bar{F}_d(\tau_k, \eta(\tau_k))x_{u_c u_d}(\tau_k))^\top \mathbb{R}_d(\tau_k, \bar{X}(\tau_k, \eta(\tau_k)), \eta(\tau_k))$$
$$\left. (u_d(k) - \bar{F}_d(\tau_k, \eta(\tau_k))x_{u_c u_d}(\tau_k)) \mid \eta(t_0) = j] \right\} \tag{6.22}$$

for all $(u_c(\cdot), \mathbf{u}_d) \in \mathscr{U}_{\text{ad}}$ where $\pi_j(t_0) = \mathscr{P}\{\eta(t_0) = j\}$. It is worth mentioning that in the case when $t_0 < \tau_{\kappa(t_0)}$, $\bar{X}(t_0^-, \eta(t_0))$ is replaced by $\bar{X}(t_0, \eta(t_0))$ in (6.22).

Employing (6.18)–(6.19) we deduce from (6.22) that

$$J(t_0, T, x_0; u_c(\cdot), \mathbf{u}_d) \geq \mathbb{E}[x_0^\top \bar{X}(t_0^-, \eta(t_0))x_0], \tag{6.23}$$

for all $(u_c(\cdot), \mathbf{u}_d) \in \mathscr{U}_{\text{ad}}$. Further, from the uniqueness of the solution of (6.2) for a given admissible control $(u_c(\cdot), \mathbf{u}_d)$ from $\mathscr{U}_{\text{ad}}$, we deduce, via IVP (6.21), that

$$x_{\bar{u}_c \bar{u}_d}(t) = \bar{x}(t), \quad \text{for all} \ \ t \in [t_0, T].$$

Substituting this equality in (6.20) we obtain from (6.22) that

$$J(t_0, T, x_0; \bar{u}_c(\cdot), \bar{\mathbf{u}}_d) = \mathbb{E}[x_0^\top \bar{X}(t_0^-, \eta(t_0))x_0]. \tag{6.24}$$

Combining (6.23) and (6.24) we may conclude that (6.20) provides an optimal control for the optimal control problem (6.17) and the minimal value of the cost (6.16) is given by (6.24). The unicity of the optimal control is immediately obtained from (6.18), (6.19) and (6.22). Thus the proof ends.

Remark 6.1 A necessary and sufficient condition in order to Assumption (c) from the statement of the previous theorem to be satisfied is provided by Theorem 5.2. Also, Corollary 5.1 provides a sufficient condition that guarantee that (c) from previous theorem is satisfied.

6.3 The Linear Quadratic Optimal Control Problem with Free End State Value. The Case of Infinite Time Horizon

Let us consider the quadratic functional $(u_c, \mathbf{u}_d) \to J(t_0, \infty, x_0; u_c(\cdot), \mathbf{u}_d)$: $\mathscr{L}^2_{\mathscr{H}}([t_0, \infty], \mathbb{R}^{m_c}) \times \ell^2_{\mathscr{H}}\{\kappa(t_0), \infty; \mathbb{R}^{m_d}\} \to \mathbb{R}$ defined by

$$J(t_0, \infty, x_0; u_c(\cdot), \mathbf{u}_d) = \int_{t_0}^{\infty} \mathbb{E}[\mathscr{W}_c(t, x_{u_c,u_d}(t), u_c(t), \eta(t))]dt$$
$$+ \sum_{k=\kappa(t_0)}^{\infty} \mathbb{E}[\mathscr{W}_d(\tau_k, x_{u_c,u_d}(\tau_k), u_d(k), \eta(\tau_k))] \quad (6.25)$$

where $x_{u_c,u_d}(t)$, $t \geq t_0$ is the solution of the IVP (6.2) determined by the control $(u_c(\cdot), \mathbf{u}_d)$ from $\mathscr{L}^2_{\mathscr{H}}([t_0, \infty], \mathbb{R}^{m_c}) \times \ell^2_{\mathscr{H}}\{\kappa(t_0), \infty; \mathbb{R}^{m_d}\}$.
In (6.25), $\mathscr{W}_c(t, x_{u_c,u_d}(t), u_c(t), \eta(t))$ and $\mathscr{W}_d(\tau_k, x_{u_c,u_d}(\tau_k), u_d(k), \eta(\tau_k))$ are the quadratic forms described in (6.3). In this section we assume that the weighting matrices from (6.3) satisfy the sign conditions:

$$R_c(t,i) \geq \delta_c I_{m_c} > 0 \quad (6.26a)$$
$$M_c(t,i) - L_c(t,i)R_c^{-1}(t,i)L_c^{\top}(t,i) \geq 0 \text{ for all } (t,i) \in \mathbb{R}_+ \times \mathscr{N} \quad (6.26b)$$

$$R_d(\tau_k, i) \geq \delta_d I_{m_d} > 0 \quad (6.27a)$$
$$M_d(\tau_k, i) - L_d(\tau_k, i)R_d^{-1}(\tau_k, i)L_d^{\top}(\tau_k, i) \geq 0 \text{ for all } (\tau_k, i) \in \mathfrak{T} \times \mathscr{N}. \quad (6.27b)$$

Bearing in mind (6.3), (6.26), (6.27) we may deduce via Lemma 5.5 that

$$J(t_0, \infty, x_0; u_c(\cdot), \mathbf{u}_d) \geq 0 \quad (6.28)$$

for any $u_c(\cdot) \in \mathscr{L}^2_{\mathscr{H}}([t_0, \infty], \mathbb{R}^{m_c})$ and $\mathbf{u}_d \in \ell^2_{\mathscr{H}}\{\kappa(t_0), \infty; \mathbb{R}^{m_d}\}$.
The set of the admissible controls $\mathscr{U}_{\text{ad}}(t_0, x_0)$ consists of all pairs $(u_c(\cdot), \mathbf{u}_d)$ where $u_c(\cdot) \in \mathscr{L}^2_{\mathscr{H}}([t_0, \infty], \mathbb{R}^{m_c})$ and $\mathbf{u}_d = \{u_d(k)\}_{k \geq \kappa(t_0)} \in \ell^2_{\mathscr{H}}\{\kappa(t_0), \infty; \mathbb{R}^{m_d}\}$ with the property that

$$J(t_0, \infty, x_0; u_c(\cdot), \mathbf{u}_d) < \infty,$$

$\kappa(t_0)$ being defined as in (6.15). Let $\mathbf{X}(\cdot) = (X(\cdot, 1), X(\cdot, 2), \ldots, X(\cdot, N)) : \mathbb{R}_+ \to \mathscr{S}_n^N$ be a bounded and positive semidefinite solution of the SJMRDE (5.10) associated to the pair formed by the controlled system (6.2a)–(6.2b) and the quadratic functional (6.25). We consider the control laws $u_c^{\mathbf{X}}(t)$ and $\mathbf{u}_d^{\mathbf{X}} = \{u_d^{\mathbf{X}}\}_{k \geq \kappa(t_0)}$ be defined by

$$u_c^{\mathbf{X}}(t) \triangleq F_c^{\mathbf{X}}(t, \eta(t))x^{\mathbf{X}}(t), \quad t \geq t_0 \quad (6.29a)$$
$$u_d^{\mathbf{X}}(k) \triangleq F_d^{\mathbf{X}}(\tau_k, \eta(\tau_k))x^{\mathbf{X}}(\tau_k), \quad k \geq \kappa(t_0) \quad (6.29b)$$

where $F_c^{\mathbf{X}}(t, i)$ and $F_d^{\mathbf{X}}(\tau_k, i)$ are the gain matrices associated to the solution $\mathbf{X}(\cdot)$ by (6.6) while $x^{\mathbf{X}}(t)$, $t \geq t_0$ is the solution of the IVP

$$dx(t) = (A_0(t,\eta(t)) + B_0(t,\eta(t))F_c^{\mathbf{X}}(t,\eta(t)))x(t)dt$$
$$+\sum_{l=1}^{r_1}(A_l(t,\eta(t)) + B_l(t,\eta(t))F_c^{\mathbf{X}}(t,\eta(t)))x(t)dw_l(t), \quad \tau_k < t \le \tau_{k+1} \tag{6.30a}$$
$$x(\tau_k^+) = (A_{d0}(\tau_k,\eta(\tau_k)) + B_{d0}(\tau_k,\eta(\tau_k))F_d^{\mathbf{X}}(\tau_k,\eta(\tau_k)))x(\tau_k)$$
$$+\sum_{p=1}^{r_2} w_{dp}(k)(A_{dp}(\tau_k,\eta(\tau_k)) + B_{dp}(\tau_k,\eta(\tau_k))F_d^{\mathbf{X}}(\tau_k,\eta(\tau_k)))x(\tau_k), \quad k \ge \kappa(t_0) \tag{6.30b}$$
$$x^{\mathbf{X}}(t_0) = x_0. \tag{6.30c}$$

Lemma 6.3 *Assume:*

(a) the Assumptions 1.2, 2.1 hold;
(b) the coefficient matrices of (6.2) and the weights matrices of the functional (6.25) are satisfying the assumptions from 5.2;
(c) the weights matrices from (6.25) are satisfying the sign conditions (6.26) and (6.27).

Under these conditions, the pairs of control laws $(u_c^{\mathbf{X}}(\cdot), \mathbf{u}_d^{\mathbf{X}})$ defined by (6.29), (6.30) lie in $\mathcal{U}_{\mathrm{ad}}(t_0, x_0)$.

Proof Let $\mathbf{X}(\cdot)$ be an arbitrary bounded on $\mathbb{R}_+$ and positive semidefinite solution of the SJMRDE of the form (5.10) associated to the controlled system (6.2) and the functional (6.25). Let $T > t_0$ be arbitrary but fixed. Applying Lemma 6.1 on the interval $[t_0, T]$ we obtain:

$$\int_{t_0}^{T} \mathbb{E}[\mathscr{W}_c(t, x_{u_c,u_d}(t), u_c(t), \eta(t)) \mid \eta(t_0) = j]dt$$
$$+\sum_{k=\kappa(t_0)}^{\kappa(T)-1} \mathbb{E}[\mathscr{W}_d(\tau_k, x_{u_c,u_d}(\tau_k), u_d(k), \eta(\tau_k)) \mid \eta(t_0) = j]$$
$$= x_0^\top X(t_0^-, j)x_0 - \mathbb{E}[x_{u_c,u_d}^\top(T)X(T,\eta(T))x_{u_c,u_d}(T) \mid \eta(t_0) = j]$$
$$+\int_{t_0}^{T} \mathbb{E}[(u_c(t) - F_c^{\mathbf{X}}(t,\eta(t))x_{u_c,u_d}(t))^\top \mathbb{R}_c(t, X(t,\eta(t)),\eta(t))$$
$$\cdot (u_c(t) - F_c^{\mathbf{X}}(t,\eta(t))x_{u_cu_d}(t)) \mid \eta(t_0) = j]dt \tag{6.31}$$
$$+\sum_{k=\kappa(t_0)}^{\kappa(T)-1} \mathbb{E}[(u_d(k) - F_d^{\mathbf{X}}(\tau_k,\eta(\tau_k))x_{u_c,u_d}(\tau_k))^\top \mathbb{R}_d(\tau_k, X(\tau_k,\eta(\tau_k)),\eta(\tau_k))$$
$$\cdot (u_d(k) - F_d^{\mathbf{X}}(\tau_k,\eta(\tau_k))x_{u_cu_d}(\tau_k)) \mid \eta(t_0) = j],$$

for any $u_c(\cdot) \in \mathscr{L}^2_{\mathscr{H}}([t_0, T], \mathbb{R}^{m_c})$ and $\{u_d(k)\}_{\kappa(t_0) \geq k \geq \kappa(T)-1} \in \ell^2_{\mathscr{H}}\{\kappa(t_0), \kappa(T) - 1; \mathbb{R}^{m_d}\}$. From the unicity of the solution of the IVP, we deduce that $x_{u_c{}^{\mathbf{X}}, u_d{}^{\mathbf{X}}}(\cdot)$ coincides to $x^{\mathbf{X}}(\cdot)$. Based on this observation we obtain from (6.31) that

$$\begin{aligned}
&\int_{t_0}^{T} \mathbb{E}[\mathscr{W}_c(t, x_{u_c{}^{\mathbf{X}}, u_d{}^{\mathbf{X}}}(t), u_c^{\mathbf{X}}(t), \eta(t)) \mid \eta(t_0) = j]\mathrm{d}t \qquad (6.32)\\
&\quad + \sum_{k=\kappa(t_0)}^{\kappa(T)-1} \mathbb{E}[\mathscr{W}_d(\tau_k, x_{u_c{}^{\mathbf{X}}, u_d{}^{\mathbf{X}}}(\tau_k), u_d^{\mathbf{X}}(k), \eta(\tau_k)) \mid \eta(t_0) = j]\\
&= x_0^\top X(t_0^-, j)x_0 - \mathbb{E}[x_{u_c{}^{\mathbf{X}}, u_d{}^{\mathbf{X}}}(T)^\top X(T, \eta(T))x_{u_c{}^{\mathbf{X}}, u_d{}^{\mathbf{X}}}(T) \mid \eta(t_0) = j]\\
&\leq x_0^\top X(t_0^-, j)x_0, \qquad \text{for all } T > t_0, \ \ j \in \mathscr{N}.
\end{aligned}$$

Since $\mathbf{X}(\cdot)$ is a bounded and positive semidefinite solution of (5.10) we may infer that there exists a constant $c > 0$ such that $0 \leq X(t_0, j) \leq cI_n$ and $0 \leq X(t_0^-, j) \leq cI_n$ for all $t_0 \in \mathbb{R}_+$, $j \in \mathscr{N}$.

Invoking again (6.26), (6.27), we obtain from (6.3) and (6.32) that

$$\begin{aligned}
0 &\leq \int_{t_0}^{T_1} \mathbb{E}[\mathscr{W}_c(t, x^{\mathbf{X}}(t), u_c{}^{\mathbf{X}}(t), \eta(t)) \mid \eta(t_0) = j]\mathrm{d}t\\
&\quad + \sum_{k=\kappa(t_0)}^{\kappa(T_1)-1} \mathbb{E}[\mathscr{W}_d(\tau_k, x^{\mathbf{X}}(\tau_k), u_d{}^{\mathbf{X}}(k), \eta(\tau_k)) \mid \eta(t_0) = j]\\
&\leq \int_{t_0}^{T_2} \mathbb{E}[\mathscr{W}_c(t, x^{\mathbf{X}}(t), u_c{}^{\mathbf{X}}(t), \eta(t)) \mid \eta(t_0) = j]\mathrm{d}t \qquad (6.33)\\
&\quad + \sum_{k=\kappa(t_0)}^{\kappa(T_2)-1} \mathbb{E}[\mathscr{W}_d(\tau_k, x^{\mathbf{X}}(\tau_k), u_d{}^{\mathbf{X}}(k), \eta(\tau_k)) \mid \eta(t_0) = j]\\
&\leq c|x_0|^2, \qquad \text{for all } t_0 < T_1 < T_2, \ \ j \in \mathscr{N}.
\end{aligned}$$

So, (6.33) shows that we may take the limit for $T \to \infty$ in (6.32) to obtain

$$0 \leq \int_{t_0}^{\infty} \mathbb{E}[\mathscr{W}_c(t, x^{\mathbf{X}}(t), u_c{}^{\mathbf{X}}(t), \eta(t)) \mid \eta(t_0) = j]$$

$$+ \sum_{k=\kappa(t_0)}^{\infty} \mathbb{E}[\mathscr{W}_d(\tau_k, x^{\mathbf{X}}(\tau_k), u_d{}^{\mathbf{X}}(k), \eta(\tau_k)) \mid \eta(t_0) = j] \leq x_0^\top X(t_0^-, j)x_0$$

which leads to

$$0 \leq J(t_0, \infty, x_0; u_c^{\mathbf{X}}(\cdot), \mathbf{u}_d^{\mathbf{X}}) \leq \mathbb{E}[x_0^\top X(t_0^-, \eta(t_0))x_0]. \tag{6.34}$$

Further, (6.26), (6.27) together with square completion technique allow us to deduce that

$$\int_{t_0}^{\infty} \mathbb{E}[|u_c^{\mathbf{X}}(t)|^2]\mathrm{d}t \leq \mathbb{E}[x_0^\top X(t_0^-, \eta(t_0))x_0] \tag{6.35a}$$

$$\sum_{k=\kappa(t_0)}^{\infty} \mathbb{E}[|u_d^{\mathbf{X}}(k)|^2]\mathrm{d}t \leq \mathbb{E}[x_0^\top X(t_0^-, \eta(t_0))x_0]. \tag{6.35b}$$

Hence, (6.34) and (6.35) confirm that $(u_c^{\mathbf{X}}, \mathbf{u}_d^{\mathbf{X}})$ lies in $\mathscr{U}_{\mathrm{ad}}(t_0, x_0)$, for all $(t_0, x_0) \in \mathbb{R}_+ \times \mathbb{R}^n$. Thus the proof is complete.

Remark 6.2 In (6.31)–(6.35), $X(t_0^-, j)$ coincides to $X(t_0, j)$ whenever $t_0 < \tau_{\kappa(t_0)}$, because $t \to X(t, j)$ is a continuous function on each interval (τ_k, τ_{k+1}), $k \in \mathbb{Z}_+$.

Now we are able to state the optimal control problem which address in this section:

The free end state value linear quadratic control problem:

For a given initial pair $(t_0, x_0) \in \mathbb{R}_+ \times \mathbb{R}^n$ find a control law $(\widetilde{u}_c(\cdot), \widetilde{\mathbf{u}}_d) \in \mathscr{U}_{\mathrm{ad}}(t_0, x_0)$ which satisfies the following optimality condition:

$$J(t_0, \infty, x_0; \widetilde{u}_c(\cdot), \widetilde{\mathbf{u}}_d) = \inf_{(u_c(\cdot), \mathbf{u}_d) \in \mathscr{U}_{\mathrm{ad}}(t_0, x_0)} J(t_0, \infty, x_0; u_c(\cdot), \mathbf{u}_d). \tag{6.36}$$

In this section we provide a set of conditions which guarantee the existence of a control $(\widetilde{u}_c(\cdot), \widetilde{\mathbf{u}}_d)$ in a state feedback form which is satisfying the optimality condition (6.36).

Remark 6.3 In the description of the class of admissible controls $\mathscr{U}_{\mathrm{ad}}(t_0, x_0)$ we do not impose any condition on the behavior of the trajectories $x_{u_c,u_d}(t)$ when $t \to \infty$. That is why, the optimal control problem involving the set of the admissible controls $\mathscr{U}_{\mathrm{ad}}(t_0, x_0)$ beside the quadratic performance criterion (6.25) and the controlled system (6.2) will be named *linear quadratic optimal control problem with free end state values.*

The main result of this section is stated and proved in the next theorem:

Theorem 6.2 *Assume:*

(a) the Assumptions 1.2, 2.1 hold;
(b) the matrix coefficients of the controlled system (6.2) and the weights matrices from (6.25) are satisfying the Assumption 5.2;
(c) the weights matrices from (6.25) are satisfying the sign conditions (6.26) and (6.27);

(d) the controlled JLSS (6.2) is mean-square stabilizable by linear state feedback.

Let $\mathbf{X}_{min}(\cdot) = (X_{min}(\cdot, 1), X_{min}(\cdot, 2), \ldots, X_{min}(\cdot, N))$ *be the unique minimal positive semidefinite and bounded on* $\mathbb{R}_+$ *solution of the SJMRDE (5.10) associated to the pair consisting by JLSS (6.2) and the quadratic functional (6.25). We consider the control law* $(\widetilde{u}_c(\cdot), \widetilde{\mathbf{u}}_d)$ *defined by*

$$\widetilde{u}_c(t) := \widetilde{F}_c(t, \eta(t))\widetilde{x}(t), \quad t \geq t_0 \tag{6.37a}$$

$$\widetilde{u}_d(k) := \widetilde{F}_d(\tau_k, \eta(\tau_k))\widetilde{x}(\tau_k), \quad k \geq \kappa(t_0), \tag{6.37b}$$

where $\widetilde{F}_c(t, i)$, $\widetilde{F}_d(\tau_k, i)$ *are computed as in (6.6) but with* $X(\cdot, i)$ *replaced by* $X_{min}(\cdot, i)$, $i \in \mathcal{N}$ *and* $\widetilde{x}(t)$, $t \geq t_0$ *is the solution of the following IVP:*

$$dx(t) = (A_0(t, \eta(t)) + B_0(t, \eta(t))\widetilde{F}_c(t, \eta(t)))x(t)dt \tag{6.38a}$$

$$+ \sum_{l=1}^{r_1} (A_l(t, \eta(t)) + B_l(t, \eta(t))\widetilde{F}_c(t, \eta(t)))x(t)dw_l(t), \quad \tau_k < t \leq \tau_{k+1}$$

$$x(\tau_k^+) = (A_{d0}(\tau_k, \eta(\tau_k)) + B_{d0}(\tau_k, \eta(\tau_k))\widetilde{F}_d(\tau_k, \eta(\tau_k)))x(\tau_k) \tag{6.38b}$$

$$+ \sum_{p=1}^{r_2} w_{dp}(k)(A_{dp}(\tau_k, \eta(\tau_k)) + B_{dp}(\tau_k, \eta(\tau_k))\widetilde{F}_d(\tau_k, \eta(\tau_k)))x(\tau_k), k \geq \kappa(t_0)$$

$$\widetilde{x}(t_0) = x_0. \tag{6.38c}$$

Under the considered assumptions the control law (6.37)–(6.38) satisfies the optimality condition (6.36). The minimal value of the cost function is:

$$J(t_0, \infty, x_0; \widetilde{u}_c(\cdot), \widetilde{\mathbf{u}}_d) = \mathbb{E}[x_0^\top X_{min}(t_0^-, \eta(t_0))x_0], \tag{6.39}$$

with the remark that $X_{min}(t_0^-, \eta(t_0)) = X_{min}(t_0, \eta(t_0))$, *whenever* $t_0 < \tau_{\kappa(t_0)}$.

Proof First, let us remark that according to Lemma 6.3, the control law described by (6.37)–(6.38) lies in $\mathcal{U}_{ad}(t_0, x_0)$ for any initial pair $(t_0, x_0) \in \mathbb{R}_+ \times \mathbb{R}^n$. Furthermore (6.34) gives

$$J(t_0, \infty, x_0; \widetilde{u}_c(\cdot), \widetilde{\mathbf{u}}_d) \leq \mathbb{E}[x_0^\top X_{min}(t_0^-, \eta(t_0))x_0]. \tag{6.40}$$

Let $(u_c(\cdot), \mathbf{u}_d) \in \mathcal{U}_{ad}(t_0, x_0)$ be arbitrary. We set

$$u_{cT}(t) \triangleq u_c(t), \quad t \in [t_0, T] \quad \text{and} \quad \mathbf{u}_{dT}(t) \triangleq \{u_d(k)\}_{\kappa(t_0) \leq k \leq \kappa(T)-1}.$$

Let $x_T(t)$, $t_0 \leq t \leq T$ be the solution of IVP (6.2) corresponding to the input $(u_c(\cdot), \mathbf{u}_d) \leftarrow (u_{cT}(\cdot, \omega), \mathbf{u}_{dT})$.

$$\bar{u}_{cT}(t, \omega) \triangleq \begin{cases} u_{cT}(t, \omega) & \text{if } t_0 \leq t \leq T \\ 0 & \text{if } t > T \end{cases}$$

$$\bar{\mathbf{u}}_{dt} = \{\bar{u}_{dt}(k,\omega)\}_{k\geq\kappa(t_0)},$$

where

$$\bar{u}_{dt}(k,\omega) = \begin{cases} u_d(k,\omega) & \text{if } \kappa(t_0) \leq k \leq \kappa(T)-1 \\ 0 & \text{if } k \geq \kappa(T) \end{cases}$$

$\bar{x}_T(\cdot,\omega)$ denotes the solution of the IVP (6.2) determined by the input $(u_c(\cdot),\mathbf{u}_d) \leftarrow (\bar{u}_{cT}(\cdot,\omega),\bar{\mathbf{u}}_{dt})$. The following equalities hold:

$$\lim_{T\to\infty} \bar{u}_{cT}(t,\omega) = u_c(t,\omega), \quad \text{for all } (t,\omega) \in \mathbb{R}_+ \times \Omega,$$

$$\lim_{T\to\infty} \bar{u}_{dt}(k,\omega) = u_d(k,\omega), \quad \text{for all } (k,\omega) \in \mathbb{Z}_+ \times \Omega,$$

$$\lim_{T\to\infty} \bar{x}_T(t,\omega) = x_{u_c,u_d}(t,\omega), \quad \text{for a.a. } (t,\omega) \in \mathbb{R}_+ \times \Omega.$$

Based on Lebesgue dominate convergence theorem we may infer that

$$\lim_{T\to\infty} J(t_0,\infty,x_0;\bar{u}_{cT}(\cdot),\bar{\mathbf{u}}_{dt}) = J(t_0,\infty,x_0;u_c(\cdot),\mathbf{u}_d), \tag{6.41}$$

for any $(u_c(\cdot),\mathbf{u}_d) \in \mathscr{U}_{ad}(t_0,x_0)$. Employing again the sign conditions (6.26) and (6.27) we deduce that

$$\begin{aligned} J(t_0,\infty,x_0;\bar{u}_{cT}(\cdot),\bar{\mathbf{u}}_{dt}) &\geq J(t_0,T,x_0;u_{cT}(\cdot),\mathbf{u}_{dt}) \\ &= \int_{t_0}^{T} \mathbb{E}[\mathscr{W}_c(t,x_T(t),u_{cT}(t),\eta(t))]dt + \sum_{k=\kappa(t_0)}^{\kappa(T)-1} \mathbb{E}[\mathscr{W}_d(\tau_k,x_T(\tau_k),u_{dt}(k),\eta(\tau_k))] \end{aligned} \tag{6.42}$$

for all $T > t_0$. Lat $\mathbf{X}_T(\cdot) = (X_T(\cdot,1),X_T(\cdot,2),\ldots,X_T(\cdot,N))$ be the solution of the SJMRDE of type (5.10) associated to the controlled JLSS (6.2) and the quadratic functional (6.25) which is satisfying the terminal conditions $X_T(T,i) = 0, \ \ i \in \mathscr{N}$. According to Corollary 5.1, $\mathbf{X}_T(\cdot)$ is defined on the whole interval $[t_0,T]$ and it satisfies

$$X_T(t,i) \geq 0, \quad \text{for all } (t,i) \in [t_0,T] \times \mathscr{N}, \quad T > t_0.$$

Applying Theorem 6.1 we obtain that

$$J(t_0,T,x_0;u_{cT}(\cdot),\mathbf{u}_{dt}) \geq \mathbb{E}[x_0^\top X_T(t_0^-,\eta(t_0))x_0]. \tag{6.43}$$

Combining (6.42) and (6.43) we get

$$J(t_0,\infty,x_0;\bar{u}_{cT}(\cdot),\bar{\mathbf{u}}_{dt}) \geq \mathbb{E}[x_0^\top X_T(t_0^-,\eta(t_0))x_0], \tag{6.44}$$

for all $T > t_0$ and any $(u_c(\cdot),\mathbf{u}_d) \in \mathscr{U}_{ad}(t_0,x_0)$. Further (5.184), (6.41), (6.44) give

$$J(t_0, \infty, x_0; u_c(\cdot), \mathbf{u}_d) \geq \mathbb{E}[x_0^\top X_{min}(t_0^-, \eta(t_0))x_0], \tag{6.45}$$

for any $(u_c(\cdot), \mathbf{u}_d) \in \mathcal{U}_{\text{ad}}(t_0, x_0)$. Particularly, for the control law $\widetilde{u}_c(\cdot)$, $\widetilde{\mathbf{u}}_d$ defined by (6.37)–(6.38) the following holds:

$$J(t_0, \infty, x_0; \widetilde{u}_c(\cdot), \widetilde{\mathbf{u}}_d) \geq \mathbb{E}[x_0^\top X_{min}(t_0^-, \eta(t_0))x_0].$$

Bearing in mind (6.40) and the last inequality we may conclude that

$$J(t_0, \infty, x_0; \widetilde{u}_c(\cdot), \widetilde{\mathbf{u}}_d) = \mathbb{E}[x_0^\top X_{min}(t_0^-, \eta(t_0))x_0]. \tag{6.46}$$

From (6.45) and (6.46) it follows that the control law $(\widetilde{u}_c(\cdot), \widetilde{\mathbf{u}}_d)$ defined in (6.37)–(6.38) is satisfying the optimality condition (6.36). The equality (6.46) confirms the validity of (6.39). Thus the proof ends.

6.4 Infinite Time Horizon Jump Stochastic Undefined Linear Quadratic Control Problem with Stabilization

In this section we study a new optimal control problem which asks for the minimization of the cost function (6.25) along the trajectories of the controlled jump linear stochastic system (6.2) determined by the controls from another set of admissible controls than that involved in the optimal control problem analyzed in the previous section.

So, in this section the set $\widetilde{\mathcal{U}}_{\text{ad}}(t_0, x_0)$ of the admissible controls consists of all pairs $(u_c(\cdot), \mathbf{u}_d)$ where $u_c(\cdot) \in \mathcal{L}^2_{\mathcal{H}}((t_0, \infty); \mathbb{R}^{m_c})$ and $\mathbf{u}_d = \{u_d(k)\}_{k \geq \kappa(t_0)} \in \ell^2_{\mathcal{H}}\{\kappa(t_0), \infty; \mathbb{R}^{m_d}\}$ which are satisfying the conditions

$$J(t_0, \infty, x_0; u_c(\cdot), \mathbf{u}_d) < \infty, \tag{6.47}$$

$$\lim_{t \to \infty} \mathbb{E}[|x_{u_c,u_d}(t; t_0, x_0)|^2 \mid \eta(t_0) = j] = 0, \tag{6.48}$$

$j \in \mathcal{N}$, $x_{u_c,u_d}(t; t_0, x_0)$, $t \geq t_0$ is the trajectory of IVP (6.2) determined by the control $(u_c(\cdot), \mathbf{u}_d)$. In the developments from this section the weighting matrices from (6.25) do not necessarily satisfy the sign conditions (6.26) and (6.27). From these reasons, the optimal control problem studied in this section is named *jump stochastic undefined linear quadratic control problem.*

In the absence of the sign conditions of the form (6.26)–(6.27), the existence of a finite lower bound as (6.28) of the cost functional is not guarantee. In the first part of this section we emphasize a set of conditions which guarantee that the lower bound of the cost functional (6.25) is finite. For $(t_0, x_0) \in \mathbb{R}_+ \times \mathbb{R}^n$ we define

$$\mathbf{v}(t_0, x_0) = \inf_{(u_c(\cdot), \mathbf{u}_d) \in \widetilde{\mathcal{U}}_{\text{ad}}(t_0, x_0)} J(t_0, \infty, x_0; u_c(\cdot), \mathbf{u}_d). \tag{6.49}$$

The function $\mathbf{v}(\cdot, \cdot)$ defined above is the *value function* of the optimal control problem under consideration.

Definition 6.1 The optimal control problem described by the cost functional (6.25), the controlled JLSS (6.2) and the set of admissible controls $\widetilde{\mathscr{U}}_{\mathrm{ad}}(t_0, x_0)$ is **well posed** if $-\infty < \mathbf{v}(t_0, x_0) < \infty$, for all $(t_0, x_0) \in \mathbb{R}_+ \times \mathbb{R}^n$.

Remark 6.4 In this section whenever we refer to the SJMRDE (5.10) we understand that it is described with the help of the matrix coefficients of the JLSS (6.2) and the weighting matrices arising in (6.25).

The next result provides a set of conditions which guarantee that the optimal control problem under consideration is well posed.

Theorem 6.3 *Assume:*

(a) the Assumptions 1.2, 2.1 are fulfilled;
(b) the matrix coefficients of the JLSS (6.2) and the weighting matrices from the quadratic functional (6.25) satisfy the conditions from the Assumption 5.2;
(c) the controlled JLSS (6.2) is mean-square stabilizable by linear state feedback;
(d) the set $\Gamma^{\Sigma}[\mathbb{R}_+]$ associated to the SJMRDE (5.10) is nonempty.

Under these conditions, the optimal control problem described by the cost functional (6.25), the controlled JLSS (6.2) and the set of admissible controls $\widetilde{\mathscr{U}}_{\mathrm{ad}}(t_0, x_0)$ is well posed. Moreover,

$$\mathbf{v}(t_0, x_0) = \mathbb{E}[x_0^\top X_{max}(t_0^-, \eta(t_0))x_0] \tag{6.50}$$

for all $(t_0, x_0) \in \mathbb{R}_+ \times \mathbb{R}^n$, $\mathbf{X}_{max}(\cdot) = (X_{max}(\cdot, 1), X_{max}(\cdot, 2), \ldots, X_{max}(\cdot, N))$ being the unique maximal and bounded solution of the SJMRDE (5.10).
In (6.50), $X_{max}(t_0^-, \eta(t_0))$ coincides to $X_{max}(t_0, \eta(t_0))$ whenever $t_0 < \tau_{\kappa(t_0)}$.

Proof First, let us remark that $\widetilde{\mathscr{U}}_{\mathrm{ad}}(t_0, x_0)$ is not empty for any $(t_0, x_0) \in \mathbb{R}_+ \times \mathbb{R}^n$ if the assumption (c) holds. Further, the Remark 5.4 and Theorem 5.3 together with Remark 6.4 allow us to deduce that under the considered assumptions the SJMRDE (5.10) has a maximal and bounded solution $t \to X_{max}(t) : \mathbb{R}_+ \to \mathscr{S}_n^N$ which is satisfying the sign conditions of type (5.68). Applying Lemma 6.1 with $\mathbf{X}(\cdot)$ replaced by $\mathbf{X}_{max}(\cdot)$, we obtain:

$$\begin{aligned}
&\int_{t_0}^{T} \mathbb{E}[\mathscr{W}_c(t, x_{u_c,u_d}(t), u_c(t), \eta(t)) \mid \eta(t_0) = j]\mathrm{d}t \\
&\quad + \sum_{k=\kappa(t_0)}^{\kappa(T)-1} \mathbb{E}[\mathscr{W}_d(\tau_k, x_{u_c,u_d}(\tau_k), u_d(k), \eta(\tau_k)) \mid \eta(t_0) = j] \\
&= x_0^\top X_{max}(t_0^-, j)x_0 - \mathbb{E}[x_{u_c,u_d}^\top(T)X_{max}(T, \eta(T))x_{u_c,u_d}(T) \mid \eta(t_0) = j]
\end{aligned}$$

$$+\int_{t_0}^{T}\mathbb{E}[(u_c(t)-\widetilde{\widetilde{F}}_c(t,\eta(t))x_{u_c,u_d}(t))^\top\mathbb{R}_c(t,X_{max}(t,\eta(t)),\eta(t))$$
$$\cdot(u_c(t)-\widetilde{\widetilde{F}}_c(t,\eta(t))x_{u_c,u_d}(t))\mid\eta(t_0)=j]\mathrm{d}t \tag{6.51}$$
$$+\sum_{k=\kappa(t_0)}^{\kappa(T)-1}\mathbb{E}[(u_d(k)-\widetilde{\widetilde{F}}_d(\tau_k,\eta(\tau_k))x_{u_c,u_d}(\tau_k))^\top\mathbb{R}_d(\tau_k,X_{max}(\tau_k,\eta(\tau_k),\eta(\tau_k))$$
$$\cdot(u_d(k)-\widetilde{\widetilde{F}}_d(\tau_k,\eta(\tau_k))x_{u_c,u_d}(\tau_k))\mid\eta(t_0)=j],$$

for all $T>t_0$, $j\in\mathcal{N}$, $\widetilde{\widetilde{F}}_c(t,i)$, $\widetilde{\widetilde{F}}_d(\tau_k,i)$ being computed as in (6.6), but with $X(\cdot,i)$ replaced by $X_{max}(\cdot,i)$, $i\in\mathcal{N}$. Since $\mathbf{X}_{max}(\cdot)$ is a bounded function, we deduce that there exists a constant $\widetilde{c}>0$ such that $|X_{max}(T,i)|\leq\widetilde{c}$ and $|X_{max}(T^-,i)|\leq\widetilde{c}$ for all $T\geq 0$ and $i\in\mathcal{N}$. So, we get:

$$|\mathbb{E}[x^\top_{u_c,u_d}(T)X_{max}(T,\eta(T))x_{u_c,u_d}(T)\mid\eta(t_0)=j]|\leq\widetilde{c}\,\mathbb{E}[|x_{u_c,u_d}(T)|^2\mid\eta(t_0)=j)]$$

and

$$|\mathbb{E}[x^\top_{u_c,u_d}(T)X_{max}(T^-,\eta(T))x_{u_c,u_d}(T)\mid\eta(t_0)=j]|\leq\widetilde{c}\mathbb{E}[|x_{u_c,u_d}(T)|^2\mid\eta(t_0)=j)].$$

Bearing in mind (6.48) one gets

$$\lim_{T\to\infty}\mathbb{E}[x^\top_{u_c,u_d}(T)X_{max}(T,\eta(T))u_{u_c,u_d}(T)\mid\eta(t_0)=j]=0$$

and

$$\lim_{T\to\infty}\mathbb{E}[x^\top_{u_c,u_d}(T)X_{max}(T^-,\eta(T))u_{u_c,u_d}(T)\mid\eta(t_0)=j]=0.$$

Taking the limit $T\to\infty$ in (6.51) we obtain via (6.47) that

$$J(t_0,\infty,x_0;u_c(\cdot),\mathbf{u}_d)=\mathbb{E}[x_0^\top X_{max}(t_0^-,\eta(t_0))x_0]+\int_{t_0}^{\infty}\mathbb{E}[(u_c(t)-\widetilde{\widetilde{F}}_c(t,\eta(t))x_{u_c,u_d}(t))^\top$$
$$\cdot\mathbb{R}_c(t,X_{max}(t,\eta(t)),\eta(t))(u_c(t)-\widetilde{\widetilde{F}}_c(t,\eta(t))x_{u_c,u_d}(t))]\mathrm{d}t \tag{6.52}$$
$$+\sum_{k=\kappa(t_0)}^{\infty}\mathbb{E}[(u_d(k)-\widetilde{\widetilde{F}}_d(\tau_k,\eta(\tau_k))x_{u_c,u_d}(\tau_k))^\top\mathbb{R}_d(\tau_k,X_{max}(\tau_k,\eta(\tau_k)),\eta(\tau_k))$$
$$\cdot(u_d(k)-\widetilde{\widetilde{F}}_d(\tau_k,\eta(\tau_k))x_{u_c,u_d}(\tau_k))]$$

for any $(u_c(\cdot),\mathbf{u}_d)\in\widetilde{\widetilde{\mathcal{U}}}_{\mathrm{ad}}(t_0,x_0)$. Combining (5.68), (6.47), (6.52) we get

$$\mathbb{E}[x_0^\top X_{max}(t_0^-,\eta(t_0))x_0]\leq J(t_0,\infty,x_0;u_c(\cdot),\mathbf{u}_d)<\infty$$

for any $(u_c(\cdot), \mathbf{u}_d) \in \widetilde{\mathscr{U}}_{\rm ad}(t_0, x_0)$, for all $(t_0, x_0) \in \mathbb{R}_+ \times \mathbb{R}^n$. This confirms that the optimal control problem under consideration is well posed and

$$\mathbf{v}(t_0, x_0) \geq \mathbb{E}[x_0^\top X_{max}(t_0^-, \eta(t_0))x_0] \tag{6.53}$$

for all $(t_0, x_0) \in \mathbb{R}_+ \times \mathbb{R}^n$. We have to show that in (6.53) is in fact an equality. To this end, we choose a sequence $\{\delta_q\}_{q\geq 0}$ with the property that $\delta_q > \delta_{q+1} > 0$ for all $q \geq 0$ and $\lim_{q\to\infty} \delta_q = 0$. For each $q \geq 0$ we consider the following disturbance of (6.25):

$$J_{\delta_q}(t_0, \infty, x_0; u_c(\cdot), \mathbf{u}_d) = J(t_0, \infty, x_0; u_c(\cdot), \mathbf{u}_d) + \delta_q \sum_{k=\kappa(t_0)}^{\infty} \mathbb{E}[|x_{u_c,u_d}(\tau_k)|^2] \tag{6.54}$$

$(u_c(\cdot), \mathbf{u}_d) \in \widetilde{\mathscr{U}}_{\rm ad}(t_0, x_0)$.

We consider the perturbed SJMRDE:

$$-\dot{X}(t,i) = \mathscr{R}_c(t, \mathbf{X}(t))(i), \quad \tau_k \leq t < \tau_{k+1} \tag{6.55a}$$

$$X(\tau_k^-, i) = \mathscr{R}_d(\tau_k, \mathbf{X}(\tau_k))(i) + \delta_q I_n, \quad k \in \mathbb{Z}_+, \tag{6.55b}$$

$i \in \mathcal{N}$, where $\mathscr{R}_c(t, \mathbf{X}(t))(i)$ and $\mathscr{R}_d(\tau_k, \mathbf{X}(\tau_k))(i)$ are described by right hand side of (5.10a) and (5.10b), respectively, associated to the quadratic functional (6.25) and to the controlled JLSS (6.2). Let Σ_{δ_q} be the quintuple of the form (5.15) associated to the SJMRDE (6.55). One check via (5.159)–(5.162) that if $\mathbf{Z}(\cdot) = (Z(\cdot, 1), Z(\cdot, 2), \ldots, Z(\cdot, N)) \in \Gamma^{\Sigma}[\mathbb{R}_+]$ then $\mathbf{Z}(\cdot) \in \widetilde{\Gamma}^{\Sigma_{\delta_q}}[\mathbb{R}_+]$. Invoking Corollary 5.7 in the case of the pair formed by the controlled system (6.2) and the perturbed SJMRDE (6.55) we deduce that (6.55) has a bounded and stabilizing solution $\mathbf{X}_s^{\delta_q}(\cdot) = (X_s^{\delta_q}(\cdot, 1), X_s^{\delta_q}(\cdot, 2), \ldots, X_s^{\delta_q}(\cdot, N)) : \mathbb{R}_+ \to \mathscr{S}_n^N$ satisfying the sign conditions of the form (5.165). According to Proposition 5.2 for each $q \geq 0$, $\mathbf{X}_s^{\delta_q}(\cdot)$ coincides to the maximal solution of the SJMRDE (6.55). Thus, Theorem 5.4 applied in the case of the JSMRDE (6.55) gives

$$X_s^{\delta_q}(t_0, i) \geq X_s^{\delta_{q+1}}(t_0, i) \geq X_{max}(t_0, i)$$

and

$$\lim_{q\to\infty} X_s^{\delta_q}(t_0, i) = X_{max}(t_0, i) \tag{6.56a}$$

$$\lim_{q\to\infty} X_s^{\delta_q}(t_0^-, i) = X_{max}(t_0^-, i) \tag{6.56b}$$

for all $(t_0, i) \in \mathbb{R}_+ \times \mathcal{N}$. Let

$$u_c^{\delta_q}(t) \triangleq F_{cs}^{\delta_q}(t, \eta(t))x_{\delta_q}(t), \quad t \in \mathbb{R}_+$$

and

$$\mathbf{u}_d^{\delta_q}(k) \triangleq F_{ds}^{\delta_q}(\tau_k, \eta(\tau_k))x_{\delta_q}(\tau_k), \quad k \in \mathbb{Z}_+,$$

where $F_{cs}^{\delta_q}(t, i)$ and $F_{ds}^{\delta_q}(\tau_k, i)$ are computed as in (5.130) but with $X_s(\cdot, i)$ replaced by $X_s^{\delta_q}(\cdot, i)$, $i \in \mathcal{N}$ and $X_{\delta_q}(t)$ is a solution of the corresponding JSLDE of the form (5.131). Since $x_{u_c^{\delta_q}, u_d^{\delta_q}}(t)$ coincides to $x_{\delta_q}(t)$, we may infer that

$$\lim_{t \to \infty} \mathbb{E}[|x_{u_c^{\delta_q}, u_d^{\delta_q}}(t)|^2 \mid \eta(t_0) = j] = 0.$$

This allows us to obtain $J_{\delta_q}(t_0, \infty, x_0; u_c^{\delta_q}(\cdot), \mathbf{u}_d^{\delta_q}) = \mathbb{E}[x_0^\top X_s^{\delta_q}(t_0^-, \eta(t_0))x_0]$. Then (6.49) and (6.54) yield

$$\mathbf{v}(t_0, x_0) \leq J(t_0, \infty, x_0; u_c^{\delta_q}(\cdot), \mathbf{u}_d^{\delta_q}) = \mathbb{E}[x_0^\top X_s^{\delta_q}(t_0^-, \eta(t_0))x_0]. \tag{6.57}$$

Taking the limit for $q \to \infty$ in (6.57) we obtain via (6.56) that

$$\mathbf{v}(t_0, x_0) \leq \mathbb{E}[x_0^\top X_{max}(t_0^-, \eta(t_0))x_0]. \tag{6.58}$$

From (6.53) and (6.58) we conclude that (6.50) holds for all initial pairs $(t_0, x_0) \in \mathbb{R}_+ \times \mathbb{R}^n$. Thus the proof ends.

Definition 6.2 (a) We say that the optimal control problem described by the performance criterion (6.25), the controlled JLSS (6.2) and the set of the admissible control $\widetilde{\mathcal{U}}_{\mathrm{ad}}(t_0, x_0)$ is *attainable* at $(t_0, x_0) \in \mathbb{R}_+ \times \mathbb{R}^n$ if there exists an admissible control $(\widetilde{\widetilde{u}}_c(\cdot), \widetilde{\widetilde{\mathbf{u}}}_d)$ with the property that

$$\mathbf{v}(t_0, x_0) = J(t_0, \infty, x_0; \widetilde{\widetilde{u}}_c(\cdot), \widetilde{\widetilde{\mathbf{u}}}_d). \tag{6.59}$$

(b) We say that the optimal control problem under consideration is *attainable* if it is attainable at any $(t_0, x_0) \in \mathbb{R}_+ \times \mathbb{R}^n$.

The control $(\widetilde{\widetilde{u}}_c(\cdot), \widetilde{\widetilde{\mathbf{u}}}_d)$ satisfying (6.59) will be named *optimal control* of the optimal control problem under consideration.

The next result provides a set of conditions which are guaranteeing the attainability of the considered optimal control problem.

Proposition 6.1 *Under the assumptions (a)–(c) from Theorem 6.3, if the set* $\widetilde{\Gamma}^{\Sigma}[\mathbb{R}_+]$ *associated to the SJMRDE (5.10) is not empty, then the optimal control problem described by the performance criterion (6.25), the controlled JLSS (6.2) and the set of admissible controls* $\widetilde{\mathcal{U}}_{\mathrm{ad}}(t_0, x_0)$ *is attainable. More precisely, under the considered assumptions, the optimal control* $(\widetilde{\widetilde{u}}_c(\cdot), \widetilde{\widetilde{\mathbf{u}}}_d)$ *is given by:*

$$\widetilde{\widetilde{u}}_c(t) = F_{cs}(t, \eta(t))\widetilde{x}(t), \quad t \in \mathbb{R}_+ \tag{6.60a}$$

$$\widetilde{\widetilde{u}}_d(k) = F_{ds}(\tau_k, \eta(\tau_k))\widetilde{x}(\tau_k), \quad k \in \mathbb{Z}_+ \tag{6.60b}$$

where $F_{cs}(t,i)$ *and* $F_{ds}(\tau_k,i)$ *are the gain matrices associated via (5.130) to the stabilizing solution* $X_s(\cdot)$ *of the SJMRDE (5.10) and* $\widetilde{x}(t)$, $t \ge t_0$ *is the solution of the closed-loop JSLDE (5.131) which is satisfying the initial condition* $\widetilde{x}(t_0) = x_0$.

Proof Invoking Proposition 5.2 and Corollary 5.7, we may infer that under the considered assumptions, the SJMRDE (5.10) associated to the control JLSS (6.2) and the cost functional (6.25) has a stabilizing and bounded solution $t \to \mathbf{X}_s(t) = (X_s(t,1), X_s(t,2), \dots, X_s(t,N)) : \mathbb{R}_+ \to \mathcal{S}_n^N$ and additionally

$$X_s(t,i) = X_{max}(t,i), \quad \text{for all } \ (t,i) \in \mathbb{R}_+ \times \mathcal{N}. \tag{6.61}$$

Writing (6.4) in the case $X(\cdot,i) \leftarrow X_s(\cdot,i)$ we obtain via (6.60), (6.61) that:

$$\begin{aligned} &\int_{t_0}^{T} \mathbb{E}[\mathcal{W}_c(t,\widetilde{x}(t),\widetilde{\widetilde{u}}_c(t),\eta(t)) \mid \eta(t_0)=j]\mathrm{d}t \\ &\quad + \sum_{k=\kappa(t_0)}^{\kappa(T)-1} \mathbb{E}[\mathcal{W}_d(\tau_k,\widetilde{x}(\tau_k),\widetilde{\widetilde{u}}_d(k),\eta(\tau_k)) \mid \eta(t_0)=j] \\ &= x_0^\top X_{max}(t_0^-,j)x_0 - \mathbb{E}[\widetilde{x}^\top(T)X_{max}(T,\eta(T))\widetilde{x}(T) \mid \eta(t_0)=j], \end{aligned} \tag{6.62}$$

for all $T > t_0$, $j \in \mathcal{N}$, $(t_0,x_0) \in \mathbb{R}_+ \times \mathbb{R}^n$. As in the proof of the previous theorem we deduce that there exists $\widetilde{c} > 0$ such that

$$|\mathbb{E}[\widetilde{x}^\top(T)X_{max}(T,\eta(T))\widetilde{x}(T) \mid \eta(t_0)=j]| \le \widetilde{c}\,\mathbb{E}[|\widetilde{x}(T)|^2 \mid \eta(t_0)=j].$$

This allows us to deduce that

$$\lim_{T\to\infty} \mathbb{E}[\widetilde{x}^\top(T)X_{max}(T,\eta(T))\widetilde{x}(T) \mid \eta(t_0)=j] = 0$$

for all $j \in \mathcal{N}$. Taking the limit for $T \to \infty$ in (6.62) we get

$$J(t_0,\infty,x_0;\widetilde{\widetilde{u}}_c(\cdot),\widetilde{\widetilde{\mathbf{u}}}_d) = \mathbb{E}[x_0^\top X_{max}(t_0^-,\eta(t_0))x_0]. \tag{6.63}$$

From (6.50) and (6.63) we may conclude that the control law (6.60) satisfies (6.59) for all $(t_0,x_0) \in \mathbb{R}_+ \times \mathbb{R}^n$. Thus the proof is complete.

Further, we shall emphasize a set of necessary and sufficient conditions for the attainability of the optimal control problem under consideration.

Theorem 6.4 *Under the assumptions of Theorem 6.3, the following are equivalent:*

(i) the optimal control problem described by the performance criterion (6.25), the controlled JLSS (6.2) and the set of the admissible controls $\widetilde{\mathcal{U}}_{\rm ad}(t_0, x_0)$ is attainable;
(ii) the jump stochastic linear differential equation

$$dx(t) = (A_0(t, \eta(t)) + B_0(t, \eta(t))\widetilde{\widetilde{F}}_c(t, \eta(t)))x(t)dt \tag{6.64a}$$
$$+ \sum_{l=1}^{r_1}(A_l(t, \eta(t)) + B_l(t, \eta(t))\widetilde{\widetilde{F}}_c(t, \eta(t)))x(t)dw_l(t), \quad \tau_k < t \le \tau_{k+1}$$

$$x(\tau_k^+) = (A_{d0}(\tau_k, \eta(\tau_k)) + B_{d0}(\tau_k, \eta(\tau_k))\widetilde{\widetilde{F}}_d(\tau_k, \eta(\tau_k)))x(\tau_k) \tag{6.64b}$$
$$+ \sum_{p=1}^{r_2} w_{dp}(k)(A_{dp}(\tau_k, \eta(\tau_k)) + B_{dp}(\tau_k, \eta(\tau_k))\widetilde{\widetilde{F}}_d(\tau_k, \eta(\tau_k)))x(\tau_k), \quad k \in \mathbb{Z}_+$$

is asymptotically stable in mean square (ASMS), where $\widetilde{\widetilde{F}}_c(t, i)$, $\widetilde{\widetilde{F}}_d(\tau_k, i)$ are computed as in (6.6) with $X(\cdot, i)$ replaced by $X_{max}(\cdot, i)$.

If one of the assumptions (i) or (ii) holds, the optimal control law is unique and it is given by

$$\widetilde{\widetilde{u}}_c(t) = \widetilde{\widetilde{F}}_c(t, \eta(t))\widetilde{\widetilde{x}}(t; t_0, x_0), \quad t \ge t_0 \tag{6.65a}$$
$$\widetilde{\widetilde{u}}_d(k) = \widetilde{\widetilde{F}}_d(\tau_k, \eta(\tau_k))\widetilde{\widetilde{x}}(\tau_k; t_0, x_0), \quad k \ge \kappa(t_0) \tag{6.65b}$$

$\widetilde{\widetilde{x}}(\cdot; t_0, x_0)$ being the solution of the JSLDE (6.64) satisfying $\widetilde{\widetilde{x}}(t_0; t_0, x_0) = x_0$.

Proof Let $(t_0, x_0) \in \mathbb{R}_+ \times \mathbb{R}^n$ arbitrary but fixed. If (i) holds, there exists $(\widehat{u}_c(\cdot), \widehat{\mathbf{u}}_d) \in \widetilde{\mathcal{U}}_{\rm ad}(t_0, x_0)$ which is satisfying

$$\mathbf{v}(t_0, x_0) = J(t_0, \infty, x_0; \widehat{u}_c(\cdot), \widehat{\mathbf{u}}_d). \tag{6.66}$$

In this case (6.52) gives

$$J(t_0, \infty, x_0; \widehat{u}_c(\cdot), \widehat{\mathbf{u}}_d) = \mathbb{E}[x_0^\top X_{max}(t_0^-, \eta(t_0))x_0] + \int_{t_0}^{\infty} \mathbb{E}[(\widehat{u}_c(t) - \widetilde{\widetilde{F}}_c(t, \eta(t))x_{\widehat{u}_c,\widehat{u}_d}(t))^\top$$
$$\cdot \mathbb{R}_c(t, X_{max}(t, \eta(t)), \eta(t))(\widehat{u}_c(t) - \widetilde{\widetilde{F}}_c(t, \eta(t))x_{\widehat{u}_c,\widehat{u}_d}(t))]dt$$
$$+ \sum_{k=\kappa(t_0)}^{\infty} \mathbb{E}[(\widehat{u}_d(k) - \widetilde{\widetilde{F}}_d(\tau_k, \eta(\tau_k))x_{\widehat{u}_c,\widehat{u}_d}(\tau_k))^\top \mathbb{R}_d(\tau_k, X_{max}(\tau_k, \eta(\tau_k)), \eta(\tau_k))$$
$$\cdot (\widehat{u}_d(k) - \widetilde{\widetilde{F}}_d(\tau_k, \eta(\tau_k))x_{\widehat{u}_c,\widehat{u}_d}(\tau_k))]. \tag{6.67}$$

Employing (6.50) and (6.66) we deduce that (6.67) holds if and only if

$$\int_{t_0}^{\infty} \mathbb{E}[(\widehat{u}_c(t) - \tilde{\tilde{F}}_c(t, \eta(t))x_{\widehat{u}_c,\widehat{u}_d}(t))^\top \mathbb{R}_c(t, X_{max}(t, \eta(t)), \eta(t))(\widehat{u}_c(t)$$

$$- \tilde{\tilde{F}}_c(t, \eta(t))x_{\widehat{u}_c,\widehat{u}_d}(t))]dt + \sum_{k=\kappa(t_0)}^{\infty} \mathbb{E}[(\widehat{u}_d(k) - \tilde{\tilde{F}}_d(\tau_k, \eta(\tau_k))x_{\widehat{u}_c,\widehat{u}_d}(\tau_k))^\top$$

$$\cdot \mathbb{R}_d(\tau_k, X_{max}(\tau_k, \eta(\tau_k)), \eta(t))(\widehat{u}_d(k) - \tilde{\tilde{F}}_d(\tau_k, \eta(\tau_k))x_{\widehat{u}_c,\widehat{u}_d}(\tau_k))] = 0. \quad (6.68)$$

Further, (6.5) together with (5.68) allow us to deduce that (6.68) holds if and only if

$$\widehat{u}_c(t) = \tilde{\tilde{F}}_c(t, \eta(t))x_{\widehat{u}_c,\widehat{u}_d}(t), \quad \text{a. s. } t \geq t_0 \quad (6.69a)$$

$$\widehat{u}_d(k) = \tilde{\tilde{F}}_d(\tau_k, \eta(\tau_k))x_{\widehat{u}_c,\widehat{u}_d}(\tau_k), \quad \text{a. s. } k \geq \kappa(t_0). \quad (6.69b)$$

We recall that $x_{\widehat{u}_c,\widehat{u}_d}(t),\ t \geq t_0$ is the solution of the IVP (6.2) determined by the control law $(\widehat{u}_c(\cdot), \widehat{\mathbf{u}}_d)$ and satisfies

$$\lim_{t\to\infty} \mathbb{E}[|x_{\widehat{u}_c,\widehat{u}_d}(t)|^2 \mid \eta(t_0) = j] = 0, \quad j \in \mathcal{N}. \quad (6.70)$$

Substituting (6.69) in (6.2) and taking into account the unicity of the solution of an IVP we may infer that

$$x_{\widehat{u}_c,\widehat{u}_d}(t) = \tilde{\tilde{x}}(t; t_0, x_0) \text{ a. s.} \quad (6.71)$$

So, (6.70) becomes $\lim_{t\to\infty} \mathbb{E}[|x(t; t_0, x_0)|^2 \mid \eta(t_0) = j] = 0$, for all $(t_0, x_0) \in \mathbb{R}_+ \times \mathbb{R}^n$, $j \in \mathcal{N}$, which shows that (6.66) is ASMS. Thus we have shown that (ii) holds if and only if (i) is true. The formulae (6.65) of the optimal control are obtained substituting (6.71) in (6.69). Thus the proof ends.

It is worth noticing that the asymptotic stability in mean square of the closed-loop JSLDE (6.64) does not guarantee that the maximal solution $\mathbf{X}_{max}(\cdot)$ of the SJMRDE is a stabilizing solution in the sense of the Definition 5.6.

In the sequel, we show that under the assumptions of Theorem 6.1, if the matrix coefficients of the weights matrices from performance criterion (6.25) are time periodic or time invariant, then the property of attainability of the optimal control problem under consideration is equivalent to the fact that maximal solution $\mathbf{X}_{max}(\cdot)$ of the corresponding SJMRDE of the form (5.10) is a stabilizing solution.

Theorem 6.5 *Assume:*

(a) the assumptions from Theorem 6.3 are fulfilled;
(b) the matrix coefficients of the controlled JLSS (6.2) and the weights matrices from the performance criterion (6.25) are satisfying the conditions from Definition 5.8 (i).

Under these conditions, the following are equivalent:

(*i*) *the optimal control problem described by the controlled JLSS (6.2), the performance criterion (6.25) and the set of the admissible controls* $\widetilde{\widetilde{\mathscr{U}}}_{ad}(t_0, x_0)$ *is attainable;*
(*ii*) *the maximal and bounded solution* $\mathbf{X}_{\max}(\cdot)$ *of the corresponding SJMRDE (5.10) is stabilizing solution.*

Proof First, according to Theorem 6.4, the optimal control problem under consideration is attainable if and only if the JSLDE (6.64) with $\tau_k = kh$, $k \in \mathbb{Z}_+$, is ASMS. On the other hand, if the assumption (b) holds, then (5.12) allows us to deduce that the corresponding SJMRDE (5.10) satisfies the conditions from the Definition 5.2 (a). Applying Corollary 5.2 (i), we may infer that the maximal solution $\mathbf{X}_{max}(\cdot)$ of (5.10) associated to the optimal control problem under consideration is a periodic function of period $\tilde{\theta}$, where $\tilde{\theta}$ is computed as in (5.166a) employing the numbers θ_c, θ_d and h, involved in Definition 5.8 (i). Hence, $\widetilde{\widetilde{F}}_c(\cdot, i)$ defined by (6.6a) with $X(\cdot, i)$ replaced by $X_{max}(\cdot, i)$ is a periodic function of period $\tilde{\theta}$, while, $\widetilde{\widetilde{F}}_d(kh, i)$, $k \in \mathbb{Z}_+$, defined by (6.68) with $X(\tau_k, i)$ replaced by $X_{max}(kh, i)$ is a periodic sequence of period $\tilde{\mathbf{k}}$ with $\tilde{\mathbf{k}}$ introduced in (5.166b). This means that under the assumption (b) from the statement, the JSLDE (6.64) with $\tau_k = kh$, $k \in \mathbb{Z}_+$, satisfies the conditions from Definition 3.2, because $\frac{\tilde{\theta}}{\tilde{\mathbf{k}}h}$ is a rational number. The implication $(iv) \Rightarrow (i)$ from Theorem 3.1 allows us to deduce that the system (6.64) is SESMS. This means that the maximal solution $\mathbf{X}_{max}(\cdot)$ of (5.10) is stabilizing solution in the sense of Definition 5.6. Thus we have shown that the implication $(i) \Rightarrow (ii)$ holds. The converse implication follows from Proposition 6.1. Thus the proof is complete.

Consider now the JLSS:

$$\begin{aligned} dx(t) &= (A_0(\eta(t))x(t) + B_0(\eta(t))u_c(t))dt \\ &\quad + \sum_{l=1}^{r_1}(A_l(\eta(t))x(t) + B_l(\eta(t))u_c(t))dw_l(t), \quad kh < t \leq (k+1)h \end{aligned} \tag{6.72a}$$

$$\begin{aligned} x(kh^+) &= A_{d0}(\eta(kh))x(kh) + B_{d0}(\eta(kh))u(k) \\ &\quad + \sum_{p=1}^{r_2} w_{dp}(k)(A_{dp}(\eta(kh))x(kh) + B_{dp}(\eta(kh))u_d(k)) \end{aligned} \tag{6.72b}$$

$$x(t_0) = x_0 \tag{6.72c}$$

which is the time-invariant version of (6.2).

We also consider the following time-invariant version of (6.25):

$$\begin{aligned} J(t_0, \infty, x_0; u_c(\cdot), \mathbf{u}_d) &= \int_{t_0}^{\infty} \mathbb{E}[x_{u_c,u_d}^{\top}(t)M_c(\eta(t))x_{u_c,u_d}(t) \\ &\quad + 2x_{u_c,u_d}^{\top}(t)L_c(\eta(t))u_c(t) + u_c^{\top}(t)R_c(\eta(t))u_c(t)]dt \end{aligned}$$

$$+ \sum_{k=\kappa(t_0)}^{\infty} \mathbb{E}[x_{u_c,u_d}^{\top}(kh)M_d(\eta(kh))x_{u_c,u_d}(kh) \tag{6.73}$$
$$+ 2x_{u_c,u_d}^{\top}(kh)L_d(\eta(kh))u_d(k) + u_d^{\top}(k)R_d(\eta(kh))u_d(k)]$$

where $M_c(i),\ M_d(i) \in \mathcal{S}_n,\ R_c(i) \in \mathcal{S}_{m_c},\ R_d(i) \in \mathcal{S}_{m_d},\ i \in \mathcal{N}$. The set of the admissible controls, in the case of the optimal control problem which is asking for the minimization of the quadratic functional (6.73) along the trajectories of the controlled system (6.72), coincides to $\widetilde{\mathcal{U}}_{\text{ad}}(t_0, x_0)$ specifying that in this case (6.47) refers to the performance criterion (6.73) and (6.48) describes the behavior of the trajectories of the JLSS (6.72).

The stochastic jump matrix Riccati differential equation (SJMRDE) involved in the solution of the optimal control problem described by the performance criterion (6.73) and the controlled system (6.72) is:

$$-\dot{X}(t,i) = \mathcal{R}_c(\mathbf{X}(t))(i),\quad kh \le t < (k+1)h \tag{6.74a}$$
$$X(kh^-, i) = \mathcal{R}_d(\mathbf{X}(kh))(i),\quad k \in \mathbb{Z}_+, \tag{6.74b}$$

$i \in \mathcal{N}$, $\mathcal{R}_c(\mathbf{X}(t))(i)$ and $\mathcal{R}_d(\mathbf{X}(kh))(i)$ being described by the right hand side of (5.10a) and (5.10b), respectively, with the updates:

$$A_l(t,i) \leftarrow A_l(i),\quad B_l(t,i) \leftarrow B_l(i),\quad 0 \le l \le r_1$$

$$M_c(t,i) \leftarrow M_c(i),\quad L_c(t,i) \leftarrow L_c(i),\quad R_c(t,i) \leftarrow R_c(i),\quad t \in \mathbb{R}_+$$

$$A_{dp}(\tau_k,i) \leftarrow A_{dp}(i),\quad B_{dp}(\tau_k,i) \leftarrow B_{dp}(i),\quad 0 \le p \le r_2,$$

$$M_d(\tau_k,i) \leftarrow M_d(i),\quad L_d(\tau_k,i) \leftarrow L_d(i),\quad R_d(\tau_k,i) \leftarrow R_d(i),$$

for all $\tau_k \in \mathfrak{T}$, $i \in \mathcal{N}$.

Theorem 6.6 *Assume:*

(a) the Assumptions 1.2, 2.1 hold;
(b) the controlled JLSS (6.72) is mean-square stabilizable by linear state feedback;
(c) the set $\Gamma^{\Sigma}[\mathbb{R}_+]$ associated to the SJMRDE (6.74) is nonempty.

Under these conditions the following are equivalent:

(i) the optimal control problem described by the performance criterion (6.73), the controlled JLSS (6.72) and the set of the admissible controls $\widetilde{\mathcal{U}}_{\text{ad}}(t_0, x_0)$ is attainable;
(ii) the maximal and bounded solution $\mathbf{X}_{max}(\cdot)$ of the SJMRDE (6.74) is stabilizing solution. If one of the above conditions hold, then the optimal control law is unique and is given by:

$$\tilde{\tilde{u}}_c(t) = F_{cs}(t, \eta(t))\tilde{\tilde{x}}(t),\quad t \ge t_0 \tag{6.75a}$$

$$\widetilde{\widetilde{u}}_d(k) = F_{ds}(\eta(kh))\widetilde{\widetilde{x}}(kh), \quad k \geq \kappa(t_0) \tag{6.75b}$$

$\widetilde{\widetilde{x}}(t)$, $t \geq t_0$, being the solution of the IVP obtained substituting (6.75) in (6.72) and the gain matrices $F_{cs}(t, i)$, $F_{ds}(i)$ are defined by:

$$F_{cs}(t, i) \triangleq -\left(\sum_{l=1}^{r_1} B_l^\top(i)X_{max}(t, i)B_l(i) + R_c(i)\right)^{-1}$$

$$\cdot\left(B_0^\top(i)X_{max}(t, i) + \sum_{l=1}^{r_1} B_l^\top(i)X_{max}(t, i)A_l(i) + L_c^\top(i)\right) \tag{6.76a}$$

$$F_{ds}(i) \triangleq -\left(\sum_{p=0}^{r_2} B_{dp}^\top(i)X_{max}(0, i)B_{dp}(i) + R_d(i)\right)^{-1}$$

$$\cdot\left(\sum_{p=0}^{r_2} B_{dp}^\top(i)X_{max}(0, i)A_{dp}(i) + L_d^\top(i)\right), \quad i \in \mathcal{N}. \tag{6.76b}$$

Proof may be done following the line of the proof of Theorem 6.5. The details are left to the reader.

It is worth noticing that according to Corollary 5.2 the maximal solution $\mathbf{X}_{max}(\cdot)$ of the SJMRDE (6.74) is a periodic function of period h. Consequently, $F_{cs}(\cdot, i)$ are periodic functions of period h, too. That is why, it is sufficient to compute the gain matrices $F_{cs}(\cdot, i)$ on the interval $[0, h]$ only.

6.5 The Linear Quadratic Optimal Control Problems for Sampled-Data Linear Stochastic Systems

Let us consider the stochastic linear controlled system described by:

$$\begin{aligned} dx(t) &= (A_0(t, \eta(t))x(t) + B_0(t, \eta(t)))dt \\ &+ \sum_{l=1}^{r}(A_l(t, \eta(t))x(t) + B_l(t, \eta(t)))dw_l(t), \quad t \geq t_0 \\ x(t_0) &= x_0 \end{aligned} \tag{6.77}$$

where $x(t) \in \mathbb{R}^n$ are the state parameters, while $u(t) \in \mathbb{R}^m$ are the control parameters. In (6.77), $\{w(t)\}_{t\geq 0}$ $(w(t) = (w_1(t)\ w_2(t)\ \ldots\ w_{r_2}(t))^\top)$ are $r-$dimensional standard Wiener process and $\{\eta(t)\}_{t\geq 0}$ is a standard homogeneous Markov process with the state space $\mathcal{N} = \{1, 2 \ldots, N\}$. We assume that $\{w(t)\}_{t\geq 0}$, $\{\eta(t)\}_{t\geq 0}$ are independent stochastic processes defined on the probability space $(\Omega, \mathfrak{F}, \mathcal{P})$.

We consider the performance criterion:

$$\mathbf{J}(t_0, \infty, x_0; u(\cdot)) = \int_{t_0}^{\infty} \mathbb{E}[x_u^\top(t)M(t)x_u(t) + u^\top(t)R(t)u(t)]dt \tag{6.78}$$

where $x_u(t)$, $t \geq t_0$ is the solution of the IVP (6.77) determined by the control $u(\cdot)$.

We assume that $t \to (A_l(t,i), B_l(t,i)) : \mathbb{R}_+ \to \mathbb{R}^{n\times n} \times \mathbb{R}^{n\times m}$, $0 \leq l \leq r$ and $t \to (M(t), R(t)) : \mathbb{R}_+ \to \mathcal{S}_n \times \mathcal{S}_m$ are bounded and continuous matrix-valued functions.

In this chapter from Drăgan et al. [1] was described a framework which guarantee the existence of a control law in a state feedback form

$$u_{\text{opt}}(t) = F_{\text{opt}}(t, \eta(t))x(t) \tag{6.79}$$

which minimize the cost functional (6.78) over a suitable set of admissible controls. One sees that to implement a control law as (6.79) it is required that the state parameters to be available for measurements in any instance time $t \in \mathbb{R}_+$.

In this section, we want to solve the problem of the minimizing of the cost functional (6.78) along of the trajectories of the controlled system (6.76) under the assumption that the state $x(\cdot)$ of the controlled system are measured at discrete instances time τ_k, $k \in \mathbb{Z}_+$, where $0 = \tau_0 < \tau_1 < \ldots < \tau_k < \ldots$. We assume that between two successive time moments τ_k, τ_{k+1} the value of the control is preserved constant. Note $\mathfrak{T} = \{\tau_k\}_{k\in\mathbb{Z}_+}$. The set of the admissible controls $\mathcal{U}_{\text{ad}}(t_0, x_0; \mathfrak{T})$ associated to the initial pair $(t_0, x_0) \in \mathbb{R}_+ \times \mathbb{R}^n$ and to the partition $\mathfrak{T}$ consists of all stochastic processes

$$u(t) = u(k), \quad \tau_k \leq t < \tau_{k+1}, \quad k \geq \kappa(t_0), \tag{6.80}$$

$\kappa(t_0)$ being defined in (6.15), where $u(k) : \Omega \to \mathbb{R}^m$ are random vectors which are measurable with respect to the $\sigma-$algebra generated by the random vectors $w(s)$, $\eta(s)$, $0 \leq s \leq \tau_k$ and satisfy the conditions

$$\sum_{k=\kappa(t_0)}^{\infty} \mathbb{E}[|u(k)|^2] < \infty \tag{6.81a}$$

$$\int_{t_0}^{\infty} \mathbb{E}[|x_u(t; t_0, x_0)|^2]dt < \infty \tag{6.81b}$$

$$\lim_{t\to\infty} \mathbb{E}[|x_u(t; t_0, x_0)|^2] = 0. \tag{6.81c}$$

It is worth mentioning that if the stochastic controlled system (6.77) is mean-square stabilizable by sampled state measurements (in the sense of Definition 4.5), then for any initial pair (t_0, x_0), the set $\mathcal{U}_{\text{ad}}(t_0, x_0; \mathfrak{T})$ is not empty.

The optimal control problem which we address in this section is asking for the finding of an admissible control $u_{\mathrm{opt}}(\cdot) \in \mathscr{U}_{\mathrm{ad}}(t_0, x_0; \mathfrak{T})$ which satisfies

$$\mathbf{J}(t_0, \infty, x_0; u_{\mathrm{opt}}(\cdot)) \leq \mathbf{J}(t_0, \infty, x_0; u(\cdot)), \tag{6.82}$$

for any $u(\cdot) \in \mathscr{U}_{\mathrm{ad}}(t_0, x_0; \mathfrak{T})$. An admissible control $u_{\mathrm{opt}}(\cdot)$ which is satisfying (6.82) will be named *optimal control.*

In this section, we shall highlight a set of conditions which guarantee the existence of an optimal control in a state feedback form. To this end, we transform the optimal control problem described by the controlled system (6.77), the performance criterion (6.78) and the set of the admissible controls $\mathscr{U}_{\mathrm{ad}}(t_0, x_0; \mathfrak{T})$ into an optimal control problem for a linear system controlled by impulses. Then, we shall use some of the results derived in the previous section, in order to obtain the existence of a control which is satisfying (6.82). Plugging a control (6.80) in (6.76) and (6.78), respectively, we obtain

$$\begin{aligned} dx(t) &= (A_0(t, \eta(t))x(t) + B_0(t, \eta(t))u(k))dt \\ &\quad + \sum_{l=1}^{r}(A_l(t, \eta(t))x(t) + B_l(t, \eta(t))u(k))dw_l(t), \quad \tau_k \leq t < \tau_{k+1} \\ x(t_0) &= x_0 \end{aligned} \tag{6.83}$$

$$\mathbf{J}(t_0, \infty, x_0; u(\cdot)) = \int_{t_0}^{\infty} \mathbb{E}[x_u^{\top}(t)M(t)x_u(t)]dt + \sum_{k=\kappa(t_0)}^{\infty} \mathbb{E}[u^{\top}(k)\int_{\tau_k}^{\tau_{k+1}} R(t)dt\, u(k)]. \tag{6.84}$$

Let $t \to \varphi(t) : \mathbb{R}_+ \to \mathbb{R}^m$ be defined by

$$\varphi(t) := u(k), \quad \tau_k < t \leq \tau_{k+1}, \quad k \in \mathbb{Z}_+. \tag{6.85}$$

One sees that $\varphi(\cdot)$ is differentiable in any $t \neq \tau_k$ and we have $\dot{\varphi}(t) = 0$ and $\varphi(\tau_k^+) = u(k), \quad k \in \mathbb{Z}_+$. If $x(t)$ is a solution of the IVP (6.83) and $\varphi(t)$ is defined by (6.85) then $\mathbf{x}(t) \stackrel{\Delta}{=} (x^{\top}(t), \ \varphi^{\top}(t))^{\top}$ is a solution of the following IVP:

$$d\mathbf{x}(t) = \mathscr{A}_0(t, \eta(t))\mathbf{x}(t)dt + \sum_{l=1}^{r} \mathscr{A}_l(t, \eta(t))\mathbf{x}(t)dw_l(t), \quad \tau_k < t \leq \tau_{k+1} \tag{6.86a}$$

$$\mathbf{x}(\tau_k^+) = A_d\mathbf{x}(\tau_k) + B_d u(k) \tag{6.86b}$$

$$\mathbf{x}(t_0) = \mathbf{x}_0 = (x_0^{\top}, \ 0^{\top})^{\top}, \tag{6.86c}$$

where

$$\mathscr{A}_l(t,i) = \begin{pmatrix} A_l(t,i) & B_l(t,i) \\ 0_{mn} & 0_m \end{pmatrix}, \quad 0 \le l \le r, \tag{6.87}$$
$$A_d = \begin{pmatrix} I_n & 0_{nm} \\ 0_{mn} & 0_m \end{pmatrix}, \quad B_d = \begin{pmatrix} 0_{nm} \\ I_m \end{pmatrix}.$$

We recall that 0_{pq} (0_p) denotes the matrix of size $p \times q$, $(p \times p,$, respectively) with all elements zero. Furthermore, one may check via (6.86), (6.87) that if $\mathbf{x}(t)$ is a solution of the IVP (6.86) determined by the control $\mathbf{u} = \{u(k)\}_{k \ge \kappa(t_0)}$ then the vector $x(t)$ formed with the first n components of $\mathbf{x}(t)$ is a solution of the IVP (6.83). The performance criterion (6.84) may be rewritten in the following equivalent form:

$$\mathscr{J}(t_0, \infty, \mathbf{x}_0; \mathbf{u}) = \int_{t_0}^{\infty} \mathbb{E}[\mathbf{x}_{\mathbf{u}}^\top(t)\mathscr{M}(t)\mathbf{x}_{\mathbf{u}}(t)]dt + \sum_{k=\kappa(t_0)}^{\infty} \mathbb{E}[u^\top(k)\mathscr{R}(k)u(k)]. \tag{6.88}$$

$\mathbf{x}_{\mathbf{u}}(t)$, $t \ge t_0$, being the solution of the IVP (6.86)–(6.87) determined by the input $\mathbf{u} = \{u(k)\}_{k \ge \kappa(t_0)}$. In (6.88) we have denoted:

$$\mathscr{M}(t) = \begin{pmatrix} M(t) & 0_{nm} \\ 0_{nm}^\top & 0_m \end{pmatrix}, \quad \mathscr{R}(k) = \int_{\tau_k}^{\tau_{k+1}} R(t)dt. \tag{6.89}$$

In order to be able to state the new optimal control problem, we introduce a new set of admissible controls associated to the pair formed by the controlled system (6.86) and the performance criterion (6.88). So, for a given initial pair $(t_0, \mathbf{x}_0) \in \mathbb{R}_+ \times \mathbb{R}^{n+m}$ and for the sequence $\mathfrak{T} = \{\tau_k\}_{k \in \mathbb{Z}_+} \subset \mathbb{R}_+$, we denote $\mathscr{U}_{sd}(t_0, \mathbf{x}_0; \mathfrak{T})$ the set of all sequences $\mathbf{u} = \{u(k)\}_{k \ge \kappa(t_0)}$ with the properties:

(a) $u(k) : \Omega \to \mathbb{R}^m$ are arbitrary random vectors which are measurable with respect to the $\sigma-$algebra generated by the random variables $w(s)$, $\eta(s)$, $0 \le s \le \tau_k$;

(b)

$$\sum_{k=\kappa(t_0)}^{\infty} \mathbb{E}[|u(k)|^2] < \infty; \tag{6.90a}$$

$$\int_{t_0}^{\infty} \mathbb{E}[|\mathbf{x}_{\mathbf{u}}(t, t_0; \mathbf{x}_0)|^2]dt < \infty; \tag{6.90b}$$

$$\lim_{t \to \infty} \mathbb{E}[|\mathbf{x}_{\mathbf{u}}(t, t_0; \mathbf{x}_0)|^2] = 0. \tag{6.90c}$$

The problem: *The optimal control problem described by the performance criterion (6.88), the controlled system (6.86) and the set of the admissible controls*

$\mathscr{U}_{sd}(t_0, \mathbf{x}_0; \mathfrak{T})$ ask for the finding an admissible control $\widehat{\mathbf{u}} = \{\widehat{u}(k)\}_{k \geq \kappa(t_0)}$ which satisfies the optimality condition:

$$\mathscr{J}(t_0, \infty, \mathbf{x}_0; \widehat{\mathbf{u}}) \leq \mathscr{J}(t_0, \infty, \mathbf{x}_0; \mathbf{u}), \tag{6.91}$$

for any $\mathbf{u} = \{u(k)\}_{k \geq \kappa(t_0)} \in \mathscr{U}_{sd}(t_0, \mathfrak{x}_0; \mathfrak{T})$.

Further we emphasize a correspondence one to one between the sets $\mathscr{U}_{ad}(t_0, x_0, \mathfrak{T})$ and $\mathscr{U}_{sd}(t_0, \mathbf{x}_0; \mathfrak{T})$ in the case when $\mathbf{x}_0 = (x_0^\top,\ 0^\top)^\top \in \mathbb{R}^{n+m}$.

Lemma 6.4 *Assume that the partition $\mathfrak{T}$ satisfies the conditions from the Assumption 2.1 and the initial state $\mathbf{x}_0$ of the IVP (6.86) is of the form $\mathbf{x}_0 = (x_0^\top,\ 0^\top)^\top$. Let $\gamma : \mathscr{U}_{ad}(t_0, x_0; \mathfrak{T}) \to \mathscr{U}_{sd}(t_0, \mathbf{x}_0; \mathfrak{T})$ be defined by*

$$\gamma(u(\cdot)) = \mathbf{u} = \{u(k)\}_{k \geq \kappa(t_0)}$$

where for each $k \geq \kappa(t_0)$, $u(k) = u(t)$, $\tau_k < t \leq \tau_{k+1}$. Then the following hold:

(i) $\gamma(\cdot)$ is a one to one mapping;
(ii) the performance criterion (6.78) and (6.88) are connected by

$$\mathbf{J}(t_0, \infty, x_0; u(\cdot)) = \mathscr{J}(t_0, \infty, \mathbf{x}_0; \gamma(u(\cdot)))$$

for any $u(\cdot) \in \mathscr{U}_{ad}(t_0, x_0, \mathfrak{T})$;
(iii) if $u_{opt}(\cdot)$ is satisfying the optimality condition (6.82) and $\widehat{\mathbf{u}}$ is satisfying the optimality condition (6.91) then $u_{opt}(\cdot) = \gamma^{-1}(\widehat{\mathbf{u}})$.

Proof Let us show that the mapping $\gamma(\cdot)$ is well defined. To this end, it is sufficient to show that if $u(\cdot)$ is satisfying conditions (6.81) then $\mathbf{u} = \gamma(u(\cdot))$ is satisfying the conditions (6.90). Indeed, (6.90a) is satisfied because it coincides to (6.81a). Further, from $\mathbf{x_u}(t; t_0, \mathbf{x}_0) = (x_u^\top(t; t_0, x_0) \quad \varphi^\top(t; t_0, 0))^\top$ we get

$$\int_{t_0}^{\infty} \mathbb{E}[|\mathbf{x_u}(t; t_0, \mathbf{x}_0)|^2]\mathrm{d}t = \int_{t_0}^{\infty} \mathbb{E}[|x_u(t; t_0, x_0)|^2]\mathrm{d}t + \int_{t_0}^{\infty} \mathbb{E}[|\varphi(t; t_0, 0)|^2]\mathrm{d}t. \tag{6.92}$$

On the other hand

$$\int_{t_0}^{\infty} \mathbb{E}[|\varphi(t; t_0, 0)|^2]\mathrm{d}t = \sum_{k=\kappa(t_0)}^{\infty} (\tau_{k+1} - \tau_k)\mathbb{E}[|u(k)|^2] \leq \bar{h} \sum_{k=\kappa(t_0)}^{\infty} \mathbb{E}[|u(k)|^2], \tag{6.93}$$

where $\bar{h}$ is provided by (2.26). Employing (6.81a), (6.81b) and (6.93) we may deduce from (6.92) that (6.90b) holds. Further from (6.85) and (6.90a) we get

$$\lim_{t \to \infty} \mathbb{E}[|\varphi(t; t_0, 0)|^2] = 0. \tag{6.94}$$

Bearing in mind that $\mathbb{E}[|\mathbf{x}_\mathbf{u}(t; t_0, \mathbf{x}_0)|^2] = \mathbb{E}[|x_u(t; t_0, x_0)|^2] + \mathbb{E}[|\varphi(t; t_0, 0)|^2]$, we obtain from (6.81c) and (6.94) that (6.90c) holds too.

Let $\widehat{\gamma} : \mathscr{U}_{\rm sd}(t_0, \mathbf{x}_0; \mathfrak{T}) \to \mathscr{U}_{\rm ad}(t_0, x_0; \mathfrak{T})$ be defined by

$$u(t) = \widehat{\gamma}(\mathbf{u})(t) = u(k)$$

if $\tau_k \leq t < \tau_{k+1}$, $k \geq \kappa(t_0)$, when $\mathbf{u} = \{u(k)\}_{k \geq \kappa(t_0)}$. One may check that $u(t) = \widehat{\gamma}(\mathbf{u})(t)$, $t \geq t_0$, satisfies conditions (6.81) if $\mathbf{u}$ satisfies (6.90). Additionally we have $\widehat{\gamma}(\gamma(u(\cdot))) = u(\cdot)$ for any $u(\cdot) \in \mathscr{U}_{\rm ad}(t_0, x_0, \mathfrak{T})$ and $\gamma(\widehat{\gamma}(\mathbf{u})) = \mathbf{u}$ for all $\mathbf{u} \in \mathscr{U}_{\rm sd}(t_0, \mathbf{x}_0, \mathfrak{T})$. This means that $\gamma(\cdot)$ is a one to one mapping which confirms the validity of the assertion (i). The validity of (ii) is obtained from (6.78), (6.84), (6.88) together with the definition of the mapping $\gamma(\cdot)$. The equality (iii) is obtained using (ii) in (6.91). Thus the proof ends.

The equality from (iii) in the statement of Lemma 6.4 allows us to obtain the optimal control $u_{\rm opt}(\cdot)$ for the optimal control problem (6.82) looking for a control $\hat{\mathbf{u}}$ from $\mathscr{U}_{\rm sd}(t_0, \mathbf{x}_0, \mathfrak{T})$ which is satisfying the optimality condition (6.91).

To this end, we shall use the results derived in the previous section. In the particular case of the IVP (6.86) and the quadratic functional (6.88), the SJMRDE (5.10) reduces to the following backward jump matrix linear differential equation (BJMLDE) with Riccati-type jumping operator:

$$\begin{aligned}
\dot{X}(t, i) &= \mathscr{A}_0^\top(t, i)X(t, i) + X(t, i)\mathscr{A}_0(t, i) + \sum_{l=1}^{r} \mathscr{A}_l^\top(t, i)X(t, i)\mathscr{A}_l(t, i) \\
&\quad + \sum_{j=1}^{N} q_{ij}X(t, j) + \mathscr{M}(t), \quad \tau_k \leq t < \tau_{k+1} \qquad (6.95a) \\
X(\tau_k^-, i) &= A_d^\top X(\tau_k, i)A_d - A_d^\top X(\tau_k, i)B_d(\mathscr{R}(k) + B_d^\top X(\tau_k, i)B_d)^{-1} \\
&\quad \cdot B_d^\top X(\tau_k, i)A_d, \quad k \in \mathbb{Z}_+. \qquad (6.95b)
\end{aligned}$$

According to Definition 5.9 a solution $t \to \mathbf{X}_s(t) = (X_s(t, 1), X_s(t, 2), \ldots, X_s(t, N)) : \mathbb{R}_+ \to \mathscr{S}_{n+m}^N$ is a stabilizing solution of (6.95) if the matrices $\mathscr{R}(k) + B_d^\top X_s(\tau_k, i)B_d$, $k \in \mathbb{Z}_+$ are invertible and the closed-loop system

$$d\mathbf{x}(t) = \mathscr{A}_0(t, \eta(t))\mathbf{x}(t)dt + \sum_{l=1}^{r} \mathscr{A}_l(t, \eta(t))\mathbf{x}(t)dw_l(t), \quad \tau_k < t \leq \tau_{k+1} \qquad (6.96a)$$

$$\mathbf{x}(\tau_k^+) = (A_d + B_d F_s(\tau_k, \eta(\tau_k)))\mathbf{x}(\tau_k), \quad k \in \mathbb{Z}_+ \qquad (6.96b)$$

is strongly exponentially stable in mean square, where

$$F_s(\tau_k, i) = -(\mathscr{R}(k) + B_d^\top X_s(\tau_k, i)B_d)^{-1} B_d^\top X_s(\tau_k, i)A_d, \quad (\tau_k, i) \in \mathfrak{T} \times \mathcal{N}. \tag{6.97}$$

A necessary and sufficient condition for the existence of a bounded and stabilizing solution of (6.95) which is satisfying the sign conditions

$$\mathscr{R}(k) + B_d^\top X_s(\tau_k, i)B_d \geq \nu I_m > 0, \quad \text{for all } (\tau_k, i) \in \mathfrak{T} \times \mathcal{N}, \tag{6.98}$$

can be obtained adapting Corollary 5.7 to the case of the BJMLDE (6.95).
The set $\widetilde{\Gamma}^\Sigma[\mathbb{R}_+]$ associated to the BJMLDE (6.95) consists from all bounded functions $\mathbf{Z}(\cdot) = (Z(\cdot, 1), Z(\cdot, 2), \ldots, Z(\cdot, N)) : \mathbb{R}_+ \to \mathscr{S}_{n+m}^N$ which are continuously differentiable on each interval $[\tau_k, \tau_{k+1})$, right continuous in each $t = \tau_k$, $k \in \mathbb{Z}_+$, and satisfying the backward linear matrix differential inequalities with jumps:

$$\dot{Z}(t, i) + \mathscr{A}_0^\top(t, i)Z(t, i) + Z(t, i)\mathscr{A}_0(t, i) + \sum_{l=1}^{r} \mathscr{A}_l^\top(t, i)Z(t, i)\mathscr{A}_l(t, i)$$
$$+ \sum_{j=1}^{N} q_{ij} Z(t, j) + \mathscr{M}(t) \geq 0, \quad t \in [\tau_k, \tau_{k+1}) \tag{6.99a}$$

$$\begin{pmatrix} -Z(\tau_k^-, i) & 0 \\ 0 & \mathscr{R}(k) \end{pmatrix} + (A_d \ \ B_d)^\top Z(\tau_k, i)(A_d \ \ B_d) \geq \tilde{\nu} I_{n+m} > 0, \tag{6.99b}$$

$(\tau_k, i) \in \mathfrak{T} \times \mathcal{N}$. Corollary 5.7 specialized to the case of the BJMLDE (6.95) becomes:

Corollary 6.1 *Under the Assumption 2.1 the following are equivalent:*

(i) the BJMLDE (6.95) has a bounded and stabilizing solution $\mathbf{X}_s(\cdot)$ satisfying the sign condition (6.98);
(ii) the impulsive controlled linear stochastic system (6.86)–(6.87) is mean-square stabilizable by impulses in a state feedback form and the set $\widetilde{\Gamma}^\Sigma[\mathbb{R}_+]$ is not empty.

Remark 6.5 The result stated in Corollary 5.9 (ii) shows that if $A_l(t, i) = A_l(i)$, $B_l(t, i) = B_l(i)$, $0 \leq l \leq r$, $i \in \mathcal{N}$, $M(t) = M \in \mathscr{S}_n$, $R(t) = R \in \mathscr{S}_m$, $t \in \mathbb{R}_+$, $\tau_k = kh$, $k \in \mathbb{Z}_+$ then the set $\widetilde{\Gamma}^\Sigma[\mathbb{R}_+]$ of the functions $\mathbf{Z}(\cdot)$ which are satisfying (6.99) is not empty if and only if it contains periodic functions of period h. Hence, in the time-invariant case it is sufficient to look for periodic functions with period h to test if the set $\widetilde{\Gamma}^\Sigma[\mathbb{R}_+]$ is or not empty.

The next result provides a set of conditions which guarantee the existence of a control $\hat{\mathbf{u}} \in \mathscr{U}_{sd}(t_0, \mathbf{x}_0; \mathfrak{T})$ which is satisfying the optimality condition (6.91).

Proposition 6.2 *Assume:*

(a) the Assumption 2.1 holds;

(b) the BJMLDE (6.95) has a bounded and stabilizing solution $\mathbf{X}_s(\cdot)$ *which is satisfying the sign condition (6.98).*

Under these assumptions, the control $\widehat{\mathbf{u}} = \{\widehat{u}(k)\}_{k \ge \kappa(t_0)}$ *with*

$$\widehat{u}(k) = F_s(\tau_k, \eta(\tau_k))\widehat{\mathbf{x}}(\tau_k; t_0, \mathbf{x}_0) \tag{6.100}$$

lies in $\mathscr{U}_{\mathrm{sd}}(t_0, \mathbf{x}_0; \mathfrak{T})$ *and satisfies the optimality condition (6.91) if* $F_s(\tau_k, i)$ *is defined as in (6.97) and* $\widehat{\mathbf{x}}(t; t_0, \mathbf{x}_0)$ *is the solution of the IVP (6.96) with the initial condition* $\mathbf{x}(t_0) = \mathbf{x}_0$. *The minimal value of the cost functional (6.88) is given by:*

$$\mathscr{J}(t_0, \infty, \mathbf{x}_0; \widehat{\mathbf{u}}) = \mathbb{E}[\mathbf{x}_0^\top X_s(t_0^-, \eta(t_0))\mathbf{x}_0] \tag{6.101}$$

with the convention that $X_s(t_0^-, \eta(t_0))$ *is replaced by* $X_s(t_0, \eta(t_0))$ *whenever* $t_0 < \tau_{\kappa(t_0)}$.

Proof Employing (6.14) in the case of the pair formed by ICLSS (6.86)–(6.87) and the performance criterion (6.88)–(6.89) we obtain via (6.80) that

$$\mathscr{J}(t_0, \infty, \mathbf{x}_0; \mathbf{u}) = \mathbb{E}[\mathbf{x}_0^\top X_s(t_0^-, \eta(t_0))\mathbf{x}_0] + \sum_{k=\kappa(t_0)}^{\infty} \mathbb{E}[(u(k) - F_s(\tau_k, \eta(\tau_k))\mathbf{x}_{\mathbf{u}}(\tau_k; t_0, \mathbf{x}_0))^\top$$
$$(B_d^\top X_s(\tau_k, \eta(\tau_k))B_d + \mathscr{R}(k))(u - F_s(\tau_k, \eta(\tau_k))\mathbf{x}_{\mathbf{u}}(\tau_k; t_0, \mathbf{x}_0))]. \tag{6.102}$$

From the unicity of the solution of an IVP, we may infer that $\mathbf{x}_{\widehat{\mathbf{u}}}(t; t_0, \mathbf{x}_0)$ coincides to $\widehat{\mathbf{x}}(t; t_0, \mathbf{x}_0)$. Thus, in the case of the control (6.100), the equality (6.102) reduces to (6.101). Then, (6.98) together with (6.101) and (6.102) give

$$\mathscr{J}(t_0, \infty, \mathbf{x}_0; \mathbf{u}) \ge \mathscr{J}(t_0, \infty, \mathbf{x}_0; \widehat{\mathbf{u}})$$

for any $\mathbf{u} \in \mathscr{U}_{\mathrm{sd}}(t_0, \mathbf{x}_0; \mathfrak{T})$ which confirms that (6.100) provides an optimal control law. Finally, (6.98) together with (6.102) shows that the control $\widehat{\mathbf{u}}$ described by (6.100) is the unique optimal control for the optimal control problem under consideration. Thus the proof ends.

Let $\begin{pmatrix} X_{s1}(t) & X_{s2}(t) \\ X_{s2}^\top(t) & X_{s3}(t) \end{pmatrix}$ be the partition of $X_s(t)$ according to the partition of the coefficients of (6.95) given in (6.87) and (6.89). With this notation the sign condition (6.98) becomes

$$\mathscr{R}(k) + X_{s3}(\tau_k, i) \ge \nu I_m > 0, \quad \text{for all } (\tau_k, i) \in \mathfrak{T} \times \mathscr{N}. \tag{6.103}$$

Further, (6.97) becomes:

$$F_s(\tau_k, i) = (F_{s1}(\tau_k, i) \quad 0), \tag{6.104a}$$
$$F_{s1}(\tau_k, i) := -(\mathscr{R}(k) + X_{s3}(\tau_k, i))^{-1} X_{s2}^\top(\tau_k, i), \quad \text{for all } (\tau_k, i) \in \mathfrak{T} \times \mathscr{N}. \tag{6.104b}$$

From $\widehat{\mathbf{x}}(t; t_0, \mathbf{x}_0) = (\widehat{x}^\top(t; t_0, x_0) \;\; \widehat{\varphi}^\top(t; t_0, 0))^\top$ and (6.104) we deduce that (6.100) reduces to

$$\widehat{u}(k) = F_{s1}(\tau_k, \eta(\tau_k))\widehat{x}(\tau_k; t_0, x_0), \quad k \geq \kappa(t_0). \tag{6.105}$$

From (6.96) we obtain

$$\widehat{\varphi}(t; t_0, 0) = F_{s1}(\tau_k, \eta(\tau_k))\widehat{x}(\tau_k; t_0, x_0), \quad \tau_k \leq t < \tau_{k+1}, \quad k \geq \kappa(t_0). \tag{6.106}$$

Substituting this equality in (6.96a) we may infer that $\widehat{x}(t; t_0, x_0)$, $t \geq t_0$ is the solution of the following IVP:

$$\begin{aligned} dx(t) &= (A_0(t, \eta(t))x(t) + B_0(t, \eta(t))F_{s1}(\tau_k, \eta(\tau_k))x(\tau_k))dt \\ &\quad + \sum_{l=1}^{r}(A_l(t, \eta(t))x(t) + B_l(t, \eta(t))F_{s1}(\tau_k, \eta(\tau_k))x(\tau_k))dw_l(t), \\ &\quad \tau_k \leq t < \tau_{k+1}, \quad k \geq \kappa(t_0), \\ &\quad x(t_0) = x_0. \end{aligned} \tag{6.107}$$

Comparing the first system to (6.83) we may deduce that (6.107) is obtained plugging the piecewise constant control

$$\widehat{u}(t) = F_{s1}(\tau_k, \eta(\tau_k))x(\tau_k), \quad \tau_k \leq t < \tau_{k+1} \tag{6.108}$$

in (6.83). Since $\widehat{u}(t) = \gamma^{-1}(\widehat{\mathbf{u}})(t)$ we may conclude via Lemma 6.4 that the control $\widehat{u}(\cdot)$ defined in (6.108) is just the optimal control $u_{\text{opt}}(\cdot)$ which satisfies the optimality condition (6.82).

Now we are in position to state the main result of this section:

Theorem 6.7 *Assume:*

(a) the terms of the sequence $\mathfrak{T}$ are satisfying the conditions from the Assumption 2.1;
(b) the BJMLDE with Riccati-type jumping operator (6.95) has a bounded and stabilizing solution $\mathbf{X}_s(\cdot) : \mathbb{R}_+ \to \mathcal{S}_{n+m}^N$ which is satisfying the sign condition (6.98).

Under these conditions the control $u_{\text{opt}}(\cdot)$ satisfying the optimality condition (6.82) is given by:

$$u_{\text{opt}}(t) = F_{s1}(\tau_k, \eta(\tau_k))x(\tau_k), \quad \tau_k < t \leq \tau_{k+1}, \quad k \geq \kappa(t_0), \tag{6.109}$$

where $F_{s1}(\tau_k, i)$ is defined in (6.104b), $x(\tau_k)$ being the value measured at instances time τ_k of the states $x(t)$ of the closed-loop system obtained when the control (6.109) is coupled to the system (6.76). The minimal value of the cost (6.78) is:

$$\mathcal{J}(t_0, \infty, x_0; u_{\text{opt}}(\cdot)) = \mathbb{E}[x_0^\top X_{s1}(t_0^-, \eta(t_0))x_0]$$

with the remark that $X_{s1}(t_0^-, \eta(t_0))$ *is replaced by* $X_{s1}(t_0, \eta(t_0))$ *when* $t_0 < \tau_{\kappa(t_0)}$.

Proof The proof is a consequence of the previous developments.

6.6 Notes and References

Regrading the linear quadratic optimal control for MJLSs, we mention the pioneering works in [2, 3] which initiated a very fertile research axe. Without any intention of being exhaustive, we mention, for instance, Blair and Sworder [4], Mariton [5], Morozan [6], Griffiths and Loparo [7], Chizeck [8], Ji and Chizeck [9, 10], Ji et al. [11], Ji and Chizeck [12], Morozan [13], Abou-Kandil et al. [14], Costa and Boukas [15], Czornik and A.Swierniak [16], Costa and do Val [17] as a representative sample of some works dealing with linear quadratic optimal control of MJLSs. For a historical perspective on the different contributions on this topic, we refer the interested reader to the monograph Costa et al. [18].

For stochastic linear systems with multiplicative noise we cite here Hausmann [19], Willems [20], Willems and Willems [21], Bismut [22], Ichikawa [23–25], Sasagawa [26], Haurie and Leizarowitz [27], Ait-Rami and El-Ghaoui [28], Chen et al. [29], Rami and Zhou [30]. In the case of stochastic systems subject to both Markovian perturbations as well as multiplicative noise, we refer the reader to the book Drăgan et al. [1] and the references therein.

For the class of Itô-type impulsive stochastic systems, we refer the reader to the recent works Qi et al. [31], Jenson et al. [32] where some linear quadratic impulsive control problems have been addressed in the finite horizon case. The authors in [31] followed a maximum principle approach while in [32] a dynamic programming algorithm has been proposed.

To the best of the authors' knowledge, linear quadratic optimal control problems for impulsive stochastic systems with Markovian jump parameters have not yet been addressed in the literature. Some material related to this chapter has been already published in [33, 34].

References

1. Drăgan, V., Morozan, T., Stoica, A.M.: Mathematical Methods in Robust Control of Linear Stochastic Systems. Springer-Verlag, New York (2013)
2. Sworder, D.: Feedback control of a class of linear systems with jump parameters. IEEE Trans. Autom. Control **14**, 9–14 (1969)
3. Wonham, W.: Random differential equations in control theory. Probab. Methods Appl. Math. (Academic, New York) **2**, 131–212 (1970)
4. Blair, W., Sworder, D.: Feedback control of a class of linear discrete systems with jump parameters and quadratic cost criteria. Int. J. Control **21**, 833–844 (1975)
5. Mariton, M.: Jump Linear Systems in Automatic Control. CRC Press (1990)

6. Morozan, T.: Optimal stationary control for dynamic systems with Markova perturbations. Stoch. Anal. Appl. **1**(3), 299–325 (1983)
7. Griffiths, B., Loparo, K.: Optimal control of jump linear quadratic gaussian systems. Int. J. Cont. **42**, 791–819 (1985)
8. Chizeck, H.J., Ji, Y.: Optimal quadratic control of jump linear systems with gaussian noise in discrete-time. In: Proceedings of the 27th Conference on Decision and Control, pp. 1989–1992 (1988)
9. Ji, Y., Chizeck, H.: Optimal quadratic control of jump linear systems with separately controlled transition probabilities. Int. J. Control **49**, 481–491 (1989)
10. Ji, Y., Chizeck, H.J.: Controllability, stabilizability, and jump linear quadratic control. IEEE Trans. Automat. Contr. **35**, 777–788 (1990)
11. Ji, Y., Chizeck, H.J., Feng, X., Loparo, K.A.: Stability and control of discrete-time jump linear systems. Control Theor. Adv. Technol. **7**, 247–270 (1991)
12. Ji, Y., Chizeck, H.: Jump linear quadratic gaussian control: steady-state solution and testable conditions. Contr. Theor. Adv. Technol. **7**, 247–270 (1991)
13. Morozan, T.: Stability and control for linear systems with jump Markov perturbations. Stoch. Anal. Appl. **13**, 91–110 (1995)
14. Abou-Kandil, H., Freiling, G., Jank, G.: On the solution of discrete-time Markovian jump linear quadratic control problems. Automatica **31**, 765–768 (1995)
15. Costa, O.L.V., Boukas, E.K.: Necessary and sufficient condition for robust stability and stabilizability of continuous-time linear systems with markovian jumps. J. Optim. Theory Appl. **99**, 359–379 (1998)
16. Czornik, A., Swierniak, A.: On the discrete jlq and jlqg problems. Nonlinear Anal. **47**, 423–434 (2001)
17. Costa, E., do Val, J.: Weak detectability and the linear-quadratic control problem of discrete-time Markov jump linear systems. Int. J. Control 1989-1992 (2002)
18. Costa, O.L.V., Fragoso, M.D., Todorov, M.G.: Continuous-Time Markov Jump Linear Systems. Springer-Verlag, Berlin Heidelberg (2013)
19. Hausmann, U.: Optimal stationary control with state and control dependent noise. SIAM J. Contr. Optim. **9**, 184–198 (1971)
20. Willems, J.: Mean square stability criteria for linear white noise stochastic systems. Probl. Contr. Inf. Theor. **2**, 199–217 (1973)
21. Willems, J.L., Willems, J.C.: Feedback stabilizability for stochastic systems with state and control dependent noise. Automatica **12**, 277–283 (1976)
22. Bismut, J.: Linear quadratic optimal control with random coefficients. SIAM J. Contr. Optim. **14**, 419–444 (1976)
23. Ichikawa, A.: Optimal control of linear stochastic evolution equation with state and control dependent noise. In: Proceedings of IMA Conference on Recent Theoretical Developments in Control, pp. 383–401 (1978)
24. Ichikawa, A.: Dynamic programming approach to stochastic evolution equation. SIAM J. Contr. Optim. **17**, 162–174 (1979)
25. Ichikawa, A.: Filtering and control of stochastic differential equations with unbounded coefficients. Stoch. Anal. Appl. **4**, 187–212 (1986)
26. Sasagawa, T.: Lp-stabilization problem for linear-stochastic control systems with multiplicative noise. J. Optim. Theor. Appl. **61**, 451–471 (1989)
27. Haurie, A., Leizarowitz, A.: Overtracking optimal regulation and tracking of piecewise diffusion linear systems. SIAM J. Contr. Optim. **30**, 816–832 (1992)
28. Ait-Rami, M., El-Ghaoui, L.: Robust stabilization of jump linear systems using linear matrix inequalities. In: Proceedings of IFAC Symposium on Robust Control Design Rio de Janeiro, pp. 91–110 (1994)
29. Chen, S., Li, X., Zhou, Y.: Stochastic linear quadratic regulators with indefinite control weight costs. SIAM J. Contr. Optim. **36**, 816–832 (1998)
30. Rami, M.A., Zhou, X.Y.: Linear matrix inequalities, riccati equations, and inde8nite stochastic linear quadratic control. IEEE Trans. Automat. Control **45**, 1131–1142 (2000)

31. Qi, Q., Qiu, Z., Ji, Z.: Optimal continuous/impulsive lq control with quadratic constraints. IEEE Access **7**, 955–963 (2019)
32. Jenson, E.L., Chen, X., Scheeres, D.J.: Optimal spacecraft guidance with asynchronous measurements and noisy impulsive controls. IEEE Control Syst. Lett. **5**, 1813–1818 (2021)
33. Dragan, V., Popa, I.L., Aberkane, S.: On the stochastic linear quadratic optimal control problem by piecewise constant controls: The infinite horizon time case. Math. Methods Appl. Sci. **47**, 3734–3762 (2024)
34. Dragan, V., Popa, I. L.: The linear quadratic optimal control problem for stochastic systems controlled by impulses. Symmetry 16 (2024)

Chapter 7
H_2 Optimal Control of an Impulsive Controlled Linear Stochastic System

Abstract In this chapter the problem of H_2 optimal control of an impulsive controlled linear stochastic system subject to Markovian jumping and independent multiplicative and additive white noise perturbations is considered. In Sect. 7.1 we begin by defining a so-called H_2 norm of jump linear stochastic systems. Such a norm serves as a measure of the effect of additive white noise perturbations over an output of the controlled system, and thus it will be used as a performance criterion for the considered optimal control problem in this chapter. Section 7.2 is devoted to the computation of the value of the proposed H_2 norm. Such a value is characterized via solutions of some suitably defined non-homogeneous forward/backward generalized Lyapunov-type jump linear differential equations on the spaces of symmetric matrices (see Theorem 7.1). A second H_2 norm definition is then introduced at the end of this section and also characterized via solutions of some suitably defined non-homogeneous forward/backward generalized Lyapunov-type jump linear differential equations (see Theorem 7.2). An interesting *dwell-time* relation between the two proposed H_2 norms is then highlighted in Remark 7.6. By using the H_2 norms' characterization results, we then address the control synthesis part. Firstly (Sect. 7.3), the problem of optimization of H_2 norms is solved under the assumption that full-state vector is available for measurements. One shows that among all stabilizing controllers of arbitrary dimension, the best performance is achieved by a zero-order controller. The corresponding feedback gain of the optimal controller is constructed based on the stabilizing solution of an adequately defined system of backward jump matrix linear differential equations with Riccati-type jumping operator (see Theorem 7.4). Secondly, the H_2 optimization problem is solved under the assumption that only an output is available for measurements. The state-space realization of the H_2-optimal controller is then obtained in Theorem 7.6. The main tool used in the construction of the optimal controller is here also a Riccati-type equation. We finally apply the obtained results to the case of sampled-data linear stochastic systems (Sect. 7.5).

V. Drăgan et al., *Robust Control of Jump Linear Stochastic Systems*, Lecture Notes in Control and Information Sciences 497, https://doi.org/10.1007/978-3-031-84070-8_7

7.1 H_2 Norm of a Jump Linear Stochastic System

We consider the JLSS $\mathbb{G}$ having the state-space representation described by

$$
\begin{aligned}
dx(t) &= A_0(\eta(t))x(t)dt + \sum_{l=1}^{r_1} A_l(\eta(t))x(t)dw_l(t) + G_c(\eta(t))d\xi(t), \\
&\quad kh < t \leq (k+1)h && (7.1a) \\
x(kh^+) &= A_{d0}(\eta(kh))x(kh) + \sum_{p=1}^{r_2} w_{dp}(k)A_{dp}(\eta(kh))x(kh) + G_d(\eta(kh))\zeta_d(k), \\
&\quad k \in \mathbb{Z}_+ && (7.1b) \\
z_c(t) &= C_{z_c}(\eta(t))x(t), \quad t \in \mathbb{R}_+ && (7.1c) \\
z_d(kh) &= C_{z_d}(\eta(kh))x(kh) + G_{z_d}(\eta(kh))\zeta_d(kh), \quad k \in \mathbb{Z}_+, && (7.1d)
\end{aligned}
$$

where $h > 0$ is fixed and $x(t) \in \mathbb{R}^n$ is the vector of the state parameters at instance time t, $z_c(t) \in \mathbb{R}^{n_{z_c}}$ is a system output often named quality output which evolves in continuous time, while $z_d(kh) \in \mathbb{R}^{n_{z_d}}$ is an output evolving in discrete time. In (7.1), $\{w(t)\}_{t\geq 0}$ ($w(t) = (w_1(t),\ w_2(t),\ \ldots,\ w_{r_1}(t))^\top$) is a r_1-dimensional standard Wiener process defined on a given probability space $(\Omega, \mathfrak{F}, \mathscr{P})$, $\{w_d(k)\}_{k\in\mathbb{Z}_+}$, $(w_d(k) = (w_{d1}(k),\ w_{d2}(k), \ldots,\ w_{dr_2}(k))^\top)$ is a sequence of r_2-dimensional independent random vectors, while $\{\eta(t)\}_{t\geq 0}$ is a homogeneous standard Markov process with the state set $\mathcal{N} = \{1, 2, \ldots, N\}$ and the transition semigroup $P(t) = e^{Qt}$, $t \geq 0$. We assume that the stochastic processes $\{w(t)\}_{t\geq 0}$, $\{w_d(k)\}_{k\in\mathbb{Z}_+}$, $\{\eta(t)\}_{t\geq 0}$ satisfy Assumption 1.2. From the state-space representation (7.1) one sees that the JLSS $\mathbb{G}$ is affected by the additive disturbances $\{\xi(t)\}_{t\geq 0}$ and $\{\zeta_d(k)\}_{k\in\mathbb{Z}_+}$. In the rest of this chapter, we will assume that the additive disturbances satisfy the assumption:

Assumption 7.1

(a) $\{\xi(t)\}_{t\geq 0}$ is a m_ξ-dimensional Wiener process defined on the probability space $(\Omega, \mathfrak{F}, \mathscr{P})$ and has the properties:

$\xi(0) = 0$; $\quad \mathbb{E}[\xi(t)] = 0$;
$\mathbb{E}[(\xi(s) - \xi(t))(\xi(s) - \xi(t))^\top] = \boldsymbol{\Lambda}_c(s-t)$, for all $s \geq t \geq 0$, where $\boldsymbol{\lambda}_c \geq 0$ is a known matrix;

(b) $\{\zeta_d(k)\}_{k\in\mathbb{Z}_+}$ is a sequence of m_{ζ_d}-dimensional independent random vectors defined on the probability space $(\Omega, \mathfrak{F}, \mathscr{P})$ having zero mean and the covariance $\mathbb{E}[\zeta_d(k)\zeta_d^\top(k)] = \boldsymbol{\Lambda}_d$, for all $k \in \mathbb{Z}_+$, where $\boldsymbol{\Lambda}_d \geq 0$ is a known matrix;

(c) $\{w(t)\}_{t\geq 0}$, $\{w_d(k)\}_{k\in\mathbb{Z}_+}$, $\{\eta(t)\}_{t\geq 0}$, $\{\xi(t)\}_{t\geq 0}$, $\{\zeta_d(k)\}_{k\in\mathbb{Z}_+}$ are mutually independent stochastic processes.

For each $t \geq 0$, $\widetilde{\mathscr{F}}_t \subset \mathfrak{F}$ stands for the σ-algebra generated by the random vectors $\xi(s)$ and $\zeta_d(k)$, $0 \leq s \leq t$ and $0 \leq k < \frac{t}{h}$, respectively, augmented by all subsets

$\mathscr{A} \in \mathfrak{F}$ with $\mathscr{P}(\mathscr{A}) = 0$. According to Assumption 7.1(c) we may infer that the σ-algebra $\widetilde{\mathscr{F}}_t$ is independent of the σ-algebra $\mathscr{F}_t \subset \mathfrak{F}$ generated by the random vectors $w(s)$ and $w_d(k)$, $0 \leq s \leq t$ and $0 \leq k < \frac{t}{h}$, respectively. For each $t \geq 0$ we set

$$\widehat{\mathscr{F}_t} \triangleq \mathscr{F}_t \vee \widetilde{\mathscr{F}_t} \quad \text{and} \quad \widehat{\mathscr{H}_t} \triangleq \widehat{\mathscr{F}_t} \vee \mathscr{G}_t$$

where $\mathscr{G}_t$ is the σ-algebra defined in Sect. 1.2, being associated to the Markov process $\{\eta(t)\}_{t\geq 0}$. Based on Assumption 7.1(c) we may conclude that similar results to those proved in Proposition 3.4, Lemma 3.1, Theorem 3.5 and Corollary 3.6 remain true in the case of the non-homogeneous JLSS (7.1a) and (7.1b) and the family of σ-algebras $\widehat{\mathscr{H}} = \{\widehat{\mathscr{H}_t}\}_{t\geq 0}$.

Remark 7.1 Unlike the stochastic processes $\{w(t)\}_{t\geq 0}$ and $\{w_d(k)\}_{k\in\mathbb{Z}_+}$ that we supposed to be *normed* (in the sense that their corresponding covariance matrices are identity matrices), in the case of the stochastic processes $\{\xi(t)\}_{t\geq 0}$ and $\{\zeta_d(k)\}_{k\in\mathbb{Z}_+}$, we kept the values $\boldsymbol{\Lambda}_c$ and $\boldsymbol{\Lambda}_d$, of the covariance matrices in order to analyze their effect on the corresponding H_2-norms.

Let consider the JSLDE:

$$dx(t) = A_0(\eta(t))x(t)dt + \sum_{l=1}^{r_1} A_l(\eta(t))x(t)dw_l(t), \quad kh < t \leq (k+1)h \tag{7.2a}$$

$$x(kh^+) = A_{d0}(\eta(kh))x(kh) + \sum_{p=1}^{r_2} w_{dp}(k)A_{dp}(\eta(kh))x(kh), \ k \in \mathbb{Z}_+, \tag{7.2b}$$

obtained from (7.1a) and (7.1b) taking $G_c(i) = 0$ and $G_d(i) = 0$, for all $i \in \mathscr{N}$.

In order to introduce a H_2-type norm of a dynamical system $\mathbb{G}$ of type (7.1), let us separately analyze the influence of the additive disturbances $\{\xi(t)\}_{t\geq 0}$ and $\{\zeta_d(k)\}_{k\in\mathbb{Z}_+}$, respectively, on the trajectories and the outputs of $\mathbb{G}$.

Let $(t_0, x_0) \in \mathbb{R}_+ \times \mathbb{R}^n$ be arbitrary but fixed. We denote $x(\cdot; t_0, x_0, \xi)$ the solution of the following IVP:

$$dx(t) = A_0(\eta(t))x(t)dt + \sum_{l=1}^{r_1} A_l(\eta(t))x(t)dw_l(t) + G_c(\eta(t))d\xi(t),$$
$$kh < t \leq (k+1)h \tag{7.3a}$$

$$x(kh^+) = A_{d0}(\eta(kh))x(kh) + \sum_{p=1}^{r_2} w_{dp}(k)A_{dp}(\eta(kh))x(kh), \quad k \geq \kappa(t_0), \tag{7.3b}$$

$$x(t_0) = x_0 \tag{7.3c}$$

where $\kappa(t_0)$ is the integer number related to t_0 by

$$(\kappa(t_0) - 1)h < t_0 \leq \kappa(t_0)h. \tag{7.4}$$

We set

$$z_c(t; t_0, x_0, \xi) := C_{z_c}(\eta(t))x(t; t_0, x_0, \xi), \quad t \geq t_0 \tag{7.5a}$$
$$z_d(kh; t_0, x_0, \xi) := C_{z_d}(\eta(kh))x(kh; t_0, x_0, \xi), \quad k \geq \kappa(t_0). \tag{7.5b}$$

It is obvious that the equations from (7.3) and the outputs (7.5) are obtained from (7.1) taking $G_d(\eta(kh)) = 0$ and $G_{z_d}(\eta(kh)) = 0$, $k \geq \kappa(t_0)$.

Let us consider $x(\cdot; t_0, 0, \zeta_d)$ the solution of the following IVP:

$$dx(t) = A_0(\eta(t))x(t)dt + \sum_{l=1}^{r_1} A_l(\eta(t))x(t)dw_l(t), \quad kh < t \leq (k+1)h \tag{7.6a}$$

$$x(kh^+) = A_{d0}(\eta(kh))x(kh) + \sum_{p=1}^{r_2} w_{dp}(k)A_{dp}(\eta(kh))x(kh)$$
$$+ G_d(\eta(kh))\zeta_d(k), \quad k \geq \kappa(t_0), \tag{7.6b}$$
$$x(t_0) = 0. \tag{7.6c}$$

The outputs determined by the trajectory $x(\cdot; t_0, 0, \zeta_d)$ are:

$$z_c(t; t_0, 0, \zeta_d) := C_{z_c}(\eta(t))x(t; t_0, 0, \zeta_d), \quad t \geq t_0 \tag{7.7a}$$
$$z_d(kh; t_0, 0, \zeta_d) := C_{z_d}(\eta(kh))x(kh; t_0, 0, \zeta_d) + G_{z_d}(\eta(kh))\zeta_d(k), k \geq \kappa(t_0). \tag{7.7b}$$

If $x(t; t_0, x_0)$ is a solution of the JSLDE (7.1a) and (7.1b) satisfying the initial condition $x(t_0; t_0, x_0) = x_0$ then from the unicity of the solution of an IVP, we may infer that $x(t; t_0, x_0) = x(t; t_0, x_0, \xi) + x(t; t_0, 0, \zeta_d)$, $t \geq t_0$ *a.s.*

Bearing in mind that both (7.3) as well as (7.6) are special cases of a non-homogeneous JSLDE of the form (3.51), we obtain:

Proposition 7.1 *Assume:*

(a) Assumptions 1.2 and 7.1 are satisfied;
(b) the JSLDE (7.2) is ESMS.

Let $(t_0, x_0) \in \mathbb{R}_+ \times \mathbb{R}^n$ *be arbitrary and* $x(\cdot; t_0, x_0, \xi)$, $x(\cdot; t_0, 0, \zeta_d)$, *respectively, be the solutions of the IVPs (7.3) and (7.6), respectively. Under the considered assumptions, the following hold:*

(i)

$$\mathbb{E}[|x(t; t_0, x_0, \xi)|^2 \mid \eta(t_0) = j] \leq \beta(e^{-\alpha(t-t_0)}|x_0|^2 + \tag{7.8}$$
$$+ \sum_{l=1}^{m_\xi} \|\boldsymbol{\Lambda}_c\| \int_0^t e^{-\alpha(t-s)}\mathbb{E}[|G_{c,l}(\eta(t))|^2 \mid \eta(t_0) = j]ds), \quad \textit{for all} \ t \geq t_0;$$

$$\mathbb{E}[|x(t;t_0,0,\zeta_d)|^2 \mid \eta(t_0)=j] \leq \sum_{l=1}^{m_{\zeta_d}} \beta \|\boldsymbol{\Lambda}_d\| \sum_{k=\kappa(t_0)}^{\kappa(t)} e^{-\alpha(t-kh)} \cdot \mathbb{E}[|G_{dl}(\eta(kh))|^2 \mid \eta(t_0)=j], \ \textit{for all} \ j \in \mathcal{N}, \ t > \kappa(t_0), \tag{7.9}$$

$\kappa(t)$ being the natural number with the property that $t \in [\kappa(t), \kappa(t)+1)$. In (7.8) and (7.9), $\alpha > 0$, $\beta \geq 1$ are constants not depending upon t, t_0, x_0;

(ii) Let $z_c(\cdot;t_0,x_0,\xi)$, $z_d(\cdot;t_0,x_0,\xi)$, be the outputs defined in (7.5) and $z_c(\cdot;t_0,0,\zeta_d)$, $z_d(\cdot;t_0,0,\zeta_d)$ be the outputs defined in (7.7). We have that $z_c(\cdot;t_0,x_0,\xi)$ and $z_c(\cdot;t_0,0,\zeta_d)$ lie in $\mathcal{L}^2_{\widehat{\mathcal{H}}}([t_0,T],\mathbb{R}^{n_{z_c}})$, for all $t > t_0$, while $z_d(\cdot;t_0,x_0,\xi)$ and $z_d(\cdot;t_0,0,\zeta_d)$, respectively, lie in $\ell^2_{\widehat{\mathcal{H}}}\{[\kappa(t_0),\widehat{\kappa}],\mathbb{R}^{n_{z_d}}\}$, for all $\widehat{\kappa} > \kappa(t_0)$. Additionally:

$$\sup_{t\geq t_0} \mathbb{E}[|z_c(t;t_0,x_0,\xi)|^2|\eta(t_0)=j] + \sup_{t\geq t_0} \mathbb{E}[|z_c(t;t_0,0,\zeta_d)|^2|\eta(t_0)=j] < \infty \tag{7.10a}$$

$$\sup_{k\geq\kappa(t_0)} \mathbb{E}[|z_d(kh;t_0,x_0,\xi)|^2|\eta(t_0)=j] + \sup_{k\geq\kappa(t_0)} \mathbb{E}[|z_d(kh;t_0,0,\zeta_d)|^2|\eta(t_0)=j] < \infty, \tag{7.10b}$$

for all $j \in \mathcal{N}$.

Proof (i) is immediately obtained by repeating step by step the proof of part (i) of Theorem 3.5 in the case of the non-homogeneous JSLDEs (7.3) and (7.6) together with the family of σ-algebras $\{\mathcal{H}_t\}_{t\geq 0}$ replaced by the family of σ-algebras $\{\widehat{\mathcal{H}}_t\}_{t\geq 0}$. The inequalities (7.10) are obtained from (7.5), (7.7) together with (7.8) and (7.9), which confirms (ii).

Now we introduce a norm of the JLSS $\mathbb{G}$ with state representation (7.1), which is denoted by $\|\mathbb{G}\|_2$, and it is defined by

$$\|\mathbb{G}\|_2 = \left[(\widehat{\|\mathbb{G}\|}_2)^2 + (\check{\|\mathbb{G}\|}_2)^2\right]^{1/2}, \tag{7.11}$$

where

$$\widehat{\|\mathbb{G}\|}_2^2 := \limsup_{T\to\infty} \frac{1}{T}\left(\int_{t_0}^{t_0+T} \mathbb{E}[|z_c(t;t_0,x_0,\xi)|^2]dt + \sum_{k=\kappa(t_0)}^{\left[\frac{t_0+T}{h}\right]} \mathbb{E}[|z_d(kh;t_0,x_0,\xi)|^2]\right) \tag{7.12a}$$

$$\check{\|\mathbb{G}\|}_2^2 := \limsup_{p\to\infty} \frac{1}{p}\left(\int_{t_0}^{t_0+(p+1)h} \mathbb{E}[|z_c(t;t_0,0,\zeta_d)|^2]dt + \sum_{k=\kappa(t_0)}^{\kappa(t_0)+p} \mathbb{E}[|z_d(kh;t_0,0,\zeta_d)|^2]\right) \tag{7.12b}$$

$\kappa(t_0)$ being introduced via (7.4).

In Sect. 7.2 we shall show that under the assumptions of Proposition 7.1 we can derive explicit formulae for the value of the limits from the right hand sides of (7.12). We will show that the values of these limits do not depend neither on the initial time instances t_0 nor on the initial state x_0.

From (7.8), (7.9) together with (7.7b) one sees that if $G_c(i) = 0$, $G_d(i) = 0$, $G_{z_d}(i) = 0$, for all $i \in \mathcal{N}$ (that is no additive white noise perturbations affect the evolution of the dynamical system $\mathbb{G}$,) then $z_c(\cdot, t_0, x_0, \xi) \in \mathcal{L}^2_{\mathcal{H}}([t_0, \infty); \mathbb{R}^{n_{z_c}})$ and $z_d(\cdot, t_0, x_0, \xi) \in \ell^2_{\mathcal{H}}\{[\kappa(t_0), \infty); \mathbb{R}^{n_{z_d}}\}$ while $z_c(t; t_0, 0, \zeta_d) = 0$, for all $t \geq t_0$, and $z_d(kh; t_0, 0, \zeta_d) = 0$, for all $k \geq \kappa(t_0)$, if the assumptions from Proposition 7.1 are satisfied. Under these conditions, the right hand sides of (7.12a) and (7.12b) are equal to zero. So, we can interpret the norm introduced in (7.11) and (7.12) as being a measure of the effect of some additive white noise-type perturbations on some outputs named quality outputs of the dynamical system $\mathbb{G}$.

In the sequel, we will call the norm introduced in (7.11) and (7.12), the H_2-norm of the JLSS $\mathbb{G}$.

7.2 The Computation of the H_2-Norm of a Jump Linear Stochastic System

7.2.1 Several Intermediate Results

Employing the coefficients of the JLSS (7.1) we introduce the following non-homogeneous backward generalized Lyapunov-type jump linear differential equation (BGLJLDE):

$$-\dot{X}(t,i) = A_0^\top(i)X(t,i) + X(t,i)A_0(i) + \sum_{l=1}^{r_1} A_l^\top(i)X(t,i)A_l(i)$$
$$+ \sum_{j=1}^{N} q_{ij}X(t,j) + C_{z_c}^\top(i)C_{z_c}(i), \quad kh \leq t < (k+1)h, \tag{7.13a}$$

$$X(kh^-, i) = \sum_{p=0}^{r_2} A_{d0}^\top(i)X(kh,i)A_{d0}(i) + C_{z_d}^\top(i)C_{z_d}(i), \quad k \in \mathbb{Z}, \quad i \in \mathcal{N}. \tag{7.13b}$$

According to the notation introduced in (2.114b) and (2.116b), we may rewrite (7.13) as:

$$-\dot{X}(t,i) = \mathcal{L}_i^*[\mathbf{X}(t)] + C_{z_c}^\top(i)C_{z_c}(i), \quad kh \leq t < (k+1)h, \tag{7.14a}$$
$$X(kh^-, i) = \mathcal{J}_i^*[\mathbf{X}(kh)] + C_{z_d}^\top(i)C_{z_d}(i), \quad k \in \mathbb{Z}, \quad i \in \mathcal{N}. \tag{7.14b}$$

The implication $(iii) \Rightarrow (i)$ from Theorem 3.1 combined with Theorem 2.13 applied in the case of (7.13) yields:

Corollary 7.1 *Under Assumption 1.2, if the JSLDE (7.2) is ESMS then the BGLJLDE (7.13) has a unique solution* $t \to \widetilde{\mathbf{X}}(t) = (\widetilde{X}(t,1), \widetilde{X}(t,2), \dots, \widetilde{X}(t,N)) : \mathbb{R} \to \mathcal{S}_n^N$ *which has the properties:*

(a) $\widetilde{\mathbf{X}}(\cdot)$ *is a differentiable function on each interval* $(kh, (k+1)h)$, *right continuous in any* $\tau_k = kh$, $k \in \mathbb{Z}$;
(b) $\widetilde{\mathbf{X}}(\cdot)$ *is a periodic function of period* h;
(c) there exists a constant $c > 0$ *such that*

$$0 \leq \widetilde{X}(t,i) \leq cI_n, \quad \textit{for all} \ (t,i) \in \mathbb{R} \times \mathcal{N}.$$

Beside the non-homogeneous BGLJLDE (7.13) we consider the non-homogeneous forward generalized Lyapunov-type jump linear differential equation (FGLJLDE):

$$\dot{Y}(t,i) = \mathcal{L}_i[\mathbf{Y}(t)] + \frac{\pi_{i\infty}}{h} G_c(i)\boldsymbol{\Lambda}_c G_c^\top(i), \quad kh < t \leq (k+1)h, \tag{7.15a}$$

$$Y(kh^+,i) = \mathcal{J}_i[\mathbf{Y}(kh)] + \pi_{i\infty} G_d(i)\boldsymbol{\Lambda}_d G_d^\top(i), \tag{7.15b}$$

$k \in \mathbb{Z}$, $i \in \mathcal{N}$, where for each $i \in \mathcal{N}$, $\mathbf{Y} \to \mathcal{L}_i[\mathbf{Y}] : \mathcal{S}_n^N \to \mathcal{S}_n$ and $\mathbf{Y} \to \mathcal{J}_i[\mathbf{Y}] : \mathcal{S}_n^N \to \mathcal{S}_n$ are the time-invariant versions of the operators defined in (2.113b) and (2.115b), respectively. In (7.15), $\pi_{i\infty} \geq 0$, $1 \leq i \leq N$ are constants whose meaning will be clarified later. $\boldsymbol{\Lambda}_c \geq 0$ and $\boldsymbol{\Lambda}_d \geq 0$ are the covariance matrices of the noises $\{\xi(t)\}_{t\geq 0}$ and $\{\zeta_d(k)\}_{k\in\mathbb{Z}_+}$, respectively, which are assumed to be known matrices. The implication $(iii) \Rightarrow (i)$ from Theorem 3.1 combined with Theorem 2.14 applied in the case of Eq. (7.15) yields:

Corollary 7.2 *Under Assumption 1.2, if the JSLDE (7.2) is ESMS, then the non-homogeneous FGLJLDE (7.15) has a unique solution* $t \to \widetilde{\mathbf{Y}}(t) = (\widetilde{Y}(t,1), \widetilde{Y}(t,2), \dots, \widetilde{Y}(t,N)) : \mathbb{R} \to \mathcal{S}_n^N$ *which has the properties:*

(a) $\widetilde{\mathbf{Y}}(\cdot)$ *is a differentiable function on each interval* $(kh, (k+1)h)$, *left continuous in any* $\tau_k = kh$, $k \in \mathbb{Z}$;
(b) $\widetilde{\mathbf{Y}}(\cdot)$ *is a periodic function of period* h;
(c) there exists a constant $\widehat{c} > 0$ *such that*

$$0 \leq \widetilde{Y}(t,i) \leq \widehat{c}I_n, \quad \textit{for all} \ (t,i) \in \mathbb{R} \times \mathcal{N}.$$

7.2.2 An Auxiliary Lemma

Before deriving explicit formulae of the limits from the right hand sides of (7.12), we prove the following preliminary results:

Lemma 7.1 *Under Assumptions 1.2 and 7.1, if the JSLDE (7.2) is ESMS, then the following hold:*

(i) if $z_c(t; t_0, x_0, \xi)$, $t \geq t_0$ and $z_d(kh; t_0, x_0, \xi)$, $kh \geq t_0$ are the outputs of type (7.5) determined by the solution $x(t; t_0, x_0, \xi)$, $t \geq t_0$ of the IVP (7.3), then we have

$$\int_{t_0}^{t_0+T} \mathbb{E}[|z_c(t; t_0, x_0, \xi)|^2 | \eta(t_0) = j]\mathrm{d}t + \sum_{k=\mathrm{K}(t_0)}^{\left[\frac{t_0+T}{h}\right]} \mathbb{E}[|z_d(kh; t_0, x_0, \xi)|^2 | \eta(t_0) = j]$$

$$= \Psi_j(t_0, T, x_0) + \sum_{i=1}^{N} \int_{t_0}^{t_0+T} \mathrm{Tr}[\widetilde{X}(t, i) G_c(i) \boldsymbol{\Lambda}_c G_c^\top(i)] p_{ji}(t - t_0)\mathrm{d}t, \tag{7.16}$$

for all $j \in \mathcal{N}$, $t_0 \in \mathbb{R}_+$, $T > \kappa(t_0)h - t_0$, $x_0 \in \mathbb{R}^n$ where

$$\lim_{T\to\infty} \frac{1}{T} \Psi_j(t_0, T, x_0) = 0; \tag{7.17}$$

(ii) if $z_c(t; t_0, x_0, \zeta_d)$, $t \geq t_0$, and $z_d(kh; t_0, 0, \zeta_d)$, $kh \geq t_0$, are the outputs of the form (7.7) generated by the solution $x(\cdot; t_0, 0, \zeta_d)$ of the IVP (7.6) then we have

$$\int_{t_0}^{t_0+(p+1)h} \mathbb{E}[|z_c(t; t_0, 0, \zeta_d)|^2 | \eta(t_0) = j]\mathrm{d}t + \sum_{k=\kappa(t_0)}^{\kappa(t_0)+p} \mathbb{E}[|z_d(kh; t_0, 0, \zeta_d)|^2 | \eta(t_0) = j]$$

$$= \widehat{\Psi}_j(t_0, p) + \sum_{i=1}^{N} \sum_{k=\kappa(t_0)}^{\kappa(t_0)+p} \mathrm{Tr}\left\{[\widetilde{X}(kh, i) G_d(i) \boldsymbol{\Lambda}_d G_d^\top(i)]\right. \tag{7.18}$$

$$\left. + \mathrm{Tr}[G_{z_d}(i) \boldsymbol{\Lambda}_d G_{z_d}^\top(i)]\right\} p_{ji}(kh - t_0),$$

for all $j \in \mathcal{N}$, $t_0 \geq 0$, $p \in \mathbb{Z}_+$ with

$$\lim_{p\to\infty} \frac{1}{p} \widehat{\Psi}_j(t_0, p) = 0. \tag{7.19}$$

In (7.16) and (7.18), $\widetilde{\mathbf{X}}(\cdot) = (\widetilde{X}(\cdot, 1), \ldots, \widetilde{X}(\cdot, N))$ is the unique bounded on $\mathbb{R}$ solution of the BGLJLDE (7.13) and $p_{ji}(\tau)$ with $\tau = t - t_0$ or $\tau = kh - t_0$ are the components of the transition probability matrix $P(\tau)$ of the Markov process.

Proof We start to prove that (7.16) and (7.17) hold. Let $(t_0, x_0) \in \mathbb{R}_+ \times \mathbb{R}^n$ be arbitrary but fixed. Denote $\widehat{x}(t) := x(t; t_0, x_0, \xi)$. Let $\{a_p\}_{p\in\mathbb{Z}_+}$, $\{b_p\}_{p\in\mathbb{Z}_+}$ be sequences of real numbers such that $kh < a_p < b_p < (k+1)h$, for all $p \in \mathbb{Z}_+$, where $k \in \mathbb{Z}_+$ is such that $k \geq \kappa(t_0)$, $\kappa(t_0)$ is associated to t_0 via (7.4).

Applying Itô's formula to the function $v(t, x, i) \stackrel{\Delta}{=} x^\top \widetilde{X}(t, i)x$ and to the stochastic process $\widehat{x}(t)$, $t \geq t_0$ on each interval $[a_p, b_p] \subset (kh, (k+1)h)$ one obtains:

$$\mathbb{E}[v(b_p,\widehat{x}(b_p),\eta(b_p)) \mid \eta(t_0)=j]-\mathbb{E}[v(a_p,\widehat{x}(a_p),\eta(a_p)) \mid \eta(t_0)=j]$$
$$=\int_{a_p}^{b_p}\mathbb{E}[\widehat{x}^\top(t)(\dot{\widetilde{X}}(t,\eta(t))+\mathscr{L}^*_{\eta(t)}[\widetilde{\mathbf{X}}(t)])\widehat{x}(t) \mid \eta(t_0)=j]\mathrm{d}t \tag{7.20}$$
$$+\int_{a_p}^{b_p}\mathbb{E}[\mathrm{Tr}[\widetilde{X}(t,\eta(t))G_c(\eta(t))\boldsymbol{\Lambda}_c G_c^\top(\eta(t))] \mid \eta(t_0)=j]\mathrm{d}t.$$

Letting $a_p\to kh$, $b_p\to(k+1)h$ and employing Lemma 3.1 together with the linear differential equation (7.14a) we get:

$$\int_{kh}^{(k+1)h}\mathbb{E}[|C_{z_c}(\eta(t))\widehat{x}(t;t_0,x_0,\xi)|^2 \mid \eta(t_0)=j]\mathrm{d}t$$
$$=\mathbb{E}[\widehat{x}^\top(kh^+)\widetilde{X}(kh,\eta(kh))\widehat{x}(kh^+) \mid \eta(t_0)=j]$$
$$-\mathbb{E}[\widehat{x}^\top((k+1)h)\widetilde{X}((k+1)h^-,\eta((k+1)h))\widehat{x}((k+1)h) \mid \eta(t_0)=j] \tag{7.21}$$
$$+\int_{kh}^{(k+1)h}\mathbb{E}[\mathrm{Tr}[\widetilde{X}(t,\eta(t))G_c(\eta(t))\boldsymbol{\Lambda}_c G_c^\top(\eta(t))] \mid \eta(t_0)=j]\mathrm{d}t.$$

Further, (7.3b), (7.14b) together with Assumption 7.1 allow us to deduce that

$$\mathbb{E}[\widehat{x}^\top(kh^+)\widetilde{X}(kh,\eta(kh))\widehat{x}(kh^+)|\eta(t_0)=j]=\mathbb{E}[\widehat{x}^\top(kh)\mathscr{I}^*_{\eta(kh)}[\widetilde{\mathbf{X}}(kh)]\widehat{x}(kh)|\eta(t_0)=j]$$
$$=\mathbb{E}[\widehat{x}^\top(kh)\widetilde{X}(kh^-,\eta(kh))\widehat{x}(kh) \mid \eta(t_0)=j]-\mathbb{E}[|C_{z_d}(\eta(kh))\widehat{x}(kh)|^2 \mid \eta(t_0)=j].$$

Substituting this equality in (7.21) we obtain via (7.5) that:

$$\int_{kh}^{(k+1)h}\mathbb{E}[|z_c(t;t_0,x_0,\xi)|^2 \mid \eta(t_0)=j]\mathrm{d}t+\mathbb{E}[|z_d(kh;t_0,x_0,\xi) \mid \eta(t_0)=j]$$
$$=\mathbb{E}[\widehat{x}^\top(kh)\widetilde{X}(kh^-,\eta(kh)\widehat{x}(kh)) \mid \eta(t_0)=j] \tag{7.22}$$
$$-\mathbb{E}[\widehat{x}^\top((k+1)h)\widetilde{X}((k+1)h^-,\eta((k+1)h))\widehat{x}((k+1)h) \mid \eta(t_0)=j]$$
$$+\int_{kh}^{(k+1)h}\mathbb{E}[\mathrm{Tr}[\widetilde{X}(t,\eta(t))G_c(\eta(t))\boldsymbol{\Lambda}_c G_c^\top(\eta(t))] \mid \eta(t_0)=j]\mathrm{d}t$$

for all $j\in\mathcal{N}$, $x_0\in\mathbb{R}^n$, $t_0\in\mathbb{R}_+$, $k\geq\kappa(t_0)$. Let $T>\kappa(t_0)h-t_0$. One can easily verify that $\kappa(t_0)\leq\left[\frac{t_0+T}{h}\right]$. Summing (7.22) from $k=\kappa(t_0)$ up to $k=\left[\frac{t_0+T}{h}\right]$ (with the convention that when $k=[\frac{t_0+T}{h}]$, $(k+1)h$ is replaced by t_0+T), we obtain:

$$\int_{\kappa(t_0)h}^{t_0+T} \mathbb{E}[|z_c(t;t_0,x_0,\xi)|^2 \mid \eta(t_0)=j]\mathrm{d}t + \sum_{k=\kappa(t_0)}^{\left[\frac{t_0+T}{h}\right]} \mathbb{E}[|z_d(kh;t_0,x_0,\xi)|^2 \mid \eta(t_0)=j]$$
$$= \mathbb{E}[\widehat{x}^\top(\kappa(t_0)h)\widetilde{X}(\kappa(t_0)h^-,\eta(\kappa(t_0)h))\widehat{x}(\kappa(t_0)h) \mid \eta(t_0)=j] \tag{7.23}$$
$$- \mathbb{E}[\widehat{x}^\top(t_0+T)\widetilde{X}(t_0+T,\eta(t_0+T))\widehat{x}(t_0+T) \mid \eta(t_0)=j]$$
$$+ \int_{\kappa(t_0)h}^{t_0+T} \mathbb{E}[\mathrm{Tr}[\widetilde{X}(t,\eta(t))G_c(\eta(t))\boldsymbol{\Lambda}_c G_c^\top(\eta(t))] \mid \eta(t_0)=j]\mathrm{d}t$$

for all $j \in \mathcal{N}$, $t_0 \in \mathbb{R}_+$, $T > \kappa(t_0)h - t_0$, $x_0 \in \mathbb{R}^n$. If t_0 is such that $(\kappa(t_0)-1)h < t_0 < \kappa(t_0)h$ one obtains based on Itô's formula together with Lemma 3.1 and the linear differential equation (7.14a) that:

$$\int_{t_0}^{\kappa(t_0)h} \mathbb{E}[|z_c(t;t_0,x_0,\xi)|^2 \mid \eta(t_0)=j]\mathrm{d}t = x_0^\top \widetilde{X}(t_0,j)x_0$$
$$- \mathbb{E}[\widehat{x}^\top(\kappa(t_0)h)\widetilde{X}(\kappa(t_0)h^-,\eta(\kappa(t_0)h))\widehat{x}(\kappa(t_0)h) \mid \eta(t_0)=j]$$
$$+ \int_{t_0}^{\kappa(t_0)h} \mathbb{E}[\mathrm{Tr}[\widetilde{X}(t,\eta(t))G_c(\eta(t))\boldsymbol{\Lambda}_c G_c^\top(\eta(t))] \mid \eta(t_0)=j]\mathrm{d}t.$$

Adding this equality to (7.23) one gets:

$$\int_{t_0}^{t_0+T} \mathbb{E}[|z_c(t;t_0,x_0,\xi)|^2 \mid \eta(t_0)=j]\mathrm{d}t + \sum_{k=\kappa(t_0)}^{\left[\frac{t_0+T}{h}\right]} \mathbb{E}[|z_d(kh;t_0,x_0,\xi)|^2 \mid \eta(t_0)=j]$$
$$= \Psi_j(t_0,T,x_0) + \int_{t_0}^{t_0+T} \mathbb{E}[\mathrm{Tr}[\widetilde{X}(t,\eta(t))G_c(\eta(t))\boldsymbol{\Lambda}_c G_c^\top(\eta(t))]|\eta(t_0)=j]\mathrm{d}t \tag{7.24}$$

where

$$\Psi_j(t_0,T,x_0) := x_0^\top \widetilde{X}(t_0,j)x_0 - \mathbb{E}[x^\top(t_0+T;t_0,x_0,\xi)\widetilde{X}(t_0+T,\eta(t_0+T))$$
$$\cdot x(t_0+T;t_0,x_0,\xi) \mid \eta(t_0)=j]. \tag{7.25}$$

Based on Corollary 7.1 (c) together with (7.10a) we may conclude that $t \to \Psi_j(t_0;T,x_0) : \mathbb{R}_+ \to \mathbb{R}$ is a bounded function which confirms the validity of (7.17). Further, we write:

$$\int_{t_0}^{t_0+T} \mathbb{E}[\mathrm{Tr}[\widetilde{X}(t,\eta(t))G_c(\eta(t))\boldsymbol{\Lambda}_c G_c^\top(\eta(t))] \mid \eta(t_0)=j]\mathrm{d}t$$
$$= \sum_{i=1}^{N} \int_{t_0}^{t_0+T} \mathrm{Tr}[\widetilde{X}(t,i)G_c(i)\boldsymbol{\Lambda}_c G_c^\top(i)]\mathbb{E}[\chi_{\{\eta(t)=i\}} \mid \eta(t_0)=j]\mathrm{d}t,$$

where $\chi_{\{\eta(t)=i\}}$ denotes the indicator function of the event $\{\eta(t)=i\}$. Since $\mathbb{E}[\chi_{\{\eta(t)=i\}} \mid \eta(t_0)=j] = \mathscr{P}[\eta(t)=i \mid \eta(t_0)=j] = p_{ji}(t-t_0)$, for all $t \geq t_0$ we deduce that

$$\int_{t_0}^{t_0+T} \mathbb{E}[\mathrm{Tr}[\widetilde{X}(t,\eta(t))G_c(\eta(t))\boldsymbol{\Lambda}_c G_c^\top(\eta(t))] \mid \eta(t_0)=j]\mathrm{d}t \tag{7.26}$$
$$= \sum_{i=1}^{N} \int_{t_0}^{t_0+T} \mathrm{Tr}[\widetilde{X}(t,i)G_c(i)\boldsymbol{\Lambda}_c G_c^\top(i)]p_{ji}(t-t_0)\mathrm{d}t.$$

Substituting (7.26) in (7.24) we obtain (7.16) which completes the proof of part (i). To prove that (7.18) and (7.19) hold, let us consider $\breve{x}(t) \triangleq x(t; t_0, 0, \zeta_d)$ be the solution of the IVP (7.6). Applying Itô's formula on the intervals $[a_p, b_p] \subset (kh, (k+1)h)$ to the function $v(t,x,i) = x^\top \widetilde{X}(t,i)x$ and to the stochastic process $\breve{x}(t)$, $t \geq 0$, we obtain:

$$\mathbb{E}[v(b_p, \breve{x}(b_p), \eta(b_p)) \mid \eta(t_0)=j] - \mathbb{E}[v(a_p, \breve{x}(a_p), \eta(a_p)) \mid \eta(t_0)=j]$$
$$= \int_{a_p}^{b_p} \mathbb{E}\left[\breve{x}^\top(t)\left(\frac{\mathrm{d}}{\mathrm{d}t}\widetilde{X}(t,\eta(t)) + \mathscr{L}^*_{\eta(t)}[\widetilde{\mathbf{X}}(t)]\right)\breve{x}(t) \mid \eta(t_0)=j\right]\mathrm{d}t.$$

Letting $a_p \to kh$ and $b_p \to (k+1)h$ we obtain via Lemma 3.1 and the linear differential equation (7.14a), that:

$$\int_{kh}^{(k+1)h} \mathbb{E}[|C_{z_c}(\eta(t))x(t; t_0, 0, \zeta_d)|^2 \mid \eta(t_0)=j]\mathrm{d}t$$
$$= \mathbb{E}[\breve{x}^\top(kh^+)\widetilde{X}(kh, \eta(kh)\breve{x}(kh^+) \mid \eta(t_0)=j] \tag{7.27}$$
$$- \mathbb{E}[\breve{x}^\top((k+1)h)\widetilde{X}((k+1)h^-, \eta((k+1)h))\breve{x}((k+1)h) \mid \eta(t_0)=j].$$

Next, (7.6b), (7.14b) together with Assumption 7.1 lead to:

$$
\begin{aligned}
&\mathbb{E}[\breve{x}^\top(kh^+)\widetilde{X}(kh,\eta(kh))\breve{x}(kh^+) \mid \eta(t_0)=j]\\
&=\mathbb{E}[\breve{x}^\top(kh)\mathscr{I}^*_{\eta(kh)}[\widetilde{\mathbf{X}}(\eta(kh))]\breve{x}(kh) \mid \eta(t_0)=j]\\
&+\mathbb{E}[\mathrm{Tr}[\widetilde{X}(kh,\eta(kh))G_d(\eta(kh))\boldsymbol{\Lambda}_d G_d^\top(\eta(kh))] \mid \eta(t_0)=j]\\
&=\mathbb{E}[\breve{x}^\top(kh)\widetilde{X}(kh^-,\eta(kh))\breve{x}(kh) \mid \eta(t_0)=j]-\mathbb{E}[|C_{z_d}(\eta(kh))\breve{x}(kh)|^2 \mid \eta(t_0)=j]\\
&+\mathbb{E}[\mathrm{Tr}[\widetilde{X}(kh,\eta(kh))G_d(\eta(kh))\boldsymbol{\Lambda}_d G_d^\top(\eta(kh))] \mid \eta(t_0)=j].
\end{aligned}
$$

Substituting this equality in (7.27) we obtain via (7.7) that

$$
\begin{aligned}
&\int_{kh}^{(k+1)h}\mathbb{E}[|z_c(t;t_0,0,\zeta_d)|^2 \mid \eta(t_0)=j]dt+\mathbb{E}[|z_d(kh;t_0,0,\zeta_d)|^2 \mid \eta(t_0)=j]\\
&=\mathbb{E}[\breve{x}^\top(kh)\widetilde{X}(kh^-,\eta(kh))\breve{x}(kh) \mid \eta(t_0)=j] \qquad (7.28)\\
&-\mathbb{E}[\breve{x}^\top((k+1)h)\widetilde{X}((k+1)h^-,\eta((k+1)h))\breve{x}((k+1)h) \mid \eta(t_0)=j]\\
&+\mathbb{E}[(\mathrm{Tr}[\widetilde{X}(kh,\eta(kh))G_d(\eta(kh))\boldsymbol{\Lambda}_d G_d^\top(\eta(kh))]\\
&+\mathrm{Tr}[G_{z_d}(\eta(kh))\boldsymbol{\Lambda}_d G_{z_d}^\top(\eta(kh))]) \mid \eta(t_0)=j]
\end{aligned}
$$

for all $t_0 \in \mathbb{R}_+$, $\kappa(t_0) \le k \le \kappa(t_0)+p$, with $p \ge 1$, $j \in \mathcal{N}$. Let us remark that from (7.4) we get $(\kappa(t_0)+p)h < t_0+(p+1)h \le (\kappa(t_0)+p+1)h$, for any natural number $p \ge 1$. This remark allows us to rewrite the use of the Itô formula on the interval $[(\kappa(t_0)+p)h, t_0+(p+1)h]$ when $k=\kappa(t_0)+p$. Summing (7.28) from $k=\kappa(t_0)$ up to $k=\kappa(t_0)+p$ yields:

$$
\begin{aligned}
&\int_{\kappa(t_0)h}^{t_0+(p+1)h}\mathbb{E}[|z_c(t;t_0,0,\zeta_d)|^2 \mid \eta(t_0)=j]dt+\sum_{k=\kappa(t_0)}^{\kappa(t_0)+p}\mathbb{E}[|z_d(kh;t_0,0,\zeta_d)|^2 \mid \eta(t_0)=j]\\
&=\mathbb{E}[\breve{x}^\top(\kappa(t_0)h)\widetilde{X}(\kappa(t_0)h^-,\eta(\kappa(t_0)h))\breve{x}(\kappa(t_0)h) \mid \eta(t_0)=j] \qquad (7.29)\\
&-\mathbb{E}[\breve{x}^\top(t_0+(p+1)h)\widetilde{X}((t_0+(p+1)h)^-,\eta(t_0+(p+1)h))\breve{x}(t_0+(p+1)h) \mid \eta(t_0)=j]\\
&+\sum_{k=\kappa(t_0)}^{\kappa(t_0)+p}\mathbb{E}[\varphi(kh,\eta(kh)) \mid \eta(t_0)=j]
\end{aligned}
$$

where

$$
\begin{aligned}
\varphi(kh,\eta(kh)) &\triangleq \mathrm{Tr}[\widetilde{X}(kh,\eta(kh))G_d(\eta(kh))\boldsymbol{\Lambda}_d G_d^\top(\eta(kh))]\\
&+\mathrm{Tr}[G_{z_d}(\eta(kh))\boldsymbol{\Lambda}_d G_{z_d}^\top(\eta(kh))]. \qquad (7.30)
\end{aligned}
$$

If $(\kappa(t_0)-1)h < t_0 < \kappa(t_0)h$ one applies Itô's formula to the function $v(t,x,i)$ on the interval $[t_0,\kappa(t_0)h]$ to obtain via the differential equation (7.14a):

$$\int_{t_0}^{\kappa t_0 h} \mathbb{E}[|z_c(t; t_0, 0, \zeta_d)|^2 \mid \eta(t_0) = j]\mathrm{d}t$$
$$= -\mathbb{E}[\breve{x}^\top(\kappa(t_0)h)\widetilde{X}(\kappa(t_0)h^-, \eta(\kappa(t_0)h))\breve{x}(\kappa(t_0)h) \mid \eta(t_0) = j].$$

Adding this equality to (7.29) we obtain:

$$\int_{t_0}^{t_0+(p+1)h} \mathbb{E}[|z_c(t; t_0, 0, \zeta_d)|^2 \,|\eta(t_0) = j]\mathrm{d}t + \sum_{k=\kappa(t_0)}^{\kappa(t_0)+p} \mathbb{E}[|z_d(kh; t_0, 0, \zeta_d)|^2 \,|\eta(t_0) = j]$$
$$= \widehat{\Psi}_j(t_0, p) + \sum_{k=\kappa(t_0)}^{\kappa(t_0)+p} \mathbb{E}[\varphi(kh, \eta(kh)) \mid \eta(t_0) = j] \tag{7.31}$$

where $\widehat{\Psi}_j(t_0, p) \stackrel{\Delta}{=} -\mathbb{E}[\breve{x}^\top(t_0 + (p+1)h)\widetilde{X}((t_0 + (p+1)h)^-, \eta(t_0 + (p+1)h))\breve{x}$ $(t_0 + (p+1)h) \mid \eta(t_0) = j]$. Employing (7.9) together with Corollary 7.1 (*c*) we may conclude that (7.19) holds. Then, proceeding as in the derivation of (7.26) we may infer that

$$\sum_{k=\kappa(t_0)}^{\kappa(t_0)+p} \mathbb{E}[\varphi(kh, \eta(kh)) \mid \eta(t_0) = j] = \sum_{i=1}^{N} \sum_{k=\kappa(t_0)}^{\kappa(t_0)+p} \varphi(kh, i) p_{ji}(kh - t_0).$$

Substituting this equality in (7.31) we obtain via (7.30) that (7.17) holds. Thus the proof is complete.

Remark 7.2 Repeating step by step the proof of part (*ii*) from Lemma 7.1 it can be obtained the following version of (7.18):

$$\int_{t_0}^{t_0+T} \mathbb{E}[|z_c(t; t_0, 0, \zeta_d)|^2 \mid \eta(t_0) = j]\mathrm{d}t + \sum_{k=\kappa(t_0)}^{\left[\frac{t_0+T}{h}\right]} \mathbb{E}[|z_d(kh; t_0, 0, \zeta_d)|^2 \mid \eta(t_0) = j]$$
$$= \breve{\Psi}_j(t_0, T) + \sum_{i=1}^{N} \sum_{k=\kappa(t_0)}^{\left[\frac{t_0+T}{h}\right]} \Big(\mathrm{Tr}[\widetilde{X}(kh, i)G_d(i)\boldsymbol{\Lambda}_d G_d^\top(i)] \tag{7.32}$$
$$+ \mathrm{Tr}[G_{z_d}(i)\boldsymbol{\Lambda}_d G_{z_d}^\top(i)]\Big)\, p_{ji}(kh - t_0)$$

with

$$\lim_{T\to\infty} \frac{1}{T}\breve{\Psi}_j(t_0, T) = 0. \tag{7.33}$$

This technical result will be used in the next subsection (please see Remark 7.3(b)).

7.2.3 The Computation of the Value of the H_2-Norm

Let $\pi(t) = (\pi_1(t), \pi_2(t), \ldots, \pi_N(t))$ be the probability distribution of the random variable $\eta(t)$, i.e., $\pi_i(t) = \mathscr{P}\{\eta(t) = i\}$, $1 \le i \le N$, $t \in \mathbb{R}_+$. Let $\widetilde{P} = (\widetilde{p}_{ij})_{(i,j)\in\mathcal{N}\times\mathcal{N}}$ be the matrix defined by $\widetilde{P} = \lim_{t\to\infty} P(t)$. We set:

$$\pi_{i\infty} \stackrel{\Delta}{=} \sum_{j=1}^{N} \pi_j(0)\widetilde{p}_{ji}, \quad i \in \mathcal{N}. \tag{7.34}$$

The next result provides explicit formulae for the computation of the H_2-norm of a JLSS $\mathbb{G}$ introduced in (7.11) and (7.12).

Theorem 7.1 *Assume:*

(a) Assumptions 1.2 and 7.1 are fulfilled;
(b) the JSLDE (7.2) is ESMS.

Under these conditions the value of the H_2-norm introduced by (7.11) and (7.12) is expressed by:

$$\begin{aligned}\|\mathbb{G}\|^2 &= \sum_{i=1}^{N} \frac{\pi_{i\infty}}{h} \int_0^h \mathrm{Tr}[\widetilde{X}(t,i)G_c(i)\boldsymbol{\Lambda}_c G_c^\top(i)]\mathrm{d}t \\ &\quad + \sum_{i=1}^{N} \pi_{i\infty}(\mathrm{Tr}[\widetilde{X}(0,i)G_d(i)\boldsymbol{\Lambda}_d G_d^\top(i)] + \mathrm{Tr}[G_{z_d}(i)\boldsymbol{\Lambda}_d G_{z_d}^\top(i)]) \\ &= \sum_{i=1}^{N} \Bigg(\int_0^h \mathrm{Tr}[C_{z_c}(i)\widetilde{Y}(t,i)C_{z_c}^\top(i)]\mathrm{d}t \\ &\quad + \mathrm{Tr}[C_{z_d}(i)\widetilde{Y}(0,i)C_{z_d}^\top(i)] + \pi_{i\infty}\,\mathrm{Tr}[G_{z_d}(i)\boldsymbol{\Lambda}_d G_{z_d}^\top(i)]\Bigg)\end{aligned} \tag{7.35}$$

where $\widetilde{\mathbf{X}}(\cdot) = (\widetilde{X}(\cdot,1), \widetilde{X}(\cdot,2), \ldots, \widetilde{X}(\cdot,N))$ is the unique bounded on $\mathbb{R}$ solution of the BGLJLDE (7.14) while $\widetilde{\mathbf{Y}}(\cdot) = (\widetilde{Y}(\cdot,1), \widetilde{Y}(\cdot,2), \ldots, \widetilde{Y}(\cdot,N))$ is the unique bounded on $\mathbb{R}$ solution of the FGLJLDE (7.15) with $\pi_{i\infty}$ computed as in (7.34).

Proof Involving the equalities (7.16) we can write

$$\int_{t_0}^{t_0+T} \mathbb{E}[|z_c(t; t_0, x_0, \xi)|^2]\mathrm{d}t + \sum_{k=\kappa(t_0)}^{\left[\frac{t_0+T}{h}\right]} \mathbb{E}[|z_d(kh; t_0, x_0, \xi)|^2] \tag{7.36}$$

$$= \sum_{j=1}^{N} \pi_j(t_0) \left\{ \int_{t_0}^{t_0+T} \mathbb{E}[|z_c(t; t_0, x_0, \xi)|^2 | \eta(t_0) = j]\mathrm{d}t \right.$$

$$+ \sum_{k=\kappa(t_0)}^{\left[\frac{t_0+T}{h}\right]} \mathbb{E}[|z_d(kh; t_0, x_0, \xi)|^2 | \eta(t_0) = j] \Bigg\}$$

$$= \sum_{j=1}^{N} \pi_j(t_0)\Psi_j(t_0, T, x_0) + \sum_{j=1}^{N}\sum_{i=1}^{N} \pi_j(t_0)$$
$$\times \int_{t_0}^{t_0+T} \mathrm{Tr}[\widetilde{X}(t,i)G_c(i)\boldsymbol{\Lambda}_c G_c^\top(i)] p_{ji}(t-t_0)\mathrm{d}t.$$

According to (7.17), we get

$$\lim_{T\to\infty} \frac{1}{T}\left(\sum_{j=1}^{N} \pi_j(t_0)\Psi_j(t_0, T, x_0)\right) = 0, \tag{7.37}$$

for all $(t_0, x_0) \in \mathbb{R}_+ \times \mathbb{R}^n$. From $\pi_j(t_0) = \sum_{l=1}^{N} \pi_l(0) p_{lj}(t_0)$ we deduce that $\sum_{j=1}^{N} \pi_j(t_0) p_{ji}(t-t_0) = \sum_{l=1}^{N} \pi_l(0) p_{li}(t)$. This allows us to write

$$\begin{aligned}
&\sum_{j=1}^{N}\sum_{i=1}^{N} \pi_j(t_0) \int_{t_0}^{t_0+T} \mathrm{Tr}[\widetilde{X}(t,i)G_c(i)\boldsymbol{\Lambda}_c G_c^\top(i)] p_{ji}(t-t_0)\mathrm{d}t \\
&= \sum_{i=1}^{N} \pi_{i\infty} \int_{t_0}^{t_0+T} \mathrm{Tr}[\widetilde{X}(t,i)G_c(i)\boldsymbol{\Lambda}_c G_c^\top(i)]\mathrm{d}t \\
&+ \sum_{i=1}^{N}\sum_{l=1}^{N} \pi_l(0) \int_{t_0}^{t_0+T} \mathrm{Tr}[\widetilde{X}(t,i)G_c(i)\boldsymbol{\Lambda}_c G_c^\top(i)](p_{li}(t) - \widetilde{p}_{li})\mathrm{d}t.
\end{aligned} \tag{7.38}$$

Bearing in mind that from the definition of the matrix $\widetilde{P}$ we know that $\lim_{t\to\infty}(p_{li}(t) - \widetilde{p}_{li}) = 0$, for all $(l, i) \in \mathcal{N} \times \mathcal{N}$, we may infer that

$$\lim_{T\to\infty} \frac{1}{T} \int_{t_0}^{t_0+T} \mathrm{Tr}[\widetilde{X}(t,i)G_c(i)\boldsymbol{\Lambda}_c G_c^\top(i)](p_{li}(t) - \widetilde{p}_{li})\mathrm{d}t = 0, \tag{7.39}$$

for all $(l, i) \in \mathcal{N} \times \mathcal{N}$. Further, from (7.12a) combined by (7.36)–(7.39) we may conclude that

$$\widehat{\|\mathbb{G}\|}_2^2 = \limsup_{T\to\infty} \frac{1}{T}\left(\sum_{i=1}^{N} \pi_{i\infty} \int_{t_0}^{t_0+T} \mathrm{Tr}[\widetilde{X}(t,i)G_c(i)\boldsymbol{\Lambda}_c G_c^\top(i)]\mathrm{d}t\right). \tag{7.40}$$

On the other hand,

$$\limsup_{T\to\infty}\frac{1}{T}\left(\sum_{i=1}^{N}\pi_{i\infty}\int_{t_0}^{t_0+T}\mathrm{Tr}[\widetilde{X}(t,i)G_c(i)\boldsymbol{\Lambda}_c G_c^{\top}(i)]\mathrm{d}t\right)$$
$$=\limsup_{T\to\infty}\frac{1}{T}\left(\sum_{i=1}^{N}\pi_{i\infty}\int_{0}^{T}\mathrm{Tr}[\widetilde{X}(t,i)G_c(i)\boldsymbol{\Lambda}_c G_c^{\top}(i)]\mathrm{d}t\right) \tag{7.41}$$

because the integrand is a bounded function. Further,

$$\limsup_{T\to\infty}\frac{1}{T}\left(\sum_{i=1}^{N}\pi_{i\infty}\int_{0}^{T}\mathrm{Tr}[\widetilde{X}(t,i)G_c(i)\boldsymbol{\Lambda}_c G_c^{\top}(i)]\mathrm{d}t\right)$$
$$=\sum_{i=1}^{N}\frac{\pi_{i\infty}}{h}\int_{0}^{h}\mathrm{Tr}[\widetilde{X}(t,i)G_c(i)\boldsymbol{\Lambda}_c G_c^{\top}(i)]\mathrm{d}t \tag{7.42}$$

because the integrand is a periodic function of period h. So, (7.40)–(7.42) yield

$$\widehat{\|\mathbb{G}\|}_2^2=\sum_{i=1}^{N}\frac{\pi_{i\infty}}{h}\int_{0}^{h}\mathrm{Tr}[\widetilde{X}(t,i)G_c(i)\boldsymbol{\Lambda}_c G_c^{\top}(i)]\mathrm{d}t. \tag{7.43}$$

Now, involving (7.18) we may write for each natural number $p\geq 1$, the equalities:

$$\int_{t_0}^{t_0+(p+1)h}\mathbb{E}[|z_c(t;t_0,x_0,\zeta_d)|^2]\mathrm{d}t+\sum_{k=\kappa(t_0)}^{\kappa(t_0)+p}\mathbb{E}[|z_d(kh;t_0,0,\zeta_d)|^2]$$
$$=\sum_{j=1}^{N}\pi_j(t_0)\left(\int_{t_0}^{t_0+(p+1)h}\mathbb{E}[|z_c(t;t_0,0,\zeta_d)|^2\mid\eta(t_0)=j]\mathrm{d}t\right.$$
$$\left.+\sum_{k=\kappa(t_0)}^{\kappa(t_0)+p}\mathbb{E}[|z_d(kh;t_0,0,\zeta_d)|^2\mid\eta(t_0)=j]\right) \tag{7.44}$$
$$=\sum_{j=1}^{N}\pi_j(t_0)\widehat{\Psi}_j(t_0,p)+\sum_{j=1}^{N}\sum_{i=1}^{N}\sum_{k=\kappa(t_0)}^{\kappa(t_0)+p}\pi_j(t_0)\varphi(kh,i)p_{ji}(kh-t_0)$$

where $\varphi(k,i)$ is defined in (7.30). Based on (7.19) we deduce that

$$\lim_{p\to\infty}\frac{1}{p}\left(\sum_{j=1}^{N}\pi_j(t_0)\widehat{\Psi}_j(t_0,p)\right)=0, \tag{7.45}$$

for all $t_0\in\mathbb{R}_+$. On the other hand, the notation introduced in the first part of the proof allow us to write

$$\sum_{j=1}^{N}\sum_{i=1}^{N}\sum_{k=\kappa(t_0)}^{\kappa(t_0)+p}\pi_j(t_0)\varphi(kh,i)p_{ji}(kh-t_0) \tag{7.46}$$
$$=\sum_{i=1}^{N}\sum_{k=\kappa(t_0)}^{\kappa(t_0)+p}\pi_{i\infty}\varphi(kh,i)+\sum_{i=1}^{N}\sum_{l=1}^{N}\sum_{k=\kappa(t_0)}^{\kappa(t_0)+p}\pi_l(0)\varphi(kh,i)(p_{li}(kh)-\widetilde{p}_{li}).$$

One can check that

$$\lim_{p\to\infty}\frac{1}{p}\left(\sum_{k=\kappa(t_0)}^{\kappa(t_0)+p}\varphi(kh,i)(p_{li}(kh)-\widetilde{p}_{li})\right)=0, \tag{7.47}$$

for all $(l,i)\in\mathcal{N}\times\mathcal{N}$, $t_0\in\mathbb{R}_+$, because the sequences $\{\varphi(kh,i)\}_{k\in\mathbb{Z}_+}$ are bounded and $\lim_{k\to\infty}(p_{li}(kh)-\widetilde{p}_{li})=0$. Now, (7.12b) together with (7.44)–(7.47) give

$$\breve{\|}\mathbb{G}\breve{\|}_2^2=\limsup_{p\to\infty}\frac{1}{p}\left(\sum_{i=1}^{N}\sum_{k=\kappa(t_0)}^{\kappa(t_0)+p}\pi_{i\infty}\varphi(kh,i)\right).$$

From (7.30) and Corollary 7.1 (b) we deduce that for each mode i of the Markov process we have

$$\breve{\|}\mathbb{G}\breve{\|}_2^2=\sum_{i=1}^{N}\pi_{i\infty}\varphi(0,i)=\sum_{i=1}^{N}\pi_{i\infty}(\mathrm{Tr}[\widetilde{X}(0,i)G_c(i)\boldsymbol{\Lambda}_cG_c^{\top}(i)]+\mathrm{Tr}[G_{z_d}(i)\boldsymbol{\Lambda}_dG_{z_d}^{\top}(i)]). \tag{7.48}$$

Plugging (7.43) and (7.48) in (7.11) we obtain the first equality from (7.35). In order to prove the validity of the second equality from (7.35) we rewrite (7.43) successively (involving (7.15a) written for $\pi_{i\infty}$ computed via (7.34)):

$$\widehat{\|\mathbb{G}\|}_2^2 = \int_0^h \left(\sum_{i=1}^{N} \mathrm{Tr}\left[\widetilde{X}(t,i) \left(\frac{\pi_{i\infty}}{h} G_c(i) \boldsymbol{\Lambda}_c G_c^\top(i) \right) \right] \right) \mathrm{d}t$$

$$= \int_0^h \langle \widetilde{\mathbf{X}}(t), \frac{\mathrm{d}}{\mathrm{d}t} \widetilde{\mathbf{Y}}(t) \rangle \mathrm{d}t - \int_0^h \langle \widetilde{\mathbf{X}}(t), \mathscr{L}[\widetilde{\mathbf{Y}}(t)] \rangle \mathrm{d}t.$$

Further, the use of (7.14a) yields:

$$\widehat{\|\mathbb{G}\|}_2^2 = \int_0^h \langle \widetilde{\mathbf{X}}(t), \frac{\mathrm{d}}{\mathrm{d}t} \widetilde{\mathbf{Y}}(t) \rangle \mathrm{d}t - \int_0^h \langle \mathscr{L}^*[\widetilde{\mathbf{X}}(t)], \widetilde{\mathbf{Y}}(t) \rangle \mathrm{d}t$$

$$= \int_0^h \langle \widetilde{\mathbf{X}}(t), \frac{\mathrm{d}}{\mathrm{d}t} \widetilde{\mathbf{Y}}(t) \rangle \mathrm{d}t + \int_0^h \langle \frac{\mathrm{d}}{\mathrm{d}t} \widetilde{\mathbf{X}}(t), \widetilde{\mathbf{Y}}(t) \rangle \mathrm{d}t + \sum_{i=1}^{N} \int_0^h \mathrm{Tr}[C_{z_c}^\top(i) C_{z_c}(i) \widetilde{Y}(t,i)] \mathrm{d}t$$

$$= \int_0^h \frac{\mathrm{d}}{\mathrm{d}t} \langle \widetilde{\mathbf{X}}(t), \widetilde{\mathbf{Y}}(t) \rangle \mathrm{d}t + \sum_{i=1}^{N} \int_0^h \mathrm{Tr}[C_{z_c}(i) \widetilde{Y}(t,i) C_{z_c}^\top(i)] \mathrm{d}t.$$

Bearing in mind the properties of the functions $\widetilde{\mathbf{X}}(\cdot)$ and $\widetilde{\mathbf{Y}}(\cdot)$ stated in Corollary 7.1 and Corollary 7.2, respectively, we can write:

$$\widehat{\|\mathbb{G}\|}_2^2 = \langle \widetilde{\mathbf{X}}(h^-), \widetilde{\mathbf{Y}}(h) \rangle - \langle \widetilde{\mathbf{X}}(0), \widetilde{\mathbf{Y}}(0^-) \rangle + \sum_{i=1}^{N} \int_0^h \mathrm{Tr}[C_{z_c}(i) \widetilde{Y}(t,i) C_{z_c}^\top(i)] \mathrm{d}t. \tag{7.49}$$

Let $\mathbf{C}_{z_d} \stackrel{\Delta}{=} \left(C_{z_d}^\top(1) C_{z_d}(1), C_{z_d}^\top(2) C_{z_d}(2), \ldots, C_{z_d}^\top(N) C_{z_d}(N) \right)$. With this, Eq. (7.14b) may be written in a compact form as: $\widetilde{\mathbf{X}}(kh^-) = \mathscr{I}^*[\widetilde{\mathbf{X}}(kh)] + \mathbf{C}_{z_d}$. Using this equation for $k = 1$ we obtain: $\langle \widetilde{\mathbf{X}}(h^-), \widetilde{\mathbf{Y}}(h) \rangle = \langle \mathscr{I}^*[\widetilde{\mathbf{X}}(h)], \widetilde{\mathbf{Y}}(h) \rangle + \langle \mathbf{C}_{z_d}, \widetilde{\mathbf{Y}}(h) \rangle = \langle \widetilde{\mathbf{X}}(h), \mathscr{I}[\widetilde{\mathbf{Y}}(h)] \rangle + \langle \mathbf{C}_{z_d}, \widetilde{\mathbf{Y}}(h) \rangle$ or equivalently

$$\langle \widetilde{\mathbf{X}}(h^-), \widetilde{\mathbf{Y}}(h) \rangle = \langle \widetilde{\mathbf{X}}(h), \widetilde{\mathbf{Y}}(h^+) \rangle - \langle \widetilde{\mathbf{X}}(h), \widehat{\mathbf{G}}_d \rangle + \langle \mathbf{C}_{z_d}, \widetilde{\mathbf{Y}}(h) \rangle, \tag{7.50}$$

where $\widehat{\mathbf{G}}_d \stackrel{\Delta}{=} \left(\pi_{1\infty} G_d(1) \boldsymbol{\Lambda}_d G_d^\top(1), \pi_{2\infty} G_d(2) \boldsymbol{\Lambda}_d G_d^\top(2), \ldots, \pi_{N\infty} G_d(N) \boldsymbol{\Lambda}_d G_d^\top (N) \right)$. Substituting (7.50) in (7.49) and taking into account that $\widetilde{\mathbf{X}}(h) = \widetilde{\mathbf{X}}(0)$, $\widetilde{\mathbf{Y}}(h) = \widetilde{\mathbf{Y}}(0)$ and $\widetilde{\mathbf{Y}}(h^+) = \widetilde{\mathbf{Y}}(0^+)$ we deduce that

$$\widehat{\|\mathbb{G}\|}_2^2 = \langle \mathbf{C}_{z_d}, \widetilde{\mathbf{Y}}(0) \rangle - \langle \widetilde{\mathbf{X}}(0), \widehat{\mathbf{G}}_d \rangle + \sum_{i=1}^{N} \int_0^h \mathrm{Tr}[C_{z_c}(i) \widetilde{Y}(t,i) C_{z_c}^\top(i)] \mathrm{d}t$$

or equivalently,

$$\widehat{\|\mathbb{G}\|}_2^2 = \sum_{i=1}^{N} \mathrm{Tr}[C_{z_d}(i)\widetilde{Y}(0,i)C_{z_d}^{\top}(i)] + \sum_{i=1}^{N} \int_0^h \mathrm{Tr}[C_{z_c}(i)\widetilde{Y}(t,i)C_{z_c}^{\top}(i)]\mathrm{d}t$$
$$- \sum_{i=1}^{N} \pi_{i\infty} \mathrm{Tr}[\widetilde{X}(0,i)G_d(i)\boldsymbol{\Lambda}_d G_d^{\top}(i)]. \tag{7.51}$$

Finally, (7.11) together with (7.48) and (7.51) give the second equality from (7.35). Thus the proof is complete.

Remark 7.3

(a) comparing the result proved in Theorem 7.1 with the one from Theorem 4.1 from Ichikawa and Katayama [1] one sees that the second and third parts of (7.35) reduce to the second and third parts, respectively, from (4.8) in the aforementioned reference, when the stochastic component from the mathematical model (7.1) vanishes. This may clarify the choice of the two kinds of limits in (7.12a) and (7.12b). At the end of this section we shall discuss the case when in the definition of the norm $\breve{\|\cdot\|}_2$ the limit for $p \to \infty$ (with p natural number) is replaced with the limit for $T \to \infty$ with T real number;
(b) following step by step the derivation of the equality (7.48), but replacing (7.18) by (7.32), we obtain that

$$\limsup_{T\to\infty} \frac{1}{T}\left(\int_{t_0}^{t_0+T} \mathbb{E}[|z_c(t;t_0,0,\zeta_d)|^2]\mathrm{d}t + \sum_{k=\kappa(t_0)}^{\left[\frac{t_0+T}{h}\right]} \mathbb{E}[|z_d(kh;t_0,0,\zeta_d)|^2] \right)$$
$$= \sum_{i=1}^{N} \frac{\pi_{i\infty}}{h}(\mathrm{Tr}[\widetilde{X}(0,i)G_d(i)\boldsymbol{\Lambda}_d G_d^{\top}(i)] + \mathrm{Tr}[G_{z_d}(i)\boldsymbol{\Lambda}_d G_{z_d}^{\top}(i)]) = \frac{1}{h}\breve{\|\mathbb{G}\|}_2^2. \tag{7.52}$$

This technical result will be used at the end of the section for the definition of the norm $\breve{\|\cdot\|}_2$.

Remark 7.4 From (7.43) and (7.48) one sees that the values of the limits from the right hand side of (7.12) do not depend neither the initial time t_0 nor the initial state x_0. Hence, the H_2-norm of the JLSS (7.1) is not affected by the initial pair (t_0, x_0) of its trajectories. The H_2-norm only depends on the so-called **permanent regime** generated by additive perturbations $\{\xi(t)\}_{t\geq 0}$ and $\{\zeta_d(k)\}_{k\in\mathbb{Z}_+}$. Hence, without loss of generality, in (7.12) we can take $(t_0, x_0) = (0, 0) \in \mathbb{R}_+ \times \mathbb{R}^n$,:

$$\widehat{\|\mathbb{G}\|}_2^2 = \limsup_{T\to\infty} \frac{1}{T} \left(\int_0^T \mathbb{E}[|z_c(t;0,0,\xi)|^2]dt + \sum_{k=0}^{[\frac{T}{h}]} \mathbb{E}[|z_d(kh;0,0,\xi)|^2] \right) \tag{7.53a}$$

$$\begin{aligned} \check{\|\mathbb{G}\|}_2^2 &= \limsup_{p\to\infty} \frac{1}{p} \left(\int_0^{(p+1)h} \mathbb{E}[|z_c(t;0,0,\zeta_d)|^2]dt + \sum_{k=0}^{p} \mathbb{E}[|z_d(kh;0,0,\zeta_d)|^2] \right) \\ &= h \limsup_{T\to\infty} \frac{1}{T} \left(\int_0^T \mathbb{E}[|z_c(t;0,0,\zeta_d)|^2]dt + \sum_{k=0}^{[\frac{T}{h}]} \mathbb{E}[|z_d(kh;0,0,\zeta_d)|^2] \right) \end{aligned} \tag{7.53b}$$

$$\|\mathbb{G}\|_2 = \left(\widehat{\|\mathbb{G}\|}_2^2 + \check{\|\mathbb{G}\|}_2^2 \right)^{1/2}. \tag{7.53c}$$

7.2.4 *Some Observations Regarding Robustness*

From (7.35) one sees that in the case when the behavior of the JLSS $\mathbb{G}$ is subject to some Markovian switchings with $N \geq 2$, then value of the H_2-norm $\|\mathbb{G}\|_2$ depends upon the initial probability distribution $\pi(0) = (\pi_1(0), \pi_2(0), \ldots, \pi_N(0))$ of the Markov process $\{\eta(t)\}_{t\geq 0}$. To overcome this inconvenient one can replace the H_2-norm defined by (7.11) and (7.12) with the following norm:

$$\begin{aligned} \widehat{|||\mathbb{G}|||}_2^2 = \limsup_{T\to\infty} \frac{1}{T} \Bigg(\sum_{j=1}^{N} \Bigg[\int_0^T \mathbb{E}[|z_c(t;0,0,\xi)|^2 \mid \eta(0)=j]\ dt \\ + \sum_{k=0}^{[\frac{T}{h}]} \mathbb{E}[|z_d(kh;0,0,\xi)|^2 \mid \eta(0)=j]\Bigg]\Bigg) \end{aligned} \tag{7.54a}$$

$$\begin{aligned} \check{|||\mathbb{G}|||}_2^2 = \limsup_{p\to\infty} \frac{1}{p} \Bigg(\sum_{j=1}^{N} \Bigg[\int_0^{(p+1)h} \mathbb{E}[|z_c(t;0,0,\zeta_d)|^2 \mid; \eta(0)=j]dt \\ + \sum_{k=0}^{p} \mathbb{E}[|z_d(kh;0,0,\zeta_d)|^2 \mid \eta(0)=j]\Bigg]\Bigg) \end{aligned} \tag{7.54b}$$

$$|||\mathbb{G}|||_2 = \left(\widehat{|||\mathbb{G}|||}_2^2 + \check{|||\mathbb{G}|||}_2^2 \right)^{1/2}. \tag{7.54c}$$

The value of the norm introduced by (7.54) is given in the next theorem:

Theorem 7.2 *Under the assumptions from Theorem 7.1 the following equalities hold:*

$$
\begin{aligned}
|||\mathbb{G}|||_2^2 &= \sum_{i=1}^{N} \frac{\nu_i}{h} \int_0^h \mathrm{Tr}[\widetilde{X}(t,i)G_c(i)\Lambda_c G_c^{\top}(i)]\mathrm{d}t \\
&\quad + \sum_{i=1}^{N} \nu_i (\mathrm{Tr}[\widetilde{X}(0,i)G_d(i)\Lambda_d G_d^{\top}(i)] + \mathrm{Tr}[G_{z_d}(i)\Lambda_d G_{z_d}(i)]) \qquad (7.55)\\
&= \sum_{i=1}^{N} \left(\int_0^h \mathrm{Tr}[C_{z_c}(i)\widetilde{Y}_\mu(t,i)C_{z_c}^{\top}(i)]\mathrm{d}t + \mathrm{Tr}[C_{z_d}(i)\widetilde{Y}_\nu(0,i)C_{z_d}^{\top}(i)] + \nu_i \,\mathrm{Tr}[G_{z_d}\Lambda_d G_{z_d}^{\top}(i)] \right)
\end{aligned}
$$

where

$$
\nu_i = \sum_{j=1}^{N} \widetilde{p}_{ji}, \quad i \in \mathcal{N}, \tag{7.56}
$$

and $\widetilde{\mathbf{X}}(\cdot) = (\widetilde{X}(\cdot,1), \widetilde{X}(\cdot,2), \ldots, \widetilde{X}(\cdot,N))$ *is the unique bounded on* $\mathbb{R}$ *solution of the BGLJLDE (7.13), while* $\widetilde{\mathbf{Y}}_\nu(\cdot) = (\widetilde{Y}_\nu(\cdot,1), \widetilde{Y}_\nu(\cdot,2), \ldots, \widetilde{Y}_\nu(\cdot,N))$ *is the unique and bounded on* $\mathbb{R}$ *solution of the FGLJLDE (7.15) with* $\pi_{i\infty}$ *replaced by* ν_i, $1 \le i \le N$.

Proof The proof may be done following step by step the proof of Theorem 7.1. The details are left to the reader.

Remark 7.5 From (7.34) and (7.56) one sees that $\pi_{i\infty} \le \nu_i$, $1 \le i \le N$ for all initial probability distribution $\pi(0)$ of the Markov process. Hence, (7.35) and (7.55) give

$$
\|\mathbb{G}\|_2 \le |||\mathbb{G}|||_2.
$$

Let us introduce the notation $\mathbb{G}(\Lambda_c, \Lambda_d)$ to mark the dependence of the behavior of the trajectories of the JLSS (7.1) with respect to the covariance matrices of the additive white noise perturbations $\{\xi(t)\}_{t\ge 0}$ and $\{\zeta_d(k)\}_{k\in\mathbb{Z}_+}$.

From (7.35) and (7.55) one obtains:

Corollary 7.3 *Under the assumptions from Theorem 7.1 we have:* $\|\mathbb{G}(\Lambda_{c1}, \Lambda_{d1})\|_2 \le \|\mathbb{G}(\Lambda_{c2}, \Lambda_{d2})\|_2$ *and* $|||\mathbb{G}(\Lambda_{c1}, \Lambda_{d1})|||_2 \le |||\mathbb{G}(\Lambda_{c2}, \Lambda_{d2})|||_2$ *if* $\Lambda_{c1} \le \Lambda_{c2}$ *and* $\Lambda_{d1} \le \Lambda_{d2}$.

Beside the norm $\|\mathbb{G}\|_2$ introduced via formulae (7.11)–(7.12), or equivalently by (7.53) we define a new type H_2-norm of the dynamical system $\mathbb{G}$ having the state-space representation (7.1). This new norm is defined by:

$$
\begin{aligned}
\widetilde{|||\mathbb{G}|||}_2^2 &\triangleq \limsup_{T\to\infty} \frac{1}{T} left(\int_0^T \mathbb{E}[|z_c(t;0,0,\xi)|^2]\mathrm{d}t + \sum_{k=0}^{\left[\frac{T}{h}\right]} \mathbb{E}[|z_d(kh;0,0,\xi)|^2] right) \\
&\quad + \limsup_{T\to\infty} \frac{1}{T} \left(\int_0^T \mathbb{E}[|z_c(t;0,0,\zeta_d)|^2]\mathrm{d}t + \sum_{k=0}^{\left[\frac{T}{h}\right]} \mathbb{E}[|z_d(kh;0,0,\zeta_d)|^2] \right) \qquad (7.57)
\end{aligned}
$$

where $z_c(\cdot; 0, 0, \xi)$ and $z_d(\cdot; 0, 0, \xi)$ are the same as in (7.53a) and $z_c(\cdot; 0, 0, \zeta_d)$, $z_d(\cdot; 0, 0, \zeta_d)$ are as in (7.53b). Explicit formulae of the value of the norm introduced by (7.57) are given in the next theorem:

Theorem 7.3 *Under the assumptions of Theorem 7.1 we have:*

$$\begin{aligned}\widetilde{\|\mathbb{G}\|}_2^2 &= \sum_{i=1}^{N} \frac{\pi_{i\infty}}{h}\left(\int_0^h \mathrm{Tr}[\widetilde{X}(t,i)G_c(i)\boldsymbol{\Lambda}_c G_c^\top(i)]\mathrm{d}t \right.\\ &\quad \left. + \mathrm{Tr}[\widetilde{X}(0,i)G_d(i)\boldsymbol{\Lambda}_d G_d^\top(i)] + \mathrm{Tr}[G_{z_d}(i)\boldsymbol{\Lambda}_d G_{z_d}^\top(i)]\right) \\ &= \sum_{i=1}^{N} \frac{1}{h}\left(\int_0^h \mathrm{Tr}[C_{z_c}(i)\tilde{\tilde{Y}}(t,i)C_{z_c}^\top(i)]\mathrm{d}t + \mathrm{Tr}[C_{z_d}(i)\tilde{\tilde{Y}}(0,i)C_{z_d}^\top(i)] \right.\\ &\quad \left. + pi_{i\infty}\,\mathrm{Tr}[G_{z_d}(i)\boldsymbol{\Lambda}_d G_{z_d}^\top(i)]\right)\end{aligned} \tag{7.58}$$

where $\widetilde{\mathbf{X}}(\cdot) = (\widetilde{X}(\cdot,1), \widetilde{X}(\cdot,2), \ldots, \widetilde{X}(\cdot,N))$ is the unique bounded on $\mathbb{R}$ solution of the BGLJLDE (7.13) and $\tilde{\tilde{\mathbf{Y}}}(\cdot) = (\tilde{\tilde{Y}}(\cdot,1), \tilde{\tilde{Y}}(\cdot,2), \ldots, \tilde{\tilde{Y}}(\cdot,N))$ is the unique bounded on $\mathbb{R}$ solution of the following FGLJLDE:

$$\begin{aligned}\dot{Y}(t,i) &= A_0(i)Y(t,i) + Y(t,i)A_0^\top(i) + \sum_{l=1}^{r_1} A_l(i)Y(t,i)A_l^\top(i) \\ &\quad + \sum_{j=1}^{N} q_{ji}Y(t,j) + \pi_{i\infty}G_c(i)\boldsymbol{\Lambda}_c G_c^\top(i), \quad kh < t \le (k+1)h\end{aligned} \tag{7.59a}$$

$$Y(kh^+,i) = \sum_{p=0}^{r_2} A_{dp}(i)Y(kh,i)A_{dp}^\top(i) + \pi_{i\infty}G_d(i)\boldsymbol{\Lambda}_d G_d^\top(i), \quad k \in \mathbb{Z}. \tag{7.59b}$$

Proof The first equality from (7.58) follows combining (7.57) with (7.43). To check the validity of the second equality from (7.58) one may proceed as in the last part of the proof of Theorem 7.1. In this case the FGLJLDE (7.59) will be involved instead of (7.15). The details are left to the reader.

Remark 7.6 Comparing the first equality from (7.35) with the first equality from (7.58) we conclude that $\|\mathbb{G}\|_2 \le \widetilde{\|\mathbb{G}\|}_2$ if $0 < h \le 1$ and $\widetilde{\|\mathbb{G}\|}_2 \le \|\mathbb{G}\|_2$, respectively, if $h \ge 1$. The above result can be used as a guideline when considering an H_2 optimal controller synthesis problem. Indeed, when the so-called dwell time h is less than 1, then an H_2 optimal controller constructed on the basis of the norm $\|\cdot\|_2$ introduced via (7.53) will provide a better disturbance attenuation when compared to the one constructed based on the norm $\widetilde{\|\cdot\|}_2$ introduced via (7.57).
If $h \ge 1$, the best attenuation of the effect of some white noise-type additive disturbances will be achieved by a H_2 optimal controller constructed based on the norm introduced via (7.57).

7.3 H_2 Optimal Control of an Impulsive Controlled Linear Stochastic System Under Perfect State Measurements

Consider an ICLSS ($\mathbb{G}$) having the state-space representation described by:

$$\begin{aligned}
\mathrm{d}x(t) &= A_0(\eta(t))x(t)\mathrm{d}t + \sum_{l=1}^{r_1} A_l(\eta(t))x(t)\mathrm{d}w_l(t) \\
&\quad + G_c(\eta(t))\mathrm{d}\xi(t), \quad kh < t \leq (k+1)h && (7.60a) \\
x(kh^+) &= A_{d0}(\eta(kh))x(kh) + B_{d0}(\eta(kh))u(k) + \sum_{p=1}^{r_2} w_{dp}(k)(A_{dp}(\eta(kh))x(kh) \\
&\quad + B_{dp}(\eta(kh))u(k)) + G_d(\eta(kh))\zeta_d(k), \quad k \in \mathbb{Z}_+ && (7.60b) \\
z_c(t) &= C_c(\eta(t))x(t), \quad t \in \mathbb{R}_+ && (7.60c) \\
z_d(kh) &= C_d(\eta(kh))x(kh)) + D_d(\eta(kh))u(k), \quad k \in \mathbb{Z}_+ && (7.60d)
\end{aligned}$$

where $x(t) \in \mathbb{R}^n$ are the vectors of the state parameters, $z_c(t) \in \mathbb{R}^{n_{z_c}}$ is a quality output evolving in continuous time, $z_d(kh) \in \mathbb{R}^{n_{z_d}}$ is a quality output evolving in discrete-time and $u(k) \in \mathbb{R}^m$ are the control parameters.

In this section we consider the case when the controls act on the dynamical system under consideration only in the impulsive instances $\tau_k = kh$, $k \in \mathbb{Z}_+$.

The stochastic processes $\{w_l(t)\}_{t\geq 0}$, $\{\xi(t)\}_{t\geq 0}$, $\{\zeta_d(k)\}_{k\in\mathbb{Z}_+}$ which affect the behavior of the dynamic of $\mathbb{G}$ satisfy Assumptions 1.2 and 7.1. For each $i \in \mathcal{N}$, $A_l(i)$, $0 \leq l \leq r_1$, $A_{dp}(i)$, $B_{dp}(i)$, $0 \leq p \leq r_2$, $C_c(i)$, $D_d(i)$, $G_c(i)$, $G_d(i)$, $D_d(i)$ are known matrices of compatible dimensions.

In this section we assume that at each instance time $\tau_k = kh$, $k \in \mathbb{Z}_+$, the exact values of the state vectors $x(kh)$ are available to the controllers involved in the control process of the ICLSS under consideration. So, the measured output which will be used in the control process of the ICLSS is:

$$y(kh) = x(kh), \quad k \in \mathbb{Z}_+. \tag{7.61}$$

To control an ICLSS having the state-space representation (7.60) and the measured output (7.61) we will use controllers $\mathbb{K}$ having the state-space representation:

$$\begin{aligned}
d\widehat{x}(t) &= \widehat{A}_0(\eta(t))\widehat{x}(t)\mathrm{d}t + \sum_{l=1}^{r_1} \widehat{A}_l(\eta(t))\widehat{x}(t)\mathrm{d}w_l(t), \quad kh < t \leq (k+1)h && (7.62a) \\
\widehat{x}(kh^+) &= \widehat{A}_{d0}(\eta(kh))\widehat{x}(kh) + \widehat{B}_{d0}(\eta(kh))\widehat{u}(k) + \sum_{p=1}^{r_2} w_{dp}(k)(\widehat{A}_{dp}(\eta(kh))\widehat{x}(kh) \\
&\quad + \widehat{B}_{dp}(\eta(kh))\widehat{u}(k)), \quad k \in \mathbb{Z}_+ && (7.62b) \\
\widehat{y}(kh) &= \widehat{C}(\eta(kh))\widehat{x}(kh) + \widehat{D}(\eta(kh))\widehat{u}(k), \quad k \in \mathbb{Z}_+, && (7.62c)
\end{aligned}$$

where $\widehat{x}(t) \in \mathbb{R}^{n_c}$ are the vectors of the state parameters of the controller, $\widehat{u}(k) \in \mathbb{R}^n$ are the inputs of the controller, while $\widehat{y}(kh) \in \mathbb{R}^m$ are the outputs of the controller. In our approach, the dimension n_c of the state space of the controllers of the form (7.62) is not prefixed. In other words, for each integer number $n_c \geq 0$, it may exist controllers $\mathbb{K}$ of the form (7.62) having the state space of dimension n_c. The dimension of the state space of a controller is often named "the order of the controller." Particularly, a zero-order controller reduces to an equality of the form:

$$\widehat{y}(kh) = \widehat{D}(\eta(kh))\widehat{u}(k), \quad k \in \mathbb{Z}_+, \tag{7.63}$$

where $\widehat{D}(i) \in \mathbb{R}^{m\times n}$, $i \in \mathcal{N}$, are arbitrary matrices. When coupling a controller $\mathbb{K}$ of the form (7.62) to the ICLSS (7.60) with the measured output (7.61) taking $u(k) = \widehat{y}(kh)$ and $\widehat{u}(k) = y(kh)$, $k \in \mathbb{Z}_+$, one obtains a closed-loop system denoted $\mathbb{G}_{cl}(\mathbb{K})$ having the state-space representation:

$$\begin{aligned} \mathrm{d}x_{cl}(t) &= A_{0,cl}(\eta(t))x_{cl}(t)\mathrm{d}t + \sum_{l=1}^{r_1} A_{l,cl}(\eta(t))x_{cl}(t)\mathrm{d}w_l(t) \\ &\quad + G_{c,cl}(\eta(t))\mathrm{d}\xi(t), \quad kh < t \leq (k+1)h \end{aligned} \tag{7.64a}$$

$$\begin{aligned} x_{cl}(kh^+) &= A_{d0,cl}(\eta(kh))x_{cl}(kh) + \sum_{p=1}^{r_2} w_{dp}(k)A_{dp,cl}(\eta(kh))x_{cl}(kh) \\ &\quad + G_{d,cl}(\eta(kh))\zeta_d(k), \quad k \in \mathbb{Z}_+ \end{aligned} \tag{7.64b}$$

$$z_{c,cl}(t) = C_{c,cl}(\eta(t))x_{cl}(t), \quad t \in \mathbb{R}_+ \tag{7.64c}$$

$$z_{d,cl}(kh) = C_{d,cl}(\eta(kh))x_{cl}(kh)), \tag{7.64d}$$

where

$$A_{l,cl}(i) = \begin{pmatrix} A_l(i) & 0 \\ 0 & \widehat{A}_l(i) \end{pmatrix}, \quad 0 \leq l \leq r_1, \tag{7.65a}$$

$$A_{dp,cl}(i) = \begin{pmatrix} A_{dp}(i) + B_{dp}(i)\widehat{D}(i) & B_{dp}(i)\widehat{C}(i) \\ \widehat{B}_{dp}(i) & \widehat{A}_{dp}(i) \end{pmatrix}, \quad 0 \leq p \leq r_2 \tag{7.65b}$$

$$C_{c,cl}(i) = (C_c(i) \quad 0), \tag{7.65c}$$

$$C_{d,cl}(i) = (C_d(i) + D_d(i)\widehat{D}(i) \quad D_d(i)\widehat{C}(i)) \tag{7.65d}$$

$$G_{c,cl}(i) = \begin{pmatrix} G_c(i) \\ 0 \end{pmatrix} \tag{7.65e}$$

$$G_{d,cl}(i) = \begin{pmatrix} G_d(i) \\ 0 \end{pmatrix} \tag{7.65f}$$

$$x_{cl}(t) = (x^\top(t) \quad \widehat{x}^\top(t))^\top \in \mathbb{R}^{n+n_c} \tag{7.65g}$$

if $n_c \geq 1$ and

$$A_{l,cl}(i) = A_l(i), \quad 0 \le l \le r_1, \tag{7.66a}$$
$$A_{dp,cl}(i) = A_{dp}(i) + B_{dp}(i)\widehat{D}(i), \quad 0 \le p \le r_2 \tag{7.66b}$$
$$C_{c,cl}(i) = C_c(i), \tag{7.66c}$$
$$C_{d,cl}(i) = C_d(i) + D_d(i)\widehat{D}(i), \tag{7.66d}$$
$$G_{c,cl}(i) = G_c(i) \tag{7.66e}$$
$$G_{d,cl}(i) = G_d(i) \tag{7.66f}$$
$$x_{cl}(t) = x(t) \tag{7.66g}$$

if $n_c = 0$, $i \in \mathcal{N}$.

Definition 7.1 A controller $\mathbb{K}$ with the state-space representation (7.62) or (7.63), respectively, is named **stabilizing controller** for the ICLSS (7.60) with the measured output (7.61) if the JSLDE:

$$dx_{cl}(t) = A_{0,cl}(\eta(t))x_{cl}(t)dt + \sum_{l=1}^{r_1} A_{l,cl}(\eta(t))x_{cl}(t)dw_l(t), \; kh < t \le (k+1)h, \tag{7.67a}$$

$$x_{cl}(kh^+) = [A_{d0,cl}(\eta(kh)) + \sum_{p=1}^{r_2} w_{dp}(k)A_{dp,cl}(\eta(kh))]x_{cl}(kh), \quad k \in \mathbb{Z}_+ \tag{7.67b}$$

is ESMS.

In the sequel we denote $\mathfrak{K}_{stab}(\mathbb{G})$ the set of all stabilizing controllers of the form (7.62) or (7.63), respectively. To introduce the H_2-norm of a closed-loop system $\mathbb{G}_{cl}(\mathbb{K})$, let us split its state-space representation (7.64) as:

$$dx_{cl}(t) = A_{0,cl}(\eta(t))x_{cl}(t)dt + \sum_{l=1}^{r_1} A_{l,cl}(\eta(t))x_{cl}(t)dw_l(t) + G_{c,cl}(\eta(t))d\xi(t), \quad kh < t \le (k+1)h \tag{7.68a}$$

$$x_{cl}(kh^+) = [A_{d0,cl}(\eta(kh)) + \sum_{p=1}^{r_2} w_{dp}(k)A_{dp,cl}(\eta(kh))]x_{cl}(kh), \quad k \in \mathbb{Z}_+ \tag{7.68b}$$

$$z_{c,cl}(t) = C_{c,cl}(\eta(t))x_{cl}(t), \quad t \in \mathbb{R}_+ \tag{7.68c}$$
$$z_{d,cl}(kh) = C_{d,cl}(\eta(kh))x_{cl}(kh), \quad k \in \mathbb{Z}_+ \tag{7.68d}$$

and

$$dx_{cl}(t) = A_{0,cl}(\eta(t))x_{cl}(t)dt + \sum_{l=1}^{r_1} A_{l,cl}(\eta(t))x_{cl}(t)dw_l(t), \; kh < t \le (k+1)h \tag{7.69a}$$

$$x_{cl}(kh^+) = [A_{d0,cl}(\eta(kh)) + \sum_{p=1}^{r_2} w_{dp}(k)A_{dp,cl}(\eta(kh))]x_{cl}(kh) + G_{d,cl}(\eta(kh))\zeta_d(k), \quad k \in \mathbb{Z}_+ \tag{7.69b}$$

$$z_{c,cl}(t) = C_{c,cl}(\eta(t))x_{cl}(t), \quad t \in \mathbb{R}_+ \tag{7.69c}$$
$$z_{d,cl}(kh) = C_{d,cl}(\eta(kh))x_{cl}(kh)), \quad k \in \mathbb{Z}_+. \tag{7.69d}$$

If $x_{cl}(\cdot)$ is the solution of the closed-loop system with the initial condition $x_{cl}(0) = 0$ and $x_{cl}(\cdot; \xi)$, $x_{cl}(\cdot; \zeta_d)$ are the solutions of (7.68) and (7.69), respectively, which are satisfying $x_{cl}(0; \xi) = 0$, $x_{cl}(0; \zeta_d) = 0$ then we have $x_{cl}(t) = x_{cl}(t; \xi) + x_{cl}(t; \zeta_d)$, for all $t \geq 0$. We set

$$z_{c,cl}(t; \xi) \triangleq C_{c,cl}(\eta(t))x_{cl}(t; \xi), \quad t \geq 0 \tag{7.70a}$$
$$z_{d,cl}(kh; \xi) \triangleq C_{d,cl}(\eta(kh))x_{cl}(kh; \xi), \quad k \geq 0 \tag{7.70b}$$

$$z_{c,cl}(t; \zeta_d) \triangleq C_{c,cl}(\eta(t))x_{cl}(t; \zeta_d), \quad t \geq 0 \tag{7.71a}$$
$$z_{d,cl}(kh; \zeta_d) \triangleq C_{d,cl}(\eta(kh))x_{cl}(kh; \zeta_d), \quad k \geq 0. \tag{7.71b}$$

The outputs $z_{c,cl}(\cdot; \xi)$, $z_{d,cl}(\cdot; \xi)$ introduced in (7.70) describe the effect of the additive white noise perturbations $\{\xi(t)\}_{t\geq 0}$ on the quality outputs of the closed-loop system, while $z_{c,cl}(\cdot; \zeta_d)$ and $z_{d,cl}(\cdot; \zeta_d)$ introduced in (7.71) describe the effect of the additive noises $\{\zeta_d(k)\}_{k\in\mathbb{Z}_+}$ on the quality output of the closed-loop system (7.64).

Now we can introduce the H_2-norm of the closed-loop system $\mathbb{G}_{cl}(\mathbb{K})$.

We define:

$$\|\mathbb{G}_{cl}(\mathbb{K})\|_2 = \left(\widehat{\|\mathbb{G}_{cl}(\mathbb{K})\|}_2^2 + \widecheck{\|\mathbb{G}_{cl}(\mathbb{K})\|}_2^2\right)^{1/2} \tag{7.72a}$$

$$\widehat{\|\mathbb{G}_{cl}(\mathbb{K})\|}_2^2 \triangleq \limsup_{T\to\infty} \frac{1}{T}\left(\int_0^T \mathbb{E}[|z_{c,cl}(t;\xi)|^2]dt + \sum_{k=0}^{[\frac{T}{h}]} \mathbb{E}[|z_{d,cl}(kh,\xi)|^2]\right) \tag{7.72b}$$

$$\widecheck{\|\mathbb{G}_{cl}(\mathbb{K})\|}_2^2 \triangleq \limsup_{p\to\infty} \frac{1}{p}\left(\int_0^{(p+1)h} \mathbb{E}[|z_{c,cl}(t;\zeta_d)|^2]dt + \sum_{k=0}^{p} \mathbb{E}[|z_{d,cl}(kh,\zeta_d)|^2]\right). \tag{7.72c}$$

The first H_2 optimal control problem we address in this section asks for finding a controller $\widetilde{\mathbb{K}}$ from $\mathfrak{K}_{stab}(\mathbb{G})$ which is satisfying the optimality condition:

$$\|\mathbb{G}_{cl}(\widetilde{\mathbb{K}})\|_2 = \inf\{\|\mathbb{G}_{cl}(\mathbb{K})\|_2 \mid \mathbb{K} \in \mathfrak{K}_{stab}(\mathbb{G})\}. \tag{7.73}$$

The second H_2 optimal control problem considered in this section asks for finding a controller $\widetilde{\mathbb{K}} \in \mathfrak{K}_{stab}(\mathbb{G})$ which is satisfying the following optimality condition:

$$\widetilde{\|\mathbb{G}_{cl}(\widetilde{\mathbb{K}})\|}_2 = \inf\{\widetilde{\|\mathbb{G}_{cl}(\mathbb{K})\|}_2 \mid \mathbb{K} \in \mathfrak{K}_{stab}(\mathbb{G})\}, \tag{7.74}$$

where

$$\widetilde{\|\mathbb{G}_{cl}(\mathbb{K}\|}_2^2 \overset{\Delta}{=} \limsup_{T\to\infty} \frac{1}{T}\left(\int_0^T \mathbb{E}[|z_{c,cl}(t;\xi)|^2]\mathrm{d}t + \sum_{k=0}^{[\frac{T}{h}]} \mathbb{E}[|z_{d,cl}(kh;\xi)|^2]\right)$$
$$+ \limsup_{T\to\infty} \frac{1}{T}\left(\int_0^T \mathbb{E}[|z_{c,cl}(t;\zeta_d|^2]\mathrm{d}t + \sum_{k=0}^{[\frac{T}{h}]} \mathbb{E}[|z_{d,cl}(kh;\zeta_d)|^2]\right) \quad (7.75)$$

where $z_{c,cl}(\cdot;\xi)$, $z_{d,cl}(\cdot;\xi)$ are the same outputs as in (7.72b) and $z_{c,cl}(\cdot;\zeta_d)$, $z_{d,cl}$ $(\cdot;\zeta_d)$ are similar to the ones involved in (7.72c).

Employing the matrix coefficients of the ICLSS (7.60) we introduce the following backward jump matrix linear differential equation (BJMLDE) with Riccati-type jumping operators:

$$-\dot{X}(t,i) = A_0^\top(t)X(t,i) + X(t,i)A_0(i) + \sum_{l=1}^{r_1} A_l^\top(i)X(t,i)A_l(i)$$
$$+ \sum_{j=1}^{N} q_{ij}X(t,j) + C_c^\top(i)C_c(i), \quad kh \le t < (k+1)h \quad (7.76a)$$

$$X(kh^-,i) = \sum_{p=0}^{r_2} A_{dp}^\top(i)X(kh,i)A_{dp}(i) - \left(\sum_{p=0}^{r_2} A_{dp}^\top(i)X(kh,i)B_{dp}(i) + C_d^\top(i)D_d(i)\right)$$
$$\cdot\left(\sum_{p=0}^{r_2} B_{dp}^\top(i)X(kh,i)B_{dp}(i) + D_d^\top(i)D_d(i)\right)^{-1}\left(\sum_{p=0}^{r_2} B_{dp}^\top(i)X(kh,i)A_{dp}(i) + D_d^\top(i)C_d(i)\right)$$
$$+ C_d^\top(i)C_d(i), \quad k\in\mathbb{Z}, \quad i\in\mathcal{N}. \quad (7.76b)$$

A globally defined solution of the BJMLDE (7.76) is a function $t \to \mathbf{X}(t) = (X(t,1), X(t,2), \dots, X(t,N)) : \mathbb{R} \to \mathcal{S}_n^N$ differentiable on each interval $(kh,(k+1)h)$ right continuous in any $\tau_k = kh$ for which the matrices $\sum_{p=0}^{r_2} D_{dp}^\top(i)X(kh,i)B_{dp}(i) + D_d^\top(i)D_d(i)$ are invertible for all $k\in\mathbb{Z}$ and $i\in\mathcal{N}$.

A globally defined solution $\mathbf{X}_s(\cdot) = (X_s(\cdot,1), X_s(\cdot,2), \dots, X_s(\cdot,N))$ of (7.76) is named "stabilizing solution" of the BJMLDE with Riccati-type jumping operator (7.76) if its corresponding closed-loop system

$$\mathrm{d}x(t) = A_0(\eta(t))x(t)\mathrm{d}t + \sum_{l=1}^{r_1} A_l(\eta(t))x(t)\mathrm{d}w_l(t), \quad kh < t \le (k+1)h \quad (7.77a)$$

$$x(kh^+) = [A_{d0}(\eta(kh)) + B_{d0}(\eta(kh))F_s(\eta(kh)) + \sum_{p=1}^{r_2} w_{dp}(k)$$
$$(A_{dp}(\eta(kh)) + B_{dp}(\eta(kh))F_s(\eta(kh)))]x(kh), \quad k\in\mathbb{Z}_+ \quad (7.77b)$$

is ESMS, where

$$F_s(i) \triangleq -\left(\sum_{p=0}^{r_2} B_{dp}^\top(i)X_s(0,i)B_{dp}(i) + D_d^\top(i)D_d(i)\right)^{-1} \times \left(\sum_{p=0}^{r_2} B_{dp}^\top(i)X_s(0,i)A_{dp}(i) + D_d^\top(i)C_d(i)\right), \quad i \in \mathcal{N}. \tag{7.78}$$

The problem of the existence of the stabilizing solution of the BJMLDE (7.76) was studied in Sect. 5.4.3. Here, we recall the most important properties of the stabilizing solution of a BJMLDE of type (7.76) which will be involved in the developments from this chapter.

Corollary 7.4 *The BJMLDE (7.76) has at most one bounded on $\mathbb{R}$ and stabilizing solution $\mathbf{X}_s(\cdot) = (X_s(\cdot,1), X_s(\cdot,2), \ldots, X_s(\cdot,N))$. The bounded and stabilizing solution, if any, is a periodic function of period h and satisfies $X_s(t,i) \geq 0$ for any $(t,i) \in \mathbb{R} \times \mathcal{N}$.*

The next result provides the state-space representation of a stabilizing controller $\widetilde{\mathbb{K}}$ which satisfies the optimality condition (7.73).

Theorem 7.4 *Assume:*

(a) Assumptions 1.2 and 7.1 are fulfilled;
(b) the BJMLDE with Riccati-type jumping operator (7.76) has a bounded on $\mathbb{R}$ and stabilizing solution $\mathbf{X}_s(\cdot) = (X_s(\cdot,1), X_s(\cdot,2), \ldots, X_s(\cdot,N))$ satisfying the sign conditions

$$\sum_{p=0}^{r_2} B_{dp}^\top(i)X_s(kh,i)B_{dp}(i) + D_d^\top(i)D_d(i) > 0, \quad i \in \mathcal{N}. \tag{7.79}$$

We consider the zero-order controller $\widetilde{\mathbb{K}}$:

$$u(k) = F_s(\eta(kh))x(kh), \quad k \in \mathbb{Z}_+ \tag{7.80}$$

where $F_s(i)$, $i \in \mathcal{N}$ are the gain matrices introduced by (7.78). Under the considered assumptions, the controller (7.80) lies in $\mathfrak{K}_{stab}(\mathbb{G})$ and it satisfies the optimality condition (7.73). The minimal value of the H_2-norm of a closed-loop system $\mathbb{G}_{cl}(\mathbb{K})$ achieved by a stabilizing controller is

$$\|\mathbb{G}_{cl}(\widetilde{\mathbb{K}})\|_2^2 = \sum_{i=1}^{N} \frac{\pi_{i\infty}}{h} \int_0^h \mathrm{Tr}[X_s(t,i)G_c(i)\boldsymbol{\Lambda}_c G_c^\top(i)]\mathrm{d}t$$
$$+ \sum_{i=1}^{N} \pi_{i\infty} \mathrm{Tr}[X_s(0,i)G_d(i)\boldsymbol{\Lambda}_d G_d^\top(i)], \qquad (7.81)$$

$\pi_{i\infty}$ being introduced by (7.34).

Proof By coupling the controller (7.80) to the ICLSS (7.60) one obtains the following closed-loop system $\mathbb{G}_{cl}(\widetilde{\mathbb{K}})$

$$\mathrm{d}x_{cl}(t) = A_0(\eta(t))x_{cl}(t)\mathrm{d}t + \sum_{l=1}^{r_1} A_l(\eta(t))x_{cl}(t)\mathrm{d}w_l(t)$$
$$+ G_c(\eta(t))\mathrm{d}\xi(t), \quad kh < t \le (k+1)h \qquad (7.82a)$$
$$x_{cl}(kh^+) = [A_{d0}(\eta(kh)) + B_{d0}(\eta(kh))F_s(\eta(kh)) + \sum_{p=1}^{r_2} w_{dp}(k) \times (A_{dp}(\eta(kh))$$
$$+ B_{dp}(\eta(kh))F_s(\eta(kh)))]x_{cl}(kh) + G_d(\eta(kh))\zeta_d(k), \quad k \in \mathbb{Z}_+ \qquad (7.82b)$$
$$z_{c,cl}(t) = C_c(\eta(t))x_{cl}(t), \quad t \in \mathbb{R}_+ \qquad (7.82c)$$
$$z_{d,cl}(kh) = (C_d(\eta(kh)) + D_d(\eta(kh))F_s(\eta(kh)))\,x_{cl}(kh), \quad k \in \mathbb{Z}_+. \qquad (7.82d)$$

The JSLDE of type (7.67) associated to the closed-loop system (7.82) is just (7.77) which is ESMS. Hence, the zero-order controller (7.80) lies in $\mathfrak{K}_{stab}(\mathbb{G})$. The analogous of the BGLJLDE (7.13) written in the case of (7.82) is given by:

$$-\dot{X}(t,i) = A_0^\top(t)X(t,i) + X(t,i)A_0(i) + \sum_{l=1}^{r_1} A_l^\top(i)X(t,i)A_l(i)$$
$$+ \sum_{j=1}^{N} q_{ij}X(t,j) + C_c^\top(i)C_c(i), \quad kh \le t < (k+1)h \qquad (7.83a)$$
$$X(kh^-,i) = \sum_{p=0}^{r_2} \left(A_{dp}(i) + B_{dp}(i)F_s(i)\right)^\top X(kh,i)\left(A_{dp}(i) + B_{dp}(i)F_s(i)\right)$$
$$+ (C_d(i) + D_d(i)F_s(i))^\top(C_d(i) + D_d(i)F_s(i)), \quad k \in \mathbb{Z}, \quad i \in \mathcal{N}. \qquad (7.83b)$$

Applying Lemma 5.1 (ii) in the case of (7.76b) taking $w_d(i) = F_s(i)$, $i \in \mathcal{N}$, we obtain that the unique bounded on $\mathbb{R}$ solution of (7.83) is just the bounded and stabilizing solution of (7.76). Thus, Theorem 7.1 applied in the case of the JLSS (7.82) confirms the validity of (7.81). It remains to show that

$$\|\mathbb{G}_{cl}(\widetilde{\mathbb{K}})\|_2 \le \|\mathbb{G}_{cl}(\mathbb{K})\|_2 \qquad (7.84)$$

for all $\mathbb{K} \in \mathfrak{K}_{stab}(\mathbb{G})$. Let us consider first that $\mathbb{K}$ is an arbitrary controller of the form (7.63). In this case, the BGLJLDE (7.14) associated to the corresponding closed-loop system $\mathbb{G}_{cl}(\mathbb{K})$ is:

$$-\dot{X}(t,i) = \mathcal{L}_i^*[\mathbf{X}(t)] + C_c^\top(i)C_c(i), \quad kh \le t < (k+1)h, \tag{7.85a}$$

$$\begin{aligned} X(kh^-,i) &= \sum_{p=0}^{r_2}(A_{dp}(i) + B_{dp}(i)\widehat{D}(i))^\top X(kh,i)(A_{dp}(i) + B_{dp}(i)\widehat{D}(i)) \\ &\quad + (C_d(i) + D_d(i)\widehat{D}(i))^\top (C_d(i) + D_d(i)\widehat{D}(i)), \quad k \in \mathbb{Z}, \ i \in \mathcal{N}. \end{aligned} \tag{7.85b}$$

On the other hand, applying again Lemma 5.1 in the case of Eq. (7.76b) satisfied by $\mathbf{X}_s(\cdot)$ we obtain that:

$$-\dot{X}(t,i) = \mathcal{L}_i^*[\mathbf{X}_s(t)] + C_c^\top(i)C_c(i), \quad kh \le t < (k+1)h, \tag{7.86a}$$

$$\begin{aligned} X(kh^-,i) &= \sum_{p=0}^{r_2}(A_{dp}(i) + B_{dp}(i)\widehat{D}(i))^\top X_s(kh,i)(A_{dp}(i) + B_{dp}(i)\widehat{D}(i)) \\ &\quad - (\widehat{D}(i) - F_s(i))^\top \left(\sum_{p=0}^{r_2} B_{dp}^\top(i)X_s(kh,i)B_{dp}(i) + D_d^\top(i)D_d(i)\right)(\widehat{D}(i) - F_s(i)) \\ &\quad + (C_d(i) + D_d(i)\widehat{D}(i))^\top (C_d(i) + D_d(i)\widehat{D}(i)), \quad k \in \mathbb{Z}, \ i \in \mathcal{N}. \end{aligned} \tag{7.86b}$$

Substracking (7.86) from (7.85) satisfied by its unique bounded on $\mathbb{R}$ solution $\widetilde{\mathbf{X}}(\cdot)$, we obtain that $\widetilde{\mathbf{X}}(\cdot) - \mathbf{X}_s(\cdot)$ is a bounded on $\mathbb{R}$ solution of the following non-homogeneous BGLJLDE:

$$-\dot{Z}(t,i) = \mathcal{L}_i^*[\mathbf{Z}(t)], \quad kh \le t < (k+1)h, \tag{7.87a}$$

$$\begin{aligned} Z(kh^-,i) &= \sum_{p=0}^{r_2}(A_{dp}(i) + B_{dp}(i)\widehat{D}(i))^\top Z(kh,i)(A_{dp}(i) + B_{dp}(i)\widehat{D}(i)) \\ &\quad + H(kh,i), \quad k \in \mathbb{Z}, \ i \in \mathcal{N}, \end{aligned} \tag{7.87b}$$

where $H(kh,i) := (\widehat{D}(i) - F_s(i))^\top(\sum_{p=0}^{r_2} B_{dp}^\top(i)X_s(kh,i)B_{dp}(i) + D_d^\top(i)D_d(i))$ $(\widehat{D}(i) - F_s(i))$. Based on (7.79) we deduce that $H(kh,i) \ge 0$, for all $(k,i) \in \mathbb{Z} \times \mathcal{N}$. So, Theorem 2.13 applied in the case of Eq. (7.87) allows us to deduce that this equation has a unique bounded on $\mathbb{R}$ solution and additionally that solution is a positive semidefinite-valued function. This means that $\widetilde{X}(t,i) \ge X_s(t,i)$, for all $(t,i) \in \mathbb{R} \times \mathcal{N}$. On the other hand from Theorem 7.1, applied in the case of the system $\mathbb{G}_{cl}(\mathbb{K})$, one obtains:

$$\|\mathbb{G}_{cl}(\mathbb{K})\|_2^2 = \sum_{i=1}^{N} \frac{\pi_{i\infty}}{h} \int_0^h \mathrm{Tr}[\widetilde{X}(t,i)G_c(i)\boldsymbol{\Lambda}_c G_c^\top(i)]\mathrm{d}t$$
$$+ \sum_{i=1}^{N} \pi_{i\infty} \mathrm{Tr}[\widetilde{X}(0,i)G_d(i)\boldsymbol{\Lambda}_d G_d^\top(i)].$$

Comparing this equality to (7.81) we may conclude that (7.84) holds for any zero-order stabilizing controller.

Let us consider now a stabilizing controller $\mathbb{K}$ of order $n_c \geq 1$. In this case, Theorem 7.1 applied in the case of the JLSS (7.64) yields:

$$\|\mathbb{G}_{cl}(\mathbb{K})\|_2^2 = \sum_{i=1}^{N} \frac{\pi_{i\infty}}{h} \int_0^h \mathrm{Tr}[\widetilde{X}_{cl}(t,i)G_{c,cl}(i)\boldsymbol{\Lambda}_c G_{c,cl}^\top(i)]\mathrm{d}t$$
$$+ \sum_{i=1}^{N} \pi_{i\infty} \mathrm{Tr}[X_{cl}(0,i)G_{d,cl}(i)\boldsymbol{\Lambda}_d G_{d,cl}^\top(i)], \tag{7.88}$$

where $\mathbf{X}_{cl}(\cdot) = (X_{cl}(\cdot, 1), X_{cl}(\cdot, 2), \dots, X_{cl}(\cdot, N))$ is the unique bounded on $\mathbb{R}$ solution of the following non-homogeneous BGLJLDE:

$$-\dot{X}_{cl}(t,i) = A_{0,cl}^\top(i)X_{cl}(t,i) + X_{cl}(t,i)A_{0,cl}(i) + \sum_{l=1}^{r_1} A_{l,cl}^\top(i)X_{cl}(t,i)A_{l,cl}(i)$$
$$+ \sum_{j=1}^{N} q_{ij}X_{cl}(t,j) + C_{c,cl}^\top(i)C_{c,cl}(i), \quad kh \leq t < (k+1)h, \tag{7.89a}$$

$$X_{cl}(kh^-, i) = \sum_{p=0}^{r_2} A_{dp,cl}^\top(i)X_{cl}(kh,i)A_{dp,cl}(i) + C_{d,cl}^\top(i)C_{d,cl}(i), \quad k \in \mathbb{Z}, \tag{7.89b}$$

$i \in \mathcal{N}$. Let $\widehat{X}_{cl}(t,i)$ be defined by

$$\widehat{X}_{cl}(t,i) \triangleq X_{cl}(t,i) - \begin{pmatrix} X_s(t,i) & 0 \\ 0 & 0 \end{pmatrix}, \quad (t,i) \in \mathbb{R} \times \mathcal{N}.$$

By direct calculation, involving (7.65), (7.76), (7.89) we deduce that $\widehat{\mathbf{X}}_{cl}(\cdot) = (\widehat{X}_{cl}(\cdot, 1), \widehat{X}_{cl}(\cdot, 2), \dots, \widehat{X}_{cl}(\cdot, N))$ solves the non-homogeneous BGLJLDE on the Hilbert space $\mathcal{S}_{n+n_c}^N$:

$$-\dot{\widehat{X}}_{cl}(t,i) = A_{0,cl}^{\top}(i)\widehat{X}_{cl}(t,i) + \widehat{X}_{cl}(t,i)A_{0,cl}(i) + \sum_{l=1}^{r_1} A_{l,cl}^{\top}(i)\widehat{X}_{cl}(t,i)A_{l,cl}(i)$$

$$+\sum_{j=1}^{N} q_{ij}\widehat{X}_{cl}(t,j), \quad kh \leq t < (k+1)h, \tag{7.90a}$$

$$\widehat{X}_{cl}(kh^-,i) = \sum_{p=0}^{r_2} A_{dp,cl}^{\top}(i)\widehat{X}_{cl}(kh,i)A_{dp,cl}(i) + H_d(kh,i), \quad k \in \mathbb{Z}, \tag{7.90b}$$

$i \in \mathcal{N}$, where $H_d(kh,i) := \left(\widehat{D}(i) - F_s(i) \quad \widehat{C}(i)\right)^{\top} (\sum_{p=0}^{r_2} B_{dp}^{\top}(i)X_s(kh,i)B_{dp}(i) + D_d^{\top}(i)D_d(i)) \left(\widehat{D}(i) - F_s(i) \quad \widehat{C}(i)\right)$. From (7.79) we deduce that $H_d(kh,i) \geq 0$, for all $(k,i) \in \mathbb{Z} \times \mathcal{N}$. Since the JSLDE (7.67) is ESMS, we may conclude via Theorem 2.13 applied in the case of the differential equation with jumps (7.90) that

$$\widehat{X}_{cl}(t,i) \geq 0, \tag{7.91}$$

for all $(t,i) \in \mathbb{R} \times \mathcal{N}$. Employing (7.81) we rewrite (7.88) as:

$$\|\mathbb{G}_{cl}(\mathbb{K})\|_2^2 = \|\mathbb{G}_{cl}(\widetilde{\mathbb{K}})\|_2^2 + \sum_{i=1}^{N} \frac{\pi_{i\infty}}{h} \int_0^h \mathrm{Tr}[\widehat{X}_{cl}(t,i)G_{c,cl}(i)\boldsymbol{\Lambda}_c G_{c,cl}^{\top}(i)]\mathrm{d}t$$

$$\pi_{i\infty}\,\mathrm{Tr}[\widehat{X}_{cl}(0,i)G_{d,cl}(i)\boldsymbol{\Lambda}_d G_{d,cl}^{\top}(i)].$$

This equality together with (7.91) show that (7.84) holds for any controller $\mathbb{K} \in \mathfrak{K}_{stab}(\mathbb{G})$. Thus the proof is complete.

Exercise 7.1 Find the state-space representation of a controller $\widetilde{\mathbb{K}}$ from $\mathfrak{K}_{stab}(\mathbb{G})$ which satisfies the optimality condition (7.74).

An important special case of the ICLSS ($\mathbb{G}$) is the one in which it is not subject to a Markovian jumping process. In this case, its state-space representation is:

$$\mathrm{d}x(t) = A_0 x(t)\mathrm{d}t + \sum_{l=1}^{r_1} A_l x(t)\mathrm{d}w_l(t) + G_c\mathrm{d}\xi(t), \; kh < t \leq (k+1)h \tag{7.92a}$$

$$x(kh^+) = A_{d0}x(kh) + B_{d0}u(k) + \sum_{p=1}^{r_2} w_{dp}(k)(A_{dp}x(kh) + B_{dp}u(k)) + G_d\zeta_d(k), \tag{7.92b}$$

$$z_c(t) = C_c x(t), \quad t \in \mathbb{R}_+ \tag{7.92c}$$

$$z_d(kh) = C_d x(kh)) + D_d u(k), \tag{7.92d}$$

$$y(kh) = x(kh), \quad k \in \mathbb{Z}_+ \tag{7.92e}$$

where $x(\cdot)$, $z_c(\cdot)$, $z_d(\cdot)$, $u(\cdot)$, $y(\cdot)$ are similar signals as the one defined for the ICLSS (7.60). The admissible controllers are both of the form (7.62) as well as (7.63), but without dependence upon the Markov process. As in the case analyzed before, the dimension n_c of the state space of an admissible controller is not prefixed. Following step by step the derivations from the first part of this section, we introduce the H_2-norm of the closed-loop system $\mathbb{G}_{cl}(\mathbb{K})$ as in (7.72) and we may state the problem of finding a stabilizing controller which provides the closed-loop system with the smallest H_2-norm. Instead of (7.76) one considers the non-homogeneous GBJLDE with Riccati-type jumping operators:

$$-\dot{X}(t) = A_0^\top X(t) + X(t)A_0 + \sum_{l=1}^{r_1} A_l^\top X(t)A_l + C_c^\top C_c, \quad kh \leq t < (k+1)h \tag{7.93a}$$

$$X(kh^-) = \sum_{p=0}^{r_2} A_{dp}^\top X(kh)A_{dp} - \left(\sum_{p=0}^{r_2} A_{dp}^\top X(kh)B_{dp} + C_d^\top D_d\right)\left(\sum_{p=0}^{r_2} B_{dp}^\top X(kh)B_{dp} + D_d^\top D_d\right)^{-1}\left(\sum_{p=0}^{r_2} B_{dp}^\top X(kh)A_{dp} + D_d^\top C_d\right) + C_d^\top C_d, k \in \mathbb{Z}. \tag{7.93b}$$

A solution $t \to X_s(t) : \mathbb{R} \to \mathcal{S}_n$ of (7.93) is named stabilizing solution if the closed-loop JSLDE:

$$dx(t) = A_0 x(t)dt + \sum_{l=1}^{r_1} A_l x(t)dw_l(t), \quad kh < t \leq (k+1)h \tag{7.94a}$$

$$x(kh^+) = [A_{d0} + B_{d0}F_s + \sum_{p=1}^{r_2} w_{dp}(k)(A_{dp} + B_{dp}F_s)]x(kh), \quad k \in \mathbb{Z}_+ \tag{7.94b}$$

is ESMS, where

$$F_s \triangleq -\left(\sum_{p=0}^{r_2} B_{dp}^\top X_s(0)B_{dp} + D_d^\top D_d\right)^{-1}\left(\sum_{p=0}^{r_2} B_{dp}^\top X_s(0)A_{dp} + D_d^\top C_d\right). \tag{7.95}$$

In the case of the ICLSS (7.92) the solution of the H_2 optimal control problem under perfect state measurements is given in the next theorem.

Theorem 7.5 *Assume:*

(a) Assumptions 1.1 and 7.1 are fulfilled;
(b) the BGLJLDE with Riccati-type jumping operator (7.93) has a bounded on $\mathbb{R}$ and stabilizing solution $X_s(\cdot)$ satisfying the sign conditions

$$\sum_{p=0}^{r_2} B_{dp}^\top X_s(kh) B_{dp} + D_d^\top D_d > 0, \quad k \in \mathbb{Z}. \tag{7.96}$$

We consider the zero-order controller $\widetilde{\mathbb{K}}$*:*

$$u(s) = F_s x(kh), \quad k \in \mathbb{Z}_+ \tag{7.97}$$

where F_s *is the gain matrix introduced in (7.95). Under the considered assumptions, the controller (7.97) lies in* $\mathfrak{K}_{stab}(\mathbb{G})$ *and additionally it satisfies*

$$\|\mathbb{G}_{cl}(\widetilde{\mathbb{K}})\|_2 \le \|\mathbb{G}_{cl}(\mathbb{K})\|_2,$$

for any stabilizing controller $\mathbb{K}$*. The minimal value of the* H_2*-norm of a closed-loop system* $\mathbb{G}_{cl}(\mathbb{K})$ *is given by:*

$$\|\mathbb{G}_{cl}(\widetilde{\mathbb{K}}\|_2^2 = \frac{1}{h}\int_0^h \mathrm{Tr}[X_s(t) G_c \boldsymbol{\Lambda}_c G_c^\top]\mathrm{d}t + \mathrm{Tr}[X_s(0) G_d \boldsymbol{\Lambda}_d G_d^\top].$$

Remark 7.7 If in (7.60), $D_d^\top(i) D_d(i) > 0$ for all $i \in \mathcal{N}$, then the sign conditions (7.79) are satisfied because the stabilizing solution of (7.76), if any, is positive semidefinite matrix-valued functions. Similarly, (7.96) is satisfied if $D_d^\top D_d > 0$.

7.4 Output-Based H_2 Optimal Control Problem for an Impulsive Controlled Linear Stochastic System Subject to Markov PerTurBations

7.4.1 *The Problem*

Let us consider the impulsive controlled linear stochastic system (ICLSS) $(\widehat{\mathbb{G}})$ having the following state-space representation:

$$\mathrm{d}x(t) = A_0(\eta(t))x(t)\mathrm{d}t + G_c(\eta(t))\mathrm{d}\xi(t), \quad kh < t \le (k+1)h \tag{7.98a}$$

$$x(kh^+) = A_d(\eta(kh))x(kh) + B_d(\eta(kh))u(k) + G_d(\eta(kh))\zeta_d(k), \quad k \in \mathbb{Z}_+ \tag{7.98b}$$

$$z_c(t) = C_c(\eta(t))x(t), \quad t \in \mathbb{R}_+ \tag{7.98c}$$

$$z_d(kh) = C_d(\eta(kh))x(kh)) + D_d(\eta(kh))u(k), \tag{7.98d}$$

$$y(kh) = C_y(\eta(kh))x(kh)) + D_y(\eta(kh))\zeta_d(k), \quad k \in \mathbb{Z}_+. \tag{7.98e}$$

In (7.98), $x(t) \in \mathbb{R}^n$, $z_c(t) \in \mathbb{R}^{n_{z_c}}$, $z_d(kh) \in \mathbb{R}^{n_{z_d}}$, $y(kh) \in \mathbb{R}^{n_y}$, $u(k) \in \mathbb{R}^m$ have the same meaning as in the case of the ICLSS (7.60) and (7.61). The stochastic processes $\{\xi(t)\}_{t\geq 0}$, $\{\zeta_d(k)\}_{k\in\mathbb{Z}_+}$ and $\{\eta(t)\}_{t\geq 0}$ are as in (7.60).

To control the ICLSS (7.98) we use controllers $(\widehat{\mathbb{K}})$ of the form:

$$d\widehat{x}(t) = \widehat{A}(\eta(t))\widehat{x}(t)dt, \quad kh < t \leq (k+1)h \tag{7.99a}$$

$$\widehat{x}(kh^+) = \widehat{A}_d(\eta(kh))\widehat{x}(kh) + \widehat{B}_d(\eta(kh))\widehat{u}(k) \tag{7.99b}$$

$$\widehat{y}(kh) = \widehat{C}(\eta(kh))\widehat{x}(kh) + \widehat{D}(\eta(kh))\widehat{u}(k), \quad k \in \mathbb{Z}_+, \tag{7.99c}$$

where $\widehat{x}(t) \in \mathbb{R}^{n_c}$ are the vectors of the state parameters of the controller $(\widehat{\mathbb{K}})$, $\widehat{u}(k) \in \mathbb{R}^{n_y}$ are the inputs of the controller, while $\widehat{y}(kh) \in \mathbb{R}^n$ is the output of the controller. As in the case of the controllers (7.62), the dimension n_c of the state space of the controller of type (7.99) is not prefixed. By coupling a controller $(\widehat{\mathbb{K}})$ of the form (7.99) to the ICLSS (7.98) taking $u(k) = \widehat{y}(kh)$ and $\widehat{u}(k) = y(kh)$, $k \in \mathbb{Z}_+$ we obtain the following closed-loop system $\mathbb{G}_{cl}(\widehat{\mathbb{K}})$:

$$dx_{cl}(t) = A_{cl}(\eta(t))x_{cl}(t)dt + G_{c,cl}(\eta(t))d\xi(t), \quad kh < t \leq (k+1)h \tag{7.100a}$$

$$x_{cl}(kh^+) = A_{d,cl}(\eta(kh))x_{cl}(kh) + G_{d,cl}(\eta(kh))\zeta_d(k), \quad k \in \mathbb{Z}_+ \tag{7.100b}$$

$$z_{c,cl}(t) = C_{c,cl}(\eta(t))x_{cl}(t), \quad t \in \mathbb{R}_+ \tag{7.100c}$$

$$z_{d,cl}(kh) = C_{d,cl}(\eta(kh))x_{cl}(kh) + D_{d,cl}(\eta(kh))\zeta_d(k), \quad k \in \mathbb{Z}_+ \tag{7.100d}$$

where $x_{cl}(t) = (x^\top(t) \quad \widehat{x}^\top(t))^\top \in \mathbb{R}^{n+n_c}$,

$$A_{cl}(i) = \begin{pmatrix} A_0(i) & 0 \\ 0 & \widehat{A}(i) \end{pmatrix}, \tag{7.101a}$$

$$A_{d,cl}(i) = \begin{pmatrix} A_d(i) + B_d(i)\widehat{D}(i)C_y(i) & B_d(i)\widehat{C}(i) \\ \widehat{B}_d(i)C_y(i) & \widehat{A}_d(i) \end{pmatrix}, \tag{7.101b}$$

$$C_{c,cl}(i) = (C_c(i) \quad 0), \tag{7.101c}$$

$$C_{d,cl}(i) = (C_d(i) + D_d(i)\widehat{D}(i)C_y(i) \quad D_d(i)\widehat{C}(i)), \tag{7.101d}$$

$$G_{c,cl}(i) = \begin{pmatrix} G_c(i) \\ 0 \end{pmatrix}, \tag{7.101e}$$

$$G_{d,cl}(i) = \begin{pmatrix} G_d(i) + B_d(i)\widehat{D}(i)D_y(i) \\ \widehat{B}_d(i)D_y(i) \end{pmatrix}, \tag{7.101f}$$

$$D_{d,cl}(i) = D_d(i)\widehat{D}(i)D_y(i). \tag{7.101g}$$

Definition 7.2 A controller $\widehat{\mathbb{K}}$ with the state-space representation of the form (7.99) is named stabilizing controller of the ICLSS (7.98) if the JSLDE:

$$\dot{x}_{cl}(t) = A_{cl}(\eta(t))x_{cl}(t), \quad kh < t \leq (k+1)h \tag{7.102a}$$

$$x_{cl}(kh^+) = A_{d,cl}(\eta(kh))x_{cl}(kh), \quad k \in \mathbb{Z}_+ \tag{7.102b}$$

is ESMS.

In the sequel, we denote, $\mathfrak{K}_{\text{stab}}(\widehat{\mathbb{G}})$ the set of all $\widehat{\mathbb{K}}$ of the form (7.100) for which the corresponding JSLDE (7.102) is ESMS. In order to introduce the H_2-norm of a closed-loop system $\mathbb{G}_{cl}(\widehat{\mathbb{K}})$ let us split its state-space representation (7.100) as:

$$\mathrm{d}x_{cl}(t) = A_{cl}(\eta(t))x_{cl}(t)\mathrm{d}t + G_{c,cl}(\eta(t))\mathrm{d}\xi(t), \quad kh < t \leq (k+1)h \tag{7.103a}$$
$$x_{cl}(kh^+) = A_{d,cl}(\eta(kh))x_{cl}(kh), \quad k \in \mathbb{Z}_+ \tag{7.103b}$$
$$z_{c,cl}(t) = C_{c,cl}(\eta(t))x_{cl}(t), \quad t \in \mathbb{R}_+ \tag{7.103c}$$
$$z_{d,cl}(kh) = C_{d,cl}(\eta(kh))x_{cl}(kh), \quad k \in \mathbb{Z}_+ \tag{7.103d}$$

and

$$dx_{cl}(t) = A_{cl}(\eta(t))x_{cl}(t) \quad kh < t \leq (k+1)h \tag{7.104a}$$
$$x_{cl}(kh^+) = A_{d,cl}(\eta(kh))x_{cl}(kh) + G_{d,cl}(\eta(kh))\zeta_d(k), \quad k \in \mathbb{Z}_+ \tag{7.104b}$$
$$z_{c,cl}(t) = C_{c,cl}(\eta(t))x_{cl}(t), \quad t \in \mathbb{R}_+ \tag{7.104c}$$
$$z_{d,cl}(kh) = C_{d,cl}(\eta(kh))x_{cl}(kh) + D_{d,cl}(\eta(kh))\zeta_d(k), \quad k \in \mathbb{Z}_+. \tag{7.104d}$$

The JLSS (7.103) reflects the influence of the additive white noise $\{\xi(t)\}_{t\in\mathbb{R}_+}$ on the trajectories and the outputs of the closed-loop system (7.100), and the JLSS (7.104) reflects the influence of the additive white noise $\{\zeta_d(k)\}_{k\in\mathbb{Z}_+}$ on the trajectories and the outputs of the closed-loop system (7.100).

Let $x_{cl}(\cdot; \xi)$ and $x_{cl}(\cdot; \zeta_d)$ be the solutions of the JSLDE (7.103a) and (7.103b) and (7.104a) and (7.104b), respectively, satisfying $x_{cl}(0; \xi) = 0$ and $x_{cl}(0; \zeta_d) = 0$. We set

$$z_{c,cl}(t; \xi) := C_{c,cl}(\eta(t))x_{cl}(t; \xi), \quad t \in \mathbb{R}_+ \tag{7.105a}$$
$$z_{d,cl}(kh; \xi) := C_{d,cl}(\eta(kh))x_{cl}(kh; \xi), \quad k \in \mathbb{Z}_+, \tag{7.105b}$$

and

$$z_{c,cl}(t; \zeta_d) := C_{c,cl}(\eta(t))x_{cl}(t; \zeta_d), \quad t \in \mathbb{R}_+ \tag{7.106a}$$
$$z_{d,cl}(kh; \zeta_d) := C_{d,cl}(\eta(kh))x_{cl}(kh; \zeta_d) + D_{d,cl}(\eta(kh))\zeta_d(k), \quad k \in \mathbb{Z}_+. \tag{7.106b}$$

Now, we are able to introduce the H_2-norm of a closed-loop system $\mathbb{G}_{cl}(\widehat{\mathbb{K}})$ associated to a stabilizing controller $\widehat{\mathbb{K}}$:

$$\|\mathbb{G}_{cl}(\mathbb{K})\|_2 = \left(\widehat{\|\mathbb{G}_{cl}(\widehat{\mathbb{K}})\|}_2^2 + \widecheck{\|\mathbb{G}_{cl}(\widehat{\mathbb{K}})\|}_2^2\right)^{1/2} \tag{7.107a}$$

$$\widehat{\|\mathbb{G}_{cl}(\widehat{\mathbb{K}})\|}_2^2 \triangleq \limsup_{T\to\infty} \frac{1}{T}\left(\int_0^T \mathbb{E}[|z_{c,cl}(t;\xi)|^2]\mathrm{d}t + \sum_{k=0}^{[\frac{T}{h}]} \mathbb{E}[|z_{d,cl}(kh,\xi)|^2]\right) \tag{7.107b}$$

$$\breve{\|\mathbb{G}_{cl}(\widehat{\mathbb{K}})\|}_2^2 \triangleq \limsup_{p\to\infty} \frac{1}{p}\left(\int_0^{(p+1)h} \mathbb{E}[|z_{c,cl}(t;\zeta_d)|^2]\mathrm{d}t + \sum_{k=0}^{p} \mathbb{E}[|z_{d,cl}(kh,\zeta_d)|^2]\right). \tag{7.107c}$$

The first H_2 optimal control problem which we want to solve in this section is asking for the finding of a stabilizing controller $\widetilde{\widetilde{\mathbb{K}}}$ of the form (7.99) which satisfies the following optimality condition:

$$\|\mathbb{G}_{cl}(\widetilde{\widetilde{\mathbb{K}}})\|_2 = \inf\{\|\mathbb{G}_{cl}(\widehat{\mathbb{K}})\|_2 \mid \widehat{\mathbb{K}} \in \mathfrak{K}_{stab}(\widehat{\mathbb{G}})\}. \tag{7.108}$$

The second H_2 optimal control problem which we address in this section is asking for finding a stabilizing controller $\widetilde{\widetilde{\mathbb{K}}}$ of the form (7.99) which have to satisfy the optimality condition:

$$\widetilde{\|\mathbb{G}_{cl}(\widetilde{\widetilde{\mathbb{K}}})\|}_2 = \inf\{\widetilde{\|\mathbb{G}_{cl}(\mathbb{K})\|}_2 \mid \mathbb{K} \in \mathfrak{K}_{stab}(\mathbb{G})\} \tag{7.109}$$

where

$$\begin{aligned}\widetilde{\|\mathbb{G}_{cl}(\mathbb{K})\|}_2^2 \triangleq\ & \limsup_{T\to\infty} \frac{1}{T}\left(\int_0^T \mathbb{E}[|z_{c,cl}(t;\xi)|^2]\mathrm{d}t + \sum_{k=0}^{[\frac{T}{h}]} \mathbb{E}[|z_{d,cl}(kh,\xi)|^2]\right)\\ &+ \limsup_{T\to\infty} \frac{1}{T}\left(\int_0^T \mathbb{E}[|z_{c,cl}(t;\zeta_d)|^2]\mathrm{d}t + \sum_{k=0}^{[\frac{T}{h}]} \mathbb{E}[|z_{d,cl}(kh,\zeta_d)|^2]\right)\end{aligned} \tag{7.110}$$

where $z_{c,cl}(\cdot;\xi)$, $z_{d,cl}(\cdot;\xi)$ were introduced by (7.105), while $z_{c,cl}(\cdot;\zeta_d)$, $z_{d,cl}(\cdot;\zeta_d)$ are introduced by (7.106).

7.4.2 Several Intermediate Results

Before stating the main result of this section, let us recall several auxiliary issues which will be involved in the next developments.

Let $\mathbf{X} \to \mathscr{L}_{cl}[\mathbf{X}] : \mathscr{S}_{n+n_c}^N \to \mathscr{S}_{n+n_c}^N$ be the linear operator defined by:

$$\mathscr{L}_{cl}[\mathbf{X}] = (\mathscr{L}_{cl,1}[\mathbf{X}], \mathscr{L}_{cl,2}[\mathbf{X}], \ldots, \mathscr{L}_{cl,N}[\mathbf{X}]) \tag{7.111a}$$

$$\mathcal{L}_{cl,i}[\mathbf{X}] = A_{cl}(i)X(i) + X(i)A_{cl}^{\top}(i) + \sum_{j=1}^{N} q_{ji}X(j), \quad i \in \mathcal{N} \tag{7.111b}$$

for all $\mathbf{X} = (X(1), X(2), \ldots, X(N)) \in \mathcal{S}_{n+n_c}^N$. Further, $\mathcal{L}_{cl}^*[\mathbf{X}] = (\mathcal{L}_{cl,1}^*[\mathbf{X}], \mathcal{L}_{cl,2}^*[\mathbf{X}], \ldots, \mathcal{L}_{cl,N}^*[\mathbf{X}])$ stands for the adjoint of the operator $\mathcal{L}_{cl}[\mathbf{X}]$ with respect to the inner product (2.111) defined on the linear space $\mathcal{S}_{n+n_c}^N$. The linear operators $\mathcal{L}_{cl}[\mathbf{X}]$, $\mathcal{L}_{cl}^*[\mathbf{X}]$ are Lyapunov-type operators associated to the stochastic linear differential equation (7.102a).

We consider the backward generalized Lyapunov-type jump linear differential equation (GBLJLDE) on the Hilbert space $\mathcal{S}_{n+n_c}^N$:

$$-\dot{X}_{cl}(t,i) = \mathcal{L}_{cl,i}^*[\mathbf{X}_{cl}(t)] + H_c(i), \quad kh \le t < (k+1)h \tag{7.112a}$$

$$X_{cl}(kh^-, i) = A_{d,cl}^{\top}(i)X_{cl}(kh,i)A_{d,cl}(i) + H_d(i), \quad k \in \mathbb{Z}_+, \quad i \in \mathcal{N}, \tag{7.112b}$$

where $H_c(i), H_d(i) \in \mathcal{S}_{n+n_c}$ are known matrices.

The implication $(iii) \Rightarrow (i)$ from Theorem 3.1 together with Theorem 2.13 applied in the case of BGLJLDE (7.112) yields:

Corollary 7.5 *If the JSLDE (7.102) is ESMS, then the non-homogeneous BGLJLDE (7.112) has a unique solution* $t \to \widetilde{\mathbf{X}}_{cl}(t) = (\widetilde{X}_{cl}(t,1), \widetilde{X}_{cl}(t,2), \ldots, \widetilde{X}_{cl}(t,N)) : \mathbb{R} \to \mathcal{S}_{n+n_c}^N$ *with the properties:*

(a) $\widetilde{\mathbf{X}}_{cl}(\cdot)$ *is differentiable on each interval* $(kh,(k+1)h)$, *right continuous in any* $\tau_k = kh$, $k \in \mathbb{Z}$;
(b) $\widetilde{\mathbf{X}}_{cl}(\cdot)$ *is a periodic function of period* h;
(c) $\sup_{t\in\mathbb{R}} \|\widetilde{\mathbf{X}}_{cl}(t)\|_\infty < \infty$, $\|\cdot\|_\infty$ *being the norm of type (2.112) defined on the space* $\mathcal{S}_{n+n_c}^N$;
(d) if $H_c(i) \ge 0$, $H_d(i) \ge 0$, $i \in \mathcal{N}$, *then* $\widetilde{X}_{cl}(t,i) \ge 0$ *for all* $(t,i) \in \mathbb{R} \times \mathcal{N}$.

Beside (7.112) we consider the following forward generalized Lyapunov-type jump linear differential equation (FGLJLDE) on the Hilbert space $\mathcal{S}_{n+n_c}^N$:

$$\dot{Y}_{cl}(t,i) = \mathcal{L}_{cl,i}[\mathbf{Y}_{cl}(t)] + \Gamma_c(i), \quad kh < t \le (k+1)h \tag{7.113a}$$

$$Y_{cl}(kh^+, i) = A_{d,cl}(i)Y_{cl}(kh,i)A_{d,cl}^{\top}(i) + \Gamma_d(i), \quad k \in \mathbb{Z}, \quad i \in \mathcal{N}, \tag{7.113b}$$

where $\Gamma_c(i), \Gamma_d(i) \in \mathcal{S}_{n+n_c}$ are known matrices.

The implication $(iii) \Rightarrow (i)$ from Theorem 3.1 combined by Theorem 2.14 applied in the case of BGLJLDE (7.113) yields:

Corollary 7.6 *If the JSLDE (7.102) is ESMS, then the non-homogeneous BGLJLDE (7.113) has a unique solution* $t \to \widetilde{\mathbf{Y}}_{cl}(t) = (\widetilde{Y}_{cl}(t,1), \widetilde{Y}_{cl}(t,2), \ldots, \widetilde{Y}_{cl}(t,N)) : \mathbb{R} \to \mathcal{S}_{n+n_c}^N$ *which has the properties:*

(a) $\widetilde{\mathbf{Y}}_{cl}(\cdot)$ *is differentiable on each interval* $(kh,(k+1)h)$, *left continuous in each* $\tau_k = kh$, $k \in \mathbb{Z}$;

(b) $\widetilde{\mathbf{Y}}_{cl}(\cdot)$ *is a periodic function of period* h;

(c) $\sup_{t\in\mathbb{R}} \|\widetilde{\mathbf{Y}}_{cl}(t)\|_\infty < \infty$;

(d) if $\Gamma_c(i) \geq 0,\ \Gamma_d(i) \geq 0,\ 1 \leq i \leq N$, *then* $\widetilde{Y}_{cl}(t,i) \geq 0$ *for all* $(t,i) \in \mathbb{R} \times \mathcal{N}$.

The equality from the next lemma can be proved following similar lines as in the proof of the second equality from (7.35) employing the properties of the solutions $\widetilde{\mathbf{X}}_{cl}(\cdot)$ and $\widetilde{\mathbf{Y}}_{cl}(\cdot)$ stated in Corollary 7.5 and Corollary 7.6, respectively.

Lemma 7.2 *The following equality holds:*

$$\sum_{i=1}^{N}\left(\frac{\gamma_i}{h}\int_0^h \mathrm{Tr}[\widetilde{X}_{cl}(t,i)\widetilde{\Gamma}_c(i)]\mathrm{d}t + \gamma_i \,\mathrm{Tr}[\widetilde{X}_{cl}(0,i)\widetilde{\Gamma}_d(i)]\right)$$
$$= \sum_{i=1}^{N}\left(\int_0^h \mathrm{Tr}[\widetilde{Y}_{cl}(t,i)H_c(i)]\mathrm{d}t + \mathrm{Tr}[\widetilde{Y}_{cl}(0,i)H_d(i)]\right)$$

where $\gamma_i \geq 0,\ i \in \mathcal{N}$ *are given real numbers and* $\widetilde{\mathbf{X}}_{cl}(\cdot) = (\widetilde{X}_{cl}(\cdot,1), \widetilde{X}_{cl}(\cdot,2), \ldots, \widetilde{X}_{cl}(\cdot,N))$ *is the unique bounded on* $\mathbb{R}$ *solution of BGLJLDE (7.112) while* $\widetilde{\mathbf{Y}}_{cl}(\cdot) = (\widetilde{Y}_{cl}(\cdot,1), \widetilde{Y}_{cl}(\cdot,2), \ldots, \widetilde{Y}_{cl}(\cdot,N))$ *is the unique bounded on* $\mathbb{R}$ *solution of the FGLJLDE (7.113) with* $\Gamma_c(i) := \frac{\gamma_i}{h}\widetilde{\Gamma}_c(i)$ *and* $\Gamma_d(i) = \gamma_i\widetilde{\Gamma}_d(i)$, $i \in \mathcal{N}$.

In the derivation of the state-space representation of a stabilizing controller which satisfies the optimality condition (7.108), the stabilizing solutions of the following jump matrix differential equation with Riccati-type jumping operators are involved. First, based on the coefficients of (7.98) we introduce the following BJMLDE with Riccati-type jumping operator:

$$-\dot{X}(t,i) = A_0^\top(i)X(t,i) + X(t,i)A_0(i) + \sum_{j=1}^{N} q_{ij}X(t,j) + C_c^\top(i)C_c(i),$$
$$kh \leq t < (k+1)h \tag{7.114a}$$
$$X(kh^-,i) = A_d^\top(i)X(kh,i)A_d(i) - \left(A_d^\top(i)X(kh,i)B_d(i) + C_d^\top(i)D_d(i)\right)$$
$$\cdot\left(B_d^\top(i)X(kh,i)B_d(i) + D_d^\top(i)D_d(i)\right)^{-1}\left(B_d^\top(i)X(kh,i)A_d(i) + D_d^\top(i)C_d(i)\right)$$
$$+ C_d^\top(i)C_d(i),\quad k \in \mathbb{Z},\ \ i \in \mathcal{N}. \tag{7.114b}$$

By analogy with the BJMLDE (7.76), a globally defined solution of (7.114) is a function $t \to \mathbf{X}(t) = (X(t,1), X(t,2), \ldots, X(t,N)) : \mathbb{R} \to \mathcal{S}_n^N$ which is differentiable on each interval $(kh, (k+1)h)$, right continuous in any $\tau_k = kh$ and the matrices $B_d^\top(i)X(kh,i)B_d(i) + D_d^\top(i)D_d(i)$ are invertible for all $i \in \mathcal{N}$ and $k \in \mathbb{Z}$ and solve Eq. (7.114) on $\mathbb{R}$.

A globally defined solution $\mathbf{X}_s(\cdot) = (X_s(\cdot,1), X_s(\cdot,2), \ldots, X_s(\cdot,N))$ of (7.114) is named stabilizing solution if the closed-loop JSLDE:

$$\dot{x}(t) = A_0(\eta(t))x(t), \quad kh < t \leq (k+1)h \tag{7.115a}$$
$$x(kh^+) = (A_d(\eta(kh)) + B_d(\eta(kh))F_s(\eta(kh)))x(kh), \quad k \in \mathbb{Z}_+ \tag{7.115b}$$

is ESMS, where for each $i \in \mathcal{N}$

$$F_s(i) = -(B_d^\top(i)X_s(0,i)B_d(i) + D_d^\top(i)D_d(i))^{-1}(B_d^\top(i)X_s(0,i)A_d(i) + D_d^\top(i)C_d(i)). \tag{7.116}$$

Remark 7.8 Comparing (7.114)–(7.116) to (7.76)–(7.78) we may conclude that the BJMLDE (7.114) is a special case of the BJMLDE (7.76), and, consequently, the Corollary 7.4 is applicable in the case of the stabilizing solution of (7.114) too.

Besides the BJMLDE (7.114), in the developments from this section, an important role will be played by the following forward jump matrix linear differential equation (FJMLDE) with a Riccati-type jumping operator:

$$\dot{Y}(t,i) = A_0(i)Y(t,i) + Y(t,i)A_0^\top(i) + \sum_{j=1}^{N} q_{ji}Y(t,j) + \frac{\pi_{i\infty}}{h}G_c(i)\boldsymbol{\Lambda}_c G_c^\top(i),$$
$$kh < t \leq (k+1)h \tag{7.117a}$$
$$Y(kh^+,i) = A_d(i)Y(kh,i)A_d^\top(i) - (A_d(i)Y(kh,i)C_y^\top(i) + \pi_{i\infty}G_d(i)\boldsymbol{\Lambda}_d D_y^\top(i))\times$$
$$(C_y(i)Y(kh,i)C_y^\top(i) + \pi_{i\infty}D_y(i)\boldsymbol{\Lambda}_d D_y^\top(i))^{-1}(C_y(i)Y(kh,i)A_d^\top(i)$$
$$+ \pi_{i\infty}D_y(i)\boldsymbol{\Lambda}_d G_d^\top(i)) + \pi_{i\infty}G_d(i)\boldsymbol{\Lambda}_d G_d^\top(i), \quad k \in \mathbb{Z}, \quad i \in \mathcal{N}. \tag{7.117b}$$

A globally defined solution of the FJMLDE (7.117) is a function $t \to \mathbf{Y}(t) = (Y(t,1), Y(t,2), \ldots, Y(t,N)) : \mathbb{R} \to \mathcal{S}_n^N$ which is differentiable on each interval $(kh, (k+1)h)$, left continuous in any $\tau_k = kh$ and such that the matrices

$$C_y(i)Y(kh,i)C_y^\top(i) + \pi_{i\infty}D_y(i)\boldsymbol{\Lambda}_d D_y^\top(i)$$

are invertible for all $k \in \mathbb{Z}$ and $i \in \mathcal{N}$ and satisfies Eq. (7.117) on the whole $\mathbb{R}$. A globally defined solution $\mathbf{Y}_s(\cdot) = (Y_s(\cdot,1), Y_s(\cdot,2), \ldots, Y_s(\cdot,N))$ is the stabilizing solution of (7.117) if the closed-loop JSLDE:

$$\dot{x}(t) = A_0(\eta(t))x(t), \quad kh < t \leq (k+1)h \tag{7.118a}$$
$$x(kh^+) = (A_d(\eta(kh)) + K_s(\eta(kh))C_y(\eta(kh)))x(kh), \quad k \in \mathbb{Z}_+ \tag{7.118b}$$

is ESMS, where

$$K_s(i) \triangleq -(A_d(i)Y_s(0,i)C_y^\top(i) + \pi_{i\infty}G_d(i)\boldsymbol{\Lambda}_d D_y^\top(i))$$
$$\times (C_y(i)Y_s(0,i)C_y^\top(i) + \pi_{i\infty}D_y(i)\boldsymbol{\Lambda}_d D_y^\top(i))^{-1} \quad i \in \mathcal{N}. \tag{7.119}$$

The problem of the existence of a stabilizing solution of a FJMLDE of type (7.117) was studied in Sect. 5.6. Here, we recall the most important properties of the bounded and stabilizing solution of (7.117) which will be involved in the developments of this section.

Corollary 7.7 *The FJMLDE with Riccati-type jumping operator (7.117) has at most a bounded on $\mathbb{R}$ and stabilizing solution. The bounded on $\mathbb{R}$ and stabilizing solution $\mathbf{Y}_s(\cdot) = (Y_s(\cdot, 1), Y_s(\cdot, 2), \dots, Y_s(\cdot, N))$ of (7.117) if any, is a periodic function of period h and satisfies*

$$Y_s(t, i) \geq 0,$$

for all $(t, i) \in \mathbb{R} \times \mathcal{N}$.

7.4.3 The State-Space Representation of the Optimal Controller

Now we are able to state and prove the main result of this section. It provides the state-space representation of a stabilizing controller which satisfies the optimality condition (7.108).

Theorem 7.6 *Assume:*

(a) the stochastic processes $\{\xi(t)\}_{t\geq 0}$ and $\{\zeta_d(k)\}_{k\in\mathbb{Z}_+}$ have the properties from 7.1 (a) and 7.1 (b);
(b) $\{\eta(t)\}_{t\geq 0}$, $\{\xi(t)\}_{t\geq 0}$, $\{\zeta_d(k)\}_{k\in\mathbb{Z}_+}$ are mutually independent stochastic processes;
(c) the BJMLDE (7.114) has a bounded on $\mathbb{R}$ and stabilizing solution $\mathbf{X}_s(\cdot) = (X_s(\cdot, 1), X_s(\cdot, 2), \dots, X_s(\cdot, N))$ satisfying the sign conditions

$$B_d^\top(i)X_s(0, i)B_d(i) + D_d^\top(i)D_d(i) > 0, \quad i \in \mathcal{N}; \tag{7.120}$$

(d) the FJMLDE (7.117) has a bounded on $\mathbb{R}$ and stabilizing solution $\mathbf{Y}_s(\cdot) = (Y_s(\cdot, 1), Y_s(\cdot, 2), \dots, Y_s(\cdot, N))$ satisfying the sign conditions

$$C_y(i)Y_s(0, i)C_y^\top(i) + \pi_{i\infty}D_y(i)\boldsymbol{\Lambda}_d D_y^\top(i) > 0, \quad i \in \mathcal{N}. \tag{7.121}$$

We associate the controller $\stackrel{\approx}{\mathbb{K}}$ having the state-space representation

$$d\widehat{x}(t) = \widehat{\mathcal{A}}(\eta(t))\widehat{x}(t)dt, \quad kh < t \leq (k+1)h \tag{7.122a}$$

$$\widehat{x}(kh^+) = \widehat{\mathcal{A}_d}(\eta(kh))\widehat{x}(kh) + \widehat{\mathcal{B}}_d(\eta(kh))\widehat{u}(k) \tag{7.122b}$$

$$\widehat{y}(kh) = \widehat{\mathcal{C}}(\eta(kh))\widehat{x}(kh) + \widehat{\mathcal{D}}(\eta(kh))\widehat{u}(k), \quad k \in \mathbb{Z}_+ \tag{7.122c}$$

where

$$\widehat{\mathscr{A}}(i) = A_0(i) \tag{7.123a}$$
$$\widehat{\mathscr{A}_d}(i) = A_d(i) + B_d(i)F_s(i) + K_s(i)C_y(i) - B_d(i)\widehat{\mathscr{D}}(i)C_y(i) \tag{7.123b}$$
$$\widehat{\mathscr{B}}_d(i) = B_d(i)\widehat{\mathscr{D}}(i) - K_d(i) \tag{7.123c}$$
$$\widehat{\mathscr{C}}(i) = F_s(i) - \widehat{\mathscr{D}}(i)C_y(i) \tag{7.123d}$$
$$\widehat{\mathscr{D}}(i) = \Delta_d^{-1}(0,i)(B_d^\top(i)X_s(0,i)K_s(i) - D_d^\top(i)C_d(i)Y_s(0,i)C_y^\top(i)\widehat{\Delta}_d^{-1}(0,i)) \tag{7.123e}$$
$$\Delta_d(0,i) = B_d^\top(i)X_s(0,i)B_d(i) + D_d^\top(i)D_d(i) \tag{7.123f}$$
$$\widehat{\Delta}_d(0,i) = C_y(i)Y_s(0,i)C_y^\top(i) + \pi_{i\infty}D_y(i)\boldsymbol{\Lambda}_d D_y^\top(i). \tag{7.123g}$$

Under the considered assumptions, the controller $\widetilde{\widetilde{\mathbb{K}}}$ described in (7.122) and (7.123) lies in $\mathfrak{K}_{stab}(\widehat{\mathbb{G}})$ and satisfies the optimality condition (7.108). The minimal value of the H_2-norm of a closed-loop system $\mathbb{G}_{cl}(\widehat{\mathbb{K}})$ is given by:

$$\begin{aligned}\|\mathbb{G}_{cl}(\widetilde{\widetilde{\mathbb{K}}})\|_2^2 = \sum_{i=1}^{N}\Bigg(&\frac{\pi_{i\infty}}{h}\int_0^h \mathrm{Tr}[X_s(t,i)G_c(i)\boldsymbol{\Lambda}_c G_c^\top(i)]dt \\ &+ \pi_{i\infty}\,\mathrm{Tr}[X_s(0,i)G_d(i)\boldsymbol{\Lambda}_d G_d^\top(i)] + \mathrm{Tr}[F_s(i)Y_s(0,i)F_s^\top(i)\Delta_d(0,i)] \\ &- \mathrm{Tr}[(D_d^\top(i)C_d(i)Y_s(0,i)C_y^\top(i) - B_d^\top(i)X_s(0,i)K_s(i)\widehat{\Delta}(0,i))\widehat{\Delta}_d^{-1}(0,i) \\ &\cdot (D_d^\top(i)C_d(i)Y_s(0,i)C_y^\top(i) - B_d^\top(i)X_s(0,i)K_s(i)\widehat{\Delta}_d(0,i))^\top \Delta_d^{-1}(0,i)]\Bigg).\end{aligned} \tag{7.124}$$

Proof Let $\widehat{\mathbb{K}}$ be an arbitrary stabilizing controller. For the closed-loop system $\mathbb{G}_{cl}(\widehat{\mathbb{K}})$ the equality (7.35) becomes:

$$\begin{aligned}\|\mathbb{G}_{cl}(\widehat{\mathbb{K}})\|_2^2 = &\sum_{i=1}^{N}\frac{\pi_{i\infty}}{h}\int_0^h \mathrm{Tr}[\widetilde{X}_{cl}(t,i)G_{c,cl}(i)\boldsymbol{\Lambda}_c G_{c,cl}^\top(i)]dt \\ &+ \sum_{i=1}^{N}\pi_{i\infty}\,\mathrm{Tr}[\widetilde{X}_{cl}(0,i)G_{d,cl}(i)\boldsymbol{\Lambda}_d G_{d,cl}^\top(i)] + \sum_{i=1}^{N}\mathrm{Tr}[D_{d,cl}(i)\boldsymbol{\Lambda}_d D_{d,cl}^\top(i)]\end{aligned} \tag{7.125}$$

where $\widetilde{\mathbf{X}}_{cl}(\cdot) = (\widetilde{X}_{cl}(\cdot,1), \widetilde{X}_{cl}(\cdot,2), \ldots, \widetilde{X}_{cl}(\cdot,N))$ is the unique bounded on $\mathbb{R}$ solution of the BGLJLDE (7.112) with $H_c(i) = C_{c,cl}^\top(i)C_{c,cl}(i)$ and $H_d(i) = C_{d,cl}^\top(i)C_{d,cl}(i)$. By direct calculation one obtains that the BJMLDE (7.114) written for $\mathbf{X}(\cdot)$ replaced with its stabilizing solution $\mathbf{X}_s(\cdot)$ takes the form:

$$\begin{aligned}&-\dot{X}_s(t,i) = A_0^\top(i)X_s(t,i) + X_s(t,i)A_0(i) + \sum_{j=1}^{N} q_{ij}X_s(t,j) + C_c^\top(i)C_c(i) \\ &kh \le t < (k+1)h\end{aligned} \tag{7.126a}$$

$$
\begin{aligned}
X_s(kh^-, i) &= (A_d(i) + B_d(i)\widehat{D}(i)C_y(i))^\top X_s(kh, i)(A_d(i) + B_d(i)\widehat{D}(i)C_y(i)) \\
&\quad - (F_s(i) - \widehat{D}(i)C_y(i))^\top \Delta_d(0, i)(F_s(i) - \widehat{D}(i)C_y(i)) \qquad (7.126b) \\
&\quad + (C_d(i) + D_d(i)\widehat{D}(i)C_y(i))^\top (C_d(i) + D_d(i)\widehat{D}(i)C_y(i)), \quad k \in \mathbb{Z}, \quad i \in \mathcal{N}.
\end{aligned}
$$

We set

$$
\widehat{X}_{cl}(t, i) \triangleq \widetilde{X}_{cl}(t, i) - \begin{pmatrix} X_s(t, i) & 0 \\ 0 & 0 \end{pmatrix} \tag{7.127}
$$

for all $(t, i) \in \mathbb{R} \times \mathcal{N}$. Employing (7.101), (7.112) written for $H_c(i) = C_{c,cl}^\top(i)C_{c,cl}(i)$ and $H_d(i) = C_{d,cl}^\top(i)C_{d,cl}(i)$ together with (7.126) we obtain that $\widehat{\mathbf{X}}_{cl}(\cdot) = (\widehat{X}_{cl}(\cdot, 1), \widehat{X}_{cl}(\cdot, 2), \ldots, \widehat{X}_{cl}(\cdot, N))$ solves the following BGLJLDE:

$$
-\frac{\mathrm{d}}{\mathrm{d}t}\widehat{X}_{cl}(t, i) = \mathcal{L}_{cl,i}^*[\widehat{\mathbf{X}}_{cl}(t)], \quad kh \le t < (k+1)h \tag{7.128a}
$$

$$
\widehat{X}_{cl}(kh^-, i) = A_{d,cl}^\top(i)\widehat{X}_{cl}(kh, i)A_{d,cl}(i) + \Theta_d^\top(i)\Delta_d(0, i)\Theta_d(i), \quad k \in \mathbb{Z}, \tag{7.128b}
$$

where we have denoted $\Theta_d(i) \triangleq (F_s(i) - \widehat{D}(i)C_y(i) \quad - \widehat{C}_y(i))$.
Substituting (7.127) in (7.125) yields:

$$
\begin{aligned}
\|\mathbb{G}_{cl}(\widehat{\mathbb{K}})\|_2^2 &= \sum_{i=1}^{N} \Bigg(\frac{\pi_{i\infty}}{h} \int_0^h \mathrm{Tr}[X_s(t, i)G_c(i)\boldsymbol{\Lambda}_c G_c^\top(i)]\mathrm{d}t \\
&\quad + \pi_{i\infty}\, \mathrm{Tr}[X_s(0, i)G_d(i)\boldsymbol{\Lambda}_d G_d^\top(i)] + \mathrm{Tr}[D_{d,cl}\boldsymbol{\Lambda}_d D_{d,cl}^\top(i)]\Bigg) \\
&\quad + \sum_{i=1}^{N} \Bigg(\frac{\pi_{i\infty}}{h} \int_0^h \mathrm{Tr}[\widehat{X}_{cl}(t, i)G_{c,cl}(i)\boldsymbol{\Lambda}_c G_{c,cl}^\top(i)]\mathrm{d}t \\
&\quad + \pi_{i\infty}\, \mathrm{Tr}[\widehat{X}_{cl}(0, i)G_{d,cl}(i)\boldsymbol{\Lambda}_d G_{d,cl}^\top(i)]\Bigg).
\end{aligned}
$$

Using the equality from Lemma 7.2 for $\widetilde{X}_{cl}(\cdot, i)$ replaced by $\widehat{X}_{cl}(\cdot, i)$, and $\widetilde{\Gamma}_c(i)$, $\widetilde{\Gamma}_d(i)$ replaced by $G_{c,cl}(i)\boldsymbol{\Lambda}_c G_{c,cl}^\top(i)$ and $G_{d,cl}(i)\boldsymbol{\Lambda}_d G_{d,cl}^\top(i)$, respectively, $G_c(i) = 0$, $H_d(i) = \Theta_d^\top(i)\Delta_d(0, i)\Theta_d(i)$, $\gamma_i = \pi_{i\infty}$ we get:

$$
\begin{aligned}
\|\mathbb{G}_{cl}(\widehat{\mathbb{K}})\|_2^2 &= \sum_{i=1}^{N} \Bigg(\frac{\pi_{i\infty}}{h} \int_0^h \mathrm{Tr}[X_s(t, i)G_c(i)\boldsymbol{\Lambda}_c G_c^\top(i)]\mathrm{d}t \qquad (7.129) \\
&+ \pi_{i\infty}\, \mathrm{Tr}[X_s(0, i)(G_d(i) + B_d(i)\widehat{D}(i)D_y(i))\boldsymbol{\Lambda}_d(G_d(i) + B_d(i)\widehat{D}(i)D_y(i))^\top] \\
&+ \mathrm{Tr}[D_{d,cl}\boldsymbol{\Lambda}_d D_{d,cl}^\top(i)]\Bigg) + \sum_{i=1}^{N} \mathrm{Tr}[\widetilde{Y}_{cl}(0, i)\Theta_d(i)\Delta_d(0, i)\Theta_d(i)]
\end{aligned}
$$

where $\widetilde{\mathbf{Y}}_{cl}(\cdot)$ is the unique solution of the FGLJLDE (7.113) written for

$$\Gamma_c(i) = \frac{\pi_{i\infty}}{h} G_c(i)\boldsymbol{\Lambda}_c G_c^\top(i) \quad \text{and} \quad \Gamma_d(i) = \pi_{i\infty} G_d(i)\boldsymbol{\Lambda}_d G_d^\top(i).$$

On the other hand, the FJMLDE (7.117) satisfied by its stabilizing solution $\mathbf{Y}_s(\cdot)$ is rewritten in the form

$$\dot{Y}_s(t,i) = A_0(i)Y_s(t,i) + Y_s(t,i)A_0^\top(i) + \sum_{j=1}^{N} q_{ji}Y_s(t,j) + \frac{\pi_{i\infty}}{h}G_c(i)\boldsymbol{\Lambda}_c G_c^\top(i),$$
$$kh < t \leq (k+1)h \tag{7.130a}$$
$$\begin{aligned} Y_s(kh^+, i) &= (A_d(i) + B_d(i)\widehat{D}(i)C_y(i))Y_s(kh,i)(A_d(i) + B_d(i)\widehat{D}(i)C_y(i))^\top \\ &\quad - (K_s(i) - B_d(i)\widehat{D}(i))\widehat{\Delta}_d(0,i)(K_s(i) - B_d(i)\widehat{D}(i))^\top \\ &\quad + \pi_{i\infty}(G_d(i) + B_d(i)\widehat{D}(i)D_y(i))\boldsymbol{\Lambda}_d(G_d(i) + B_d(i) + \widehat{D}(i)D_y(i))^\top, \\ &\quad k \in \mathbb{Z},\ i \in \mathcal{N}. \end{aligned} \tag{7.130b}$$

Let $\widehat{\mathbf{Y}}_{cl}(\cdot) = (\widehat{Y}_{cl}(\cdot,1), \widehat{Y}_{cl}(\cdot,2), \ldots, \widehat{Y}_{cl}(\cdot,N))$ be defined by

$$\widehat{Y}_{cl}(t,i) = \widetilde{Y}_{cl}(t,i) - \begin{pmatrix} Y_s(t,i) & 0 \\ 0 & 0 \end{pmatrix} \tag{7.131}$$

for all $(t,i) \in \mathbb{R} \times \mathcal{N}$.

Employing (7.101), (7.113) written for $\Gamma_c(i) = \frac{\pi_{i\infty}}{h}G_c(i)\boldsymbol{\Lambda}_c G_c^\top(i)$ and $\Gamma_d(i) = \pi_{i\infty}G_d(i)\boldsymbol{\Lambda}_d G_d^\top(i)$ together with (7.130) and (7.131) we deduce that $\widehat{\mathbf{Y}}_{cl}(\cdot)$ solves the following FGLJLDE:

$$\frac{\mathrm{d}}{\mathrm{d}t}\widehat{Y}_{cl}(t,i) = \mathcal{L}_{cl,i}[\widehat{\mathbf{Y}}_{cl}(t)], \quad kh < t \leq (k+1)h \tag{7.132a}$$
$$\widehat{Y}_{cl}(kh^+, i) = A_{d,cl}(i)\widehat{Y}_{cl}(kh,i)A_{d,cl}^\top(i) + \Xi_d(i)\widehat{\Delta}_d(0,i)\Xi_d^\top(i) \quad k \in \mathbb{Z}, \tag{7.132b}$$

where $\Xi_d(i) = \begin{pmatrix} K_s(i) - B_d(i)\widehat{D}(i) \\ -B_d(i) \end{pmatrix}$. Substituting (7.131) in (7.129) we obtain

$$\|\mathbb{G}_{cl}(\widehat{\mathbb{K}})\|_2^2 = \sum_{i=1}^{N} \rho_i(\widehat{D}(i)) + \sum_{i=1}^{N} \mathrm{Tr}[\widehat{Y}_{cl}(0,i)\Theta_d^\top(i)D_d(0,i)\Theta_d(i)] \tag{7.133}$$

where

$$\rho_i(\widehat{D}(i)) := \frac{\pi_{i\infty}}{h}\int_0^h \mathrm{Tr}[X_s(t,i)G_c(i)\boldsymbol{\Lambda}_c G_c^\top(i)]\mathrm{d}t \\ + \pi_{i\infty}\,\mathrm{Tr}[X_s(0,i)(G_d(i)+B_d(i)\widehat{D}(i)D_y(i))\boldsymbol{\Lambda}_d(G_d(i) \\ + B_d(i)\widehat{D}(i)D_y(i))^\top] + \mathrm{Tr}[D_d(i)\widehat{D}(i)D_y(i)\boldsymbol{\Lambda}_d D_y^\top(i)\widehat{D}^\top(i)D_d^\top(i)] \\ + \mathrm{Tr}[(F_s(i)-\widehat{D}(i)C_y(i))Y_s(0,i)(F_s(i)-\widehat{D}(i)C_y(i))^\top\Delta_d(0,i)]. \tag{7.134}$$

Since the FGLJLDE (7.132) is a special case of (7.113) we may infer via Corollary 7.6 that $\widehat{Y}_{cl}(t,i) \geq 0$, for all $(t,i) \in \mathbb{R}\times\mathcal{N}$. This allows us to conclude via (7.133) that

$$\|\mathbb{G}_{cl}(\widehat{\mathbb{K}})\|_2^2 \geq \sum_{i=1}^{N}\rho_i(\widehat{D}(i)) \tag{7.135}$$

for any stabilizing controller $\widehat{\mathbb{K}} \in \mathfrak{K}_{\text{stab}}(\widehat{\mathbb{G}})$, $\widehat{D}(i), 1 \leq i \leq N$ being matrices involved in the state-space representation of the controller $\widehat{\mathbb{K}}$. Further, we display a subset $\widehat{\mathfrak{K}}(\widehat{\mathbb{G}}) \subset \mathfrak{K}_{\text{stab}}(\widehat{\mathbb{G}})$, which consist of the stabilizing controllers $\widehat{\mathbb{K}}$ of the form (7.99) for which the inequality (7.135) becomes equality. Let $\widehat{\mathfrak{K}}(\widehat{\mathbb{G}})$ be the subset of all controllers of the form (7.99) whose coefficients are described by:

$$\widehat{A}(i) := A_0(i) \tag{7.136a}$$
$$\widehat{A}_d(i) := A_d(i) + B_d(i)F_s(i) + K_s(i)C_y(i) - B_d(i)\widehat{D}_d(i)C_y(i) \tag{7.136b}$$
$$\widehat{B}_d(i) := B_d(i)\widehat{D}(i) - K_s(i) \tag{7.136c}$$
$$\widehat{C}(i) := F_s(i) - \widehat{D}(i)C_y(i), \tag{7.136d}$$

$i \in \mathcal{N}$, where $\widehat{D}(i) \in \mathbb{R}^{m\times n_y}$, $1 \leq i \leq N$, are arbitrary matrices. Comparing (7.123a)–(7.123d) by (7.136a)–(7.136d) it follows that the controller $\widetilde{\widetilde{\mathbb{K}}}$ with the state-space representation (7.122) lies in $\widehat{\mathfrak{K}}(\widehat{\mathbb{G}})$. To show that the inclusion $\widehat{\mathfrak{K}}(\widehat{\mathbb{G}}) \subset \mathfrak{K}_{\text{stab}}(\widehat{\mathbb{G}})$ holds, we have to check that all controllers $\widehat{\mathbb{K}}$ of the form (7.99) and (7.136) are stabilizing controllers. To this end, we have to show that the corresponding JSLDE of the form (7.102) is ESMS. Setting $x_{cl}(t) = (x^\top(t) \quad \widehat{x}^\top(t))^\top \in \mathbb{R}^n \times \mathbb{R}^n$ we obtain via (7.101a) - (7.101b) the following partition of the corresponding JSLDE (7.102):

$$d\begin{pmatrix} x(t) \\ \widehat{x}(t)\end{pmatrix} = \begin{pmatrix} A_0(\eta(t)) & 0 \\ 0 & A_0(\eta(t))\end{pmatrix}\begin{pmatrix} x(t) \\ \widehat{x}(t)\end{pmatrix}\mathrm{d}t, \quad kh < t \leq (k+1)h \tag{7.137a}$$

$$\begin{pmatrix} x(kh^+) \\ \widehat{x}(kh^+)\end{pmatrix} = \begin{pmatrix} A_d(\eta(kh)) + B_d(\eta(kh))\widehat{D}(\eta(kh))C_y(\eta(kh)) & B_d(\eta(kh))\widehat{C}(\eta(kh)) \\ \widehat{B}_d(\eta(kh))C_y(\eta(kh)) & \widehat{A}_d(\eta(kh))\end{pmatrix} \\ \cdot\begin{pmatrix} x(kh) \\ \widehat{x}(kh)\end{pmatrix}, \quad k \in \mathbb{Z}_+. \tag{7.137b}$$

If $(x(t), \widehat{x}(t))$ is a solution of (7.137) we define $e(t) = x(t) - \widehat{x}(t)$, $t \geq 0$. By direct calculation, involving (7.136) and (7.137) one obtains that $t \to (e(t), \widehat{x}(t))$ solves the following JSLDE:

$$de(t) = A_0(\eta(t))e(t)dt, \quad kh < t \leq (k+1)h \tag{7.138a}$$
$$e(kh^+) = (A_d(\eta(kh)) + K_s(\eta(kh))C_y(\eta(kh)))e(kh), \quad k \in \mathbb{Z}_+ \tag{7.138b}$$

$$d\widehat{x}(t) = A_0(\eta(t))\widehat{x}(t)dt, \quad kh < t \leq (k+1)h \tag{7.139a}$$
$$\begin{aligned}\widehat{x}(kh^+) &= (A_d(\eta(kh)) + B_d(\eta(kh))F_s(\eta(kh)))\widehat{x}(kh) \\ &\quad + \widehat{B}_d(\eta(kh))C_y(\eta(kh))e(kh), \quad k \in \mathbb{Z}_+.\end{aligned} \tag{7.139b}$$

Since the JSLDE (7.138) coincides to the JSLDE (7.118) we may infer that there exist $\beta_1 \geq 1$, $\alpha_1 > 0$ such that the solutions of (7.138) to satisfy

$$\mathbb{E}[|e(t)|^2] \leq \beta_1 e^{-\alpha_1(t-t_0)}\mathbb{E}[|e(t_0)|^2], \text{ for all } \quad t \geq t_0 \geq 0.$$

On the other hand (7.139) is a non-homogeneous JSLDE whose linear part is ESMS and the forcing term has an exponential decay when $t \to \infty$. Applying Theorem 3.5 in the case of (7.139) we deduce that the solutions of this JSLDE are satisfying $\lim_{t\to\infty} \mathbb{E}[|\widehat{x}(t)|^2] = 0$. Hence, the solutions of the JSLDE (7.137) satisfy

$$\lim_{t\to\infty} \mathbb{E}[|x_{cl}(t)|^2] = \lim_{t\to\infty} \mathbb{E}[|x(t)|^2 + |\widehat{x}(t)|^2] = 0.$$

Finally, invoking the implication $(iv) \Rightarrow (iii)$ from Theorem 3.1 we may conclude that for an arbitrary controller $\widehat{\mathbb{K}}$ having the state-space representation (7.99), (7.136) the corresponding JSLDE (7.102) is ESMS. In this way the inclusion $\widehat{\mathfrak{K}}(\widehat{\mathbb{G}}) \subset \mathfrak{K}_{\text{stab}}(\widehat{\mathbb{G}})$ has been certified. According to (7.136d) we deduce from (7.133) that:

$$\begin{aligned}\|\mathbb{G}_{cl}(\widehat{\mathbb{K}})\|_2^2 &= \sum_{i=1}^{N} \rho_i(\widehat{D}(i)) + \sum_{i=1}^{N} \text{Tr}[(F_s(i) - \widehat{D}(i)C_y(i))\check{Y}_{11}(0,i)(F_s(i) \\ &\quad - \widehat{D}(i)C_y(i))^\top \Delta_d(0,i)],\end{aligned} \tag{7.140}$$

where $\check{Y}_{11}(0,i) = (I_n \quad -I_n)\widehat{Y}_{cl}(0,i)(I_n \quad -I_n)^\top$. One sees that $\check{Y}_{11}(0,i)$ is the 11-block of the matrix $\check{Y}_{cl}(0,i) \stackrel{\Delta}{=} \mathscr{T}\widehat{Y}_{cl}(0,i)\mathscr{T}^\top$ with $\mathscr{T} = \begin{pmatrix} I_n & -I_n \\ 0 & I_n \end{pmatrix}$. By direct calculation involving (7.132) and (7.136) one obtains that $(\check{Y}_{11}(\cdot,1), \check{Y}_{11}(\cdot,2), \ldots, \check{Y}_{11}(\cdot,N))$ is a bounded on $\mathbb{R}$ solution of the following FGLJLDE

$$\frac{d}{dt}\check{Y}_{11}(t,i) = A_0(i)\check{Y}_{11}(t,i) + \check{Y}_{11}(t,i)A_0^\top(i) + \sum_{j=1}^{N} q_{ji}\check{Y}_{11}(t,j), \quad kh < t \leq (k+1)h \tag{7.141a}$$

$$\check{Y}_{11}(kh^+, i) = (A_d(i) + K_s(i)C_y(i))\check{Y}_{11}(kh, i)(A_d(i) + K_s(i)C_y(i))^\top \quad k \in \mathbb{Z}, \quad i \in \mathcal{N}. \tag{7.141b}$$

One sees that (7.141) is the FGLJLDE defined by the JSLDE (7.118). Hence, (7.141) may have only one bounded on $\mathbb{R}$ solution. This allows us to conclude that $\check{Y}_{11}(t, i) = 0$, for all $(t, i) \in \mathbb{R} \times \mathcal{N}$. So, (7.140) reduces to

$$\|\mathbb{G}_{cl}(\widehat{\mathbb{K}})\|_2^2 = \sum_{i=1}^{N} \rho_i(\widehat{D}(i)), \tag{7.142}$$

whenever $\widehat{\mathbb{K}} \in \widehat{\mathfrak{K}}(\widehat{\mathbb{G}})$. Further, we rewrite (7.134) as:

$$\rho_i(\widehat{D}(i)) = \widetilde{\rho}_i + \mathrm{Tr}[(\widehat{D}(i) - \widehat{\mathscr{D}}(i))\widehat{\Delta}_d(0, i)(\widehat{D}(i) - \widehat{\mathscr{D}}(i))^\top \Delta_d(0, i)] \tag{7.143}$$

where $\widehat{\mathscr{D}}_d(i)$ are defined in (7.123e) and

$$\begin{aligned}
\widetilde{\rho}_i := & \frac{\pi_{i\infty}}{h} \int_0^h \mathrm{Tr}[X_s(t, i)G_c(i)\boldsymbol{\Lambda}_c G_c^\top(i)]\mathrm{d}t \\
& + \pi_{i\infty} \mathrm{Tr}[X_s(0, i)G_d(i)\boldsymbol{\Lambda}_d G_d^\top(i)] + \mathrm{Tr}[F_s(i)Y_s(0, i)F_s^\top(i)\Delta_d(0, i)] \\
& - \mathrm{Tr}[(D_d^\top(i)C_d(i)Y_s(0, i)C_y^\top(i) - B_d^\top(i)X_s(0, i)K_s(i)\widehat{\Delta}_d(0, i)) \\
& \cdot \widehat{\Delta}_d^{-1}(0, i)(D_d^\top(i)C_d(i)Y_s(0, i)C_y^\top(i) \\
& - B_d^\top(i)X_s(0, i)K_s(i)\widehat{\Delta}_d(0, i))^\top \Delta_d^{-1}(0, i)].
\end{aligned}$$

Invoking (7.120), (7.121) together with the properties of the trace operator, we deduce from (7.143) that

$$\|\mathbb{G}_{cl}(\widehat{\mathbb{K}})\|_2^2 \geq \sum_{i=1}^{N} \widetilde{\rho}_i = \sum_{i=1}^{N} \rho_i(\widehat{\mathscr{D}}(i)) = \|\mathbb{G}_{cl}(\widetilde{\widetilde{\mathbb{K}}})\|_2^2 \tag{7.144}$$

for all $\widehat{\mathbb{K}} \in \widehat{\mathfrak{K}}(\widehat{\mathbb{G}})$. Combining (7.135), (7.142), (7.144) we may conclude that

$$\|\mathbb{G}_{cl}(\widehat{\mathbb{K}})\|_2 \geq \|\mathbb{G}_{cl}(\widetilde{\widetilde{\mathbb{K}}})\|_2 \tag{7.145}$$

for any stabilizing controller $\widehat{\mathbb{K}}$ of the form (7.99). Thus, we have shown that the controller $\widetilde{\widetilde{\mathbb{K}}}$ with the state-space representation (7.122) and (7.123) achieves the minimal value of the H_2-norm of the closed-loop systems (7.100)–(7.102). The minimal value of the H_2-norm given in (7.124) is obtained from (7.144). Thus the proof ends.

Exercise 7.2 Find the state-space representation of a stabilizing controller $\widetilde{\widetilde{\mathbb{K}}}$ of the form (7.99) which is satisfying the optimality condition (7.109).

7.4.4 A Discussion of the Robustness of the Optimal Controller

The additive white noises $\{\xi(t)\}_{t\geq 0}$ and $\{\zeta_d(k)\}_{k\in\mathbb{Z}_+}$ which are affecting the dynamical system $\widehat{\mathbb{G}}$ are characterized by the covariance matrices $\boldsymbol{\Lambda}_c$ and $\boldsymbol{\Lambda}_d$, respectively.

Let us denote $\mathbb{G}_{cl}^{\boldsymbol{\Lambda}_c\boldsymbol{\Lambda}_d}(\widehat{\mathbb{K}})$ the closed-loop system of the form (7.100) in the case when the additive white noise have the covariance matrices $\boldsymbol{\Lambda}_c$ and $\boldsymbol{\Lambda}_d$, respectively. According to Corollary 7.3 applied in the case of the closed-loop system, we may infer that

$$\|\mathbb{G}_{cl}^{\boldsymbol{\Lambda}_c\boldsymbol{\Lambda}_d}(\widehat{\mathbb{K}})\|_2 \leq \|\mathbb{G}_{cl}^{\widetilde{\boldsymbol{\Lambda}}_c\widetilde{\boldsymbol{\Lambda}}_d}(\widehat{\mathbb{K}})\|_2, \tag{7.146}$$

whenever $\boldsymbol{\Lambda}_c \leq \widetilde{\boldsymbol{\Lambda}}_c$ and $\boldsymbol{\Lambda}_d \leq \widetilde{\boldsymbol{\Lambda}}_d$, respectively. The inequality (7.146) allows us to propose a procedure for the design of a stabilizing controller $\widehat{\mathbb{K}}$ which achieves an acceptable level of the attenuation of the effect of the additive white noise type without needing prior knowledge of precise values of the covariance matrices of these additive perturbations. So, if the designer wants to protect the behavior of trajectories of a dynamical system $\widehat{\mathbb{G}}$ by the effect of some additive white noise perturbations whose intensity does not exceed a certain prescribed level, he/she will choose positive semidefinite matrices $\widetilde{\boldsymbol{\Lambda}}_c$, $\widetilde{\boldsymbol{\Lambda}}_d$ and will use these matrices to obtain the state-space representation of a H_2 optimal controller $\widetilde{\widetilde{\mathbb{K}}}$ using, for example, (7.123) with $\boldsymbol{\Lambda}_c$ replaced by $\widetilde{\boldsymbol{\Lambda}}_c$ and $\boldsymbol{\Lambda}_d$ replaced by $\widetilde{\boldsymbol{\Lambda}}_d$. From (7.145) and (7.146) it follows that:

$$\|\mathbb{G}_{cl}^{\boldsymbol{\Lambda}_c\boldsymbol{\Lambda}_d}(\widetilde{\widetilde{\mathbb{K}}})\|_2 \leq \|\mathbb{G}_{cl}^{\widetilde{\boldsymbol{\Lambda}}_c\widetilde{\boldsymbol{\Lambda}}_d}(\widetilde{\widetilde{\mathbb{K}}})\|_2 \leq \|\mathbb{G}_{cl}^{\widetilde{\boldsymbol{\Lambda}}_c\widetilde{\boldsymbol{\Lambda}}_d}(\widehat{\mathbb{K}})\|_2$$

for all $0 \leq \boldsymbol{\Lambda}_c \leq \widetilde{\boldsymbol{\Lambda}}_c$, $0 \leq \boldsymbol{\Lambda}_d \leq \widetilde{\boldsymbol{\Lambda}}_d$ and an arbitrary $\widehat{\mathbb{K}} \in \mathfrak{K}_{\text{stab}}(\widehat{\mathbb{G}})$.

This means that in the presence of some additive perturbations $\{\xi(t)\}_{t\geq 0}$ and $\{\zeta_d(k)\}_{k\in\mathbb{Z}_+}$ having the covariance matrices $\boldsymbol{\Lambda}_c \leq \widetilde{\boldsymbol{\Lambda}}_c$ and $\boldsymbol{\Lambda}_d \leq \widetilde{\boldsymbol{\Lambda}}_d$, respectively, the optimal controller $\widetilde{\widetilde{\mathbb{K}}}$ will achieve a level of attenuation of the additive perturbations better than the level achieved by the same controller in the presence of some white noise-type additive perturbations having the covariance matrices $\widetilde{\boldsymbol{\Lambda}}_c$, $\widetilde{\boldsymbol{\Lambda}}_d$, respectively.

7.5 H_2-Optimal Control of Sampled-Data Linear Stochastic Systems

7.5.1 H_2-Optimal Control of Sampled-Data Linear Stochastic System Under Perfect State Measurements

Let us consider the linear stochastic system $\mathbb{G}$ described by

$$\begin{aligned} dx(t) =&(A_0(\eta(t))x(t) + B_0(\eta(t))u(t))dt \\ &+ \sum_{l=1}^{r}(A_l(\eta(t))x(t) + B_l(\eta(t))u(t))dw_l(t) + G(\eta(t))d\xi(t), \quad t \in \mathbb{R}_+ \end{aligned} \tag{7.147}$$

where $x(t) \in \mathbb{R}^n$ are the vectors of the state parameters and $u(t) \in \mathbb{R}^m$ are vectors of the control parameters. In (7.147), $\{w(t)\}_{t\geq 0}$ $(w(t) = (w_1(t)\ w_2(t)\ \dots\ w_r(t))^\top)$ is a r-dimensional standard Wiener process and $\{\eta(t)\}_{t\geq 0}$ is a standard homogeneous Markov process with the set of the states $\mathcal{N} = \{1, 2, \dots, N\}$ while $\{\xi(t)\}_{t\geq 0}$ is a m_ξ-dimensional stochastic process satisfying Assumption 7.1(a).

We assume that $\{w(t)\}_{t\geq 0}$, $\{\eta(t)\}_{t\geq 0}$ and $\{\xi(t)\}_{t\geq 0}$ are defined on the same probability space $(\Omega, \mathfrak{F}, \mathcal{P})$ and are mutually independent stochastic processes. We assume also that the states $x(\cdot)$ of (7.147) are available for measurements, only at discrete-time instances $\tau_k = kh$, $k = 0, 1, 2, \dots$ and $h > 0$ is fixed. The admissible controls take constant values $u(k)$ on each interval $(kh, (k+1)h]$. Roughly speaking, our aim is to design a controller activated by the measurements $y(kh) = x(kh), k = 0, 1, \dots,$ which minimize the effect of the additive noises $\{\xi(t)\}_{t\geq 0}$ on the quality outputs:

$$z_c(t) = C_{z_c}(\eta(t))x(t), \quad t \in \mathbb{R}_+ \tag{7.148a}$$

$$z_d(kh) = C_{z_d}(\eta(kh))x(kh) + D_{z_d}(\eta(kh))u(kh), \quad k \in \mathbb{Z}_+. \tag{7.148b}$$

Proceeding as in the case of the system (1.31) (with $G_j(\cdot, i) = 0$, $i \in \mathcal{N}$) one may show that if $x(t)$ is a solution of (7.147) when $u(t) = u(k)$ for $t \in (kh, (k+1)h]$, $k \in \mathbb{Z}_+$ and if $\varphi(\cdot)$ is defined by $\varphi(t) = u(k)$, $kh < t \leq (k+1)h$, $k \in \mathbb{Z}_+$, then $\mathbf{x}(t) \stackrel{\Delta}{=} (x^\top(t) \quad \varphi^\top(t))^\top$ solves the following impulsive controlled linear stochastic system (ICLSS):

$$\begin{aligned} d\mathbf{x}(t) = \mathbf{A}_0(\eta(t))\mathbf{x}(t)dt + \sum_{l=1}^{r}\mathbf{A}_l(\eta(t))\mathbf{x}(t)dw_l(t) + \mathbf{G}(\eta(t))d\xi(t),& \\ kh < t \leq (k+1)h& \end{aligned} \tag{7.149a}$$

$$\mathbf{x}(kh^+) = \mathbf{A}_d\mathbf{x}(kh) + \mathbf{B}_d u(k), \quad k \in \mathbb{Z}_+ \tag{7.149b}$$

$$\mathbf{z}_c(t) = \mathbf{C}_c(\eta(t))\mathbf{x}(t), \quad t \in \mathbb{R}_+ \tag{7.149c}$$

$$\mathbf{z}_d(kh) = \mathbf{C}_d(\eta(kh))\mathbf{x}(kh) + \mathbf{D}_d(\eta(kh))u(k), \quad k \in \mathbb{Z}_+ \tag{7.149d}$$

$$\mathbf{y}(kh) = \mathbf{x}(kh), \quad k \in \mathbb{Z}_+ \tag{7.149e}$$

where for each $i \in \mathcal{N}$ we have denoted

$$\mathbf{A}_l(i) = \begin{pmatrix} A_l(i) & B_l(i) \\ 0 & 0 \end{pmatrix}, \quad 0 \le l \le r, \quad \mathbf{A}_d = \begin{pmatrix} I_n & 0 \\ 0 & 0 \end{pmatrix}, \quad \mathbf{B}_d = \begin{pmatrix} 0 \\ I_m \end{pmatrix}, \tag{7.150}$$

$$\mathbf{G}(i) = \begin{pmatrix} G(i) \\ 0 \end{pmatrix}, \quad \mathbf{C}_c(i) = \begin{pmatrix} C_{z_c}(i) & 0 \end{pmatrix}, \quad \mathbf{C}_d(i) = \begin{pmatrix} C_{z_d}(i) & 0 \end{pmatrix}, \quad \mathbf{D}_d(i) = D_{z_d}(i).$$

The ICLSS (7.149) is a special case of (7.60) and (7.61). To control the system (7.149) we shall use controllers of the form (7.62) and (7.63). One can see that by coupling a controller $\mathbb{K}$ of the form (7.62) or (7.63) (but with the inputs $\widehat{u}(k)$ of dimension $n + m$) to the ICLSS (7.149) one obtains a closed-loop system $\mathbb{G}_{cl}(\mathbb{K})$ of the form (7.64) with $G_{d,cl}(i) = 0$, $i \in \mathcal{N}$. This means that the outputs $\mathbf{z}_{c,cl}(\cdot, \zeta_d)$ and $\mathbf{z}_{d,cl}(\cdot, \zeta_d)$ of the form (7.71) are satisfying

$$\mathbf{z}_{c,cl}(t, \zeta_d) = 0, \quad \text{for all } \ t \in \mathbb{R}_+$$

$$\mathbf{z}_{d,cl}(kh, \zeta_d) = 0, \quad \text{for all } \ k \in \mathbb{Z}_+.$$

This means that

$$\mathbf{z}_{c,cl}(t) = \mathbf{z}_{c,cl}(t, \xi), \quad t \in \mathbb{R}_+$$

$$\mathbf{z}_{d,cl}(kh) = \mathbf{z}_{d,cl}(kh, \xi), \quad k \in \mathbb{Z}_+.$$

Hence, in this case, (7.72) reduces to:

$$\|\mathbb{G}_{cl}(\mathbb{K})\|_2^2 = \limsup_{T\to\infty} \frac{1}{T} \left(\int_0^T \mathbb{E}[|\mathbf{z}_{c,cl}(t)|^2]\mathrm{d}t + \sum_{k=0}^{[\frac{T}{h}]} \mathbb{E}[|\mathbf{z}_{d,cl}(kh)|^2] \right). \tag{7.151}$$

To obtain the state-space representation of a controller which minimizes the H_2-norm of a closed-loop system obtained by coupling an admissible controller $\mathbb{K}$ to the ICLSS (7.149) we apply Theorem 7.4. In this case, the BJMLDE with Riccati-type jumping operator (7.76) becomes:

$$-\dot{X}(t,i) = \mathbf{A}_0^\top(i)X(t,i) + X(t,i)\mathbf{A}_0(i) + \sum_{l=1}^{r} \mathbf{A}_l^\top(i)X(t,i)\mathbf{A}_l(i)$$
$$+ \sum_{j=1}^{N} q_{ij}X(t,j) + \mathbf{C}_c^\top(i)\mathbf{C}_c(i), \quad kh \le t < (k+1)h \tag{7.152a}$$

$$X(kh^-, i) = \mathbf{A}_d^\top X(kh,i)\mathbf{A}_d - (\mathbf{A}_d^\top X(kh,i)\mathbf{B}_d + \mathbf{C}_d^\top(i)\mathbf{D}_d(i))$$

$$\cdot (\mathbf{B}_d^\top X(kh,i)\mathbf{B}_d + \mathbf{D}_d^\top(i)\mathbf{D}_d(i))^{-1}(\mathbf{B}_d^\top X(kh,i)\mathbf{A}_d + \mathbf{D}_d^\top(i)\mathbf{C}_d(i))$$
$$+ \mathbf{C}_d^\top(i)\mathbf{C}_d(i), \quad k \in \mathbb{Z}_+, \quad i \in \mathcal{N}. \tag{7.152b}$$

A global solution $t \to \mathbf{X}_s(t) = (X_s(t,1), X_s(t,2), \dots, X_s(t,N)) : \mathbb{R} \to \mathcal{S}_{n+m}^N$ of the BJMLDE (7.152) is **stabilizing solution** if its corresponding closed-loop system on $\mathbb{R}^{n+m}$

$$\mathrm{d}\mathbf{x}(t) = \mathbf{A}_0(\eta(t))\mathbf{x}(t)\mathrm{d}t + \sum_{l=1}^{r} \mathbf{A}_l(\eta(t))\mathbf{x}(t)\mathrm{d}w_l(t), \quad kh < t \le (k+1)h \tag{7.153a}$$

$$\mathbf{x}(kh^+) = (\mathbf{A}_d + \mathbf{B}_d\mathbf{F}_s(\eta(kh))\mathbf{x}(kh), \quad k \in \mathbb{Z}_+ \tag{7.153b}$$

is ESMS, where

$$\mathbf{F}_s(i) \triangleq - (\mathbf{D}_d^\top(i)\mathbf{D}_d(i) + \mathbf{B}_d^\top X_s(0,i)\mathbf{B}_d)^{-1}$$
$$(\mathbf{B}_d^\top X_s(0,i)\mathbf{A}_d + \mathbf{D}_d^\top(i)\mathbf{C}_d(i)), \quad i \in \mathcal{N}. \tag{7.154}$$

According to Corollary 7.4 applied in the case of the BJMLDE (7.152) we have that $X_s(t,i) \ge 0$, for all $(t,i) \in \mathbb{R}_+ \times \mathcal{N}$. Let

$$X_s(t,i) = \begin{pmatrix} X_{s,11}(t,i) & X_{s,12}(t,i) \\ X_{s,12}^\top(t,i) & X_{s,22}(t,i) \end{pmatrix} \tag{7.155}$$

be the partition of the stabilizing solution of the BJMLDE (7.152) such that $X_{s,11}(t,i) \in \mathcal{S}_n$ and $X_{s,22}(t,i) \in \mathcal{S}_m$, $(t,i) \in \mathbb{R}_+ \times \mathcal{N}$.
Employing (7.150), (7.154) and (7.155) we deduce that if $i \in \mathcal{N}$ we have:

$$\mathbf{F}_s(i) = (F_s(i) \quad 0) \tag{7.156a}$$

$$F_s(i) \triangleq -(X_{s,22}(0,i) + D_{z_d}^\top(i)D_{z_d}(i))^{-1}(X_{s,12}^\top(0,i) + D_{z_d}^\top(i)C_{z_d}(i)). \tag{7.156b}$$

The main result of this subsection is:

Theorem 7.7 *Assume that the BJMLDE with Riccati-type jumping operator (7.152) has a bounded and stabilizing solution $t \to \mathbf{X}_s(t) = (X_s(t,1), X_s(t,2), \dots, X_s(t,N)) : \mathbb{R} \to \mathcal{S}_{n+m}^N$ satisfying the sign condition*

$$X_{s,22}(kh,i) + D_{z_d}^\top(i)D_{z_d}(i) > 0, \tag{7.157}$$

for all $(k,i) \in \mathbb{Z} \times \mathcal{N}$. Consider the zero-order controller $\widetilde{\mathbb{K}}$

$$u(k) = \mathbf{F}_s(\eta(kh))\mathbf{x}(kh), \quad k \in \mathbb{Z}, \tag{7.158}$$

where $\mathbf{F}_s(i)$ is in (7.154) and $\mathbf{x}(kh)$ are the values of the solution of the closed-loop system obtained substituting (7.158) in (7.149) measured at time instances $\tau_k = kh$.

Under the considered assumptions, the controller (7.158) stabilizes the ICLSS (7.149) and satisfies $\|\mathbb{G}_{cl}(\widetilde{\mathbb{K}})\|_2 \le \|\mathbb{G}_{cl}(\mathbb{K})\|_2$, where $\mathbb{K}$ is an arbitrary stabilizing controller of type (7.62) or (7.63) (with outputs $\widehat{u}(k)$ of dimension $n+m$). The minimal value of the H_2-norm of the closed-loop system $\mathbb{G}_{cl}(\mathbb{K})$ is:

$$\|\mathbb{G}_{cl}(\widetilde{\mathbb{K}})\|_2 = \left(\sum_{i=1}^{N} \frac{\pi_{i\infty}}{h} \int_0^h \mathrm{Tr}[X_{s,11}(t,i)G(i)\boldsymbol{\Lambda}_c G^\top(i)]\mathrm{d}t\right)^{1/2}.$$

Proof The proof follows applying Theorem 7.4 in the case of the ICLSS (7.149) and (7.150).

Remark 7.9 Based on Corollary 7.4 applied in the case of the BJMLDE (7.152) we may infer that a sufficient condition to (7.157) holds is $D_{z_d}^\top(i)D_{z_d}(i) > 0,\ i \in \mathcal{N}$.

Let $\mathbf{x}(t) = (x^\top(t)\ \ \varphi^\top(t))^\top$ be a solution of the closed-loop system (7.153) with $x(t) \in \mathbb{R}^n$ and $\varphi(t) \in \mathbb{R}^m$. Employing (7.150), (7.156) we may infer that the component $x(t)$ of the solution of (7.153) satisfies:

$$\begin{aligned}\mathrm{d}x(t) &= (A_0(\eta(t))x(t) + B_0(\eta(t))F_s(\eta(t))x(kh))\mathrm{d}t + \sum_{l=1}^{r}(A_l(\eta(t))x(t)\\ &\quad + B_l(\eta(t))F_s(\eta(t))x(kh))\mathrm{d}w_l(t),\quad kh < t \le (k+1)h,\quad k \in \mathbb{Z}_+.\end{aligned} \tag{7.159}$$

Since $\mathbf{F}_s(i)$ introduced via (7.154) is the stabilizing feedback gain, we deduce via (7.156) and (7.158) that the solution of (7.159) satisfies

$$\mathbb{E}[|x(t)|^2] \le \beta e^{-\alpha(t-t_0)}\mathbb{E}[|x(t_0)|^2],\quad \text{for all } t \ge t_0 \ge 0,$$

where $\alpha > 0$, $\beta \ge 1$ are constants not depending upon t, t_0 and $x(t_0)$. Also, (7.150) and (7.156) allow us to deduce that the component $x_{cl}(\cdot)$ of the solution $\mathbf{x}_{cl}(\cdot)$ of the closed-loop system obtained when the control (7.158) is coupled to the system (7.149) satisfies

$$\begin{aligned}\mathrm{d}x_{cl}(t) &= (A_0(\eta(t))x_{cl}(t) + B_0(\eta(t))F_s(\eta(kh))x_{cl}(kh))\mathrm{d}t\\ &\quad + \sum_{l=1}^{r}(A_l(\eta(t))x_{cl}(t) + B_l(\eta(t))F_s(\eta(kh))x_{cl}(kh))\mathrm{d}w_l(t)\\ &\quad + G_c(\eta(t))\mathrm{d}\xi(t),\quad kh < t \le (k+1)h\end{aligned} \tag{7.160a}$$

$$z_{c,cl}(t) = C_{z_c}(\eta(t))x_{cl}(t),\quad t \in \mathbb{R}_+ \tag{7.160b}$$

$$z_{d,cl}(kh) = (C_{z_d}(\eta(kh)) + D_{z_d}(\eta(kh))F_s(\eta(kh)))x_{cl}(kh),\quad k \in \mathbb{Z}_+. \tag{7.160c}$$

The stochastic differential equation (7.160a) is obtained by coupling to the controlled linear stochastic system (7.147) the piecewise constant control

$$u(t) = F_s(\eta(kh))x(kh), \quad kh \le t < (k+1)h. \tag{7.161}$$

Based on (7.156), we may conclude that (7.160b) and (7.160c) are just the outputs of the closed-loop system obtained when (7.158) is coupled to (7.149). This means that the piecewise constant control (7.161) minimizes the effect of additive perturbations $\{\xi(t)\}_{t\ge 0}$ on the quality outputs (7.148) of the controlled system (7.147).

Example 7.1 Let us consider the dynamic system from Example 4.3 with the additional numerical parameters:

$$G(i) = I_n, \ C_{z_c}(i) = I_n, \ C_{z_d}(i) = \begin{pmatrix} I_n \\ 0_{m\times n} \end{pmatrix}, \ D_{z_d}(i) = \begin{pmatrix} 0_{n\times m} \\ I_m \end{pmatrix}, i \in \mathcal{N}.$$

We have fixed $r = 1$ and generated randomly the matrices $A_1(i)$ and $A_1(i)$, $i \in \mathcal{N}$. The sampling period is $h = 0.1$ and $\boldsymbol{\lambda}_c = \boldsymbol{\lambda}_d = 0.01$. Applying Theorem 7.7, we obtain the following feedback gains:

$$\begin{aligned} F_s(1) &= \begin{bmatrix} -0.8382 & -1.8539 & 3.4339 & -1.2810 \end{bmatrix} \\ F_s(2) &= \begin{bmatrix} -0.7291 & -2.0774 & 2.3272 & -1.4268 \end{bmatrix} \\ F_s(3) &= \begin{bmatrix} -0.9544 & -1.9208 & 3.4961 & -1.1637 \end{bmatrix} \\ F_s(4) &= \begin{bmatrix} -0.7196 & -1.9264 & 3.3150 & -1.3725 \end{bmatrix}. \end{aligned}$$

The corresponding value of the H_2-norm of the closed-loop system is $\|\mathbb{G}_{cl}(\widetilde{\mathbb{K}})\|_2 = 4.3898$.

Figure 7.1 shows the evolution of the closed-loop system states corresponding to a single realization of the stochastic processes.

For the second part of the simulation, we would like to illustrate the robustness property raised in Corollary 7.3. To this end, we choose $\boldsymbol{\lambda}_c = \boldsymbol{\lambda}_d = 0.05$. The corresponding value of the H_2-norm of the closed-loop system is $\|\mathbb{G}_{cl}(\widetilde{\mathbb{K}})\|_2 = 9.8159$. Figure 7.2 shows the evolution of the closed-loop system states corresponding to a single realization of the stochastic processes.

Figure 7.3 shows the evolution of $\mathbb{E}[|z_{c,cl}(t)|^2]$ for the two different values of $(\boldsymbol{\lambda}_c, \boldsymbol{\lambda}_d)$. The result in Corollary 7.3 is then confirmed by the simulation example.

7.5.2 *Output-Based H_2-Optimal Control of a Class of Sampled-Data Linear Stochastic System Subject to Jump Markov Perturbations*

We consider the controlled system:

$$dx(t) = (A(\eta(t))x(t) + B(\eta(t))u(t))dt + G(\eta(t))d\xi(t), \quad t \in \mathbb{R}_+, \tag{7.162}$$

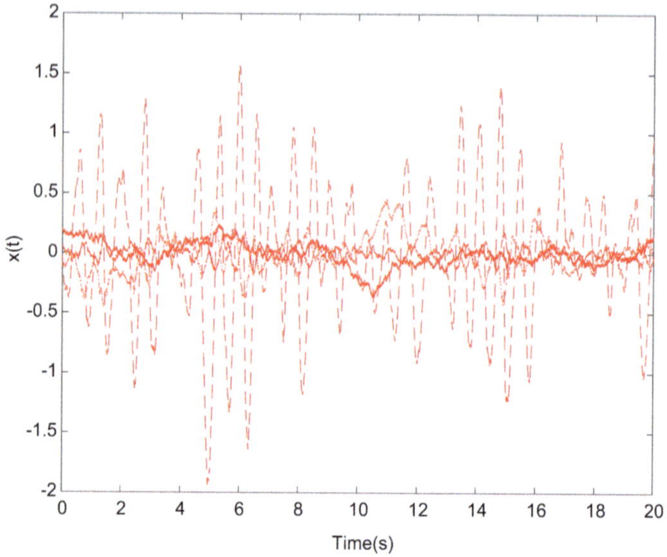

Fig. 7.1 Closed-loop states evolution: $\boldsymbol{\lambda}_c = \boldsymbol{\lambda}_d = 0.01$

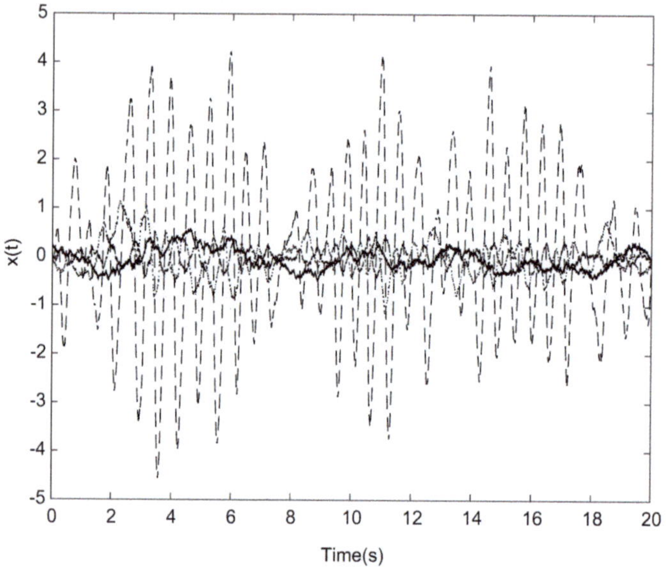

Fig. 7.2 Closed-loop states evolution: $\boldsymbol{\lambda}_c = \boldsymbol{\lambda}_d = 0.05$

where $x(t) \in \mathbb{R}^n$ are the vectors of the state parameters and $u(t) \in \mathbb{R}^m$ are the vectors of the control parameters. In (7.162), $\{\eta(t)\}_{t\geq 0}$ is a standard homogeneous continuous-time Markov process defined on a given probability space $(\Omega, \mathfrak{F}, \mathscr{P})$

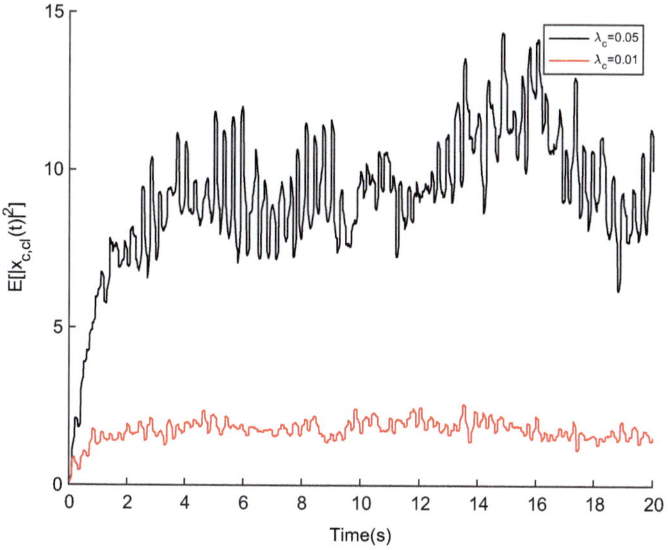

Fig. 7.3 Evolution of $\mathbb{E}[|z_{c,cl}|^2]$ for $\boldsymbol{\lambda}_c = 0.05$ and $\boldsymbol{\lambda}_c = 0.01$

taking values in the set $\mathcal{N} = \{1, 2, \ldots, N\}$ and having the transition semigroup $P(t) = e^{Qt}$, $t \geq 0$, and $\{\xi(t)\}_{t\geq 0}$ is a m_ξ-dimensional stochastic process satisfying Assumption 7.1 (a). We assume that only sampled measurements of the form:

$$y(kh) = C_y(\eta(kh))x(kh) + D_y(\eta(kh))\zeta_d(k), \quad k \in \mathbb{Z}_+ \tag{7.163}$$

are available for activating a controller. In (7.163), $\{\zeta_d(k)\}_{k\in\mathbb{Z}_+}$ is a sequence of m_{ζ_d}-dimensional independent random vectors which are satisfying Assumption 7.1 (b). As in the case of the system (7.147), we assume that to control the system (7.162), we shall use controls taking constant values $u(k)$ on each interval $(kh, (k+1)h]$, $k = 0, 1, \ldots$.

Our aim is to design a controller activated by the measurements $y(kh)$ of the form (7.163) which minimize the effect of the additive perturbations $\{\xi(t)\}_{t\geq 0}$ and $\{\zeta_d(k)\}_{k\in\mathbb{Z}_+}$ on the quality outputs:

$$z_c(t) = C_c(\eta(t))x(t), \quad t \in \mathbb{R}_+ \tag{7.164a}$$

$$z_d(kh) = C_d(\eta(kh))x(kh) + D_d(\eta(kh))u(kh), \quad k \in \mathbb{Z}_+. \tag{7.164b}$$

Proceeding as in the previous subsection we may show that if $x(t)$ is a solution of the controlled system (7.162) controlled by the piecewise constant control $u(t) = u(k)$, $kh < t \leq (k+1)h$, $k \in \mathbb{Z}_+$ and if $\varphi(t)$ is defined by $\varphi(t) \stackrel{\Delta}{=} u(k)$, $kh < t \leq (k+1)h$, $k \in \mathbb{Z}_+$, then $\mathbf{x}(t) := (x^\top(t) \quad \varphi^\top(t))^\top$ solves the impulsive controlled linear stochastic system $\widehat{\mathbb{G}}$:

$$d\mathbf{x}(t) = \mathbf{A}(\eta(t))\mathbf{x}(t)dt + \mathbf{G}(\eta(t))d\xi(t), \quad kh < t \le (k+1)h \tag{7.165a}$$
$$\mathbf{x}(kh^+) = \mathbf{A}_d\mathbf{x}(kh) + \mathbf{B}_d u(k), \quad k \in \mathbb{Z}_+ \tag{7.165b}$$
$$\mathbf{z}_c(t) = \mathbf{C}_c(\eta(t))\mathbf{x}(t), \quad t \in \mathbb{R}_+ \tag{7.165c}$$
$$\mathbf{z}_d(kh) = \mathbf{C}_d(\eta(kh))\mathbf{x}(kh) + \mathbf{D}_d(\eta(kh))u(k), \quad k \in \mathbb{Z}_+ \tag{7.165d}$$
$$\mathbf{y}(kh) = \mathbf{C}_y(\eta(kh))\mathbf{x}(kh) + \mathbf{D}_y(\eta(kh))u(k), \quad k \in \mathbb{Z}_+ \tag{7.165e}$$

where

$$\begin{aligned}
&\mathbf{A}(i) = \begin{pmatrix} A(i) & B(i) \\ 0 & 0 \end{pmatrix}, \quad \mathbf{A}_d = \begin{pmatrix} I_n & 0 \\ 0 & 0 \end{pmatrix}, \quad \mathbf{B}_d = \begin{pmatrix} 0 \\ I_m \end{pmatrix}, \mathbf{G}(i) = \begin{pmatrix} G(i) \\ 0 \end{pmatrix}, \\
&\mathbf{C}_c(i) = \begin{pmatrix} C_c(i) & 0 \end{pmatrix}, \quad \mathbf{C}_d(i) = \begin{pmatrix} C_d(i) & 0 \end{pmatrix}, \quad \mathbf{D}_d(i) = D_d(i), \\
&\mathbf{C}_y(i) = \begin{pmatrix} C_y(i) & 0 \end{pmatrix}, \quad \mathbf{D}_y(i) = D_y(i), \quad i \in \mathcal{N}.
\end{aligned} \tag{7.166}$$

One sees that (7.165) is an ICLSS of the form (7.98), but having the state space of dimension $n+m$. By coupling a controller $\widehat{\mathbb{K}}$ of the form (7.99) to the ICLSS (7.165) one obtains a closed-loop system $\mathbb{G}_{cl}(\widehat{\mathbb{K}})$ of the form of (7.100) whose coefficients have the structure:

$$A_{cl}(i) = \begin{pmatrix} \mathbf{A}(i) & 0 \\ 0 & \widehat{A}(i) \end{pmatrix}, \quad A_{d,cl}(i) = \begin{pmatrix} \mathbf{A}_d + \mathbf{B}_d\widehat{D}(i)\mathbf{C}_y(i) & \mathbf{B}_d\widehat{C}(i) \\ \widehat{B}_d(i)\mathbf{C}_y(i) & \widehat{A}_d(i) \end{pmatrix},$$

$$C_{c,cl}(i) = \begin{pmatrix} \mathbf{C}_c(i) & 0, \end{pmatrix} \quad C_{d,cl}(i) = \begin{pmatrix} \mathbf{C}_d(i) + \mathbf{D}_d(i)\widehat{D}(i)\mathbf{C}_y(i) & \mathbf{D}_d(i)\widehat{C}(i) \end{pmatrix},$$

$$G_{c,cl}(i) = \begin{pmatrix} \mathbf{G}(i) \\ 0 \end{pmatrix}, \quad G_{d,cl}(i) = \begin{pmatrix} \mathbf{B}_d\widehat{D}(i)\mathbf{D}_y(i) \\ \widehat{B}_d(i)\mathbf{D}_y(i) \end{pmatrix}, \quad D_{d,cl}(i) = \mathbf{D}_d(i)\widehat{D}(i)\mathbf{D}_y(i).$$

To obtain the state-space representation of a stabilizing controller for the ICLSS (7.165) which minimizes the H_2-norm of the closed-loop system we shall apply Theorem 7.6 in the special case of the ICLSS (7.165) and (7.166).

In this case, the BJMLDE with Riccati-type jumping operator (7.114) becomes:

$$-\dot{X}(t,i) = \mathbf{A}^\top(i)X(t,i) + X(t,i)\mathbf{A}(i) + \sum_{j=1}^{N} q_{ij}X(t,j) + \mathbf{C}_c^\top(i)\mathbf{C}_c(i),$$
$$kh \le t < (k+1)h \tag{7.167a}$$

$$\begin{aligned}
X(kh^-,i) &= \mathbf{A}_d^\top X(kh,i)\mathbf{A}_d - (\mathbf{A}_d^\top X(kh,i)\mathbf{B}_d + \mathbf{C}_d^\top(i)\mathbf{D}_d(i)) \\
&\quad \cdot (\mathbf{B}_d^\top X(kh,i)\mathbf{B}_d + \mathbf{D}_d^\top(i)\mathbf{D}_d(i))^{-1}(\mathbf{B}_d^\top X(kh,i)\mathbf{A}_d + \mathbf{D}_d(i)^\top\mathbf{C}_d(i)) \\
&\quad + \mathbf{C}_d^\top(i)\mathbf{C}_d(i), \quad k \in \mathbb{Z}, \quad i \in \mathcal{N}.
\end{aligned} \tag{7.167b}$$

A global solution $t \to \widetilde{\mathbf{X}}(t) = (\widetilde{X}(t,1), \widetilde{X}(t,2), \ldots, \widetilde{X}(t,N)) : \mathbb{R} \to \mathcal{S}_{n+m}^N$ of the BJMLDE (7.167) is called stabilizing solution if the closed-loop system

$$\dot{\mathbf{x}}(t) = \mathbf{A}(\eta(t))\mathbf{x}(t), \quad kh < t \le (k+1)h \tag{7.168a}$$

$$\mathbf{x}(kh^+) = (\mathbf{A}_d + \mathbf{B}_d\widetilde{\mathbf{F}}(\eta(kh)))\mathbf{x}(kh), \quad k \in \mathbb{Z}_+ \tag{7.168b}$$

is ESMS, where

$$\widetilde{\mathbf{F}}(i) \triangleq -(\mathbf{B}_d^\top \widetilde{X}(0,i)\mathbf{B}_d + \mathbf{D}_d(i)^\top \mathbf{D}_d(i))^{-1}(\mathbf{B}_d^\top \widetilde{X}(0,i)\mathbf{A}_d + \mathbf{D}_d^\top(i)\mathbf{C}_d(i)). \tag{7.169}$$

Let

$$\widetilde{X}(t,i) = \begin{pmatrix} \widetilde{X}_{11}(t,i) & \widetilde{X}_{12}(t,i) \\ \widetilde{X}_{12}^\top(t,i) & \widetilde{X}_{22}(t,i) \end{pmatrix} \quad (t,i) \in \mathbb{R} \times \mathcal{N}, \tag{7.170}$$

be the partition of the stabilizing solution of (7.167) such that $\widetilde{X}_{11}(t,i) \in \mathcal{S}_n$ and $\widetilde{X}_{22}(t,i) \in \mathcal{S}_m$. Employing (7.166), (7.169), (7.170) we deduce that

$$\widetilde{\mathbf{F}}(i) = \begin{pmatrix} \widetilde{F}(i) & 0 \end{pmatrix} \tag{7.171a}$$

$$\widetilde{F}(i) = -(\widetilde{X}_{22}(0,i) + D_d^\top(i)D_d(i))^{-1}(\widetilde{X}_{12}^\top(0,i) + D_d^\top(i)C_d(i)), \quad i \in \mathcal{N}. \tag{7.171b}$$

Remark 7.10 Comparing (7.167)–(7.169) to (7.76)–(7.78) we may conclude that the BJMLDE (7.167) is a special case of (7.76). Hence, Corollary 7.4 is applicable also in the case of the stabilizing solution $\widetilde{\mathbf{X}}(\cdot)$ of the BJMLDE (7.167). So $\widetilde{\mathbf{X}}(\cdot)$ is a periodic function of period h and $\widetilde{X}(t,i) \ge 0$, for all $(t,i) \in \mathbb{R} \times \mathcal{N}$.

In the special case of the ICLSS (7.165) and (7.166) the FJMLDE (7.117) becomes:

$$\dot{Y}(t,i) = \mathbf{A}(i)Y(t,i) + Y(t,i)\mathbf{A}^\top(i) + \sum_{j=1}^{N} q_{ji}Y(j) + \frac{\pi_{i\infty}}{h}\mathbf{G}(i)\boldsymbol{\Lambda}_c\mathbf{G}^\top(i),$$

$$kh < t \le (k+1)h \tag{7.172a}$$

$$\begin{aligned} Y(kh^+,i) = {} & \mathbf{A}_d Y(kh,i)\mathbf{A}_d^\top - \mathbf{A}_d Y(kh,i)\mathbf{C}_y^\top(i)(\mathbf{C}_y(i)Y(kh,i)\mathbf{C}_y^\top(i) \\ & + \pi_{i\infty}\mathbf{D}_y(i)\boldsymbol{\Lambda}_d\mathbf{D}_y^\top(i))^{-1}\mathbf{C}_y(i)Y(kh,i)\mathbf{A}_d^\top, \quad k \in \mathbb{Z},\ i \in \mathcal{N}, \end{aligned} \tag{7.172b}$$

where $\pi_{i\infty}$ are introduced in (7.34). A global solution $t \to \mathbf{Y}_s(t) = (Y_s(t,1), Y_s(t,2), \ldots, Y_s(t,N)) : \mathbb{R} \to \mathcal{S}_{n+m}^N$ of the FJMLDE (7.172) is stabilizing solution of its corresponding closed-loop FJLDE:

$$\dot{\mathbf{x}}(t) = \mathbf{A}(\eta(t))\mathbf{x}(t), \quad kh < t \le (k+1)h \tag{7.173a}$$

$$\mathbf{x}(kh^+) = (\mathbf{A}_d + \mathbf{K}_s(\eta(kh)\mathbf{C}_c(\eta(kh)))\mathbf{x}(kh), \quad k \in \mathbb{Z}_+ \tag{7.173b}$$

is ESMS, where

$$\mathbf{K}_s(i) \triangleq -\mathbf{A}_d Y_s(0,i)\mathbf{C}_y^\top(i)(\mathbf{C}_y(i)Y_s(0,i)\mathbf{C}_y^\top(i) + \pi_{i\infty}\mathbf{D}_y(i)\boldsymbol{\Lambda}_d\mathbf{D}_y^\top(i))^{-1}. \tag{7.174}$$

From Corollary 7.7 applied in the case of the FJMLDE (7.172) we deduce that the unique bounded on $\mathbb{R}$ and stabilizing solution $\mathbf{Y}_s(\cdot) = (Y_s(\cdot,1), Y_s(\cdot,2), \ldots, Y_s(\cdot,N))$ of (7.172), if any, is a periodic function of period h and satisfies $Y_s(t,i) \geq 0$, for all $(t,i) \in \mathbb{R} \times \mathcal{N}$. We set $Y_s(t,i) = \begin{pmatrix} \widetilde{Y}(t,i) & \widetilde{Y}_{12}(t,i) \\ \widetilde{Y}_{12}^\top(t,i) & \widetilde{Y}_{22}(t,i) \end{pmatrix}$ such that $\widetilde{Y}(t,i) \in \mathcal{S}_n$ and $\widetilde{Y}_{22}(y,i) \in \mathcal{S}_m$ for all $(t,i) \in \mathbb{R} \times \mathcal{N}$. According to (7.166) we obtain the following partition of the FJMLDE (7.172) written for its stabilizing solution

$$\begin{aligned}
\frac{\mathrm{d}}{\mathrm{d}t}\widetilde{Y}(t,i) &= A(i)\widetilde{Y}(t,i) + \widetilde{Y}(t,i)A^\top(i) + B(i)\widetilde{Y}_{12}(t,i) + \widetilde{Y}_{12}^\top(t,i)B^\top(i) \\
&\quad + \sum_{j=1}^{N} q_{ji}\widetilde{Y}(t,j) + \frac{\pi_{i\infty}}{h}G(i)\boldsymbol{\Lambda}_c G^\top(i) && (7.175a) \\
\frac{\mathrm{d}}{\mathrm{d}t}\widetilde{Y}_{12}(t,i) &= A(i)\widetilde{Y}_{12}(t,i) + B(i)\widetilde{Y}_{22}(t,i) + \sum_{j=1}^{N} q_{ji}\widetilde{Y}_{12}(t,j) && (7.175b) \\
\frac{\mathrm{d}}{\mathrm{d}t}\widetilde{Y}_{22}(t,i) &= \sum_{j=1}^{N} q_{ji}\widetilde{Y}_{22}(t,j) && (7.175c) \\
\widetilde{Y}(kh^+,i) &= \widetilde{Y}(kh,i) - \widetilde{Y}(kh,i)C_y^\top(i)(C_y(i)\widetilde{Y}(kh,i)C_y^\top(i) \\
&\quad + \pi_{i\infty}D_y(i)\boldsymbol{\Lambda}_d D_y^\top(i))^{-1}C_y(i)\widetilde{Y}(kh,i) && (7.175d) \\
\widetilde{Y}_{12}(kh^+,i) &= 0, && (7.175e) \\
\widetilde{Y}_{22}(kh^+,i) &= 0, \quad k \in \mathbb{Z}, \quad i \in \mathcal{N}. && (7.175f)
\end{aligned}$$

From (7.175b), (7.175c) together with (7.175e) and (7.175f) one sees that on each interval $(kh,(k+1)h]$ the function $(\widetilde{\mathbf{Y}}_{12}(\cdot), \widetilde{\mathbf{Y}}_{22}(\cdot))$ is the solution of a system of linear differential equations with zero initial values at $t = kh$. This means that $\widetilde{\mathbf{Y}}_{12}(t) = 0$, $\widetilde{\mathbf{Y}}_{22}(t) = 0$, for all $t \in (kh,(k+1)h]$, $k \in \mathbb{Z}$. Hence,

$$Y_s(t,i) = \begin{pmatrix} \widetilde{Y}(t,i) & 0 \\ 0 & 0 \end{pmatrix}, \tag{7.176}$$

for all $(t,i) \in \mathbb{R} \times \mathcal{N}$, where $t \to \widetilde{\mathbf{Y}}(t) = (\widetilde{Y}(t,1), \widetilde{Y}(t,2), \ldots, \widetilde{Y}(t,N)) : \mathbb{R} \to \mathcal{S}_n^N$ solves the following FJMLDE on $\mathcal{S}_n^N$:

$$\begin{aligned}
\dot{Y}(t,i) &= A(i)Y(t,i) + Y(t,i)A^\top(i) + \sum_{j=1}^{N} q_{ji}Y(j) + \frac{\pi_{i\infty}}{h}G(i)\boldsymbol{\Lambda}_c G^\top(i), \\
&\quad kh < t \leq (k+1)h && (7.177a)
\end{aligned}$$

$$Y(kh^+, i) = Y(kh, i) - Y(kh, i)C_y^\top(i)(C_y(i)Y(kh, i)C_y^\top(i) + \pi_{i\infty} D_y(i)\boldsymbol{\Lambda}_d D_y^\top(i))^{-1} \cdot C_y(i)Y(kh, i), \quad k \in \mathbb{Z}, \quad i \in \mathcal{N}. \tag{7.177b}$$

Further, (7.166), (7.174), (7.176) give

$$\mathbf{K}_s(i) = \begin{pmatrix} \widetilde{K}(i) \\ 0 \end{pmatrix} \tag{7.178a}$$

$$\widetilde{K}(i) = -\widetilde{Y}(0, i)C_y^\top(i)(C_y(i)\widetilde{Y}(0, i)C_y^\top(i) + \pi_{i\infty} D_y(i)\boldsymbol{\Lambda}_d D_y^\top(i))^{-1}. \tag{7.178b}$$

If $\mathbf{x}(t) = (x^\top(t) \quad \varphi^\top(t))^\top$ is a solution of (7.173) then according to (7.166), (7.178) we obtain the partition

$$\begin{aligned} &\dot{x}(t) = A(\eta(t))x(t) + B(\eta(t))\varphi(t), \\ &\dot{\varphi}(t) = 0, \quad kh < t \le (k+1)h, \end{aligned} \tag{7.179a}$$

$$\begin{aligned} &x(kh^+) = (I_n + \widetilde{K}(\eta(kh))C_y(\eta(kh)))x(kh), \\ &\varphi(kh^+) = 0, \quad k \in \mathbb{Z}_+. \end{aligned} \tag{7.179b}$$

From (7.179) we may deduce that if $\mathbf{x}(t; t_0, \mathbf{x}_0) = (x^\top(t; t_0, \mathbf{x}_0) \quad \varphi^\top(t; t_0, \mathbf{x}_0))^\top$ is the solution of (7.173) with the initial condition $\mathbf{x}(t_0; t_0, \mathbf{x}_0) = \mathbf{x}_0 = (x_0^\top \quad \varphi_0^\top)^\top$ then

$$\varphi(t) = \begin{cases} \varphi_0, & \text{if } t = t_0 \\ 0, & \text{if } t > t_0. \end{cases} \tag{7.180}$$

From (7.179) and (7.180) we deduce that if $\mathbf{x}(t; t_0, \mathbf{x}_0)$ is the solution of the JSLDE (7.173) satisfying $\mathbf{x}(t_0; t_0, \mathbf{x}_0) = \mathbf{x}_0 = (x_0^\top \quad \varphi_0^\top)^\top$ with $x_0 \in \mathbb{R}^n$, then

$$\mathbf{x}(t; t_0, \mathbf{x}_0) = \begin{cases} \mathbf{x}_0, & \text{if } \ t = t_0 \\ (x^\top(t; t_0, x_0) \quad 0^\top)^\top, & \text{if } \ t > t_0, \end{cases} \tag{7.181}$$

where $\mathbf{x}(t; t_0, x_0)$ solves

$$\dot{x}(t) = A(\eta(t))x(t), \quad kh < t \le (k+1)h \tag{7.182a}$$

$$x(kh^+) = (I_n + \widetilde{K}(\eta(kh))C_y(\eta(kh)))x(kh), \quad k = k_0, k_0 + 1, \ldots \tag{7.182b}$$

$$x(t_0) = x_0 \tag{7.182c}$$

where k_0 is such that $(k_0 - 1)h < t_0 \le k_0 h$.

Now we may prove:

Lemma 7.3 *Let* $\mathbf{Y}_s(\cdot) = (Y_s(\cdot, 1), \ldots, Y_s(\cdot, N))$ *be a global solution of the FJMLDE (7.172) and* $\widetilde{\mathbf{Y}}(\cdot) = (\widetilde{Y}(\cdot, 1), \ldots, \widetilde{Y}(\cdot, N))$ *be a global solution of the FJMLDE (7.177). If* $\mathbf{Y}_s(\cdot)$ *and* $\widetilde{\mathbf{Y}}(\cdot)$ *are connected by (7.176) then the following are equivalent:*

(i) $\mathbf{Y}_s(\cdot)$ *is the stabilizing solution of the FJMLDE (7.172);*
(ii) $\widetilde{\mathbf{Y}}(\cdot)$ *is the stabilizing solution of the FJMLDE (7.177).*

Proof From (7.181) we deduce that the JSLDE (7.173) is ESMS if and only if the JSLDE (7.182) is ESMS. This means that the equivalence statement holds.

The next result provides the state-space representation of a sampled-data controller which achieves the smallest value of the H_2-norm of the closed-loop system.

Theorem 7.8 *Assume:*

(a) the BJMLDE with Riccati-type jumping operator (7.167) has a bounded on $\mathbb{R}$ *and stabilizing solution* $\widetilde{\mathbf{X}}(\cdot) = (\widetilde{X}(\cdot, 1), \ldots, \widetilde{X}(\cdot, N))$ *which is satisfying the sign condition:*

$$\widetilde{X}_{22}(kh, i) + D_d^\top(i) D_d(i) > 0, \quad \textit{for all} \ \ (k, i) \in \mathbb{Z} \times \mathcal{N}, \tag{7.183}$$

$\widetilde{X}_{22}(kh, i) \in \mathcal{S}_m$ *being the 22-block of the partition (7.170) of the matrices* $\widetilde{X}(kh, i)$;

(b) the FJMLDE (7.177) has a bounded on $\mathbb{R}$ *and stabilizing solution* $\widetilde{\mathbf{Y}}(\cdot) = (\widetilde{Y}(\cdot, 1), \ldots, \widetilde{Y}(\cdot, N))$ *which satisfies the sign condition*

$$C_y(i)\widetilde{Y}(kh, i)C_y^\top(i) + \pi_{i\infty} D_y(i) \Lambda_d D_y^\top(i) > 0, \quad \textit{for all} \ \ (k, i) \in \mathbb{Z} \times \mathcal{N}. \tag{7.184}$$

Consider the controller $\widetilde{\widetilde{\mathbb{K}}}$ *having the state-space representation:*

$$\frac{\mathrm{d}}{\mathrm{d}t}\widehat{x}(t) = A(\eta(t))\widehat{x}(t) + B(\eta(t))\widehat{\varphi}(t), \quad kh < t \le (k+1)h \tag{7.185a}$$

$$\widehat{x}(kh^+) = (I_n + \widetilde{K}(\eta(kh))C_y(\eta(kh)))\widehat{x}(kh) - \widetilde{K}(\eta(kh))\widehat{u}(k), \tag{7.185b}$$

$$\begin{aligned}\widehat{y}(kh) = {} & \widetilde{F}(\eta(kh))(I_n + \widetilde{K}(\eta(kh))C_y(\eta(kh)))\widehat{x}(kh) \\ & - \widetilde{F}(\eta(kh))\widetilde{K}(\eta(kh))\widehat{u}(k), \quad k \in \mathbb{Z}_+\end{aligned} \tag{7.185c}$$

where $\widehat{\varphi}(\cdot)$ *is defined by:*

$$\begin{aligned}\widehat{\varphi}(t) \stackrel{\Delta}{=} {} & \widetilde{F}(\eta(kh))(I_n + \widetilde{K}(\eta(kh))C_y(\eta(kh)))\widehat{x}(kh) - \widetilde{F}(\eta(kh))\widetilde{K}(\eta(kh))\widehat{u}(k), \\ & kh \le t < (k+1)h, \quad k \in \mathbb{Z}_+,\end{aligned} \tag{7.186}$$

where $\widetilde{F}(i)$ *and* $\widetilde{K}(i)$, $i \in \mathcal{N}$, *are stabilizing gain matrices introduced by (7.171b) and (7.178b), respectively.*

Under the considered assumptions, if the controller (7.185) and (7.186) is coupled to the controlled system (7.162) and (7.163) taking $u(t) = \widehat{y}(kh), \ \ kh < t \le (k+1)h$ *and* $\widehat{u}(k) = y(kh), \ \ k \in \mathbb{Z}_+$, *we obtain a jump linear stochastic system* $\mathbb{G}_{cl}(\widetilde{\mathbb{K}})$ *which in the absence of the system disturbances* $\{\xi(t)\}_{t\ge0}$ *and* $\{\zeta_d(k)\}_{k\in\mathbb{Z}_+}$ *is ESMS and in the presence of these additive perturbations will achieve the smallest value* *of the H_2-norm. The value of the H_2-norm of the closed-loop system determined by the controller (7.185) and (7.186) is*

$$\|\mathbb{G}_{cl}(\tilde{\tilde{\mathbb{K}}})\|_2^2 = \sum_{i=1}^{N} \left(\frac{\pi_{i\infty}}{h} \int_0^h \mathrm{Tr}[\widetilde{X}_{11}(t,i)G(i)\boldsymbol{\Lambda}_c G^\top(i)]dt \right.$$
$$+ \mathrm{Tr}[\widetilde{F}(i)\widetilde{Y}(0,i)\widetilde{F}^\top(i)(\widetilde{X}_{22}(0,i) + D_d^\top(i)D_d(i))] - \mathrm{Tr}[(\widetilde{X}_{22}(0,i)$$
$$\left. + D_c^\top(i)D_d(i))\widetilde{F}(i)\widetilde{K}(i)(C_y(i)\widetilde{Y}(0,i)C_y^\top(i) + \pi_{i\infty}D_y(i)\boldsymbol{\Lambda}_d D_y^\top(i))\widetilde{K}^\top(i)\widetilde{F}(i)] \right). \tag{7.187}$$

Proof The conclusion is obtained applying Theorem 7.6 in the special case of the ICLSS (7.165) and (7.166). The controller (7.185) and (7.186) is the special case of (7.122) and (7.123) obtained employing (7.166), (7.171), (7.178). The details are omitted.

Remark 7.11 (a) bearing in mind the periodicity property of the stabilizing solution $\widetilde{\mathbf{X}}(\cdot)$ and $\widetilde{\mathbf{Y}}(\cdot)$ we deduce that it is enough that (7.183) and (7.184) to be satisfied for $k = 0$, $i \in \mathcal{N}$;

(b) if $D_d^\top(i)D_d(i) > 0$ and $\pi_{i\infty}D_y(i)\boldsymbol{\Lambda}_d D_y^\top(i) > 0$ for all $i \in \mathcal{N}$, then the sign conditions (7.183) and (7.184), respectively, are fulfilled.

7.6 Notes and References

It seems that the $\mathcal{H}_2$ control problem for MJLSs was first introduced in the literature in Costa et al. [2] for the discrete-time case and in Costa et al. [3] for the continuous-time case. Since then, a variety of results have been reported in the literature de Farias et al. [4]; Dragan et al. [5, 6]; Aberkane et al. [7]; Dong and Yang [8]; Xiong and Lam [9]; Shen and Yang [10]. In the last decade, two main research directions seem to be prioritized: $\mathcal{H}_2$ control with hidden Markov process Costa et al. [11]; Stadtmann and Costa [12]; de Oliveira et al. [13]; Romero et al. [14], $\mathcal{H}_2$ control for MJLSs where the Markov process is in a general Borel-measurable state space Figueiredo and Costa [15]; Ma and Wang [16]. Regarding the $\mathcal{H}_2$ control problem for stochastic systems with multiplicative white noise we cite Prato and Ichikawa [17]; Dragan et al. [18].

For impulsive MJLSs, to the best of the authors' knowledge, there are very few results that have been reported in the literature. We cite here Geromel and Gabriel [19]; Souza et al. [20]; Geromel [21] and we recall that an analysis similar to the one we have conducted at the end of Chap. 3 about the behavior of the Markov process at the impulse times applies also in this context.

References

1. Ichikawa, A., Katayama, H.: Linear Time Varying Systems and Sampled-data Systems. Springer, London (2001)
2. Costa, O., do Val J, Geromel J,: A convex programming approach to h_2-control of discrete-time markovian jump linear systems. International Journal of Control **66**, 557–579 (1997)
3. Costa, O., do Val J, Geromel J,: Continuous-time state-feedback h_2-control of markovian jump linear systems via convex analysis. Automatica **35**, 259–268 (1999)
4. de Farias, D., Geromel, J., do Val J, Costa O,: Output feedback control of markov jump linear systems in continuous-time. IEEE Trans Automat Contr **45**, 944–948 (2000)
5. Dragan, V., Morozan, T., Stoica, A.: Optimal h_2 state-feedback control for linear systems with markovian jumps. Rev Roum Sci Tech **48**, 523–537 (2003)
6. Dragan, V., Morozan, T., Stoica, A.: h_2 optimal control for linear stochastic systems. Automatica **40**, 1103–1113 (2004)
7. Aberkane, S., Ponsart, J., Rodrigues, M., Sauter, D.: Output feedback control of a class of stochastic hybrid systems. Automatica **44**, 1325–1332 (2008)
8. Dong, J., Yang, G.: Robust h_2 control of continuous-time markov jump linear systems. Automatica **44**, 1431–1436 (2008)
9. Xiong, J., Lam, J.: Robust h_2 control of markovian jump systems with uncertain switching probabilities. International Journal of Systems Science **40**, 255–265 (2009)
10. Shen, M., Yang, G.: h_2 state feedback controller design for continuous markov jump linear systems with partly known information. International Journal of Systems Science **43**, 255–265 (2012)
11. Costa O, Fragoso M, Todorov M (2014) A new approach for the h_2 control of markov jump linear systems with partial information. IFAC Proceedings Volumes 47:11,099–11,104
12. Stadtmann, F., Costa, O.L.V.: h_2-control of continuous-time hidden markov jump linear systems. IEEE Transactions on Automatic Control **62**, 4031–4037 (2017)
13. de Oliveira, A.M., do Valle Costa OL, Daafouz J,: Suboptimal and static output feedback control of hidden markov jump linear systems. European Journal of Control **51**, 10–18 (2020)
14. Romero, L.H., Ribeiro, J.R., Costa, E.F.: On the h_2 control of hidden markov jump linear systems. IEEE Control Systems Letters **7**, 1315–1320 (2023)
15. Figueiredo DZ, Costa OLV (2017) h_2-control and the separation principle for discrete-time jump systems with the markov chain in a general state space. International Journal of Systems Science 48
16. Ma H, Wang Y (2022) Full information h_2 control of borel-measurable markov jump systems with multiplicative noises. Mathematics 10
17. Prato, G.D., Ichikawa, A.: Quadratic control for linear periodic systems. Appl Math Optim **18**, 39–66 (1988)
18. Dragan, V., Halanay, A., Morozan, T.: Optimal stabilizing compensator for linear systems with state-dependent noise. Stoch Anal Appl **10**, 557–572 (1992)
19. Geromel, J.C., Gabriel, G.W.: Optimal h_2 state feedback sampled-data control design of markov jump linear systems. Automatica **54**(1), 182–188 (2015)
20. Souza, M., Fioravanti, A.R., Araujo, V.S.P.: Impulsive markov jump linear systems: Stability analysis and h_2 control. Nonlinear Analysis: Hybrid Systems **42**(101), 089 (2021)
21. Geromel, J.C.: Differential Linear Matrix Inequalities. Springer, In Sampled-Data Systems Filtering and Control (2023)

Chapter 8
H_2-Optimal Filtering

Abstract Estimating a remote signal from the measurements of another observed signal is a classical problem in the systems theory with numerous applications in different real-life problems. Such a problem is usually named in the literature a signal filtering problem or signal estimation/observation problem. In this chapter, we will address the problem of optimal filtering for a general class of systems described by impulsive stochastic linear differential equations. The metric used as an optimality measure of the proposed estimation scheme belongs to the H_2-type norm setting. More specifically, we will consider the two H_2-type norms defined in Chap. 7. Our aim is to find the state-space representation of a dynamical system named *filter*, which fed at its input with the measured signals, provides at its output a signal which is the best estimate (in the sense of H_2 norm minimization) of the signal of interest. The dimension of the state space of the considered filter is not prefixed. The state-space representations of the H_2-optimal filters corresponding to the two different considered norms are obtained in Theorems 8.1 and 8.3, respectively. We show that the dimensions of the optimal filters are equal to the dimension of the system under consideration. The corresponding feedback gains of the optimal filters are constructed based on the stabilizing solution of an adequately defined system of forward jump matrix linear differential equations with Riccati-type jumping operators. We finally apply the obtained results to the case of sampled-data linear stochastic systems (Theorems 8.4 and 8.5).

8.1 The Problem Setting

We consider the dynamical system $\mathbb{G}$ having the state-space representation described by:

$$\begin{aligned} dx(t) &= A_0(\eta(t))x(t)dt + \sum_{l=1}^{r_1} A_l(\eta(t))x(t)dw_l(t) \\ &\quad + G_c(\eta(t))d\xi(t), \quad kh < t \le (k+1)h \end{aligned} \tag{8.1a}$$

V. Drăgan et al., *Robust Control of Jump Linear Stochastic Systems*, Lecture Notes in Control and Information Sciences 497, https://doi.org/10.1007/978-3-031-84070-8_8

$$x(kh^+) = A_{d0}(\eta(kh))x(kh) + \sum_{p=1}^{r_2} w_{dp}(k)A_{dp}(\eta(kh))x(kh) + G_d(\eta(kh))\zeta_d(k), \quad k \in \mathbb{Z}_+ \tag{8.1b}$$

$$y(kh) = C_0(\eta(kh))x(kh) + \sum_{p=1}^{r_2} w_{dp}(k)C_p(\eta(kh))x(kh) + D(\eta(kh))\zeta_d(k), \quad k \in \mathbb{Z}_+ \tag{8.1c}$$

$$z(t) = C_z(\eta(t))x(t), \quad t \in \mathbb{R}_+ \tag{8.1d}$$

where $x(t) \in \mathbb{R}^n$ are the vectors of the state parameters while $y(kh) \in \mathbb{R}^{n_y}$ are the measurements available at the instance time $\tau_k = kh$, $k \in \mathbb{Z}_+$ and $z(t) \in \mathbb{R}^{n_z}$ is the remote signal which must be estimated.

In (8.1), $\{w(t)\}_{t\geq 0}$ $(w(t) = (w_1(t) \;\; w_2(t) \;\; \dots w_{r_1}(t))^\top)$ is a r_1-dimensional standard Wiener process; $\{w_d(k)\}_{k\in\mathbb{Z}_+}$ $(w_d(k) = (w_{d1}(k) \;\; w_{d2}(k) \;\; \dots w_{dr_2}(k))^\top)$ is a sequence of r_2-dimensional independent random vectors and $\{\eta(t)\}_{t\geq 0}$ is a standard homogeneous continuous-time Markov process taking values in the set $\mathcal{N} = \{1, 2, \dots, N\}$. We assume that the stochastic processes $\{w(t)\}_{t\geq 0}$, $\{\eta(t)\}_{t\geq 0}$, $\{w_d(k)\}_{k\in\mathbb{Z}_+}$ are defined on the given probability space $(\Omega, \mathfrak{F}, \mathcal{P})$ and satisfy the Assumption 1.2.

The additive disturbances $\{\xi(t)\}_{t\geq 0}$ and $\{\zeta_d(k)\}_{k\in\mathbb{Z}_+}$ which are affecting the dynamical system $(\mathbb{G})$ are assumed to be stochastic processes which are satisfying the Assumption 7.1. For each $i \in \mathcal{N}$, $A_l(i) \in \mathbb{R}^{n\times n}$, $0 \leq l \leq r_1$, $A_{dp}(i) \in \mathbb{R}^{n\times n}$, $C_p(i) \in \mathbb{R}^{n_y\times n}$, $0 \leq p \leq r_2$, $C_z(i) \in \mathbb{R}^{n_z\times n}$, $G_c(i) \in \mathbb{R}^{n\times n_\xi}$, $G_d(i) \in \mathbb{R}^{n\times n_{\zeta_d}}$, $D(i) \in \mathbb{R}^{n_y\times n_{\zeta_d}}$ are known matrices.

Our aim is to find the state-space representation of a dynamical system $(\mathbb{G}_F)$ named filter, which fed at its input with the measured values $y(kh)$, $0 \leq kh \leq t$, to provide at its output a signal $z_F(t)$, $t \geq 0$, which is the best estimate of the remote signal $z(t)$, $t \geq 0$.

We consider filters $(\mathbb{G}_F)$ having the state-space representation of the form:

$$dx_F(t) = A_F(\eta(t))x_F(t)dt, \quad kh < t \leq (k+1)h \tag{8.2a}$$

$$x_F(kh^+) = A_{Fd}(\eta(kh))x_F(kh) + B_{Fd}(\eta(kh))u(k), \quad k \in \mathbb{Z}_+ \tag{8.2b}$$

$$z_F(t) = C_F(\eta(t))x_F(t), \quad t \in \mathbb{R}_+ \tag{8.2c}$$

where $x_F(t) \in \mathbb{R}^{n_F}$ are the vectors of the state parameters of the filter, $u(k) \in \mathbb{R}^{n_y}$, $k \in \mathbb{Z}_+$ are the inputs of the filter, while $z_F(t) \in \mathbb{R}^{n_z}$ is the output which provides the estimation of the output $z(t)$ of $(\mathbb{G})$. The dimension $n_F \geq 1$ of the state space of a filter of the form (8.2) is not prefixed. In (8.2) $A_F(i) \in \mathbb{R}^{n_F\times n_F}$, $A_{Fd}(i) \in \mathbb{R}^{n_F\times n_F}$, $B_{Fd}(i) \in \mathbb{R}^{n_F\times n_y}$, $C_F(i) \in \mathbb{R}^{n_z\times n_F}$ are the arbitrary matrices subject only to the condition that the corresponding jump stochastic linear differential equation (JSLDE) on $\mathbb{R}^{n_F}$:

$$dx_F(t) = A_F(\eta(t))x_F(t)dt, \quad kh < t \leq (k+1)h \tag{8.3a}$$

$$x_F(kh^+) = A_{Fd}(\eta(kh))x_F(kh), \quad k \in \mathbb{Z}_+ \tag{8.3b}$$

is exponentially stable in mean square (ESMS).

In the sequel, we denote $\mathscr{F}_{\text{adm}}(\mathbb{G})$ the set of dynamical systems $(\mathbb{G}_F)$ of the form of (8.2) with arbitrary dimension of their state space subject only to the constraints that the corresponding JSLDE of the form (8.3) is ESMS. A filter $\mathbb{G}_F$ from $\mathscr{F}_{\text{adm}}(\mathbb{G})$ will be named admissible filter. Coupling an arbitrary admissible filter $\mathbb{G}_F$ to the dynamical system $\mathbb{G}$, taking $u(k) = y(kh)$, $k \in \mathbb{Z}_+$, one obtains the following resulting system $(\mathbb{G}_{cl}(\mathbb{G}_F))$:

$$\begin{aligned} \mathrm{d}x_{cl}(t) &= A_{0,cl}(\eta(t))x_{cl}(t)\mathrm{d}t + \sum_{l=1}^{r_1} A_{l,cl}(\eta(t))x_{cl}(t)\mathrm{d}w_l(t) \\ &\quad + G_{c,cl}(\eta(t))d\xi(t), \quad kh < t \le (k+1)h \end{aligned} \tag{8.4a}$$

$$\begin{aligned} x_{cl}(kh^+) &= (A_{d0,cl}(\eta(kh)) + \sum_{p=1}^{r_2} w_{dp}(k)A_{dp,cl}(\eta(kh)))x_{cl}(kh) \\ &\quad + G_{d,cl}(\eta(kh))\zeta_d(k), \quad k \in \mathbb{Z}_+ \end{aligned} \tag{8.4b}$$

$$z_{cl}(t) = C_{cl}(\eta(t))x_{cl}(t), \quad t \in \mathbb{R}_+ \tag{8.4c}$$

where $x_{cl}(t) = \begin{pmatrix} x^\top(t) & x_F^\top(t) \end{pmatrix}^\top \in \mathbb{R}^n \times \mathbb{R}^{n_F}$ and for each $i \in \mathcal{N}$ we have denoted:

$$A_{0,cl}(i) = \begin{pmatrix} A_0(i) & 0 \\ 0 & A_F(i) \end{pmatrix} \tag{8.5a}$$

$$A_{l,cl}(i) = \begin{pmatrix} A_l(i) & 0 \\ 0 & 0 \end{pmatrix}, \quad 1 \le l \le r_1 \tag{8.5b}$$

$$A_{d0,cl}(i) = \begin{pmatrix} A_{d0}(i) & 0 \\ B_{Fd}(i)C_0(i) & A_{Fd}(i) \end{pmatrix} \tag{8.5c}$$

$$A_{dp,cl}(i) = \begin{pmatrix} A_{dp}(i) & 0 \\ B_{Fd}(i)C_p(i) & 0 \end{pmatrix}, \quad 1 \le p \le r_2 \tag{8.5d}$$

$$G_{c,cl}(i) = \begin{pmatrix} G_c(i) \\ 0 \end{pmatrix} \tag{8.5e}$$

$$G_{d,cl}(i) = \begin{pmatrix} G_d(i) \\ B_{Fd}(i)D(i) \end{pmatrix} \tag{8.5f}$$

$$C_{cl}(i) = \begin{pmatrix} C_z(i) & -C_F(i) \end{pmatrix}. \tag{8.5g}$$

The quality of the estimation achieved by an admissible filter $(\mathbb{G}_F)$ will be determined by the H_2−norm of the corresponding resulting system, that is $\|\mathbb{G}_{cl}(\mathbb{G}_F)\|_2$ or $\widetilde{\|}\mathbb{G}_{cl}(\mathbb{G}_F)\widetilde{\|}_2$, respectively.

In the next section we shall derive the state-space representation of a filter $(\widetilde{\mathbb{G}}_F)$ which achieves the best estimation of the signal $z(t)$. More precisely, we shall provide explicit formulae of the state-space representation of an admissible filter ($\widetilde{\mathbb{G}}_F$) which is satisfying one of the following two optimality conditions

$$\|\mathbb{G}_{cl}(\widetilde{\mathbb{G}}_F)\|_2 \leq \|\mathbb{G}_{cl}(\mathbb{G}_F)\|_2, \tag{8.6}$$

$$\widetilde{\|}\mathbb{G}_{cl}(\widetilde{\mathbb{G}}_F)\widetilde{\|}_2 \leq \widetilde{\|}\mathbb{G}_{cl}(\mathbb{G}_F)\widetilde{\|}_2, \tag{8.7}$$

for any $\mathbb{G}_F \in \mathscr{F}_{\text{adm}}(\mathbb{G})$. In (8.6), $\|\mathbb{G}_{cl}(\widetilde{\mathbb{G}}_F)\|_2$ is introduced as in (7.72) and $\widetilde{\|}\mathbb{G}_{cl}(\widetilde{\mathbb{G}}_F)\widetilde{\|}_2$ is defined as in (7.74) for $z_{d,cl}(kh;\xi) = 0$, $z_{d,cl}(kh;\zeta_d) = 0$ for all $k \in \mathbb{Z}_+$.

8.2 H_2-Optimal Filtering

8.2.1 Several Auxiliary Results

The developments in this section will be done under the following assumption:

Assumption 8.1 The JSLDE:

$$\mathrm{d}x(t) = A_0(\eta(t))x(t)\mathrm{d}t + \sum_{l=1}^{r_1} A_l(\eta(t))x(t)\mathrm{d}w_l(t), \quad kh < t \leq (k+1)h \tag{8.8a}$$

$$x(kh^+) = (A_{d0}(\eta(kh)) + \sum_{p=1}^{r_2} w_{dp}(k)A_{dp}(\eta(kh)))x(kh), \quad k \in \mathbb{Z}_+ \tag{8.8b}$$

is ESMS.

The JSLDE (8.8) is obtained from (8.1a)–(8.1b) taking $G_c(\eta(t)) \equiv 0$ and $G_d(\eta(kh)) \equiv 0$. A first consequence of the Assumption 8.1 is:

Proposition 8.1 *Under the Assumption 8.1 for any filter $\mathbb{G}_F$ from $\mathscr{F}_{\text{adm}}(\mathbb{G})$, the corresponding resulting JSLDE:*

$$dx_{cl}(t) = A_{0,cl}(\eta(t))x_{cl}(t)dt + \sum_{l=1}^{r_1} A_{l,cl}(\eta(t))x_{cl}(t)dw_l(t), kh < t \leq (k+1)h \tag{8.9a}$$

$$x_{cl}(kh^+) = (A_{d0,cl}(\eta(kh)) + \sum_{p=1}^{r_2} w_{dp}(k)A_{dp,cl}(\eta(kh)))x_{cl}(kh), \quad k \in \mathbb{Z}_+ \tag{8.9b}$$

is ESMS.

Proof We have to show that under the conditions from the statement there exist the constants $\beta \geq 1$ and $\alpha > 0$ with the property that the solutions of the JSLDE (8.9) are satisfying:

$$\mathbb{E}[|x_{cl}(t; t_0, x_{0,cl}|^2] \leq \beta e^{-\alpha(t-t_0)}|x_{0,cl}|^2, \tag{8.10}$$

for all $t \geq t_0 \geq 0$, $x_{0,cl} = (x_0^\top \ \ x_{F0}^\top)^\top \in \mathbb{R}^{n+n_F}$. We set $(\widehat{x}^\top(t) \ \ \widehat{x}_F^\top(t))^\top$ the partition of the solution $x_{cl}(t; t_0, x_{0,cl})$. According to (8.5a)–(8.5d) and (8.9) we obtain the equations

$$d\widehat{x}(t) = A_0(\eta(t))\widehat{x}(t)dt + \sum_{l=1}^{r_1} A_l(\eta(t))\widehat{x}(t)dw_l(t), \quad kh < t \leq (k+1)h \tag{8.11a}$$

$$\widehat{x}(kh^+) = (A_{d0}(\eta(kh)) + \sum_{p=1}^{r_2} w_{dp}(k)A_{dp}(\eta(kh)))\widehat{x}(kh), \quad k \in \mathbb{Z}_+ \tag{8.11b}$$

$$d\widehat{x}_F(t) = A_F(\eta(t))\widehat{x}_F(t)dt, \quad kh < t \leq (k+1)h \tag{8.12a}$$

$$\widehat{x}_F(kh^+) = A_{Fd}(\eta(kh)) + g_0(k) + \sum_{p=1}^{r_2} w_{dp}(k)g_p(k), \quad k \in \mathbb{Z}_+ \tag{8.12b}$$

where

$$g_p(k) := B_{Fd}(\eta(kh))C_p(\eta(kh))\widehat{x}(kh), \quad 0 \leq p \leq r_2, \quad k \in \mathbb{Z}_+. \tag{8.13}$$

Comparing (8.11) with (8.8) we deduce from 8.1 that there exist constants $\widehat{\beta} \geq 1$, $\widehat{\alpha} > 0$ such that

$$\mathbb{E}[|\widehat{x}(t)|^2] \leq \widehat{\beta} e^{-\widehat{\alpha}(t-t_0)}|\widehat{x}(t_0)|^2, \tag{8.14}$$

for all $t \geq t_0 \geq 0$, $\widehat{x}(t_0) \in \mathbb{R}^n$ and any initial probability distribution of the Markov process. From (8.13) and (8.14) we deduce via Corollary 3.6 that $\{g_p(k)\}_{k \in \mathbb{Z}_+} \in \ell^2_{\mathcal{H}}(\mathbb{Z}_{\kappa_0}; \mathbb{R}^{n_F})$ where $\kappa_0 \in \mathbb{Z}$ is such that $(\kappa_0 - 1)h < t_0 \leq \kappa_0 h$ and $\mathbb{Z}_{\kappa_0} = \{\kappa \in \mathbb{Z} \mid \kappa \geq \kappa_0\}$. Applying Theorem 3.5 (iii) in the case of the non-homogeneous JSLDE (8.12) we deduce that $\lim_{t \to \infty} \mathbb{E}[|\widehat{x}_F(t)|^2] = 0$, for all $(t_0, x_{F0}) \in \mathbb{R}_+ \times \mathbb{R}^{n_F}$ and any initial probability distribution of the Markov process. Thus we have obtained that

$$\lim_{t \to \infty} \mathbb{E}[|x_{cl}(t; t_0, x_{0,cl})|^2] = \lim_{t \to \infty} \mathbb{E}[|\widehat{x}(t)|^2] + \lim_{t \to \infty} \mathbb{E}[|\widehat{x}_F(t)|^2] = 0,$$

for all $x_{0,cl} = \left(x_0^\top \ \ x_{F0}^\top\right)^\top \in \mathbb{R}^{n+n_F}$ and any initial probability distribution of the Markov process. So, the JSLDE (8.9) is asymptotic stable in mean square (ASMS).

Finally, the equivalence $(iii) \Leftrightarrow (iv)$ from Theorem 3.1 applied in the case of the JSLDE (8.9) allows us to conclude that (8.10) holds for some $\alpha > 0$, $\beta \geq 1$ not depending upon t, t_0, $x_{0,cl}$ and any initial probability distribution of the Markov process. Thus the proof ends.

Let us consider the JSLDE on $\mathbb{R}^n \times \mathbb{R}^{n_F}$:

$$dx_{cl}(t) = A_{0,cl}(\eta(t))x_{cl}(t)dt, \quad kh < t \leq (k+1)h \tag{8.15a}$$
$$x_{cl}(kh^+) = A_{d0,cl}(\eta(kh))x_{cl}(kh), \quad k \in \mathbb{Z}_+ \tag{8.15b}$$

obtained from (8.9) when the state multiplicative white noise perturbations vanish.

Corollary 8.1 *Under the Assumptions 1.2, 7.1, 8.1 for each admissible filter $\mathbb{G}_F$ the corresponding JSLDE of the form (8.15) is ESMS.*

Proof Based on Proposition 8.1 we may infer that under the considered assumptions the JSLDE (8.9) is ESMS. Further, the equivalence $(i) \Leftrightarrow (v)$ from Theorem 3.4 applied in the case of the JSLDE (8.9) guarantee that the non-homogeneous FGLJLDE on $\mathscr{S}^N_{n+n_F}$:

$$\dot{X}_{cl}(t,i) = A_{0,cl}(i)X_{cl}(t,i) + X_{cl}(t,i)A_{0,cl}^{\top}(i) + \sum_{l=1}^{r_1} A_{l,cl}(i)X_{cl}(t,i)A_{l,cl}^{\top}(i)$$
$$+ \sum_{j=1}^{N} q_{ji}X_{cl}(t,j) + I_{n+n_F}, \quad kh < t \leq (k+1)h \tag{8.16a}$$

$$X_{cl}(kh^+,i) = A_{d0,cl}(i)X_{cl}(kh,i)A_{d0,cl}^{\top}(i) + \sum_{p=1}^{r_2} A_{dp,cl}(i)X_{cl}(kh,i)A_{dp,cl}^{\top}(i)$$
$$+ I_{n+n_F}, \quad k \in \mathbb{Z}, \tag{8.16b}$$

$i \in \mathcal{N}$, has a bounded on $\mathbb{R}$ solution $\widetilde{\mathbf{X}}_{cl}(t) = (\widetilde{X}_{cl}(t,1), \ldots, \widetilde{X}_{cl}(t,N))$ which is a periodic function of period h and satisfy $\widetilde{X}_{cl}(t,i) \geq cI_{n+n_F} > 0$, for all $(t,i) \in \mathbb{R} \times \mathcal{N}$. So, (8.16) yields

$$-\dot{\widetilde{X}}_{cl}(t,i) + A_{0,cl}(i)\widetilde{X}_{cl}(t,i) + \widetilde{X}_{cl}(t,i)A_{0,cl}^{\top}(i) + \sum_{j=1}^{N} q_{ji}\widetilde{X}_{cl}(t,j) \leq -I_{n+n_F},$$
$$kh < t \leq (k+1)h$$
$$\widetilde{X}_{cl}(kh^+,i) \geq A_{d0,cl}(i)\widetilde{X}_{cl}(kh,i)A_{d0,cl}^{\top}(i) + I_{n+n_F}, \quad k \in \mathbb{Z}.$$

The conclusion follows from the implication $(vii) \Rightarrow (i)$ from Theorem 3.4 combined by Remark 3.4 applied in the case of the JSLDE (8.15).

The result proved in Proposition 8.1 allows us to use the second equality from (7.35) and the second equality from (7.58), respectively, to compute the value of the norm

$\|\mathbb{G}_{cl}(\mathbb{G}_F)\|_2$ and $\widetilde{\|}\mathbb{G}_{cl}(\mathbb{G}_F)\widetilde{\|}_2$, respectively, for an admissible filter $\mathbb{G}_F$. To this end, we take into account that the resulting system (8.4) is a jump linear stochastic system (JLSS) of the form (7.1) without the outputs evolving in the discrete time. Thus, applying Theorem 7.1 and Theorem 7.3, respectively, in the case of the resulting system $\mathbb{G}_{cl}(\mathbb{G}_F)$ corresponding to an admissible filter $\mathbb{G}_F$, we obtain:

Corollary 8.2 *Under the Assumptions 1.2, 7.1 and 8.1 for each admissible filter $\mathbb{G}_F$ we have:*

(a)

$$\|\mathbb{G}_{cl}(\mathbb{G}_F)\|_2^2 = \sum_{i=1}^{N} \frac{1}{h} \int_0^h \mathrm{Tr}[C_{cl}(i)\widetilde{Y}_{cl}(t,i)C_{cl}^{\top}(i)]dt \tag{8.17}$$

where $\widetilde{\mathbf{Y}}_{cl}(\cdot) = \big(\widetilde{Y}_{cl}(\cdot,1), \widetilde{Y}_{cl}(\cdot,2), \ldots, \widetilde{Y}_{cl}(\cdot,N)\big)$ is the unique bounded on $\mathbb{R}$ solution of the following FGLJLDE:

$$\dot{Y}_{cl}(t,i) = A_{0,cl}(i)Y_{cl}(t,i) + Y_{cl}(t,i)A_{0,cl}^{\top}(i) + \sum_{l=1}^{r_1} A_{l,cl}(i)Y_{cl}(t,i)A_{l,cl}^{\top}(i)$$
$$+ \sum_{j=1}^{N} q_{ji}Y_{cl}(t,j) + \pi_{i\infty}G_{c,cl}(i)\mathbf{\Lambda}_c G_{c,cl}^{\top}(i), \quad kh < t \le (k+1)h \tag{8.18a}$$

$$Y_{cl}(kh^+,i) = \sum_{p=0}^{r_2} A_{dp,cl}(i)Y_{cl}(kh,i)A_{dp,cl}^{\top}(i) + h\pi_{i\infty}G_{d,cl}(i)\mathbf{\Lambda}_d G_{d,cl}^{\top}(i). \tag{8.18b}$$

(b)

$$\widetilde{\|}\mathbb{G}_{cl}(\mathbb{G}_F)\widetilde{\|}_2^2 = \sum_{i=1}^{N} \frac{1}{h} \int_0^h \mathrm{Tr}[C_{cl}(i)\approx{Y}_{cl}(t,i)C_{cl}^{\top}(i)]dt \tag{8.19}$$

where $\approx{\mathbf{Y}}_{cl}(\cdot) = \big(\approx{Y}_{cl}(\cdot,1), \approx{Y}_{cl}(\cdot,2), \ldots, \approx{Y}_{cl}(\cdot,N)\big)$ is the unique bounded on $\mathbb{R}$ solution of the following FGLJLDE:

$$\dot{Y}_{cl}(t,i) = A_{0,cl}(i)Y_{cl}(t,i) + Y_{cl}(t,i)A_{0,cl}^{\top}(i) + \sum_{l=1}^{r_1} A_{l,cl}(i)Y_{cl}(t,i)A_{l,cl}^{\top}(i)$$
$$+ \sum_{j=1}^{N} q_{ji}Y_{cl}(t,j) + \pi_{i\infty}G_{c,cl}(i)\mathbf{\Lambda}_c G_{c,cl}^{\top}(i), \quad kh < t \le (k+1)h \tag{8.20a}$$

$$Y_{cl}(kh^+, i) = \sum_{p=0}^{r_2} A_{dp,cl}(i) Y_{cl}(kh, i) A_{dp,cl}^\top(i) + \pi_{i\infty} G_{d,cl}(i) \mathbf{\Lambda}_d G_{d,cl}^\top(i). \tag{8.20b}$$

In (8.18) and (8.20), $\pi_{i\infty}$ *being defined in (7.34) while* $\mathbf{\Lambda}_c \geq 0$, $\mathbf{\Lambda}_d \geq 0$ *are the covariance matrices introduced via the Assumption 7.1.*

Remark 8.1 (a) Employing Proposition 8.1 we may deduce via Corollary 7.2 that under the Assumptions 1.2, 7.1 and 8.1, the FGLJLDE (8.18) has a unique bounded solution $t \to \widetilde{\mathbf{Y}}_{cl}(t) = (\widetilde{Y}_{cl}(t, 1), \widetilde{Y}_{cl}(t, 2), \ldots, \widetilde{Y}_{cl}(t, N)) : \mathbb{R} \to \mathcal{S}_{n+n_F}^N$. This solution is a differentiable function on each interval $(kh, (k+1)h)$, left continuous in each $\tau_k = kh$, $k \in \mathbb{Z}$. Furthermore $\widetilde{\mathbf{Y}}_{cl}(\cdot)$ is a periodic function of period h and satisfies $\widetilde{Y}_{cl}(t, i) \geq 0$, for all $(t, i) \in \mathbb{R} \times \mathcal{N}$.

(b) Under the same set of assumptions, the FGLJLDE (8.20) has a unique bounded solution $t \to \approx{\mathbf{Y}}_{cl}(t) = (\approx{Y}_{cl}(t, 1), \approx{Y}_{cl}(t, 2), \ldots, \approx{Y}_{cl}(t, N)) : \mathbb{R} \to \mathcal{S}_{n+n_F}^N$. This solution is differentiable on each interval $(kh, (k+1)h)$, left continuous in each $\tau_k = kh$, $k \in \mathbb{Z}$. Moreover $\approx{\mathbf{Y}}_{cl}(\cdot)$ is a periodic function of period h and satisfies $\approx{Y}_{cl}(t, i) \geq 0$, for all $(t, i) \in \mathbb{R} \times \mathcal{N}$.

Involving the coefficients of the state-space representation (8.1) of the dynamical system $\mathbb{G}$ we introduce the following FGLJLDE:

$$\dot{Z}(t, i) = A_0(i) Z(t, i) + Z(t, i) A_0^\top(i) + \sum_{l=1}^{r_1} A_l(i) Z(t, i) A_l^\top(i)$$

$$+ \sum_{j=1}^{N} q_{ji} Z(t, j) + \pi_{i\infty} G_c(i) \mathbf{\Lambda}_c G_c^\top(i), \quad kh < t \leq (k+1)h \tag{8.21a}$$

$$Z(kh^+, i) = \sum_{p=0}^{r_2} A_{dp}(i) Z(kh, i) A_{dp}^\top(i) + h\pi_{i\infty} G_d(i) \mathbf{\Lambda}_d G_d^\top(i), \quad k \in \mathbb{Z}. \tag{8.21b}$$

Along with (8.21) we consider the following FGLJLDE

$$\dot{Z}(t, i) = A_0(i) Z(t, i) + Z(t, i) A_0^\top(i) + \sum_{l=1}^{r_1} A_l(i) Z(t, i) A_l^\top(i)$$

$$+ \sum_{j=1}^{N} q_{ji} Z(t, j) + \pi_{i\infty} G_c(i) \mathbf{\Lambda}_c G_c^\top(i), \quad kh < t \leq (k+1)h \tag{8.22a}$$

$$Z(kh^+, i) = \sum_{p=0}^{r_2} A_{dp}(i) Z(kh, i) A_{dp}^\top(i) + \pi_{i\infty} G_d(i) \mathbf{\Lambda}_d G_d^\top(i), \quad k \in \mathbb{Z}. \tag{8.22b}$$

Remark 8.2 Invoking Corollary 7.2, we may infer that under the Assumption 8.1 the following hold:

(a) the FGLJLDE (8.21) has a unique bounded solution $t \to \widetilde{\mathbf{Z}}(t) = (\widetilde{Z}(t, 1), \widetilde{Z}(t, 2), \dots, \widetilde{Z}(t, N)) : \mathbb{R} \to \mathcal{S}_{\tilde{n}}^N$. Additionally, this solution is a periodic function of period h and satisfies $\widetilde{Z}(t, i) \geq 0$, for all $(t, i) \in \mathbb{R} \times \mathcal{N}$.

(b) the FGLJLDE (8.22) has a unique solution $t \to \widetilde{\widetilde{\mathbf{Z}}}(t) = (\widetilde{\widetilde{Z}}(t, 1), \widetilde{\widetilde{Z}}(t, 2), \dots, \widetilde{\widetilde{Z}}(t, N)) : \mathbb{R} \to \mathcal{S}_n^N$. This solution is a periodic function of period h and satisfies $\widetilde{\widetilde{Z}}(t, i) \geq 0$, for all $(t, i) \in \mathbb{R} \times \mathcal{N}$.

We introduce the following notation:

$$\widetilde{M}_c(t,i) := \pi_{i\infty} G_c(i)\mathbf{\Lambda}_c G_c^\top(i) + \sum_{l=1}^{r_1} A_l(i)\widetilde{Z}(t,i)A_l^\top(i), \ t \in \mathbb{R} \tag{8.23a}$$

$$\begin{aligned}\widetilde{M}_d(i) &:= h\pi_{i\infty} G_d(i)\mathbf{\Lambda}_d G_d^\top(i) + \sum_{p=1}^{r_2} A_{dp}(i)\widetilde{Z}(0,i)A_{dp}^\top(i)\\ &= h\pi_{i\infty} G_d(i)\mathbf{\Lambda}_d G_d^\top(i) + \sum_{p=1}^{r_2} A_{dp}(i)\widetilde{Z}(kh,i)A_{dp}^\top(i), \quad k \in \mathbb{Z}\end{aligned} \tag{8.23b}$$

$$\begin{aligned}\widetilde{L}_d(i) &:= h\pi_{i\infty} G_d(i)\mathbf{\Lambda}_d D^\top(i) + \sum_{p=1}^{r_2} A_{dp}(i)\widetilde{Z}(0,i)C_p^\top(i)\\ &= h\pi_{i\infty} G_d(i)\mathbf{\Lambda}_d D^\top(i) + \sum_{p=1}^{r_2} A_{dp}(i)\widetilde{Z}(kh,i)C_p^\top(i), \quad k \in \mathbb{Z}\end{aligned} \tag{8.23c}$$

$$\begin{aligned}\widetilde{R}_d(i) &:= h\pi_{i\infty} D(i)\mathbf{\Lambda}_d D^\top(i) + \sum_{p=1}^{r_2} C_p(i)\widetilde{Z}(0,i)C_p^\top(i)\\ &= h\pi_{i\infty} D(i)\mathbf{\Lambda}_d D^\top(i) + \sum_{p=1}^{r_2} C_p(i)\widetilde{Z}(kh,i)C_p^\top(i), \quad k \in \mathbb{Z}, \quad i \in \mathcal{N}.\end{aligned} \tag{8.23d}$$

$$\widetilde{\widetilde{M}}_c(t,i) := \pi_{i\infty} G_c(i)\mathbf{\Lambda}_c G_c^\top(i) + \sum_{l=1}^{r_1} A_l(i)\widetilde{\widetilde{Z}}(t,i)A_l^\top(i), \ t \in \mathbb{R} \tag{8.24a}$$

$$\begin{aligned}\widetilde{\widetilde{M}}_d(i) &:= \pi_{i\infty} G_d(i)\mathbf{\Lambda}_d G_d^\top(i) + \sum_{p=1}^{r_2} A_{dp}(i)\widetilde{\widetilde{Z}}(0,i)A_{dp}^\top(i)\\ &= \pi_{i\infty} G_d(i)\mathbf{\Lambda}_d G_d^\top(i) + \sum_{p=1}^{r_2} A_{dp}(i)\widetilde{\widetilde{Z}}(kh,i)A_{dp}^\top(i), \quad k \in \mathbb{Z}\end{aligned} \tag{8.24b}$$

$$\begin{aligned}\widetilde{\widetilde{L}}_d(i) &:= \pi_{i\infty} G_d(i)\mathbf{\Lambda}_d D^\top(i) + \sum_{p=1}^{r_2} A_{dp}(i)\widetilde{\widetilde{Z}}(0,i)C_p^\top(i)\\ &= \pi_{i\infty} G_d(i)\mathbf{\Lambda}_d D^\top(i) + \sum_{p=1}^{r_2} A_{dp}(i)\widetilde{\widetilde{Z}}(kh,i)C_p^\top(i),\quad k\in\mathbb{Z} \qquad (8.24c)\\ \widetilde{\widetilde{R}}_d(i) &:= \pi_{i\infty} D(i)\mathbf{\Lambda}_d D^\top(i) + \sum_{p=1}^{r_2} C_p(i)\widetilde{\widetilde{Z}}(0,i)C_p^\top(i)\\ &= \pi_{i\infty} D(i)\mathbf{\Lambda}_d D^\top(i) + \sum_{p=1}^{r_2} C_p(i)\widetilde{\widetilde{Z}}(kh,i)C_p^\top(i),\quad k\in\mathbb{Z},\quad i\in\mathcal{N}. \qquad (8.24d)\end{aligned}$$

The validity of the second equality from (8.23b)–(8.23d) and (8.24b)–(8.24d) is based on the periodicity property of the bounded solution $\widetilde{Z}(\cdot)$ of the FGLJLDE (8.21) and $\widetilde{\widetilde{Z}}(\cdot)$ of (8.22). To obtain the state-space representation of a filter $\widetilde{\mathbb{G}}_F$ which satisfies the optimality condition (8.6) we need the following forward jump matrix linear differential equation (FJMLDE) with Riccati-type jumping operator:

$$\begin{aligned}\dot{Y}(t,i) &= A_0(i)Y(t,i) + Y(t,i)A_0^\top(i) + \sum_{j=1}^{N} q_{ji}Y(t,j)\\ &\quad + \widetilde{M}_c(t,i),\quad kh < t \le (k+1)h \qquad (8.25a)\\ Y(kh^+,i) &= A_{d0}(i)Y(kh,i)A_{d0}^\top(i) - \left(A_{d0}(i)Y(kh,i)C_0^\top(i) + \widetilde{L}_d(i)\right)\\ &\quad\cdot\left(C_0(i)Y(kh,i)C_0^\top(i) + \widetilde{R}_d(i)\right)^{-1}\left(C_0(i)Y(kh,i)A_{d0}^\top(i) + \widetilde{L}_d^\top(i)\right)\\ &\quad + \widetilde{M}_d(i), k\in\mathbb{Z}, \qquad (8.25b)\end{aligned}$$

$i \in \mathcal{N}$. A globally defined solution $t \to \widetilde{\mathbf{Y}}_s(t) = (\widetilde{Y}_s(t,1), \widetilde{Y}_s(t,2), \ldots, \widetilde{Y}_s(t,N)) : \mathbb{R} \to \mathcal{S}_n^N$ of (8.25) is named stabilizing solution if the matrices $\widetilde{R}_d(i) + C_0(i)\widetilde{Y}_s(kh,i)C_0^\top(i)$ are invertible for any $(k,i) \in \mathbb{Z}\times\mathcal{N}$ and the corresponding closed-loop system:

$$\begin{aligned}dx(t) &= A_0(\eta(t))x(t)dt,\quad kh < t \le (k+1)h \qquad (8.26a)\\ x(kh^+) &= \left(A_{d0}(\eta(kh)) + \widetilde{K}_s(\eta(kh))C_0(\eta(kh))\right)x(kh),\quad k\in\mathbb{Z}_+ \qquad (8.26b)\end{aligned}$$

is ESMS, where

$$\widetilde{K}_s(i) \stackrel{\Delta}{=} -(A_{d0}(i)\widetilde{Y}_s(0,i)C_0^\top(i) + \widetilde{L}_d(i))(\widetilde{R}_d(i) + C_0(i)\widetilde{Y}_s(0,i)C_0^\top(i))^{-1}, \qquad (8.27)$$

$i \in \mathcal{N}$. Along with (8.25) we consider the following FJMLDE with Riccati-type jumping operator

$$\dot{Y}(t,i) = A_0(i)Y(t,i) + Y(t,i)A_0^\top(i)$$

$$+\sum_{j=1}^{N} q_{ji}Y(t,j)+\widetilde{\widetilde{M}}_c(t,i),\quad kh<t\le (k+1)h \tag{8.28a}$$

$$\begin{aligned}Y(kh^+,i) &= A_{d0}(i)Y(kh,i)A_{d0}^{\top}(i)-\left(A_{d0}(i)Y(kh,i)C_0^{\top}(i)+\widetilde{\widetilde{L}}_d(i)\right)\\ &\cdot\left(C_0(i)Y(kh,i)C_0^{\top}(i)+\widetilde{\widetilde{R}}_d(i)\right)^{-1}\left(C_0(i)Y(kh,i)A_{d0}^{\top}(i)+\widetilde{\widetilde{L}}_d^{\top}(i)\right)\\ &+\widetilde{\widetilde{M}}_d(i),\quad k\in\mathbb{Z},\end{aligned} \tag{8.28b}$$

$i\in\mathcal{N}$. A globally defined solution $t\to\widetilde{\widetilde{\mathbf{Y}}}_s(t)=(\widetilde{\widetilde{Y}}_s(t,1),\widetilde{\widetilde{Y}}_s(t,2),\dots,\widetilde{\widetilde{Y}}_s(t,N)):\mathbb{R}\to\mathcal{S}_n^N$ of (8.28) is named stabilizing solution if the matrices $\widetilde{\widetilde{R}}_d(i)+C_0(i)\cdot\widetilde{\widetilde{Y}}_s(kh,i)C_0^{\top}(i)$ are invertible for any $(k,i)\in\mathbb{Z}\times\mathcal{N}$ and the corresponding closed-loop system:

$$dx(t)=A_0(\eta(t))x(t)dt,\quad kh<t\le(k+1)h \tag{8.29a}$$

$$x(kh^+)=\left(A_{d0}(\eta(kh))+\widetilde{\widetilde{K}}_s(\eta(kh))C_0(\eta(kh))\right)x(kh),\quad k\in\mathbb{Z}_+ \tag{8.29b}$$

is ESMS, where for each $i\in\mathcal{N}$

$$\widetilde{\widetilde{K}}_s(i)\triangleq-\left(A_{d0}(i)\widetilde{\widetilde{Y}}_s(0,i)C_0^{\top}(i)+\widetilde{\widetilde{L}}_d(i)\right)\left(C_0(i)\widetilde{\widetilde{Y}}_s(0,i)C_0^{\top}(i)+\widetilde{\widetilde{R}}_d(i)\right)^{-1}. \tag{8.30}$$

The problem of the existence of a bounded and stabilizing solution of the FJMLDE with Riccati-type jumping operator (8.25) and (8.28), respectively, was discussed in a more general framework in Sect. 5.6. Here, we summarize the most important properties of the bounded on $\mathbb{R}$ and stabilizing solution $\widetilde{\mathbf{Y}}_s(\cdot)$ of (8.25) and (8.28), respectively, which is satisfying some imposed sign conditions.

Corollary 8.3 *The FJMLDE (8.25) and (8.28) have at most one bounded on $\mathbb{R}$ and stabilizing solution $\widetilde{\mathbf{Y}}_s(\cdot)=(\widetilde{Y}_s(\cdot,1),\widetilde{Y}_s(\cdot,2),\dots,\widetilde{Y}_s(\cdot,N))$ which satisfies the sign conditions*

$$\widetilde{R}_d(i)+C_0(i)\widetilde{Y}_s(kh,i)C_0^{\top}(i)>0, \tag{8.31}$$

and $\widetilde{\widetilde{\mathbf{Y}}}_s(\cdot)=(\widetilde{\widetilde{Y}}_s(\cdot,1),\widetilde{\widetilde{Y}}_s(\cdot,2),\dots,\widetilde{\widetilde{Y}}_s(\cdot,N))$ which satisfies the sign conditions

$$\widetilde{\widetilde{R}}_d(i)+C_0(i)\widetilde{\widetilde{Y}}_s(kh,i)C_0^{\top}(i)>0, \tag{8.32}$$

for all $(k,i)\in\mathbb{Z}\times\mathcal{N}$. These solutions are differentiable on each interval $(kh,(k+1)h)$, left continuous in each $\tau_k=kh$, $k\in\mathbb{Z}$. Moreover, $\widetilde{\mathbf{Y}}_s(\cdot)$, $\widetilde{\widetilde{\mathbf{Y}}}_s(\cdot)$ are periodic functions of period h and satisfy $\widetilde{Y}_s(t,i)\ge 0$, and $\widetilde{\widetilde{Y}}_s(t,i)\ge 0$, for all $(t,i)\in\mathbb{R}\times\mathcal{N}$.

8.2.2 The State-Space Representation of the H_2-Optimal Filter

Now we are in position to provide explicit formulae of the state-space representation of a filter $\widetilde{\mathbb{G}}_F$ which satisfies the optimality condition (8.6).

Theorem 8.1 *Assume:*

(a) the Assumptions 1.2, 7.1 and 8.1 are fulfilled;
(b) the FJMLDE with Riccati-type jumping operator (8.25) has a bounded on $\mathbb{R}$ and stabilizing solution $\widetilde{\mathbf{Y}}_s(\cdot)$ which satisfies the sign conditions (8.31).

Consider the filter $\widetilde{\mathbb{G}}_F$ having the dimension of the state space $n_F = n$ and the state-space representation given by

$$d\widetilde{x}_F(t) = A_0(\eta(t))\widetilde{x}_F(t)dt, \quad kh < t \leq (k+1)h \tag{8.33a}$$

$$\widetilde{x}_F(kh^+) = (A_{d0}(\eta(kh))+\widetilde{K}_s(\eta(kh))C_0(\eta(kh)))\widetilde{x}_F(kh)-\widetilde{K}_s(\eta(kh))y(kh), k \in \mathbb{Z}_+ \tag{8.33b}$$

$$z_F(t) = C_z(\eta(t))\widetilde{x}_F(t), \quad t \in \mathbb{R}_+, \tag{8.33c}$$

where $\widetilde{K}_s(i)$, $i \in \mathcal{N}$ are computed via (8.27). Under the considered assumptions $\widetilde{\mathbb{G}}_F$ lies in $\mathscr{F}_{\mathrm{adm}}(\mathbb{G})$ and satisfies the optimality condition

$$\sum_{i=1}^{N}\frac{1}{h}\int_0^h \mathrm{Tr}[C_z(i)\widetilde{Y}_s(t,i)C_z^\top(i)]dt = \|\mathbb{G}_{cl}(\widetilde{\mathbb{G}})\|_2^2 \leq \|\mathbb{G}_{cl}(\mathbb{G}_F)\|_2^2, \tag{8.34}$$

for any admissible filter $\mathbb{G}_F$ of arbitrary dimension $n_F \geq 1$ of its state space.

Proof The fact that the filter $\widetilde{\mathbb{G}}_F$ lies in $\mathscr{F}_{\mathrm{adm}}(\mathbb{G})$ is obvious because in the special case of (8.33) the corresponding JSLDE (8.3) coincides to (8.26) which is ESMS.

Let now $\mathbb{G}_F$ be an arbitrary admissible filter with the dimension of its state space $n_F \geq 1$. The value of the performance achieved by this filter is given in (8.17) being computed based on the unique bounded on $\mathbb{R}$ solution of the non-homogeneous FGLJLDE (8.18). Let

$$\widetilde{Y}_{cl}(t,i) = \begin{pmatrix} Y_{11}(t,i) & Y_{12}(t,i) \\ Y_{12}^\top(t,i) & Y_{22}(t,i) \end{pmatrix} \tag{8.35}$$

for all $(t,i) \in \mathbb{R} \times \mathcal{N}$, be the partition of the bounded solution of (8.18) such that $Y_{11}(t,i) \in \mathscr{S}_n$ and $Y_{22}(t,i) \in \mathscr{S}_{n_F}$. According to the structure of the coefficients of the resulting system (8.4) given in (8.5), we obtain the following partitions of the FGLJLDE (8.18):

$$\dot{Y}_{11}(t,i) = A_0(i)Y_{11}(t,i) + Y_{11}(t,i)A_0^\top(i) + \sum_{l=1}^{r_1} A_l(i)Y_{11}(t,i)A_l^\top(i) + \sum_{j=1}^{N} q_{ji}Y_{11}(t,j) + \pi_{i\infty}G_c(i)\mathbf{\Lambda}_c G_c^\top(i) \tag{8.36a}$$

$$\dot{Y}_{12}(t,i) = A_0(i)Y_{12}(t,i) + Y_{12}(t,i)A_F^\top(i) + \sum_{j=1}^{N} q_{ji}Y_{12}(t,j) \tag{8.36b}$$

$$\dot{Y}_{22}(t,i) = A_F(i)Y_{22}(t,i) + Y_{22}(t,i)A_F^\top(i) + \sum_{j=1}^{N} q_{ji}Y_{22}(t,j), \tag{8.36c}$$

$$Y_{11}(kh^+,i) = \sum_{p=0}^{r_2} A_{dp}(i)Y_{11}(kh,i)A_{dp}^\top(i) + h\pi_{i\infty}G_d(i)\mathbf{\Lambda}_d G_d^\top(i) \tag{8.36d}$$

$$Y_{12}(kh^+,i) = A_{d0}(i)Y_{11}(kh,i)C_0^\top(i)B_{Fd}^\top(i) + A_{d0}(i)Y_{12}(kh,i)A_{Fd}^\top(i) + \sum_{p=1}^{r_2} A_{dp}(i)Y_{11}(kh,i)C_p^\top(i)B_{Fd}^\top(i) + h\pi_{i\infty}G_d(i)\mathbf{\Lambda}_d D^\top(i)B_{Fd}^\top(i) \tag{8.36e}$$

$$\begin{aligned} Y_{22}(kh^+,i) &= B_{Fd}(i)C_0(i)Y_{11}(kh,i)C_0^\top(i)B_{Fd}^\top(i) + A_{Fd}(i)Y_{12}^\top(kh,i)C_0^\top(i)B_{Fd}^\top(i) \\ &+ B_{Fd}(i)C_0(i)Y_{12}(kh,i)A_{Fd}^\top(i) + A_{Fd}(i)Y_{22}(kh,i)A_{Fd}^\top(i) \\ &+ \sum_{p=1}^{r_2} B_{Fd}C_p(i)Y_{11}(kh,i)C_p^\top(i)B_{Fd}^\top(i) + h\pi_{i\infty}B_{Fd}(i)D(i)\mathbf{\Lambda}_d D^\top(i)B_{F_d}(i) \end{aligned} \tag{8.36f}$$

$k \in \mathbb{Z}$, $i \in \mathcal{N}$. From (8.36a) and (8.36d) we may infer that $\mathbf{Y}_{11}(\cdot) = (Y_{11}(\cdot,1), Y_{11}(\cdot,2), \ldots, Y_{11}(\cdot,N))$ is a bounded on $\mathbb{R}$ solution of the FGLJLDE (8.21). From the unicity of the bounded on $\mathbb{R}$ solution of (8.21) we may conclude that $Y_{11}(t,i) = \widetilde{Z}(t,i)$, for all $(t,i) \in \mathbb{R} \times \mathcal{N}$. Hence, (8.35) becomes

$$\widetilde{Y}_{cl}(t,i) = \begin{pmatrix} \widetilde{Z}(t,i) & Y_{12}(t,i) \\ Y_{12}^\top(t,i) & Y_{22}(t,i) \end{pmatrix}, \tag{8.37}$$

for all $(t,i) \in \mathbb{R} \times \mathcal{N}$. We set

$$U(t,i) \triangleq \begin{pmatrix} \widetilde{Z}(t,i) - \widetilde{Y}_s(t,i) & Y_{12}(t,i) \\ Y_{12}^\top(t,i) & Y_{22}(t,i) \end{pmatrix}, \tag{8.38}$$

for all $(t,i) \in \mathbb{R} \times \mathcal{N}$. By direct calculation involving (8.5), (8.18)–(8.25), (8.27), (8.35)–(8.38) we deduce that $t \to \mathbf{U}(t) = (U(t,1), U(t,2), \ldots, U(t,N)) : \mathbb{R} \to \mathcal{S}_{n+n_F}^N$ is a bounded on $\mathbb{R}$ solution of the following FGLJLDE:

$$\dot{U}(t,i) = A_{0,cl}(i)U(t,i) + U(t,i)A_{0,cl}^\top(i) + \sum_{j=1}^{N} q_{ji}U(t,j), kh < t \le (k+1)h \tag{8.39a}$$

$$U(kh^+,i) = A_{d0,cl}(i)U(kh,i)A_{d0,cl}^\top(i) + \Theta_d(i)(R_d(i) + C_0(i)Y_s(0,i)C_0^\top(i))\Theta_d^\top(i), \tag{8.39b}$$

$k \in \mathbb{Z}$, $i \in \mathcal{N}$ where $\Theta_d(i) \triangleq \begin{pmatrix} -\widetilde{K}_s(i) \\ B_{Fd}(i) \end{pmatrix}$. Since (8.39) is a special case of the FGLJLDE (7.15) we deduce via Corollary 7.2 and Corollary 8.1 that the FGLJLDE (8.39) has a unique bounded on $\mathbb{R}$ solution and the values of that solution lie in $\mathcal{S}_{n+n_F}^{N+}$. So,

$$U(t,i) \geq 0, \tag{8.40}$$

for all $(t,i) \in \mathbb{R} \times \mathcal{N}$. Further, employing (8.38) we rewrite (8.17) as:

$$\|\mathbb{G}_{cl}(\mathbb{G}_F)\|_2^2 = \sum_{i=1}^N \frac{1}{h} \int_0^h \mathrm{Tr}[C_z(i)\widetilde{Y}_s(t,i)C_z^\top(i)]\mathrm{d}t + \sum_{i=1}^N \int_0^h \mathrm{Tr}[C_{cl}(i)U(t,i)C_{cl}^\top(i)]\mathrm{d}t, \tag{8.41}$$

for all $\mathbb{G}_F \in \mathcal{F}_{\mathrm{adm}}(\mathbb{G})$. Combining (8.40) and (8.41) we deduce that

$$\|\mathbb{G}_{cl}(\mathbb{G}_F)\|_2^2 \geq \sum_{i=1}^N \frac{1}{h} \int_0^h \mathrm{Tr}[C_z(i)\widetilde{Y}_s(t,i)C_z^\top(i)]\mathrm{d}t, \tag{8.42}$$

for any admissible filter $\mathbb{G}_F$. To complete the proof we show that (8.42) becomes equality when the filter $\mathbb{G}_F$ has the state-space representation given by (8.33). If $C_F(i) = C_z(i)$, $i \in \mathcal{N}$, we have:

$$\begin{aligned} C_{cl}(i)U(t,i)C_{cl}^\top(i) &= C_z(i)(I_n \quad - I_n)U(t,i)(I_n \quad - I_n)^\top C_z^\top(i) \\ &= C_z(i)\widehat{U}_{11}(t,i)C_z^\top(i), \end{aligned} \tag{8.43}$$

for all $(t,i) \in \mathbb{R} \times \mathcal{N}$, where $\widehat{U}_{11}(t,i)$ is the 11-block of the matrix

$$\widehat{U}(t,i) \triangleq \mathcal{I}U(t,i)\mathcal{I}^\top \qquad \mathcal{I} = \begin{pmatrix} I_n & -I_n \\ 0 & I_n \end{pmatrix}. \tag{8.44}$$

Here we took into account that in the case of the filter described by (8.33) we have $n_F = n$. By direct calculation, involving (8.39) and (8.44) we deduce that in the case when the controller $\mathbb{G}_F$ is described by (8.33) then $\widehat{\mathbf{U}}_{11}(\cdot) = (\widehat{U}_1 1(\cdot,1), \widehat{U}_{11}(\cdot,2), \ldots, \widehat{U}_{11}(\cdot,N))$ is a bounded on $\mathbb{R}$ solution of the following FGLJLDE:

$$\dot{U}_{11}(t,i) = A_0(i)U_{11}(t,i) + U_{11}(t,i)A_0^\top(i) + \sum_{j=1}^N q_{ji}U_{11}(t,j),\ kh < t \leq (k+1)h \tag{8.45a}$$

$$U_{11}(kh^+, i) = (A_{d0}(i) + \widetilde{K}_s(i)C_0(i))U_{11}(kh, i)(A_{d0}(i) + \widetilde{K}_s(i)C_0(i))^\top, \quad (8.45b)$$

$k \in \mathbb{Z}$, $i \in \mathcal{N}$. Bearing in mind that the JSLDE (8.26) is ESMS we may conclude via Corollary 7.2 applied in the case of the FGLJLDE (8.45) that this equation has an unique bounded on $\mathbb{R}$ solution. Hence, $\widehat{U}_{11}(t, i) = 0$, for all $(t, i) \in \mathbb{R} \times \mathcal{N}$. So, in the case of the filter $\widetilde{\mathbb{G}}_F$ described by (8.33) we have that $C_{cl}(i)U(t, i)C_{cl}^\top(i) = 0$, for all $(t, i) \in \mathbb{R} \times \mathcal{N}$. In this case (8.41) becomes

$$\|\mathbb{G}_{cl}(\widetilde{\mathbb{G}}_F)\|_2^2 = \sum_{i=1}^{N} \frac{1}{h} \int_0^h \mathrm{Tr}[C_z(i)\widetilde{Y}_s(t, i)C_z^\top(i)]\mathrm{d}t. \quad (8.46)$$

The validity of (8.34) is confirmed combining (8.42) and (8.46).
Thus the proof ends.

The next result provides the state-space representation of a filter from $\mathcal{F}_{\text{adm}}(\mathbb{G})$ which satisfies the optimality condition (8.7).

Theorem 8.2 *Assume:*

(a) the Assumptions 1.2, 7.1 and 8.1 are fulfilled;
(b) the FJMLDE with Riccati-type jumping operator (8.28) has a bounded on $\mathbb{R}$ and stabilizing solution $\widetilde{\mathbf{Y}}_s(\cdot)$ which is satisfying the sign conditions (8.32).

We consider the filter $\approx{\mathbb{G}}_F$ having the dimension of the state space $n_F = n$ and the state-space representation given by:

$$d\widetilde{\widetilde{x}}_F(t) = A_0(\eta(t))\widetilde{\widetilde{x}}_F(t)dt, \quad kh < t \leq (k+1)h \quad (8.47a)$$

$$\widetilde{\widetilde{x}}_F(kh^+) = (A_{d0}(\eta(kh)) + \widetilde{\widetilde{K}}_s(\eta(kh))C_0(\eta(kh)))\widetilde{\widetilde{x}}_F(kh) - \widetilde{\widetilde{K}}_s(\eta(kh))u(k),\ k \in \mathbb{Z}_+ \quad (8.47b)$$

$$z_F(t) = C_z(\eta(t))\widetilde{\widetilde{x}}_F(t), \quad t \in \mathbb{R}_+. \quad (8.47c)$$

Under the considered assumptions the filter $\widetilde{\widetilde{\mathbb{G}}}_F$ lies in $\mathcal{F}_{\text{adm}}(\mathbb{G})$ and satisfies the optimality condition $\|\!|\mathbb{G}_{cl}(\mathbb{G}_F)\|\!|_2^2 \geq \|\!|\mathbb{G}_{cl}(\widetilde{\widetilde{\mathbb{G}}}_F)\|\!|_2^2 = \sum_{i=1}^N \frac{1}{h}\int_0^h \mathrm{Tr}[C_z(i)\widetilde{\widetilde{Y}}_s(t, i)C_z^\top(i)]dt$ for any filter $\mathbb{G}_F \in \mathcal{F}_{\text{adm}}(\mathbb{G})$.

Proof Can be done following the line of the proof of Theorem 8.1, but starting from (8.19) instead of (8.17). The details are left to the reader.

8.2.3 The Case of a Jump Linear Stochastic System Without the Effect of Markovian Switching

Now we deal with the problem of estimation of a remote signal generated by a dynamical system $\mathbb{G}$ having the state-space representation described by:

$$dx(t) = A_0 x(t)dt + \sum_{l=1}^{r_1} A_l x(t)dw_l(t) + G_c d\xi(t), \quad kh < t \le (k+1)h \tag{8.48a}$$

$$x(kh^+) = (A_{d0} + \sum_{p=1}^{r_2} w_{dp}(k)A_{dp})x(kh) + G_d\zeta_d(k), \quad k \in \mathbb{Z}_+ \tag{8.48b}$$

$$y(kh) = (C_0 + \sum_{p=1}^{r_2} w_{dp}(k)C_p)x(kh) + D\zeta_d(k), \quad k \in \mathbb{Z}_+ \tag{8.48c}$$

$$z(t) = C_z x(t), \quad t \in \mathbb{R}_+ \tag{8.48d}$$

where $x(t)$, $z(t)$, $y(kh)$ have the same meaning as in the case of (8.1). The state multiplicative noises $\{w(t)\}_{t\ge 0}$, $\{w_d(k)\}_{k\in\mathbb{Z}_+}$ are stochastic processes which are satisfying the Assumption 1.1 while the additive noises $\{\xi(t)\}_{t\ge 0}$, $\{\zeta_d(k)\}_{k\in\mathbb{Z}_+}$ are stochastic processes which satisfy the Assumption 7.1. The set of the admissible filters consist of all jump linear systems $\mathbb{G}_F$ with the state-space representation:

$$dx_F(t) = A_F x_F(t)dt, \quad kh < t \le (k+1)h \tag{8.49a}$$
$$x_F(kh^+) = A_{Fd}x_F(kh) + D_{Fd}u(k), \quad k \in \mathbb{Z}_+ \tag{8.49b}$$
$$z_F(t) = C_F x_F(t), \quad t \in \mathbb{R}_+ \tag{8.49c}$$

subject to the condition that the corresponding JLDE:

$$dx_F(t) = A_F x_F(t)dt, \quad kh < t \le (k+1)h \tag{8.50a}$$
$$x_F(kh^+) = A_{Fd}x_F(kh), \quad k \in \mathbb{Z}_+ \tag{8.50b}$$

is exponentially stable (ES). As in the previous case, the problem of optimal filtering in the case of the signal $z(t)$ generated by the dynamical system (8.48) require the finding of an admissible filter $\tilde{\tilde{\mathbb{G}}}_F$ of the form (8.49) which minimizes the H_2-norm of the resulting system obtained when the inputs $u(k)$ are replaced by the measurements $y(kh)$, $k \in \mathbb{Z}_+$.

The value of the H_2-norms $\|\mathbb{G}_{cl}(\mathbb{G}_F)\|_2$ and $\widetilde{\|}\mathbb{G}_{cl}(\mathbb{G}_F)\widetilde{\|}_2$ of the resulting system determined by an admissible filter $\mathbb{G}_F$ is computed by

$$\|\mathbb{G}_{cl}(\mathbb{G}_F)\|_2^2 = \frac{1}{h}\int_0^h \mathrm{Tr}[C_{cl}\widetilde{Y}_{cl}(t)C_{cl}^\top]dt \tag{8.51}$$

where $t \to \widetilde{Y}_{cl}(t) : \mathbb{R} \to \mathcal{S}_{n+n_F}$ is the unique bounded solution of the following non-homogeneous FGLJLDE:

$$\dot{Y}_{cl}(t) = A_{0,cl}Y_{cl}(t) + Y_{cl}(t)A_{0,cl}^\top + \sum_{l=1}^{r_1} A_{l,cl}Y_{cl}(t)A_{l,cl}^\top$$

$$+ G_{c,cl}\mathbf{\Lambda}_c G_{c,cl}^\top,\ kh < t \leq (k+1)h \tag{8.52a}$$

$$Y_{cl}(kh^+) = \sum_{p=0}^{r_2} A_{dp,cl} Y_{cl}(kh) A_{dp,cl}^\top + hG_{d,cl}\mathbf{\Lambda}_d G_{d,cl}^\top,\ k \in \mathbb{Z} \tag{8.52b}$$

whose coefficients are computed as in (8.5) suppressing the dependence with respect to the node i.

In this new framework (8.19) becomes:

$$\widetilde{\|}\mathbb{G}_{cl}(\mathbb{G}_F)\widetilde{\|}_2^2 = \frac{1}{h}\int_0^h \mathrm{Tr}[C_{cl}\widetilde{\widetilde{Y}}_{cl}(t)C_{cl}^\top]\mathrm{d}t \tag{8.53}$$

where $t \to \widetilde{\widetilde{Y}}_{cl}(t) : \mathbb{R} \to \mathscr{S}_{n+n_F}$ is the unique bounded solution of the following non-homogeneous FGLJLDE:

$$\dot{Y}_{cl}(t) = A_{0,cl}Y_{cl}(t) + Y_{cl}(t)A_{0,cl}^\top + \sum_{l=1}^{r_1} A_{l,cl}Y_{cl}(t)A_{l,cl}^\top$$
$$+ G_{c,cl}\mathbf{\Lambda}_c G_{c,cl}^\top,\ kh < t \leq (k+1)h \tag{8.54a}$$

$$Y_{cl}(kh^+) = \sum_{p=0}^{r_2} A_{dp,cl} Y_{cl}(kh) A_{dp,cl}^\top + G_{d,cl}\mathbf{\Lambda}_d G_{d,cl}^\top,\ k \in \mathbb{Z}. \tag{8.54b}$$

The properties highlighted in Remark 8.1 for the bounded solution $\widetilde{\mathbf{Y}}_{cl}(\cdot)$ of the FGLJLDE (8.18) and $\widetilde{\widetilde{\mathbf{Y}}}_{cl}(\cdot)$ of (8.28) are transferred mutatis-mutandis to the bounded on $\mathbb{R}$ solution $\widetilde{Y}_{cl}(\cdot)$ of (8.52) and $\widetilde{\widetilde{Y}}_{cl}(\cdot)$ of (8.54). Involving the coefficients of (8.48a), (8.48b) we introduce the following non-homogeneous FGLJLDEs:

$$\dot{Z}(t) = A_0 Z(t) + Z(t)A_0^\top + \sum_{l=1}^{r_1} A_l Z(t) A_l^\top + G_c\mathbf{\Lambda}_c G_c^\top,\ kh < t \leq (k+1)h \tag{8.55a}$$

$$Z(kh^+) = \sum_{p=0}^{r_2} A_{dp} Z(kh) A_{dp}^\top + hG_d\mathbf{\Lambda}_d G_d^\top,\ k \in \mathbb{Z}, \tag{8.55b}$$

$$\dot{Z}(t) = A_0 Z(t) + Z(t)A_0^\top + \sum_{l=1}^{r_1} A_l Z(t) A_l^\top + G_c\mathbf{\Lambda}_c G_c^\top,\ kh < t \leq (k+1)h \tag{8.56a}$$

$$Z(kh^+) = \sum_{p=0}^{r_2} A_{dp} Z(kh) A_{dp}^\top + G_d\mathbf{\Lambda}_d G_d^\top,\ k \in \mathbb{Z}. \tag{8.56b}$$

Let $t \to \widetilde{Z}(t) : \mathbb{R} \to \mathscr{S}_n$ and $t \to \widetilde{\widetilde{Z}}(t) : \mathbb{R} \to \mathscr{S}_n$ be the bounded solutions of (8.55) and (8.56), respectively. Specializing to the case $\mathscr{N} = \{1\}$, $q_{11} = 0$, the properties of the solutions $\widetilde{\mathbf{Z}}(\cdot)$ of (8.21) and $\widetilde{\widetilde{\mathbf{Z}}}(\cdot)$ of (8.22) highlighted in Remark 8.2 we obtain useful properties of the solutions $\widetilde{Z}(\cdot)$ and $\widetilde{\widetilde{Z}}(\cdot)$.

In this context the Assumption 8.1 becomes:

Assumption 8.2 The JSLDE:

$$\mathrm{d}x(t) = A_0 x(t)\mathrm{d}t + \sum_{l=1}^{r_1} A_l x(t)\mathrm{d}w_l(t), \quad kh < t \le (k+1)h \tag{8.57a}$$

$$x(kh^+) = (A_{d0} + \sum_{p=1}^{r_2} w_{dp}(k) A_{dp}) x(kh), \quad k \in \mathbb{Z}_+ \tag{8.57b}$$

is ESMS.

In the new framework, the notation (8.23) take the form

$$\widetilde{M}_c(t) := G_c \mathbf{\Lambda}_c G_c^\top + \sum_{l=1}^{r_1} A_l \widetilde{Z}(t) A_l^\top, \quad t \in \mathbb{R} \tag{8.58a}$$

$$\widetilde{M}_d \coloneqq hG_d \mathbf{\Lambda}_d G_d^\top + \sum_{p=1}^{r_2} A_{dp} \widetilde{Z}(0) A_{dp}^\top = hG_d \mathbf{\Lambda}_d G_d^\top + \sum_{p=1}^{r_2} A_{dp} \widetilde{Z}(kh) A_{dp}^\top, \quad k \in \mathbb{Z} \tag{8.58b}$$

$$\widetilde{L}_d \coloneqq hG_d \mathbf{\Lambda}_d D^\top + \sum_{p=1}^{r_2} A_{dp} \widetilde{Z}(0) C_p^\top = hG_d \mathbf{\Lambda}_d D^\top + \sum_{p=1}^{r_2} A_{dp} \widetilde{Z}(kh) C_p^\top, \quad k \in \mathbb{Z} \tag{8.58c}$$

$$\widetilde{R}_d := hD \mathbf{\Lambda}_d D^\top + \sum_{p=1}^{r_2} C_p \widetilde{Z}(0) C_p^\top = hD \mathbf{\Lambda}_d D^\top + \sum_{p=1}^{r_2} C_p \widetilde{Z}(kh) C_p^\top, \quad k \in \mathbb{Z}. \tag{8.58d}$$

Similar, (8.24) reduces to:

$$\widetilde{\widetilde{M}}_c(t) := G_c \mathbf{\Lambda}_c G_c^\top + \sum_{l=1}^{r_1} A_l \widetilde{\widetilde{Z}}(t) A_l^\top, \quad t \in \mathbb{R} \tag{8.59a}$$

$$\widetilde{\widetilde{M}}_d \coloneqq G_d \mathbf{\Lambda}_d G_d^\top + \sum_{p=1}^{r_2} A_{dp} \widetilde{\widetilde{Z}}(0) A_{dp}^\top = G_d \mathbf{\Lambda}_d G_d^\top + \sum_{p=1}^{r_2} A_{dp} \widetilde{\widetilde{Z}}(kh) A_{dp}^\top, \quad k \in \mathbb{Z} \tag{8.59b}$$

$$\tilde{\tilde{L}}_d := G_d\Lambda_d D^\top + \sum_{p=1}^{r_2} A_{dp}\tilde{\tilde{Z}}(0)C_p^\top = G_d\Lambda_d D^\top + \sum_{p=1}^{r_2} A_{dp}\tilde{\tilde{Z}}(kh)C_p^\top,\ k \in \mathbb{Z} \tag{8.59c}$$

$$\tilde{\tilde{R}}_d := D\Lambda_d D^\top + \sum_{p=1}^{r_2} C_p\tilde{\tilde{Z}}(0)C_p^\top = D\Lambda_d D^\top + \sum_{p=1}^{r_2} C_p\tilde{\tilde{Z}}(kh)C_p^\top,\ \ k \in \mathbb{Z}. \tag{8.59d}$$

The validity of the second equality from (8.58b)–(8.58d) and (8.59b)–(8.59d), respectively, is guaranteed by the periodicity property of the solution $\tilde{Z}(\cdot)$ of (8.55) and $\tilde{\tilde{Z}}(\cdot)$ of (8.56), respectively.

We consider the FJMLDE with Riccati-type jumping operator:

$$\dot{Y}(t) = A_0Y(t) + Y(t)A_0^\top + \tilde{M}_c(t),\ \ kh < t \le (k+1)h \tag{8.60a}$$

$$Y(kh^+) = A_{d0}Y(kh)A_{d0}^\top - (A_{d0}Y(kh)C_0^\top + \tilde{L}_d)(C_0Y(kh)C_0^\top + \tilde{R}_d)^{-1} \cdot (C_0Y(kh)A_{d0}^\top + \tilde{L}_d^\top) + \tilde{M}_d,\ \ k \in \mathbb{Z}. \tag{8.60b}$$

A globally defined solution $t \to Y_s(t) : \mathbb{R} \to \mathcal{S}_n$ is a stabilizing solution of (8.60) if the matrices $C_0Y_s(kh)C_0^\top + R_d$ are invertible for all $k \in \mathbb{Z}$ and the corresponding jump linear differential equation (JLDE):

$$\dot{x}(t) = A_0x(t),\ \ kh < t \le (k+1)h \tag{8.61a}$$

$$x(kh^+) = (A_{d0} + \tilde{K}_sC_0)x(kh),\ \ k \in \mathbb{Z}, \tag{8.61b}$$

is exponentially stable (ES), where

$$\tilde{K}_s \triangleq -(A_{d0}Y_s(0)C_0^\top + \tilde{L}_d)(C_0Y_s(0)C_0^\top + \tilde{R}_d)^{-1}. \tag{8.62}$$

The FJMLDE with Riccati-type jumping operator (8.28) reduces to

$$\dot{Y}(t) = A_0Y(t) + Y(t)A_0^\top + \tilde{\tilde{M}}_c(t),\ \ kh < t \le (k+1)h \tag{8.63a}$$

$$Y(kh^+) = A_{d0}Y(kh)A_{d0}^\top - (A_{d0}Y(kh)C_0^\top + \tilde{\tilde{L}}_d)(C_0Y(kh)C_0^\top + \tilde{\tilde{R}}_d)^{-1} \cdot (C_0Y(kh)A_{d0}^\top + \tilde{\tilde{L}}_d^\top) + \tilde{\tilde{M}}_d,\ \ k \in \mathbb{Z}. \tag{8.63b}$$

A globally defined solution $t \to \tilde{\tilde{Y}}_s(t) : \mathbb{R} \to \mathcal{S}_n$ is a stabilizing solution of (8.63) if the matrices $\tilde{\tilde{R}}_d + C_0\tilde{\tilde{Y}}_s(kh)C_0^\top$ are invertible for all $k \in \mathbb{Z}$ and the jump linear system:

$$\dot{x}(t) = A_0x(t),\ \ kh < t \le (k+1)h \tag{8.64a}$$

$$x(kh^+) = (A_{d0} + \widetilde{\widetilde{K}}_s C_0)x(kh), \quad k \in \mathbb{Z}, \tag{8.64b}$$

is exponentially stable, where $\widetilde{\widetilde{K}}_s \stackrel{\Delta}{=} -(A_{d0}\widetilde{\widetilde{Y}}_s(0)C_0^\top + \widetilde{\widetilde{L}}_d)(C_0\widetilde{\widetilde{Y}}_s(0)C_0^\top + \widetilde{\widetilde{R}}_d)^{-1}$. The properties of the bounded and stabilizing solution of the FJMLDEs (8.25) and (8.28), respectively, stated in Corollary 8.3 remain valid in the case of the bounded and stabilizing solution $\widetilde{Y}_s(\cdot)$ of the FJMLDE (8.60), and $\widetilde{\widetilde{Y}}_s(\cdot)$ of (8.63), respectively, because (8.60) and (8.63) are particular cases of (8.25) and (8.28) for $\mathcal{N} = \{1\}$ and $q_{11} = 0$.

The next result provides the state-space representation of an admissible filter of type (8.49) which satisfies condition (8.50) and achieves the smallest value of the right hand side of (8.51) with respect to all admissible filters.

Theorem 8.3 *Assume:*

(a) the Assumptions 1.1, 7.1 and 8.2 are fulfilled;
(b) the FJMLDE with Riccati-type jumping operator (8.60) has a bounded on $\mathbb{R}$ and stabilizing solution $\widetilde{Y}_s(\cdot)$ which satisfies the sign conditions

$$C_0\widetilde{Y}_s(kh)C_0^\top + \widetilde{R}_d > 0 \ \textit{for all} \ k \in \mathbb{Z}. \tag{8.65}$$

We consider the filter $\widetilde{\widetilde{\mathbb{G}}}_F$ having the dimension of the state space $n_F = n$ and the state-space representation given by

$$dx_F(t) = A_0 x_F(t)dt, \quad kh < t \le (k+1)h \tag{8.66a}$$

$$x_F(kh^+) = (A_{d0} + \widetilde{K}_s C_0)x_F(kh) - \widetilde{K}_s u(k), \quad k \in \mathbb{Z}_+ \tag{8.66b}$$

$$z_F(t) = C_z x_F(t), \quad t \in \mathbb{R}_+. \tag{8.66c}$$

Under the considered assumptions the filter $\widetilde{\widetilde{\mathbb{G}}}_F$ satisfies the constraints (8.50) as well as the optimality condition

$$\|\mathbb{G}_{cl}(\mathbb{G}_F)\|_2^2 \ge \|\mathbb{G}_{cl}(\widetilde{\widetilde{\mathbb{G}}}_F)\|_2^2 = \frac{1}{h}\int_0^h \mathrm{Tr}[C_z\widetilde{Y}_s(t)C_z^\top]dt,$$

for any filter $\mathbb{G}_F$ having the state-space representation (8.49) subject to the constraints (8.50) and having arbitrary dimension of the state space. In (8.66), $\widetilde{K}_s$ is the gain matrix associated to the bounded and stabilizing solution $\widetilde{Y}_s(\cdot)$ via (8.62).

Proof The proof can be done following step by step the proof of Theorem 8.1. The details are left to the reader.

Remark 8.3 Due to the periodicity property of the stabilizing solutions $\widetilde{\mathbf{Y}}_s(\cdot)$ and $\widetilde{Y}_s(\cdot)$, respectively, it is enough that the sign conditions (8.31) and (8.65), respectively, to be satisfied for $k = 0$.

Exercise 8.1 Find the state-space representation of a filter $\tilde{\tilde{\mathbb{G}}}_F$ from $\mathscr{F}_{\text{adm}}(\mathbb{G})$ which satisfies the optimality condition

$$\tilde{\|}\mathbb{G}_{cl}(\tilde{\tilde{\mathbb{G}}}_F)\tilde{\|}_2 = \min\{\tilde{\|}\mathbb{G}_{cl}(\mathbb{G}_F)\tilde{\|}_2 \mid \mathbb{G}_F \in \mathscr{F}_{\text{adm}}(\mathbb{G})\}.$$

Hint: One may proceed as in the proof of the previous theorem, but employing the stabilizing solution $\tilde{\tilde{Y}}_s(\cdot)$ of the FJMLDE with Riccati-type jumping operator (8.63).

8.3 H_2-Optimal Filtering Under Sampled Measurements

Let us consider the dynamical system $\mathbb{G}$ with the state-space representation described by:

$$dx(t) = A_0(\eta(t))x(t)dt + \sum_{l=1}^{r_1} A_l(\eta(t))x(t)dw_l(t) + G_c(\eta(t))d\xi(t), \tag{8.67a}$$

$$z(t) = C_z(\eta(t))x(t), \quad t \in \mathbb{R}_+ \tag{8.67b}$$

$$y(kh) = (C_0(\eta(kh)) + \sum_{p=1}^{r_2} w_{dp}(k)C_p(\eta(kh)))x(kh) + D(\eta(kh))\zeta_d(k), \, k \in \mathbb{Z}_+ \tag{8.67c}$$

where $x(t) \in \mathbb{R}^n$ are the states of the dynamical system, $z(t) \in \mathbb{R}^{n_z}$ describes the remote signal which have to be estimated and $y(kh) \in \mathbb{R}^{n_y}$, $k \in \mathbb{Z}_+$ are the measurements available to activate a filter $\mathbb{G}_F$ which must generates a signal $z_F(t)$ that is the best estimation of the remote signal $z(t)$. In (8.67), $\{w_l(t)\}_{t\geq 0}$, $1 \leq l \leq r_1$, $\{w_{dp}(k)\}_{k\in\mathbb{Z}_+}$, $1 \leq p \leq r_2$ and $\{\eta(t)\}_{t\geq 0}$ are stochastic processes satisfying the Assumption 1.2, while $\{\xi(t)\}_{t\geq 0}$ and $\{\zeta_d(k)\}_{k\in\mathbb{Z}_+}$ are stochastic processes satisfying the Assumption 7.1.

Since the filters involved in the filtering problem addressed in this section are activated at their inputs by discrete-time signals and must generate at their outputs signals which are evolving in continuous time, we consider that the most suitable types of admissible filters are the filters $\mathbb{G}_F$ having the state-space representation described by (8.2) and satisfying the stability condition (8.3).

The developments from this section are done under the following assumptions:

Assumption 8.3 The stochastic linear differential equation (SLDE):

$$dx(t) = A_0(\eta(t))x(t)dt + \sum_{l=1}^{r_1} A_l(\eta(t))x(t)dw_l(t), \quad t \in \mathbb{R}_+ \tag{8.68}$$

is ESMS.

The SLDE (8.68) is obtained from (8.67a) when $G(\eta(t)) \equiv 0$.

Remark 8.4 (a) The SLDE (8.68) is equivalent to the following jump stochastic linear differential equation (JSLDE):

$$dx(t) = A_0(\eta(t))x(t)dt + \sum_{l=1}^{r_1} A_l(\eta(t))x(t)dw_l(t), \quad kh < t \le (k+1)h \tag{8.69a}$$

$$x(kh^+) = A_{d0}(\eta(kh))x(kh), \quad k \in \mathbb{Z}_+ \tag{8.69b}$$

where $A_{d0}(i) = I_n$ and $A_l(i)$ are those from (8.68). One sees that $x(\cdot)$ is a solution of (8.68) if and only if it is a solution of the JSLDE (8.69). Hence, the SLDE (8.68) is ESMS if and only if the JSLDE (8.69) is ESMS. So, the Assumption 8.3 holds for the SLDE (8.68) if and only if the Assumption 8.1 is fulfilled in the case of the JSLDE (8.69).

(b) The non-homogeneous stochastic linear differential equation (8.67a) is equivalent to the non-homogeneous jump stochastic linear differential equation

$$\begin{aligned} dx(t) &= A_0(\eta(t))x(t)dt + \sum_{l=1}^{r_1} A_l(\eta(t))x(t)dw_l(t) \\ &\quad + G(\eta(t))d\xi(t), \quad kh < t \le (k+1)h \qquad (8.70a) \\ x(kh^+) &= A_{d0}(\eta(kh))x(kh), \quad k \in \mathbb{Z}_+. \qquad (8.70b) \end{aligned}$$

It is obvious that (8.70) is the particular case of (8.1a)–(8.1b) for $A_{d0}(\eta(kh)) = I_n$, $A_{dp}(\eta(kh)) = 0$, $1 \le p \le r_2$, $G_c(\eta(t)) = G(\eta(t))$, $G_d(\eta(kh)) = 0$.

Coupling a filter $\mathbb{G}_F$ of the form (8.2) to the dynamical system (8.67), taking $u(k) = y(kh)$, $k \in \mathbb{Z}_+$, we obtain the following resulting system $\mathbb{G}_{cl}(\mathbb{G}_F)$:

$$\begin{aligned} dx_{cl}(t) &= A_{0,cl}(\eta(t))x_{cl}(t)dt + \sum_{l=1}^{r_1} A_{l,cl}(\eta(t))x_{cl}(t)dw_l(t) \\ &\quad + G_{c,cl}(\eta(t))d\xi(t), \quad kh < t \le (k+1)h \qquad (8.71a) \\ x_{cl}(kh^+) &= \left(A_{d0,cl}(\eta(kh)) + \sum_{p=1}^{r_2} w_{dp}(k)A_{dp,cl}(\eta(kh)) \right) x_{cl}(kh) \\ &\quad + G_{d,cl}(\eta(kh))\zeta_d(k), \quad k \in \mathbb{Z}_+ \qquad (8.71b) \\ z_{cl}(t) &= C_{cl}(\eta(t))x_{cl}(t), \quad t \in \mathbb{R}_+, \qquad (8.71c) \end{aligned}$$

where $x_{cl}(t) = (x^\top(t) \;\; x_F^\top)^\top \in \mathbb{R}^{n+n_F}$ and for each $i \in \mathcal{N}$ we have denoted:

$$A_{0,cl}(i) = \begin{pmatrix} A_0(i) & 0 \\ 0 & A_F(i) \end{pmatrix} \tag{8.72a}$$

$$A_{l,cl}(i) = \begin{pmatrix} A_l(i) & 0 \\ 0 & 0 \end{pmatrix}, \quad 1 \le l \le r_1 \tag{8.72b}$$

$$A_{d0,cl}(i) = \begin{pmatrix} I_n & 0 \\ B_{Fd}(i)C_0(i) & A_{Fd}(i) \end{pmatrix} \tag{8.72c}$$

$$A_{dp,cl}(i) = \begin{pmatrix} 0 & 0 \\ B_{Fd}(i)C_p(i) & 0 \end{pmatrix}, \quad 1 \le p \le r_2 \tag{8.72d}$$

$$C_{cl}(i) = \begin{pmatrix} C_z(i) & -C_F(i) \end{pmatrix} \tag{8.72e}$$

$$G_{c,cl}(i) = \begin{pmatrix} G(i) \\ 0 \end{pmatrix} \tag{8.72f}$$

$$G_{d,cl}(i) = \begin{pmatrix} 0 \\ B_{Fd}(i)D(i) \end{pmatrix}. \tag{8.72g}$$

Our aim is to derive the state-space representation of a filter $\widetilde{\mathbb{G}}_F \in \mathscr{F}_{\text{adm}}(\mathbb{G})$ which satisfies one of the optimality condition (8.6) or (8.7) written for the resulting system (8.71)–(8.72).

To this end, we shall use the results from Theorems 8.1 and 8.2 specialized to the case of the resulting systems of the form (8.71)–(8.72).

First, let us remark that in the case of the framework considered in this section, the resulting system (8.71)–(8.72) is a special case of (8.4)–(8.5) when (8.1a)–(8.1b) is replaced by (8.70). In this case, the bounded on $\mathbb{R}$ solutions $\mathbf{Z}(\cdot)$ and $\widetilde{\widetilde{\mathbf{Z}}}(\cdot)$ of (8.21) and (8.22), respectively, coincide with the unique bounded on $\mathbb{R}$ solution of the following Lyapunov-type differential equation of stochastic control:

$$\dot{Z}(t,i) = A_0(i)Z(t,i) + Z(t,i)A_0^\top(i) + \sum_{l=1}^{r_1} A_l(i)Z(t,i)A_l^\top(i) + \sum_{j=1}^{N} q_{ji}Z(t,j) + \pi_{i\infty}G_c(i)\Lambda_c G_c^\top(i), \quad i \in \mathcal{N}. \tag{8.73}$$

Applying Theorem 2.3.7 (iii) and Theorem 2.6.1 (i) from [1], we deduce that under the Assumption 8.3 the equation (8.73) has a unique bounded on $\mathbb{R}$ solution and that solution is constant. So, $\widetilde{Z}(t,i) = \widetilde{\widetilde{Z}}(t,i) = \widetilde{\mathscr{Z}}(i)$, for all $(t,i) \in \mathbb{R} \times \mathcal{N}$, where $(\widetilde{\mathscr{Z}}(1), \widetilde{\mathscr{Z}}(2), \ldots, \widetilde{\mathscr{Z}}(N))$ is the unique solution of the algebraic Lyapunov-type equation on $\mathcal{S}_n^N$:

$$A_0(i)Z(i) + Z(i)A_0^\top(i) + \sum_{l=1}^{r_1} A_l(i)Z(i)A_l^\top(i)$$

$$+\sum_{j=1}^{N} q_{ji} Z(j) + \pi_{i\infty} G(i) \mathbf{\Lambda}_c G^{\top}(i) = 0, \quad i \in \mathcal{N}. \tag{8.74}$$

In this context, (8.23) and (8.24) become:

$$\begin{aligned}
\widetilde{M}_c(i) &= \widetilde{\widetilde{M}}_c(i) \triangleq \pi_{i\infty} G_c(i) \mathbf{\Lambda}_c G_c^{\top}(i) + \sum_{l=1}^{r_1} A_l(i) \widetilde{\mathscr{Z}}(i) A_l^{\top}(i) := \mathscr{M}_c(i) \\
\widetilde{M}_d(i) &= \widetilde{\widetilde{M}}_d(i) = 0, \qquad \widetilde{L}_d(i) = \widetilde{\widetilde{L}}_d(i) = 0, \\
\widetilde{R}_d(i) &= h\pi_{i\infty} D(i) \mathbf{\Lambda}_d D^{\top}(i) + \sum_{p=1}^{r_2} C_p(i) \widetilde{\mathscr{Z}}(i) C_p^{\top}(i) \triangleq \widetilde{\mathscr{R}}_d(i) \\
\widetilde{\widetilde{R}}_d(i) &= \pi_{i\infty} D(i) \mathbf{\Lambda}_d D^{\top}(i) + \sum_{p=1}^{r_2} C_p(i) \widetilde{\widetilde{\mathscr{Z}}}(i) C_p^{\top}(i) \triangleq \widetilde{\widetilde{\mathscr{R}}}_d(i),
\end{aligned} \tag{8.75}$$

$i \in \mathcal{N}$. The FJMLDE with Riccati-type jumping operator (8.25) reduces to

$$\dot{Y}(t,i) = A_0(i) Y(t,i) + Y(t,i) A_0^{\top}(i) + \sum_{j=1}^{N} q_{ji} Y(t,j) + \widetilde{\mathscr{M}_c}(i), \quad kh < t \le (k+1)h \tag{8.76a}$$

$$\begin{aligned} Y(kh^+, i) = {} & Y(kh,i) - Y(kh,i) C_0^{\top}(i) (\widetilde{\mathscr{R}}_d(i) + C_0(i) Y(kh,i) C_0^{\top}(i))^{-1} \\ & C_0(i) Y(kh,i), \quad k \in \mathbb{Z}. \end{aligned} \tag{8.76b}$$

A globally defined solution $t \to \widetilde{\mathbf{Y}}_s(t) = (\widetilde{\mathscr{Y}}_s(t,1), \widetilde{\mathscr{Y}}_s(t,2), \ldots, \widetilde{\mathscr{Y}}_s(t,N)) : \mathbb{R} \to \mathscr{S}_n^N$ is named the stabilizing solution of (8.76) if the matrices $\widetilde{\mathscr{R}}_d(i) + C_0(i)\widetilde{\mathscr{Y}}_s(kh,i)C_0^{\top}(i)$ are invertible for any $(k,i) \in \mathbb{Z} \times \mathcal{N}$ and the closed-loop JSLDE

$$\mathrm{d}x(t) = A_0(\eta(t)) x(t) \mathrm{d}t, \quad kh < t \le (k+1)h \tag{8.77a}$$

$$x(kh^+) = (I_n + \widetilde{\mathscr{K}_s}(i) C_0(i)) x(kh), \quad k \in \mathbb{Z}_+ \tag{8.77b}$$

is ESMS, where

$$\widetilde{\mathscr{K}_s}(i) \triangleq -\widetilde{\mathscr{Y}}_s(0,i) C_0^{\top}(i) (\widetilde{\mathscr{R}}_d(i) + C_0(i) \widetilde{\mathscr{Y}}_s(0,i) C_0^{\top}(i))^{-1}, \quad i \in \mathcal{N}. \tag{8.78}$$

Besides (8.76) we consider the following FJMLDE with Riccati-type jumping operator

$$\dot{Y}(t,i) = A_0(i)Y(t,i) + Y(t,i)A_0^\top(i) + \sum_{j=1}^{N} q_{ji}Y(t,j) + \widetilde{\mathscr{M}}_c(i), \quad kh < t \le (k+1)h \tag{8.79a}$$

$$Y(kh^+,i) = Y(kh,i) - Y(kh,i)C_0^\top(i)(\widetilde{\widetilde{\mathscr{R}}}_d(i) + C_0(i)Y(kh,i)C_0^\top(i))^{-1} C_0(i)Y(kh,i), \quad k \in \mathbb{Z}, \tag{8.79b}$$

$i \in \mathscr{N}$, where $\widetilde{\mathscr{M}}_c(i)$ and $\widetilde{\widetilde{\mathscr{R}}}_d(i)$ are defined in (8.75). Analogously, a globally defined solution $t \to \widetilde{\widetilde{\mathbf{Y}}}_s(t) = (\widetilde{\widetilde{\mathscr{Y}}}_s(t,1), \widetilde{\widetilde{\mathscr{Y}}}_s(t,2), \dots, \widetilde{\widetilde{\mathscr{Y}}}_s(t,N)) : \mathbb{R} \to \mathscr{S}_n^N$ is named stabilizing solution of (8.79) if the matrices $\widetilde{\widetilde{\mathscr{R}}}(i) + C_0(i)\widetilde{\widetilde{\mathscr{Y}}}_s(kh,i)C_0^\top(i)$ are invertible for any $(k,i) \in \mathbb{Z} \times \mathscr{N}$ and the closed-loop JSLDE

$$\mathrm{d}x(t) = A_0(\eta(t))x(t)\mathrm{d}t, \quad kh < t \le (k+1)h \tag{8.80a}$$

$$x(kh^+) = (I_n + \widetilde{\widetilde{\mathscr{K}}}_s(\eta(kh))C_0(\eta(kh)))x(kh), \quad k \in \mathbb{Z}_+ \tag{8.80b}$$

is ESMS, where

$$\widetilde{\widetilde{\mathscr{K}}}_s(i) \triangleq -\widetilde{\widetilde{\mathscr{Y}}}_s(0,i)C_0^\top(i)(\widetilde{\widetilde{\mathscr{R}}}_d(i) + C_0(i)\widetilde{\widetilde{\mathscr{Y}}}_s(0,i)C_0^\top(i))^{-1}, \quad i \in \mathscr{N}. \tag{8.81}$$

Since (8.76) and (8.79) are particular cases of (8.25) and (8.28), respectively, the properties stated in Corollary 8.3 can be transferred in the case of the stabilizing solutions $\widetilde{\mathbf{Y}}_s(\cdot)$ and $\widetilde{\widetilde{\mathbf{Y}}}_s(\cdot)$, too.

The next results provide the state-space representation of an admissible filter which satisfies the optimality condition (8.6) or (8.7), respectively, written for the resulting system (8.71)–(8.72). Even if apparently the statements of these results seem to repeat the statements of Theorem 8.1 and Theorem 8.2, we prefer to present them in detail so that readers can compare the state-space representation of the optimal filters.

Theorem 8.4 *Assume:*

(a) the Assumptions 1.2, 7.1 and 8.3 are fulfilled;
(b) the FJMLDE with Riccati-type jumping operator (8.76) has a stabilizing and bounded on $\mathbb{R}$ solution $\widetilde{\mathbf{Y}}_s(\cdot)$ satisfying the sign conditions

$$C_0(i)\widetilde{\mathscr{Y}}_s(kh,i)C_0^\top(i) + \widetilde{\mathscr{R}}_d(i) > 0, \quad \textit{for all } (k,i) \in \mathbb{Z} \times \mathscr{N}.$$

We consider the filter $\widetilde{\mathbb{G}}_F$ having the dimension of the state space $n_F = n$ and the state-space representation

$$d\widetilde{x}_F(t) = A_0\eta(t)\widetilde{x}_F(t)dt, \quad kh < t \le (k+1)h \tag{8.82a}$$

$$\widetilde{x}_F(kh^+) = (I_n + \widetilde{\mathscr{K}}_s(\eta(kh))C_0(\eta(kh)))\widetilde{x}_F(kh) - \widetilde{\mathscr{K}}_s(\eta(kh))u(k), k \in \mathbb{Z}_+ \tag{8.82b}$$

where $\widetilde{\mathscr{K}}_s(i)$, $i \in \mathcal{N}$ are computed as in (8.78). Under the considered assumptions the filter (8.82) lies in $\mathscr{F}_{\text{adm}}(\mathbb{G})$ and satisfies the optimality condition

$$\|\mathbb{G}_{cl}(\mathbb{G}_F)\|_2^2 \geq \|\mathbb{G}_{cl}(\{\widetilde{\mathbb{G}}_F)\|_2^2 = \sum_{i=1}^{N} \frac{1}{h} \int_0^h \text{Tr}[C_z(i)\widetilde{\mathscr{Y}}_s(t,i)C_z^\top(i)]dt, \tag{8.83}$$

for all $\mathbb{G}_F \in \mathscr{F}_{\text{adm}}(\mathbb{G})$.

Proof The proof is similar to the proof of Theorem 8.1. The details are omitted.

Theorem 8.5 *Assume:*

(a) the Assumptions 1.2, 7.1 and 8.3 are fulfilled;
(b) the FJMLDE with Riccati-type jumping operator (8.79) has a bounded on $\mathbb{R}$ and stabilizing solution $\widetilde{\widetilde{\mathbf{Y}}}_s(\cdot)$ which satisfies the following sign conditions

$$C_0(i)\widetilde{\widetilde{\mathscr{Y}}}_s(kh,i)C_0^\top(i) + \widetilde{\widetilde{\mathscr{R}}}_d(i) > 0, \quad \textit{for all} \quad (k,i) \in \mathbb{Z} \times \mathcal{N}.$$

We consider the filter $\widetilde{\widetilde{\mathbb{G}}}_F$ having the dimension of the state space $n_F = n$ and its state-space representation described by

$$d\widetilde{\widetilde{x}}_F(t) = A_0\eta(t)\widetilde{\widetilde{x}}_F(t)dt, \quad kh < t \leq (k+1)h \tag{8.84a}$$

$$\widetilde{\widetilde{x}}_F(kh^+) = (I_n + \widetilde{\widetilde{\mathscr{K}}}_s(\eta(kh))C_0(\eta(kh)))\widetilde{\widetilde{x}}_F(kh) - \widetilde{\widetilde{\mathscr{K}}}_s(\eta(kh))u(k), \ k \in \mathbb{Z}_+ \tag{8.84b}$$

$$z_F(t) = C_z(\eta(t))\widetilde{\widetilde{x}}_F(t), \quad t \in \mathbb{R}_+. \tag{8.84c}$$

Under the considered assumptions the filter described by (8.84) lies in $\mathscr{F}_{\text{adm}}(\mathbb{G})$ and satisfies the optimality condition

$$\widetilde{\|}\mathbb{G}_{cl}(\mathbb{G}_F)\widetilde{\|}_2^2 \geq \widetilde{\|}\mathbb{G}_{cl}(\widetilde{\widetilde{\mathbb{G}}}_F)\widetilde{\|}_2^2 = \sum_{i=1}^{N} \frac{1}{h} \int_0^h \text{Tr}[C_z(i)\widetilde{\widetilde{\mathscr{Y}}}_s(t,i)C_z^\top(i)]dt,$$

for any filter $\mathbb{G}_F$ from $\mathscr{F}_{\text{adm}}(\mathbb{G})$.

In both (8.83) and (8.84), $\mathbb{G}_{cl}(\mathbb{G}_F)$ denotes the resulting system of the form (8.71)–(8.72) obtained by coupling a filter $\mathbb{G}_F$ of the form (8.2) satisfying the stability condition (8.3), to the dynamical system $\mathbb{G}$ described by (8.67).

Example 8.1 To illustrate the proposed filtering strategy, we consider a problem of field monitoring (see Liu et al. [2], Mo et al. [3], Zhang et al. [4] and the reference therein) that will be adapted to our setting. In this problem, sensors are distributed on a rectangular region in order to estimate the state of a diffusion process described by the following partial differential equation:

$$\frac{\partial \xi(s,t)}{\partial t} = \nabla^2 \xi(s,t) \tag{8.85}$$

with Dirichlet boundary conditions:

$$\xi(s,\cdot) = 0 \quad s \in \partial \mathscr{D} \tag{8.86}$$

where $\xi(s,t)$ is the state value at location s and time t, ∇^2 denotes the Laplace operator, and $\partial \mathscr{D}$ denotes the boundary of the rectangular region of interest $\mathscr{D}$.
We consider a spatially discretized approximation of (8.85) (with boundary conditions (8.86)) which can be simply generated by setting Liu et al. Liu et al. [2], Mo et al. [3], Zhang et al. [4]:

$$\begin{aligned} \nabla^2 \xi(s,t)\big|_{s=(i,j)} \approx & \frac{\xi(i+1,j,t) - 2\xi(i,j,t) + \xi(i-1,j,t)}{d^2} \\ & + \frac{\xi(i,j+1,t) - 2\xi(i,j,t) + \xi(i,j-1,t)}{d^2} \end{aligned} \tag{8.87}$$

where $i = 0, 1, \cdots, l_h$, $j = 0, 1, \cdots, l_v$, $l_h + 2$ and $l_v + 2$ are the width and length of the considered rectangular region, d is the physical distance between the lattice points and $\xi(-1, j, t) = \xi(l_h + 1, j, t) = \xi(i, -1, t) = \xi(i, l_v + 1, t) = 0$ for all position indices i, j and time t. From (8.85)–(8.87), one can easily deduce the evolution equations:

$$\frac{\mathrm{d}}{\mathrm{d}t} x(t) = A_0 x(t)$$

where $x(t) \in \mathbb{R}^{n_x}$, $n_x = (l_h + 1) \times (l_v + 1)$, denotes the state vector $x(t) = \left[\, \xi(0,0,t)\ \xi(0,1,t) \cdots \xi(l_h, l_v, t) \,\right]^T$ and A_0 can be directly computed from (8.87). In order to fit in our setting, we introduce a process noise and a state-dependent noise yielding:

$$\mathrm{d}x(t) = A_0 x(t)\mathrm{d}t + A_1 x(t)\mathrm{d}w(t) + G_c d\zeta_c(t)$$

where A_1 and G_c have been randomly generated. We assume that M sensors, $M < n_x$, are deployed to make measurements of the state. In our example, we choose $l_h = l_v = 4$ (hence $n_x = 25$) and $M = 10$. The location of the sensors as well as the state variables in the rectangular field are illustrated by Fig. 8.1.

We consider that the sensors and the filter communicate over a communication channel. We assume that the communication channel induces some communication constraints. More specifically, the constraint we deal with here is referred to in the literature as a *medium access constraint*. In this case, the shared network can only accommodate a limited number of simultaneous communications between components. In this context, it is only meaningful to specify a filter in conjunction with a communication policy which indicates the times at which the plant's sensors are to be granted medium access. This communication policy is known in the literature as

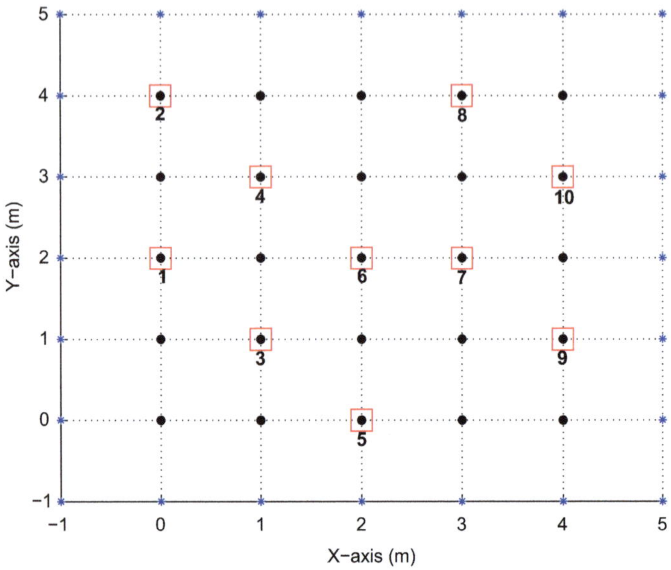

Fig. 8.1 $M = 10$ sensors deployed in a 6×6 region: $*$-Boundary points, •-Estimated points, □-Sensor locations

communication sequence. The communication sequence specifies which sensors are able to send information to the filter at each time step. To meet this requirement, we use a stochastic scheduling protocol. Such a protocol can be described as follows. Suppose that there are N sensor nodes which are indexed from 1 to N, each node contains r sensors. Due to communication limitation, at each sampling time, only one sensor node is allowed to send the measurements to the filter. The switching between sensor nodes is governed by a time-homogeneous Markovian process $\eta(t) \in \mathcal{N} = \{1, 2, \cdots, N\}$.

In our example, we consider $N = 4$. The different sensor nodes are given as follows:

$$\mathcal{N} = \left\{ \begin{pmatrix} 1 \\ 2 \\ 3 \\ 4 \end{pmatrix}, \begin{pmatrix} 3 \\ 4 \\ 5 \\ 6 \end{pmatrix}, \begin{pmatrix} 5 \\ 6 \\ 7 \\ 8 \end{pmatrix}, \begin{pmatrix} 7 \\ 8 \\ 9 \\ 10 \end{pmatrix} \right\}$$

where $\begin{pmatrix} i \\ j \\ k \\ l \end{pmatrix}$ represents the node containing the corresponding sensors i, j, k, l in Fig. 8.1. The measurement equation is then given by:

$$y(kh) = C_{d0}(\eta(kh))x(kh)$$

where $h > 0$ ($h = 0.1$ in our simulation example) is the sampling period. In order to fit in with our setting, we introduce a measurement noise and a state-dependent noise leading to:

$$y(kh) = [C_{d0}(\eta(kh)) + C_{d1}(\eta(kh))w_d(k)]x(kh) + D_d(\eta(kh))\zeta_d(k).$$

The matrices $C_1(i)$ and $D_d(i)$, $0 \leq i \leq N$, are generated randomly.

For the first experiment, the sensor scheduling protocol is defined via the following transition rate matrix:

$$Q^1 = \begin{bmatrix} -3 & 1 & 1 & 1 \\ 1 & -3 & 1 & 1 \\ 1 & 1 & -3 & 1 \\ 1 & 1 & 1 & -3 \end{bmatrix}.$$

In this case, we have: $\pi_{i\infty} = 0.25, i = 1, \cdots, N$. This corresponds to a protocol that doesn't affect any priority to the sensor nodes. A realization of the corresponding Markov process is given by Fig. 8.2. In this case, the optimal H_2 norm is

$$\|\mathbb{G}_{cl}(\widetilde{\mathbb{G}}_F)\|_2 = 76.6793.$$

In a second experiment, we choose the following transition rate matrix:

$$Q^2 = \begin{bmatrix} -7 & 1 & 1 & 5 \\ 5 & -11 & 1 & 5 \\ 5 & 1 & -11 & 5 \\ 5 & 1 & 1 & -7 \end{bmatrix}$$

for which one has:

$$\begin{cases} \pi_{1\infty} = 0.4167; \\ \pi_{2\infty} = 0.0833; \\ \pi_{3\infty} = 0.0833; \\ \pi_{4\infty} = 0.4167. \end{cases}$$

Here, a priority is affected to nodes 1 and 4, respectively. A realization of the Markov process is given by Fig. 8.3. The optimal H_2 norm in this case is

$$\|\mathbb{G}_{cl}(\widetilde{\mathbb{G}}_F)\|_2 = 73.6763.$$

This result suggests the existence of an optimal scheduling protocol which is different from the *uniformly distributed* one. Note that computing the optimal scheduling protocol is beyond the problem we are addressing here.

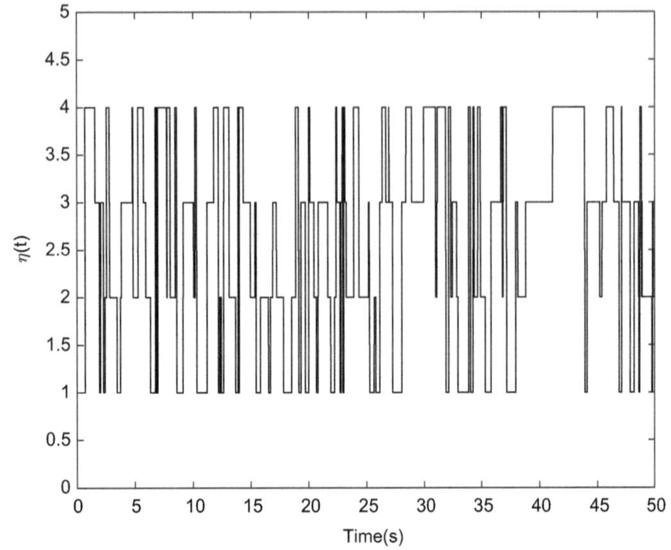

Fig. 8.2 Realization of $\eta(t)$ with transition rate matrix Π^1

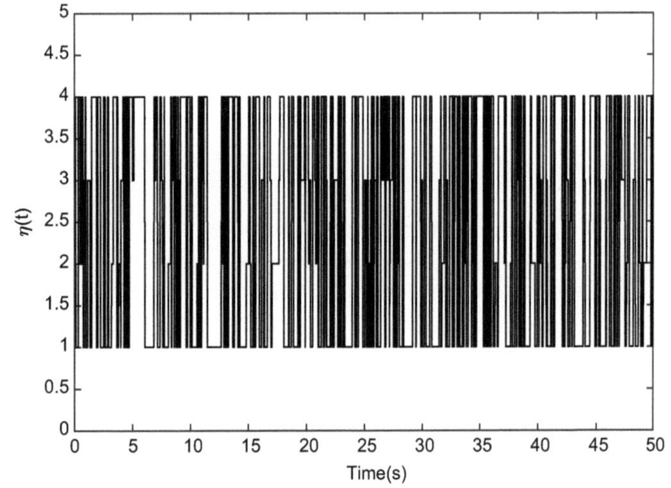

Fig. 8.3 Realization of $\eta(t)$ with transition rate matrix Π^2

Figure 8.4 shows the evolution of the states of the system and there estimates (due to the large number of state variables, we have only showed here the evolution of some components).

In the last part of this section, let us consider the particular case of a dynamical system of the form (8.67) whose coefficients are not affected by Markovian switching. In this case, (8.67) becomes:

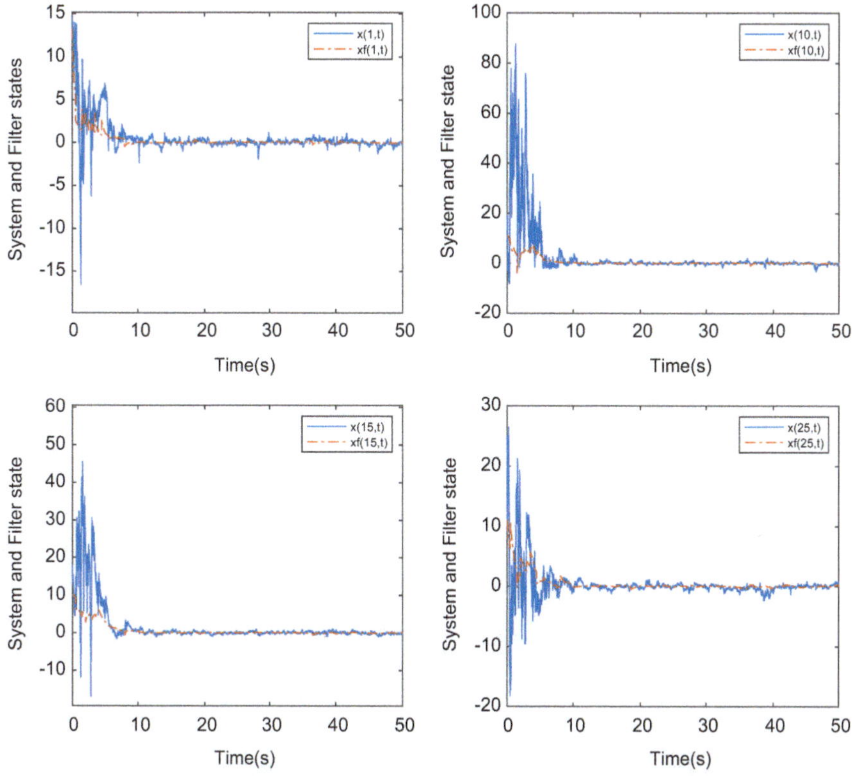

Fig. 8.4 States of the system and there estimates

$$\mathrm{d}x(t) = A_0 x(t)\mathrm{d}t + \sum_{l=1}^{r_1} A_l x(t)\mathrm{d}w_l(t) + G_c d\xi(t), \tag{8.88a}$$

$$z(t) = C_z x(t), \quad t \in \mathbb{R}_+ \tag{8.88b}$$

$$y(kh) = (C_0 + \sum_{p=1}^{r_2} w_{dp}(k)C_p)x(kh) + D\zeta_d(k), \quad k \in \mathbb{Z}_+ \tag{8.88c}$$

with $x(t)$, $z(t)$ and $y(kh)$ having the same meaning as those in (8.67) and $\{w_l(t)\}_{t\geq 0}$, $1 \leq l \leq r_1$, $\{w_{dp}(k)\}_{k\in\mathbb{Z}_+}$, $1 \leq p \leq r_2$ are stochastic processes which are satisfying the Assumption 1.1, while $\{\xi(t)\}_{t\geq 0}$, $\{\zeta_d(k)\}_{k\in\mathbb{Z}_+}$ are stochastic processes satisfying the Assumption 7.1.

Our goal is to find the state-space representation of the filter $\widetilde{\mathbb{G}}_F$ of the form (8.49) subject to the stability condition (8.50), which activated at its inputs by the measurements $y(kh)$, $k = 0, 1, \ldots, \left[\frac{t}{h}\right]$ to generate the best estimation of the remote signal $z(t)$, $t \in \mathbb{R}_+$.

The next developments are done under the following assumption:

Assumption 8.4 The stochastic linear differential equation:

$$\mathrm{d}x(t) = A_0 x(t)\mathrm{d}t + \sum_{l=1}^{r_1} A_l x(t)\mathrm{d}w_l(t), \quad t \in \mathbb{R}_+,$$

is ESMS.

Under the Assumption 8.4 the non-homogeneous algebraic Lyapunov-type equation of the stochastic control

$$A_0 Z + Z A_0^\top + \sum_{l=1}^{r_1} A_l Z A_l^\top + G_c \mathbf{\Lambda}_c G_c^\top = 0 \tag{8.89}$$

has a unique solution $\widetilde{Z} \in \mathcal{S}_n^{N+}$.

The FJMLDEs with Riccati-type jumping operators (8.76) and (8.79) reduce to

$$\dot{Y}(t) = A_0 Y(t) + Y(t) A_0^\top + \widetilde{M}_c, \quad kh < t \le (k+1)h \tag{8.90a}$$

$$Y(kh^+) = Y(kh) - Y(kh) C_0^\top \left(\widetilde{R}_d + C_0 Y(kh) C_0^\top\right)^{-1} C_0 Y(kh), \quad k \in \mathbb{Z} \tag{8.90b}$$

and

$$\dot{Y}(t) = A_0 Y(t) + Y(t) A_0^\top + \widetilde{\widetilde{M}}_c, \quad kh < t \le (k+1)h \tag{8.91a}$$

$$Y(kh^+) = Y(kh) - Y(kh) C_0^\top \left(\widetilde{\widetilde{R}}_d + C_0 Y(kh) C_0^\top\right)^{-1} C_0 Y(kh), \quad k \in \mathbb{Z} \tag{8.91b}$$

where $\widetilde{M}_c = \widetilde{\widetilde{M}}_c \triangleq G_c \mathbf{\Lambda}_c G_c^\top + \sum_{l=1}^{r_1} A_l \widetilde{Z} A_l^\top, \quad \widetilde{R}_d \triangleq h D \mathbf{\Lambda}_d D^\top + \sum_{p=1}^{r_2} C_p \widetilde{Z} C_p^\top,$

$$\widetilde{\widetilde{R}}_d = D \mathbf{\Lambda}_d D^\top + \sum_{p=1}^{r_2} C_p \widetilde{Z} C_p^\top.$$

Let $t \to \widetilde{Y}_s(t) : \mathbb{R} \to \mathcal{S}_n$ and $t \to \widetilde{\widetilde{Y}}_s(t) : \mathbb{R} \to \mathcal{S}_n$ be the bounded on $\mathbb{R}$ and stabilizing solution of (8.90) and (8.91), respectively.

This means that the following closed-loop jump linear differential equations

$$\dot{x}(t) = A_0 x(t), \quad kh < t \le (k+1)h \tag{8.92a}$$

$$x(kh^+) = (I_n + \widetilde{K}_s C_0) x(kh), \quad k \in \mathbb{Z}_+ \tag{8.92b}$$

and

$$\dot{x}(t) = A_0 x(t), \quad kh < t \le (k+1)h \tag{8.93a}$$

$$x(kh^+) = (I_n + \widetilde{\widetilde{K}}_s C_0)x(kh), \quad k \in \mathbb{Z}_+, \tag{8.93b}$$

respectively, are exponentially stable, where

$$\widetilde{K}_s \triangleq -\widetilde{Y}_s(0)C_0^\top (C_0\widetilde{Y}_s(0)C_0^\top + \widetilde{R}_d)^{-1} \tag{8.94}$$

and

$$\widetilde{\widetilde{K}}_s \triangleq -\widetilde{\widetilde{Y}}_s(0)C_0^\top (C_0\widetilde{\widetilde{Y}}_s(0)C_0^\top + \widetilde{\widetilde{R}}_d)^{-1}, \tag{8.95}$$

respectively.

If $\mathbb{G}_{cl}(\mathbb{G}_F)$ is the resulting system obtained by coupling an arbitrary filter $\mathbb{G}_F$ of the form (8.49) satisfying the stability condition (8.50) to the dynamical system (8.88) we want to obtain the state-space representation of a filter $\widetilde{\mathbb{G}}_F$ which minimize the values of the norms $\|\mathbb{G}_{cl}(\mathbb{G}_F)\|_2$ and $\widetilde{\|}\mathbb{G}_{cl}(\mathbb{G}_F)\widetilde{\|}_2$ with respect to all admissible filters.

Theorem 8.6 *Assume:*

(a) the Assumptions 1.1, 7.1 and 8.4 are fulfilled;
(b) the FJMLDE with Riccati-type jumping operator (8.90) has a bounded on $\mathbb{R}$ and stabilizing solution $\widetilde{Y}_s(\cdot)$ satisfying the sign conditions

$$C_0\widetilde{Y}_s(kh)C_0^\top + \widetilde{R}_d > 0, \quad \text{for all } \; k \in \mathbb{Z}.$$

We consider the filter $\widetilde{\mathbb{G}}_F$ having the state-space representation described by

$$\dot{\widetilde{x}}_F(t) = A_0\widetilde{x}_F(t), \quad kh < t \le (k+1)h \tag{8.96a}$$

$$\widetilde{x}_F(kh^+) = \left(I_n + \widetilde{K}_s C_0\right)\widetilde{x}_F(kh) - \widetilde{K}_s u(k), \quad k \in \mathbb{Z}_+ \tag{8.96b}$$

$$z_F(t) = C_z\widetilde{x}_F(t), \quad t \in \mathbb{R}_+. \tag{8.96c}$$

Under the considered assumptions the filter $\widetilde{\mathbb{G}}_F$ described by (8.96) is admissible and satisfies the optimality condition

$$\|\mathbb{G}_{cl}(\mathbb{G}_F)\|_2^2 \ge \|\mathbb{G}_{cl}(\widetilde{\mathbb{G}}_F)\|_2^2 = \frac{1}{h}\int_0^h \mathrm{Tr}[C_z\widetilde{Y}_s(t)C_z^\top]dt, \tag{8.97}$$

for any admissible filter $\mathbb{G}_F$ of the form (8.49) satisfying the stability condition (8.50). In (8.96) and (8.97), $\widetilde{Y}_s(\cdot)$ is the bounded on $\mathbb{R}$ and stabilizing solution of (8.90) and $\widetilde{K}_s$ is the gain matrix associated via (8.94).

Theorem 8.7 *Assume:*

(a) the Assumptions 1.1, 7.1 and 8.4 are fulfilled;

(b) the FJMLDE with Riccati-type jumping operator (8.91) has a bounded on $\mathbb{R}$ *and stabilizing solution* $\widetilde{\widetilde{Y}}_s(\cdot)$ *satisfying the sign conditions*

$$C_0\widetilde{\widetilde{Y}}_s(kh)C_0^{\top} + \widetilde{\widetilde{R}}_d > 0, \quad \textit{for all} \;\; k \in \mathbb{Z}.$$

We consider the filter $\widetilde{\widetilde{\mathbb{G}}}_F$ *having the state-space representation described by*

$$\dot{\widetilde{\widetilde{x}}}_F(t) = A_0\widetilde{\widetilde{x}}_F(t), \quad kh < t \leq (k+1)h \tag{8.98a}$$

$$\widetilde{\widetilde{x}}_F(kh^+) = \left(I_n + \widetilde{\widetilde{K}}_s C_0\right)\widetilde{\widetilde{x}}_F(kh) - \widetilde{\widetilde{K}}_s u(k), \quad k \in \mathbb{Z}_+ \tag{8.98b}$$

$$z_F(t) = C_z\widetilde{\widetilde{x}}_F(t), \quad t \in \mathbb{R}_+. \tag{8.98c}$$

Under the considered assumptions the filter (8.98) is admissible and satisfies the optimality condition

$$\widetilde{\|}\mathbb{G}_{cl}(\mathbb{G}_F)\widetilde{\|}_2^2 \geq \widetilde{\|}\mathbb{G}_{cl}(\widetilde{\widetilde{\mathbb{G}}}_F)\widetilde{\|}_2^2 = \frac{1}{h}\int_0^h \mathrm{Tr}[C_z\widetilde{\widetilde{Y}}_s(t)C_z^{\top}]dt, \tag{8.99}$$

for all admissible filter $\mathbb{G}_F$*. In (8.98) the gain matrix* $\widetilde{\widetilde{K}}_s$ *is defined by (8.95).*

At the end of this section, let us remark that the result stated in Theorem 8.7 contains as particular cases the main results from [5, 6] when $\theta_c = h$ and $\theta_d = 1$.

8.4 Notes and References

The problem of filtering is fundamental in control theory and signal processing, and several approaches have been developed in the literature. The famous results of Kalman and Bucy derived in the 1960s (see Kalman [7], Kalman and Bucy [8]) and their successful implementation in many applications including aerospace, signal processing, geophysics, etc., strongly stimulated the research in this area. A comprehensive survey of linear filtering and estimation can be found in Kalman and Bucy [9].

For the class of MJLSs, the problem of $\mathscr{H}_2$ filtering has been addressed for example in Fioravanti et al. [10] via an LMI setting. In Costa and Tuesta [11], the authors studied the so-called separation principle for the $\mathscr{H}_2$ control and filtering, where the optimal $\mathscr{H}_2$ filter is obtained through coupled Riccati equations. An important issue concerning the filtering performance is the robustness with respect to the modeling uncertainty of the system which generates the signal to be estimated. In the context of MJLSs, great efforts have been devoted from one part to the case of uncertainties affecting the transition probabilities of the underlying Markov process mainly in the discrete-time setting where several modeling frameworks have been adopted in the literature: we can cite here the polytopic setting Goncalves et al. [12], the interval (bounded transition probabilities) case Goncalves et al. [13] and the situation where there are completely unknown parts of the transition probability matrix Zhang and Boukas [14]. The multisimplex setting, a generic representation of the uncertain probabilities (that includes the bounded, polytopic and partly unknown approaches as particular cases) has been also proposed in Li et al. [15], Morais et al. [16] for the discrete-time case and in Morais et al. [17], Li et al. [15] for the continuous-time case. All the aforementioned works assume that the Markov state is available to the filter, an assumption that cannot always be fulfilled. Hence another part of robust filtering theory for MJLSs has been devoted to the issue of partial observation of the Markov process mode. The cluster case, introduced in do Val et al. [18], considers that the Markov modes can be grouped in disjoint sets called clusters so that the controller would know the cluster in which the Markov chain is currently operating. This result was extended to the $\mathscr{H}_2$ and $\mathscr{H}_\infty$ filtering problem in Goncalves et al. [12, 19]. For the case where there is no information at all regarding the underlying Markov chain, the so-called mode-independent case, we can cite Fioravanti et al. [10], Goncalves et al. [12, 19]. More recently, Costa et al. [20] introduced the so-called detector approach, or Hidden Markov model approach. In this particular model, it is assumed that the Markov chain is hidden and there is only an estimation available. Application of this setting in the filtering problem has been reported in de Oliveira and Costa [21].

For the class of Itô stochastic systems, one the robustness issues that received much attention recently is related to the problem of filtering with the system parameters being subject to random perturbations, requiring stochastic models with state-dependent noise (or multiplicative noise). One can see for example Costa and Kubrusly [22], Mangoubi [23], Gershon et al. [24] and the reference therein. It is known that in the case of linear stochastic systems with multiplicative white noise perturbations the state-space representation of the analogous of Kalman filter is hardly implementable because it incorporates the multiplicative white noise perturbations which affect the given plant. That is why, for $\mathscr{H}_2$-type control problems and H_2-type filtering problems for controlled systems with multiplicative white noise perturbations only suboptimal solutions were derived based on solvability of some systems of LMIs (see Gershon et al. [24]). Recently, the problem of $\mathscr{H}_2$ filtering for linear Itô-type systems with state-dependent noise in both the state and the output equations has been addressed in Dragan and Aberkane [25]. The authors succeed in obtaining the optimal solution to this problem by using a Riccati-type approach.

For stochastic impulsive systems, we cite here Jazwinski [26] where a filtering problem has been addressed in the context of sampled-data systems. A Kalman-type impulsive filter has been proposed where the impulsive correction gain of the filter is obtained by integrating a continuous-time Riccati equation with finite jumps. Note however that in Jazwinski [26], the author does not consider the case with state-dependent noise affecting both the state-space equation and/or the measurement equation. Hence, a direct extension of the results in Jazwinski [26] to the case of stochastic systems with state-dependent noise will lead to a filter which is hardly implementable because it incorporates the multiplicative white noise perturbations which affect the given plant. In Xu and Chen [27], the authors addressed the filtering problem with only multiplicative noise in the state equation. They obtained a suboptimal solution via a LMI formalism. The problem with multiplicative noise in both the state and the measurement equations has been recently solved in Dragan et al. [28] where the authors succeeded to obtain the optimal solution by using the theory of generalized Riccati equations with finite jumps.

For the class of MJLSs with impulsive effects, Dong et al. [29] addressed the problem of $\mathscr{H}_\infty$-filtering in a sampled-data setting. We recall here the discussion from Chapter 3 about the crucial issue of describing the behavior of the Markovian process in the neighbor of the impulse times. In Dong et al. [29], the authors overlooked this issue by imposing that the matrices involved in the impulsive part of the dynamics are deterministic and do not depend on the Markov jump process. They also obtained a suboptimal solution in terms of a LMI optimization problem.

References

1. Drăgan, V., Morozan, T., Stoica, A.M.: Mathematical Methods in Robust Control of Linear Stochastic Systems. Springer-Verlag, New York (2013)
2. Liu, S., Fardad, M., Masazade, E., Varshney, P.: Optimal periodic sensor scheduling in large-scale dynamical networks. IEEE Trans. Signal Process. **34**, 3055–3068 (2014)
3. Mo, Y., Ambrosino, R., Sinopoli, B.: Sensor selection strategies for state estimation in energy constrained wireless sensor networks. Automatica J. IFAC **47**, 1330–1338 (2011)
4. Zhang, H., Moura, J., Krogh, B.: Dynamic field estimation using wireless sensor networks: tradeoffs between estimation error and communication cost. IEEE Trans. Signal Process. **57**, 1330–1338 (2009)
5. Drăgan, V.: AS,: Optimal h_2 filtering for a class of linear stochastic systems with sampling. Automatica **18**, 2494–2501 (2012)
6. Drăgan, V., Aberkane, I.L.P.S.: Optimal h2 filtering for periodic linear stochastic systems with multiplicative white noise perturbations and sampled measurements. J. FRANKLIN I **352**(12), 5985–6010 (2016)
7. Kalman, R.: A new approach to linear filtering and prediction problems. ASME Trans. D **82**, 34–45 (1960)
8. Kalman, R., Bucy, R.: New results in linear filtering and prediction theory. ASME Trans. D **83**, 95–108 (1961)
9. Kalman, R., Bucy, R.: A view of three decades of linear filtering theory. IEEE Trans. Inform. Theory **20**, 146–181 (1974)
10. Fioravanti, A.R., Gonçalves, A.P.C., Geromel, J.C.: h_2 filtering of discrete-time Markov jump linear systems through linear matrix inequalities. Int. J. Cont. **81**, 1221–1231 (2008)
11. Costa, O.L.V., Tuesta, E.F.: h_2 control and the separation principle for discrete-time markovian jump linear systems. Math. Control, Signals Syst. **16**, 320–350 (2004)
12. Goncalves, A.P.C., Fioravanti, A.R., Geromel, J.C.: Filtering of discrete-time Markov jump linear systems with uncertain transition probabilities. Int. J. Robust Nonblinear Control **21**, 613–624 (2011)
13. Goncalves, A.P.C., Fioravanti, A.R., Geromel, J.C.: h_∞ robust and networked control of discrete-time mjls through lmis. J. Franklin Inst. **349**, 2171–2181 (2012)
14. Zhang, L., Boukas, E.K.: Mode-dependent filtering for discrete-time Markovian jump linear systems with partly unknown transition probabilities. Automatica **46**, 1462–1467 (2009)
15. Li, X.: b JL, Gao H, Xiong J,: h_∞ and h_2 filtering for linear systems with uncertain Markov transitions. Automatica **67**, 252–266 (2016)

16. Morais, C.F., Palma, J.M., Peres, P.L., Oliveira, R.C.: An lmi approach for h_2 and h_∞ reduced-order filtering of uncertain discrete-time Markov and Bernoulli jump linear systems. Automatica **69**, 463–471 (2018)
17. Morais, C.F., Braga, M.F., Lacerda, M.J., Oliveira, R.C.L.F., Peres, P.L.D.: h_2 and h_∞ filter design for polytopic continuous-time Markov jump linear systems with uncertain transition rates. Int. J. Adaptive Control Sig. Proc. **16**, 320–350 (2015)
18. do Val, J.B., Geromel, J.C., Goncalves, A.P.C.: The h_2 control for jump linear systems: Cluster observations of the Markov state. Automatica **38**, 343–349 (2002)
19. Goncalves, A.P.C., Fioravanti, A.R., Geromel, J.C.: Markov jump linear systems and filtering through network transmitted measurements. Signal Proc. **90**, 2842–2850 (2010)
20. Costa, O.L.V., Fragoso, M.D., Todorov, M.G.: A detector-based approach for the h_2 control of Markov jump linear systems with partial information. IEEE Trans. Autom. Control **60**, 1219–1234 (2015)
21. de Oliveira, A.M., Costa, O.L.V.: Mixed h_2/h_∞ filtering for Markov jump linear systems. Int. J. Syst. Sci. **49**, 3023–3036 (2018)
22. Costa, O., Kubrusly, C.: State-feedback h_∞-control for discrete-time infinite-dimensional stochastic bilinear systems. J. Math. Sys. Estim. Control **6**, 1–32 (1996)
23. Mangoubi, R.: Robust Estimation and Failure Detection. Springer (1998)
24. Gershon, E., Shaked, U., Yaesh, I.: H_∞ Control and Estimation of State- Multiplicative Linear Systems. Lecture Notes in Control and Information Sciences, vol. 318. Springer-Verlag, London Limited (2005)
25. Dragan, V., Aberkane, S.: h_2 optimal filtering for continuous-time periodic linear stochastic systems with state-dependent noise. Syst. Control Lett. **66**, 35–42 (2014)
26. Jazwinski, A.: Stochastic processes and filtering theory. Math. Sci, Eng (1970)
27. Xu, S., Chen, T.: Robust h_∞ filtering for uncertain impulsive stochastic systems under sampled measurements. Automatica **39**, 509–516 (2003)
28. Dragan, V., Aberkane, S., Popa, I.L.: Optimal h_2 filtering for periodic linear stochastic systems with multiplicative white noise perturbations and sampled measurements. J. Franklin Inst. **352**, 5985–6010 (2015)
29. Dong, Y., Sun, J., Wu, Q.: h_∞ filtering for a class of stochastic Markovian jump systems with impulsive effects. Int. J. Robust Nonlinear Control **18**, 1–13 (2008)

Chapter 9
Robust Stability and Robust Stabilization of Jump Linear Stochastic Systems Under a Class of Stochastic Uncertainties

Abstract This chapter is devoted to a the robust stability analysis, as well as to the problem of the robust stabilization of the class of impulsive Markovian jump linear systems subject to block-diagonal stochastic parameter perturbations. To the best of the authors knowledge, robust stabilization for linear impulsive stochastic systems has not been considered yet in the literature. In Sect. 9.1, a detailed description of the model of the system as well as the model of the considered class of structured stochastic uncertainties are given. The considered parametric uncertainties are of multiplicative white noise type with unknown intensity. In order to measure the level of the robustness of the stability of the considered class of impulsive systems with respect to the stochastic structured uncertainties we introduce in Sect. 9.2 the concept of stability radius. In order to effectively address the multiperturbations case, we use scaling techniques; hence in Sect. 9.3 we begin by a detailed description of such techniques when applied to our setting. By using these scaling techniques, we then succeed to obtain an estimation of the lower bound of the stability radius. A first characterization of a lower bound of the stability radius is obtained in terms of the unique bounded and positive semidefinite solutions of an adequately defined parameterized backward Lyapunov-type jump linear differential equations (Theorems 9.1 and 9.2). A second characterization is given in terms of the existence of positive definite solutions of adequately defined parameterized backward Lyapunov-type jump differential inequalities (Theorem 9.3). This second result is then exploited in order to solve a robust control synthesis problem in Sect. 9.4. A first answer to the robust control problem is obtained in Theorem 9.5 where the stabilizing feedback gains are characterized via the solution of an adequately defined nonlinear backward jumps matrix differential inequality. The nonlinear nature of the conditions obtained in Theorem 9.5 makes there numerical exploitation not tractable. Hence, a second answer (numerically tractable) to the robust control problem is proposed in Corollary 9.3 where the stabilizing feedback gains are characterized via the unique stabilizing solution of an adequately defined backward jumps matrix Lyapunov differential equation with Riccati-type jumping operator. We finally apply the obtained results to the case of sampled-data linear stochastic systems (Sect. 9.5).

V. Drăgan et al., *Robust Control of Jump Linear Stochastic Systems*, Lecture Notes in Control and Information Sciences 497, https://doi.org/10.1007/978-3-031-84070-8_9

9.1 The Problem Setting

9.1.1 The Model Description

We consider the dynamic system $\mathbf{G}_\Delta$ having the state-space representation described by the jump stochastic differential equations

$$dx(t) = A_0(\eta(t))x(t)dt + \sum_{\iota=1}^{r_1} A_\iota(\eta(t))x(t)dw_\iota(t)$$

$$+\sum_{l=1}^{r_2} D_{c,l}(\eta(t))\Delta_{c,l}(t, z_{c,l}(t), \eta(t)), \quad kh < t \le (k+1)h \tag{9.1a}$$

$$x(kh^+) = (A_{d0}(\eta(kh))x(kh) + \sum_{q=1}^{r_3} A_{dq}(\eta(kh))w_{dq}(k))x(kh)$$

$$+\sum_{p=1}^{r_4} D_{d,p}(\eta(kh))\Delta_{d,p}(k, z_{d,p}(kh), \eta(kh))\zeta(k), \quad k \in \mathbb{Z}_+ \tag{9.1b}$$

$$z_{c,l}(t) = C_{c,l}(\eta(t))x(t), \quad 1 \le l \le r_2, \quad t \in \mathbb{R}_+ \tag{9.1c}$$

$$z_{d,p}(kh) = C_{d,p}(\eta(kh))x(kh), \quad 1 \le p \le r_4, \quad k \in \mathbb{Z}_+ \tag{9.1d}$$

where $x(t) \in \mathbb{R}^n$ is the system state vector. The stochastic processes involved in (9.1) are defined on the probability space $(\Omega, \mathfrak{F}, \mathscr{P})$ and have the properties included in the following assumption:

Assumption 9.1

(a) $\{w(t)\}_{t\in\mathbb{R}_+}$, $(w(t) = (w_1(t)\ w_2(t)\ \dots w_{r_1}(t))^\top)$ is a r_1-dimensional standard Wiener process;

(b) $\{w_d(k)\}_{k\in\mathbb{Z}_+}$ $(w_d(k) = (w_{d1}(k)\ w_{d2}(k)\ \dots\ wdr_3(k))$ is a sequence of r_3-dimensional independent random vectors with zero mean and covariance $\mathbb{E}[w_d(k)w_d^\top(k)] = I_{r_3}$, $k \in \mathbb{Z}_+$;

(c) $\{\xi(t)\}_{t\in\mathbb{R}_+}$ $(\xi(t) = (\xi_1(t)\ \xi_2(t)\ \dots \xi_{r_2}(t))^\top)$ is a r_2-dimensional Wiener process with zero mean and $\mathbb{E}[(\xi_{l_1}(t) - \xi_{l_1}(s))(\xi_{l_2}(t) - \xi_{l_2}(s))] = \lambda_{c,l_1}^2 \delta_{l_1 l_2}(t-s)$, for all $t \ge s \ge 0$, $l_1, l_2 \in \{1, 2, \dots, r_2\}$ where $\lambda_{c,l} > 0$ for all $1 \le l \le r_2$ and

$$\delta_{l_1 l_2} = \begin{cases} 1 & \text{if } l_1 = l_2 \\ 0 & \text{if } l_1 \ne l_2; \end{cases}$$

(d) $\{\zeta(k)\}_{k\in\mathbb{Z}_+}$ $(\zeta(k) = (\zeta_1(k),\ \zeta_2(k)\ \dots\ \zeta_{r_4}(k))^\top)$ is a sequence of r_4–dimensional independent vectors with zero mean and $\mathbb{E}[\zeta_{p_1}(k)\zeta_{p_2}(k)] = \lambda_{d,p_1}^2 \delta_{p_1 p_2}$, for all $p_1, p_2 \in \{1, 2, \dots, r_4\}$ where $\lambda_{d,p} > 0$, $1 \le p \le r_4$ and $\delta_{p_1 p_2}$ are as in (c);

(e) $\{\eta(t)\}_{t\in\mathbb{R}_+}$ is a standard homogeneous Markov process taking values in the finite set $\mathcal{N} = \{1, 2, \dots, N\}$ and has the transition semigroup $P(t) = e^{Qt}$, for all

$t > 0$ where $Q \in \mathbb{R}^{N\times N}$ is a matrix whose elements q_{ij} satisfy conditions as (1.22);

(f) $\{w(t)\}_{t\in\mathbb{R}_+}$, $\{\eta(t)\}_{t\in\mathbb{R}_+}$, $\{\xi(t)\}_{t\in\mathbb{R}_+}$, $\{w_d(k)\}_{k\in\mathbb{Z}_+}$, $\{\zeta(k)\}_{k\in\mathbb{Z}_+}$ are independent stochastic processes.

In (9.1), $A_\iota(i) \in \mathbb{R}^{n\times n}, 0 \le \iota \le r_1$, $A_{dq}(i) \in \mathbb{R}^{n\times n}, 0 \le q \le r_3$, $D_{c,l}(i) \in \mathbb{R}^{n\times m_l}$, $C_{c,l}(i) \in \mathbb{R}^{\nu_{m_l}\times n}$, $1 \le l \le r_2$, $D_{d,p}(i) \in \mathbb{R}^{n\times m_{d,p}}$, $C_{d,p}(i) \in \mathbb{R}^{\nu_{d,p}\times m}$, $1 \le p \le r_4$, $i \in \mathcal{N}$, are given matrices whose values are known, while the $(t, z_{c,l}) \to \Delta_{c,l}(t, z_{c,l}, i) : \mathbb{R}\times\mathbb{R}^{\nu_l} \to \mathbb{R}^{m_l}$, $1 \le l \le r_2$, $z_{d,p} \to \Delta_{d,p}(k, z_{d,p}, i) : \mathbb{R}^{\nu_{d,p}} \to \mathbb{R}^{m_{d,p}}$, $1 \le p \le r_4$, $k \in \mathbb{Z}_+$, $i \in \mathcal{N}$, are arbitrary unknown functions which satisfy the assumption:

Assumption 9.2

(a) for all $i \in \mathcal{N}$, $0 \le l \le r_2$, $(t, z_{c,l}) \to \Delta_{c,l}(t, z_{c,l}, i) : \mathbb{R}_+ \times \mathbb{R}^{\nu_l} \to \mathbb{R}^{m_l}$ are measurable functions having the following additional properties:

(i)
$$\Delta_{c,l}(t, 0, i) = 0 \quad \text{for all} \quad (t, i) \in \mathbb{R}_+ \times \mathcal{N}; \tag{9.2}$$

(ii)
$$|\Delta_{c,l}(t, \widehat{z}_{c,l}, i) - \Delta_{c,l}(t, \widetilde{z}_{c,l}, i)| \le \gamma_{c,l}|\widehat{z}_{c,l} - \widetilde{z}_{c,l}|, \tag{9.3}$$

for all $\widehat{z}_{c,l}, \widetilde{z}_{c,l} \in \mathbb{R}^{\nu_l}$, where $\gamma_{c,l} > 0$ does not depend upon $(t, i) \in \mathbb{R}_+ \times \mathcal{N}$, but it may depend upon the function $\Delta_{c,l}(\cdot, \cdot, \cdot)$.

(b) For each $i \in \mathcal{N}$ and $0 \le p \le r_4$, the functions $z_{d,p} \to \Delta_{d,p}(k, z_{d,p}, i) : \mathbb{R}^{\nu_{d,p}} \to \mathbb{R}^{m_{d,p}}$ are measurable and have the additional properties:

$$\Delta_{d,p}(k, 0, i) = 0, \quad \text{for all} \quad (t, i) \in \mathbb{Z}_+ \times \mathcal{N}; \tag{9.4}$$

$$|\Delta_{d,p}(k, z_{d,p}, i)| \le \gamma_{d,p}|z_{d,p}|, \quad \text{for all} \quad (k, z_{d,p}, i) \in \mathbb{Z}_+ \times \mathbb{R}^{\nu_{d,p}} \times \mathcal{N}. \tag{9.5}$$

In (9.5) $\gamma_{d,p} > 0$ does not depend upon $(k, z_{d,p}, i)$ but it may depends upon the function $\Delta_{d,p}(k, \cdot, i)$.

Here and in the sequel $|\cdot|$ denotes the Euclidean norm of a vector.

Remark 9.1

(a) Under (9.2) a Lipschitz-type condition of the form (9.3) guarantees the existence of the stochastic differential equation (9.1a) on each interval $[kh, (k+1)h)$, $k \in \mathbb{Z}_+$. For the reader convenience we refer, for example to Theorem 1.1, Chap. 5 in Friedman [1].
(b) Under (9.4) a linear growth condition of the form (9.5) seems to be more relax than a Lipschitz-type condition as it was considered in the continuous-time case. The measurability of the functions $\Delta_{d,p}(k, \cdot, i)$ was required in Assumption 9.2 (b) in order to be sure that the stochastic process $\{x(kh^+)\}_{k\in\mathbb{Z}_+}$ is adopted to the filtration generated by the stochastic processes involved in (9.1).

Further we denote $\mathcal{NL}_{[c]}(\mathbb{R}^{\nu_l}, \mathbb{R}^{m_l})$, $1 \le l \le r_3$ the set of all measurable functions $(t, z_{c,l}) \to \Delta_{c,l}(t, z_{c,l}, i) : \mathbb{R}_+ \times \mathbb{R}^{\nu_l} \to \mathbb{R}^{m_l}$, $i \in \mathcal{N}$, which satisfy conditions of the form (9.2) and (9.3). $\mathcal{NL}_{[d]}(\mathbb{R}^{\nu_{d,p}}, \mathbb{R}^{m_{d,p}})$, $1 \le l \le r_4$ the set of all measurable functions $z_{d,p} \to \Delta_{d,p}(k, z_{d,p}, i) : \mathbb{Z}_+ \times \mathbb{R}^{\nu_{d,p}} \to \mathbb{R}^{m_{d,p}}$, $i \in \mathcal{N}$, satisfying conditions of the form (9.4) and (9.5).

To ease the presentation, we shall denote $\Delta_{c,l}(\cdot) \stackrel{\Delta}{=} (\Delta_{c,l}(\cdot,\cdot,1), \Delta_{c,l}(\cdot,\cdot,2), \dots, \Delta_{c,l}(\cdot,\cdot,N))$, the elements of the set $\mathcal{NL}_{[c]}(\mathbb{R}^{\nu_l}, \mathbb{R}^{m_l})$, $1 \le l \le r_2$, and

$$\Delta_{d,p}(\cdot) \stackrel{\Delta}{=} \left(\{\Delta_{d,p}(k,\cdot,1)\}_{k\in\mathbb{Z}_+}, \{\Delta_{d,p}(k,\cdot,2)\}_{k\in\mathbb{Z}_+}, \dots, \{\Delta_{d,p}(k,\cdot,N)\}_{k\in\mathbb{Z}_+}\right)$$

where $\Delta_{d,p}(k,\cdot,1) \in \mathcal{NL}_{[d]}(\mathbb{R}^{\nu_{d,p}}, \mathbb{R}^{m_{d,p}})$. This allows us to describe the set $\mathcal{D}$ of the admissible uncertainties as:

$$\mathcal{D} := \{\boldsymbol{\Delta}(\cdot) = (\boldsymbol{\Delta}_c(\cdot), \boldsymbol{\Delta}_d(\cdot)) \mid \boldsymbol{\Delta}_c(\cdot) \in \mathcal{D}_c,\ \boldsymbol{\Delta}_d(\cdot) \in \mathcal{D}_d\} \tag{9.6a}$$

$$\mathcal{D}_c := \{\boldsymbol{\Delta}_c(\cdot) = (\Delta_{c,1}(\cdot), \Delta_{c,2}(\cdot), \dots, \Delta_{c,r_2}(\cdot)) \mid \Delta_{c,l}(\cdot) \in \mathcal{NL}_{[c]}(\mathbb{R}^{\nu_l}, \mathbb{R}^{m_l}),\ 1 \le l \le r_2\} \tag{9.6b}$$

$$\mathcal{D}_d := \{\boldsymbol{\Delta}_d(\cdot) = (\Delta_{d,1}(\cdot), \Delta_{d,2}(\cdot), \dots, \Delta_{d,r_4}(\cdot)) \mid \Delta_{d,p}(\cdot) \in \mathcal{NL}_{[d]}(\mathbb{R}^{\nu_{d,p}}, \mathbb{R}^{m_{d,p}}),\ 1 \le p \le r_4\}. \tag{9.6c}$$

9.1.2 The Problem Setting

The problem (9.1) can be seen as a disturbance of the nominal system $\mathbf{G}_0$ having the state-space representation described by:

$$dx(t) = A_0(\eta(t))x(t)dt + \sum_{\iota=1}^{r_1} A_\iota(\eta(t))x(t)dw_\iota(t), \quad kh < t \le (k+1)h, \tag{9.7a}$$

$$x(kh^+) = (A_{d,0}(\eta(kh)) + \sum_{q=1}^{r_3} w_{dq}(k)A_{dq}(\eta(kh)))x(kh), \quad k \in \mathbb{Z}_+. \tag{9.7b}$$

The terms:

$$\sum_{l=1}^{r_2} D_{c,l}(\eta(t))\Delta_{c,l}(t, z_{c,l}(t), \eta(t))d\xi_l(t) \tag{9.8}$$

and

$$\sum_{p=1}^{r_4} D_{d,p}(\eta(kh))\Delta_{d,p}(k, z_{d,p}(kh), \eta(kh))\zeta_p(k) \tag{9.9}$$

are modeling possible parametric excitations of white noise type of the coefficients of the nominal system (9.7) with unknown intensity.

Let us assume that the nominal system (9.7) is exponentially stable in mean square (ESMS). It is known that the exponentially stable behavior of the trajectories of (9.7) is preserved under small variations of its coefficients.

Our aim is to highlight a large subset $\mathfrak{D}$ of the set $\mathscr{D}$ of the admissible parametric uncertainties described by (9.6) such that the zero state equilibrium of the perturbed system $\mathbf{G}_{\Delta}$ to be stable in mean square (in the sense of Definition 9.2) for any admissible uncertainties $\boldsymbol{\Delta}(\cdot) \in \mathfrak{D}$ if the nominal system (9.7) is ESMS.

For a precise description of the set $\mathfrak{D}$ we shall use the notion of stability radius.

Remark 9.2 Now is clear why with Assumption 9.1 the stochastic processes $\{w(t)\}_{t\geq 0}$ and $\{w_d(k)\}_{k\in\mathbb{Z}_+}$ are normed stochastic processes (in the sense of Assumption 9.1(a) and (b)). These stochastic processes affect the nominal system (9.7) which is assumed to be known. The stochastic processes $\{\xi_l(t)\}_{t\geq 0}$, $1 \leq l \leq r_2$ and $\{\zeta_p(k)\}_{k\in\mathbb{Z}_+}$, $1 \leq p \leq r_4$ which affect the perturbed system $\mathbf{G}_{\Delta}$ via (9.8) and (9.9), can have unknown intensities $\lambda_{c,l}$, $\lambda_{d,p}$, respectively. We will see that the values of the intensities of the white noise involved in the models (9.8) and (9.9), respectively, of the parametric uncertainties will be involved in the computation of the value of the stability radius.

The property of exponential stability in mean square of the nominal system (9.7) is that introduced in Definition 3.1 (c). A list of Lyapunov-type criteria for exponential stability in mean square of the nominal system is provided by Theorem 3.4.

The next definition extends to the case of the nonlinear system (9.1) the kind of the stochastic stability introduced by Definition 3.1 (d)–(f).

Definition 9.1 The perturbed system $\mathbf{G}_{\Delta}$ is:

(i) $\mathscr{L}^2$-global uniform mean-square stable with conditioning ($\mathscr{L}^2$-GUMSS-C) if its solutions $x(\cdot; t_0, x_0)$ satisfy

$$\mathbb{E}\left[\int_{t_0}^{\infty} |x(t; t_0, x_0)|^2 \mathrm{d}t \mid \eta(t_0) = j\right] \leq c|x_0|^2, \tag{9.10}$$

for any initial pair $(t_0, x_0) \in \mathbb{R}_+ \times \mathbb{R}^n$, $j \in \mathscr{N}$, and any initial probability distribution π_0 of the Markov process. In (9.10) $c > 0$ is a constant which does not depend upon $(t_0, x_0) \in \mathbb{R}_+ \times \mathbb{R}^n$, $j \in \mathscr{N}$ and π_0.

(ii) $\mathscr{L}^2$-global uniform mean-square stable ($\mathscr{L}^2$-GUMSS) if its solutions $x(\cdot; t_0, x_0)$ satisfy

$$\mathbb{E}\left[\int_{t_0}^{\infty} |x(t; t_0, x_0)|^2 \mathrm{d}t\right] \leq c|x_0|^2 \tag{9.11}$$

for any initial pair $(t_0, x_0) \in \mathbb{R}_+ \times \mathbb{R}^n$ and any initial probability distribution π_0 of the Markov process. In (9.11), $c > 0$ is a constant which does not depend upon (t_0, x_0, π_0).

Definition 9.2 The constant solution $x = 0$ of the nonlinear jump stochastic differential equation (9.1) is:

(i) uniform mean-square stable (UMSS) if for any $\varepsilon > 0$ there exists $\delta(\varepsilon) > 0$ such that for all initial time instant $t_0 \geq 0$, the solution $x(\cdot; t_0, x_0)$ satisfies $\mathbb{E}[|x(t; t_0, x_0)|^2] < \varepsilon$, for all $t \geq t_0 \geq 0$, whenever the initial state x_0 is satisfying $|x_0| < \delta(\varepsilon)$;

(ii) global uniform mean-square asymptotic stable (GUMSAS) if it is UMSS and additionally, $\lim_{t\to\infty} \mathbb{E}[|x(t; t_0, x_0)|^2] = 0$, for any initial pair $(t_0, x_0) \in \mathbb{R}_+ \times \mathbb{R}^n$.

9.2 The Stability Radius

9.2.1 *The Definition of the Stability Radius*

In order to define the concept of stability radius, we introduce a norm on the space $\mathscr{D}$ of the admissible parametric uncertainties described in (9.6). If $\Delta_{c,l}(\cdot) \in \mathscr{N}\mathscr{L}_{[c]}(\mathbb{R}^{\nu_l}, \mathbb{R}^{m_l})$ we set

$$\begin{aligned} |\Delta_{c,l}(\cdot)| := \inf\{\widetilde{\gamma}_{c,l} > 0 \mid |\Delta_{c,l}(t, z_{c,l}, i)| \leq \widetilde{\gamma}_{c,l}|z_{c,l}|, \\ \text{for all } (t, z_{c,l}, i) \in \mathbb{R}_+ \times \mathbb{R}^{\nu_l} \times \mathscr{N}\}. \end{aligned} \tag{9.12}$$

If $\Delta_{d,p}(\cdot) \in \mathscr{N}\mathscr{L}_{[d]}(\mathbb{R}^{\nu_{d,p}}, \mathbb{R}^{m_{d,p}})$ we set

$$\begin{aligned} |\Delta_{d,p}(\cdot)| := \inf\{\widetilde{\gamma}_{d,p} > 0 \mid |\Delta_{d,p}(k, z_{d,p}, i)| \leq \widetilde{\gamma}_{d,p}|z_{d,p}|, \\ \text{for all } (k, z_{d,p}, i) \in \mathbb{Z}_+ \times \mathbb{R}^{\nu_{d,p}} \times \mathscr{N}\}. \end{aligned} \tag{9.13}$$

Now we define

$$\|\boldsymbol{\Delta}_c(\cdot)\| := \max_{1\leq l\leq r_2} |\Delta_{c,l}(\cdot)| \tag{9.14a}$$

$$\|\boldsymbol{\Delta}_d(\cdot)\| := \max_{1\leq p\leq r_4} |\Delta_{d,p}(\cdot)| \tag{9.14b}$$

$$\|\boldsymbol{\Delta}(\cdot)\| := \max\{\|\boldsymbol{\Delta}_c(\cdot)\|, \|\boldsymbol{\Delta}_d(\cdot)\|\}, \tag{9.14c}$$

for all $\boldsymbol{\Delta}(\cdot) = (\boldsymbol{\Delta}_c(\cdot), \boldsymbol{\Delta}_d(\cdot)) \in \mathscr{D}$. From (9.12) and (9.14a) we deduce that

$$|\Delta_{c,l}(t, z_{c,l}, i)| \leq \|\boldsymbol{\Delta}_c(\cdot)\| \, |z_{c,l}|, \tag{9.15}$$

for all $(t, z_{c,l}, i) \in \mathbb{R}_+ \times \mathbb{R}^{\nu_l} \times \mathscr{N}$. Similarly, (9.13) and (9.14b) give

$$|\Delta_{d,p}(k, z_{d,p}, i)| \leq \|\boldsymbol{\Delta}_d(\cdot)\| \, |z_{d,p}|, \tag{9.16}$$

for all $(k, z_{d,p}, i) \in \mathbb{R}_+ \times \mathbb{R}^{\nu_{d,p}} \times \mathcal{N}$. From (9.8) and (9.9) one sees that the parametric uncertainties which affect the dynamic of the numerical system (9.7) are structured perturbations. The structure of these perturbations is modeled by the pairs of matrices $(D_{c,l}(i), C_{c,l}(i)) \in \mathbb{R}^{n\times m_l} \times \mathbb{R}^{\nu_l \times n}$, $i \in \mathcal{N}$, $1 \le l \le r_2$, in the case of the parametric uncertainties which affect the dynamic in the continuous time of the nominal system and the pairs $(D_{d,p}(i), C_{d,p}(i)) \in \mathbb{R}^{n\times m_{d,p}} \times \mathbb{R}^{\nu_{d,p}\times n}$, $i \in \mathcal{N}$, $1 \le p \le r_4$, in the case of the parametric uncertainties which appear in the case of the equations which describe the jump of the nominal system at the impulse instant. Employing the notation introduced in Sect. 5.1, we set:

$$\begin{aligned}
\mathbf{D}_{c,l} &= (D_{c,l}(1), D_{c,l}(2), \dots, D_{c,l}(N)) \in \mathcal{M}^N_{n,m_l},\\
\mathbf{C}_{c,l} &= (C_{c,l}(1), C_{c,l}(2), \dots, C_{c,l}(N)) \in \mathcal{M}^N_{\nu_l,n}, \quad 1 \le l \le r_2,\\
\mathbf{D}_{d,p} &= (D_{d,p}(1), D_{d,p}(2), \dots, D_{d,p}(N)) \in \mathcal{M}^N_{n,m_{d,p}},\\
\mathbf{C}_{d,p} &= (C_{d,p}(1), C_{d,p}(2), \dots, C_{d,p}(N)) \in \mathcal{M}^N_{\nu_{d,p},n}, \quad 1 \le p \le r_4.
\end{aligned}$$

Now we can define the concept of stability radius for the nominal system $\mathbf{G_0}$ with respect to the structured parametric uncertainties of the form (9.8) and (9.9).

Definition 9.3 The stability radius of the nominal system $\mathbf{G_0}$ with respect to the set $\mathcal{D}$ of structured parametric uncertainties having the structure described by the pair $(\mathbf{D}_{c,l}, \mathbf{C}_{c,l})$, $l \in \underline{r_2}$, and $(\mathbf{D}_{d,p}, \mathbf{C}_{d,p})$, $p \in \underline{r_4}$, is the number:

$$\begin{aligned}
&\rho_{\mathcal{D}}[\mathbf{G_0} \mid (\mathbf{D}_{c,l}, \mathbf{C}_{c,l}),\ l \in \underline{r_2},\ (\mathbf{D}_{d,p}, \mathbf{C}_{d,p}),\ p \in \underline{r_4}] \stackrel{\Delta}{=}\\
&\inf\{\rho > 0 \mid \exists\, \boldsymbol{\Delta}(\cdot) \in \mathcal{D} \text{ with } \|\boldsymbol{\Delta}(\cdot)\| \le \rho \text{ such that the zero state}\\
&\text{equilibrium of the corresponding system (9.1) is not GUMSAS}\}.
\end{aligned} \tag{9.17}$$

Here and in the sequel $\underline{r_j}$ is an abbreviation for the set $\{1, 2, \dots, r_j\}$.
From the previous definition it follows that if $\boldsymbol{\Delta}(\cdot) \in \mathcal{D}$ with

$$\|\boldsymbol{\Delta}(\cdot)\| \le \rho_{\mathcal{D}}[\mathbf{G_0} \mid (\mathbf{D}_{c,l}, \mathbf{C}_{c,l}),\ l \in \underline{r_2},\ (\mathbf{D}_{d,p}, \mathbf{C}_{d,p}),\ p \in \underline{r_4}],$$

the corresponding dynamical system $\mathbf{G}_{\boldsymbol{\Delta}}$ is GUMSAS. Particularly, if

$$\rho_{\mathcal{D}}[\mathbf{G_0} \mid (\mathbf{D}_{c,l}, \mathbf{C}_{c,l}),\ l \in \underline{r_2},\ (\mathbf{D}_{d,p}, \mathbf{C}_{d,p}),\ p \in \underline{r_4}] > 0$$

then the nominal system $\mathbf{G_0}$ is GUMSAS or equivalently it is ASMS, because it is a linear system. Then, the equivalence $(iii) \Leftrightarrow (iv)$ from Theorem 3.1 allows us to conclude that the nominal system $\mathbf{G_0}$ must be ESMS, in order to be sure that its stability radius is a positive number.

In the next section we shall provide an estimation of a lower bound of the stability radius introduced via (9.17).

9.2.2 Several Particular Cases

A. A first particular case of the perturbed system (9.1) is described by:

$$dx(t) = A_0(\eta(t))x(t)dt + \sum_{\iota=1}^{r_1} A_\iota(\eta(t))x(t)dw_\iota(t)$$
$$+ \sum_{l=1}^{r_2} D_{c,l}(\eta(t))\Delta_{c,l}(t, z_{c,l}(t), \eta(t))d\xi_l(t), \quad kh < t \le (k+1)h \quad (9.18a)$$

$$x(kh^+) = (A_{d0}(\eta(kh)) + \sum_{q=1}^{r_3} w_{dq}(k)A_{dq}(\eta(kh)))x(kh), \quad k \in \mathbb{Z}_+ \quad (9.18b)$$

$$z_{c,l}(t) = C_{c,l}(\eta(t))x(t), \quad 1 \le l \le r_2, \quad t \in \mathbb{R}_+. \quad (9.18c)$$

Systems of the form (9.18) describe the cases when the nominal system $\mathbf{G_0}$ is affected only by parametric uncertainties of the form (9.8) where only unknown functions $\boldsymbol{\Delta}_c(\cdot)$ from $\mathscr{D}_c$ are involved. That is why we shall denote $\mathbf{G}_{\boldsymbol{\Delta}_c}$ a system of the form (9.18). Let us remark that in the special case when $A_{dq}(i) = 0$, $1 \le q \le r_3$ and $A_{d0}(i) = I_n$, $i \in \mathscr{N}$, the jump stochastic differential equation (9.18) reduces to the stochastic differential equation:

$$dx(t) = A_0(\eta(t))x(t)dt + \sum_{\iota=1}^{r_1} A_\iota(\eta(t))x(t)dw_\iota(t)$$
$$+ \sum_{l=1}^{r_2} D_{c,l}(\eta(t))\Delta_{c,l}(t, z_{c,l}(t), \eta(t))d\xi_l(t), \quad (9.19a)$$

$$z_{c,l}(t) = C_{c,l}(\eta(t))x(t), \quad t \in \mathbb{R}_+, \quad (9.19b)$$

and the nominal system (9.7) reduces to:

$$dx(t) = A_0(\eta(t))x(t)dt + \sum_{\iota=1}^{r_1} A_\iota(\eta(t))x(t)dw_\iota(t), \quad t \in \mathbb{R}_+. \quad (9.20)$$

When $A_\iota(i) = 0$, $1 \le \iota \le r_1$, $i \in \mathscr{N}$, (9.19) recover the stochastic system subject to some structured parametric uncertainties studied in V. Drăgan [2].

If the Markov process has only one state, that is $\mathscr{N} = \{1\}$, the system (9.19) reduces to the system of deterministic linear differential equations subject to some structured parametric uncertainties studied by D. Hinrichsen [3]. In the particular case of the perturbed systems of the form (9.18) of the nominal system (9.7) the Definition 9.3 becomes:

Definition 9.4 The stability radius of the nominal system $\mathbf{G_0}$ with respect to the set $\mathscr{D}_c$ of the structured parametric uncertainties having the structure described by the pair $(\mathbf{D}_{c,l}, \mathbf{C}_{c,l})$, $l \in \underline{r_2}$, is the number:

$$\rho_{\mathscr{D}_c}[\mathbf{G_0} \mid (\mathbf{D}_{c,l}, \mathbf{C}_{c,l}),\ l \in \underline{r_2},\ ] \stackrel{\Delta}{=} \inf\{\rho > 0 \mid \exists\, \boldsymbol{\Delta}_c(\cdot) \in \mathscr{D}_c \text{ with } \|\boldsymbol{\Delta}_c(\cdot)\| \leq \rho \text{ such that the zero state equilibrium of the corresponding system (9.18) is not GUMSAS}\}. \tag{9.21}$$

B. Another particular case of the perturbed system (9.1) is described by:

$$dx(t) = A_0(\eta(t))x(t)dt + \sum_{\iota=1}^{r_1} A_\iota(\eta(t))x(t)dw_\iota(t), \quad kh < t \leq (k+1)h \tag{9.22a}$$

$$x(kh^+) = (A_{d0}(\eta(kh)) + \sum_{q=1}^{r_3} w_{dq}(k)A_{d,q}(\eta(kh)))x(kh) + \sum_{p=1}^{r_4} D_{d,p}(\eta(kh))\Delta_{d,p}(k, z_{d,p}(kh), \eta(kh))\zeta_p(k), \quad k \in \mathbb{Z}_+ \tag{9.22b}$$

$$z_{d,p}(kh) = C_{d,p}(\eta(kh))x(kh), \quad 1 \leq p \leq r_4, \quad k \in \mathbb{Z}_+. \tag{9.22c}$$

Systems of the form (9.22) describe the cases when the nominal system $\mathbf{G_0}$ is affected only by structured parametric uncertainties of the form (9.9) in which only unknown functions of the form $\boldsymbol{\Delta}_d(\cdot)$ from $\mathscr{D}_d$ are involved. That is why a perturbed system of the form (9.22) will be denoted $\mathbf{G}_{\boldsymbol{\Delta}_d}$.
The stability radius $\rho_{\mathscr{D}_d}[\mathbf{G_0} \mid (\mathbf{D}_{d,p}, \mathbf{C}_{d,p}),\ p \in \underline{r_4}\ ]$ of the nominal system $\mathbf{G_0}$ with respect to the set $\mathscr{D}_d$ of the structured parametric uncertainties having the structure described by the pairs $(\mathbf{D}_{d,p}, \mathbf{C}_{d,p})$, $p \in \underline{r_4}$ may be introduced as in Definition 9.4, but taking $\mathscr{D}_d$ instead of $\mathscr{D}_c$ and $\boldsymbol{\Delta}_d(\cdot)$ instead of $\boldsymbol{\Delta}_c(\cdot)$. The details are left to the reader.

C. Among the admissible perturbances of the form (9.8) and (9.9) there are those of the form:

$$\sum_{l=1}^{r_2} D_{c,l}(\eta(t))\widetilde{\Delta}_{c,l}(t, \eta(t))z_{c,l}(t)d\xi_l(t) \tag{9.23}$$

and

$$\sum_{p=1}^{r_4} D_{d,p}(\eta(kh))\widetilde{\Delta}_{d,p}(t, \eta(kh))z_{d,p}(kh)\zeta_p(k), \quad k \in \mathbb{Z}_+, \tag{9.24}$$

respectively, where $t \to \widetilde{\Delta}_{c,l}(t, i) : \mathbb{R}_+ \to \mathbb{R}^{\nu_l \times m_l}$ are arbitrary measurable and bounded matrix-valued functions and $\{\widetilde{\Delta}_{d,p}(k, i)\}_{k \in \mathbb{Z}_+} \subset \mathbb{R}^{\nu_{d,p} \times m_{d,p}}$ are arbitrary bounded sequences. In this case we have

$$\Delta_{c,l}(t, z_{c,l}, i) = \widetilde{\Delta}_{c,l}(t, i) z_{c,l} \tag{9.25a}$$

$$\Delta_{d,p}(k, z_{d,p}, i) = \widetilde{\Delta}_{d,p}(k, i) z_{d,p}. \tag{9.25b}$$

So, (9.12) becomes:

$$|\Delta_{c,l}(\cdot)| = \max_{i \in \mathcal{N}} \sup_{t \in \mathbb{R}_+} \|\widetilde{\Delta}_{c,l}(t, i)\|_2 \tag{9.26}$$

and (9.13) becomes:

$$|\Delta_{d,p}(\cdot)| = \max_{i \in \mathcal{N}} \sup_{k \in \mathbb{Z}_+} \|\widetilde{\Delta}_{d,p}(k, i)\|_2. \tag{9.27}$$

In (9.26) and (9.27), $\|\cdot\|_2$ denotes the euclidean norm of a matrix. In the sequel, based on the perturbations of the form (9.25) we may define the subsets $\mathcal{D}_L \subset \mathcal{D}$, $\mathcal{D}_{L,c} \subset \mathcal{D}_c$, $\mathcal{D}_{L,d} \subset \mathcal{D}_d$ which consist of structured linear parametric uncertainties. Restricting the Definition 9.3 and the Definition 9.4 to the structured linear uncertainties one may obtain the stability radius: $\rho_{\mathcal{D}_L}[\mathbf{G_0} \mid (\mathbf{D}_{c,l}, \mathbf{C}_{c,l}),\ l \in \underline{r_2},\ (\mathbf{D}_{d,p}, \mathbf{C}_{d,p}),\ p \in \underline{r_4}]$ and $\rho_{\mathcal{D}_{L,c}}[\mathbf{G_0} \mid (\mathbf{D}_{c,l}, \mathbf{C}_{c,l}),\ l \in \underline{r_2}]$.

9.3 An Estimate of a Lower Bound of the Stability Radius

9.3.1 The Scaling of the Perturbations

In this section we shall adapt to the case of the perturbed systems of the form (9.1) the technique of the scaling of the perturbations developed in D. Hinrichsen [3] in order to obtain an improved estimation of the stability radius of the nominal system (9.7) with respect to the set $\mathcal{D}$ of the admissible structured perturbations described in (9.6).

To this end, we rewrite the perturbed system (9.1) in the following equivalent form:

$$\begin{aligned} dx(t) &= A_0(\eta(t))x(t)dt + \sum_{\iota=1}^{r_1} A_\iota(\eta(t))x(t)dw_\iota(t) \\ &+ \sum_{l=1}^{r_2} \frac{1}{\alpha_{c,l}(\eta(t))} D_{c,l}(\eta(t))\Delta_{c,l}^{\alpha_{c,l}}(t, z_{c,l}^{\alpha_{c,l}}(t), \eta(t))d\xi_l(t),\ kh < t \leq (k+1)h \end{aligned} \tag{9.28a}$$

$$\begin{aligned} x(kh^+) &= (A_{d0}(\eta(kh)) + \sum_{q=1}^{r_3} w_{dq}(k)A_{dq}(\eta(kh)))x(kh) \\ &+ \sum_{p=1}^{r_4} \frac{1}{\alpha_{d,p}(\eta(kh))} D_{d,p}(\eta(kh))\Delta_{d,p}^{\alpha_{d,p}}(k, z_{d,p}^{\alpha_{d,p}}(kh), \eta(kh))\zeta_p(k),\ k \in \mathbb{Z}_+ \end{aligned} \tag{9.28b}$$

$$z_{c,l}^{\alpha_{c,l}}(t) = \alpha_{c,l}(\eta(t))C_{c,l}(\eta(t))x(t), \quad 1 \le l \le r_2, \quad t \in \mathbb{R}_+ \tag{9.28c}$$
$$z_{d,p}^{\alpha_{d,p}}(kh) = \alpha_{d,p}(\eta(kh))C_{d,p}(\eta(kh))x(kh), \quad 1 \le p \le r_4, \quad k \in \mathbb{Z}_+, \tag{9.28d}$$

where $\alpha_{c,l}(\eta(t)) \in (0,\infty)$, $l \in \underline{r_2}$, $t \in \mathbb{R}_+$ and $\alpha_{d,p}(\eta(kh)) \in (0,\infty)$, $p \in \underline{r_4}$, $k \in \mathbb{Z}_+$.
In (9.28a) the function $z_{c,l} \to \Delta_{c,l}^{\alpha_{c,l}}(t, z_{c,l}, \eta(t))$ is defined by

$$\Delta_{c,l}^{\alpha_{c,l}}(t, z_{c,l}, \eta(t)) = \alpha_{c,l}(\eta(t))\Delta_{c,l}(t, \alpha_{c,l}^{-1}(\eta(t))z_{c,l}, \eta(t)), \tag{9.29}$$

$l \in \underline{r_2}$, $t \in \mathbb{R}_+$, $\Delta_{c,l}(\cdot,\cdot,i)$ being an arbitrary measurable function satisfying (9.2) and (9.3).

In (9.28b) the function $z_{d,p} \to \Delta_{d,p}^{\alpha_{d,p}}(k, z_{d,p}, \eta(kh))$ is defined by

$$\Delta_{d,p}^{\alpha_{d,p}}(k, z_{d,p}, \eta(kh)) = \alpha_{d,p}(\eta(kh))\Delta_{d,p}(k, \alpha_{d,p}^{-1}(\eta(kh))z_{d,p}, \eta(kh)), \tag{9.30}$$

$\Delta_{d,p}(\cdot,\cdot,i)$ being an arbitrary measurable function satisfying (9.4) and (9.5).
From (9.2), (9.3) and (9.29) one sees that $\Delta_{c,l}^{\alpha_{c,l}}(\cdot,\cdot,i) \in \mathcal{NL}_{[c]}(\mathbb{R}^{\nu_l}, \mathbb{R}^{m_l})$ if $\Delta_{c,l}(\cdot,\cdot,i) \in \mathcal{NL}_{[c]}(\mathbb{R}^{\nu_l}, \mathbb{R}^{m_l})$, $l \in \underline{r_2}$. Similarly, from (9.4), (9.5) and (9.30) one sees that $\Delta_{d,p}^{\alpha_{d,p}}(k,\cdot,i)$ lies in $\mathcal{NL}_{[d]}(\mathbb{R}^{\nu_{d,p}}, \mathbb{R}^{m_{d,p}})$ if $\Delta_{d,p}(k,\cdot,i) \in \mathcal{NL}_{[d]}(\mathbb{R}^{\nu_{d,p}}, \mathbb{R}^{m_{d,p}})$, $p \in \underline{r_4}$. Furthermore, from (9.12) and (9.29) we may infer that

$$|\Delta_{c,l}^{\alpha_{c,l}}(\cdot)| = |\Delta_{c,l}(\cdot)|, \quad \text{for all} \quad \alpha_{c,l}(i) > 0, \quad l \in \underline{r_2}, \tag{9.31}$$

and from (9.13) and (9.30) we have that

$$|\Delta_{d,p}^{\alpha_{d,p}}(\cdot)| = |\Delta_{d,p}(\cdot)|, \quad \text{for all} \quad \alpha_{d,p}(i) > 0, \quad p \in \underline{r_4}, \tag{9.32}$$

$i \in \mathcal{N}$. These allow us to deduce via (9.14) that

$$\|\boldsymbol{\Delta}_c^{\alpha_c}(\cdot)\| = \|\boldsymbol{\Delta}_c(\cdot)\| \tag{9.33a}$$
$$\|\boldsymbol{\Delta}_d^{\alpha_d}(\cdot)\| = \|\boldsymbol{\Delta}_d(\cdot)\| \tag{9.33b}$$
$$\|\boldsymbol{\Delta}^{\alpha_c\alpha_d}(\cdot)\| = \max\{\|\boldsymbol{\Delta}_c^{\alpha_c}(\cdot)\|, \|\boldsymbol{\Delta}_d^{\alpha_d}(\cdot)\|\}, \tag{9.33c}$$

where

$$\boldsymbol{\Delta}_c^{\alpha_c}(\cdot) = (\Delta_{c,1}^{\alpha_{c,1}}(\cdot), \Delta_{c,2}^{\alpha_{c,2}}(\cdot), \ldots, \Delta_{c,r_2}^{\alpha_{c,r_2}}(\cdot)) \tag{9.34a}$$
$$\boldsymbol{\Delta}_d^{\alpha_d}(\cdot) = (\Delta_{d,1}^{\alpha_{d,1}}(\cdot), \Delta_{d,2}^{\alpha_{d,2}}(\cdot), \ldots, \Delta_{d,r_4}^{\alpha_{d,r_4}}(\cdot)) \tag{9.34b}$$
$$\boldsymbol{\Delta}^{\alpha_c\alpha_d}(\cdot) = (\boldsymbol{\Delta}_c^{\alpha_c}(\cdot), \boldsymbol{\Delta}_d^{\alpha_d}(\cdot)). \tag{9.34c}$$

From (9.29), (9.33a), (9.34a) we may infer that $\boldsymbol{\Delta}_c(\cdot) \to \boldsymbol{\Delta}_c^{\alpha_c}(\cdot) : \mathcal{D}_c \to \mathcal{D}_c$ is an isometry. Similarly, (9.30), (9.33b), (9.34b) allow us to deduce that $\boldsymbol{\Delta}_d(\cdot) \to \boldsymbol{\Delta}_d^{\alpha_d}(\cdot) : \mathcal{D}_d \to \mathcal{D}_d$ is also an isometry. Finally, (9.33), (9.34) allow us to conclude that

$(\boldsymbol{\Delta}_c(\cdot), \boldsymbol{\Delta}_d(\cdot)) \to (\boldsymbol{\Delta}_c^{\alpha_c}(\cdot), \boldsymbol{\Delta}_d^{\alpha_d}(\cdot)) : \mathscr{D} \to \mathscr{D}$ is an isometry, too. For any vector of scaling parameters

$$\alpha_c = \{\alpha_{c,l}(i)\}_{(l,i)\in \underline{r_2}\times\mathcal{N}} \in (0,\infty)^{\underline{r_2}\times\mathcal{N}}$$

and

$$\alpha_d = \{\alpha_{d,p}(i)\}_{(p,i)\in \underline{r_4}\times\mathcal{N}} \in (0,\infty)^{\underline{r_4}\times\mathcal{N}}.$$

9.3.2 Parametrized Lyapunov-Type Jump Linear Differential Equations

In the derivation of a lower bound of the value of the stability radius

$$\rho_{\mathscr{D}}[\mathbf{G_0} \mid (\mathbf{D}_{c,l}, \mathbf{C}_{c,l}), l \in \underline{r_2},\ (\mathbf{D}_{d,p}, \mathbf{C}_{d,p}),\ p \in \underline{r_4}]$$

an important role will be played by the bounded on $\mathbb{R}$ solution of the following backward Lyapunov-type jump linear differential equation (BLJLDE):

$$-\dot{X}(t,i) = A_0^\top(i)X(t,i) + X(t,i)A_0(i) + \sum_{\iota=1}^{r_1} A_\iota^\top(i)X(t,i)A_\iota(i)$$

$$+\sum_{j=1}^{N} q_{ij}X(t,j) + \sum_{l=1}^{r_2} \alpha_{c,l}^2(i)C_{c,l}^\top(i)C_{c,l}(i),\quad kh \le t < (k+1)h \tag{9.35a}$$

$$X(kh^-,i) = \sum_{q=0}^{r_3} A_{d,q}^\top(i)X(kh,i)A_{d,q}(i) + \sum_{p=1}^{r_4} \alpha_{d,p}^2(i)C_{d,p}^\top(i)C_{d,p}(i),\quad k \in \mathbb{Z}, \tag{9.35b}$$

where $\alpha_{c,l}(i) \ge 0$, $\alpha_{d,p}(i) \ge 0$ are parameters. The BLJLDE (9.35) can be written in a compact form as:

$$-\dot{X}(t,i) = \mathscr{L}_{c,i}^*[\mathbf{X}(t)] + \sum_{l=1}^{r_2} \alpha_{c,l}^2(i)C_{c,l}^\top(i)C_{c,l}(i),\quad kh \le t < (k+1)h \tag{9.36a}$$

$$X(kh^-,i) = \mathscr{L}_{d,i}^*[\mathbf{X}(kh)] + \sum_{p=1}^{r_4} \alpha_{d,p}^2(i)C_{d,p}^\top(i)C_{d,p}(i),\quad k \in \mathbb{Z}, \tag{9.36b}$$

$i \in \mathcal{N}$, where $\mathbf{X} \to \mathscr{L}_c^*[\mathbf{X}] = (\mathscr{L}_{c,1}^*[\mathbf{X}], \mathscr{L}_{c,2}^*[\mathbf{X}], \dots, \mathscr{L}_{c,N}^*[\mathbf{X}]) : \mathscr{S}_n^N \to \mathscr{S}_n^N$ is the adjoint with respect to the inner product (2.111) of the linear operator $\mathbf{X} \to \mathscr{L}_c[\mathbf{X}] = (\mathscr{L}_{c,1}[\mathbf{X}], \mathscr{L}_{c,2}[\mathbf{X}], \dots, \mathscr{L}_{c,N}[\mathbf{X}]) : \mathscr{S}_n^N \to \mathscr{S}_n^N$ defined as:

$$\mathcal{L}_{c,i}[\mathbf{X}] = A_0(i)X(i) + X(i)A_0^\top(i) + \sum_{\iota=1}^{r_1} A_\iota(i)X(i)A_\iota^\top(i) + \sum_{j=1}^{N} q_{ji}X(j), \tag{9.37}$$

$i \in \mathcal{N}$, for all $\mathbf{X} = (X(1), \dots, X(N)) \in \mathcal{S}_n^N$.

The operator $\mathbf{X} \to \mathcal{L}_d^*[\mathbf{X}] = (\mathcal{L}_{d,1}^*[\mathbf{X}], \mathcal{L}_{d,2}^*[\mathbf{X}], \dots, \mathcal{L}_{d,N}^*[\mathbf{X}]) : \mathcal{S}_n^N \to \mathcal{S}_n^N$ involved in (9.36b) is the adjoint with respect to the inner product (2.111) of the linear operator $\mathbf{X} \to \mathcal{L}_d[\mathbf{X}] = (\mathcal{L}_{d,1}[\mathbf{X}], \mathcal{L}_{d,2}[\mathbf{X}], \dots, \mathcal{L}_{d,N}[\mathbf{X}]) : \mathcal{S}_n^N \to \mathcal{S}_n^N$ given by:

$$\mathcal{L}_{d,i}[\mathbf{X}] = \sum_{q=0}^{r_3} A_{d,q}(i)X(i)A_{d,q}^\top(i), \quad i \in \mathcal{N}. \tag{9.38}$$

Based on Theorem 3.1 applied in the nominal system (9.7) we obtain:

Corollary 9.1 *If the stochastic processes $\{w(t)\}_{t\geq 0}$, $\{\eta(t)\}_{t\geq 0}$, $\{w_d(k)\}_{k\in\mathbb{Z}_+}$ involved in the nominal system (9.7) satisfy the conditions from Assumption 9.1, then the following are equivalent:*

(i) the nominal system (9.7) is ESMS;

(ii) the generalized Lyapunov-type jump linear differential equation (GLJLDE) on the Hilbert space $\mathcal{S}_n^N$:

$$\dot{\mathbf{X}}(t) = \mathcal{L}_c[\mathbf{X}(t)], \quad kh < t \leq (k+1)h \tag{9.39a}$$

$$\mathbf{X}(kh^+) = \mathcal{L}_d[\mathbf{X}(kh)], \quad k \in \mathbb{Z}, \tag{9.39b}$$

is exponentially stable (ES).

Further, applying Theorem 2.13 in the case of the BLJLDE (9.35) we obtain via Corollary 9.1:

Proposition 9.1 *Assume:*

(a) the stochastic processes $\{w(t)\}_{t\geq 0}$, $\{\eta(t)\}_{t\geq 0}$, $\{w_d(k)\}_{k\in\mathbb{Z}_+}$ which affect the nominal system (9.7) satisfy the conditions from Assumption 9.1;

(b) the nominal system (9.7) is ESMS.

Under these conditions for each pair of vectors of scaling parameters (α_c, α_d), with $\alpha_c = \{\alpha_{c,l}(i)\}_{(l,i)\in\underline{r_2}\times\mathcal{N}} \subset (0, \infty)$ and $\alpha_d = \{\alpha_{d,p}(i)\}_{(p,i)\in\underline{r_4}\times\mathcal{N}} \subset (0, \infty)$, the corresponding BLJLDE (9.35) has a unique bounded solution $t \to \mathbf{X}^{\alpha_c,\alpha_d}(t) = (X^{\alpha_c,\alpha_d}(t, 1), X^{\alpha_c,\alpha_d}(t, 2), \dots, X^{\alpha_c,\alpha_d}(t, N)) : \mathbb{R} \to \mathcal{S}_n^N$. This solution has the following additional properties:

(i) $\mathbf{X}^{\alpha_c,\alpha_d}(\cdot)$ is a periodic function of period h;

(ii) $\mathbf{X}^{\alpha_c,\alpha_d}(\cdot)$ is right continuous in each impulsive time instance $\tau_k = kh$, $k \in \mathbb{Z}$;

(iii) $X^{\alpha_c,\alpha_d}(t, i) \geq 0$ for all $(t, i) \in \mathbb{R} \times \mathcal{N}$.

Besides the BLJLDE (9.35), the following its particular forms will be involved in this section:

$$-\dot{X}(t,i) = \mathcal{L}^*_{c,i}[\mathbf{X}(t)] + \sum_{l=1}^{r_2} \alpha^2_{c,l}(i) C^\top_{c,l}(i) C_{c,l}(i), \quad kh \le t < (k+1)h \tag{9.40a}$$

$$X(kh^-, i) = \mathcal{L}^*_{d,i}[\mathbf{X}(kh)], \quad k \in \mathbb{Z}, \tag{9.40b}$$

$i \in \mathcal{N}$, and

$$-\dot{X}(t,i) = \mathcal{L}^*_{c,i}[\mathbf{X}(t)], \quad kh \le t < (k+1)h \tag{9.41a}$$

$$X(kh^-, i) = \mathcal{L}^*_{d,i}[\mathbf{X}(kh)] + \sum_{p=1}^{r_4} \alpha^2_{d,p}(i) C^\top_{d,p}(i) C_{d,p}(i), \quad k \in \mathbb{Z}, \tag{9.41b}$$

$i \in \mathcal{N}$. It is worth mentioning that in the BLJLDE (9.35) the pair of vectors of scaling parameters (α_c, α_d) is involved, while in (9.40) and (9.41) the pairs $(\alpha_c, 0)$ and $(0, \alpha_d)$, respectively, are involved, where

$$\alpha_c = \{\alpha_{c,l}(i)\}_{(l,i) \in \underline{r_2} \times \mathcal{N}} \subset [0, \infty) \tag{9.42}$$

and

$$\alpha_d = \{\alpha_{d,p}(i)\}_{(p,i) \in \underline{r_4} \times \mathcal{N}} \subset [0, \infty). \tag{9.43}$$

With this remark we may invoke Proposition 9.1 to obtain:

Corollary 9.2 *Under the assumption from Proposition 9.1, for each vector of scaling parameters α_c as in (9.42) and α_d as in (9.43), the BLJLDE (9.40) and (9.41) have unique bounded on $\mathbb{R}$ solutions $t \to \mathbf{X}^{\alpha_c,0}(t) = (X^{\alpha_c,0}(t,1), X^{\alpha_c,0}(t,2), \dots, X^{\alpha_c,0}(t,N)) : \mathbb{R} \to \mathcal{S}_n^N$ and $t \to \mathbf{X}^{0,\alpha_d}(t) = (X^{0,\alpha_d}(t,1), X^{0,\alpha_d}(t,2), \dots, X^{0,\alpha_d}(t,N)) : \mathbb{R} \to \mathcal{S}_n^N$, respectively. These solutions have the additional properties:*

(i) $\mathbf{X}^{\alpha_c,0}(\cdot)$ and $\mathbf{X}^{0,\alpha_d}(\cdot)$ are periodic functions of period h;
(ii) $\mathbf{X}^{\alpha_c,0}(\cdot)$ and $\mathbf{X}^{0,\alpha_d}(\cdot)$ are right continuous functions at each impulsive time instance $\tau_k = kh$, $k \in \mathbb{Z}$;
(iii) $X^{\alpha_c,0}(t,i) \ge 0$ and $X^{0,\alpha_d}(t,i) \ge 0$ for all $(t,i) \in \mathbb{R} \times \mathcal{N}$.

9.3.3 Lower Bound of the Stability Radius

First we prove:

Theorem 9.1 *Assume:*

(a) Assumption 9.1 hold;
(b) the nominal system (9.7) is ESMS.

Let $\sigma_k > 0$, $k = 1, 2$, be given. If there exist vectors of scaling parameters

$$\alpha_c = \{\alpha_{c,l}(i)\}_{(l,i)\in \underline{r_2}\times\mathcal{N}} \subset (0,\infty)^{\underline{r_2}\times\mathcal{N}}$$

and

$$\alpha_d = \{\alpha_{d,p}(i)\}_{(p,i)\in \underline{r_4}\times\mathcal{N}} \subset (0,\infty)^{\underline{r_4}\times\mathcal{N}}$$

such that

$$\max_{l\in\underline{r_2}}\max_{j\in\mathcal{N}}\sup_{0\le t\le h}\left(\frac{\sigma_1\lambda_{c,l}}{\alpha_{c,l}(j)}\right)^2 \|D_{c,l}^\top(j)X^{\alpha_c,\alpha_d}(t,j)D_{c,l}(j)\|_2 < 1 \tag{9.44}$$

$$\max_{p\in\underline{r_4}}\max_{j\in\mathcal{N}}\left(\frac{\sigma_2\lambda_{d,p}}{\alpha_{d,p}(j)}\right)^2 \|D_{d,p}^\top(j)X^{\alpha_c,\alpha_d}(0,j)D_{d,p}(j)\|_2 < 1 \tag{9.45}$$

then

$$\rho_{\mathcal{D}}[\mathbf{G_0} \mid (\mathbf{D}_{c,l},\mathbf{C}_{c,l}),\ l\in\underline{r_2},\ (\mathbf{D}_{d,p},\mathbf{C}_{d,p}),\ p\in\underline{r_4}] \ge \max\{\sigma_1,\sigma_2\}. \tag{9.46}$$

In (9.44) and (9.45), $\mathbf{X}^{\alpha_c,\alpha_d}(\cdot) = (X^{\alpha_c,\alpha_d}(\cdot,1), X^{\alpha_c,\alpha_d}(\cdot,2),\ldots,X^{\alpha_c,\alpha_d}(\cdot,N))$ is the unique bounded on $\mathbb{R}$ solution of the BLJLDE (9.35) corresponding to the pair (α_c,α_d) of vectors of scaling parameters under consideration and $\|\cdot\|_2$ is the euclidean norm of a matrix.

Proof Let $\boldsymbol{\Delta}(\cdot) = (\boldsymbol{\Delta}_c(\cdot), \boldsymbol{\Delta}_d(\cdot)) \in \mathcal{D}$ be arbitrary with the conditions

$$\|\boldsymbol{\Delta}_c\| \le \sigma_1, \quad \|\boldsymbol{\Delta}_d(\cdot)\| \le \sigma_2. \tag{9.47}$$

We show that the zero state equilibrium of the corresponding perturbed system $\mathbf{G}_{\boldsymbol{\Delta}}$ having the state-space representation described by (9.1) is GUMSAS.

Let $\alpha_c = \{\alpha_{c,l}(j)\}_{(l,j)\in\underline{r_2}\times\mathcal{N}}$ and $\alpha_d = \{\alpha_{d,p}(j)\}_{(p,j)\in\underline{r_4}\times\mathcal{N}}$ be the vectors of scaling parameters which satisfy (9.44) and (9.45), respectively. Based on Proposition 9.1 we deduce that under the considered assumptions the corresponding BLJLDE (9.35) has a unique bounded on $\mathbb{R}$ solution $\mathbf{X}^{\alpha_c,\alpha_d}(\cdot)$ which is a periodic function of period h and satisfies $X^{\alpha_c,\alpha_d}(t,i)\ge 0$, for all $(t,i)\in\mathbb{R}\times\mathcal{N}$. Let $(t_0,x_0)\in\mathbb{R}_+\times\mathbb{R}^n$ be an arbitrary initial pair and $x(\cdot;t_0,x_0)$ be the solution of the perturbed system (9.1) which starts from x_0 at the initial time t_0 and corresponding to the disturbances $\boldsymbol{\Delta}_c(\cdot) = (\Delta_{c,1}(\cdot),\Delta_{c,2}(\cdot),\ldots,\Delta_{c,r_2}(\cdot))$ and $\boldsymbol{\Delta}_d(\cdot) = (\Delta_{d,1}(\cdot),\Delta_{d,2}(\cdot),\ldots,\Delta_{d,r_4}(\cdot))$ which are satisfying (9.47).

Let κ_0 be the natural number satisfying

$$(\kappa_0 - 1)h < t_0 \le \kappa_0 h. \tag{9.48}$$

Applying Itô formula on the interval $[kh,(k+1)h]$, $k\ge\kappa_0$ and $[t_0,\kappa_0 h]$ (if $t_0<\kappa_0 h$) in the case of the function $v(t,x,i) = x^\top X^{\alpha_c,\alpha_d}(t,i)x$ and of the stochastic process $x(t) := x(t;t_0,x_0)$, $t\ge t_0$ (viewed as solution of the equivalent version (9.28) of (9.1)) we obtain via Lemma 3.1 and Proposition 9.1 (ii) that

$$\mathbb{E}[x^\top((k+1)h)X^{\alpha_c,\alpha_d}((k+1)h^-,\eta((k+1)h))x((k+1)h) \mid \eta(t_0)=j]$$
$$-\mathbb{E}[x^\top(kh^+)X^{\alpha_c,\alpha_d}(kh,\eta(kh))x(kh^+) \mid \eta(t_0)=j]$$
$$=\mathbb{E}\left[\int_{kh}^{(k+1)h} x^\top(t)(\dot{X}^{\alpha_c,\alpha_d}(t,\eta(t))+\mathscr{L}^*_{c,\eta(t)}[\mathbf{X}^{\alpha_c,\alpha_d}(t)])x(t)\mathrm{d}t \mid \eta(t_0)=j\right]$$
$$+\sum_{l=1}^{r_2}\mathbb{E}\left[\int_{kh}^{(k+1)h}\left(\frac{\lambda_{c,l}}{\alpha_{c,l}(\eta(t))}\right)^2\Gamma_{c,l}^{\alpha_c,\alpha_d}(t)\mathrm{d}t \mid \eta(t_0)=j\right],$$

$k\geq\kappa_0$ and

$$\mathbb{E}[x^\top(\kappa_0 h)X^{\alpha_c,\alpha_d}(\kappa_0 h^-,\eta(\kappa_0 h))x(\kappa_0 h) \mid \eta(t_0)=j]-x_0^\top X^{\alpha_c,\alpha_d}(t_0,j)x_0$$
$$=\mathbb{E}\left[\int_{t_0}^{\kappa_0 h} x^\top(t)(\dot{X}^{\alpha_c,\alpha_d}(t,\eta(t))+\mathscr{L}^*_{c,\eta(t)}[\mathbf{X}^{\alpha_c,\alpha_d}(t)])x(t)\mathrm{d}t \mid \eta(t_0)=j\right]$$
$$+\sum_{l=1}^{r_2}\mathbb{E}\left[\int_{t_0}^{\kappa_0 h}\left(\frac{\lambda_{c,l}}{\alpha_{c,l}(\eta(t))}\right)^2\Gamma_{c,l}^{\alpha_c,\alpha_d}(t)\mathrm{d}t \mid \eta(t_0)=j\right],$$

where we denoted:

$$\Gamma_{c,l}^{\alpha_c,\alpha_d}(t):=\left(\Delta_{c,l}^{\alpha_{c,l}}(t,z_{c,l}^{\alpha_{c,l}}(t),\eta(t))\right)^\top \cdot D_{c,l}^\top(\eta(t))X^{\alpha_c,\alpha_d}(t,\eta(t))D_{c,l}(\eta(t))\Delta_{c,l}^{\alpha_{c,l}}(t,z_{c,l}^{\alpha_{c,l}}(t),\eta(t)). \tag{9.49}$$

Then, (9.36a) gives via (9.28c) the equalities:

$$\mathbb{E}\left[\int_{kh}^{(k+1)h}|z_{c,l}^{\alpha_{c,l}}(t)|^2\mathrm{d}t \mid \eta(t_0)=j\right]$$
$$=\mathbb{E}[x^\top(kh^+)X^{\alpha_c,\alpha_d}(kh,\eta(kh))x(kh^+) \mid \eta(t_0)=j] \tag{9.50}$$
$$-\mathbb{E}[x^\top((k+1)h)X^{\alpha_c,\alpha_d}((k+1)h^-,\eta((k+1)h))x((k+1)h) \mid \eta(t_0)=j]$$
$$+\sum_{l=1}^{r_2}\mathbb{E}\left[\int_{kh}^{(k+1)h}\left(\frac{\lambda_{c,l}}{\alpha_{c,l}(\eta(t))}\right)^2\Gamma_{c,l}^{\alpha_c,\alpha_d}(t)\mathrm{d}t \mid \eta(t_0)=j\right],$$

for $k\geq\kappa_0$ and

$$\sum_{l=1}^{r_2}\mathbb{E}\left[\int_{t_0}^{\kappa_0 h}|z_{c,l}^{\alpha_{c,l}}(t)|^2\mathrm{d}t \mid \eta(t_0)=j\right] \tag{9.51}$$
$$= x_0^\top X^{\alpha_c,\alpha_d}(t_0,j)x_0 - \mathbb{E}\left[x^\top(\kappa_0 h)X^{\alpha_c,\alpha_d}(\kappa_0 h^-,\eta(\kappa_0 h))x(\kappa_0 h)\mid \eta(t_0)=j\right]$$
$$+\sum_{l=1}^{r_2}\mathbb{E}\left[\int_{t_0}^{\kappa_0 h}\left(\frac{\lambda_{c,l}}{\alpha_{c,l}(\eta(t))}\right)^2\Gamma_{c,l}^{\alpha_c,\alpha_d}(t)\mathrm{d}t \mid \eta(t_0)=j\right],$$

$j\in\mathcal{N}$. Further, using the same technique as in the derivation of the value of the right hand side of (3.10b), we obtain via (9.28b) that:

$$\begin{aligned}&\mathbb{E}[x^\top(kh^+)X^{\alpha_c,\alpha_d}(kh,\eta(kh)x(kh^+)\mid\eta(t_0)=j]\\&\quad=\mathbb{E}[x^\top(kh)\mathcal{L}_{d,\eta(kh)}^*[\mathbf{X}^{\alpha_c,\alpha_d}(kh)]x(kh)\mid\eta(t_0)=j]\\&\qquad+\sum_{p=1}^{r_4}\mathbb{E}[\left(\frac{\lambda_{d,p}}{\alpha_{d,p}(\eta(kh))}\right)^2\Gamma_{d,p}^{\alpha_c,\alpha_d}(kh)\mid\eta(t_0)=j],\end{aligned}$$

where

$$\begin{aligned}\Gamma_{d,p}^{\alpha_c,\alpha_d}(kh):=&(\Delta_{d,p}^{\alpha_{d,p}}(k,z_{d,p}^{\alpha_{d,p}}(kh),\eta(kh)))^\top D_{d,p}^\top(\eta(kh))\\&\cdot X^{\alpha_c,\alpha_d}(kh,\eta(kh))D_{d,p}(\eta(kh))\Delta_{d,p}^{\alpha_{d,p}}(k,z_{d,p}^{\alpha_{d,p}}(kh),\eta(kh)).\end{aligned} \tag{9.52}$$

Then, (9.36b), (9.28d) allow us to deduce that

$$\begin{aligned}&\mathbb{E}[x^\top(kh^+)X^{\alpha_c,\alpha_d}(kh,\eta(kh))x(kh^+)\mid\eta(t_0)=j]\\&=\mathbb{E}[x^\top(kh)X^{\alpha_c,\alpha_d}(kh^-,\eta(kh))x(kh)\mid\eta(t_0)=j]\\&-\sum_{p=1}^{r_4}\mathbb{E}[|z_{d,p}^{\alpha_{d,p}}(kh)|^2\mid\eta(t_0)=j]+\sum_{p=1}^{r_4}\mathbb{E}[\left(\frac{\lambda_{d,p}}{\alpha_{d,p}(\eta(kh))}\right)^2\Gamma_{d,p}^{\alpha_c,\alpha_d}(kh)\mid\eta(t_0)=j].\end{aligned}$$

Substituting this equality in (9.50) we get:

$$\begin{aligned}&\sum_{l=1}^{r_2}\mathbb{E}\left[\int_{kh}^{(k+1)h}|z_{c,l}^{\alpha_{c,l}}(t)|^2\mathrm{d}t\mid\eta(t_0)=j]+\sum_{p=1}^{r_4}\mathbb{E}[|z_{d,p}^{\alpha_{d,p}}(kh)|^2\mid\eta(t_0)=j\right]\\&=\mathbb{E}[x^\top(kh)X^{\alpha_c,\alpha_d}(kh^-,\eta(kh))x(kh)\mid\eta(t_0)=j]\\&\quad-\mathbb{E}[x^\top((k+1)h)X^{\alpha_c,\alpha_d}((k+1)h^-,\eta((k+1)h))x((k+1)h)\mid\eta(t_0)=j]\\&\quad+\sum_{l=1}^{r_2}\mathbb{E}\left[\int_{kh}^{(k+1)h}\left(\frac{\lambda_{c,l}}{\alpha_{c,l}(\eta(t))}\right)^2\Gamma_{c,l}^{\alpha_c,\alpha_d}(t)\mathrm{d}t\mid\eta(t_0)=j\right]\end{aligned} \tag{9.53}$$

$$+\sum_{p=1}^{r_4}\mathbb{E}\left[\left(\frac{\lambda_{d,p}}{\alpha_{d,p}(\eta(kh))}\right)^2\Gamma_{d,p}^{\alpha_c,\alpha_d}(kh)\mid\eta(t_0)=j\right],$$

for all $k\geq\kappa_0$, $j\in\mathcal{N}$. Let $T>\kappa_0 h$ be arbitrary. Summing (9.53) for $k=\kappa_0$ to $k=\left[\frac{T}{h}\right]$ (with the convention that when $k=\left[\frac{T}{h}\right]$, the interval $[kh,(k+1)h]$ is replaced by the interval $\left[\left[\frac{T}{h}\right]h,T\right]$) one obtains:

$$\begin{aligned}
&\sum_{l=1}^{r_2}\mathbb{E}\left[\int_{\kappa_0 h}^{T}|z_{c,l}^{\alpha_{c,l}}(t)|^2\mathrm{d}t\mid\eta(t_0)=j\right]+\sum_{p=1}^{r_4}\sum_{k=\kappa_0}^{\left[\frac{T}{h}\right]}\mathbb{E}[|z_{d,p}^{\alpha_{d,p}}(kh)|^2\mid\eta(t_0)=j]\\
&=\mathbb{E}[x^\top(\kappa_0 h)X^{\alpha_c,\alpha_d}(\kappa_0 h^-,\eta(\kappa_0 h))x(\kappa_0 h)\mid\eta(t_0)=j]\\
&\quad-\mathbb{E}[x^\top(T)X^{\alpha_c,\alpha_d}(T,\eta(T)x(T)\mid\eta(t_0)=j]\\
&\quad+\sum_{l=1}^{r_2}\int_{\kappa_0 h}^{T}\mathbb{E}\left[\left(\frac{\lambda_{c,l}}{\alpha_{c,l}(\eta(t))}\right)^2\Gamma_{c,l}^{\alpha_c,\alpha_d}(t)\mid\eta(t_0)=j\right]\mathrm{d}t\\
&\quad+\sum_{p=1}^{r_4}\sum_{k=\kappa_0}^{\left[\frac{T}{h}\right]}\mathbb{E}[\left(\frac{\lambda_{d,p}}{\alpha_{d,p}(\eta(kh))}\right)^2\Gamma_{d,p}^{\alpha_c,\alpha_d}(kh)\mid\eta(t_0)=j],
\end{aligned}\tag{9.54}$$

where $\left[\frac{T}{h}\right]$ is the integer part of the real number $\frac{T}{h}$.

If $t_0<\kappa_0 h$, we add (9.51) with (9.54) and obtain:

$$\begin{aligned}
&\sum_{l=1}^{r_2}\int_{t_0}^{T}\mathbb{E}\left[|z_{c,l}^{\alpha_{c,l}}(t)|^2\mid\eta(t_0)=j\right]\mathrm{d}t+\sum_{p=1}^{r_4}\sum_{k=\kappa_0}^{\left[\frac{T}{h}\right]}\mathbb{E}[|z_{d,p}^{\alpha_{d,p}}(kh)|^2\mid\eta(t_0)=j]\\
&=x_0^\top X^{\alpha_c,\alpha_d}(t_0,j)x_0-\mathbb{E}[x^\top(T)X^{\alpha_c,\alpha_d}(T,\eta(T))x(T)\mid\eta(t_0)=j]\\
&\quad+\sum_{l=1}^{r_2}\int_{t_0}^{T}\mathbb{E}\left[\left(\frac{\lambda_{c,l}}{\alpha_{c,l}(\eta(t))}\right)^2\Gamma_{c,l}^{\alpha_c,\alpha_d}(t)\mid\eta(t_0)=j\right]\mathrm{d}t\\
&\quad+\sum_{p=1}^{r_4}\sum_{k=\kappa_0}^{\left[\frac{T}{h}\right]}\mathbb{E}\left[\left(\frac{\lambda_{d,p}}{\alpha_{d,p}(\eta(kh))}\right)^2\Gamma_{d,p}^{\alpha_c,\alpha_d}(kh)\mid\eta(t_0)=j\right],
\end{aligned}\tag{9.55}$$

for all $T>\kappa_0 h$, $j\in\mathcal{N}$. When $t_0=\kappa_0 h$, (9.54) will coincide with (9.55). Further, (9.49) together with the property of the solution $\mathbf{X}^{\alpha_c,\alpha_d}(\cdot)$ yield:

$$\begin{aligned}
0&\leq\mathbb{E}[\left(\frac{\lambda_{c,l}}{\alpha_{c,l}(\eta(t))}\right)^2\Gamma_{c,l}^{\alpha_c,\alpha_d}(t)\mid\eta(t_0)=j]\\
&\leq\max_{i\in\mathcal{N}}\sup_{0\leq t\leq h}\left(\left(\frac{\lambda_{c,l}}{\alpha_{c,l}(i)}\right)^2\|D_{c,l}^\top(i)X^{\alpha_c,\alpha_d}(t,i)D_{c,l}(i)\|_2\right)
\end{aligned}\tag{9.56}$$

$$\cdot \mathbb{E}[|\Delta_{c,l}^{\alpha_{c,l}}(t, z_{c,l}^{\alpha_{c,l}}(t), \eta(t))|^2 \mid \eta(t_0) = j],$$

for all $t \in \mathbb{R}_+$, $j \in \mathcal{N}$, $l \in \underline{r_2}$. Then, (9.15), (9.31), (9.33a), (9.47) give:

$$\mathbb{E}[|\Delta_{c,l}^{\alpha_{c,l}}(t, z_{c,l}^{\alpha_{c,l}}(t), \eta(t)) \mid \eta(t_0) = j] \le \sigma_1^2 \mathbb{E}[|z_{c,l}^{\alpha_{c,l}}(t)|^2 \mid \eta(t_0) = j], \tag{9.57}$$

for all $t \ge t_0$, $l \in \underline{r_2}$. Combining (9.56) and (9.57) we get

$$\begin{aligned} 0 &\le \mathbb{E}[\left(\frac{\lambda_{c,l}}{\alpha_{c,l}(\eta(t))}\right)^2 \Gamma_{c,l}^{\alpha_c,\alpha_d}(t) \mid \eta(t_0) = j] \qquad (9.58) \\ &\le \max_{i \in \mathcal{N}} \sup_{0 \le t \le h} \left(\left(\frac{\sigma_1 \lambda_{c,l}}{\alpha_{c,l}(i)}\right)^2 \|D_{c,l}^{\top}(i) X^{\alpha_c,\alpha_d}(t, i) D_{c,l}(i)\|_2 \right) \mathbb{E}[|z_{c,l}^{\alpha_{c,l}}(t)|^2 \mid \eta(t_0) = j], \end{aligned}$$

for all $t \ge t_0$, $l \in \underline{r_2}$, $j \in \mathcal{N}$. On the other hand, (9.52) gives:

$$\begin{aligned} 0 &\le \mathbb{E}[\left(\frac{\lambda_{d,p}}{\alpha_{d,p}(\eta(kh))}\right)^2 \Gamma_{d,p}^{\alpha_c,\alpha_d}(kh) \mid \mid \eta(t_0) = j] \\ &\le \max_{i \in \mathcal{N}} \left(\left(\frac{\lambda_{d,p}}{\alpha_{d,p}(i)}\right)^2 \|D_{d,p}^{\top}(i) X^{\alpha_c,\alpha_d}(0, i) D_{d,p}(i)\|_2 \right) \qquad (9.59) \\ &\quad \cdot \mathbb{E}[|\Delta_{d,p}^{\alpha_{d,p}}(k, z_{d,p}^{\alpha_{d,p}}(kh), \eta(kh))|^2 \mid \eta(t_0) = j] \end{aligned}$$

for all $p \in \underline{r_4}$, $k \ge \kappa_0$, $j \in \mathcal{N}$. Employing (9.16), (9.32), (9.33b), (9.47) we deduce that:

$$\mathbb{E}[|\Delta_{d,p}^{\alpha_{d,p}}(k, z_{d,p}^{\alpha_{d,p}}(kh), \eta(kh)) \mid \eta(t_0) = j] \le \sigma_2^2 \mathbb{E}[|z_{d,p}^{\alpha_{d,p}}(kh)|^2 \mid \eta(t_0) = j], \tag{9.60}$$

for all $p \in \underline{r_4}$, $k \ge \kappa_0$, $j \in \mathcal{N}$. From (9.59), (9.60) we get

$$\begin{aligned} 0 &\le \mathbb{E}\left[\left(\frac{\lambda_{d,p}}{\alpha_{d,p}(\eta(kh))}\right)^2 \Gamma_{d,p}^{\alpha_c,\alpha_d}(kh) \mid \eta(t_0) = j\right] \qquad (9.61) \\ &\le \max_{i \in \mathcal{N}} \left(\left(\frac{\sigma_2 \lambda_{d,p}}{\alpha_{d,p}(i)}\right)^2 \|D_{d,p}^{\top}(i) X^{\alpha_c,\alpha_d}(0, i) D_{d,p}(i)\|_2 \right) \mathbb{E}[|z_{d,p}^{\alpha_{d,p}}(kh)|^2 \mid \eta(t_0) = j], \end{aligned}$$

for all $p \in \underline{r_4}$, $k \ge \kappa_0$, $j \in \mathcal{N}$. Plugging (9.58) and (9.61) in (9.53) we may infer that

$$
\sum_{l=1}^{r_2}\left(1-\max_{i\in\mathcal{N}}\sup_{0\le t\le h}\left[\left(\frac{\sigma_1\lambda_{c,l}}{\alpha_{c,l}(i)}\right)^2\|D_{c,l}^\top(i)X^{\alpha_c,\alpha_d}(t,i)D_{c,l}(i)\|_2\right]\right)
\cdot\int_{t_0}^{T}\mathbb{E}[|z_{c,l}^{\alpha_{c,l}}(t)|^2|\eta(t_0)=j]\mathrm{d}t+\sum_{p=1}^{r_4}(1-\max_{i\in\mathcal{N}}(\frac{\sigma_2\lambda_{d,p}}{\alpha_{d,p}(i)})^2\|D_{d,p}^\top(i)X^{\alpha_c,\alpha_d}(0,i)D_{d,p}(i)\|_2)
\cdot\sum_{k=\kappa_0}^{\left[\frac{T}{h}\right]}\mathbb{E}[|z_{d,p}^{\alpha_{d,p}}(kh)|^2\mid\eta(t_0)=j]
\le x_0^\top X^{\alpha_c,\alpha_d}(t_0,j)x_0-\mathbb{E}[x^\top(T)X^{\alpha_c,\alpha_d}(T,\eta(T))x(T)\mid\eta(t_0)=j], \tag{9.62}
$$

for all $T>\kappa_0 h,\ j\in\mathcal{N}$. Let

$$
\delta_1:=1-\max_{l\in\underline{r_2}}\max_{i\in\mathcal{N}}\sup_{0\le t\le h}\left(\frac{\sigma_1\lambda_{c,l}}{\alpha_{c,l}(i)}\right)^2\|D_{c,l}^\top(i)X^{\alpha_c,\alpha_d}(t,i)D_{c,l}(i)\|_2 \tag{9.63a}
$$

$$
\delta_2:=1-\max_{p\in\underline{r_4}}\max_{i\in\mathcal{N}}\left(\frac{\sigma_2\lambda_{d,p}}{\alpha_{d,p}(i)}\right)^2\|D_{d,p}^\top(i)X^{\alpha_c,\alpha_d}(0,i)D_{d,p}(i)\|_2. \tag{9.63b}
$$

Employing (9.44), (9.45), (9.63) together with Proposition 9.1 (iii) we obtain from (9.62) that

$$
\delta_1\sum_{l=1}^{r_2}\int_{t_0}^{T}\mathbb{E}[|z_{c,l}^{\alpha_{c,l}}(t)|^2|\eta(t_0)=j]\mathrm{d}t+\delta_2\sum_{p=1}^{r_4}\sum_{k=\kappa_0}^{\left[\frac{T}{h}\right]}\mathbb{E}[|z_{d,p}^{\alpha_{d,p}}(kh)|^2|\eta(t_0)=j]\le c|x_0|^2, \tag{9.64}
$$

for all $T>\kappa_0 h,\ j\in\mathcal{N}$, and any initial pair $(t_0,x_0)\in\mathbb{R}_+\times\mathbb{R}^n$, κ_0 being related to t_0 via (9.48), where $c>0$ is a constant such that $0\le X^{\alpha_c,\alpha_d}(t,i)\le cI_n$, for all $(t,i)\in\mathbb{R}\times\mathcal{N}$. In order to obtain (9.64) from (9.62) we took into account that

$$
0<\delta_1\le 1-\max_{i\in\mathcal{N}}\sup_{0\le t\le h}\left(\frac{\sigma_1\lambda_{c,l}}{\alpha_{c,l}(i)}\right)^2\|D_{c,l}^\top(i)X^{\alpha_c,\alpha_d}(t,i)D_{c,l}(i)\|_2,\quad \text{for all } l\in\underline{r_2}
$$

and $0<\delta_2\le 1-\max_{i\in\mathcal{N}}\left(\frac{\sigma_2\lambda_{d,p}}{\alpha_{d,p}(i)}\right)^2\|D_{d,p}^\top(i)X^{\alpha_c,\alpha_d}(0,i)D_{d,p}(i)\|_2$ for all $p\in\underline{r_4}$. Letting $T\to\infty$ in (9.64) we get

$$
\int_{t_0}^{\infty}\mathbb{E}[|z_{c,l}^{\alpha_{c,l}}(t)|^2\mid\eta(t_0)=j]\mathrm{d}t\le c\delta_1^{-1}|x_0|^2 \tag{9.65a}
$$

$$
\sum_{k=\kappa_0}^{\infty}\mathbb{E}[|z_{d,p}^{\alpha_{d,p}}(kh)|^2\mid\eta(t_0)=j]\le c\delta_2^{-1}|x_0|^2, \tag{9.65b}
$$

for all $l \in \underline{r_2}$, $p \in \underline{r_4}$, for any initial pair $(t_0, x_0) \in \mathbb{R}_+ \times \mathbb{R}^n$, $j \in \mathcal{N}$, κ_0 being connected by t_0 through (9.48). From (9.15), (9.65a) we obtain

$$\int_{t_0}^{\infty} \mathbb{E}[|\Delta_{c,l}^{\alpha_{c,l}}(t, z_{c,l}^{\alpha_{c,l}}(t), \eta(t))|^2 \mid \eta(t_0) = j]dt \tag{9.66}$$

$$\leq \|\boldsymbol{\Delta}_c(\cdot)\|^2 \int_{t_0}^{\infty} \mathbb{E}[|z_{c,l}^{\alpha_{c,l}}(t)|^2 \mid \eta(t_0) = j]dt \leq \|\boldsymbol{\Delta}_c(\cdot)\|^2 \, c\delta_1^{-1}|x_0|^2,$$

for all $\Delta_{c,l}^{\alpha_{c,l}}(t, \cdot, i)$ defined as in (9.29) for an arbitrary $\Delta_{c,l}(t, \cdot, i) \in \mathcal{N}\mathcal{L}_{[c]}(\mathbb{R}^{\nu_l}, \mathbb{R}^{m_l})$, $l \in \underline{r_2}$. Hence, $t \to \Delta_{c,l}^{\alpha_{c,l}}(t, z_{c,l}^{\alpha_{c,l}}(t), \eta(t)) : [t_0, \infty) \to \mathbb{R}^{m_l}$ lies in $\mathcal{L}^2_{\mathcal{H}}([t_0, \infty), \mathbb{R}^{m_l})$.

Similarly, from (9.16) and (9.65b) we get

$$\sum_{k=\kappa_0}^{\infty} \mathbb{E}[|\Delta_{d,p}^{\alpha_{d,p}}(k, z_{d,p}^{\alpha_{d,p}}(kh), \eta(kh))|^2 \mid \eta(t_0) = j] \tag{9.67}$$

$$\leq \|\boldsymbol{\Delta}_d(\cdot)\|^2 \sum_{k=\kappa_0}^{\infty} \mathbb{E}[|z_{d,p}^{\alpha_{d,p}}(kh)|^2 \,|\eta(t_0) = j] \leq \|\boldsymbol{\Delta}_d(\cdot)\|^2 \, c\delta_2^{-1}|x_0|^2,$$

for all $\Delta_{d,p}^{\alpha_{d,p}}(k,\cdot,i)$ defined as in (9.30) for an arbitrary measurable function $\Delta_{d,p}(k,\cdot,i)$ which is satisfying conditions (9.4) and (9.5). So, $\{\Delta_{d,p}^{\alpha_{d,p}}(k, z_{d,p}^{\alpha_{d,p}}(kh), \eta(kh))\}_{k\geq\kappa_0} \in \ell^2_{\mathcal{H}}\{\kappa_0, \infty; \mathbb{R}^{m_{d,p}}\}$, for all $p \in \underline{r_4}$.

Taking $f_l(t) \equiv 0$ if $0 \leq l \leq r_1$,

$$f_{r_1+l}(t) := \frac{1}{\alpha_{c,l}(\eta(t))} D_{c,l}(\eta(t))\Delta_{c,l}^{\alpha_{c,l}}(t, z_{c,l}^{\alpha_{c,l}}(t), \eta(t)) \quad \text{for} \quad 1 \leq l \leq r_2,$$

$t \in \mathbb{R}_+$, and $g_p = 0$ if $0 \leq p \leq r_3$,

$$g_{r_3+p}(k) := \frac{1}{\alpha_{d,p}(\eta(kh))} D_{d,p}(\eta(kh))\Delta_{d,p}^{\alpha_{d,p}}(k, z_{d,p}^{\alpha_{d,p}}(kh), \eta(kh)) \quad \text{for} \quad 1 \leq p \leq r_4,$$

$k \in \mathbb{Z}_+$, we obtain from Theorem 3.5 (i) that:

$$\mathbb{E}[|x(t; t_0, x_0)|^2 \mid \eta(t_0) = j] \leq \beta(e^{-\gamma(t-t_0)}|x_0|^2 + \sum_{p=1}^{r_4} \max_{i\in\mathcal{N}}\{\alpha_{d,p}^{-2}(i)\|D_{d,p}(i)\|_2^2\}$$

$$\cdot \sum_{k=\kappa_0}^{\kappa_t} e^{-\gamma(t-kh)}\mathbb{E}[|\Delta_{d,p}^{\alpha_{d,p}}(k, z_{d,p}^{\alpha_{d,p}}(kh)), \eta(kh))|^2 \mid \eta(t_0) = j] \tag{9.68}$$

$$+\sum_{l=1}^{r_2} \max_{i\in\mathcal{N}}\{\alpha_{c,l}^{-2}(i)\|D_{c,l}(i)\|_2^2\} \int_{t_0}^{t} e^{-\gamma(t-s)}\mathbb{E}[|\Delta_{c,l}^{\alpha_{c,l}}(s, z_{c,l}^{\alpha_{c,l}}(s), \eta(s))|^2|\eta(t_0) = j]ds),$$

for all $t > \kappa_0 h$, where $\gamma > 0$ is guaranteed by assumption (b) from the statement and κ_t is a natural number satisfying $\kappa_t h \le t < (\kappa_t + 1)h$. If $t \in [t_0, \kappa_0 h]$ we get

$$\mathbb{E}[|x(t; t_0, x_0)|^2 \mid \eta(t_0) = j] \le \beta(e^{-\gamma(t-t_0)}|x_0|^2 + \sum_{l=1}^{r_2} \max_{i\in\mathcal{N}}\{\alpha_{c,l}^{-2}(i)\|D_{c,l}(i)\|_2^2\}$$

$$\cdot \int_{t_0}^{t} e^{-\gamma(t-s)}\mathbb{E}[|\Delta_{c,l}^{\alpha_{c,l}}(s, z_{c,l}^{\alpha_{c,l}}(s)), \eta(s)|^2 \mid \eta(t_0) = j]ds). \quad (9.69)$$

Combining (9.66) and (9.69) we may infer that $\mathbb{E}[|x(t; t_0, x_0)|^2] \le \widetilde{c}|x_0|^2$ for all $t \ge t_0$, an arbitrary initial pair $(t_0, x_0) \in \mathbb{R}_+ \times \mathbb{R}^n$ and an initial probability distribution π_0 of the Markov process, $\widetilde{c} > 0$ being a constant which does not depend upon t_0, x_0, t, π_0. So, if $|x_0| \le \widetilde{c}^{-1}\varepsilon$ then $\mathbb{E}[|x(t; t_0, x_0)|^2] \le \varepsilon$ for all $t \ge t_0 \ge 0$. This means that (9.1) is UMSS. Finally, based on Theorem 3.5 (iii) we obtain that $\lim_{t\to\infty} \mathbb{E}[|x(t; t_0, x_0)|^2] = 0$, for all $(t_0, x_0) \in \mathbb{R}_+ \times \mathbb{R}^n$ and any initial probability distribution π_0. This allows us to conclude that the perturbed system (9.1) corresponding to some admissible parametric uncertainties which satisfies (9.47) is GUMSAS. Thus the proof ends.

The next result provides the best lower bound of the stability radius which may be obtained via the parametric scaling technique.

Theorem 9.2 *Under the assumptions of Theorem 9.1 the following holds:*

$$\rho_{\mathcal{D}}[\mathbf{G_0} \mid (\mathbf{D}_{c,l}, \mathbf{C}_{c,l}),\ l \in \underline{r_2},\ (\mathbf{D}_{d,p}, \mathbf{C}_{d,p}),\ p \in \underline{r_4}] \ge \widehat{\mu}, \quad (9.70a)$$

$$\widehat{\mu} = \sup_{(\alpha_c,\alpha_d)\in(0,\infty)^{\underline{r_2}\times\mathcal{N}} \times(0,\infty)^{\underline{r_4}\times\mathcal{N}}} \min\{\mu_c(\alpha_c, \alpha_d), \mu_d(\alpha_c, \alpha_d)\} \quad (9.70b)$$

where

$$\mu_c(\alpha_c, \alpha_d) = [\max_{l\in\underline{r_2}} \max_{i\in\mathcal{N}} \sup_{0\le t\le h} \left(\frac{\lambda_{c,l}}{\alpha_{c,l}(i)}\right)^2 \|D_{c,l}^{\top}(i)X^{\alpha_c,\alpha_d}(t, i)D_{c,l}(i)\|_2]^{-1/2} \quad (9.71a)$$

$$\mu_d(\alpha_c, \alpha_d) = [\max_{p\in\underline{r_4}} \max_{i\in\mathcal{N}} \left(\frac{\lambda_{d,p}}{\alpha_{d,p}(i)}\right)^2 \|D_{d,p}^{\top}(i)X^{\alpha_c,\alpha_d}(0, i)D_{d,p}(i)\|_2]^{-1/2} \quad (9.71b)$$

with $\mathbf{X}^{\alpha_c,\alpha_d}(\cdot) = (X^{\alpha_c,\alpha_d}(\cdot, 1), X^{\alpha_c,\alpha_d}(\cdot, 2), \ldots, X^{\alpha_c,\alpha_d}(\cdot, N))$ *is the unique bounded on* $\mathbb{R}$ *solution of the BLJLDE (9.35) corresponding to the vectors* α_c, α_d, $\alpha_c = \{\alpha_{c,l}(i)\}_{(l,i)\in\underline{r_2}\times\mathcal{N}}$, $\alpha_d = \{\alpha_{d,p}(i)\}_{(p,i)\in\underline{r_4}\times\mathcal{N}}$.

Proof Let us assume by contrary that (9.70a) is not true. This means that

$$\rho_{\mathscr{D}}[\mathbf{G_0} \mid (\mathbf{D}_{c,l}, \mathbf{C}_{c,l}),\ l \in \underline{r_2},\ (\mathbf{D}_{d,p}, \mathbf{C}_{d,p}),\ p \in \underline{r_4}] < \widehat{\mu}.$$

Let an arbitrary σ be such that

$$\rho_{\mathscr{D}}[\mathbf{G_0} \mid (\mathbf{D}_{c,l}, \mathbf{C}_{c,l}),\ l \in \underline{r_2},\ (\mathbf{D}_{d,p}, \mathbf{C}_{d,p}),\ p \in \underline{r_4}] < \sigma < \widehat{\mu}. \tag{9.72}$$

From (9.70b) it follows that for each $\sigma_1 \in (\sigma, \widehat{\mu})$ there exists $(\widehat{\alpha}_c, \widehat{\alpha}_d) \in (0, \infty)^{\underline{r_2} \times \mathscr{N}} \times (0, \infty)^{\underline{r_4} \times \mathscr{N}}$ such that

$$\sigma_1 < \min\{\mu_c(\widehat{\alpha}_c, \widehat{\alpha}_d), \mu_d(\widehat{\alpha}_c, \widehat{\alpha}_d)\}. \tag{9.73}$$

Further, (9.71) written for (α_c, α_d) replaced by $(\widehat{\alpha}_c, \widehat{\alpha}_d)$ together with (9.73) lied to

$$\sigma_1 < (\max_{l \in \underline{r_2}} \max_{i \in \mathscr{N}} \sup_{0 \le t \le h} \left(\frac{\lambda_{c,l}}{\widehat{\alpha}_{c,l}(i)}\right)^2 \|D_{c,l}^{\top}(i) X^{\widehat{\alpha}_c, \widehat{\alpha}_d}(t, i) D_{c,l}(i)\|_2)^{-1/2}$$

and

$$\sigma_1 < (\max_{p \in \underline{r_4}} \max_{i \in \mathscr{N}} \left(\frac{\lambda_{d,p}}{\widehat{\alpha}_{d,p}(i)}\right)^2 \|D_{d,p}^{\top}(i) X^{\widehat{\alpha}_c, \widehat{\alpha}_d}(0, i) D_{d,p}(i)\|_2)^{-1/2},$$

or equivalently

$$\max_{l \in \underline{r_2}} \max_{i \in \mathscr{N}} \sup_{0 \le t \le h} \left(\frac{\sigma_1 \lambda_{c,l}}{\widehat{\alpha}_{c,l}(i)}\right)^2 \|D_{c,l}^{\top}(i) X^{\widehat{\alpha}_c, \widehat{\alpha}_d}(t, i) D_{c,l}(i)\|_2 < 1 \tag{9.74a}$$

$$\max_{p \in \underline{r_4}} \max_{i \in \mathscr{N}} \left(\frac{\sigma_1 \lambda_{d,p}}{\widehat{\alpha}_{d,p}(i)}\right)^2 \|D_{d,p}^{\top}(i) X^{\widehat{\alpha}_c, \widehat{\alpha}_d}(0, i) D_{d,p}(i)\|_2 < 1. \tag{9.74b}$$

Applying Theorem 9.1 we may conclude via (9.74) that

$$\rho_{\mathscr{D}}[\mathbf{G_0} \mid (\mathbf{D}_{c,l}, \mathbf{C}_{c,l}),\ l \in \underline{r_2},\ (\mathbf{D}_{d,p}, \mathbf{C}_{d,p}),\ p \in \underline{r_4}] \ge \sigma_1 > \sigma$$

but this is in contradiction with (9.72). Hence, (9.70a) holds. Thus the proof is complete.

In the case of the stability radius $\rho_{\mathscr{D}_c}[\mathbf{G_0} \mid (\mathbf{D}_{c,l}, \mathbf{C}_{c,l}),\ l \in \underline{r_2}]$ defined in (9.21) of the nominal system $\mathbf{G_0}$ with respect to the set $\mathscr{D}_c$ of the structured parametric uncertainties having the structure described by the pairs $(\mathbf{D}_{c,l}, \mathbf{C}_{c,l})$, $l \in \underline{r_2}$ we obtain:

Theorem 9.3 *Under the assumptions of Theorem 9.1 the following holds:*

$$\rho_{\mathscr{D}_c}[\mathbf{G_0} \mid (\mathbf{D}_{c,l}, \mathbf{C}_{c,l}),\ l \in \underline{r_2}] \ge \widetilde{\mu}_c, \tag{9.75a}$$

$$\widetilde{\mu}_c = \sup_{\alpha_c \in (0,\infty)^{\underline{r_2} \times \mathcal{N}}} (\max_{l \in \underline{r_2}} \max_{i \in \mathcal{N}} \sup_{0 \le t \le h} (\frac{\lambda_{c,l}}{\alpha_{c,l}(i)})^2 \| D_{c,l}^{\top}(i) X^{\alpha_c,0}(t,i) D_{c,l}(i) \|_2)^{-1/2} \quad (9.75b)$$

$\mathbf{X}^{\alpha_c,0}(\cdot) = (X^{\alpha_c,0}(\cdot,1), X^{\alpha_c,0}(\cdot,2), \dots, X^{\alpha_c,0}(\cdot,N))$ *being the unique bounded on* $\mathbb{R}$ *solution of the BJLDE (9.40) corresponding to the vector of scaling parameters* $\alpha_c = \{\alpha_{c,l}(i)\}_{(l.i) \in \underline{r_2} \times \mathcal{N}} \in (0,\infty)^{\underline{r_2} \times \mathcal{N}}$.

Proof Is obtained from the previous theorem taking $D_{d,p}(i) = 0, \quad C_{d,p}(i) = 0$ $1 \le p \le r_4$, $i \in \mathcal{N}$. The details are left to the reader.

In the case of the stability radius $\rho_{\mathcal{D}_d}[\mathbf{G_0} \mid (\mathbf{D}_{d,p}, \mathbf{C}_{d,p}),\ l \in \underline{r_4}]$ of the nominal system $\mathbf{G_0}$ with respect to the set $\mathcal{D}_d$ of the structured parametric uncertainties having the structure described by the pairs $(\mathbf{D}_{d,p}, \mathbf{C}_{d,p})$, $p \in \underline{r_4}$, we have:

Theorem 9.4 *Under the assumptions of Theorem 9.1 the following holds:*

$$\rho_{\mathcal{D}_d}[\mathbf{G_0} \mid (\mathbf{D}_{d,p}, \mathbf{C}_{d,p}),\ p \in \underline{r_4}] \ge \widetilde{\mu}_d, \quad (9.76a)$$

$$\widetilde{\mu}_d = \sup_{\alpha_d \in (0,\infty)^{\underline{r_4} \times \mathcal{N}}} \left(\max_{p \in \underline{r_4}} \max_{i \in \mathcal{N}} \left(\frac{\lambda_{d,p}}{\alpha_{d,p}(i)} \right)^2 \| D_{d,p}^{\top}(i) X^{0,\alpha_d}(0,i) D_{d,p}(i) \|_2 \right)^{-1/2} \quad (9.76b)$$

where $\mathbf{X}^{0,\alpha_d}(\cdot) = (X^{0,\alpha_d}(\cdot,1), X^{0,\alpha_d}(\cdot,2), \dots, X^{0,\alpha_d}(\cdot,N))$ *is the unique bounded on* $\mathbb{R}$ *solution of the BLJLDE (9.41) corresponding to the vector of scaling parameters* $\alpha_d = \{\alpha_{d,p}(i)\}_{(p.i) \in \underline{r_4} \times \mathcal{N}}$.

Proof The conclusion follows from Theorem 9.2 taking into account that the perturbed system (9.22) is the particular case of (9.1) for $D_{c,l}(i) = 0$, $C_{c,l} = 0$, $l \in \underline{r_2}$, $i \in \mathcal{N}$.

In the last part of this section we present several results which will be used in the developments from the next section.

Proposition 9.2 *Assume that Assumption 9.1 holds. Let* $\widehat{\mu}$ *be introduced in (9.70b). For a given scalar* $\sigma > 0$, *the following are equivalent:*

(i) the nominal system (9.7) is ESMS and $\sigma < \widehat{\mu}$;
(ii) there exist the vectors of scaling parameters $\alpha_c = \{\alpha_{c,l}(i)\}_{(l,i) \in \underline{r_2} \times \mathcal{N}} \in (0,\infty)^{\underline{r_2} \times \mathcal{N}}$ *and* $\alpha_d = \{\alpha_{d,p}(i)\}_{(p,i) \in \underline{r_4} \times \mathcal{N}} \in (0,\infty)^{\underline{r_4} \times \mathcal{N}}$ *and the bounded and h-periodic function* $t \to \mathbf{Y}(t) = (Y(t,1), Y(t,2), \dots, Y(t,N)) : \mathbb{R} \to \mathcal{S}_n^N$ *differentiable on each interval* $(kh, (k+1)h)$, *right continuous in any* $\tau_k = kh$, $k \in \mathbb{Z}$ *solving the following backward Lyapunov-type jump linear differential inequality (BLJLDI):*

$$\dot{Y}(t,i) + \mathcal{L}_{c,i}^{*}[\mathbf{Y}(t)] + \sum_{l=1}^{r_2} \alpha_{c,l}^2(i) C_{c,l}^{\top}(i) C_{c,l}(i) \le 0, \quad kh \le t < (k+1)h \quad (9.77a)$$

$$\mathcal{L}_{d,i}^{*}[\mathbf{Y}(kh)] - Y(kh^{-}, i) + \sum_{p=1}^{r_4} \alpha_{d,p}^{2}(i) C_{d,p}^{\top}(i) C_{d,p}(i) < 0, \quad k \in \mathbb{Z}, \tag{9.77b}$$

together with the following constraints

$$Y(t, i) > 0 \tag{9.78a}$$

$$\alpha_{c,l}^{2}(i) I_{m_l} - \sigma^2 \lambda_{c,l}^{2} D_{c,l}^{\top}(i) Y(t, i) D_{c,l}(i) > 0, \quad l \in \underline{r_2}, \quad t \in \mathbb{R}, \tag{9.78b}$$

$$\alpha_{d,p}^{2} I_{m_{d,p}} - \sigma^2 \lambda_{d,p}^{2} D_{d,p}^{\top}(i) Y(0, i) D_{d,p}(i) > 0, \quad p \in \underline{r_4}, \quad i \in \mathcal{N}. \tag{9.78c}$$

Proof We start to prove the implication $(i) \Rightarrow (ii)$. If (i) holds, we take an arbitrary $\sigma_1 \in (\sigma, \widehat{\mu})$. From (9.70b) and (9.71) we may infer that there exist vectors of scaling parameters $\alpha_c = \{\alpha_{c,l}(i)\}_{(l,i) \in \underline{r_2} \times \mathcal{N}} \in (0, \infty)^{\underline{r_2} \times \mathcal{N}}$ and $\alpha_d = \{\alpha_{d,p}(i)\}_{(p,i) \in \underline{r_4} \times \mathcal{N}} \in (0, \infty)^{\underline{r_4} \times \mathcal{N}}$ such that

$$\sigma_1^2 < [\max_{(l,i) \in \underline{r_2} \times \mathcal{N}} \sup_{0 \le t \le h} \left(\frac{\lambda_{c,l}}{\alpha_{c,l}(i)}\right)^2 \|D_{c,l}^{\top}(i) X^{\alpha_c, \alpha_d}(t, i) D_{c,l}(i)\|_2]^{-1}$$

and

$$\sigma_1^2 < [\max_{(p,i) \in \underline{r_4} \times \mathcal{N}} \left(\frac{\lambda_{d,p}}{\alpha_{d,p}(i)}\right)^2 \|D_{d,p}^{\top}(i) X^{\alpha_c, \alpha_d}(0, i) D_{d,p}(i)\|_2]^{-1}$$

where $\mathbf{X}^{\alpha_c, \alpha_d}(\cdot) = (X^{\alpha_c, \alpha_d}(\cdot, 1), X^{\alpha_c, \alpha_d}(\cdot, 2), \ldots, X^{\alpha_c, \alpha_d}(\cdot, N))$ is the unique bounded on $\mathbb{R}$ solution of the BLJLDE (9.36) corresponding to the pair of scaling parameter (α_c, α_d). The previous inequalities are equivalent to:

$$\| \left(\frac{\sigma_1 \lambda_{c,l}}{\alpha_{c,l}(i)}\right)^2 D_{c,l}^{\top}(i) X^{\alpha_c, \alpha_d}(t, i) D_{c,l}(i)\|_2 < 1 \tag{9.79}$$

for all $(l, i) \in \underline{r_2} \times \mathcal{N}$, $0 \le t \le h$ and

$$\| \left(\frac{\sigma_1 \lambda_{d,p}}{\alpha_{d,p}(i)}\right)^2 D_{d,p}^{\top}(i) X^{\alpha_c, \alpha_d}(0, i) D_{d,p}(i)\|_2 < 1, \tag{9.80}$$

for all $(p, i) \in \underline{r_4} \times \mathcal{N}$. Further, (9.79) and (9.80), respectively, allow us to obtain $\left(\frac{\sigma_1 \lambda_{c,l}}{\alpha_{c,l}(i)}\right)^2 D_{c,l}^{\top}(i) X^{\alpha_c, \alpha_d}(t, i) D_{c,l}(i) < I_{m_l}$ for all $(l, i) \in \underline{r_2} \times \mathcal{N}$, $0 \le t \le h$ and $\left(\frac{\sigma_1 \lambda_{d,p}}{\alpha_{d,p}(i)}\right)^2 D_{d,p}^{\top}(i) X^{\alpha_c, \alpha_d}(0, i) D_{d,p}(i) < I_{m_{d,p}}$, for all $(p, i) \in \underline{r_4} \times \mathcal{N}$. Hence

$$\left(\frac{\sigma \lambda_{c,l}}{\alpha_{c,l}(i)}\right)^2 D_{c,l}^{\top}(i) X^{\alpha_c, \alpha_d}(t, i) D_{c,l}(i) < I_{m_l} \tag{9.81}$$

for all $(l, i) \in \underline{r_2} \times \mathcal{N}$, $0 \le t \le h$ and

$$\left(\frac{\sigma\lambda_{d,p}}{\alpha_{d,p}(i)}\right)^2 D_{d,p}^\top(i) X^{\alpha_c,\alpha_d}(0,i) D_{d,p}(i) < I_{m_{d,p}}, \tag{9.82}$$

for all $(p, i) \in \underline{r_4} \times \mathcal{N}$, because $\sigma < \sigma_1$. On the other hand, for an arbitrary $\varepsilon > 0$ we denote $\mathbf{Y}_\varepsilon(\cdot) = (Y_\varepsilon(\cdot, 1), Y_\varepsilon(\cdot, 2), \ldots, Y_\varepsilon(\cdot, N))$ the bounded on $\mathbb{R}$ solution of the following BLJLDE:

$$-\dot{Y}_\varepsilon(t,i) = \mathcal{L}_{c,i}^*[\mathbf{Y}_\varepsilon(t)] + \sum_{l=1}^{r_2} \alpha_{c,l}^2(i) C_{c,l}^\top(i) C_{c,l}(i), \quad kh \le t < (k+1)h \tag{9.83a}$$

$$Y_\varepsilon(kh^-, i) = \mathcal{L}_{d,i}^*[\mathbf{Y}_\varepsilon(kh)] + \sum_{p=1}^{r_4} \alpha_{d,p}^2(i) C_{d,p}^\top(i) C_{d,p}(i) + \varepsilon^2 I_n, \quad k \in \mathbb{Z}. \tag{9.83b}$$

Based on Corollary 9.1 and Theorem 2.13 (iii), we may infer that if the nominal system (9.7) is ESMS, the BLJLDE (9.83) has a unique solution $t \to \mathbf{Y}_\varepsilon(t) : \mathbb{R} \to \mathcal{S}_n^N$ and additionally this solution is a periodic function of period h. Moreover, Theorem 2.13 (v) allows us to conclude that $Y_\varepsilon(t, i) \ge cI_n$, for all $(t, i) \in \mathbb{R} \times \mathcal{N}$, where $c > 0$ does not depend upon t, i. From (9.36) and (9.83) we deduce that $t \to \mathbf{Z}(t) \triangleq \mathbf{Y}_\varepsilon(t) - \mathbf{X}^{\alpha_c,\alpha_d}(t)$ is a bounded on $\mathbb{R}$ solution of the following non-homogeneous backward jump linear differential equation on the Hilbert space $\mathcal{S}_n^N$:

$$-\dot{Z}(t,i) = \mathcal{L}_c^*[\mathbf{Z}(t)], \quad kh \le t < (k+1)h \tag{9.84a}$$

$$Z(kh^-) = \mathcal{L}_d^*[\mathbf{Z}(kh)] + \varepsilon^2 \mathbf{I}, \quad k \in \mathbb{Z}, \tag{9.84b}$$

where $\mathbf{I} = (I_n \ I_n \ \ldots \ I_n) \in \operatorname{Int} \mathcal{S}_n^{N+}$. The representation formula (2.87) applied in the special case of the solution $\mathbf{Z}(\cdot)$ gives:

$$\mathbf{Z}(t) = \varepsilon^2 \sum_{k=\kappa_t}^{\infty} \mathbf{T}^*(kh^-, t)\mathbf{I}, \tag{9.85}$$

for all $t \in \mathbb{R}$, where $\mathbf{T}^*(kh^-, t)$ is the adjoint of the linear evolution operator defined on $\mathcal{S}_n^N$, by the FJLDE (9.39) and κ_t is the integer linked by the current time t, by $(\kappa_t - 1)h \le t < \kappa_t h$. Based again on Corollary 9.1 we may infer that the nominal system (9.7) is ESMS, which means that

$$\|\mathbf{T}^*(kh^-, t)\|_\infty \le \beta e^{-\alpha(kh-t)}, \tag{9.86}$$

for all $kh \ge t$, $t \in \mathbb{R}$, $\beta \ge 1$, $\alpha > 0$ are constants and $\|\cdot\|_\infty$ is the norm on the space $\mathcal{S}_n^N$ introduced in (2.112). Employing (9.86) we may deduce from (9.85) that

$$\|\mathbf{Y}_\varepsilon(t) - \mathbf{X}^{\alpha_c,\alpha_d}(t)\|_\infty \leq \delta\varepsilon^2, \tag{9.87}$$

for all $t \in \mathbb{R}$, $\delta > 0$ being a constant not depending upon t. Let us remark that if the nominal system (9.7) is ESMS, then for each $\varepsilon > 0$, the unique bounded on $\mathbb{R}$ solution $\mathbf{Y}_\varepsilon(\cdot)$ of the BLJLDE (9.83) is also a solution of the BLJLDI (9.77), satisfying the constraint (9.78a). Further, (9.87) and (9.81), (9.82) show that the function $\mathbf{Y}_\varepsilon(\cdot)$ satisfies the constraints (9.78b) and (9.78c) if $\varepsilon > 0$ is small enough. So, we have proven that (ii) holds if (i) is true.

To prove the implication $(ii) \Rightarrow (i)$, let us first remark that if (9.77) has a solution which satisfies (9.78a) then the implication $(vii) \Rightarrow (i)$ from Theorem 3.4 allows to conclude that the nominal system (9.7) is ESMS. Under these considerations, the corresponding BLJLDE of the form (9.36) has a unique bounded solution $\mathbf{X}^{\alpha_c,\alpha_d}(\cdot) : \mathbb{R} \to \mathscr{S}_n^N$ and additionally, for any solution $\mathbf{Y}(\cdot) = (Y(\cdot, 1), Y(\cdot, 2), \ldots, Y(\cdot, N))$ of (9.77) corresponding to the pair (α_c, α_d) of scaling parameters will satisfy

$$Y(t, i) \geq X^{\alpha_c,\alpha_d}(t, i), \quad \text{for all} \quad (t, i) \in \mathbb{R} \times \mathcal{N}. \tag{9.88}$$

Combining (9.81), (9.82) and (9.88) we may infer that $\left(\frac{\sigma\lambda_{c,l}}{\alpha_{c,l}(i)}\right)^2 D_{c,l}^\top(i) X^{\alpha_c,\alpha_d}(t, i) D_{c,l}(i) < I_{m_l}$ for all $(l, i) \in \underline{r_2} \times \mathcal{N}$, $0 \leq t \leq h$ and $\left(\frac{\sigma\lambda_{d,p}}{\alpha_{d,p}(i)}\right)^2 D_{d,p}^\top(i) X^{\alpha_c,\alpha_d}(0, i) D_{d,p}(i) < I_{m_{d,p}}$, for all $(p, i) \in \underline{r_4} \times \mathcal{N}$, or equivalently,

$$\sigma < \left[\max_{l\in\underline{r_2}} \max_{i\in\mathcal{N}} \max_{0\leq t\leq h} \left(\frac{\lambda_{c,l}}{\alpha_{c,l}(i)}\right)^2 \|D_{c,l}^\top(i) X^{\alpha_c,\alpha_d}(t, i) D_{c,l}(i)\|_2\right]^{-1/2} \tag{9.89a}$$

$$\sigma < \left[\max_{p\in\underline{r_4}} \max_{i\in\mathcal{N}} \left(\frac{\lambda_{d,p}}{\alpha_{d,p}(i)}\right)^2 \|D_{d,p}^\top(i) X^{\alpha_c,\alpha_d}(0, i) D_{d,p}(i)\|_2\right]^{-1/2}. \tag{9.89b}$$

With (9.89) and (9.71) we deduce that $\sigma < \min\{\mu_c(\alpha_c, \alpha_d), \mu_d(\alpha_c, \alpha_d)\}$. Finally, we obtain via (9.70b) that $\sigma < \widehat{\mu}$, which complete the proof.

In the case of the stability radius of the nominal system (9.7) with respect to the set $\mathscr{D}_c$ of the structured parametric uncertainties having the structure described by the pairs $(\mathbf{D}_{c,l}, \mathbf{C}_{c,l})_{l\in\underline{r_2}}$, one obtains the following version of Proposition 9.2:

Proposition 9.3 *Assume that Assumption 9.1 holds. Let $\widetilde{\mu}_c$ be the lower bound of the stability radius $\rho_{\mathscr{D}_c}[\mathbf{G_0} \mid (\mathbf{D}_{c,l}, \mathbf{C}_{c,l}),\ l \in \underline{r_2}]$ provided by Theorem 9.3. Then, for a given $\sigma > 0$ the following are equivalent:*

- *(i) the nominal system (9.7) is ESMS and $\sigma < \widetilde{\mu}_c$;*
- *(ii) there exists a vector of scaling parameters $\alpha_c = \{\alpha_{c,l}(i)\}_{(l,i)\in\underline{r_2}\times\mathcal{N}} \in (0, \infty)^{\underline{r_2}\times\mathcal{N}}$, a bounded and h-periodic function $t \to \widetilde{\mathbf{Y}}(t) = (\widetilde{Y}(t, 1), \widetilde{Y}(t, 2), \ldots, \widetilde{Y}(t, N)) : \mathbb{R} \to \mathscr{S}_n^N$ differentiable on each interval $(kh, (k + 1)h)$, right continuous in any $\tau_k = kh$, satisfying the following BLJLDI:*

$$\dot{\widetilde{Y}}(t,i) + \mathscr{L}_{c,i}^{*}[\widetilde{\mathbf{Y}}(t)] + \sum_{l=1}^{r_2} \alpha_{c,l}^2(i) C_{c,l}^{\top}(i) C_{c,l}(i) \leq 0, \quad kh \leq t < (k+1)h \tag{9.90a}$$

$$\mathscr{L}_{d,i}^{*}[\widetilde{\mathbf{Y}}(kh)] - \widetilde{Y}(kh^-, i) < 0, \quad k \in \mathbb{Z}, \tag{9.90b}$$

together with the constrains

$$\widetilde{Y}(t,i) > 0 \tag{9.91a}$$

$$\alpha_{c,l}^2(i) I_{m_l} - \sigma^2 \lambda_{c,l}^2 D_{c,l}^{\top}(i) \widetilde{Y}(t,i) D_{c,l}(i) > 0, \quad l \in \underline{r_2}, \quad (t,i) \in \mathbb{R} \times \mathscr{N}. \tag{9.91b}$$

Proof The proof may be done following the line of the proof of Proposition 9.2. The details are omitted.

9.4 Robust Stabilization by Impulses in a State Feedback Form

We consider the impulsive controlled linear stochastic systems (ICLSS):

$$\mathrm{d}x(t) = A_0(\eta(t))x(t)\mathrm{d}t + \sum_{\iota=1}^{r_1} A_\iota(\eta(t))x(t)\mathrm{d}w_\iota(t) \tag{9.92a}$$
$$+ \sum_{l=1}^{r_2} D_{c,l}(\eta(t))\Delta_{c,l}(t, z_{c,l}(t), \eta(t))\mathrm{d}\xi_l(t), \quad kh < t \leq (k+1)h$$

$$x(kh^+) = A_{d,0}(\eta(kh))x(kh) + B_{d,0}(\eta(kh))u(k)$$
$$+ \sum_{q=1}^{r_3} w_{d,q}(k)(A_{d,q}(\eta(kh)x(kh) + B_{d,q}(\eta(kh))u(k)) \tag{9.92b}$$
$$+ \sum_{p=1}^{r_4} D_{d,p}(\eta(kh))\Delta_{d,p}(k, z_{d,p}(kh), \eta(kh))\zeta_p(k), \quad k \in \mathbb{Z}_+,$$

$$z_{c,l}(t) = C_{c,l}(\eta(t))x(t), \quad t \in \mathbb{R}_+ \tag{9.92c}$$

$$z_{d,p}(kh) = C_{d,p}(\eta(kh))x(kh), \quad k \in \mathbb{Z}_+, \tag{9.92d}$$

where $x(t) \in \mathbb{R}^n$, $t \in \mathbb{R}_+$ are state vectors and $u(k) \in \mathbb{R}^{n_u}$, $k \in \mathbb{Z}_+$ are the vectors of the control parameters. The stochastic processes $\{w(t)\}_{t\geq 0}$, $(w(t) = (w_1(t) \;\; w_2(t) \;\; \dots \;\; w_{r_1}(t))^{\top})$, $\{w_d(k)\}_{k\in\mathbb{Z}_+}$, $(w_d(k) = (w_{d1}(k) \;\; w_{d2}(k) \;\; \dots \;\; w_{dr_3}(k))^{\top})$, $\{\xi(t)\}_{t\geq 0}$, $(\xi(t) = (\xi_1(t) \;\; \xi_2(t) \;\; \dots \;\; \xi_{r_2}(t))^{\top})$, $\{\zeta(k)\}_{k\in\mathbb{Z}_+}$, $(\zeta(k) = (\zeta_1(k) \;\; \zeta_2(k) \;\; \dots \;\; \zeta_{r_4}(k))^{\top})$, $\{\eta(t)\}_{t\geq 0}$ which affect the system (9.92) satisfy Assumption 9.1.

For each $i \in \mathcal{N}$, in (9.92) $A_\iota(i) \in \mathbb{R}^{n\times n}$, $0 \le \iota \le r_1$, $A_{d,q}(i) \in \mathbb{R}^{n\times m}$, $B_{d,q} \in \mathbb{R}^{n\times n_u}$, $0 \le q \le r_3$, $D_{c,l}(i) \in \mathbb{R}^{n\times m_l}$, $C_{c,l}(i) \in \mathbb{R}^{\nu_l\times n}$, $1 \le l \le r_2$, $D_{d,p}(i) \in \mathbb{R}^{n\times m_{d,p}}$, $C_{d,p}(i) \in \mathbb{R}^{\nu_{d,p}\times n}$, $1 \le p \le r_4$, are known matrices, while $(t, z_{c,l}) \to \Delta_{c,l}(t, z_{c,l}, i) : \mathbb{R}_+ \times \mathbb{R}^{\nu_l} \to \mathbb{R}^{m_l}$, $1 \le l \le r_2$, and $z_{d,p} \to \Delta_{d,p}(k, z_{d,p}, i) : \mathbb{R}^{\nu_{d,p}} \to \mathbb{R}^{m_{d,p}}$, $1 \le p \le r_4$, $k \in \mathbb{Z}_+$ are arbitrary unknown functions which satisfy the conditions from Assumption 9.2.

The system (9.92) may be viewed as a perturbation of the following impulsive controlled linear stochastic system (ICLSS):

$$dx(t) = A_0(\eta(t))x(t)dt + \sum_{\iota=1}^{r_1} A_\iota(\eta(t))x(t)dw_\iota(t), \quad kh < t \le (k+1)h \tag{9.93a}$$

$$\begin{aligned} x(kh^+) &= A_{d,0}(\eta(kh))x(kh) + B_{d,0}(\eta(kh))u(k) \\ &\quad + \sum_{q=1}^{r_3} w_{d,q}(k)(A_{d,q}(\eta(kh)x(kh) + B_{d,q}(\eta(kh))u(k)), \quad k \in \mathbb{Z}_+, \end{aligned} \tag{9.93b}$$

which will be named nominal ICLSS. To control the ICLSS (9.93) we shall use control laws of the form:

$$u(k) = F(\eta(kh))x(kh), \quad k \in \mathbb{Z}_+ \tag{9.94}$$

where $F(i) \in \mathbb{R}^{n_u\times n}$, $i \in \mathcal{N}$.

Coupling a control of the form (9.94) to the system (9.92) one obtains the following closed-loop system:

$$\begin{aligned} dx(t) &= A_0(\eta(t))x(t)dt + \sum_{\iota=1}^{r_1} A_\iota(\eta(t))x(t)dw_\iota(t) \qquad (9.95a) \\ &\quad + \sum_{l=1}^{r_2} D_{c,l}(\eta(t))\Delta_{c,l}(t, z_{c,l}(t), \eta(t))d\xi_l(t), \quad kh < t \le (k+1)h \\ x(kh^+) &= (A_{d,0}(\eta(kh)) + B_{d,0}(\eta(kh))F(\eta(kh)))x(kh) \\ &\quad + \sum_{q=1}^{r_3} w_{d,q}(k)(A_{d,q}(\eta(kh)) + B_{d,q}(\eta(kh))F(\eta(kh)))x(kh) \qquad (9.95b) \\ &\quad + \sum_{p=1}^{r_4} \zeta_p(k)D_{d,p}(\eta(kh))\Delta_{d,p}(k, z_{d,p}(kh), \eta(kh)), \quad k \in \mathbb{Z}_+, \end{aligned}$$

$$z_{c,l}(t) = C_{c,l}(\eta(t))x(t), \quad t \in \mathbb{R}_+, \quad 1 \le l \le r_2, \tag{9.95c}$$

$$z_{d,p}(kh) = C_{d,p}(\eta(kh))x(kh), \quad k \in \mathbb{Z}_+, \quad 1 \le p \le r_4. \tag{9.95d}$$

The closed-loop system (9.95) may be interpreted as a perturbation of the following jump linear stochastic system $\mathbf{G_{0,F}}$ (JLSS):

$$dx(t) = A_0(\eta(t))x(t)dt + \sum_{\iota=1}^{r_1} A_\iota(\eta(t))x(t)dw_\iota(t), \quad kh < t \le (k+1)h \tag{9.96a}$$

$$\begin{aligned} x(kh^+) &= (A_{d,0}(\eta(kh)) + B_{d,0}(\eta(kh))F(\eta(kh)))x(kh) \\ &+ \sum_{q=1}^{r_3} w_{d,q}(k)(A_{d,q}(\eta(kh)) + B_{d,q}(\eta(kh))F(\eta(kh)))x(kh), \quad k \in \mathbb{Z}_+. \end{aligned} \tag{9.96b}$$

The JLSS $\mathbf{G_{0,F}}$ is the closed-loop system obtained by coupling a control law of type (9.94) (having the gain matrices $\mathbf{F} = (F(1),\ F(2),\ \ldots,\ F(N)))$ to the nominal system (9.93). That is why the JLSS (9.96) will be named **the closed-loop nominal system**.

Our aim is to find necessary and sufficient conditions which guarantee the existence of a control law of type (9.94) that stabilizes the closed-loop nominal system (9.96) and additionally, its stability radius with respect to the set $\mathscr{D}$ of the structured parametric uncertainties having the structure described by the pairs $(\mathbf{D}_{c,l}, \mathbf{C}_{c,l})$, $l \in \underline{r_2}$ and $(\mathbf{D}_{d,p}, \mathbf{C}_{d,p})$, $p \in \underline{r_4}$, defined as in (9.17) to satisfy:

$$\rho_{\mathscr{D}}[\mathbf{G_{0,F}} \mid (\mathbf{D}_{c,l}, \mathbf{C}_{c,l}),\ l \in \underline{r_2},\ (\mathbf{D}_{d,p}, \mathbf{C}_{d,p}),\ p \in \underline{r_4}] > \sigma \tag{9.97}$$

for a prefixed level of robustness $\sigma > 0$.

Remark 9.3 In order to obtain a stabilizing control law (9.94) that achieves a given level of robustness as in (9.97) we'll look for necessary and sufficient conditions which guarantee the existence of a control law of type (9.94) which stabilizes the nominal system (9.93) and additionally to satisfy the robustness condition:

$$\widehat{\mu}_{\mathbf{F}} > \sigma, \tag{9.98}$$

where

$$\widehat{\mu}_{\mathbf{F}} = \sup_{(\alpha_c,\alpha_d)\in(0,\infty)^{\underline{r_2}\times\mathcal{N}}\times(0,\infty)^{\underline{r_4}\times\mathcal{N}}} (\min\{\mu_{c,\mathbf{F}}(\alpha_c, \alpha_d), \mu_{d,\mathbf{F}}(\alpha_c, \alpha_d)\}) \tag{9.99}$$

where

$$\mu_{c,\mathbf{F}}(\alpha_c, \alpha_d) = [\max_{(l,i)\in\underline{r_2}\times\mathcal{N}} \max_{0\le t\le h} \left(\frac{\lambda_{c,l}}{\alpha_{c,l}(i)}\right)^2 \|D_{c,l}^\top(i) X_{\mathbf{F}}^{\alpha_c,\alpha_d}(t, i) D_{c,l}(i)\|_2]^{-1/2} \tag{9.100a}$$

$$\mu_{d,\mathbf{F}}(\alpha_c, \alpha_d) = [\max_{(p,i)\in\underline{r_4}\times\mathcal{N}} \left(\frac{\lambda_{d,p}}{\alpha_{d,p}(i)}\right)^2 \|D_{d,p}^\top(i) X_{\mathbf{F}}^{\alpha_c,\alpha_d}(0, i) D_{d,p}(i)\|_2]^{-1/2} \tag{9.100b}$$

$t \to \mathbf{X}_{\mathbf{F}}^{\alpha_c,\alpha_d}(t) = (X_{\mathbf{F}}^{\alpha_c,\alpha_d}(t,1)\ X_{\mathbf{F}}^{\alpha_c,\alpha_d}(t,2)\ \ldots\ X_{\mathbf{F}}^{\alpha_c,\alpha_d}(t,N)) : \mathbb{R} \to \mathscr{S}_n^N$ is the unique bounded solution of the following BLJLDE:

$$- \dot{X}(t,i) = \mathcal{L}_{c,i}^{*}[\mathbf{X}(t)] + \sum_{l=1}^{r_2} \alpha_{c,l}^2(i) C_{c,l}^\top(i) C_{c,l}(i), \quad kh \le t < (k+1)h \tag{9.101a}$$

$$X(kh^-,i) = \sum_{q=0}^{r_3} (A_{d,q}(i) + B_{d,q}(i)F(i))^\top X(kh,i)(A_{d,q}(i) + B_{d,q}(i)F(i))$$

$$+ \sum_{p=1}^{r_4} \alpha_{d,p}^2(i) C_{d,p}^\top(i) C_{d,p}(i), \quad k \in \mathbb{Z}, \tag{9.101b}$$

$i \in \mathcal{N}$. According to Theorem 9.2 applied in the case of the closed-loop system (9.96) we obtain that

$$\rho_{\mathcal{D}}[\mathbf{G}_{0,\mathbf{F}} \mid (\mathbf{D}_{c,l}, \mathbf{C}_{c,l}),\ l \in \underline{r_2},\ (\mathbf{D}_{d,p}, \mathbf{C}_{d,p}),\ p \in \underline{r_4}] \ge \widehat{\mu}_{\mathbf{F}} > \sigma \tag{9.102}$$

if $\mathbf{F} = (F(1),\ F(2), \ldots, F(N))$ are designed such that (9.98) to be satisfied for a prefixed $\sigma > 0$.

The next theorem provides a first answer to the above problem.

Theorem 9.5 *Assume:*

(a) Assumption 9.1 is fulfilled;
(b) rank $B_{d,q}(i) = n_u$ *for any* $i \in \mathcal{N}$, $0 \le q \le r_3$.

For a given level of robustness $\sigma > 0$, *the following are equivalent:*

(i) there exists a control law in a state feedback form of type (9.94) which stabilizes the nominal system (9.93) such that the stability radius of the closed-loop nominal system is satisfying (9.102);
(ii) there exist vectors of scaling parameters $\alpha_c = \{\alpha_{c,l}(i)\}_{(l.i) \in \underline{r_2} \times \mathcal{N}} \in (0,\infty)^{\underline{r_2} \times \mathcal{N}}$*,,*
$\alpha_d = \{\alpha_{d,p}(i)\}_{(p.i) \in \underline{r_4} \times \mathcal{N}} \in (0,\infty)^{\underline{r_4} \times \mathcal{N}}$ *and a bounded and h-periodic function* $t \to \mathbf{X}(t) = (X(t,1),\ X(t,2),\ \ldots\ , X(t,N)) : \mathbb{R} \to \mathcal{S}_n^N$ *differentiable on each interval* $(kh,(k+1)h)$, *right continuous in any* $\tau_k = kh,\ k \in \mathbb{Z}$ *and satisfying the following backward jumps matrix differential inequality (BJMDI):*

$$\dot{X}(t,i) + A_0^\top(i)X(t,i) + X(t,i)A_0(i) + \sum_{\iota=1}^{r_1} A_\iota^\top(i)X(t,i)A_\iota(i) \tag{9.103a}$$

$$+ \sum_{j=1}^{N} q_{ij}X(t,j) + \sum_{l=1}^{r_2} \alpha_{c,l}^2(i) C_{c,l}^\top(i) C_{c,l}(i) \le 0, \quad kh \le t < (k+1)h$$

$$\sum_{q=0}^{r_3} A_{d,q}^\top(i) X(kh,i) A_{d,q}(i) - X(kh^-,i) + \sum_{p=1}^{r_4} \alpha_{d,p}^2(i) C_{d,p}^\top(i) C_{d,p}(i)$$

$$
-\left(\sum_{q=0}^{r_3} A_{d,q}^{\top}(i)X(kh,i)B_{d,q}(i)\right)\left(\sum_{q=0}^{r_3} B_{d,q}^{\top}(i)X(kh,i)B_{d,q}(i)\right)^{-1}
$$

$$
\cdot\left(\sum_{q=0}^{r_3} A_{d,q}^{\top}(i)X(kh,i)B_{d,q}(i)\right)^{\top} < 0, \quad k \in \mathbb{Z}, , i)B_{d,q}(i))^{\top} < 0, \quad k \in \mathbb{Z}, \tag{9.103b}
$$

together with the constraints:

$$
X(t,i) \geq \gamma^2 I_n, \quad t \in \mathbb{R}, \tag{9.104a}
$$

$$
\alpha_{c,l}^2(i)I_{m_l} - \sigma^2\alpha_{c,l}^2 D_{c,l}^{\top}(i)X(t,i)D_{c,l}(i) \geq \gamma_l^2 I_{m_l}, \quad 1 \leq l \leq r_2, \quad t \in \mathbb{R}, \tag{9.104b}
$$

$$
\alpha_{d,p}^2(i)I_{m_{d,p}} - \sigma^2\lambda_{d,p}^2 D_{d,p}^{\top}(i)X(0,i)D_{d,p}(i) > 0, \tag{9.104c}
$$

$i \in \mathcal{N}$. Moreover, if $\mathbf{X}(t)$, $t \in \mathbb{R}$ is a solution of the BJMDI (9.103) satisfying the constraints (9.104) then, a control law of type (9.94) which robustly stabilizes the nominal system (9.93) achieving a stability radius which satisfies (9.102) has the gain matrices given by

$$
\widehat{F}(i) = -\left(\sum_{q=0}^{r_3} B_{d,q}^{\top}(i)X(0,i)B_{d,q}(i)\right)^{-1}\left(\sum_{q=0}^{r_3} B_{d,q}^{\top}(i)X(0,i)A_{d,q}(i)\right). \tag{9.105}
$$

Proof First we prove the implication $(i) \Rightarrow (ii)$. Let us assume that there exists a control law of type (9.94) which stabilizes the nominal system (9.93) such that the achieved stability radius to satisfy (9.102). Applying $(i) \Rightarrow (iii)$ from Proposition 9.2 in the case of the closed-loop systems (9.95) and (9.96), we deduce that there exist the scalars $\alpha_{c,l}(i) > 0, \ \ (l,i) \in \underline{r_2} \times \mathcal{N}, \ \ \alpha_{d,p}(i) > 0, \ \ (p,i) \in \underline{r_2} \times \mathcal{N}$, and the bounded and h-periodic function $t \to \mathbf{X}(t) = (X(t,1), \ X(t,2), \ldots, \ X(t,N))$ differentiable on each interval $(kh, (k+1)h)$, right continuous in any $\tau_k = kh$, $k \in \mathbb{Z}$ and satisfying the following BLJLDI:

$$
\dot{X}(t,i) + \mathcal{L}_{c,i}^*[\mathbf{X}] + \sum_{l=1}^{r_2} \alpha_{c,l}^2(i)C_{c,l}^{\top}(i)C_{c,l}(i) \leq 0, \quad kh < t \leq (k+1)h \tag{9.106a}
$$

$$
\sum_{q=0}^{r_3}(A_{d,q}(i) + B_{d,q}(i)F(i))^{\top}X(kh,i)(A_{d,q}(i) + B_{d,q}(i)F(i))
$$

$$
- X(kh^-,i) + \sum_{p=1}^{r_4} \alpha_{d,p}^2(i)C_{d,p}^{\top}(i)C_{d,p}(i) < 0, \quad k \in \mathbb{Z}, \tag{9.106b}
$$

$i \in \mathcal{N}$, together with the constraints:

$$X(t,i) \geq \gamma^2 I_n, \quad t \in \mathbb{R}, \tag{9.107a}$$

$$\alpha_{c,l}^2(i) I_{m_l} - \sigma^2 \alpha_{c,l}^2 D_{c,l}^\top(i) X(t,i) D_{c,l}(i) \geq \gamma_l^2 I_{m_l}, \quad l \in \underline{r_2}, \quad t \in \mathbb{R}, \tag{9.107b}$$

$$\alpha_{d,p}^2(i) I_{m_{d,p}} - \sigma^2 \lambda_{d,p}^2 D_{d,p}^\top(i) X(0,i) D_{d,p}(i) > 0, \tag{9.107c}$$

$i \in \mathcal{N}$, $F(i)$, $i \in \mathcal{N}$ being the gain matrices of the control (9.94) under consideration. The existence of the uniform lower bounds in (9.107a) and (9.107b) is guaranteed by the periodicity property of the functions involved in (9.106a) and (9.106b).

The assumption (b) from the statement and (9.107a) gives:

$$\sum_{q=0}^{r_3} B_{d,q}^\top(i) X(kh,i) B_{d,q}(i) > 0, \tag{9.108}$$

for all $k \in \mathbb{Z}$. This allows us to rewrite (9.106b) in the form:

$$\begin{aligned}
&\sum_{q=0}^{r_3} A_{d,q}^\top(i) X(kh,i) A_{d,q}(i) - X(kh^-,i) + \sum_{p=1}^{r_4} \alpha_{d,p}^2(i) C_{d,p}^\top(i) C_{d,p}(i) \\
&- \left(\sum_{q=0}^{r_3} A_{d,q}^\top(i) X(kh,i) B_{d,q}(i)\right) \left(\sum_{q=0}^{r_3} B_{d,q}^2(i) X(kh,i) B_{d,q}(i)\right)^{-1} \\
&\times \left(\sum_{q=0}^{r_3} A_{d,q}^\top(i) X(kh,i) B_{d,q}(i)\right) + (F(i) - \widehat{F}(i))^\top \\
&\times \left(\sum_{q=0}^{r_3} B_{d,q}^\top(i) X(kh,i) B_{d,q}(i))(F(i) - \widehat{F}(i)\right) < 0,
\end{aligned} \tag{9.109}$$

$k \in \mathbb{Z}$, $i \in \mathcal{N}$, where $\widehat{F}(i)$ are those defined in (9.105). Bearing in mind (9.108) we deduce that (9.106b) may be true if $X(kh,i)$, $k \in \mathbb{Z}$ is satisfying (9.103b). Hence, the solution $\mathbf{X}(\cdot)$ of (9.106) which satisfies the constraints (9.107) solves also the BJMDI (9.103) and satisfies the constraints (9.104). Thus, we have shown that (ii) holds if (i) is true.

To prove the implication $(ii) \Rightarrow (i)$ let us assume that (9.103) has a solution $\mathbf{X}(\cdot)$ satisfying the constraints (9.104). Since (9.104a) holds, it follows that (9.108) holds too. Hence, the corresponding gain matrices $\widehat{F}(i)$ are well defined via (9.105). Further, we may rewrite the BJMDI (9.103) as:

$$\dot{X}(t,i) + \mathscr{L}^*_{c,i}[\mathbf{X}(t)] + \sum_{l=1}^{r_2} \alpha^2_{c,l}(i) C^\top_{c,l}(i) C_{c,l}(i) \leq 0, \quad kh \leq t < (k+1)h \tag{9.110a}$$

$$\sum_{q=0}^{r_3} (A_{d,q}(i) + B_{d,q}(i)\widehat{F}(i))^\top X(kh,i)(A_{d,q}(i) + B_{d,q}(i)\widehat{F}(i))$$

$$- X(kh^-, i) + \sum_{p=1}^{r_4} \alpha^2_{d,p}(i) C^\top_{d,p}(i) C_{d,p}(i) < 0, \quad k \in \mathbb{Z}, \tag{9.110b}$$

$i \in \mathscr{N}$. Since the solution $\mathbf{X}(\cdot)$ of (9.110) satisfies also (9.104a) we may conclude via the implication $(vii) \Rightarrow (i)$ from Theorem 3.4 that the closed-loop JLSS:

$$dx(t) = A_0(\eta(t))x(t)dt + \sum_{\iota=1}^{r_1} A_\iota(\eta(t))x(t)dw_\iota(t), \quad kh < t \leq (k+1)h \tag{9.111a}$$

$$\begin{aligned} x(kh^+) &= (A_{d,0}(\eta(kh)) + B_{d,0}(\eta(kh))\widehat{F}(\eta(kh)))x(kh) \\ &+ \sum_{q=1}^{r_3} w_{d,q}(k)(A_{d,q}(\eta(kh)) + B_{d,q}(\eta(kh))\widehat{F}(\eta(kh)))x(kh), \ k \in \mathbb{Z}_+ \end{aligned} \tag{9.111b}$$

is ESMS. This means that the control

$$u(k) = \widehat{F}(\eta(kh))x(kh), \quad k \in \mathbb{Z}_+ \tag{9.112}$$

stabilizes the nominal system (9.93). If we take into account that the solution $\mathbf{X}(\cdot)$ of (9.110) satisfies the conditions (9.104), too, we may conclude via the implication $(ii) \Rightarrow (i)$ from Proposition 9.2 applied in the case of the closed-loop system (9.111) that

$$\sigma < \widehat{\mu}_{\widehat{\mathbf{F}}} \tag{9.113}$$

$\widehat{\mu}_{\widehat{\mathbf{F}}}$ being computed as in (9.99) and (9.100) for $F(i)$ replaced by $\widehat{F}(i)$ provided by (9.105). Further, Theorem 9.2 applied to the case of the closed-loop system (9.111) yields:

$$\rho_{\mathscr{D}}[\mathbf{G}_{\mathbf{0},\widehat{\mathbf{F}}} \mid (\mathbf{D}_{c,l}, \mathbf{C}_{c,l}),\ l \in \underline{r_2},\ (\mathbf{D}_{d,p}, \mathbf{C}_{d,p}),\ p \in \underline{r_4}] \geq \widehat{\mu}_{\widehat{\mathbf{F}}}. \tag{9.114}$$

Finally, (9.113) and (9.114) show that (9.102) is satisfied in the case of the control law (9.112). Thus the proof ends.

Remark 9.4 One sees that due to (9.108), the nonlinear matrix inequality (9.103b) cannot be converted in an obvious way into a LMI. For these reasons, the numerical computation of a solution of the BJMRDI (9.103) is difficult, and consequently, the numerical computation by (9.105) of the gain matrices $\widehat{F}(i)$ is also difficult. To overcome this inconvenient, we propose to replace the solution of BJMDI (9.103) by the stabilizing solution of a suitable backward jump matrix linear differential equation (BJMLDE) with Riccati-type jumping operator associated to the nominal system (9.93).

Let us consider the following BJMLDE with Riccati-type jumping operator associated to the nominal system (9.93):

$$-\dot{X}(t,i) = A_0^\top(i)X(t,i) + X(t,i)A_0(i) + \sum_{\iota=1}^{r_1} A_\iota^\top(i)X(t,i)A_\iota(i)$$
$$+\sum_{j=1}^{N} q_{ij}X(t,j) + \sum_{l=1}^{r_2} \alpha_{c,l}^2(i)C_{c,l}^\top(i)C_{c,l}(i) = 0, \quad kh \le t < (k+1)h \tag{9.115a}$$

$$X(kh^-,i) = \sum_{q=0}^{r_3} A_{d,q}^\top(i)X(kh,i)A_{d,q}(i) - \left(\sum_{q=0}^{r_3} A_{d,q}^\top(i)X(kh,i)B_{d,q}(i)\right)$$
$$\cdot\left(\sum_{q=0}^{r_3} B_{d,q}^\top(i)X(kh,i)B_{d,q}(i)\right)^{-1}\left(\sum_{q=0}^{r_3} A_{d,q}^\top(i)X(kh,i)B_{d,q}(i)\right)^\top$$
$$+\sum_{p=1}^{r_4} \alpha_{d,p}^2(i)C_{d,p}^\top(i)C_{d,p}(i), \quad k \in \mathbb{Z}, \tag{9.115b}$$

$i \in \mathcal{N}$. One may see that (9.115) is a special case of the BJMLDE (5.168). The following definition is an adaptation to the case of the BJMLDE (9.115) of the Definition 5.9.

Definition 9.5 A globally defined solution $t \to \mathbf{X}_s(t) = (X_s(t,1), X_s(t,2), \dots, X_s(t,N)) : \mathbb{R} \to \mathcal{S}_n^N$ of (9.115) is named stabilizing solution if the matrices $\sum_{p=0}^{r_3} B_{d,q}^\top(i)X_s(kh,i)B_{d,q}(i)$, $(k,i) \in \mathbb{Z} \times \mathcal{N}$, are invertible and the corresponding closed-loop system:

$$dx(t) = (A_0(\eta(t))x(t)dt + \sum_{l=1}^{r_1} A_\iota(\eta(t))x(t)dw_\iota(t), \quad kh < t \le (k+1)h \tag{9.116a}$$

$$x(kh^+) = (A_{d0}(\eta(kh)) + B_{d,0}(\eta(kh))F_s(\eta(kh)))x(kh) \tag{9.116b}$$
$$+\sum_{q=1}^{r_3} w_{d,q}(k)\left(A_{d,q}(\eta(kh)) + B_{d,q}(\eta(kh))F_s(\eta(kh))\right)x(kh), \quad k \in \mathbb{Z}_+,$$

is SESMS, when

$$F_s(i) := -\left(\sum_{q=0}^{r_3} B_{d,q}^\top(i)X_s(0,i)B_{d,q}(i)\right)^{-1}\left(\sum_{q=0}^{r_3} B_{d,q}^\top(i)X_s(0,i)A_{d,q}(i)\right). \tag{9.117}$$

Customizing the result proved in Theorem 5.6 to the case of the BJMLDE (5.127) we obtain:

Corollary 9.3 *Under Assumption 9.1 the following are equivalent:*

(i) the BJMLDE with Riccati-type jumping operator has a unique stabilizing solution $\mathbf{X}_s(\cdot) = (X_s(\cdot,1), X_s(\cdot,2), \dots, X_s(\cdot,N))$ *satisfying the following sign conditions:*

$$\sum_{q=0}^{r_3} B_{d,q}^\top(i)X_s(kh,i)B_{d,q}(i) > 0 \text{ for all } (k,i) \in \mathbb{Z}\times\mathcal{N}; \tag{9.118}$$

(ii) (a) the nominal system (9.93) is mean-square stabilizable by impulses in a linear state feedback form;

(b) there exists a function $t \to \mathbf{Z}(t) = (Z(t,1),\ Z(t,2),\ \dots,\ Z(t,N)) : [0,h] \to \mathcal{S}_n^N$ *differentiable on* $(0,h)$*, right continuous in* $t=0$ *and satisfying the following linear matrix differential inequality with constraints:*

$$\dot{Z}(t,i) + A_0^\top(i)Z(t,i) + Z(t,i)A_0(i) + \sum_{\iota=1}^{r_1} A_\iota^\top(i)Z(t,i)A_\iota(i) + \sum_{j=1}^{N} q_{ij}Z(t,j) + \sum_{l=1}^{r_2}\alpha_{c,l}^2(i)C_{c,l}^\top(i)C_{c,l}(i) \geq 0,\quad 0\leq t<h \tag{9.119a}$$

$$\begin{pmatrix} \sum_{p=1}^{r_4}\alpha_{d,p}^2(i)C_{d,p}^\top(i)C_{d,p}(i) - Z(h^-,i) & 0 \\ 0 & 0\end{pmatrix} + \sum_{q=0}^{r_3}\begin{pmatrix}A_{d,q}(i) & B_{d,q}(i)\end{pmatrix}^\top Z(0,i)\begin{pmatrix}A_{d,q}(i) & B_{d,q}(i)\end{pmatrix} > 0 \tag{9.119b}$$

$$Z(h,i) = Z(0,i), \tag{9.119c}$$

$i \in \mathcal{N}$*. Moreover, if (9.119) is feasible, then the stabilizing solution* $\mathbf{X}_s(\cdot)$ *is a periodic function of period* h *and satisfies* $X_s(t,i)\geq 0$*, for all* $(t,i) \in \mathbb{R}\times\mathcal{N}$.

Remark 9.5 For the numerical computation of the stabilizing solution of (9.115) the iterative procedure described in (5.281) may be adapted.

The result proven in the following lemma provides a comparison between the solutions of the LMI (9.106) associated to an arbitrary stabilizing feedback gain $\mathbf{F} = (F(1),\ F(2),\ \ldots,\ F(N))$ and the stabilizing solution of the BJMLDE (9.115).

Lemma 9.1 *Assume that Assumption 9.1 holds.*

(a) If $\mathbf{X}(\cdot) = (X(\cdot,1),\ X(\cdot,2),\ \ldots,\ X(\cdot,N))$ *is a solution of the BLJLDI (9.106) corresponding to a set* $\mathbf{F} = (F(1), F(2), \ldots, F(N))$ *of the stabilizing gain matrices for the nominal system (9.93) and to a pair* $(\alpha_c, \alpha_d) \in (0,\infty)^{\underline{r_2} \times \mathcal{N}} \times (0,\infty)^{\underline{r_4} \times \mathcal{N}}$ *of vectors of scaling parameters, then,* $X(t,i) > X_s(t,i)$*, for all* $(t,i) \in \mathbb{R} \times \mathcal{N}$*, where* $\mathbf{X}_s(\cdot) = (X_s(\cdot,1), X_s(\cdot,2), \ldots, X_s(\cdot,N))$ *is the unique bounded and stabilizing solution which is satisfying the sign condition (9.118) of the BJMLDE with Riccati-type jumping operator of the form (9.115) corresponding to the same pair* (α_c, α_d) *of vectors of scaling parameters.*

(b) If $\mathbf{X}(\cdot) = (X(\cdot,1),\ X(\cdot,2),\ \ldots,\ X(\cdot,N))$ *is a solution of the BJMDI (9.103) satisfying (9.108) then* $X(t,i) > X_s(t,i)$*, for all* $(t,i) \in \mathbb{R} \times \mathcal{N}$.

Proof (a) The BLJLDI (9.106) corresponding to the stabilizing gain matrices $\mathbf{F} = (F(1),\ F(2),\ \ldots,\ F(N))$ and to the pair (α_c, α_d) of vectors of scaling parameters may be rewritten in the form of a BLJLDE as:

$$-\dot{X}(t,i) = \mathcal{L}^*_{c,i}[\mathbf{X}(t)] + \sum_{l=1}^{r_2} \alpha^2_{c,l}(i) C^\top_{c,l}(i) C_{c,l}(i) + M_c(t,i), \quad kh < t \le (k+1)h \tag{9.120a}$$

$$X(kh^-,i) = \sum_{q=0}^{r_3} (A_{d,q}(i) + B_{d,q}(i)F(i))^\top X(kh,i)(A_{d,q}(i) + B_{d,q}(i)F(i)) + \sum_{p=1}^{r_4} \alpha^2_{d,p}(i) C^\top_{d,p}(i) C_{d,p}(i) + M_d(kh,i), \quad k \in \mathbb{Z}, \tag{9.120b}$$

where $M_c(t,i) \ge 0$, for all $(t,i) \in \mathbb{R} \times \mathcal{N}$ and $M_d(kh,i) > 0$, for all $(k,i) \in \mathbb{Z} \times \mathcal{N}$, $\mathbf{X} \to \mathcal{L}^*_c[\mathbf{X}] = (\mathcal{L}^*_{c,1}[\mathbf{X}],\ \mathcal{L}^*_{c,2}[\mathbf{X}],\ \ldots, \mathcal{L}^*_{c,N}[\mathbf{X}]) : \mathcal{S}^N_n \to \mathcal{S}^N_n$ is the adjoint of the operator $\mathcal{L}_c[\cdot]$ defined in (9.37).

Further, the BJMLDE with Riccati-type jumping operator (9.115) written for its stabilizing solution $\mathbf{X}_s(\cdot)$ may be rewritten in the following form:

$$-\dot{X}_s(t,i) = \mathcal{L}^*_{c,i}[\mathbf{X}_s(t)] + \sum_{l=1}^{r_2} \alpha^2_{c,l}(i) C^\top_{c,l}(i) C_{c,l}(i), \quad kh \le t < (k+1)h \tag{9.121a}$$

$$X_s(kh^-,i) = \sum_{q=0}^{r_3} (A_{d,q}(i) + B_{d,q}(i)F(i))^\top X_s(kh,i)(A_{d,q}(i) + B_{d,q}(i)F(i))$$

$$+\sum_{p=1}^{r_4}\alpha_{d,p}^2(i)C_{d,p}^\top(i)C_{d,p}(i) \tag{9.121b}$$

$$-(F_s(i)-F(i))^\top(\sum_{q=0}^{r_3}B_{d,q}^\top(i)X_s(kh,i)B_{d,q}(i))(F_s(i)-F(i)),\ \ k\in\mathbb{Z},$$

$i \in \mathcal{N}$. Subtracting (9.121) from (9.120) we may infer that $t \to \mathbf{Y}(t) := \mathbf{X}(t) - \mathbf{X}_s(t) : \mathbb{R} \to \mathcal{S}_n^N$ is a bounded solution of the following backward generalized Lyapunov-type jump linear differential equation GLJLDE:

$$-\dot{Y}(t,i) = \mathcal{L}_{c,i}^*[\mathbf{Y}(t)] + M_c(t,i),\ \ kh \le t < (k+1)h \tag{9.122a}$$

$$Y(kh^-,i) = \sum_{q=0}^{r_3}(A_{d,q}(i)+B_{d,q}(i)F(i))^\top Y(kh,i)(A_{d,q}(i)+B_{d,q}(i)F(i))$$
$$+\widetilde{M}_d(kh,i),\ \ k\in\mathbb{Z}, \tag{9.122b}$$

$i \in \mathcal{N}$, where

$$\widetilde{M}_d(kh,i) \coloneqq M_d(kh,i) + (F_s(i)-F(i))^\top \times \left(\sum_{q=0}^{r_3}B_{d,q}^\top(i)X_s(kh,i)B_{d,q}(i)\right)(F_s(i)-F(i)).$$

Bearing in mind the sign conditions (9.118), we deduce that $\widetilde{M}_d(kh,i) \ge M_d(kh,i) > 0$. Applying Theorem 2.13 in the case of the backward GLJLDE (9.122) we deduce that this equation has a unique bounded on $\mathbb{R}$ solution and that solution is positive definite. So, we may conclude that $X(t,i) > X_s(t,i)$, for all $(t,i) \in \mathbb{R}\times\mathcal{N}$, which confirms the validity of (a).

(b) If $\mathbf{X}(\cdot)$ is a solution of the BJMDI (9.103) which is satisfying (9.108) it is also a solution of (9.110). The conclusion is obtained from (a) because (9.110) is a particular case of (9.106). Thus the proof ends.

Now, we introduce a notation which ease the statement of the following result. So, we denote

$$\mathbf{\Sigma}[\mathbf{X}(\cdot),\alpha_c,\alpha_d] \triangleq \{\sigma>0 \mid \sigma^2\lambda_{c,l}^2D_{c,l}^\top(i)X(t,i)D_{c,l}(i) < \alpha_{c,p}^2(i)I_{m_l},\ l\in\underline{r_2},\ 0\le t\le h,$$
$$\text{and } \sigma^2\lambda_{d,p}^2D_{d,p}^\top(i)X(0,i)D_{d,p}(i) < \alpha_{d,p}^2(i)I_{m_{d,p}},\ \ p\in\underline{r_4},\ i\in\mathcal{N}\} \tag{9.123}$$

where $\mathbf{X}(\cdot) = (X(\cdot,1),X(\cdot,2),\ldots,X(\cdot,N)) : [0,h] \to \mathcal{S}_n^N$ is a given function and $(\alpha_c,\alpha_d) \in (0,\infty)^{\underline{r_2}\times\mathcal{N}} \times (0,\infty)^{\underline{r_4}\times\mathcal{N}}$ are vectors of scaling parameters.

Theorem 9.6 *Assume:*

(a) Assumption 9.1 is fulfilled;
(b) the nominal system (9.93) is mean-square stabilizable by impulses in a linear state feedback form;
(c) there exists a pair $(\alpha_c, \alpha_d) \in (0, \infty)^{\underline{r_2} \times \mathcal{N}} \times (0, \infty)^{\underline{r_4} \times \mathcal{N}}$ *of vectors of scaling parameters for which there exists a function* $\mathbf{Z}(\cdot) : [0, h] \to \mathcal{S}_n^N$ *satisfying the conditions from (ii) (b) of Corollary 9.3.*

Let $F_s(i)$, $i \in \mathcal{N}$ *be the gain matrices associated to the stabilizing solution* $\mathbf{X}_s(\cdot)$ *of the BJMLDE with Riccati-type jumping operator (9.115).*

Under these conditions the control:

$$u(k) = F_s(\eta(kh))x(kh), \quad k \in \mathbb{Z}_+ \tag{9.124}$$

robustly stabilizes the nominal system (9.93). The stability radius achieved by this control law is satisfying:

$$\begin{aligned}\rho_{\mathcal{D}}[\mathbf{G_{0,F_s}} \mid (\mathbf{D}_{c,l}, \mathbf{C}_{c,l}),\ l \in \underline{r_2},\ (\mathbf{D}_{d,p}, \mathbf{C}_{d,p}),\ p \in \underline{r_4}]\\ > \sup\{\sigma > 0 \mid \sigma \in \mathbf{\Sigma}[\mathbf{X}_s(\cdot), \alpha_c, \alpha_d]\}.\end{aligned} \tag{9.125}$$

Proof Under the considered assumptions the BJMLDE with Riccati-type jumping operator (9.115) has a unique bounded on $\mathbb{R}$ and stabilizing solution $\mathbf{X}_s(\cdot) = (X_s(\cdot, 1),\ X_s(\cdot, 2),\ \ldots,\ X_s(\cdot, N))$ which is satisfying the sign conditions (9.118). The BJMLDE (9.115) written for its stabilizing solution may be rewritten as:

$$-\dot{X}_s(t, i) = \mathcal{L}_{c,i}^*[\mathbf{X}_s(t)] + \sum_{l=1}^{r_2} \alpha_{c,l}^2(i) C_{c,l}^\top(i) C_{c,l}(i), \quad kh \le t < (k+1)h \tag{9.126a}$$

$$X(kh^-, i) = \sum_{q=0}^{r_3} (A_{d,q}(i) + B_{d,q}(i)F_s(i))^\top X_s(kh, i)(A_{d,q}(i) + B_{d,q}(i)F_s(i))$$
$$+ \sum_{p=1}^{r_4} \alpha_{d,p}^2(i) C_{d,p}^\top(i) C_{d,p}(i), \quad k \in \mathbb{Z}, \tag{9.126b}$$

$i \in \mathcal{N}$. Let $\sigma \in \mathbf{\Sigma}[\mathbf{X}_s(\cdot), \alpha_c, \alpha_d]$ be arbitrary. From (9.123) written for $\mathbf{X}(\cdot)$ replaced by $\mathbf{X}_s(\cdot)$, we obtain: $\left(\frac{\sigma\lambda_{c,l}}{\alpha_{c,l}(i)}\right)^2 D_{c,l}^\top(i) X_s(t, i) D_{c,l}(i) < I_{m_l}$ for all $l \in \underline{r_2}$, $0 \le t \le h$ and $\left(\frac{\sigma\lambda_{d,p}}{\alpha_{d,p}(i)}\right)^2 D_{d,p}^\top(i) X_s(0, i) D_{d,p}(i) < I_{m_{d,p}}$ for all $p \in \underline{r_4}$ and $i \in \mathcal{N}$. Hence

$$\sigma < \left[\max_{(l,i)\in \underline{r_2}\times\mathcal{N}} \max_{0\le t\le h} \left\| \left(\frac{\lambda_{c,l}}{\alpha_{c,l}(i)}\right)^2 D_{c,l}^\top(i) X_s(t, i) D_{c,l}(i) \right\|_2\right]^{-1/2}$$

and

$$\sigma < \left[\max_{(p,i)\in \underline{r_4}\times \mathcal{N}} \| \left(\frac{\lambda_{d,p}}{\alpha_{d,p}(i)} \right)^2 D_{d,p}^{\top}(i) X_s(0,i) D_{d,p}(i) \|_2 \right]^{-1/2}.$$

Since (9.122) is a special case of (9.101) we may conclude, via Theorem 9.2 applied in the case of the closed-loop system (9.116) when (9.70) and (9.71) take the form from (9.99) and (9.100) written for $F(i)$ replaced by $F_s(i)$, that

$$\sigma < \widehat{\mu}_{\mathbf{F}_s}$$

which leads to

$$\widehat{\mu}_{\mathbf{F}_s} \geq \sup\{\sigma > 0 \mid \sigma \in \mathbf{\Sigma}[\mathbf{X}_s(\cdot), \alpha_c\alpha_d]\} \tag{9.127}$$

then from (9.70a) applied in the case of the closed-loop system (9.116) we deduce that

$$\rho_{\mathscr{D}}[\mathbf{G}_{\mathbf{0},\widehat{\mathbf{F}}_s} \mid (\mathbf{D}_{c,l}, \mathbf{C}_{c,l}, \ l \in \underline{r_2}, \ (\mathbf{D}_{d,p}, \mathbf{C}_{d,p}), \ p \in \underline{r_4}] \geq \widehat{\mu}_{\mathbf{F}_s}. \tag{9.128}$$

Finally, (9.125) is obtained from (9.127) and (9.128). Thus the proof ends.

Remark 9.6 From Lemma 9.1 we may deduce that if $\mathbf{X}(\cdot) = (X(\cdot, 1), X(\cdot, 2), \ldots, X(\cdot, N))$ is a solution of the BLJLDE (9.106) corresponding to a control (9.94) which is stabilizing the nominal system (9.93) and to the vectors α_c, α_d of scaling parameters, then,

$$\mathbf{\Sigma}[\mathbf{X}(\cdot), \alpha_c\alpha_d] \subset \mathbf{\Sigma}[\mathbf{X}_s(\cdot), \alpha_c\alpha_d]. \tag{9.129}$$

From (9.102) and (9.129) we may conclude that for a pair of vectors of scaling parameters (α_c, α_d) for which the conditions from (ii) (b) from Corollary 9.3 are fulfilled, among all controls (9.94) which are stabilizing the nominal system (9.93), the largest lower bound of the stability radius is achieved by the control (9.124).

In the last part of this section, let investigate the problem of robust stabilization of the nominal system (9.93) with respect to the set $\mathscr{D}_c$ of the structured parametric uncertainties having the structure described by the pairs $(\mathbf{D}_{c,l}, \mathbf{C}_{c,l})$, $l \in \underline{r_2}$. More precisely, let us consider the case when the system (9.92) reduces to:

$$\begin{aligned} dx(t) &= A_0(\eta(t))x(t)dt + \sum_{\iota=1}^{r_1} A_\iota(\eta(t))x(t)dw_\iota(t) \\ &\quad + \sum_{l=1}^{r_2} D_{c,l}(\eta(t))\Delta_{c,l}(t, z_{c,l}(t), \eta(t))d\xi_l(t), \quad kh < t \leq (k+1)h \end{aligned} \tag{9.130a}$$

$$x(kh^+) = A_{d,0}(\eta(kh))x(kh) + B_{d,0}(\eta(kh))u(k) + \sum_{q=1}^{r_3} w_{d,q}(k)(A_{d,q}(\eta(kh))x(kh) + B_{d,q}(\eta(kh))u(k)), \quad k \in \mathbb{Z}_+, \tag{9.130b}$$

$$z_{c,l}(t) = C_{c,l}(\eta(t))x(t), \quad t \in \mathbb{R}_+ \tag{9.130c}$$

where

$$(t, z_{c,l}) \to \Delta_{c,l}(t, z_{c,l}, i) : \mathbb{R}_+ \times \mathbb{R}^{\nu_l} \to \mathbb{R}^{m_l}$$

are arbitrary functions which are satisfying conditions from Assumption 9.2 (*a*). In this case, the BJMLDE with Riccati-type jumping operator (9.115) becomes:

$$-\dot{X}(t,i) = \mathscr{L}^*_{c,i}[\mathbf{X}(t)] + \sum_{l=1}^{r_2} \alpha^2_{c,l}(i) C^\top_{c,l}(i) C_{c,l}(i), \quad kh \le t < (k+1)h \tag{9.131a}$$

$$X(kh^-, i) = \sum_{q=0}^{r_3} (A^\top_{d,q}(i) X(kh,i) A_{d,q}(i) - \left(\sum_{q=0}^{r_3} (A^\top_{d,q}(i) X(kh,i) B_{d,q}(i)\right) \cdot \left(\sum_{q=0}^{r_3} B^\top_{d,q}(i) X(kh,i) B_{d,q}(i)\right)^{-1} \left(\sum_{q=0}^{r_3} (A^\top_{d,q}(i) X(kh,i) B_{d,q}(i)\right)^\top, \quad k \in \mathbb{Z}, \tag{9.131b}$$

$i \in \mathscr{N}$. Since (9.131) is a particular case of (9.115), the concept of stabilizing solution of BJMLDE (9.131) is that introduced via Definition 9.5. The analogous of Corollary 9.3 is:

Corollary 9.4 *Under Assumption 9.1 the following are equivalent:*

(i) the BJMLDE with Riccati-type jumping operator has a unique bounded and stabilizing solution $\mathbf{X}_s(\cdot)$ *satisfying the following sign conditions:*

$$\sum_{q=0}^{r_3} B^\top_{d,q}(i) \widehat{X}_s(kh,i) B_{d,q}(i) > 0, \tag{9.132}$$

for all $(k,i) \in \mathbb{Z} \times \mathscr{N}$;

(ii) (a) the nominal system (9.93) is mean-square stabilizable by impulses in a state feedback form;

(b) there exists a function

$$t \to \mathbf{Z}(t) = (Z(t,1),\ Z(t,2),\ \ldots,\ Z(t,N)) : [0,h] \to \mathscr{S}^N_n$$

differentiable on $(0, h)$, *right continuous in* $t = 0$, *satisfying the following linear matrix differentiable inequality with constraints:*

$$\dot{Z}(t,i) + \mathscr{L}_{l,i}^{*}[\mathbf{Z}(t)] + \sum_{l=1}^{r_2} \alpha_{c,l}^2(i) C_{c,l}^{\top}(i) C_{c,l}(i) \geq 0, \quad 0 \leq t < h \tag{9.133a}$$

$$\begin{pmatrix} -Z(h^-, i) & 0 \\ 0 & 0 \end{pmatrix} + \sum_{q=0}^{r_3} \left(A_{d,q}(i) \; B_{d,q}(i)\right)^{\top} Z(0,i) \left(A_{d,q}(i) \; B_{d,q}(i)\right) > 0 \tag{9.133b}$$

$$Z(h,i) = Z(0,i), \tag{9.133c}$$

$i \in \mathcal{N}$. *Moreover, if (9.133) is feasible, then the unique bounded and stabilizing solution* $\widehat{\mathbf{X}}_s(\cdot)$ *is a periodic function of period h and satisfies*

$$\widehat{X}_s(t,i) \geq 0,$$

for all $(t,i) \in \mathbb{R} \times \mathcal{N}$.

Let

$$\widehat{F}_s(i) \triangleq -\left(\sum_{q=0}^{r_3} B_{d,q}^{\top}(i) \widehat{X}_s(0,i) B_{d,q}(i)\right)^{-1} \left(\sum_{q=0}^{r_3} B_{d,q}^{\top}(i) \widehat{X}_s(0,i) A_{d,q}(i)\right), \tag{9.134}$$

$i \in \mathcal{N}$, be the gain matrices associated to the stabilizing solution $\widehat{X}_s(\cdot)$ of (9.131). In the case of the stability radius $\rho_{\mathscr{D}_c}[\mathbf{G_{0,F}} \mid (\mathbf{D}_{c,l}, \mathbf{C}_{c,l}), l \in \underline{r_2}]$ of the closed-loop nominal system (9.96) with respect to the set $\mathscr{D}_c$ of the admissible uncertainties which affect the continuous-time dynamics of the nominal system, we obtain:

Theorem 9.7 *Assume:*

(a) Assumption 9.1 holds;
(b) the nominal system (9.93) is mean-square stabilizable by impulses in a linear state feedback form;
(c) there exists a vector of scaling parameters $\alpha_c, \in (0, \infty)^{\underline{r_2} \times \mathcal{N}}$ *for which there exists a function* $\mathbf{Z}(\cdot) : [0, h] \to \mathscr{S}_n^N$ *satisfying the conditions from (ii) (b) in Corollary 9.4.*

Let $\widehat{F}_s(i)$, $i \in \mathcal{N}$ *be the gain matrices associated via (9.134) to the stabilizing solution* $\widehat{\mathbf{X}}_s(\cdot)$ *of (9.131). Under these conditions the control:*

$$u(k) = \widehat{F}_s(\eta(kh)) x(kh), \quad k \in \mathbb{Z}_+ \tag{9.135}$$

robustly stabilizes the nominal system (9.93). The stability radius with respect to the set $\mathscr{D}_c$ *of the admissible parametric uncertainties is satisfying:*

$$\rho_{\mathscr{D}_c}[\mathbf{G}_{\mathbf{0},\widehat{\mathbf{F}}_{\mathbf{s}}} \mid (\mathbf{D}_{c,l}, \mathbf{C}_{c,l}, \ l \in \underline{r_2}] \geq \sup\{\sigma > 0 \mid \left(\frac{\sigma\lambda_{c,l}}{\alpha_{c,l}(i)}\right)^2 D_{c,l}^{\top}(i)\widehat{X}_s(t,i)D_{c,l}(i) < I_{m_l},$$
$$\textit{for all} \ \ (t,i) \in \underline{r_2} \times \mathscr{N}, \ \ 0 \leq t \leq h\}. \tag{9.136}$$

Proof The proof may be done following step by step the proof of Theorem 9.6.

Remark 9.7 From (9.125) and (9.136) one can see how the intensities $\lambda_{c,l}$ and $\lambda_{d,p}$, respectively, of the white noise-type additive perturbations $\{\xi_l(t)\}_{t\geq 0}$ and $\{\zeta_p(k)\}_{k\in\mathbb{Z}_+}$, respectively, is affecting the level of robustness achieved by the stabilizing controllers (9.124) and (9.135), respectively.

9.5 Robust Stabilization by Sampled State Measurements

Let us consider the controlled linear stochastic system

$$\begin{aligned} \mathrm{d}x(t) = (A_0(\eta(t))x(t) + B_0(\eta(t))u(t))\mathrm{d}t + \sum_{\iota=1}^{r_1}(A_\iota(\eta(t))x(t) \\ + B_\iota(\eta(t))u(t))\mathrm{d}w_\iota(t) \\ + \sum_{l=1}^{r_2} D_{c,l}(\eta(t))\Delta_{c,l}(t, z_{c,l}(t), \eta(t))\mathrm{d}\xi_l(t), \end{aligned} \tag{9.137a}$$

$$z_{c,l}(t) = C_{c,l}(\eta(t))x(t), \ \ t \in \mathbb{R}_+, \ \ 1 \leq l \leq r_2, \tag{9.137b}$$

where $x(t) \in \mathbb{R}^n$ are the state parameters and $u(t) \in \mathbb{R}^{n_u}$ are the control parameters, while $\{w(t)\}_{t\geq 0}$, $(w(t) = (w_1(t), \ w_2(t), \ \dots, w_{r_1}(t))^\top)$, $\{\xi(t)\}_{t\geq 0}$, $(\xi(t) = (\xi_1(t), \ \xi_2(t), \ \dots, \ \xi_{r_2}(t))^\top)$ and $\{\eta(t)\}_{t\geq 0}$ are stochastic processes which are satisfying the conditions from Assumption 9.1.

In (9.137), $A_\iota(i) \in \mathbb{R}^{n\times n}$, $B_\iota(i) \in \mathbb{R}^{n\times n_u}$, $0 \leq \iota \leq r_1$, $D_{c,l}(i) \in \mathbb{R}^{n\times m_l}$, $C_{c,l}(i) \in \mathbb{R}^{\nu_l\times n}$, $1 \leq l \leq r_2$, are known matrices, while, $(t, z_{c,l}) \to \Delta_{c,l}(t, z_{c,l}, i) : \mathbb{R}_+ \times \mathbb{R}^{\nu_l} \to \mathbb{R}^{m_l}$, $1 \leq l \leq r_2$, are arbitrary functions which are satisfying the conditions from Assumption 9.2. This allows us to regard the system as a disturbance of the following linear controlled system named nominal system

$$\begin{aligned} \mathrm{d}x(t) =&(A_0(\eta(t))x(t) + B_0(\eta(t))u(t))\mathbf{d}t \\ &+ \sum_{\iota=1}^{r_1}(A_\iota(\eta(t))x(t) + B_\iota(\eta(t))u(t))\mathbf{d}w_\iota(t), \ \ t \in \mathbb{R}_+. \end{aligned} \tag{9.138}$$

The term $\sum_{l=1}^{r_2} D_{c,l}(\eta(t))\Delta_{c,l}(t, z_{c,l}(t), \eta(t))\mathrm{d}\xi_l(t)$ model some random parametric uncertainties which may affect the mathematical model of the dynamical system which must be stabilized. To stabilize the dynamical system modeled by (9.137) and (9.138) we use control laws in a linear state feedback form. To implement these kinds of controls we need by the measured values of the state parameters $X(t)$ of the system under consideration. In this section, we assume that the states $x(\cdot)$ of the system (9.137) and (9.138), respectively, are available for measurements only to some discrete-time instants $\tau_k = kh$, $k = 0, 1, \ldots,$ where $h > 0$ is the sampling period. We assume also that on the intervals $[kh, (k+1)h)$ (between two instances time when the states measurements are possible) the controls have constant values u_k, $k \geq 0$.

Our aim is to design a control law of the form

$$u(t) = F(\eta(kh))x(kh), \quad kh \leq t < (k+1)h, \quad k = 0, 1, 2, \ldots \tag{9.139}$$

which stabilizes the nominal system (9.138) **together with all its disturbances of the form** (9.137) **corresponding to all parametric uncertainties** $\Delta(\cdot) \in \mathscr{D}_c$ **with** $\|\Delta_c(\cdot)\| < \sigma$ **for a prescribed level of robustness** $\sigma > 0$.

The controls of the form (9.139) belong to a more general class of piecewise constant controls of the form:

$$u(t) = u_k, \quad t \in [kh, (k+1)h), \quad k \in \mathbb{Z}_+ \tag{9.140}$$

where $u_k : \Omega \to \mathbb{R}^{n_u}$ are random vectors measurable with respect to the σ-algebra generated by the random variables $w(s)$, $\xi(s)$, $\eta(s)$, $0 \leq s \leq kh$ and have finite second moments.

Plugging a control (9.140) in the controlled system (9.137) and in the nominal system (9.138) we get:

$$\begin{aligned} \mathrm{d}x(t) &= (A_0(\eta(t))x(t) + B_0(\eta(t))u_k)\mathrm{d}t + \sum_{\iota=1}^{r_1}(A_\iota(\eta(t))x(t) + B_\iota(\eta(t))u_k)\mathrm{d}w_\iota(t) \\ &\quad + \sum_{l=1}^{r_2} D_{c,l}(\eta(t))\Delta_{c,l}(t, z_{c,l}(t), \eta(t))\mathrm{d}\xi_l(t), \quad kh \leq t < (k+1)h, \quad k \in \mathbb{Z}_+ \end{aligned} \tag{9.141a}$$

$$z_{c,l}(t) = C_{c,l}(\eta(t))x(t), \quad t \in \mathbb{R}_+, \tag{9.141b}$$

and

$$\begin{aligned} \mathbf{d}x(t) &= (A_0(\eta(t))x(t) + B_0(\eta(t))u_k)\mathbf{d}t + \sum_{\iota=1}^{r_1}(A_\iota(\eta(t))x(t) + B_\iota(\eta(t))u_k)\mathrm{d}w_\iota(t) \\ &\quad kh \leq t < (k+1)h, \quad k \in \mathbb{Z}_+, \end{aligned} \tag{9.142}$$

respectively.

Let $t \to \varphi(t) : \mathbb{R}_+ \to \mathbb{R}^{n_u}$ be defined by

$$\varphi(t) = u_k, \quad t \in [kh, (k+1)h), \quad k \in \mathbb{Z}_+. \tag{9.143}$$

Setting $\mathbf{x}(t) = (x^\top(t) \;\; \varphi^\top(t))^\top$ by direct calculation we deduce that if $x(\cdot)$ is a solution of (9.141) and $\varphi(\cdot)$ is that from (9.143), we obtain that

$$\begin{aligned} \mathrm{d}\mathbf{x}(t) &= \mathscr{A}_0(\eta(t))\mathbf{x}(t)\mathrm{d}t + \sum_{\iota=1}^{r_1} \mathscr{A}_\iota(\eta(t))\mathbf{x}(t)\mathrm{d}w_\iota(t) \\ &\quad + \sum_{l=1}^{r_2} \widehat{D}_{c,l}(\eta(t))\Delta_{c,l}(t, \widehat{z}_{c,l}(t), \eta(t))\mathrm{d}\xi_l(t), \quad kh < t \le (k+1)h \quad (9.144a) \\ \mathbf{x}(kh^+) &= \mathscr{A}_d\mathbf{x}(kh) + \mathscr{B}_d u_k, \quad k \in \mathbb{Z}_+ \quad (9.144b) \\ \widehat{z}_{c,l}(t) &= \widehat{C}_{c,l}(\eta(t))\mathbf{x}(t), \quad t \in \mathbb{R}_+, \quad t \in \mathbb{R}_+, \quad (9.144c) \end{aligned}$$

where we denote:

$$\begin{aligned} \mathscr{A}_\iota(i) &= \begin{pmatrix} A_\iota(i) & B_\iota(i) \\ 0 & 0 \end{pmatrix} \in \mathbb{R}^{(n+n_u)\times(n+n_u)}, \quad 0 \le \iota \le r_1, \\ \mathscr{A}_d &= \begin{pmatrix} I_n & 0 \\ 0 & 0 \end{pmatrix} \in \mathbb{R}^{(n+n_u)\times(n+n_u)}, \quad (9.145) \\ \mathscr{B}_d &= \begin{pmatrix} 0 \\ I_{n_u} \end{pmatrix} \in \mathbb{R}^{(n+n_u)\times n_u}, \quad \widehat{D}_{c,l}(i) = \begin{pmatrix} D_{c,l}(i) \\ 0 \end{pmatrix} \in \mathbb{R}^{(n+n_u)\times m_l} \\ \widehat{C}_{c,l}(i) &= \begin{pmatrix} C_{c,l}(i) & 0 \end{pmatrix} \in \mathbb{R}^{\nu_l\times(n+n_u)}, \quad 1 \le l \le r_2. \end{aligned}$$

Conversely, if $\mathbf{x}(t; \mathbf{u})$ is a solution of (9.144) and (9.145) determined by the control $\mathbf{u} = \{\widetilde{u}_k\}_{k\in\mathbb{Z}_+}$, then the vector $x(t; \mathbf{u})$ formed by the first n-scalar components of $\mathbf{x}(t; \mathbf{u})$ is a solution of (9.137) determined by the control law of the form (9.140) constructed based on the vectors $\widetilde{u}_k$, $k \in \mathbb{Z}_+$.

Proceeding in the same manner we may equivalently transform the nominal system (9.138) in the following impulsive controlled linear stochastic system (ICLSS):

$$\begin{aligned} \mathrm{d}\mathbf{x}(t) &= \mathscr{A}_0(\eta(t))\mathbf{x}(t)\mathrm{d}t + \sum_{\iota=1}^{r_1} \mathscr{A}_\iota(\eta(t))\mathbf{x}(t)\mathrm{d}w_\iota(t), \quad kh < t \le (k+1)h \quad (9.146a) \\ \mathbf{x}(kh^+) &= \mathscr{A}_d\mathbf{x}(kh) + \mathscr{B}_d u_k, \quad k \in \mathbb{Z}_+. \quad (9.146b) \end{aligned}$$

In the following, the ICLSS (9.146) is named nominal system and (9.144) is regarded as a disturbance of this nominal system.

To obtain an answer to the robust stabilization problem stated before, we shall apply Theorem 9.7 in the case of the systems (9.144), (9.146). In this case, the BJMLDE with Riccati-type jumping operator (9.131) becomes:

$$\dot{X}(t,i) + \mathscr{A}_0^\top(i)X(t,i) + X(t,i)\mathscr{A}_0(i) + \sum_{\iota=1}^{r_1} \mathscr{A}_\iota^\top(i)X(t,i)\mathscr{A}_\iota(i)$$

$$+\sum_{j=1}^{N} q_{ij} X(t,j) + \sum_{l=1}^{r_2} \alpha_{c,l}^2(i) \widehat{C}_{c,l}^\top(i) \widehat{C}_{c,l}(i) = 0, \quad kh \le t < (k+1)h \tag{9.147a}$$

$$X(kh^-, i) = \mathscr{A}_d^\top X(kh,i)\mathscr{A}_d - \mathscr{A}_d^\top X(kh,i)\mathscr{B}_d(\mathscr{B}_d^\top X(kh,i)\mathscr{B}_d)^{-1} \cdot \mathscr{B}_d^\top X(kh,i)\mathscr{A}_d, \quad k \in \mathbb{Z}. \tag{9.147b}$$

Definition 9.6 A globally defined solution $t \to \mathbf{X}_s(t) = (X_s(t,1), X_s(t,2), \ldots, X_s(t,N)) : \mathbb{R} \to \mathscr{S}_{n+n_u}^N$ is a stabilizing solution of the BJMLDE with Riccati-type jumping operator (9.147) if $\det(\mathscr{B}_d^\top X_s(kh,i)\mathscr{B}_d) \neq 0$, for all $(k,i) \in \mathbb{Z} \times \mathscr{N}$ and the closed-loop system

$$d\mathbf{x}(t) = \mathscr{A}_0(\eta(t))\mathfrak{x}(t)dt + \sum_{\iota=1}^{r_1} \mathscr{A}_\iota(\eta(t))\mathbf{x}(t)dw_\iota(t), \quad kh < t \le (k+1)h \tag{9.148a}$$

$$\mathbf{x}(kh^+) = (\mathscr{A}_d + \mathscr{B}_d \mathbf{F}_s(\eta(kh)))\mathbf{x}(kh), \quad k \in \mathbb{Z}_+. \tag{9.148b}$$

is ESMS, where

$$\mathbf{F}_s(i) := -(\mathscr{B}_d^\top X_s(0,i)\mathscr{B}_d)^{-1}\mathscr{B}_d^\top X_s(0,i)\mathscr{A}_d. \tag{9.149}$$

The analogous of Corollary 9.4 is:

Corollary 9.5 *Under Assumption 9.1 the following are equivalent:*

(i) the BJMLDE with Riccati-type jumping operator (9.147) has a unique bounded and stabilizing solution $\mathbf{X}_s(\cdot) = (X_s(\cdot,1), X_s(\cdot,2), \ldots, X_s(\cdot,N))$ which is satisfying the following sign conditions:

$$\mathscr{B}_d^\top X_s(kh,i)\mathscr{B}_d > 0, \tag{9.150}$$

for all $(k,i) \in \mathbb{Z} \times \mathscr{N}$;

(ii) (a) the nominal system (9.138) is mean-square stabilizable by sampled state measurements;

(b) there exists a function $t \to \mathbf{Z}(t) = (Z(t,1), Z(t,2), \ldots, Z(t,N)) : [0,h] \to \mathscr{S}_{n+n_u}^N$ differentiable on $(0,h)$, right continuous in $t = 0$ and satisfying the linear matrix differential inequality with constraints:

$$\dot{Z}(t,i) + \mathscr{A}_0^\top(i)Z(t,i) + Z(t,i)\mathscr{A}_0(i) + \sum_{\iota=1}^{r_1} \mathscr{A}_\iota^\top(i)Z(t,i)\mathscr{A}_\iota(i) + \sum_{j=1}^{N} q_{ij} Z(t,j) + \sum_{l=1}^{r_2} \alpha_{c,l}^2(i)\widehat{C}_{c,l}^\top(i)\widehat{C}_{c,l}(i) \ge 0, \quad 0 \le t < h \tag{9.151a}$$

$$\begin{pmatrix} -Z(h^-, i) & 0 \\ 0 & 0 \end{pmatrix} + \left(\mathscr{A}_d \ \mathscr{B}_d\right)^\top Z(0, i) \left(\mathscr{A}_d \ \mathscr{B}_d\right) > 0 \tag{9.151b}$$

$$Z(h, i) = Z(0, i), \tag{9.151c}$$

$i \in \mathcal{N}$. If (9.151) hold then the unique bounded and stabilizing solution of (9.147) is a periodic function of period h and satisfies $X_s(t, i) \geq 0$, for all $(t, i) \in \mathbb{R}_+ \times \mathcal{N}$.

Let $\begin{pmatrix} X_{s,1}(t, i) & X_{s,2}(t, i) \\ X_{s,2}^\top(t, i) & X_{s,3}(t, i) \end{pmatrix}$, $(t, i) \in \mathbb{R} \times \mathcal{N}$, be the partition of the matrix components of $X(t, i)$ of the stabilizing solution $\mathbf{X}_s(t)$ of the BJMLDE (9.147) where $X_{s,1}(t, i) \in \mathcal{S}_n$ and $X_{s,3}(t, i) \in \mathcal{S}_{n_u}$.

Bearing in mind (9.145) together with the periodicity property we may infer that the sign conditions (9.150) reduces to:

$$X_{s,3}(k, i) = X_{s,3}(0, i) > 0, \quad \text{for all} \quad (k, i) \in \mathbb{Z} \times \mathcal{N}. \tag{9.152}$$

The stabilizing gain matrices (9.149) become:

$$\mathbf{F}_s(i) = \left(\widetilde{F}_s(i) \quad 0\right) \tag{9.153a}$$

$$\widetilde{F}_s(i) = -X_{s,3}^{-1}(0, i) X_{s,2}^\top(0, i), \tag{9.153b}$$

$i \in \mathcal{N}$. By direct calculation, involving (9.145), (9.153) we obtain that if the closed-loop system (9.148) is ESMS, then the solutions $x(t; t_0, x_0)$ of the stochastic differential equation

$$\begin{aligned} dx(t) =& (A_0(\eta(t))x(t) + B_0(\eta(t))\widetilde{F}_s(\eta(kh))x(kh))dt \\ &+ \sum_{\iota=1}^{r_1} (A_\iota(\eta(t))x(t) + B_\iota(\eta(t))\widetilde{F}_s(\eta(kh))x(kh))dw_\iota(t), \\ & kh < t \leq (k+1)h, \quad k \in \mathbb{Z}_+, \end{aligned} \tag{9.154}$$

are satisfying

$$\mathbb{E}[|x(t; t_0, x_0)|^2] \leq \beta e^{-\alpha(t-t_0)}|x_0|^2,$$

for all $t \geq t_0 \geq 0$, where $\beta \geq 1$, $\alpha > 0$ do not depend upon t, t_0, x_0 and by the initial probability distribution π_0 of the Markov process.
One sees that (9.154) is obtained by coupling the piecewise constant control

$$u(t) = \widetilde{F}_s(\eta(kh))x(kh), \quad kh \leq t < (k+1)h, \quad k \in \mathbb{Z}_+, \tag{9.155}$$

to the nominal system (9.138).

To obtain a lower bound of the robustness level $\rho_{\mathscr{D}_c}[\mathbf{G}_{\mathbf{0},\widetilde{\mathbf{F}}_\mathbf{s}} \mid (\mathbf{D}_{c,l}, \mathbf{C}_{c,l}),\ l \in \underline{r_2}]$ achieved by the control law (9.155) when it is coupled to the perturbed system (9.137) we apply Theorem 9.7 in the case of the system (9.144) and (9.146).

Theorem 9.8 *Assume:*

(a) Assumption 9.1 holds;
(b) the nominal system (9.138) is mean-square stabilizable by sampled state measurements;
(c) there exists a vector $\alpha_c \in (0, \infty)^{\underline{r_2} \times \mathcal{N}}$ *of scaling parameters for which there exists a function* $\mathbf{Z}(\cdot) : [0, h] \to \mathcal{S}_{n+n_u}^N$ *satisfying the conditions from (ii) (b) in Corollary 9.5.*

Let $\widetilde{F}_s(i) \in \mathbb{R}^{n_u \times n}$ *be the stabilizing feedback gain matrices defined in (9.153) based on the bounded and stabilizing solution of the BJMLDE with Riccati-type jumping operator (9.147). Under these conditions the control law (9.155) robustly stabilizes the nominal system (9.138). The stability radius with respect to the set* $\mathcal{D}_c$ *of the admissible parametric uncertainties is satisfying:*

$$\rho_{\mathcal{D}_c}[\mathbf{G}_{\mathbf{0},\widetilde{\mathbf{F}}_\mathbf{s}} | (\mathbf{D}_{c,l}, \mathbf{C}_{c,l}), l \in \underline{r_2}] \geq \sup\{\sigma > 0 \mid \left(\frac{\sigma \lambda_{c,l}}{\alpha_{c,l}(i)}\right)^2 D_{c,l}^\top(i) X_{s,1}(t,i) D_{c,l}(i) < I_{m_l},$$
$$\textit{for all } (t,i) \in \underline{r_2} \times \mathcal{N}, \ 0 \leq t \leq h\}, \tag{9.156}$$

where $X_{s,1}(t,i)$ *is the* $1,1$*-block of the matrix component* $X_s(t,i)$ *of the stabilizing solution* $\mathbf{X}_s(\cdot)$ *of the BJMLDE with Riccati-type jumping operator (9.147).*

9.6 Notes and References

Robustness with respect to stochastic parametric uncertainties has attracted a lot of interest from the research community. This is partly due to the engineering applications of such a modeling paradigm. One of the most efficient and elegant tools that has been developed to tackle robustness issues is the concept of the stability radii which, as coined by stability radii which, as coined by Costa et al. [4], lie at the heart of robustness analysis problems and thus constitute an area of central interest. Hence, since the seminal work of D. Hinrichsen and A. J. Pritchard [5, 6], this has been a very active research domain.

In the MJLS context, two seminal references are Bouhtouri and Hadri [7, 8]. For recent developments on the topic we refer to Todorov and Fragoso [9] and Dragan and Aberkane [10]. For Itô-type stochastic systems, we cite here Bouhtouri et al. [11] and Bouhtouri and Hadri [12]. Very recent references such us Kada et al. [13] illustrates that this domain remains active.

To the best of the authors knowledge, the problem of robust stability/stabilization of impulsive MJLSs via the stability radii approach have not been addressed yet in the literature. Hence, this chapter can be viewed as the first rigorous study of such a mathematical problem.

References

1. Friedman, A.: Stochastic Differential Equations and Applications, vol. 1. Academic (1975)
2. Drăgan, V., Aberkane, S.: Robust stability of time-varying Markov jump linear systems with respect to a class of structured, stochastic, nonlinear parametric uncertainties. Axioms **10**(3), 148 (2021)
3. Hinrichsen, D., Pritchard, A.: Stability radii of systems with stochastic uncertainty and their optimization by output feedback. SIAM J. Control Optim. **34**, 1972–1998 (1996)
4. Costa, O.L.V., Fragoso, M.D., Todorov, M.G.: Continuous-Time Markov Jump Linear Systems. Springer, Berlin, Heidelberg (2013)
5. Hinrichsen, D., Pritchard, A.: Stability radii for linear systems. Syst. Control Lett. **7**, 1–10 (1986)
6. Hinrichsen, D., Pritchard, A.: Stability radius for structured perturbations and the algebraic Riccati equation. Syst. Control Lett. **8**, 105–113 (1986)
7. Bouhtouri, A.E., Hadri, K.E.: Robust stabilization of jump linear systems subject to structured uncertainties in the state and input matrices. IMA J. Math. Control Inf. **17**, 281–293 (2000)
8. Dragan, V., Morozan, T.: Stability and robust stabilization to linear stochastic systems described by differential equations with Markovian jumping and multiplicative white noise. Stoch. Anal. Appl.. **20**, 33–92 (2002)
9. Todorov, M., Fragoso, M.: A new perspective on the robustness of Markov jump linear systems. Automatica **49**, 735–747 (2013)
10. Dragan, V., Aberkane, S.: Robust stability of time-varying Markov jump linear systems with respect to a class of structured, stochastic, nonlinear parametric uncertainties. Axioms **10** (2021)
11. Bouhtouri, A.E., Hinrichsen, D., Pritchard, A.J.: Stability radii of discrete-time stochastic systems with respect to blockdiagonal perturbations. Automatica **36**, 1033–1040 (2000)
12. Bouhtouri, A.E., Hadri, K.E.: Robust stabilization of discrete-time jump linear systems with multiplicative noise. IMA J. Math. Control Inf. **23**, 447–462 (2005)
13. Kada, M., Kameche, A., Heddar, A., Mennoun, A.: Some new results about the stability radius of infinite-dimensional systems perturbed stochastically and deterministically. Int. J. Control (2024)

Index

V. Drăgan et al., *Robust Control of Jump Linear Stochastic Systems*, Lecture Notes in Control and Information Sciences 497, https://doi.org/10.1007/978-3-031-84070-8